Machine Tool and Manufacturing Technology

Steve F. Krar

Mario Rapisarda

Albert F. Check

Delmar Publishers

an International Thomson Publishing company I(T)P

Albany • Bonn • Boston • Cincinnati • Detroit • London • Madrid
Melbourne • Mexico City • New York • Pacific Grove • Paris • San Francisco
Singapore • Tokyo • Toronto • Washington

NOTICE TO THE READER

Cover photo courtesy of: Cincinnati Milacron Inc.
Cover Design: Courtesy of Brian Deep

Delmar Staff
Publisher: Nancy K. Roberson
Acquisitions Editor: Vernon R. Anthony
Editorial Assistant: Rhonda Kreshover
Production Manager: Larry Main
Production Coordinator: Karen Smith
Art and Design Coordinator: Cheri Plasse

COPYRIGHT © 1998
By Delmar Publishers
an International Thomson Publishing company

The ITP logo is a trademark under license

Printed in the United States of America

For more information, contact:

Delmar Publishers
3 Columbia Circle, Box 15015
Albany, New York 12212-5015

International Thomson Publishing Europe
Berkshire House 168-173
High Holborn
London, WC1V7AA
England

Thomas Nelson Australia
102 Dodds Street
South Melbourne, 3205
Victoria, Australia

Nelson Canada
1120 Birchmount Road
Scarborough, Ontario
Canada M1K 5G4

International Thomson Editores
Campos Eliseos 385, Piso 7
Col Polanco
11560 Mexico D F Mexico

International Thomson Publishing Gmbh
Königswinterer Strasse 418
53227 Bonn
Germany

International Thomson Publishing Asia
221 Henderson Road
#05-10 Henderson Building
Singapore 0315

International Thomson Publishing Japan
Hirakawacho Kyowa Building, 3F
2-2-1 Hirakawacho
Chiyoda-ku, Tokyo 102
Japan

Delmar Publishers' Online Service
To access Delmar on the World WIde Web, point your browser to:
http://www.delmar.com/delmar.html
To access through Gopher: gopher://gopher.delmar.com
(Delmar Online is part of "thomson.com", an Internet site with information on more than 30 publishers of the International Thomson Publishing organization.)
For information on our products and services:
email: info @delmar.com
or call 800-347-7707

3 4 5 6 7 8 9 10 XXX 03 02 01 00 99 98

Library of Congress Cataloging-in-Publication Data

Krar, Steve F.
    Machine tool and manufacturing technology / Steve Krar, Mario
Rapisarda, Albert F. Check.
      Includes Index   p.   cm.
     ISBN 0–8273–6351–6
     1. Machining. 2. Machine-tools. 3. Manufactures—Technological innovations.
     I. Rapisarda, Mario. II. Check, Albert F. III. Title
TJ1185.K6678 1996
671.3'5—dc20
                                    95-25386
                                        CIP

Contents

About the Authors

Steve F. Krar

Steve F. Krar majored in Machine Shop Practice and spent fifteen years in the trade, first as a machinist and finally as a tool and diemaker. He then entered Teachers' College and graduated from the University of Toronto with a Specialist's Certificate in Machine Shop Practice. During his twenty years of teaching, Mr. Krar was active in Vocational and Technical education and served on the executive boards of many educational organizations. For ten years, he was on the summer staff of the College of Education, University of Toronto, involved in technical teacher training. Active in machine tool associations, Steve Krar is a Life Member of the Society of Manufacturing Engineers (SME) and former Associate Director of the GE Superabrasives Partnership for Manufacturing Productivity, an industry/education alliance.

Steve Krar's continual research in manufacturing over the past thirty-eight years, especially in the new technologies, has involved many courses with leading world manufacturers and an opportunity to study under Dr. W. Edwards Deming. Mr. Krar is co-author of over forty-five technical books, such as *Machine Shop Training*, *Technology of Machine Tools*, *Machine Tool Operations*, *CNC Technology and Programming*, and *Superabrasives: Grinding and Machining*, some of which have been translated into six languages and used throughout the world. Steve Krar has received international recognition and has been invited to China twice to share his knowledge of the latest developments in machining/manufacturing technology.

Mario Rapisarda

Mario Rapisarda is a multimedia writer and producer whose credits include developing interactive teaching programs as well as being published in two technical textbooks. His first effort with Kelmar Associates was doing research and some photography for the text *Superabrasives: Grinding and Machining*. He is the author of *Precision Metal Technology*, published by Harcourt Brace Jovanovich. Mr. Rapisarda's varied teaching experience includes working in the vocational school system of Connecticut, CETA job-training programs for the Norwalk Board of Education, and the National Tooling and Machining Association (NTMA). A member of the Society of Manufacturing Engineers (SME), his practical experiences are the result of working at several levels of engineering, beginning as an apprentice tool and diemaker. He also developed and produced a series of audiovisual programs on machine shop practices for Photocom Productions.

Albert F. Check

Albert F. Check has worked in the machine tool trade for many years and has experience in the setup and operation of NC/CNC machine tools. He holds a master of science degree in Occupational Education from Chicago State University. Al Check has been a full-time faculty member at Triton Community College for twenty-one years and served as the Coordinator for Machine Tool Technology for sixteen years. His extensive trade background makes him well suited for teaching industrial in-plant training courses through the Employee Development Institute. Mr. Check continues to keep up to date with technological developments by attending industrial training seminars offered by the Society of Manufacturing Engineers (SME), industrial machine tool manufacturers, and GE Superabrasives Grinding and Machining Technology. He has coauthered technical texts such as *Technology of Machine Tools* and *Machine Shop Fundamentals*.

Al Check is a Senior Member of SME, has been a VICA (Vocational Industrial Clubs of America) judge for the State of Illinois Precision Machining Skill Olympics, and has been an active participant in the Vocational Instructional Practicum sponsored by the State of Illinois. Mr. Check has acted as a mentor for a visiting Turkish Educator as part of a World Bank project. He has served on many college and local elementary education committees and is currently a member of the Educators' Advisory Council of the Industrial Diamond Association's Partnership for Manufacturing Productivity.

Preface

Throughout history, the most progressive and wealthiest countries in the world have all had a vibrant manufacturing base. The ability to manufacture high quality products at the lowest possible price ensures any country or manufacturer a good share of the world market. A strong manufacturing base determines the wealth of a country and is responsible for the high standard of living that its people enjoy. Therefore, it is wise to examine what makes this possible.

Ever since the beginning of time, humans have used tools in order to survive and satisfy their needs. As the development of tools evolved over the centuries from the stone axe to the numerical control machines of today, it continually became easier and faster to produce high-quality goods with less effort. The entire world is a user of tools, and in these days of rapidly changing technology, the countries or manufacturers that recognize the tools of tomorrow and learn to use them today will assure themselves of a place in tomorrow's prosperity.

Over the past thirty years, new technologies have made a dramatic impact on manufacturing, making it possible to produce goods of better quality, faster, and at lower costs. The countries which have incorporated new technology into their manufacturing processes have increased their manufacturing productivity and gained an ever-increasing share of world markets. In turn, the countries that have not taken advantage of the new technologies have seen their manufacturing productivity and their share of the world market and wealth decline.

Machine shop is the basis of all manufacturing and it is one of the most important technical subjects taught in schools, since it has a direct bearing on the productivity, the standard of living, and the economy of a country. Therefore, it is important that new manufacturing technologies are incorporated into curriculums as soon as possible. The standard machine trade curriculum of twenty-five to thirty years ago does not prepare a student to enter and compete in this technological age.

The main purpose of this book is to interest students in machine trade programs with a curriculum that introduces some of the more exciting manufacturing technologies used by industry. The need for the machinist of yesterday who was skilled in operating a variety of conventional machine tools has been declining for the past thirty years. What industry needs today is a person who has a knowledge of conventional machining theory and processes along with a basic knowledge of the processes involved in manufacturing in this technological age. In order to prepare students for this technological age, machine trade courses should be constantly modified to include new technologies and adapt to world trends. These courses should incorporate the following principles:

1. Basic metal-removal operations of conventional machines should still be taught to reinforce the related theory. Less time should be spent on developing manual skills that are rapidly being replaced by numerical control machine tools.

2. Basic computer programming skills on the most common Computer Numerical Control (CNC) machine tools must be a part of every machine shop course because almost 90 percent of the machine tools manufactured in the world today are numerically controlled.

3. All students must have a basic knowledge of all the new manufacturing technologies and processes. This will show them the important role that machine trades play and the many employment opportunities available in the related manufacturing technologies.

TEACHING RESOURCES

One of the problems facing educators is finding time to become familiar with the many new manufacturing technologies as they develop. It seems that changes in manufacturing technology are occurring every month, and in many cases, one- or two-year-old technology has already been replaced by new technology. To assist instructors in introducing new technologies, a list of videotapes and their source is included in the Teacher's Guide for this book. These videotapes are generally well done and provide a good understanding of how each process works and the applications it has in the metalworking industry. Resources such as these are available from libraries, manufacturers, Society of Manufacturing Engineers (SME), professional organizations, and publishers. Many of these are free, although in some cases there might be a modest fee for the purchase or rental by educators.

New technologies provide instructors with the opportunity to enrich their basic courses and attract more students to manufacturing programs. Students at all levels respond well to technological developments.

S. F. Krar
M. Rapisarda
A. F. Check

ACKNOWLEDGMENTS

The authors wish to express their sincere appreciation to Alice H. Krar for the countless hours she devoted to typing, proofreading, and checking the manuscript for this book. Her assistance has contributed greatly to the book's clarity and completeness. A special note of thanks must go to J.W. Oswald for his assistance and organization of the many pieces of artwork, and to Sam Fugazatto of Reynolds Machine & Tool Corp. for assistance in securing technical information and artwork sources.

We owe a special debt of gratitude to the many teachers, students, and industrial personnel who were kind enough to offer constructive criticism and suggestions. For their assistance and valuable suggestions in reviewing the manuscript and help with art design, which clarified and made the book as current as possible, we gratefully acknowledge the following: Richard Abate, AQL Corp.; William Fulkerson, John Knapp, Robert D. Wismer, David T. Wrzesinski, Deere & Company Manufacturing; Mel Sudhacker, Deckel-Maho, Inc.; Bob Kelly, Emco Maier Corp.; Ian Bradbury, GM Powertrain Div.; G. Howland Blackiston, Juran Institute, Inc.; Matthew Spiller, Makino Inc.; Joe Sarkess, Niagara Cutter, Inc.; Bob Crowl, North Central Technical College; Greg Smith, F.K. Smith Co., Inc.; Rudy Bernegger, Maurice Aube, View Engineering, Inc.

As many suggestions as possible were incorporated in this edition in order to make it a more useful reference for both the student and the instructor.

We are also grateful to the following firms, which were kind enough to supply technical information and illustrations for this text:
Acme Screw and Gear
Allen-Bradley Co.
American CNC
American Iron and Steel Institute
AMT - The Association for Manufacturing Technology
Autodesk, Inc.
Balzers Tool Coating Inc.
Bridgeport Machines, Inc.
Brown & Sharpe Manufacturing Co.
Carborundum Abrasives of North America
Charmilles Technologies Corp.
Cincinnati Milacron Inc.
Clausing Industrial, Inc.
Cleveland Twist Drill Co.

Concentric Tool Corp.
Computervision Corp.
Cushman Industries, Inc.
DEA Corp.
Deere & Company Mfg.
Deckel Maho Inc.
Delta File Works
Delta International Machinery Corp.
DoALL Company
Dorian Tool International
Electronic Industries Association
Emco Maier Corp.
Everett Industries, Inc.
Fadal Engineering Co., Inc.
Forkardt, Inc.
GE Superabrasives
Giddings & Lewis, Inc.
GM Powertrain Group
Greenfield Industries, Inc.
Grinding Wheel Institute
Hamar Laser Instruments, Inc.
Hardinge Brothers, Inc.
Hewlett-Packard Co.
Ingersoil Cutting Tool Co.
Inland Steel Co.
Jacobs© Chuck Manufacturing Co.
James Neill & Co. (Sheffield) Ltd.
Juran Institute
Kaiser Precision Tooling, Inc.
Kelmar Associates
Kennametal, Inc.
KTS Industries
Kurt Manufacturing
LaserMike Div., Techmet Co.
Makino Inc.
Manufacturing Engineering Magazine
Mazak Co.
Modern Machine Shop Magazine, Copyright 1994. Gardner Publications Inc.
Monarch Machine Tool Co.
Morse Twist Drill and Machine Co.
MTI Co.
National Tooling & Machining Association
National Twist Drill & Tool Co.

Niagara Cutter, Inc.
Nicholson File Co. Ltd.
Northwestern Tools, Inc.
Norton Co.
Nucor Corp.
Numerical Control Computer Sciences
H. Pauline & Co. Ltd
Pratt & Whitney Co., Inc.
Praxair, Inc.
Pro Link
Prohold Workholding, Inc.
Rohm Products of America
Shell Oil Co.
Society of Manufacturing Engineers
South Bend Lathe Corp.
Spectra-Physics
Standard-Modern Machine Co.
Stanley Tools, Division of the Stanley Works
L.S. Starrett Co.
Steel Company
Summagraphics
Sunbeam Equipment Corp.
Taft-Peirce Manufacturing Co.
Tecnomatix Technologies, Inc.
Toolex Systems, Inc.
Union Butterfield Corp.
United States Steel Corp.
Valenite Inc.
View Engineering
The Weldon Tool Co.
Wellsaw, Inc.
J.H. Williams & Co., Division of Snap-On Inc.
Whiteman & Barnes 3D Systems

Special thanks are to be given to the reviewers of this edition: Don W. Alexander, Wytheville Community College, Wytheville, VA; H. Allen Anthis, Columbia Basin College, Pasco, WA; Hans Boettcher, South Western Oregon Community College, Coos Bay, OR; Mohon S. Devgun, State University College at Buffalo, Buffalo, NY; David Sizemore, Monore County Technical Center, Lindside WV; John Spencer, British Columbia Institute of Technology, Burnaby British Columbia, Canada.

Evolution of Machine Tools

Through carbon dating, it has been estimated that human civilization originated millions of years ago in the valleys of the Tigris and Euphrates rivers in the Middle East. The first tools were natural implements such as sharp shells, jagged rocks, or splintered wood. The making of tools such as stone axes and spears and animal-bone or wooden clubs was responsible for the start of community living. It was the use of tools, made largely of stone, that satisfied the needs for food and shelter for primitive people during the Stone Age.

Sometime between 2500 and 2000 B.C., axes and spears made of stone were replaced by harder and more durable tools made of bronze, a mixture of copper and small amounts of tin. Tools could now be made to develop better tools, which in turn helped to improve the standard of living. During the Bronze Age, draft animals such as horses and oxen were used to supplement human power.

The Bronze Age gave way to the Iron Age, when it was discovered how to make iron by smelting iron ore. One of the earliest uses of iron was as ornamental articles for rulers and kings. As smiths unlocked the secrets of hardening and tempering, the use of iron and steel spread throughout civilization. Hard, sharp, flexible sword blades of carbon-rich, Damascus steel were made by skilled craftspeople.

Figures courtesy of (clockwise from far left): Wilkie Brothers Foundation, Wilkie Brothers Foundation, Wilkie Brothers Foundation, Cincinnati Milacron, Inc., Cincinnati Milacron, Inc., South Bend Lathe Corp.

U N I T 1

Machine Tools

Over the centuries, as new tools and machines were developed, it became possible for humans to produce better-quality goods faster and thereby improve their standard of living, Fig.1-1A. The capabilities of today's machine tools are beyond the wildest dreams of our ancestors. Today most machine tools are controlled by computers, which reduces the amount of manual skill required, but increases the need for operators capable of planning and programming. The manufacturing world of today and tomorrow will require personnel who have a good knowledge of machine tool operations and processes, the basics of computer numerical control programming, and an overview of the technological manufacturing processes.

OBJECTIVES

After completing this unit, you should be able to:

- Understand how the development of tools throughout history improved the life of humans.
- Identify the three categories of machine tools.
- Identify the space-age machine tools and processes developed during and since the second industrial revolution.

KEY TERMS

chip-producing machine tools

non-chip-producing machine tools

new-generation machine tools

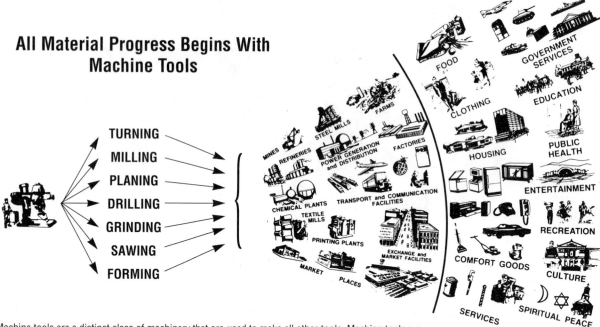

All Material Progress Begins With Machine Tools

TURNING
MILLING
PLANING
DRILLING
GRINDING
SAWING
FORMING

Machine tools are a distinct class of machinery that are used to make all other tools. Machine tools are man's "Robots" to do the first work operation, such as labeled and illustrated above.

Besides making the special machinery used for all types of manufacturing, machine tools are often employed to directly manufacture the parts used in production.

The machine tool industry is relatively small, but it is the key to all material progress.

Fig. 1-1A The development of machine tools has continually improved the standard of living of humans. (Courtesy of DoAll Co.)

● EVOLUTION OF MACHINE TOOLS

About 300 years ago, the Iron Age turned into the Machine Age and people began to develop and use sources of power other than human or animal strength. Water-power wheels were used for grinding grain, sawing wood, and even for manufacturing, as power-driven machines replaced hand-operated tools. While water, wind, and animal power continued to be used, they were gradually replaced by steam power.

The development of steam power was aided by machines that had originally been built for other uses. In order to make the cannon more accurate, the internal hole (bore) was finished with a screw-like drill to straighten and smooth the uneven cast hole. This boring machine enabled James Watt to develop the first steam engine in 1769. With the unleashing of this more-powerful energy, Watt made a new era in technology possible—the Industrial Revolution. As machines improved, new energy sources such as petroleum, Fig. 1-1B, which is the source of products such as gasoline, fuel oil, and lubricants, were developed. Electric generators as well as diesel and gasoline engines were developed. Improved

Fig. 1-1B Oil (petroleum) helped to power diesel and gasoline engines of all kinds, as depicted in this promotional drawing from the 1950s. (Courtesy of Shell Oil Co.)

generators and central power plants, designed by inventor Thomas Edison and others, supplied electricity to drive factory machines and light up office buildings and homes. Electricity did much to advance technical achievements and spur industrial growth.

Industrial growth and technical advancements resulted in increased productivity at the start of the twentieth century. During World War II, the need for greater production and more accurate machines became very important to support the armed forces.

Second Industrial Revolution

Since the 1950s, technological developments have made many changes in the way products are manufactured. Space-age computer technology made possible the development of calculators, robots, and automated machines and plants. Related fields of measurement and metallurgy became more precise and exacting sciences. Nuclear energy is now used in electrical power plants and to power submarines and ships. The ability to launch satillites into orbit and to explore outer space depends heavily on computer technology. Production machines, capable of producing parts to an accuracy of millionths of an inch, are now computer-controlled. Parts are welded, painted, and assembled by computer-controlled robots.

The computer, Fig. 1-2, plays an important role in engineering, allowing design engineers and drafters to produce complex drawings quickly and accurately with the help of computer-aided design (CAD) systems. Engineers no longer need to have parts and assemblies made before they can see if they work. "Working models" are now made on computers which can isolate, assemble, or magnify design parts for debugging and for simulating how they will function when they are manufactured.

The application of hydraulics and computerized numerical control devices has also done much to improve the standard machine tool. Through the use of new technologies, modern machine tools have become more reliable, accurate, and productive. All these factors play an important role in creating a higher standard of living.

● MACHINE TOOLS

Machine tools used in the metalworking industry generally fall into three categories:

1. **Chip-producing machine tools** which produce parts to size and shape by cutting away metal in the form of chips. Examples includes saws, lathes, and grinders. These types of machines, usually referred to as standard or conventional machine tools, can also include some types of Computer Numerical Control (CNC) machines.

2. **Non-chip-producing machine tools** which produce parts to form and size through a drawing, pressing, shearing, or punching action. Examples include punch presses and stamping machines. These types of machines are commonly used to produce parts made out of metal sheets.

3. **New-generation machine tools,** such as electro-discharge and laser machines, which form or cut parts to size and shape using either chemical or electrical energy.

● MACHINE TOOL PERFORMANCE

The performance of any machine tool is generally stated in terms of its metal-removal rate, accuracy, and repeatability, Fig. 1-3. *Metal-removal rate* depends upon the cutting speed, feed rate, and the depth of cut. *Accuracy* is determined by how precisely the machine can position the cutting tool to a given location once. *Repeatability* is the ability of the machine to return the cutting tool consistently to any given position.

Fig. 1-2 Computers assist engineers to produce complex drawings, and also prepare CNC machining programs. (Reprinted with the permission from and under the copyright of ©1995 Autodesk, Inc.)

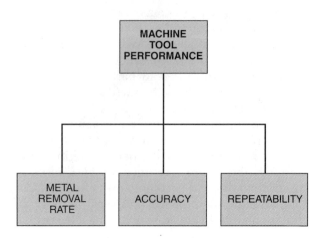

Fig. 1-3 The factors which determine the performance of any machine tool. (Courtesy of Kelmar Associates)

Fig. 1-4 Toolholders, which can hold different styles of tools, can reduce tool inventory and increase productivity. (Courtesy of Ingersoll Cutting Tool Co.)

The use of proper cutting tool technology is very important, especially on high-cost CNC machine tools. Improper or out-of-date tooling can be very expensive due to the loss of productivity that results from its use. The cutting tool is an important link in the manufacturing chain, and no one can afford a weak link in the metal-removal operation.

New cutting tools and toolholders have been developed in the last few years which are designed to:

1. **Remove metal fast and accurately.**
   The new high-positive, high-shear, indexable cutters for roughing operations and indexable insert cutters for semi-finishing provide good chip control, run at lower power, and are available in a wide range of insert grades, coatings, and shapes.

2. **Keep the tools running.**
   Use new-design tooling that combines round, square, and hexagonal inserts to reduce the number of toolholders required, Fig. 1-4. By using fewer inserts and toolholders, less time will be spent changing tools. Wherever possible, use high-speed heads, especially for finishing operations, to get more work out of the tools.

3. **Use operator labor wisely.**
   Use finishing tools that are identical so that new tool lengths and diameter offsets do not have to be reset when tools are changed. True-radius, true-diameter inserts can reduce finishing time for molds and eliminate many mistakes which could have required welding or repairing.

STANDARD MACHINE TOOLS

The most common machine tools found in machine shops are chip-producing machines. These machines can be grouped under the type of machining operations generally performed by each.

Sawing Machines

Sawing machines include machine tools which shape or cut metal by a sawing action. There are two types of metal-cutting saws in general use—bandsaws, which can be either vertical or horizontal, and cut-off saws with reciprocal motion. The vertical bandsaw, Fig. 1-5, can be used for cutting metal to size or shape. On this machine, work can be fastened to the table or guided by hand and brought into contact with a continuously revolving saw blade. The horizontal bandsaw and cut-off saw are primarily used for cutting workpieces to length.

Hole-Producing Machines

Hole-producing machines include those machines which produce straight holes in a workpiece or alter the shape of the hole through the use of various cutting tools. The most common drill press is the bench model, Fig. 1-6A, found in many machine shops and hobby shops. Other models, such as floor, multi-spindle, and CNC drill presses, find wide use for large workpieces and mass production. The most common drill press operations are drilling, reaming, coun-

Fig. 1-5 Vertical bandsaws are widely used to quickly cut material to approximate size and shape. (Courtesy of DoALL Co.)

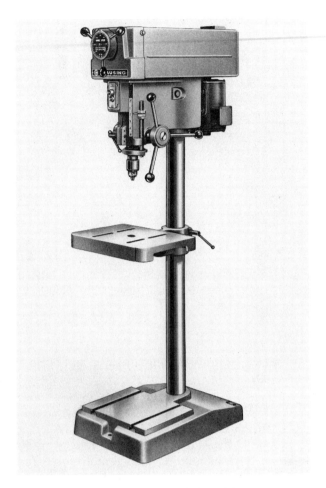

Fig. 1-6A The bench drill press is found in many machine shops and home hobby shops. (Courtesy of Clausing Industrial, Inc.)

tersinking, counterboring, spotfacing, and tapping, Fig.1-6B.

Turning Machines

Turning machines are used to produce round forms or shapes by bringing a stationary cutting tool into contact with a workpiece which is being held and revolved in a chuck or fixture. The engine lathe, Fig. 1-7A, is the most common turning machine in use today and is the fundamental tool of industry. Practically every modern mechanical invention and improvement has been developed through the use of engine lathes which produce many of the components for new machines. Other forms of turning machines include the turret lathe, single- and multiple-spindle automatic lathe, tracer lathe, vertical boring mill, and CNC turning and chucking center.

Some of the more common operations performed on lathes are turning, facing, form turn-

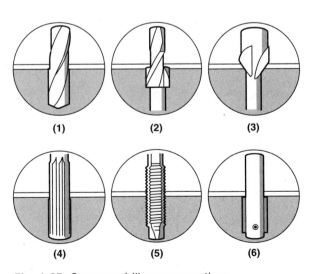

Fig. 1-6B Common drill press operations.
(1) Drilling (2) Counterboring (3) Countersinking
(4) Reaming (5) Tapping (6) Boring (Courtesy of Kelmar Associates)

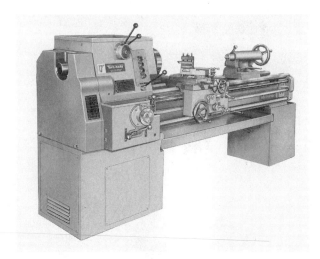

Fig. 1-7A The engine lathe is the most common turning machine that is used to produce round parts. (Courtesy of South Bend Lathe Corp.)

ing, tapering, screw-cutting, drilling, and boring, Fig. 1-7B.

Milling Machines

Milling machines are used to produce flat or formed surfaces on a workpiece held securely in some form of holding device or fixture. The machining operation is performed by using one or more revolving milling cutters that are brought into contact with the stationary workpiece. The milling machine is one of the most versatile machine tools in machine shops due to the wide va-

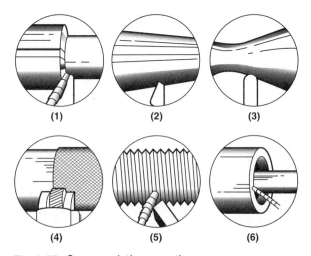

Fig.1-7B Common lathe operations.
(1) Turning (2) Tapering (3) Form turning
(4) Knurling (5) Threading (6) Boring (Courtesy of Kelmar Associates)

riety of work it can produce. The most common types of milling machines are the horizontal milling machine, Fig. 1-8A, and the vertical milling machine, Fig.1-8B. Other types of milling machines include the plain automatic, manufacturing, tracer-controlled, and the CNC machining center.

Some of the most common operations performed on milling machines are machining flat and contour surfaces, helical forms, gear teeth, drilling, boring, etc., Fig. 1-8C.

Grinding Machines

Grinding machines play a very important role in the finishing of workpieces to an accurate size and producing a high surface finish. The grinding process involves bringing a revolving abrasive wheel into contact with the surface of a workpiece and each abrasive grain on the wheel periphery removes a minute (small) chip of metal. Grinding wheels can cut very hard materials that would be difficult, if not impossible, to cut by any other process. New grinding machines and abrasives, such as superabrasives, have made it possible to consistently finish workpieces to tolerances of .0002 in. (0.005 mm) or less on high production runs.

The most common grinder found in many shops is the surface grinder, Fig. 1-9A, used for producing flat, angular, or contoured forms on a workpiece. Other types of grinders used in industry include the cylindrical, centerless, tool and cutter, and pedestal or bench. Some of the more common grinding operations include surface, cylindrical, cutter sharpening, tool grinding, and cut-off, Fig. 1-9B.

● CNC MACHINE TOOLS

Numerical control is a process where a machine tool can be controlled and guided through a series of coded instructions, consisting of letters, numbers, and symbols, to carry out a machining operation quickly and accurately. Computer Numerical Control (CNC), developed in the late 1960s and early 1970s as an offshoot of Direct Numerical Control (DNC), uses a minicomputer built into the machine's control system. CNC machine tools, Fig. 1-10, have revolutionized the metal-removal process by greatly increasing manufacturing productivity and making it possible to produce more accurate workpieces than was previously possible. These CNC machines have been able to reduce manufacturing costs

Fig. 1-8A Horizontal milling machines are used to produce flat or formed surfaces with the use of milling cutters. (Courtesy of Cincinnati Milacron Inc.)

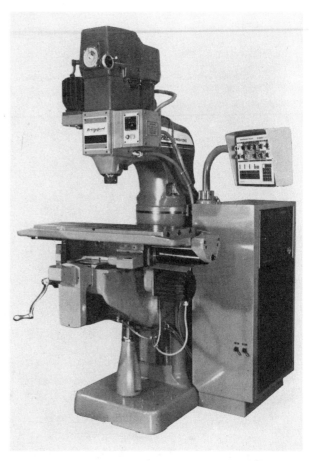

Fig. 1-8B The vertical milling machine is one of the most versatile machines used in industry. (Courtesy of Bridgeport Machines, Inc.)

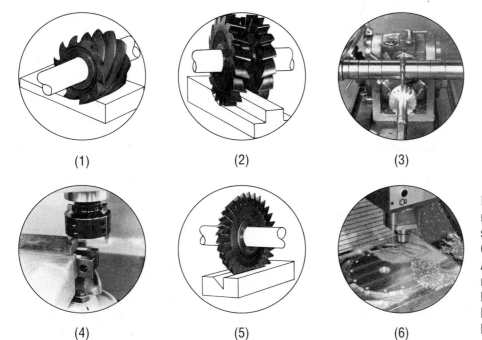

(1)

(2)

(3)

(4)

(5)

(6)

Fig. 1-8C Common milling machine operations. (1) Flat surface (2) Gang milling (3) Gear cutting (4) Flycutting (5) Angular milling (6) Circular milling (Courtesy #1, 2, 5 Niagara Cutters, Inc., #3, 4 Kelmar Associates, #6 Deckel-Maho Inc.)

and produce parts of consistent quality. The continual development of CNC machines and their computer control systems has led to wide acceptance of this technology in industry. As a result, almost 90 percent of the machine tools manufactured in the world today have some form of computer control system.

● SPECIAL MACHINE TOOLS

Special purpose machine tools are designed to perform specific operations on a single component or do all the operations to complete the part. Some examples of special purpose machine tools are: gear-generating machines; centerless, cam, and thread grinders; turret lathes; and automatic screw machines. The introduction of electro-discharge machining, electro-chemical machining, and electrolytic grinding, made it possible to machine materials and produce shapes that were difficult or impossible to produce by conventional methods.

● COMPUTER-AIDED MANUFACTURING

The flexibility and reliability of the computer have led to its application in many machining and manufacturing-related areas. Computer-Aided Design (CAD) has revolutionized the process that drafters and tool engineers use to design and test new products, Fig. 1-11. The ability of Computer-Aided Manufacturing (CAM) to accurately control the entire operation

Fig. 1-9A The surface grinder is used to bring flat and formed surfaces to size and produce a fine surface finish. (Courtesy of DoALL Co.)

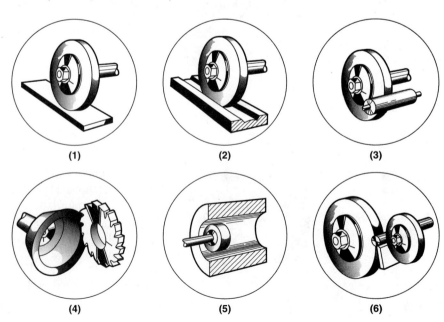

(1)   (2)   (3)

(4)   (5)   (6)

Fig. 1-9B Common grinding operations. (1) Surface (2) Form (3) Cylindrical (4) Cutter (5) Internal (6) Centerless grinding. (Courtesy of Kelmar Associates)

of machine tools has greatly improved productivity. During the 1980s, the introduction of Computer-Integrated Manufacturing (CIM) saw computers become involved in all aspects of the manufacturing process including designing the product, financing, purchasing, manufacturing, marketing, servicing, etc.

With the development of manufacturing cells and systems, better machine tools, robots, in-process gaging, and advanced cutting tool materials, productivity has increased tremendously over what was possible with standard machine tools. Many products can now be produced automatically with a continuous flow of finished parts from these numerical control systems and the use of Just-In-Time (JIT) manufacturing methods. With reliable quality control, high production rates, and reduced manufacturing costs, it is possible to enjoy the pleasures and conveniences of automobiles, power lawn mowers, automatic washers, microwave ovens, stereo and laser music systems, and scores of other products.

Fig. 1-10 A CNC machining center provides high productivity with speed and accuracy.

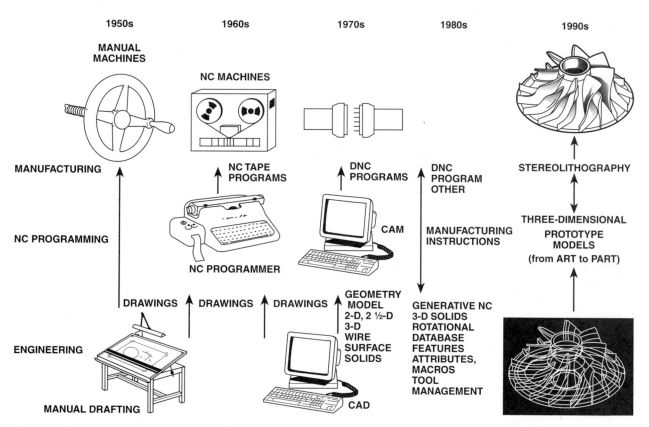

Fig. 1-11 The combination of CAD and CAM has provided industry with valuable tools for increasing productivity. (Courtesy of Ingersoll Milling Machine Co., and 3D Systems)

SUMMARY

- The gradual development of more-sophisticated tools and machines over thousands of years has made it possible for humans to greatly improve their standard of living.

- Steam energy was literally one of the driving forces of the Industrial Revolution, and its use would not have been possible without the development of tools and machines.

- The second industrial revolution, which began in the 1950s, initiated the widespread use of computers in industrial operations. Since then, technological developments have included automated machines, robots, CAD/CAM systems, and CNC devices.

- Technologies to cut metal to size or shape are developing at a rapid rate.

- Machine tool performance is measured in terms of metal-removal rate, accuracy, and repeatability.

- The three major classes of machine tools are chip-producing, non-chip-producing, and new-generation.

KNOWLEDGE REVIEW

1. What was one of the first uses of iron?

2. Name four sources of power developed during the period when machine tools evolved.

3. List the three categories of machine tools used in metalworking.

4. What three factors determine the performance of any machine tool?

5. State three purposes for which new cutting tools and toolholders were designed.

6. Name two machines that are examples of each of the following:

   a. sawing machines

   b. hole-producing machines

   c. turning machines

   d. milling machines

   e. grinding machines

7. What are three manufacturing benefits of CNC machine tools?

8. Name three areas, other than CNC, where the use of the computer has improved manufacturing processes.

REWARDING CAREER

MANUFACTURING TECHNOLOGY

PROFESSION

MANUFACTURING ENGINEER

TECHNOLOGIST

TECHNICIAN

TOOL AND DIE MAKER

MACHINIST

APPRENTICE

Careers in Manufacturing

The development of new technologies, mainly due to the use of computers in the machine tool trade, has created many opportunities for those who seek a career in the world of manufacturing. The rapidly changing technology in the machine trade will require a person to continually keep up to date with the new developments and the ever-increasing use of computers in manufacturing. Bright young people who are precise and do not hesitate to assume responsibility are always in demand by industry. One of the secrets to success is being able to do a job to the best of one's ability and to never be satisfied with shoddy or inferior workmanship. Continual improvement in product quality and manufacturing processes must be the goal of everyone associated with machining/manufacturing technology. The aim should be to always produce the best-quality product in the shortest possible time in order to be competitive with domestic and foreign products in a global economy.

U N I T

2

Machine/Manufacturing Careers

Education and training in manufacturing-related courses are the key to our survival as a manufacturing nation. The key people who will succeed in manufacturing careers require versatility and willingness to keep up with the rapidly changing technology. Manufacturing programs must be in every school curriculum to expose students to the exciting, challenging, and high-paying jobs available to them in manufacturing.

OBJECTIVES

After completing this unit, you should be able to:

- Understand the effect of a strong manufacturing base on the economy of any country.
- Identify exciting new manufacturing technologies available as careers.
- Identify requirements for careers in the machine tool trade.

KEY TERMS

apprentice   general shop   job shop
production shop

TRADE OPPORTUNITIES

The wealthiest countries in the world have always used the latest technological developments and the most modern machine tools to develop a strong manufacturing base. This allowed them to produce high-quality products at the lowest manufacturing costs and increase their share of the world market. A strong manufacturing base is the core of business, the economy, and the standard of living which people enjoy. Most third world countries do not have a strong manufacturing base, and as a result, their standard of living is lower.

SCHOOLS AND INDUSTRY

The school system has an important role to play in preparing students to become the manufacturing personnel and leaders of tomorrow. It must foster a value system that will attract the brightest and the best students into manufacturing-related courses. Schools and industry should combine their efforts to provide continual updating of courses in manufacturing technology in order to keep pace with new developments.

Machine Shop, the basis of all manufacturing, is the prerequisite that best prepares students for careers in many of the exciting manufacturing technologies available today, Fig. 2-1. A number of these technologies are listed as follows:

- **Artificial Intelligence (AI)** technology is finding many manufacturing applications in the fields of expert systems, machine vision systems, artificial vision, robotics, natural language understanding, and voice recognition.

- **Computer-Aided Design (CAD),** which has almost replaced handmade drawings, uses computers to accurately produce and revise complex engineering drawings on a video screen. The data used in making the drawing can also be used to produce the CNC program to machine the part.

- **Computer-Aided Manufacturing (CAM)** uses the technologies of CAD and CAE (Computer-Aided Engineering) to produce a computer model of the manufacturing process. It also provides the machine codes needed to control the various production machines, material handling equipment, and control systems.

- **Computer-Integrated Manufacturing (CIM)** uses a central host computer to control an entire network of computers involved in all phases of the manufacturing operation. The ability of the CIM factory to oversee all aspects of the factory results in a new concept in inventory control and manufacturing scheduling which is called Just-In-Time (JIT) manufacturing.

- **Flexible Manufacturing Systems (FMS)** generally consist of a group of CNC machines, linked together with an automated material handling system, under the supervision of one or more dedicated supervisory (executive) computers. They can randomly process a group of different parts or part families, and the computer adjusts the system automatically to changes in part production, mixes, and levels of output.

- **Group Technology (GT)** is a concept of classifying parts (workpieces) into part families on the basis of their similarities in physical characteristics and manufacturing processes. The aim of Group Technology is to maximize the use of equipment and labor, keep the work-in-process (WIP) to a minimum, and reduce manufacturing costs, even in low-volume situations.

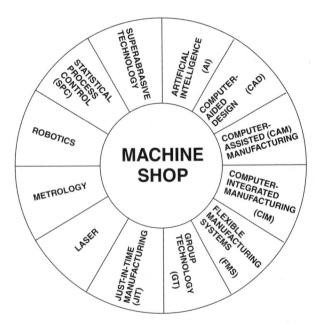

Fig. 2-1 Machine Shop provides the basic groundwork for all manufacturing technology careers. (Courtesy of Kelmar Associates)

- **Just-In-Time (JIT)** manufacturing was developed to assist manufacturers to improve productivity, reduce costs, overcome the shortage of machines, reduce finished-goods inventories and work-in-process, and utilize manufacturing space efficiently. Just-In-Time manufacturing requires precise scheduling to create a smooth, continuous flow of parts and material to the proper work stations at the time when they are needed; not too early, not too late.

- **Lasers** (**L**ight **A**mplification by **S**timulated **E**mission of **R**adiation) use an intense beam of light which can be focused to pinpoint accuracy, to perform manufacturing operations such as measurement and inspection, metal cutting and welding, metal cladding (hard facing or surfacing), drilling, hardening, and some types of machining on lathes and milling machines.

- **Metrology** or the science of precision measurement, is a vital part of the manufacturing process since it provides a means of determining the quality of the part manufactured. With modern electronic and optical measuring/ inspection tools and systems, it is possible to reduce or eliminate errors from the manufacturing process in order to produce the highest quality parts at the lowest possible cost.

- **Robotics,** a form of programmable automation made possible by numerical control technology, is a rapidly growing manufacturing technology that is used to automate many repetitive production operations. Robots are programmable, multifunction manipulators designed to move material, parts, tools, or special devices through various motions to perform an operation, or a series of operations, quickly and accurately.

- **Statistical Process Control (SPC)** procedures are used to ensure that a manufacturing process is producing parts that meet the user's specifications. *Inspection data*, which can be collected manually or through automated quality control systems, is analyzed. This data, which shows variations in the manufacturing operation, allows corrective action to be taken early so that the operation comes as close as possible to *zero defect* manufacturing.

- **Superabrasive (SA) Grinding and Machining** technology uses diamond and cubic boron nitride (CBN) cutting tools and grinding wheels to increase manufacturing productivity. These high-technology tools have revolutionized metal-removal processes, increased productivity, produced high-quality parts, and lowered manufacturing costs.

● TYPES OF MACHINE SHOPS

Machinists may qualify to work in a variety of machine shops. The three most commonly found types are general, production, and jobbing shops.

A **general shop,** sometimes called a maintenance shop, is usually associated with a manufacturing plant, a school lab, or a foundry. A machinist in a general shop must be able to operate all machine tools and is called upon to make and replace parts for all types of work-holding devices, cutting tools, and production machinery.

A **production shop,** usually associated with a large plant or factory, makes many types of machined parts. These may include shafts, pulleys, bushings, motors, and sheet metal parts. Machinists working in a production shop usually operate one type of machine tool and generally produce identical parts, Fig. 2-2.

Fig. 2-2 A production machinist usually operates only one machine producing identical parts.

A jobbing shop, often called a **job shop,** is generally equipped with a variety of standard and CNC machine tools. A job shop usually does not have a product of its own to manufacture; it does a variety of work for other companies. These jobs may include the fabrication of jigs, fixtures, dies, molds, or short-run production of special parts. Any person working in a job shop is usually a qualified machinist or toolmaker and should be able to operate most machine tools and measuring equipment.

● ENTERING THE WORKPLACE

Technical and vocational schools provide basic skills and knowledge about many areas of technology. In the past, the industrial age stressed the importance of physical skills in manufacturing. The information age of today requires more mental skills because we are encountering more and more knowledge-intensive technology. It is important for students choosing a career in manufacturing to get a good background in machine operations and processes.

Most manufacturing careers start with some form of apprenticeship, and where they lead is limited only by the initiative and determination of each individual. The person most likely to succeed in the manufacturing world is one who realizes the need for *life-long learning* in order to keep up with the ever-changing technology. A sampling of job descriptions and requirements follows:

Apprenticeships

An **apprentice,** Fig. 2-3, is one who is employed, during a learning period, under the guidance of skilled tradespeople for the purpose of learning a trade. The apprenticeship program, set up with the cooperation and approval of company and trade union, is conducted under the supervision of the U.S. Department of Labor. Such a program usually takes two to four years (4000 to 8000 work hours) and includes both on-the-job training and related theory or classroom work.

Requirements for entry into an apprenticeship program are demanding. It is to the advantage of the applicant to have graduated from high school or have an equivalent education. It is also desirable for the applicant to have good mechanical abilities and a background in science, mathematics, English, and mechanical drawing.

Fig. 2-3 An apprentice works under the guidance of a skilled machinist or tool and diemaker to learn the trade. (Courtesy of Fadal Engineering Co., Inc.)

A journeyman's certificate is granted upon completion of an apprenticeship program. This certificate should not be considered as the end of the learning process, but rather as the beginning. Opportunities in the future are limited only by the initiative and interest of each individual. It is quite possible for an apprentice to eventually become an engineer.

Machine Operator

Machine operators, Fig. 2-4, are workers who are classified as semi-skilled tradespeople and usually operate only one machine. They are limited in responsibility, and are usually rated and paid according to their skill, knowledge, and job classification. A Class "A" operator should be able to operate the machine as well as:

- interpret prints and drawings,
- use precision measuring tools,
- calculate cutting speeds and feeds,
- make machine setups,
- adjust cutting tools.

Due to continued advancements in technology and computer numerical control machines, the number of machine operators required in the future will decline. However, machine tool operators who continue to upgrade themselves with

Fig. 2-4 A machine operator is a person who is skilled at running only one machine. (Courtesy of Kelmar Associates)

Tool and Diemaker

A tool and diemaker must possess the knowledge and skills to make different types of tooling including dies, molds, cutting tools, jigs and fixtures. These tools may be used in mass production of metal, plastic, or other parts. In order to make a die, the tool and diemaker must be able to select, machine, and heat-treat the steel for the die components. In mold making, the tool and diemaker must know the type of plastic used for the part, the surface finish required, and the heat-treating process for the mold components.

In order to qualify as a tool and diemaker, a person must complete the apprenticeship program, have several years of practical experience, and have above-average mechanical aptitude. The person should also possess a good knowledge of machine shop mathematics and machining processes, and be able to interpret shop drawings and prints. Tool and diemakers should also have a good understanding of metallurgy, heat-treating, computers and computer-age manufacturing processes.

CNC Machine Operator

With the wide acceptance of CNC, machines can be programmed to make parts quickly and accurately, each part being an exact copy of the previous part. It is the job of a CNC machine

advanced technical knowledge through courses and seminars can become programmers and operators of CNC turning centers, machining centers, CNC robots, etc.

Machinist

Machinists, Fig. 2-5, are considered skilled workers who are capable of operating all standard machine tools found in a machine shop. They must be able to read technical drawings and use the latest precision measuring instruments. Machinists should also have a basic knowledge of numerical control programming and machining processes. A good technical background, which allows them to perform any bench, layout, or machine tool operation and includes knowledge about mathematics, metallurgy, and heat treating, is essential for a machinist.

Fig. 2-5 A machinist should be able to operate and perform operations on all conventional machine tools. (Courtesy of DoALL Co.)

operator to see that CNC machine tools are properly set up and used in order to achieve the maximum benefit from them, Fig. 2-6.

The duties of a CNC machine operator will vary. In some shops, a person may operate only one machine, while in other shops, a person may operate more than one machine at a time. Generally, the work assignments come in written form from a supervisor and will include:

- loading the program into the machine control unit (MCU),
- securing all the tools required,
- positioning the workpiece,
- checking the level of coolants and lubricants.

Once a job has been set up properly and the program tested and corrected if necessary, the operator should only have to monitor the machine as it operates.

CNC Programmer

Programmers for CNC machine tools must be thoroughly familiar with all types and uses of cutting tools, work fixtures and setups, and machining processes. In order to program a machine tool to produce a part, a programmer must be able to:

- read and interpret work drawings,
- select the best cutting tools for each machining operation,
- calculate speeds and feeds required for different materials,
- understand machine tool processes,
- calculate production costs.

Programming background and knowledge may be obtained by attending vocational schools, technical schools, colleges, and universities that offer CNC programming courses.

Technician

A technician, Fig. 2-7, is a person who has completed high school and has at least two years of post-secondary education at a technical institute, community college, or university. A technician, who works at a level between the machinist and engineer, must have a good knowledge of drafting, mathematics, machining processes, and technical writing.

The development of CNC machining centers, turning centers, electro-discharge machines,

Fig. 2-6 The CNC machine operator should be able to prepare the computer program, set up the machine, and produce the finished part. (Courtesy of American CNC)

Fig. 2-7 A technician works at a level between a machinist and an engineer. (Courtesy of Fadal Engineering Co., Inc.)

robots, etc., has provided many job opportunities for technicians. Technicians are usually trained in only one area of technology, which could include manufacturing, electrical, metallurgy, or machine tools. A machine tool technician requires a working knowledge of industrial machine and manufacturing processes in order to recognize and recommend the best method of manufacturing a product. A technician may qualify for a technologist's position after at least one year of on-the-job training with a technologist or an engineer.

Technologist

A technologist holds a position between that of a technician and a graduate engineer. Technologists are generally three- or four-year graduates of technical or community colleges, and their studies will have included advanced mathematics, physics, chemistry, computer programming, engineering graphics, business organization, and management.

Engineering technologists may perform many tasks previously done by an engineer. These include production planning, laboratory experiments, and supervision of technicians. Technologists may be employed in other management positions such as cost and quality control, production control, labor relations, employee training, and product analysis. After completing further university studies, a technologist can become an engineer by passing the required examinations.

Tool and Manufacturing Engineers

Tool and manufacturing engineers are generally responsible for the design and development of a new product. They are also responsible for directing and coordinating the processes required to run a manufacturing plant. These processes can include setting company standards, determining production methods, and establishing quality control. Tool and diemakers can become tool and manufacturing engineers by taking upgrading courses provided by the Society of Manufacturing Engineers (SME) or enroling in a college or university program offering industrial or mechanical engineering, or manufacturing technology. Required courses should include advanced mathematics, manufacturing processes, statistical process control, technical writing,

Fig. 2-8 The new manufacturing technologies offer an exciting, challenging, and rewarding career in teaching. (Courtesy of Robert K. Throop, *Reaching Your Potential*, 1993, Delmar Publishers)

physics, and engineering sciences. Tool engineers should be able to suggest and design the best and least-expensive method of producing a product quickly and accurately.

Professions

People having many years' experience in manufacturing may consider a career in teaching, a very rewarding and challenging profession, Fig. 2-8. To qualify for teaching, a person generally requires a number of years of experience in the machine tool trade along with graduation from a teacher training program.

Engineers are responsible for the design and development of new products produced for the world marketplace. They must also develop production methods, redesign, and improve existing products. Most engineers specialize in a certain area of engineering such as electronics, aerospace, electrical, metallurgical, or mechanical.

A degree in engineering will not only allow a person to enter the profession, but positions in educational administration are also available for those with this type of background. Some classes of engineers progress through practical experience and on-the-job training programs. This applies to tool engineers who come up through the trade and upgrade themselves through courses and seminars and qualify by passing the required examinations.

SUMMARY

- Persons in manufacturing careers will need to engage in continual learning to keep up with ongoing developments in technology.

- A strong manufacturing base is necessary for a country to have a high standard of living and to compete in world markets.

- A modern machine shop can use a wide variety of new technologies and methods to produce parts accurately and efficiently.

- The most common types of machine shops are general, production, and jobbing shops.

- Most workers will enter skilled trades through an apprenticeship, an extended period of learning, and on-the-job training under close supervision. Qualifications for an apprenticeship applicant include a high-school diploma or its equivalent. Other desirable factors include good mechanical aptitude and an educational background that includes science, and mathematics, English, and mechanical drawing.

- There are many different types of jobs available in the machine trades. Depending on his or her interests and abilities, a person can choose from a variety of career paths.

KNOWLEDGE REVIEW

1. Why is a strong manufacturing base important to any country?

2. List four of the most important manufacturing technologies for which machine shop provides the basic groundwork.

3. What type of skills were/are important in the:

   a. industrial age?

   b. information age?

4. What should a person expect after successful completion of an apprenticeship program?

5. Compare a machinist and a machine operator.

6. Describe a job shop.

7. How can a machinist qualify to become a tool and diemaker?

8. What knowledge should a person have to become a successful CNC programmer?

9. What four qualities should a technician possess?

10. How can a tool and diemaker become a tool and manufacturing engineer?

11. What qualifications are required for a person to become a technical teacher?

SECTION
3

Manufacturing Process Planning

The manufacturing world is changing so rapidly that we not only must compete with companies in our country but also with companies throughout the world. It is becoming a race between countries to see which can compete most effectively for a greater share of the world market. Companies that intend to survive must update their manufacturing processes in order to be more productive and become *world-class manufacturers.* Manufacturing processes of twenty to thirty years ago cannot compete effectively in the high-technology world in which we live today. The countries that will be the leaders in the world are those that combine the most advanced manufacturing technology with the best human resources to produce the highest quality goods at the lowest cost.

Photo courtesy of AMT–The Association for Manufacturing Technology.

UNIT 3

Technical Drawings/Prints

A *technical drawing,* sometimes referred to as a drawing or print, is the language used by a draftsperson, tool designer, and engineer to communicate the physical requirements of a part to the machinist and toolmaker. These drawings may consist of a variety of lines—thick, thin, broken, or solid—that have different meanings but help a person to read and understand the drawing. It is with the help of those lines that surfaces, edges, and contours of a workpiece are described. The addition of dimension lines, sizes, symbols and word notes allows the draftsperson to indicate the exact material, heat treatment, plating, and other specifications of each individual part. The student must have a good knowledge of this visual language in order to understand the drawing and produce parts quickly and accurately.

OBJECTIVES

After completing this unit, you should be able to:

- Recognize the various types of drawings/views and how they are used.
- Recognize the various lines and symbols used on engineering (technical) drawings.
- Read working prints in order to make parts to the exact size and shape required.

KEY TERMS

allowance	basic size	fit
limits	orthographic view	scale size
section view	surface finish	tolerance

● DRAWINGS AND PRINTS

An *assembly drawing*, Fig. 3-1, is made by a draftsperson for the purpose of showing how the individual components of a product are assembled. Assembly drawings, which have a minimum of dimension and detail, include assembly instructions, part numbers for each component, and overall dimensions, etc., of the part, unit, or machine shown. A *detail drawing*, Fig. 3-2, is used to show each single part with a complete and exact description of its shape, dimensions, tolerances, surface finish, etc. Names and part numbers of adjoining or connecting units should also be shown on the detail drawing. This is the working drawing which is used by the machinist or toolmaker to produce the parts that eventually become a part of, or, the complete product.

Dimensioning Systems

Dimensions are used on drawings to give the distance between two points, lines, planes, or some combination of points, lines, and planes. The numerical value gives the actual measurement (distance), the dimension line indicates the direction in which the value applies, and the arrowheads indicate the points between which the value applies.

There are three main systems for showing dimensions on technical drawings:

1. The *common-fraction system* has dimensional values written as units and common fractions, for example, 1/8, 3/4, 1-1/2, 4-3/16, etc.

2. The *common-fraction, decimal-fraction system* is used primarily for machine shop

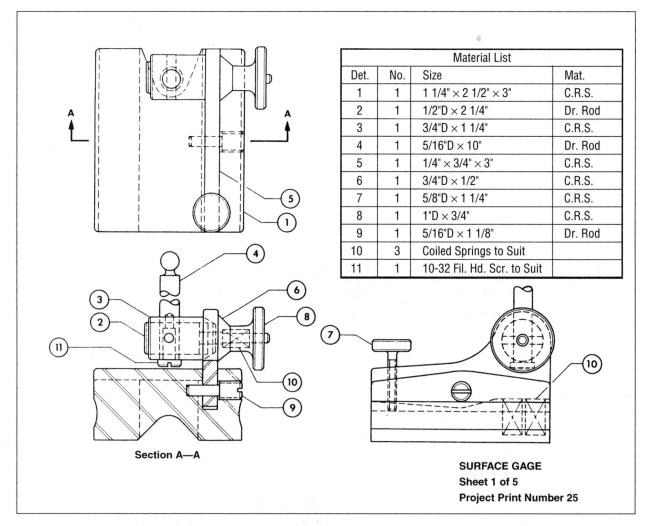

Material List			
Det.	No.	Size	Mat.
1	1	1 1/4" × 2 1/2" × 3"	C.R.S.
2	1	1/2"D × 2 1/4"	Dr. Rod
3	1	3/4"D × 1 1/4"	C.R.S.
4	1	5/16"D × 10"	Dr. Rod
5	1	1/4" × 3/4" × 3"	C.R.S.
6	1	3/4"D × 1/2"	C.R.S.
7	1	5/8"D × 1 1/4"	C.R.S.
8	1	1"D × 3/4"	C.R.S.
9	1	5/16"D × 1 1/8"	Dr. Rod
10	3	Coiled Springs to Suit	
11	1	10-32 Fil. Hd. Scr. to Suit	

Section A—A

SURFACE GAGE
Sheet 1 of 5
Project Print Number 25

Fig. 3-1 An assembly drawing shows how the individual parts or components go together to make a final product. (Courtesy of National Tooling & Machining Association)

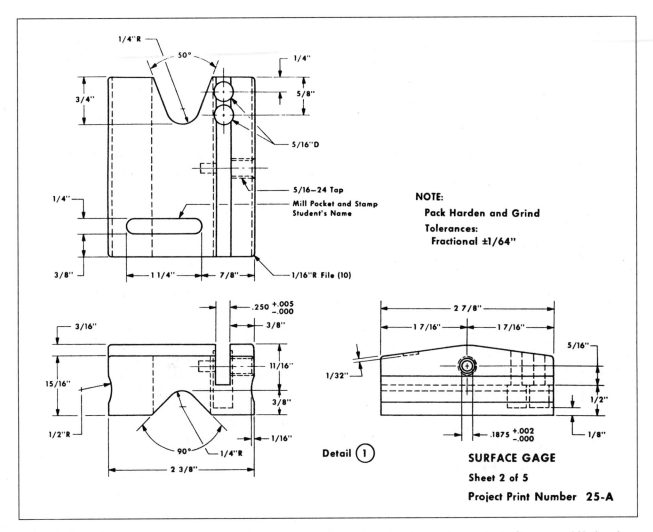

Fig. 3-2 A detail drawing shows the exact size and shape of each individual part of a product. (Courtesy of National Tooling & Machining Association)

drawings. The common fraction in units and fractions is used for distances not requiring an accuracy of less than 1/64 inch (in.). The decimal fraction, in units and decimals, is used for distances requiring greater precision, for example, 2.750, .125, etc.

3. The *complete decimal system,* used for machine shop and CNC work, uses only decimal fractions for all dimensional values. In CNC work, two types of dimensioning are used:

(a) *Incremental system* where all dimensions are given from a previously known point, Fig. 3-3.

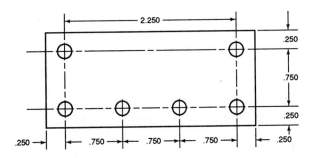

Fig. 3-3 In incremental dimensioning, all sizes are given from the previous point or location. (Courtesy of Kelmar Associates)

(b) *Absolute system* where all dimensions or positions are given from a fixed zero or origin point, Fig. 3-4.

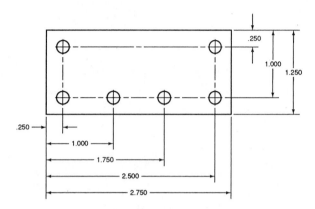

Fig. 3-4 In absolute dimensioning, all sizes are given from one zero or reference point. (Courtesy of Kelmar Associates)

Types of Drawings

One type of drawing may not be enough to adequately describe all parts or objects. Parts may have special or unusual features that are hard to illustrate. As a result, different types of drawings or views have been developed to meet these needs.

Orthographic Views

To describe the shape and size of an object accurately on a drawing or print, the draftsperson may use the orthographic view or projection method. The **orthographic view** shows the object

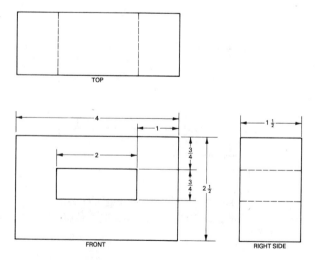

Fig. 3-5 Technical drawings, made with the orthographic projection method, generally include the top, front, and right-side views. (Courtesy of Kelmar Associates)

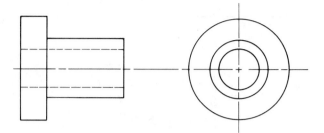

Fig. 3-6 Cylindrical parts are generally shown with the front and right-side views. (Courtesy of Kelmar Associates)

as seen from three different views—the top, front, and right-side view, Fig. 3-5. Each view may be seen by looking directly perpendicular (90°) to each of the three surfaces. These three views enable the drafter to completely describe the part or object accurately so that the machinist knows exactly what is required.

Cylindrical parts are generally drawn on prints in two views—the front and right-side views, Fig. 3-6. If a part contains many details, a third view may be used to accurately describe the part to the machinist.

Section Views

Complicated interior forms are often difficult to describe by a draftsperson in the usual manner. These interior details may be shown more clearly on a **section view.** This involves making an imaginary cut through the object to produce a view of the cross section. This imaginary cut can be made in a straight line in any direction, drawn as though part of the object were cut away to show the details of the inside surfaces, Fig. 3-7.

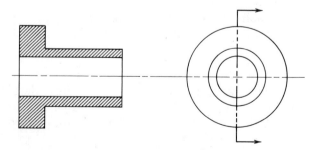

Fig. 3-7 Section views are used to show the interior form of an object that could not be shown clearly by conventional methods. (Courtesy of Kelmar Associates)

Types of Lines

Various standard line widths and styles are used on engineering drawings by the designer to precisely specify what is required. Heavy, light, medium, and broken lines are among the types of lines used on shop or engineering drawings. The standards for these lines were developed by the American National Standards Institute (ANSI) and are known as the *alphabet of lines.* See Table 3-1 for an example, description, and purpose of some of the more common lines used on technical drawings.

Drafting Terms

Common drafting terms and symbols are used on shop and engineering drawings to help the designer describe each part as accurately as possible. If these widely accepted terms, symbols, and abbreviations were not available, designers would have to use extensive notes to describe exactly what is required. These notes would make drawings difficult to read, easy to misunderstand, and could lead to errors in the machined part. Some of the more commonly used drafting terms and symbols are explained as follows:

- **Basic size** is the size from which limits of size are based when applying allowances and tolerances.
- **Allowance,** Fig. 3-8, is the intentional difference in the sizes of mating parts, such as the diameter of a shaft and the size of a hole. On a shop drawing, both the hole and shaft sizes are marked with the maximum and minimum dimensions permissible in order to produce the best fit between mating parts.
- **Fit** is indicated by the range of tightness between two mating parts. There are two general classes of fits.

Table 3-1   Standard lines used on technical drawings

Example	Name	Description	Use
A	Object lines	Thick black lines approximately $\frac{1}{32}$-in. wide (width may vary to suit drawing size).	Indicate the visible form or edges of an object.
B	Hidden lines	Medium-weight black lines ($\frac{1}{8}$-in. long dashes and $\frac{1}{16}$-in. spaces).	Indicate hidden contours of an object.
C	Center lines	Thin lines with alternating long lines and short dashes. Long lines $\frac{1}{2}$- to 3-in. long. Short dashes $\frac{1}{16}$- to $\frac{1}{8}$-in. long, spaces $\frac{1}{16}$-in. long.	Indicate centers of holes, cylindrical objects, and other sections.
D	Dimension lines	Thin black lines with arrowhead at each end and a space in the center for a dimension.	Indicate dimensions of an object.
E	Cutting-plane lines	Thick black lines, a series of one long line and two short dashes. Arrowheads show line of sight from where section is taken.	Show imagined section cut.
F	Cross-section lines	Fine evenly spaced parallel lines at 45°. Line spacing is in proportion to the part size.	Show surfaces exposed when a section is cut.

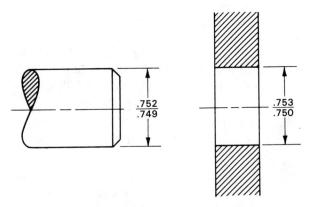

Fig. 3-8 Allowance shows the largest and smallest dimensions of mating parts. (Courtesy of Kelmar Associates)

Clearance fit, the space a part may rotate or move in relation to a mating part.

Interference fit, the condition in which the part (shaft) is larger than the internal part (hole). The two parts must be forced together in assembly.

- **Limits,** Fig. 3-9, represent the largest and smallest permissible sizes of a part (the maximum and minimum dimensions), both sizes being given on a technical drawing.

  EXAMPLE .626 largest dimension
  .624 smallest dimension

- **Scale size** is used on most shop or engineering drawings to indicate the ratio of the drawing size to the actual size of the part. It is often necessary to enlarge small parts for clarity and to have room for dimensions and other details. Large objects are often drawn at a reduced scale in order to get the necessary information to fit a convenient size sheet of paper. The scale is generally found in the title block of a drawing and indicates that the drawing is in a certain proportion to the size of the workpiece.

> The dimensions shown on the drawing give the correct size of the part required. The actual drawing should never be measured to determine the size to be machined.

Scale	Definition
1:1	Drawing made to actual size of part (full scale).
1:2	Drawing made one-half actual size of part (one-half scale).
2:1	Drawing made twice actual size of part (twice or double scale).

- **Tolerance,** Fig. 3-10, is the permissible variation of the size of a part. A drawing generally gives the basic size along with a plus or minus to show the variation allowed.

  EXAMPLE .625 +.001
  −.003

  The tolerance in this case would be .004 (the difference between +.001 oversize and −.003 undersize).

Drafting Symbols

Symbols and abbreviations are used on shop drawings to tell a machinist the surface finish, type of material, roughness, and special details of the part.

Surface Finish Symbols

The deviation from the nominal surface caused by the machining operation is indicated by a **surface finish** symbol, generally resembling a square root sign or a large check mark. The information around this symbol, which is generally located on the surface of the part, may

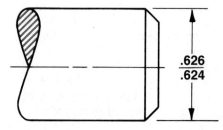

Fig. 3-9 Limits represent the largest and smallest permissible sizes of a part. (Courtesy of Kelmar Associates)

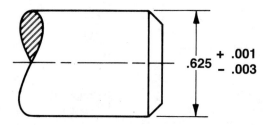

Fig. 3-10 Tolerance is the permissible variation of the size of a part. (Courtesy of Kelmar Associates)

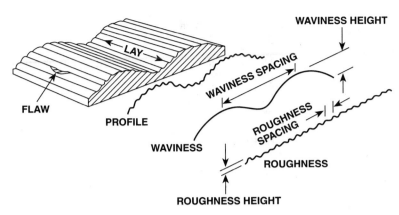

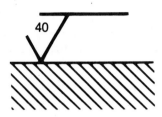

Fig. 3-11A A pictorial explanation of surface finish terminology. (Courtesy of *Manufacturing Engineering Magazine*)

Fig. 3-11B Surface finish symbols are used to indicate the type of finish required on the surface of a part. (Courtesy of Kelmar Associates)

include waviness, roughness, lay, and flaws specifications required for the work surface, Fig. 3-11A. These surface characteristics can be measured by a surface finish indicator in microinches. The number inside the symbol indicates the quality of surface finish, Fig. 3-11B. In this case, the roughness height or the measurement of the finely spaced irregularities caused by the cutting tool cannot be greater than 40 microinches, Fig. 3-11B.

When the surface of a part must be finished to exact specifications, each portion of the surface finish symbol must be complete, Fig. 3-12, as follows:

EXAMPLE

30	surface finish in microinches
.0015	waviness height in thousandths of an inch
.002	roughness width in thousandths of an inch
=	machining marks perpendicular to the surface boundary

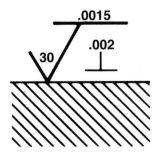

Fig. 3-12 Surface finish specifications required on the top surface of the part. (Courtesy of Kelmar Associates)

SUMMARY

- Technical drawings are used to communicate the physical requirements of a part. They include pictorial views, dimensions, symbols, data, and notes.

- A detail drawing shows an individual part and gives its complete description. An assembly drawing is less detailed and shows how a number of different parts are to be fitted together or assembled.

- Special views may be needed to accurately and completely show a part on a drawing. Examples include orthographic views and section views.

- Dimensions are used to specify sizes and distances and give the actual dimensions of the part.

- The scale size shows how the drawing size of a part relates to the part's actual size.

- Drafting terms and symbols are used to further describe parts as accurately and completely as possible. These can include basic size, allowance, fit, limits, tolerance, and surface finish.

KNOWLEDGE REVIEW

1. Name three reasons why manufacturers wishing to compete for world markets must use the best manufacturing processes available.

2. What is the purpose of an assembly drawing?

3. List three systems used on technical drawings to show dimensions.

4. What two dimensioning systems are used for numerical control work?

5. Why are section views used on technical drawings?

6. Define the following:

   a. allowance

   b. limits

   c. tolerance

7. What information is generally given around a surface finish symbol?

UNIT 4

Machining Procedures

The proper manufacturing processes must be employed in order to best use human resources, materials, and machine tools to produce high-quality parts quickly, accurately, and at the lowest possible cost. This same procedure must be used by students and apprentices when producing individual parts in a machine shop. Machine shop work consists of machining a variety of parts (round, flat, contoured, etc.) that are either assembled into a unit or used separately. To reduce the time and cost of machining any part, whether round or flat, it is very important that the **sequence of operations** be carefully planned in order to produce the part quickly and accurately. Improper planning or following wrong machining procedures can often result in scrap or spoiled work.

OBJECTIVES

After completing this unit, you should be able to:

- Understand the importance of following specific machining procedures to produce parts quickly and accurately.
- Describe the proper machining sequences for work held between lathe centers or in a chuck.
- Follow the general rules for laying out and machining flat workpieces.

KEY TERMS

finish turning
(finishing)

rough turning
(roughing)

sequence of operations

● MACHINING PROCEDURE FOR ROUND WORK

Much of the work in the machine shop is round and is machined on a lathe or a turning center. In industry, much of the round work is held in a lathe chuck; in training programs, a larger percentage of work is machined between centers due to the need to reset work more often.

General Rules for Round Work

It is good practice to begin by **rough turning** all diameters to within 1/32 in. (0.8 mm) of finish size required (Fig. 4-1A) before finishing any diameter to allow enough material for the finish turning operation. Use coolant wherever possible for all metal-removal operations and be sure that the work has been allowed to return to room temperature before **finish turning.**

1. Machine the largest diameter first, then machine the remaining diameters in sequence from largest to smallest.

- Rough turning small diameters first might allow the part to bend or distort when cutting large diameters.

2. All measurements must be taken from the end of the workpiece to avoid errors and to be sure that enough material is left on the steps and shoulders.

- Rough turn all steps and shoulders to within 1/32 in. (0.8 mm) of the length required, Fig. 4-1B.

3. Special operations, especially those involving force or pressure, such as knurling or grooving, should be done next.

4. Allow the workpiece to return to room temperature before finish turning to avoid inaccurate measurements due to metal expansion.

5. First, finish the shoulder of each step to the correct length, then cut the diameter to size.

6. Finish turn all diameters and lengths.

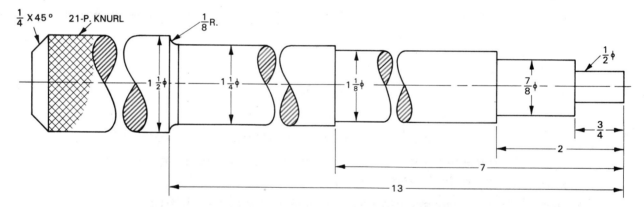

Fig. 4-1A The finish-turn dimensions required for a round part. (Courtesy of Kelmar Associates)

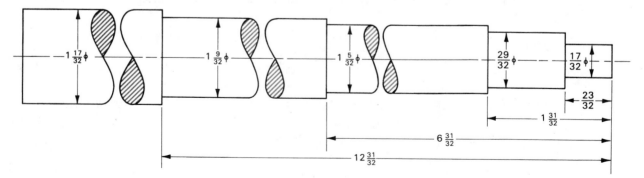

Fig. 4-1B The rough-turn dimensions required for the round part shown in Fig. 4-1A. (Courtesy of Kelmar Associates)

Workpieces with Center Holes

In training programs, it is common practice to machine workpieces between lathe centers because the work must be removed from the lathe more often than is necessary in an industrial situation. In this case, center holes are drilled in each end of the workpiece allowing the work to be taken out of the machine and replaced to the same accuracy as many times as necessary. It also allows the entire part to be machined by reversing the part on the lathe centers and machining the other end. When machining workpieces between the centers on a lathe, follow the machining procedures in the General Rules for Round Work on page 34.

Round Work Held in a Chuck

The procedure for machining round workpieces held in a lathe chuck (three-jaw, four-jaw, collet, etc.) is basically the same as that used for machining round work, Fig. 4-2. If both the external and internal surfaces must be machined on chuck-held work, the sequence of some operations is changed. However, it is important to do all rough turning first before starting any finishing operations.

Work held in a lathe chuck must be held short for purposes of rigidity and to prevent accidents. Always be sure that the chuck jaws have been tightened securely to prevent the workpiece from coming loose during machining. Never extend work beyond the chuck jaws more than three times its diameter unless it is supported by other means such as a steady rest or lathe center. For example, a 1 in. diameter workpiece should never extend more than 3 in. beyond the chuck jaw.

● MACHINING PROCEDURES FOR FLAT WORK

There are many variations in size and shape of flat workpieces, and as a result, it is difficult to give specific machining rules to follow in every case. Some general rules are listed; however, these rules may have to be modified, depending on the type of workpiece or the machining operation being performed, Fig. 4-3.

1. Cut off the proper material approximately 1/8 in. (3.17 mm) larger on all surfaces than the finish size required.

2. Machine all surfaces to size in a vertical or horizontal milling machine.

3. Coat the work surface with a layout dye.

4. Lay out the physical outline of the part according to the dimensions on the print.

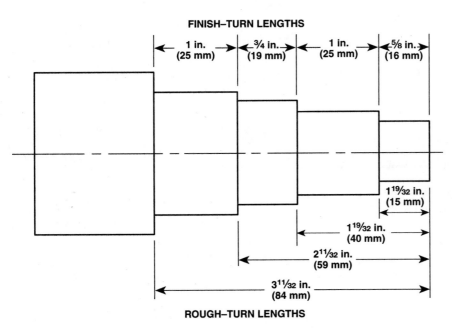

Fig. 4-2 A round part that requires external turning operations. (Courtesy of Kelmar Associates)

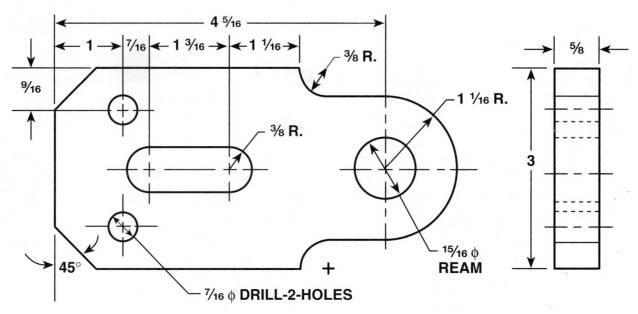

Fig. 4-3 A typical flat part that requires the operations of laying out and machining. (Courtesy of Kelmar Associates)

5. Use a sharp, fine punch to lightly prick punch the layout lines to outline the part.

6. Remove large sections of the workpiece on a contour bandsaw, staying at least 1/32 in. (0.8 mm) away from the layout lines.

7. Machine all contours such as steps, angles, radii, and grooves.

8. Lay out all hole locations and scribe the reference circle with dividers.

9. Drill all holes and tap those which require threads.

10. Ream all holes required.

11. Surface grind all surfaces indicated for grinding on the part print.

need for resetting. In industry, round work is usually held in a chuck.

- When machining round work in a lathe chuck, the workpiece should never be extended more than three times its diameter beyond the chuck jaws.

- When machining round work, rough turn all diameters first, beginning with the largest and working in sequence to the smallest.

- Finish turning should be done only after all rough turning has been completed.

- Procedures for flat work generally include layout, cutting, and machining.

SUMMARY

- Standardized machining procedures have been developed to facilitate training and to ensure that parts are produced accurately, quickly, and with minimum waste.

- In training programs, round work is usually machined between centers because of the

KNOWLEDGE REVIEW

1. List the six main steps which should be followed when rough turning diameters.

2. What precautions should be observed when holding work in a lathe chuck?

3. List the five main steps for preparing and laying out a flat workpiece.

SECTION

4

SHOW-OFF

DANGER
HIGH VOLTAGE

INEXPERIENCED

SAFETY FIRST

LAZY

TIRED

CAUSES OF ACCIDENTS

CARELESS

RULES

DISOBEDIENT

FORGETFUL

CAUTION

HOT-HEADED

Safety

SECTION 4

People learning to use hand tools or operate machine tools must first learn the safety requirements and precautions that apply to each tool or machine. They must learn how to work safely because the safe way helps to prevent accidents and is usually the most correct and efficient way of working. Far too many accidents are caused by carelessness or horseplay. It is easier and far more sensible to develop safe work habits than to suffer the consequences of an accident. Developing good safety practices should become a habit to be followed in a machine shop and should be practiced in everything one does in life.

Safety is Everyone's Business;
Safety is Everyone's Responsibility

UNIT 5

Accident Prevention

Many safety programs are started by accident prevention associations, safety councils, government agencies, and industrial safety committees that are constantly striving to reduce accidents, yet far too many accidents still occur. These result in millions of dollars' worth of lost time and production, needless pain, and suffering, loss of work time, and, unfortunately, some physical handicaps. While modern machine tools are equipped with safety features, it is still the responsibility of the operator to follow recommended safety procedures and use machines wisely and as safely as possible.

OBJECTIVES

After completing this unit, you should be able to:

- Explain why safety is very important to everyone.
- Identify and correct unsafe practices in a machine shop.
- Use shop tools and machines in a safe manner.

KEY TERMS

safety glasses   safety shoes

Accidents just don't happen; they are generally caused by carelessness on someone's part, and they can be avoided.

Any person learning the machine tool trade should first develop safe work habits which include the following:

- Being neat and tidy at all times, and being dressed safely for the job being performed.
- Observing proper operating procedures for all tools and machines.
- Developing a sense of responsibility to one's self.
- Learning to consider the welfare of fellow workers.
- Thinking and working safely at all times.

● EYE PROTECTION

The following rules should be observed to protect the eyes when working in a machine shop:

Approved **safety glasses** must always be worn in the machine shop. It is a widespread practice in industry that all visitors and employees wear safety glasses or some form of eye protection when entering a shop area. There are several types of eye protective devices for use in the machine shop area.

1. The most common to be found are the *plain safety glasses with side shields,* Fig. 5-1A. These glasses provide good eye protection for operating machines or performing bench or assembly operations. The lenses are made of shatterproof glass or similar material, and the side shields protect the sides of the eye area from flying chips and particles.

2. *Plastic safety goggles,* Fig. 5-1B, are generally worn to protect a person's regular prescription eyeglasses. These goggles are made of soft, flexible plastic and fit closely around the upper cheeks and hug the forehead. They are generally provided with ventilation holes to prevent them from fogging up.

3. *A face shield*, Fig. 5-1C, consists of a large plastic shield long enough to give full face protection. Face shields, also used by persons wearing prescription glasses, allow air to circulate between the face and shield, preventing fogging under most conditions. They are particularly useful during heat-treating operations, especially when working with hot metal or where there is any danger of hot flying particles.

(A) PLAIN SAFETY GLASSES

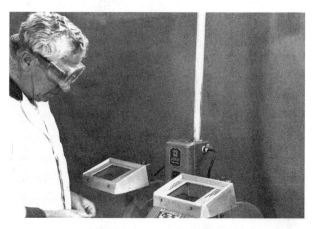

(B) PLASTIC GOGGLES

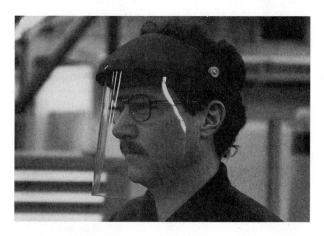

(C) FACE SHIELD

Fig. 5-1 Common types of safety glasses used in a machine shop. (A), (B) (Courtesy of Kelmar Associates)

Each person has only one set of eyes that must last a lifetime. Be sure to always use the proper safety glasses for each operation. NEVER USE ANY GLASSES ThAT DO NOT HAVE SHATTERPROOF LENSES—EYES CANNOT BE RE-PLACED.

PERSONAL GROOMING

In the machine shop, personal grooming plays a vital part in avoiding accidents. The following grooming rules should be observed:

1. Never wear loose clothes when operating machines; they can be caught in revolving parts of the machine, Fig. 5-2.

- Never wear a necktie while working in a machine shop.

- Always roll up shirtsleeves above the elbow, or wear short-sleeved clothing.

- Avoid wearing clothing made of loosely woven material such as sweaters. Material of this type can easily get caught in a machine; *always wear clothing made of hard, smooth fabric.*

- Shop coats or aprons should always be tied at the back to prevent their strings from getting caught in rotating parts of the machine, Fig. 5-3.

Fig. 5-3 Tie apron strings around the back to prevent them from being caught in machinery.

2. Remove wristwatches, rings, bracelets, necklaces, or other jewelry; they are very dangerous in a machine shop and have resulted in many serious injuries, Fig. 5-4.

3. Never wear gloves when operating machines; these can be caught in rotating parts of machines or the workpiece.

4. Long hair must be covered or protected by a hair net or approved protective shop cap, Fig. 5-5.

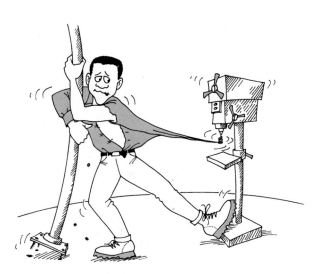

Fig. 5-2 Loose clothing is dangerous around moving parts of machinery.

Fig. 5-4 Do not wear jewelry (watches, bracelets, necklaces, rings) in a machine shop.

Fig. 5-5 Wear a shop cap or hair net to prevent hair from being caught in machinery. (Courtesy of Kelmar Associates)

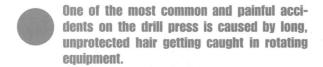

One of the most common and painful accidents on the drill press is caused by long, unprotected hair getting caught in rotating equipment.

5. Canvas shoes or open-toe sandals must never be worn in a machine shop. This type of footwear offers no protection against sharp chips or falling objects. Many companies have a policy requiring employees to wear **safety shoes** with steel-reinforced toes.

● **WORKPLACE ENVIRONMENT**

Everyone should remember a good working environment is essential to safe working conditions. There should always be a place for everything and everything should be kept in its place.

Good housekeeping is more than neatness and cleanliness; orderliness contributes to safety and efficiency.

While it is impossible to list all the possible causes of accidents in a machine shop, the following general suggestions are offered as ways to reduce and prevent accidents:

1. All shop areas should be well-lighted. Broken or improperly functioning lights and light fixtures should be repaired promptly.

2. Always keep machine and hand tools free of chips, grease, or oil.

Oily or greasy surfaces are difficult to hold, and metal chips can prevent a workpiece from being clamped securely.

3. Always use a brush or vacuum to remove any chips; never use a cloth or your hand.

4. Clean oily surfaces and tools with a dry cloth.

5. Tools and materials should not be stored on the machine. A serious accident may occur if they fall.

6. To prevent slipping or falling accidents, always clean oil, grease, or other liquids from the floor around the machine, Fig. 5-6.

7. Metal chips on the floor should be swept up frequently. Chips can become embedded in the soles or heels of shoes and can cause dangerous walking conditions and/or slippage, Fig. 5-7. Set up a shoe scraper in the shop so that shoes can be cleaned periodically.

8. Keep the floor areas close to machines free of tools or materials.

These objects can cause tripping accidents or interfere with an operator's ability to safely operate the machine.

9. After cutting off stock to the required length, return the bar stock to its proper storage rack; never leave it on the floor where it becomes a safety hazard.

Fig. 5-6 Remove oil or grease spots from the floor to prevent accidents caused by slipping.

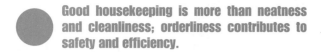

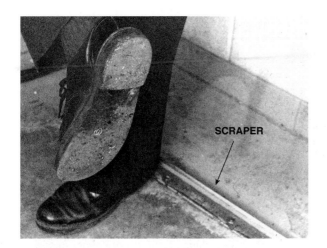

Fig. 5-7 Use a shoe scraper to remove steel chips from the soles of shoes. (Courtesy of Kelmar Associates)

10. Never use compressed air to blow chips away from a machine.

> Flying metal chips can hit someone with shrapnel-like force. Dirt and small chips can cause unnecessary wear when they get into machine slides.

SAFE WORK PRACTICES

When actually operating tools and machines, the following practices should be observed:

1. Do not attempt to operate any machine or power tool equipment before understanding how it works and how to stop it quickly if something unexpected happens.

2. Make sure all guards and safety devices are in place and operational before operating any equipment. They are there to protect the operator and should not be removed or tampered with.

3. Always disconnect the electric power and lock it off at the switch box when making repairs or adjustments to any equipment, Fig. 5-8.

> A sign should be placed on the equipment to indicate it is out of order.

4. Check to be sure that the workpiece and cutting tool are properly secured before operating the machine.

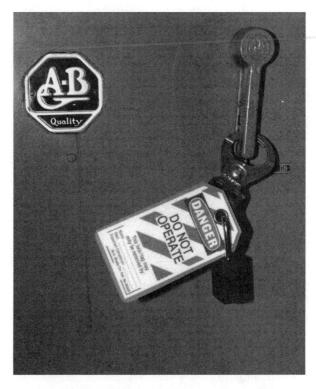

Fig. 5-8 Lock off the power switch on any machine being repaired. (Courtesy of Direct Safety Co.)

5. Keep hands away from moving parts. Never attempt to "feel" the surface of revolving work, or to stop the machine by hand.

6. Never measure, clean, or make any adjustment unless the machine has come to a complete stop.

7. Rags should never be used near moving parts as both the hand and rag can be drawn into the machine.

8. Only one person should operate a machine at any one time.

> Not knowing what the other person will do can lead to a very serious accident.

9. Every injury, regardless of how minor, should get first-aid attention immediately. Even the smallest cut should be treated to prevent infection.

10. Always remove all sharp burrs from workpieces with a file before handling.

11. Get help before attempting to lift anything too heavy or odd-shaped and follow safe lifting practices.

- Get into a squatting position with the knees bent and back straight, Fig. 5-9.
- Hold the object firmly.
- Straighten the legs and keep the back straight while lifting. This uses the leg muscles and helps prevent back injury.

12. Always make sure that the work is clamped securely before attempting a machining operation.

13. Never start a machine until you are certain that the workpiece will clear the cutting tool and machine parts, Fig. 5-10.

14. Use a proper fitting wrench for each job. An oversize wrench can lead to worn corners on nuts. Worn nuts should always be replaced.

15. Pulling a wrench is much safer than pushing it. Pushing on a wrench can be dangerous, especially if the hands or wrench are greasy.

16. During grinding, use coolant wherever possible to control the grinding dust. **NEVER INHALE GRINDING DUST.**

Fig. 5-10 Always be sure that everything will clear before starting a machine. (Courtesy of Kelmar Associates)

 Use a vacuum dust collector system for all grinding operations and, if this is not available, use a safety-approved respirator.

GENERAL SAFETY CONSIDERATIONS

1. Be aware that not only does haste make waste, it can also cause accidents. Avoid hurrying or running near machines and equipment. Be sure to take adequate time to properly set up and operate tools and machines.

2. Do not engage in horseplay or other careless activities that could result in an accident.

3. Know the locations of the nearest telephones in the event of an emergency.

4. Keep a first-aid kit on the premises.

FIRE PREVENTION

1. Keep oily rags in proper metal containers to avoid spontaneous combustion.

2. Follow the proper procedures before lighting a gas heat-treating furnace.

3. Know the location of every fire extinguisher in the shop and know how to operate this equipment.

4. Be familiar with the location of the nearest fire exit.

5. Know the location and how to operate the nearest fire alarm box.

6. Be sure to keep combustible materials away

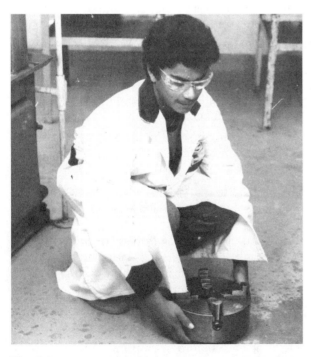

Fig. 5-9 Use correct lifting procedures to prevent back injuries. (Courtesy of Kelmar Associates)

from all hot metal operations such as welding, cutting, forging, heat treatment, etc.

7. Follow all rules and regulations regarding smoking and use of matches or open flames.

SUMMARY

- Safety is the responsibility of everyone in the workplace. Accidents are usually caused by carelessness.

- Proper eye protection should be worn at all times. This can include safety glasses with side shields, plastic safety goggles, and face shields.

- Personal grooming can affect shop safety. Follow safety rules and guidelines with respect to clothing, hair, gloves, shoes, and jewelry.

- Good housekeeping and proper work environment contribute to shop safety.

- Follow safe work practices when operating any machine or power tool.

- Fire prevention is a serious concern in a machine shop. Follow proper procedures and use common sense to avoid fire-related accidents and injuries.

KNOWLEDGE REVIEW

1. What is the first thing a person must learn before operating machine tools?

2. List the five safe work habits a person should develop before working in a machine shop.

3. Name three types of safety glasses used in a machine shop.

4. What four precautions regarding clothing should be observed for safe work practices?

5. Why are safety shoes recommended for shop work?

6. How should steel chips be removed from a machine?

7. Name three reasons why steel chips are dangerous.

8. List the proper procedure for lifting heavy work.

9. Name two precautions which should be observed when using wrenches.

10. What three methods can be used to control grinding dust?

SECTION

5

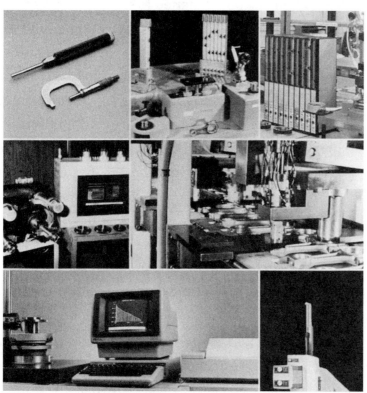

Measurement

Customers throughout the world are sending a very strong message to manufacturers that they are looking for better-quality products and services. It started in the 1960s when customers started to show their preference for quality by buying large amounts of automotive and electronic products from offshore manufacturers. What they were saying to manufacturers was that it was no longer acceptable to produce low-quality goods. To keep pace with customer demand, better-quality machine tools and measuring equipment were developed in order to produce products accurate to millionths of an inch. The race for better-quality products requires that they be inspected during the manufacturing operation, or immediately after, to ensure the quality of the product.

The Science of Precision Measurement has come a long way from the days when humans used sticks, stones, or parts of the body as standards of measurement. Measurement now has been developed to an exacting science which is known as metrology or dimensional metrology. To ensure accuracy in manufacturing operations, tools capable of measuring in millionths of an inch are available, such as:

- *digital measuring tools* that can download data to computers,
- *in-process gaging* that measures work while it is being manufactured,
- *electronic measuring systems* that use light magnification to convert small spindle movements to large needle movements on a gage,
- *laser non-contact tools* capable of 500 measurements per second.

The need for newer and better inspection and measuring tools will continue as industry searches for ways to make better-quality parts and achieve the goal of zero-defect manufacturing. These new precision tools will make precision measurements easier and help to increase manufacturing productivity.

Courtesy of Giddings and Lewis—Sheffield Measurement.

UNIT
6

Measurement Systems

The inch system, often called the English system of measurement, and the metric system are the two major systems of measurement used in the world. Over 90% of the countries in the world are presently using the metric system. The United States and Canada are two of the industrial powers still using the inch system of measurement. Even though the U.S. Congress officially adopted the metric standard as far back as 1866, the standard still has not been widely adopted.

It may take quite some time before countries using the inch system convert to the metric system. With the reality of global manufacturing continually expanding, the need for machinists to be able to work in both systems of measurement will continue to grow. To assist everyone to become familiar with both systems of measurement, this book uses dual measurements throughout, with the inch dimension given first, followed by the metric equivalent in parentheses, for example, .750 in. (19.05 mm).

OBJECTIVES

After completing this unit, you should be able to:

- Explain how important measurement is in the manufacture of high-quality products.
- Describe the relationship between the inch, decimal, and metric systems of measurement.
- Read and use inch, decimal, and metric steel rules.

KEY TERMS

inch system   metric system

● INCH SYSTEM

The **inch system** is the standard of measurement for North American industry. Unlike the metric system, there is no relationship of other linear units to the base inch unit. The values of yard, rod, mile, etc., have to be studied and kept in memory in order to use them. The inch can be divided by halves, quarters, eighths, sixteenths, thirty-seconds, sixty-fourths, tenths, hundreds, thousandths, ten-thousandths, etc. Some inch/metric equivalents are shown in Table 6-1.

Other quantities, such as weight, volume, pressure, and temperature, also do not have simple relationships with each other. This situation is further complicated by the fact that fluid measurement varies between the United States and Canada.

Decimal-Inch System

The *decimal-inch system* is used when measurements smaller than a sixty-fourth of an inch are required. The inch is divided into ten equal parts, each having a value of one hundred thousandths (.100). Each of these parts is again divided into ten equal parts, and so on, until the inch is divided into tenths, hundredths, thousandths, and ten-thousandths of an inch, Fig. 6-1.

The common decimal-inch fraction used in a machine shop is one thousandth of an inch (.001), or three numbers to the right of the decimal point. Figures to the left of the decimal point

Table 6-1   Inch—Metric Equivalents

Inch	Metric
.001 inch (in.)	0.025 millimeters (mm)
1 inch	25.4 millimeters
1 foot (ft.)	0.3048 meter (m)
1 yard (yd.)	0.9134 meter
1 ounce (oz.)	28.35 grams (g)
1 pound (lb.)	0.4536 kilograms (kg)
Metric	**Inch**
1 millimeter	.039 inches
1 meter	39.37 inches
1 kilometer (km)	.6213 mile
1 gram	15.432 grains (gr)
1 kilogram	2.204 pounds
1 liter (l)	1.0567 quarts (qt.)

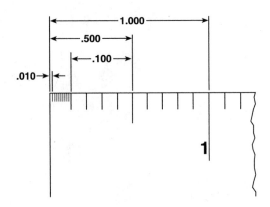

Fig. 6-1 Decimal rules generally have .010 in. graduations, with every tenth line numbered for easy reading. (Courtesy of Kelmar Associates)

represent whole numbers, while those to the right represent the fraction (less than one inch). Using a size of 1.375 in. as an example, the dimension would be one inch plus 375 thousandths of an inch.

There are times when dimensions on prints are shown without three numbers to the right of the decimal point. In this case, it is good practice to always add enough zeros (0) to make three numbers to the right of the decimal to avoid reading errors. For example, for the dimension .2 in., add two zeros (00) to make it .200 or 200 thousandths of an inch.

The addition of zeros to the right of the decimal point does not change the value of the dimension. Some decimal fractions and their value are given as examples:

- .001 represents one thousandth of an inch.
- .020 means twenty thousandths.
- .075 means seventy-five thousandths.
- .150 means one hundred and fifty thousandths.
- 1.312 means one inch and three hundred and twelve thousandths.

● METRIC SYSTEM

The **metric system** uses the meter and linear units based on the meter as its standards of measure. The meter, originally introduced by the French Academy of Sciences toward the end of the eighteenth century, has become the standard of length for most countries. Over the years, a number of different physical standards were used to define the meter, starting with the platinum-iridium end bar. At the General Conference on Weights and Measures in October, 1983, the meter, defined as the distance traveled by light in a vacuum during 1/299,792,458 of a second, was approved as a world standard. This was a real breakthrough for measurement science, because the meter was defined ten times more accurately than ever before.

All multiples and subdivisions of the meter are directly related to the meter by a factor of ten. This makes it easy to use the decimal system for calculations involving metric units. In order to convert from a smaller to a larger unit, or vice versa, it is necessary only to multiply or divide by 10, 100, 1000, etc. For example:

prefix	meaning	multiplier	symbol
micro	one millionth	.000 001	μ
milli	one thousandth	.001	m
centi	one hundredth	.01	c
deci	one tenth	.1	d
deca	ten	10	da
hecto	one hundred	100	h
kilo	one thousand	1 000	k
mega	one million	1 000 000	M

● MEASURING EQUIPMENT

A modern machine shop could not function without precise measuring equipment. The parts produced by machine tools that cut and form are useless if they are not made to the exact sizes specified.

Care

Proper care of measuring tools and instruments is very important to maintain the accuracy and quality of these tools. Precision measuring tools and instruments are expensive and should be treated with care, otherwise their accuracy can be destroyed. Inaccurate tools are worse than no tool at all, because inaccurate work will be produced when they are used. A skilled craftsperson is judged by the care and condition of the tools that he or she is using. Skilled machinists or toolmakers take a great deal of pride in keeping their measuring tools accurate and in good condition by observing the following:

1. Never drop a measuring tool.

2. Keep measuring tools away from chips and dirt.

3. Never place measuring tools on oily or dirty surfaces.

4. Store measuring tools in separate boxes to avoid scratches, nicks, or dents.

5. Clean the tools and apply a light film of oil on the handling surfaces before putting them away.

Steel Rules

Steel rules are the most common linear measuring tools and are available with graduations in the inch or metric system. Inch rules are graduated in fractional parts of an inch, while metric rules are graduated in both millimeters and half-millimeters. Some rules are available with both inch and millimeter graduations, for example, one edge may be graduated in thirty-seconds of an inch, while the other edge is graduated in millimeters. The back of the same rule could have sixty-fourths of an inch on one edge, while the other edge is graduated in half-millimeters.

Inch Rules

The steel rule most commonly used in the machine shop is a spring-tempered, quick-reading, six in. rule, which is incorrectly called a scale in the trade. The most common fraction graduations on the six in. rule are 1/64, 1/32, 1/16, and 1/8 of an inch, Fig. 6-2. Rules are used for measuring dimensions within an accuracy of 1/64 in. Dimensions of 1/64 of an inch are about as small as can be seen accurately without the aid of a magnifying glass.

Decimal-inch dimensions require the use of precision measuring instruments such as micrometers, verniers, and dial calipers.

Metric Rules

Metric rules, Fig. 6-3, are usually graduated in millimeters and half-millimeter. They are

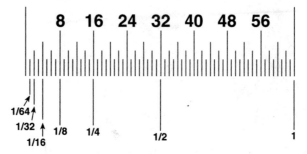

Fig. 6-2 The most common fractions found on inch rules. (Courtesy of Kelmar Associates)

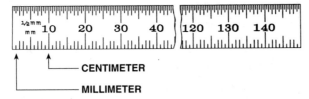

Fig. 6-3 Metric rules are usually graduated in millimeters and half millimeters. (Courtesy of The L.S. Starrett Co.)

available in lengths ranging from 150 mm to 1 m, the most common being between 150 mm and 300 mm in length.

Types of Rules

A wide variety of rules are available to help the machinist measure various diameters and lengths.

- *Spring tempered rules,* (quick-reading) 6 in. or 15 cm rules, Figs. 6-2 and 6-3, are the most common rules used in machine shop work.

- *Hook rules,* Fig. 6-4, are used to take accurate measurements from an edge, step, or shoulder of a workpiece. They may also be used for setting inside calipers to a dimension.

- *Short-length rules,* Fig. 6-5, are used for measuring narrow spaces or hard-to-reach surfaces that are too difficult to reach by ordinary means. Short-length rules come in a set of five, ranging in sizes between 1/4 in. (6.35 mm) and 1 in. (25.4 mm), which can be interchanged in a holder.

Fig. 6-4 Hook rules are handy for measuring from a shoulder, step, or an edge. (Courtesy of Kelmar Associates)

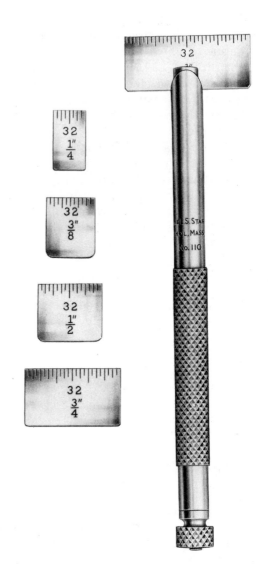

Fig. 6-5 Short-length rules are used for measuring narrow spaces or hard-to-reach surfaces. (Courtesy of The L.S. Starrett Co.)

- *Decimal rules,* Fig. 6-6, are often used where the linear measurement is less than 1/64 in. Divisions on decimal rules are graduated to measurements of .010 (1/100 in.), .020 (1/50 in.), .050 (1/20 in.), and .100 (1/10 in.). Fig. 6-6 shows a 6 in. decimal rule.

● MEASURING LENGTHS

The length of common rules used in a machine shop will range from 1 to 72 in. The accuracy of the measurements taken with rules

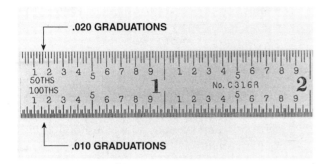

Fig. 6-6 Decimal rules are used where measurements smaller than 1/64 in. are required. (Courtesy of Kelmar Associates)

depends upon the skill of the person using them. Due to neglect or wear, the end of a rule may become rounded or inaccurate. To avoid inaccurate measurements, proceed as follows:

1. Place the center of the 1 in. or 1 cm graduation line on the edge of the work.

2. Hold the rule parallel to the edge of the workpiece or across the center line on round work.

3. Take the measurement, Fig. 6-7, and then subtract the 1 in. or 1 cm from the reading.

Whenever possible, butt the end of the rule against a step or shoulder on the workpiece to get the most accurate measurements, Fig.6-8.

Since the edge of a steel rule is ground flat, it may be used as a straightedge to test the flatness of a work surface. This is done by placing the edge of the rule on the work surface, and then holding it up to some light. If there is any unevenness of the surface, even as small as a few thousandths of an inch (0.05 mm), light will show through.

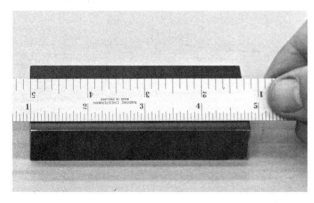

Fig. 6-7 For accurate measurements, start at a 1 in. or 1 cm line, especially when using a rule with a worn end. (Courtesy of Kelmar Associates)

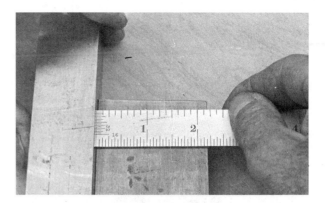

Fig. 6-8 Butting the end of a rule against a shoulder or step gives the most accurate measurements. (Courtesy of Kelmar Associates)

● OUTSIDE CALIPERS

Outside calipers are comparison tools used to make approximate measurements of the outside diameter of round work. *They are not meant to be used on work that requires accuracy.* The spring joint caliper, Fig. 6-9, consisting of two curved legs, a spring, and an adjusting nut, is sometimes used in machine shop work to take approximate measurements. The caliper cannot be read directly and its setting must be checked with a rule or a standard-size gage.

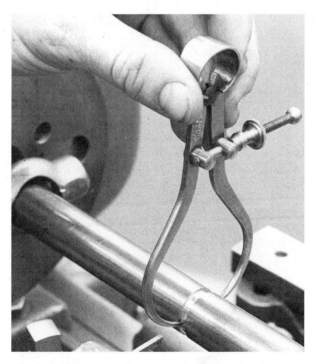

Fig. 6-9 Checking a diameter with an outside caliper. (Courtesy of Kelmar Associates)

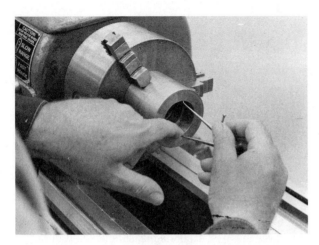

Fig. 6-10 Check an inside caliper setting with a micrometer for greater accuracy. (Courtesy of Kelmar Associates

Checking the Work Size

Although checking work size with the caliper setting is not an accurate method of measurement, reasonably accurate measurements can be made after the machinist develops a sense of touch or "feel." Calipers may also be used to measure narrow grooves where a micrometer cannot be used.

 Never attempt to measure the work while it is revolving or moving. It is a very dangerous practice, and the measurement will not be accurate.

● INSIDE CALIPERS

Inside calipers should only be used for taking internal measurements if no precision tools are available and the approximate measurement is acceptable for the part being produced. When using inside calipers, always check their setting with a micrometer or internal gage, Fig. 6-10.

SUMMARY

- Extremely precise measuring systems are required to meet increasing customer demands for high-quality products.

- The two major systems of measurement are the inch (English) system and the metric, or SI, system.

- Because they are both based on powers of ten, the metric system of measurement and the decimal system of numbers are closely related. The decimal system is also used with the inch system of measurement when high precision is required.

- Measuring tools and instruments need proper care to maintain their accuracy and quality.

- Commonly used measuring tools include steel rules and outside and inside calipers.

KNOWLEDGE REVIEW

1. Why were better-quality machine tools and measuring equipment developed after the late 1960s?

2. List four tools capable of measurements in millionths of an inch.

Measurement Systems and Tools

3. What is the world standard for the meter since 1983?

4. Name the advantage of the SI system over the inch system.

5. Why should measuring tools be treated with care?

6. Why is an inaccurate tool worse than no tool at all?

7. What is the most common inch rule? What is the smallest dimension that can be measured accurately?

8. Name the common graduations found on a metric rule.

9. What type of rule should be used to measure the following?
   a. accurate measurements from a step
   b. narrow, hard-to-reach surfaces
   c. divisions smaller than 1/64 in.

10. Explain how an accurate measurement can be made using a rule with a worn end.

U N I T
7

Micrometers

Micrometers are the most commonly used precision instrument in the machine tool industry and is available in both inch and metric types. The standard inch micrometer, shown in a cutaway view in Fig. 7-1A, can measure accurately to within .001 in. Fig. 7-1B shows the graduations on a standard inch micrometer. Inch micrometers with a vernier scale on the sleeve make it possible to measure as small as .0001 in.

The standard metric micrometer, Fig. 7-2, measures in hundredths of a millimeter. Equipped with a vernier scale, the metric micrometer is accurate to 0.002 mm.

OBJECTIVES

After completing this unit, you should be able to:

- Identify and use various types of outside micrometers.
- Measure workpieces to within an accuracy of .001 in. (0.02 mm)using a micrometer.
- Read vernier micrometers to within .0001 in. (0.002 mm) accuracy.

KEY TERMS

inch micrometer   metric micrometer   vernier micrometer

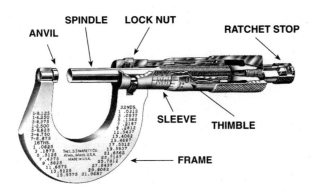

Fig. 7-1A The main operative parts of a standard inch micrometer. (Courtesy of The L.S. Starrett Co.)

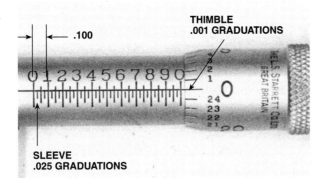

Fig. 7-1B The graduations on a standard inch micrometer. (Courtesy of Kelmar Associates)

● MICROMETER PARTS

The inch micrometer and metric micrometer are similar in construction. The only difference between them is the pitch of the spindle screw and the graduations on the sleeve and thimble. The most common parts and their functions are:

- Frame—the U-shaped body which holds all other parts together.
- Anvil—the fixed measuring face.
- Spindle—the movable measuring face; the spindle can be moved to measure various sizes of workpieces.
- Sleeve (barrel)—graduated in .025 in. divisions for inch micrometers (0.5 mm divisions for metric micrometers).
- Thimble—graduated around the circumference in .001 in. divisions for inch micrometers (0.01 mm divisions for metric micrometers).

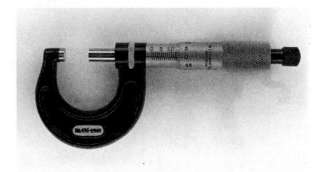

Fig. 7-2 The metric micrometer has a 0.5 mm pitch on its spindle thread. (Courtesy of The L.S. Starrett Co.)

- Friction thimble—assures that the same pressure is applied when measuring objects; it prevents damage to the micrometer.

● PRINCIPLES OF THE INCH MICROMETER

There are 40 threads per inch on the micrometer spindle, and it is this principle which must be understood to read **inch micrometers.**

- It takes 40 complete turns of the micrometer spindle to open the measuring faces one inch.
- There are 40 graduations on the micrometer sleeve with every fourth line numbered for easy reading.
- The numbered graduations each have a value of .100, for example, #2 = .200 in., #3 = .300 in., etc.
- One complete turn of the micrometer will open or close the measuring faces 1/40 in. or .025 in.
- When the micrometer is completely closed, the zero line on the thimble matches the center line on the sleeve.
- One complete turn of the thimble counter-clockwise will expose one line on the sleeve or .025 in.
- There are 25 equal divisions around the thimble, and each division equals .001 in.

To Read a Standard Inch Micrometer

To read a standard inch micrometer, follow the steps listed below:

1. Note the last numbered line showing on the sleeve and multiply it by .100 in.

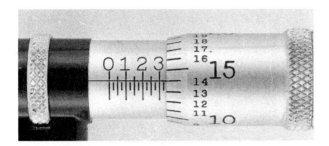

Fig. 7-3 A standard inch micrometer reading of .339 in. (Courtesy of Kelmar Associates)

2. Count the number of small lines after the last number and multiply it by .025 in.

3. To these readings, add the number of the thimble division that aligns with the index line.

In Fig. 7-3, the reading is obtained as follows:

- #3 shown on the
  sleeve $3 \times .100 = .300$
- 1 line past #3 $1 \times .025 = .025$
- #14 line on the
  thimble aligns
  with the index line $14 \times .001 = \underline{.014}$

  Total reading = .339 in.

● INCH VERNIER MICROMETER

The inch **vernier micrometer** is basically the same as the standard inch micrometer with the addition of a vernier scale on the sleeve.

To Read a Vernier Micrometer

To read an inch vernier micrometer, use the following steps:

1. Read the micrometer setting to the nearest thousandth.

2. Find the line on the vernier scale that matches any line on the thimble. This line will indicate the number of ten-thousandths that must be added to make the reading a four-place decimal figure.

In Fig. 7-4, the inch vernier micrometer reading is obtained as follows:

- #4 shown on the
  sleeve $4 \times .100 = .400$
- No full line after #4 $0 \times .025 = .000$

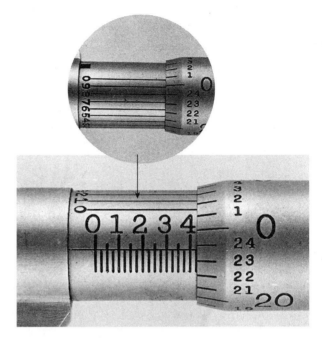

Fig. 7-4 A vernier inch micrometer reading of .4238 in. (Courtesy of Kelmar Associates)

- #23 on thimble is
  past the index line $23 \times .001 = .023$
- #8 line on vernier
  scale matches a
  thimble line $8 \times .0001 = \underline{.0008}$

  Reading = .4238 in.

● METRIC MICROMETER

The standard **metric micrometer,** Fig. 7-5, is used to make measurements accurate to within one-hundredth of a millimeter (0.01 mm). There are 25 graduations above the index line, each having a value of 1 mm (1 mm), and 25 graduations below, which have a value of 1/2 mm (0.5 mm) each. The pitch of the micrometer spindle

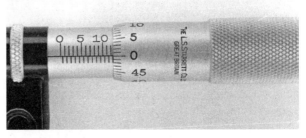

Fig. 7-5 The graduations usually found on a standard metric micrometer. (Courtesy of Kelmar Associates)

screw is 0.5 mm, and one complete turn of the thimble moves the measuring faces 0.5 mm.

Principles of the Metric Micrometer

- It takes 50 complete turns of the thimble to open the measuring faces 25 mm.
- There are 25 graduations *above the sleeve index line,* each having a value of 1 mm, with every fifth one numbered for easy reading.
- The numbered graduations represent millimeters, Fig. 7-6. e.g., #5 = 5 mm, #20 = 20 mm
- There are 25 graduations *below the sleeve index line,* each having a value of 0.5 mm.
- When the micrometer faces are closed, the zero line on the thimble aligns with the index line on the sleeve.
- There are 50 equal divisions on the micrometer thimble, each having a value of 0.01 mm.
- Each complete turn of the thimble will show one line on the sleeve or 0.5 mm.

To Read a Metric Micrometer

To read a metric micrometer, use the following steps:

1. Note the last division number above the sleeve index line and multiply this number by 1 mm.
2. Count the number of lines (above and below the index line) which show past the numbered line and multiply them by 0.5 mm.

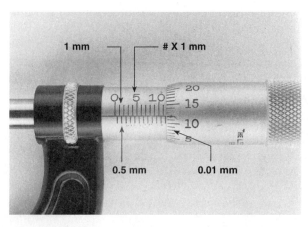

Fig. 7-6 A metric micrometer reading of 11.62 mm. (Courtesy of Kelmar Associates)

3. To the sleeve reading, add the number of the line on the thimble which lines up with the index line.

In Fig. 7-6, the metric micrometer reading is obtained as follows:

- 11 lines above the index line $11 \times 1 = 11$
- 1 line past the 11th line below the index line $1 \times 0.5 = 0.5$
- 12th line on thimble aligns with the index line $12 \times 0.01 = \underline{0.12}$

$$\text{Reading} = 11.62 \text{ mm}$$

● METRIC VERNIER MICROMETERS

The metric **vernier micrometer,** Fig. 7-7, is basically the same as the standard metric micrometer with the addition of a vernier scale on the sleeve. Each division of the vernier scale on the sleeve has a value of 0.002 mm, making metric readings possible in thousandths of a millimeter.

To Read a Metric Vernier Micrometer

To read a metric vernier micrometer, use the following steps:

1. Read to the nearest hundredth millimeter as with the standard metric micrometer.

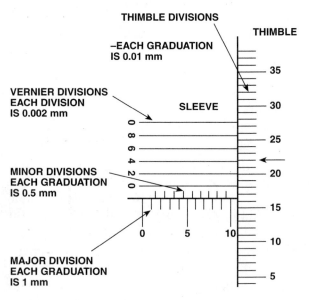

Fig. 7-7 A metric vernier micrometer reading of 10.168 mm. (Courtesy of Kelmar Associates)

2. Find the line on the vernier scale that aligns with any line on the thimble. This thimble line will indicate the number of two-thousandths of a millimeter (0.002 mm) that must be added to the standard metric reading.

In Fig. 7-7, the metric vernier micrometer reading is obtained as follows:

- 10 major divisions *below* the index line $10 \times 1 = 10$

- No minor divisions *above* the index line $0 \times 0.5 = 0$

- 16 thimble divisions past the index line $16 \times 0.01 = 0.16$

- #4 vernier line matches a thimble line $4 \times 0.002 = 0.008$

Total Reading = 10.168 mm

METRIC READING
6.28 mm

SLEEVE

INCH READING
.2472 in.

THIMBLE

(A)

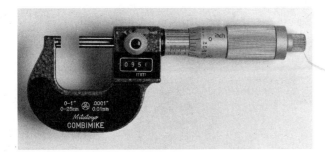

(B)

Fig. 7-8 Combination inch/metric micrometers. (A) Dual scales on sleeve and thimble. (B) Inch on sleeve and thimble, metric on readout. (Courtesy of MTI Corp.)

COMBINATION INCH-METRIC MICROMETER

Combination inch-metric micrometers are shown in Fig. 7-8. Available in several styles, they are useful where both inch and metric measurements are required. They give inch readings as with a standard inch micrometer, and a metric digital reading in a boxed window on the micrometer frame.

SUMMARY

- The micrometer is the most common precision measuring instrument used in machine shops.

- The inch micrometer can measure accurately to within one-thousandth of an inch.

- The metric micrometer can measure accurately to within one-hundredth of a millimeter.

- When vernier scales are added to standard micrometers, they increase the measuring precision even further.

- Combination micrometers give readings in both inch and metric measurements.

KNOWLEDGE REVIEW

1. Compare the inch and metric micrometers with respect to:

   a. pitch of the spindle thread

   b. value of sleeve graduations

   c. value of thimble graduations

2. What are the readings for the following inch micrometer settings?

   a.

Courtesy of Kelmar Associates

b.

Courtesy of Kelmar Associates

b.

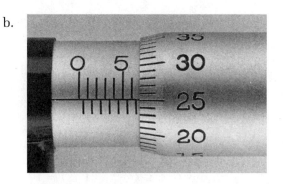

Courtesy of Kelmar Associates

c.

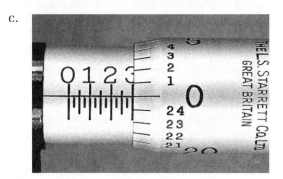

Courtesy of Kelmar Associates

c.

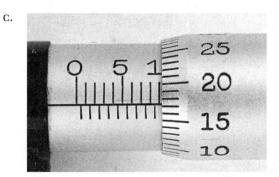

Courtesy of Kelmar Associates

d.

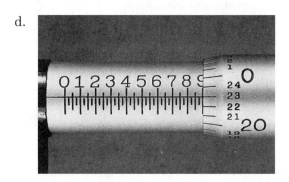

Courtesy of Kelmar Associates

d.

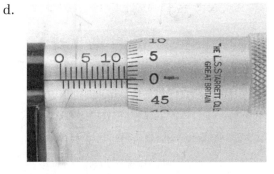

Courtesy of Kelmar Associates

3. What are the readings for the following metric micrometer settings?

a.

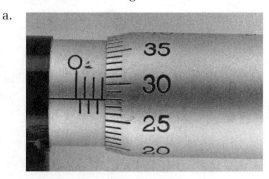

Courtesy of Kelmar Associates

UNIT
8

Vernier Calipers

Vernier calipers, Fig. 8-1, are precision measuring instruments used to make internal and external measurements that would be difficult to make with a micrometer or other measuring instrument. Vernier calipers are capable of readings within .001 in. for inch tools (0.02 mm for metric tools). The most common vernier calipers have 50 divisions on the movable (sliding) jaw.

OBJECTIVES

After completing this unit, you should be able to:

- Read and use a 50-division vernier caliper.
- Measure workpieces to an accuracy of plus or minus .001 in.
- Measure with a metric vernier caliper to an accuracy of plus or minus 0.02 mm.

KEY TERMS

dial caliper   vernier caliper

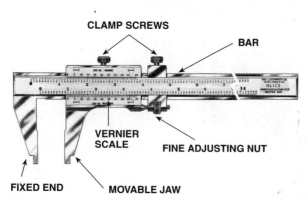

Fig. 8-1 The main parts of an inch vernier caliper. (Courtesy of The L.S. Starrett Co.)

● **VERNIER CALIPER PARTS**

Inch and metric vernier calipers are available for measurements in either system. Some styles of vernier calipers provide inch readings on one side and metric on the other side. All vernier calipers have the same basic parts:

- Frame—an L-shaped bar consisting of the bar and fixed jaw.
- Bar—contains the main scale, with the bottom edge graduated for outside measurements and the top for inside measurements.
- Movable jaw—contains the vernier scale for either inch or metric measurements.
- Clamp screws—lock the movable jaw in position for taking measurements.
- Adjusting nut—used for fine adjustment when taking a measurement.

Outside measurements are taken on the inside of the fixed and movable jaws and read on the bottom scale of the bar. Inside measurements are taken on the outside of the nibs on the fixed and movable jaws and read on the top scale of the bar.

● **PRINCIPLES OF THE 50-DIVISION VERNIER INCH CALIPER**

Most vernier calipers have 50 divisions on the vernier scale of the sliding jaw.

- The large numbers, usually in the center of the bar, represent inches, for example, #3 = 3 × 1 or 3 in.
- The small numbers on the bar represent .100 inch, for example, #6 = 6 × .100 or .600 in.

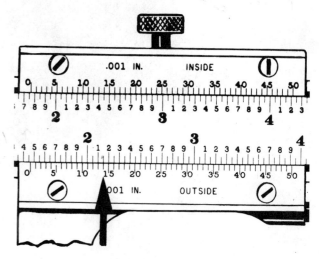

Fig. 8-2 A 50-division inch vernier caliper reading of 1.464 in. (Courtesy of The L.S. Starrett Co.)

- Each line on the bar represents a value of .050 in.
- The vernier scale on the sliding jaw has 50 equal divisions, each having a value of .001 in.
- The 50 vernier scale divisions occupy the same space as 49 divisions on the bar; therefore only one line of the vernier scale will line up exactly with a line on the bar at any setting.

To read the 50-division vernier setting in Fig. 8-2, first note in sequence how many inches (1.00 in.), hundred thousandths (.100), and fifty thousandths (.050) the zero on the movable jaw is past the zero on the bar. To this total, add the number of the vernier scale line in thousandths (.001) which exactly aligns with a line on the bar.

In Fig. 8-2:

- The large #1 on the bar   1 × 1.000 = 1.000
- The small #4 past the large #1   4 × .100 = .400
- 1 line showing past the small #4   1 × .050 = .050
- The 14th vernier scale line aligns with a bar line   14 × .001 = .014

Total reading = 1.464 in.

● **METRIC VERNIER CALIPERS**

Metric measuring tools are becoming more common throughout the world because of the

Fig. 8-3 Inch vernier calipers are available with the 50-division scale and the direct-reading type. (Courtesy of MTI Corp.)

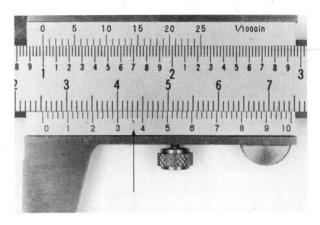

Fig. 8-4 A 50-division metric vernier caliper reading of 25.36 mm. (Courtesy of Kelmar Associates)

global manufacturing concept where goods can be manufactured anywhere in the world and assembled somewhere else. The most common metric vernier calipers are the *50*-division and the direct-reading types, Fig. 8-3.

- The main scale on the bar of a 50-division metric vernier caliper is graduated in millimeters, and every tenth division is numbered.

- Each numbered division has a value of 10 millimeters; for example, #1 represents 10 mm, #2 represents 20 mm, etc.

- There are 50 graduations on the sliding or vernier scale, with every fifth one being numbered.

- Each division on the vernier scale has a value of 0.02 mm.

- The 50 vernier scale graduations occupy the same space as 49 main scale graduations (49 mm).

Therefore 1 vernier division $= \dfrac{49}{50}$
$= 0.98$ mm

The difference between 1
main scale division and 1
vernier scale division $= 1 - 0.98$
$= 0.02$ mm

To Read a Metric Vernier Caliper

To read a metric vernier caliper, follow the suggested steps:

1. The last numbered division on the bar to the left of the zero on the vernier scale represents the number of millimeters multiplied by 10.

2. Note how many graduations are showing between this numbered division and the zero on the vernier scale; multiply this number by 1 mm.

3. Locate the line on the vernier scale which aligns with a bar line; multiply this number by 0.02 mm.

In Fig. 8-4:

- The large #2
  graduation
  on the bar 2×10 mm $= 20$

- Five lines past
  the #2 bar
  graduation 5×1 mm $=\ \ 5$

- The 18th vernier
  scale line aligns
  with a bar line $18 \times 0.02 = \underline{\ \ 0.36}$

  Total reading $= \overline{25.36}$ mm

● DIRECT-READING DIAL CALIPERS

The direct-reading **dial caliper,** Fig. 8-5, has a dial indicator mounted on the movable jaw which provides a direct reading in inches or millimeters.

Inch Dial Calipers

- The edge of the bar is usually graduated into 100 equal divisions with every tenth line, represented by the large numbers, indicating inches.

- The indicator dial is graduated into 100 equal divisions, each having a value of .001 in.

Fig. 8-5 Metric direct-reading dial calipers can be used to measure sizes in millimeters. (Courtesy of The L.S. Starrett Co.)

Metric Dial Calipers

- The edge of the bar is usually graduated in millimeters, with every tenth line numbered for easy reading.
- The indicator dial is usually graduated into 100 equal divisions, each having a value of 0.05 mm, or as fine as 0.002 mm.

Dial calipers take measurements quickly and are easier to read than standard vernier calipers. Both types of dial calipers have a narrow sliding blade attached to the movable jaw which can be used as a depth gage.

SUMMARY

- Vernier calipers are used to make precision internal and external measurements that would be difficult to make with other measuring instruments.
- Vernier calipers can be used to make either inch or metric measurements.
- Inch vernier calipers can make measurements to within one-thousandth of an inch.
- On vernier calipers, outside measurements are read on the bottom scale of the bar and inside measurements are read on the top scale of the bar.
- Dial calipers make measurements quickly and are easier to read than vernier calipers.

KNOWLEDGE REVIEW

1. For what purpose are vernier calipers used in machine shop work?
2. Compare the inch and metric vernier calipers with respect to:
   a. each division on the bar
   b. each division on the vernier scale
3. What are the readings for the following 50-division inch vernier caliper?

a.

(Courtesy of Kelmar Associates)

b.

(Courtesy of Kelmar Associates)

c.

(Courtesy of Kelmar Associates)

d.

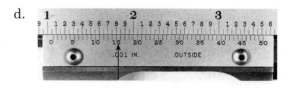

(Courtesy of Kelmar Associates)

U N I T
9

Inspection Tools

The development of new measuring tools and gaging systems, such as electronic, optical, digital, and laser instruments, have made it possible to measure to an accuracy of .000 050 in. or less. It is even possible to measure the accuracy of a part or dimension while it being machined, saving valuable time spent on inspection and eliminating scrap work.

OBJECTIVES

After completing this unit, you should be able to:
- Explain the importance of electro-optical measuring systems in manufacturing processes.
- Explain the importance of gage blocks as a standard of measurement and make a gage block for a given dimension.
- Describe the use of laser and digital measuring systems.

KEY TERMS

electro-optical measuring tools   gage blocks

● GAGE BLOCKS

Global manufacturing can only be possible if everyone involved in manufacturing throughout the world has an acceptable standard to which all measuring tools are set. **Gage blocks**, Fig. 9-1, the accepted world physical standard of measurement, give the inch or millimeter physical form and allow them to be used as a means of calibration or measurement. Some gage block sets are accurate within 2 millionths of an inch (.000002 in.), and for metric blocks, 50 millionths of a millimeter (0.00005 mm). Gage blocks are used to calibrate (set) precision measuring instruments, set sine bars to accurate angles, position machine tool components, set cutting tools, and for quality control (inspection) purposes.

Gage Block Manufacture

Gage blocks, consisting of hardened and ground alloy steel, chrome-plated, cemented carbide, and ceramics, are used for various gaging purposes. The hardened and ground alloy steel blocks are the most common sets used in a machine shop. They are stabilized, through alternate cycles of extreme cold and heat to remove all internal stresses and strains from the metal's microstructure. After this, the two measuring surfaces are ground, lapped, and polished, with the temperature kept at 68°F (20°C) to provide an optically flat surface. The size of each block is marked on one of its surfaces.

Gage Block Accuracy

Gage blocks are used in the machine tool trade as physical standards against which all

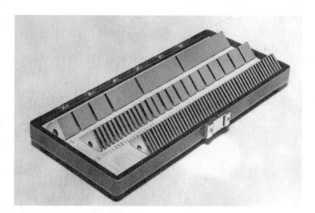

Fig. 9-1 An 83-piece set of inch gage blocks. (Courtesy of DoALL Co.)

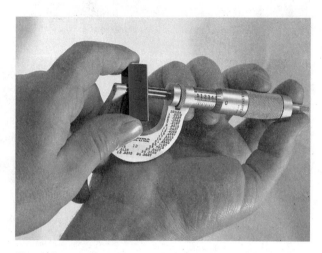

Fig. 9-2 Checking the accuracy of a micrometer using a gage block. (Courtesy of Kelmar Associates)

gages and measuring tools are checked for accuracy. All types of measuring tools, such as micrometers, can be checked for accuracy by measuring a gage block, Fig. 9-2; the micrometer reading should be the same as the size on the gage block. Any micrometer that is not accurate should be adjusted by a qualified person.

Gage blocks in the inch and metric systems are manufactured to three common degrees of accuracy, depending on the purpose for which they are used.

1. The *Class AA set*, commonly called a laboratory or master set, is accurate to ±.000 002 in. in the inch system, and ±0.000 05 mm in the metric system. These gage blocks are used in temperature- and humidity-controlled laboratories as references to compare or check the accuracy of working gages.

2. The *Class A set* is used for inspection purposes and is accurate to ±.000 004 in. in the inch system and ±0.000 1 mm in the metric system.

3. The *Class B set*, commonly called the working set, is accurate to ±.000 008 in. in the inch system and ±0.000 2 mm in the metric system. These blocks are used in the machine shop for machine tool setups, layout work, and measurement.

Gage Block Buildups

It is possible to make over 120,000 different measurements with an 83-piece set of gage blocks. When wrung together properly, gage blocks adhere (stick) so well to each other that they can withstand a 200-pound (lb.) [890 newton (N)] pull. Scientists think this adhesion may be the result of atmospheric pressure, molecular attraction, or the extremely flat surfaces of the blocks. A combination of any or all of these could be responsible.

Wringing Procedure

When wringing blocks together, it is important not to damage them. The procedure for wringing blocks together correctly is shown in Fig. 9-3.

1. Clean the blocks with a clean, soft cloth.

2. Wipe each of the contacting surfaces on the clean palm of the hand or on the wrist. This will remove any dust particles left by the cloth and also applies a light film of oil.

3. Place the end of one block over the end of another block as shown in Fig. 9-3.

4. While applying down pressure on the two blocks, slide one block over the other.

5. If the blocks do not adhere to each other, it is generally because the blocks have not been properly cleaned.

Making a Gage Block Buildup

The 83-piece set of gage blocks consists of four series of block sizes.

1st series— 9 blocks from 1.0001 to 1.0009 in increments of .0001 in.

2nd series—49 blocks from .101 to .149 in increments of .001 in.

3rd series— 19 blocks from .050 to .950 in increments of .050 in.

4th series— 4 blocks from 1.000 to 4.000 in increments of 1.000 in.

For example, to make a gage block for 2.6753 in., proceed as follows:

		Procedure Column	Proof Column
1.	Write the dimension required	2.6753	
2.	Select a block to eliminate the right-hand digit	.1003 ———— 2.5750	.1003
3.	Select a block to eliminate the right-hand digit and bring the next number to a five (5) or zero (0).	.125 ———— 2.450	.125
4.	Continue to eliminate the digits right to left until the remainder is at zero (0)	.450 ———— 2.000	.450
5.	Use a 2.000 block to finish	2.000	2.000
6.	Total the check column to be sure it matches the dimension required.	.000	2.6753

The DoAll vernier gage block, Fig. 9-4, makes it possible to vary dimensions in incre-

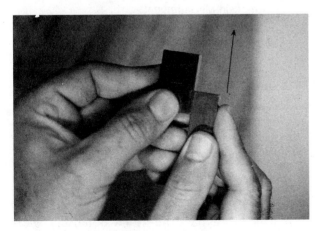

Fig. 9-3 Keep down pressure on the top block when wringing gage blocks together. (Courtesy of Kelmar Associates)

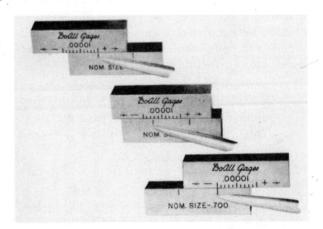

Fig. 9-4 Vernier gage blocks make it possible to vary dimensions in increments of .000 010 in. (Courtesy of DoALL Co.)

ments of .000 010 in. The block is made in two halves, at a very slight angle, and has a nominal size of .700 in. There are 10 vernier graduations on the top block, and if the top half is slid from left to right, the measurement is increased by .000 010 in. The graduations are large enough to be able to make measurements as accurate as .000 002 in.

● ELECTRO-OPTICAL INSPECTION

As the competition for customers becomes keener throughout the world, the need for quality control of the manufacturing process becomes very important. Industry has realized that quality must be manufactured into a product, it cannot be added later. Therefore, careful inspection must become part of the manufacturing operation. As a result, new, improved inspection systems are being used for manufacturing process control to prevent errors and produce better-quality products at lower prices. The ultimate goal of improved quality must be customer satisfaction.

Electro-optical measuring tools and systems are used for a wide variety of workpiece measurements such as dimensions, angles, contours, surface conditions, etc. These tools can be of two types—contact measuring tools and non-contact measuring tools.

Contact Measuring Tools

Contact measuring tools are those that come into physical contact with the part during the measuring or inspection process. These tools are generally electronic or a combination of electro-mechanical. They are generally used when the machining operation is stopped or after the machining is completed to inspect the dimensions of a part or component. The most common electronic measuring tools are:

1. Electronic calipers, micrometers, height gages, and indicators, which supply measurement data to a miniprocessor for statistical process or quality control purposes, Fig. 9-5A.

2. Geometry profile systems, for linear, rotary, and surface finish measurements, collect inspection data quickly, easily, and display statistical process control data on a screen or provide a hard copy, Fig. 9-5B.

3. In-Process gaging is used to control the accuracy of an operation by measuring the

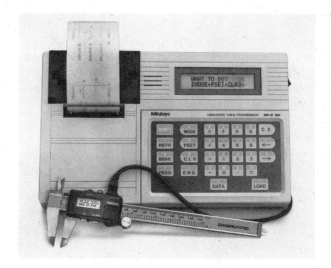

Fig. 9-5A Electronic indicators can supply measurement data for statistical process control. (Courtesy of MTI Corp.)

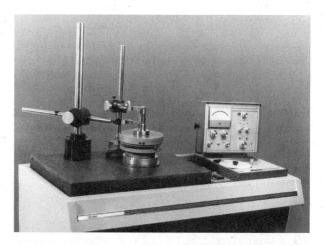

Fig. 9-5B Rotary and linear profile systems are used to collect inspection data. (Courtesy of Giddings & Lewis—Sheffield Measurement)

size of the part while it is being manufactured and providing continuous feedback to the control of the machine to stop the operation when the part is to size.

4. Coordinate Measuring Machines (CMMs), Fig. 9-5C, are advanced multi-purpose quality control systems which replace long, complex, and inefficient conventional inspection methods with simple accurate, and much faster procedures. They can be used to inspect the dimensional and geometric accuracy of a wide variety of parts such as engine blocks, circuit boards, machine components, etc.

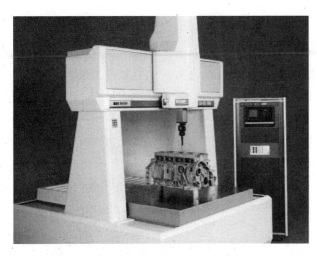

Fig. 9-5C The coordinate measuring machine is an accurate multi-purpose quality control system. (Courtesy of Giddings & Lewis—Sheffield Measurement)

Fig. 9-5D Autometrology systems are used in production where 100% of the part or component must be inspected. (Courtesy of Giddings & Lewis—Sheffield Measurement)

5. Autometrology systems, Fig. 9-5D, are automated systems designed for production operations which require precise tolerances, select fits, or 100% inspection of the part or component. These systems can provide improved quality and faster inspection to high-, medium-, and low-volume manufacturing operations.

Non-Contact Measuring Tools

Non-contact measuring tools are those that do not come into physical contact with the part in the manufacturing or inspection process. These generally consist of some form of optical gaging system which focuses a light from a controlled source onto the object to be inspected. The light is either reflected as a measure of that object's distance or changed in some way by the object's presence. This modification is noted and the dimension is recorded and displayed.

The following are some of the optical gaging systems used in manufacturing:

1. Scanning laser system, Fig. 9-6, consists of a thin beam of laser light which goes through a focusing mirror and scans the part or measurement area at a constant speed. The object being inspected interrupts this beam for a period of time which determines the diameter or thickness of the part. This system is very fast, measuring as many as 500 dimensions in a second, and accurate to within 10 millionths (.000 010) in. or 0.000 250 mm. It is finding wide use in industry to measure moving parts and parts in the process of being manufactured.

2. Linear array consists of a parallel light source which is on one side of the object to be measured and a photo-optical diode array (scale) on the other side. The size of the object is measured by the number of elements in the array which are blocked off. This high-speed system can measure to 50 millionths of an inch, and is only limited by how fast the objects can be electronically scanned and the data processed by the microcomputer.

3. Triangular optical scanners are used for CNC and coordinate measuring machine probes to measure the distance from it to the surface of the part. The signal from the controlled light source of the probe is picked up by a detector. When the part position and the probe position are known, the signal from the optical sensor completes the three-point dimensioning which results in the accurate measurement of the part.

Fig. 9-6 Scanning laser systems can measure as many as 500 parts per second to an accuracy of 10 millionths of an inch. (Courtesy of Hewlett-Packard Co.)

4. Television cameras are being used for some forms of non-contact measurement. The TV camera digitizes the image it sees and compares it to a similar picture of the object which is stored in memory. From this comparison, it is possible to determine the size and shape of the part being measured.

It is quite possible that optical gaging systems will be widely used in the future, because they provide a means of measuring a part during the machining operation and controlling the machine tool so that only accurate parts will be produced.

SUMMARY

- Gage blocks are the tools accepted as the world standard of measurement of both the inch and metric systems.

- Gage blocks are used to calibrate other precision measuring instruments.

- One set of gage blocks can be used to make more than 120,000 different measurements.

- Gage blocks are wrung together in order to construct buildups to provide a wide variety of measurements.

- Electro-optical inspection systems include both contact and non-contact measuring tools.

- Electro-optical inspection systems use electronics and light to perform workplace

measurements with high precision. They can be used to inspect and measure parts while the parts are in the process of being manufactured.

KNOWLEDGE REVIEW

1. List two purposes for gage blocks.

2. Explain how gage blocks are manufactured to ensure their accuracy and stability.

3. What is the accuracy of the following sets of inch gage blocks?
   a. Class AA
   b. Class B

4. Make a gage block buildup for 3.3333 in.

Electro-Optical Inspection

5. Name two types of electro-optical measuring tools.

6. Name four of the most common electronic measuring tools used in industry.

7. Which of the tools in question 6 can be used to measure work while it is being machined?

8. Briefly describe how a scanning laser system measures a part.

9. Name two other non-contact measuring systems used by industry.

SECTION

6

Layout Tools and Procedures

Laying out is the operation of scribing (drawing) center locations, straight lines, arcs, circles, or contour lines on the surface of a piece of metal to show the machinist the finished size and shape of the part to be manufactured. The information regarding the size and shape of the part is taken from a technical drawing, or print, prepared by a draftsperson or tool designer. The care and accuracy of the layout plays an important part in determining the accuracy of the finished part, since the machinist uses these layout lines as a guide for machining. A machinist is expected to be able to read prints, select and use the proper layout tools, and transfer the size and shape of the part from the print to the metal workpiece.

Figures courtesy of (clockwise from far left): Kelmar Associates, Kelmar Associates, The L.S. Starrett Co., The L.S. Starrett Co., Kelmar Associates, Kelmar Associates.

UNIT 10

Laying Out

Layouts are generally classified under two types, *semi-precision* and *precision*. The accuracy required on the finished part determines which type is used. For layouts to be as accurate as possible, all lines and measurements must be made from a machined edge, baseline, or square surface. Regardless of what type of layout is used, the machinist should use the layout lines only as a guide while machining, and should check the actual size of the part with the proper measuring tools.

OBJECTIVES

After completing this unit, you should be able to:

• Prepare various work surfaces for the laying-out operation.
• Select and use the tools for semi-precision layouts.
• Select and use the tools for precision layouts.

KEY TERMS

height gage   laying out
surface gage   surface plate

● LAYING OUT

Laying out is an important first step before any machining operations take place. The following sections describe types of layouts and detailed layout procedures.

The type of layout operation to be performed depends on the degree of accuracy that is required in the finished product. The two main types are the semi-precision layout and the precision layout.

Semi-Precision Layout

This type of layout is generally used on work that does not require an accuracy of less than 1/64 in. (0.38 mm). The tools used for semi-precision layout could include a rule, dividers, adjustable square, center head, bevel protractor, and surface gage, Fig. 10-1. *Witness marks* are placed at intervals along the layout lines with a prick punch, in case the lines are accidentally rubbed off.

Precision Layout

Precision layout is generally used on work that requires an accuracy of less than 1/64 in. (0.38 mm). This layout generally requires the use of electronic, digital, optical, and laser tools, which can be set to an accuracy of .001 in. (0.025 mm) or less. The tools used for precision layout could include vernier, electronic, and digital height gages, gage blocks, optical and laser tools, and layout machines.

Surface Preparation

In order to see layout lines clearly, the work surface must be coated with some type of layout

Fig. 10-2 Layout dye is the most common material used to coat work surfaces. (Courtesy of Kelmar Associates)

material. The most common layout material used in a machine shop is a dark blue liquid called layout dye, Fig. 10-2. Layout dye is fast-drying and provides a sharp contrast so that scribed lines can be clearly seen. A light coating should be applied to a clean, dry work surface with a brush or from a spray can. Care should be used when applying layout dye because it can stain clothing, workbench tops, or tools if not used with care. It is good practice to place the workpiece on a paper towel or cardboard to prevent the dye from staining other surfaces.

Other materials and methods that are sometimes used for coating surfaces for layout are:

- Blue vitriol, a copper sulfate solution, can be used for coating clean *ferrous* machined surfaces.
- Chalk may be rubbed into the rough surface of castings.
- Some metals may be heated only until their surface turns a bluish color.

Layout Equipment and Tools

Many types of equipment and tools are required to produce an accurate, usable layout. Some of the most common layout tools are as follows:

Surface Plates

Surface plates, Fig. 10-3, with their accurate, flat surfaces, provide a good reference surface or base for the starting point of any layout. They may be made of cast iron, granite, or ceramic materials, and some of their surfaces are ground and lapped flat within .0001 in. (0.0025 mm). Granite surface plates are the most common type used for

Fig. 10-1 Common tools used for semi-precision layout work. (Courtesy of Kelmar Associates)

Fig. 10-3 Surface plates provide flat, accurate surfaces for layout and inspection work. (Courtesy of The Taft-Peirce Mfg. Co.)

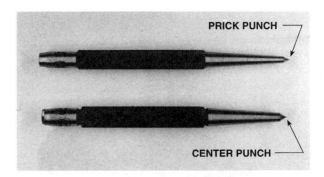

Fig. 10-5 The prick punch and center punch are used for layout work. (Courtesy of Kelmar Associates)

inspection purposes and jig and fixture work. To maintain the accuracy of any surface plate, the top should be covered to keep it clean and free from damage when not in use.

Scriber

A scriber is a layout tool used for drawing layout lines on a work surface. Scribers are made of 3/16 in. (5 mm) diameter tool steel with hardened and tempered points. It is important that the point of the scriber be as sharp as possible to produce clear, thin, layout lines. It is a good habit to occasionally hone the scriber point to keep it sharp. When using a scriber, it is good practice to hold the scriber at an angle so that the point is against the edge of the rule or work edge, Fig. 10-4.

Prick Punch and Center Punch

The prick punch and center punch are very similar layout tools, the only difference being in the angle of their points, Fig. 10-5. The prick punch point is ground to an angle of 30 to 60°, and is used to place witness marks on layout lines. The center punch point is ground to an angle of 90° and is used to enlarge prick punch marks so that a drill may be started easily at the correct hole location. It is important that the point on both of these punches be sharp to maintain the accuracy of the layout or the location for drilled holes. Automatic center punches, Fig. 10-6, have a striking mechanism in the handle, and a downward pressure on the handle releases the striking mechanism to make an impression on the work surface.

The following points are recommended when using a center or prick punch:

1. Make sure that the point of the punch is sharp before starting.

2. Hold the punch at a 45° angle and place the point carefully on the layout line.

3. Tilt the punch to a vertical position and strike it gently with a light hammer.

4. If the punch mark is not in the proper position, correct it as necessary.

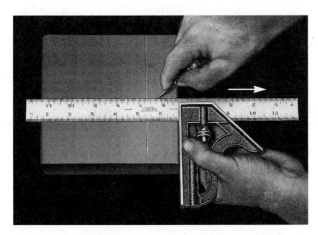

Fig. 10-4 For accurate layout lines, hold the scriber at a slight angle and keep the point against the edge of the rule. (Courtesy of The L.S. Starrett Co.)

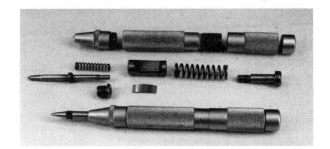

Fig. 10-6 The automatic center punch has a striking mechanism inside the handle. (Courtesy of Kelmar Associates)

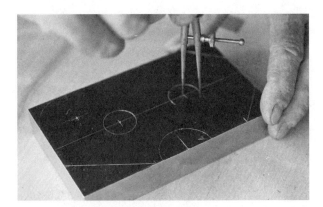

Fig. 10-7 To set dividers to an accurate size, start at a 1 in. or 1 cm line on a rule. (Courtesy of Kelmar Associates)

Divider

A divider, Fig. 10-7, is a layout tool which is used for scribing arcs and circles, transferring measurements, and comparing distances. Spring dividers are adjustable to any size within the maximum opening between the two pointed legs.

To set a divider to a specific size:

1. Place one of the divider points in the 1 in. or 1 cm graduation line of a steel rule.

2. Turn the adjusting nut until the point of the other leg splits the desired graduation line.

3. For accurate layouts, check the setting with a magnifying glass.

Solid Square

The solid square, Fig. 10-8, often called the master square, is used to compare the squareness of workpieces or other squares for accuracy. It consists of two main parts—the beam and the blade, both of which are hardened and ground to ensure the solid square's accuracy. Another type of solid square has a beveled-edge blade which allows it to make line contact with the work for greater accuracy. Three common uses of a solid square are to:

a. check a work surface for flatness,

b. check if two adjacent surfaces are at 90° to each other,

c. compare other squares against its accuracy.

Combination Set

The combination set, Fig. 10-9, is one of the most useful layout and measuring items used in

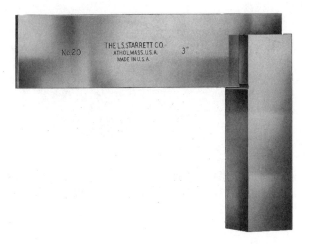

Fig. 10-8 The knife-edge solid square provides a line contact with workpieces or other tools for high accuracy. (Courtesy of The L.S. Starrett Co.)

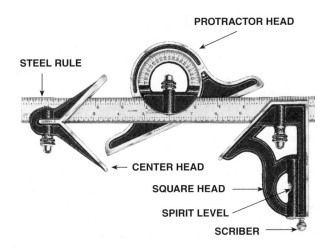

Fig. 10-9 The combination set has many uses in layout and inspection work. (Courtesy of The L.S. Starrett Co.)

a machine shop. The set consists of four main tools—steel rule, square head, bevel protractor, and a center head.

Steel Rule

The *steel rule* is grooved along its length which allows other parts such as the square head, bevel protractor, or center head to be fastened to it. It can also be used as a straightedge or for measuring purposes. Inch combination set rules are generally graduated in eighths and sixteenths on one side and thirty-seconds and sixty-fourths on the other side. Metric combination set rules are generally graduated in millimeters and half millimeters.

Square Head

The *square head* can be fastened to any position along the rule and is used for laying out lines parallel or at right angles to an edge. The combination square can also be used for checking 45° and 90° angles, as well as a depth gage.

Laying out Lines Parallel and at Right Angles

To ensure accuracy in any layout, be sure to remove all burrs from the workpiece and start the layout from a square machined surface or work edge. Proceed with the layout using the following steps:

1. Clean the work surface and apply a light coat of layout dye.

2. Hold the workpiece in a vise, or by any other means which will not interfere with the layout tools.

3. Lay out all lines *parallel* to an edge first, starting with the dimension closest to the work edge.

4. Extend the rule beyond the body of the square until the desired dimension line is split in half.

5. Tighten the locknut and recheck the rule setting; reset if necessary.

6. Hold the body of the square tightly against the machined edge, keeping the rule flat on the work surface.

7. Hold a sharp scriber at an angle, keeping the point against the rule edge, and scribe a line, Fig. 10-10.

8. Continue to move the square and scribe lines until the layout line is the required length.

9. Set the rule in the square to the dimensions for other parallel lines, being sure that only half the graduation line is showing.

10. With a sharp scriber, draw all horizontal lines, keeping the body of the square firmly against the machined edge.

11. For the lines or dimensions at right angles, use the squared end of the workpiece, right angle to the other edge, as the starting surface.

12. Follow the same procedure as in steps 3 to 10 to scribe the lines at right angles, Fig. 10-11.

13. Compare the accuracy of all layout lines with the dimensions on the part drawing.

14. Correct dimensions, if necessary, lightly prick-punch the part outline, and center punch the hole locations.

Bevel Protractor

When the *bevel protractor* is fitted to the rule, it can be used to lay out and check angles from 0 to 180° to an accuracy of ± 0.5° (30°). On some bevel protractor models, the scale is graduated from 0 to 90° in both directions from the center.

Center Head

The included angle of a center head is 90°, and when it is fastened on the rule, it forms a center square. The *center head* may be used for locating centers on the ends of round, square, and octagonal stock. It is more accurate for locat-

Fig. 10-10 Using the combination square head to lay out lines parallel to a machined edge. (Courtesy of Kelmar Associates)

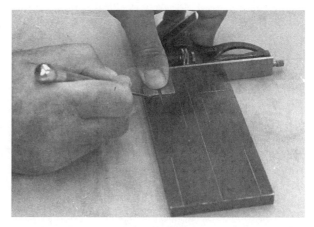

Fig. 10-11 Laying out lines at right angles with a combination square head. (Courtesy of Kelmar Associates)

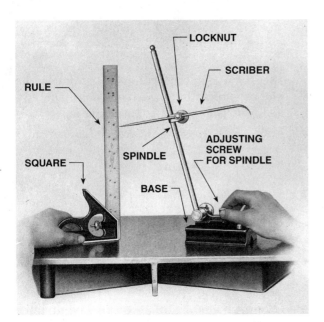

Fig. 10-12 Use the combination square to set a surface gage to dimensions which do not require greater accuracy than 1/64 in. (0.38 mm). (Courtesy of Brown & Sharpe Mfg. Co.)

ing center when the work is true in shape and the ends have been machined square.

Surface Gage

The **surface gage** is a useful layout tool that can be used on a surface plate or any flat surface for scribing layout lines on a workpiece. It consists of a heavy base, a movable upright spindle, and an adjustable scriber which can be clamped to any position along the spindle length.

The surface gage must be set to a required dimension using a combination square, gage, or other type of tool, Fig. 10-12. The type of comparison instrument used to set the surface gage will depend upon the accuracy required for the workpiece. When a workpiece is clamped to an angle plate, the surface gage may be used to scribe vertical parallel lines on the workpiece. When the angle plate is turned on its side, without removing the workpiece, horizontal parallel lines can be laid out at 90° to the vertical lines. Therefore, both horizontal and vertical workpiece lines can be drawn by using a surface gage in one setup if the workpiece is accurately clamped to the angle plate.

The base of the surface gage has a V-groove which allows it to be used on cylindrical surfaces. The base also contains two pins which,

when pushed down and held against the edge of a surface plate or a workpiece, can be used for scribing parallel lines.

Height Gages

Height gages, which include the vernier height gage and dial height gage are precision instruments used in toolrooms and inspection rooms to lay out and measure vertical distances to .001 in. (0.025 mm) accuracy. They generally consist of a hardened and ground base, a graduated vertical beam, a sliding jaw containing the vernier scale, and a scriber. The sliding jaw assembly can be raised or lowered to any dimension within the range of the beam. Fine measurement adjustments are made by means of an adjusting nut. The scale of the standard vernier height gage is read the same as the scale of a vernier caliper.

Accessories such as offset scriber, depth gage attachment, and dial indicator, make the *vernier height gage* suitable for layout and inspection work. With a scriber fastened to the movable jaw, Fig. 10-13A, accurate layout lines can be scribed parallel to the top of the surface plate, while an offset scriber allows setting of heights from the face of the surface plate. A depth gage attachment mounted on the movable jaw makes the measurement of heights and steps, and the depths of holes and slots an easy operation.

A dial indicator may be fastened to the sliding jaw to make the vernier height gage an excellent inspection tool. With this setup, all dimensions such as holes or surfaces can be checked within an accuracy of .001 in. (0.02 mm) on the vernier scale. Greater accuracy is possible if the indicator setting is checked with gage blocks.

Dial and digital height gages, Fig. 10-13B, have a big advantage over standard height gages because they are easier to read and their accuracy is usually .001 in. (0.02 mm) or less. Measurements can be read directly from a dial or digital display, eliminating many reading errors which have been common with standard height gages. The attachments that fit standard height gages can also be used on dial and digital height gages, making them very accurate and useful layout and inspection tools.

● LAYOUT ACCESSORIES

There are many accessories available for layout work which can help the operator to make accurate layouts easier, Fig. 10-14.

(A) VERNIER HEIGHT GAGE

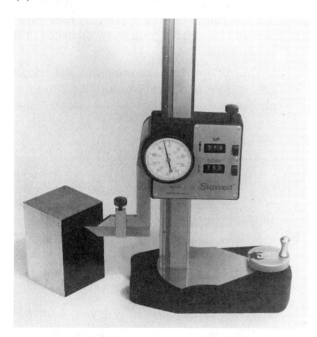

(B) DIGITAL HEIGHT GAGE

Fig. 10-13 Types of height gages which can be used for layout work. (Courtesy of The L.S. Starrett Co.)

a.   An angle plate is a precision L-shaped tool usually made of hardened steel. All its surfaces are ground to an accurate 90° angle and are square and parallel. Toolmaker's clamps, C-clamps, or bolt holes, are used to

(A) ANGLE PLATE

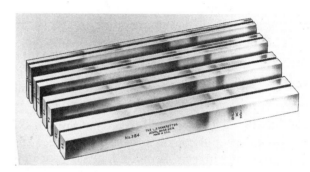

(B) PARALLELS

(C) V-BLOCKS

(D) RADIUS GAGES

Fig. 10-14 Accessories that can be useful in layout work. (A), (C) (Courtesy of Kelmar Associates) (B), (D) (Courtesy of The L.S. Starrett Co.)

hold work vertically on the face of an angle plate for layout purposes. Since all adjacent sides on an angle plate are at 90° to each other, vertical and horizontal layout lines can be scribed on a work surface by turning the angle plate on its side.

b. Parallels are hardened square or rectangular steel bars with surfaces that have been ground square and parallel. They are usually available in matched pairs and used in layout work to raise the workpiece height while keeping the machined edge or baseline parallel with the top of the surface plate.

c. V-blocks, generally made of hardened and ground steel, are usually available in matched pairs. They have an accurate 90° V-slot which is used to hold round work for layout, inspection, drilling, or grinding purposes. Some V-blocks may be turned 90° to allow layout lines to be scribed at right angles to other lines without removing the workpiece from the V-block.

d. Radius gages are used to lay out and check concave and convex forms. These gages are available in a set of individual gages or in a series of radius leaves mounted in one holder. Radius gages are available in a wide range of inch and metric sizes, with the most common inch set consisting of gages from 1/64 to 1/2 in. varying in steps of 1/64 in.

e. A template is a master pattern made of a thin piece of metal that has been machined to the exact shape and size of the finished workpiece. A template can be used as a guide or master to lay out a number of workpieces which have the same form. After the surface of the work is coated with a layout dye, the outline of the template can be scribed (traced) on each workpiece using a scriber with a sharp point. Templates may also be used to check the accuracy of forms, shapes, or other special contours.

● CNC MACHINE LAYOUT

CNC machine tools do not require extensive layout work because the cutting path of the tool is controlled by the computer program and not by the machine operator. However, on the first part to be machined, it is considered good practice to lay out the starting point of the cut and three other key points on the job as a visual

check of the program. If the cut does not start at the correct point of the layout, it is wise to stop the machine to check the layout and, if necessary, the computer program.

SUMMARY

- Laying out is the process of marking the surface of a piece of metal to show the machinist the finished size and shape of a part.
- Layout information is obtained from a technical drawing or print.
- The choice of whether to use precision layout or semi-precision layout depends on the degree of accuracy required for the finished product.
- Tools and equipment typically used in layout include surface plate, layout dye, scriber, divider, punches, surface and height gages, etc.
- Layout lines are only to be used as a rough guide; the part size should be carefully checked using more precise measuring tools.

KNOWLEDGE REVIEW

1. Describe the process of laying out.
2. a. When should semi-precision layout be used?
   b. What layout tools are used for this layout?
3. a. When should precision layout be used?
   b. What layout tools are used for this layout?
4. Name three precautions that should be observed when using layout dye.
5. What type of surface plate is commonly used for inspection and jig and fixture work?
6. Compare the point angle of a prick punch and a center punch and state where each is used.
7. How can a divider be set fairly accurately to size?

8. Which type of solid square allows it to make line contact with the work surface?

9. Name the four main parts of a combination set.

10. Which part of the combination set can be used to lay out lines parallel and at right angles?

11. What tools can be used to set a surface gage to a dimension?

12. Name three height gage accessories and state the purpose of each.

13. What advantages do dial and digital height gages have over standard height gages?

14. List four layout accessories and give one purpose for each.

15. Why does work machined on CNC tools not require extensive layout work?

SECTION

7

PRE-REDUCTION

Metals and Their Properties

Iron is one of the most abundant elements found on the earth and makes up approximately 5% of the earth's crust. For at least 5000 years, humans have been removing this valuable raw material from the earth to create products that have helped them improve their productivity and their standard of living. From the first crude lump of metal that was created fifty centuries ago in a primitive campfire, literally thousands of different kinds of irons and steels have been developed. Iron and steel products are found in, or used to produce, almost every product used by humans.

Steel, the most versatile of metals[1], can be made hard enough to cut glass, pliable enough to use in paper clips, flexible as the steel in springs, strong enough to stand stresses up to 500,000 psi (3445 MPa), or heat-resistant enough to withstand the searing heat of rocket exhaust engines. It can be drawn into wire .001 in. (0.02 mm) thick, or formed into giant girders for buildings and bridges. With the addition of certain alloying elements, it can be made to resist heat, cold, rust, and chemical action. The age of space exploration and nuclear energy requires newer and tougher steels many of which have already been developed. It seems that there will be no end to the development of new and better steels to serve the special needs of humans.

The need to preserve our environment and reduce pollution throughout the world has resulted in many changes in the iron and steelmaking industries. The Bessemer furnace is no longer used, and the open-hearth furnace, long an industry standard, has rarely been used since the early 1960s. It seems only a matter of time before blast furnaces and coke-making ovens will be replaced by new furnace techniques, direct ironmaking, direct steelmaking, and minimills that create less pollution and are far more efficient.

In order to use metals productively, it is necessary to understand their special properties and to develop methods for enhancing those properties. Because each metal is unique, it may be perfect to use in one application but totally unsuitable for another. Thus, knowledge of metallic properties is important in selecting and using machine tools.

[1]*Although steel, brass, and bronze are metal alloys in common usage, they are frequently referred to as metals. This convention will also be followed in this textbook.*

Figures courtesy of (clockwise from far left): Kaiser steel Corp., Nucor Corp., American Iron and Steel Institute, American Iron and Steel Institute.

UNIT 11

Manufacture of Iron and Steel

The study of metals, which includes the manufacture of iron and steel and the modification of their structures to give them certain qualities and properties, should include:

- The mining and processing of ore into metals.
- The improvement of metals and the development of alloy steels.
- The heat treatment of metals and alloys to improve their qualities.

In ancient times, iron was a rare and precious metal. Steel, a purified form of iron, has become one of our most useful servants. Nature supplied the basic raw products of iron ore, coal, and limestone, and our ingenuity has converted them into countless numbers of products that improve our living standards.

OBJECTIVES

After completing this unit, you should be able to:

- Explain the procedure for producing iron and steel.
- Describe the various types of furnaces used for steelmaking.
- Explain the processes involved in direct iron and steelmaking and describe minimills.

KEY TERMS

basic oxygen furnace
direct steelmaking
minimills

cast iron
electric-arc furnace
wustite

continuous casting
ladle metallurgical station

● RAW MATERIALS

The raw materials used for steelmaking, iron ore, coal, and limestone must be brought together, often from great distances, and smelted in a blast furnace to produce the pig iron that is used to make steel.

Iron Ore

Iron ore is the chief raw material used in the manufacture of iron and steel. The main sources of iron ore in the United States are the Great Lakes states of Michigan, Minnesota, and Wisconsin. In Canada, large ore deposits are found in the Steep Rock and Michipicoten districts on the north shore of Lake Superior and in the Ungava district near the Quebec-Labrador border.

Mining Iron Ore

When layers of iron ore are near the earth's surface, a process called open-pit mining is used to mine the ore. In this process, the surface material, consisting of sand, gravel, and boulders, is first removed. Next, the iron ore is scooped up by power shovels and transported by trucks, railway cars, or conveyors to the steel mills, Fig. 11-1A. About 75% of the ore mined is removed by open-pit mining.

When layers of iron ore are too deep in the earth to make open pit mining economical, underground or shaft mining is used, Fig. 11-1B. Shafts are sunk into the earth, and passageways are cut into the ore body. The ore is blasted loose or dug out and brought to the surface by shuttle cars or conveyor belts.

Fig. 11-A Shallow iron ore deposits are mined by the open-pit method. (Courtesy of U.S. Steel Corp.)

Fig. 11-1B Deep iron ore deposits are mined by the underground or shaft mining method. (Courtesy Of U.S. Steel Corp.)

Types of Iron Ore

Some of the most important types of iron ore are:

Hematite, a rich ore containing about 70% iron. It ranges in color from gray to bright red.

Limonite, a high-grade brown ore containing water which must be removed before the ore can be shipped to steel mills.

Magnetite, a rich, gray to black magnetic ore containing over 70% iron.

Taconite, a low-grade ore containing only about 20 to 30% iron, which is uneconomical to use without further treatment.

Pelletizing Process

Low-grade iron ores are uneconomical to use in the blast furnace, and as a result, go through a pelletizing process where most of the rock is removed and the ore is brought to a higher iron concentration. Some steelmaking firms are now pelletizing most of their ores to reduce transportation costs and the problems of pollution and slag disposal at the steel mills.

The crude ore is crushed and ground into a powder and passed through magnetic separators where the iron content is increased to about 65%, Fig. 11-2A. This high-grade material is mixed with clay and formed into pellets about 1/2 to 3/4 in. (12 to 20 mm) in diameter in a pelletizer. The pellets are then covered with coal dust and sintered (baked) at 2354°F (1290°C), Fig. 11-2B. The resultant hard, highly concentrated pellets will remain intact during transportation to the steel mills and unloading into the blast furnace.

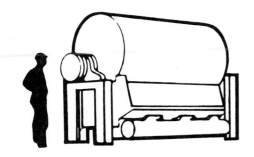

Fig. 11-2A Iron ore is separated from the rock in a magnetic separator. (Courtesy of American Iron and Steel Institute)

Fig. 11-2B Iron ore pellets are sintered (hardened) in a pellet hardening furnace. (Courtesy of American Iron and Steel Institute)

Coal

Coke, which is used as a source of heat in the blast furnace, is made from a special grade of soft coal that contains small amounts of phosphorus and sulfur. The main sources of this coal are mines in West Virginia, Pennsylvania, Kentucky, and Alabama. Coal may be mined by either strip or underground mining. Before being converted into coke, the coal is crushed and washed. It is then loaded into the top of long, high, narrow ovens which are tightly sealed to exclude air. The coal is baked at 2150°F (1150°C) for about eighteen hours, then dumped into railcars and quenched with water, Fig. 11-3.

The gases formed during the coking process are distilled into valuable by-products such as tar, ammonia, and light oils. These by-products are used in the manufacture of fertilizers, nylon, synthetic rubber, dyes, plastics, explosives, aspirin, and sulfa drugs.

Limestone

Limestone, a gray rock consisting mainly of calcium carbonate, is used in the blast furnace as a flux to fuse and remove the impurities from the

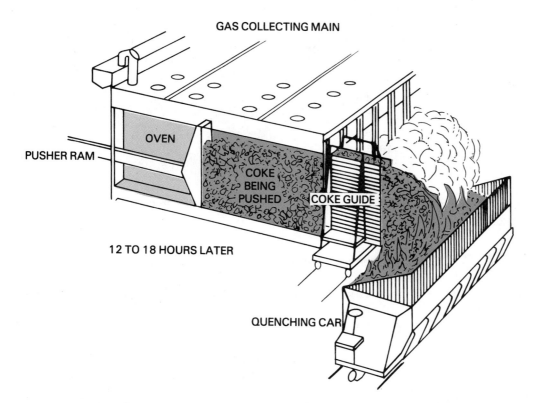

Fig. 11-3 Coal is converted into fast-burning coke in long, narrow coking ovens. (Courtesy of American Iron and Steel Institute)

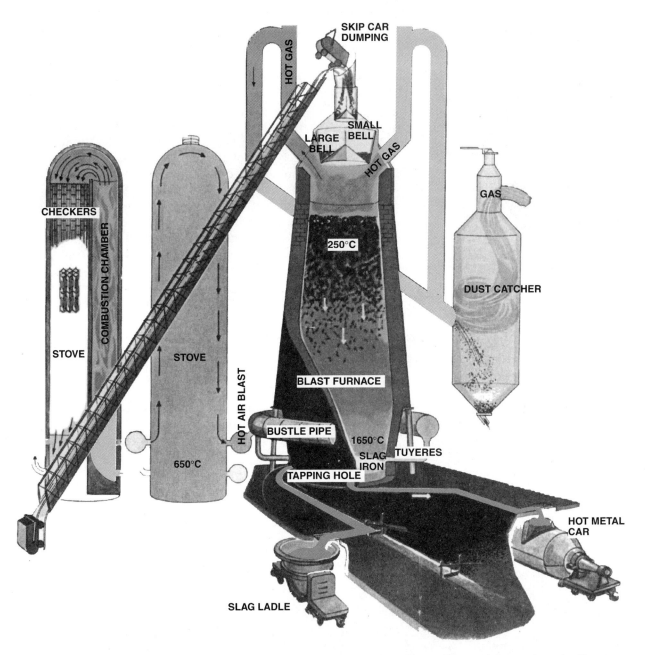

Fig. 11-4 A blast furnace produces pig iron that is the first step in the production of most irons and steels. (Courtesy of American Iron and Steel Institute)

iron ore. It is also used as a purifier in the steel-making furnaces. Limestone is usually found fairly close to steelmaking centers. It is generally mined by open-pit quarrying where the rock is removed by blasting. It is crushed to size before shipment to the steel mill.

● MANUFACTURE OF PIG IRON

The first step in the manufacture of any iron or steel is the production of pig iron in the blast furnace. The blast furnace, Fig. 11-4, about 130 ft. (40 m) high, is a huge steel shell lined with heat-resistant brick. Once started, the blast furnace runs continuously until the brick lining needs renewal or the demand for the pig iron drops.

Iron ore, coke, and limestone are measured out carefully in proper proportions and carried to the top of the furnace in a skip car. Each ingredient is dumped separately into the furnace through the bell system, forming layers of coke, limestone, and iron ore in the top of the furnace.

A continuous blast of hot air from the stoves at 1200°F (650°C) passes through the bustle pipe and tuyeres causing the coke to burn vigorously. The temperature at the bottom of the furnace reaches about 3000°F (1650°C) or higher. The carbon of the coke unites with the oxygen of the air to form carbon monoxide, which removes the oxygen from the iron ore and liberates the metallic iron. The molten iron trickles through the charge and collects in the bottom of the furnace.

The intense heat also melts the limestone which combines with the impurities from the iron ore and coke to form a scum called slag. The slag also seeps down to the bottom of the charge and floats on top of the molten pig iron.

Every four to five hours, the furnace is tapped and the molten iron, up to 350 tons (318 t), flows into a hot metal or bottle car and is taken to the steelmaking furnaces. Sometimes the pig iron is cast directly into pigs, which are used in foundries for making cast iron. At more frequent intervals, the slag is drawn off into a slag car or ladle and is later used for making mineral wood insulation, building blocks, and other products.

Direct Ironmaking

As with other technologies which are constantly changing, the ironmaking processes are undergoing changes which could possibly eliminate the need for the blast furnace and coking ovens in the future. In the late 1980s, iron and steel manufacturers, in conjunction with the American Iron and Steel Institute, started a five-year developmental program to improve the efficiency of ironmaking. This has the full support of the U.S. Department of Energy which has encouraged manufacturers to improve their steelmaking capabilities and also reduce the pollution caused by former methods. The aim of the direct ironmaking process, Fig. 11-5, is to produce iron in a one-step process, instead of the normal three steps, which would eliminate the need for the blast furnace and the coking

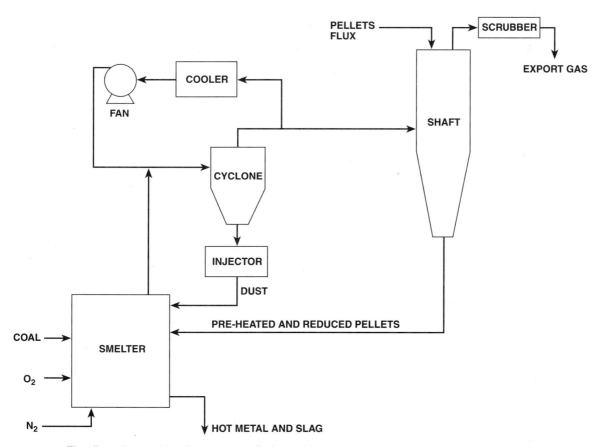

Fig. 11-5 The direct ironmaking furnace can eliminate the need for the blast furnace and coking ovens. (Courtesy of American Iron and Steel Institute)

ovens. It is based on the smelting of iron ore, pulverized coal, fluxes, and oxygen in a liquid bath.

This process, called bath smelting or coal-based ironmaking, should be available to industry by the mid- to late-1990s. It has the potential of a continuous, less-polluting process, requiring 27% less energy, and lowering manufacturing costs by $10 per ton. The direct ironmaking process reuses the high-temperature heat from the postcombustion of the process offgas in the smelting process. The goal of this completely enclosed smelting process is to make the steel industry more competitive with offshore manufacturers and reduce its capital and operating costs.

MANUFACTURE OF CAST IRON

Most of the pig iron manufactured in a blast furnace is used to make steel. However, some is used to manufacture cast iron products. **Cast iron** is manufactured in a cupola furnace which resembles a huge stovepipe, Fig. 11-6.

Layers of coke, solid pig iron, scrap iron, and limestone, are charged into the top of the furnace. After the furnace is charged, the fuel is ignited and air is forced in near the bottom to assist combustion. When the iron is melted, it settles to the bottom of the furnace and is then tapped into ladles.

The molten iron is poured into sand molds of the required shape, and the metal assumes the shape of the mold. After the metal has cooled, the castings are removed from the molds.

The principal types of iron castings are:

Gray iron castings, made from a mixture of pig iron and steel scrap, are the most widely used. They are made into a wide variety of products, including bathtubs, sinks, parts for automobiles, locomotives, and machinery.

Chilled iron castings are made by pouring molten metal into metal molds so that the surface cools very rapidly. The surface of such castings becomes very hard. These castings are used for crusher rolls and for other products requiring a hard, wear-resistant surface.

Alloyed castings contain certain amounts of alloying elements such as chromium, molybdenum, and nickel. Castings of this type are used extensively by the automobile industry.

Malleable castings are made from a special grade of pig iron and foundry scrap. After these castings have solidified, they are an-

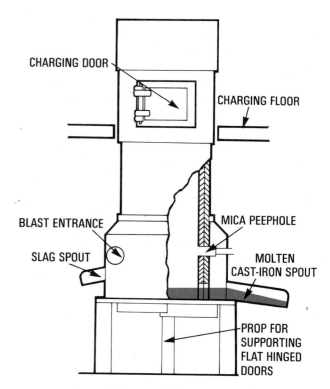

Fig. 11-6 A cupola furnace is used to manufacture various types of cast irons. (Courtesy of Kelmar Associates)

nealed in special furnaces. This makes the iron malleable and resistant to shock.

MANUFACTURE OF STEEL

In order to improve productivity and reduce manufacturing costs, the methods of manufacturing steel have undergone tremendous changes. The open-hearth furnace and the Bessemer converter, little used since the 1960s, have been largely replaced by more efficient direct-current, electric-arc, and basic oxygen furnaces. New, smaller steel plants, called minimills, are now producing steel faster and at lower costs. New iron and steelmaking processes such as direct ironmaking, direct steelmaking, post-combustion, use of iron carbide, strip casting, and coal-based production of liquid iron have been developed, or are in their late developmental stages. Steel manufacturers throughout the world are faced with the problems of reducing pollution, lowering manufacturing costs, and increasing the quality of their products.

Manufacturing Process

Before molten pig iron from the blast furnace can be converted into steel, some of its impurities must be burned out. This is done in one of two different types of furnaces: the basic oxygen furnace, or the DC or AC electric-arc furnace.

For many years, about 90% of all steel produced in North American was made by open-hearth furnaces. The remainder of the steel was produced by Bessemer converters and electric furnaces.

With the introduction of the basic oxygen process in 1955, the emphasis on steelmaking processes shifted. Steelmakers found that the addition of oxygen to any steelmaking process speeded production. The basic oxygen furnace was developed as a result. Today, about 50% of all steel is made in basic oxygen furnaces and in open-hearth furnaces modified by the addition of oxygen lances. Special tool steels are still made by electric furnaces, but the production of steel by the Bessemer process has become practically non-existent.

Basic Oxygen Furnace

The **basic oxygen furnace**, Fig. 11-7, is a cylindrical, brick-lined furnace with a dished bottom and cone-shaped top. It may be tilted in two directions for charging and tapping, but is kept in the vertical position during the steelmaking process.

With the furnace tilted forward, scrap steel (30 to 40% of the total charge) is loaded into the furnace, Fig. 11-7A. Molten pig iron (60 to 70% of the total charge) is added, Fig. 11-7B. The furnace is then moved to the vertical position where the fluxes (mainly burnt lime) are added, Fig. 11-7C. When the furnace is still in the upright position, a water-cooled oxygen lance is lowered into the furnace until the top of the lance is the required height above the molten metal. High-pressure oxygen is blown into the furnace, causing a churning, turbulent action during which the undesirable elements are burned out of the steel, Fig. 11-7D. The oxygen blow lasts for about 20 minutes. The lance is then removed and the furnace is tilted to a horizontal position. The temperature of the metal is taken and samples of the

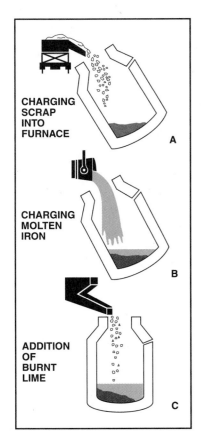

CHARGING SCRAP INTO FURNACE — A

CHARGING MOLTEN IRON — B

ADDITION OF BURNT LIME — C

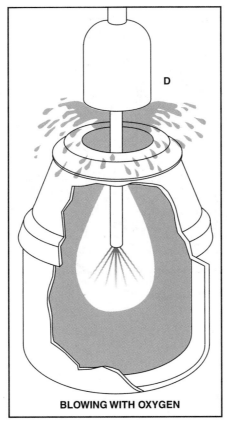

BLOWING WITH OXYGEN

D

TRAPPING THE FURNACE — E

POURING THE SLAG — F

Fig. 11-7 The key steps in the operation of a basic oxygen furnace. (Courtesy of Inland Steel Co.)

metal are tested. If the temperature and the samples are correct, the furnace is tilted to the tapping position, Fig. 11-7E, and the metal is tapped into a ladle. Alloying elements are added at this time to give the steel its desired properties. An alloy is a combination of two or more metals designed to give desired properties. After tapping, the furnace is tilted in the opposite direction to an almost vertical position to dump the slag into a slag pot, Fig. 11-7F. About 300 tons (272 t) of steel can be produced in 1 hour in a large basic oxygen furnace.

Electric Arc Furnace

Either DC or AC **electric-arc furnaces**, Fig. 11-8, are used to make fine alloy and tool steels. Because the heat, the amount of oxygen, and the atmospheric conditions can be accurately controlled in the electric furnaces, they are used to make steels that cannot readily be produced in any other way.

Carefully selected steel scrap, containing smaller amounts of the alloying elements than are required in the finished steel, is loaded into the furnace. Three carbon electrodes are lowered until an arc strikes from them to the scrap. The heat generated by the electric arcs gradually melts all the steel scrap. Alloying materials, such as chromium, nickel, tungsten, etc., are then added to make the type of alloy steel required. Depending on the size of the furnace, it takes from 4 to 12 hours to make a heat of steel. When the metal is ready to be tapped, the furnace is tilted forward and the steel flows into a large ladle. The steel is teemed (poured) into ingots from the ladle.

● DIRECT STEELMAKING

In the late 1980s, the American Iron and Steel Institute, along with most of the iron and steel manufacturers, initiated a developmental program, with U. S. Department of Energy funding, to improve the efficiency of steelmaking and reduce manufacturing costs. The process was designed to bypass the blast furnace and the coking ovens and to make steel directly from iron ore. Fig. 11-9 shows the **direct steelmaking** process which involves four major components or steps:

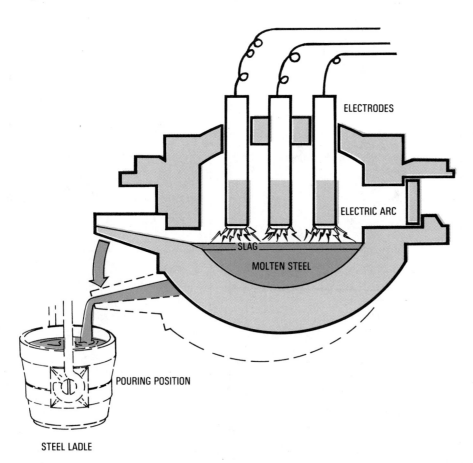

ELECTRODES

ELECTRIC ARC

SLAG

MOLTEN STEEL

POURING POSITION

STEEL LADLE

Fig. 11-8 Electric furnaces are used to manufacture fine alloy and tool steels. (Courtesy of U.S. Steel Corp.)

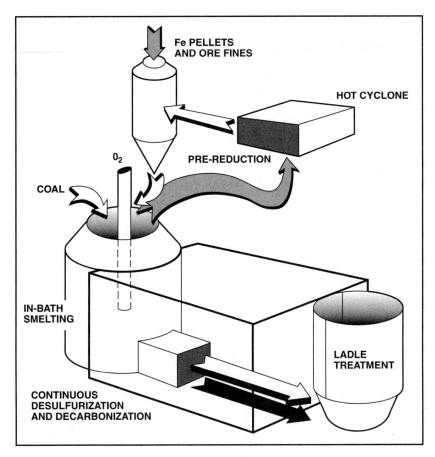

Fig. 11-9 The direct-steelmaking process involves four steps; smelting, pre-reduction, offgas cleaning, and refining. (Courtesy of American Iron and Steel Institute)

1. Smelting

In-bath smelting is the heart of the direct steelmaking process. Oxygen, iron pellets or pre-reduced iron ore, pulverized coal, and flux are gravity-fed into a molten iron-slag bath. Oxygen partially burns the coal to produce some of the heat required to drive the process. The iron oxide melts in the slag and is reduced to molten iron by the carbon in the coal. The rest of the energy required comes from the partial combustion of the carbon monoxide which is produced during the pre-reduction of the iron oxide.

2. Pre-Reduction

This process is used to preheat and pre-reduce the iron ore fed to the smelter using the gases given off during the smeltering operation. Removal of 30% of the oxygen in the iron ore pellets reduces them to **wustite**, an iron oxide. The reducing capabilities of the smelter offgases limit the reduction of the iron ore to wustite only, thereby producing a stable feedstock to the smelter.

3. Offgas Cleaning and Handling

The hot, dust-laden gas produced in the smelter is cooled with a recirculated gas stream and then passed through the hot cyclone separator to remove most of the dust. The dust, which consists mostly of carbon and iron oxide, is recycled back into the smelter.

The cleaned, cooled gas exiting from the cyclone is split into two streams. One passes through a water scrubber that cools it for mixing with the hot smelter gases. The other stream goes to a vertical shaft furnace to heat and partially reduce the iron ore pellets before they are charged into the smelter.

4. Refining

The refining process (desulfurization and decarburization) produces liquid steel which is suitable for ladle metallurgy treatment (addition of desirable chemical elements). From this point, the liquid steel is processed into steel products by continuous casting, rolling, etc.

The direct steelmaking system is entirely enclosed for the refining stage, making it almost pollution-free. The only release into the atmosphere is from the combustion of the off-gas from the shaft-furnace to capture its residual fuel value. This fuel gas has been reduced to less than 40 parts/million (ppm) hydrogen sulfide. This process is very efficient because the only energy lost is the difference in heat energy from the smelter offgas and the cooler shaft furnace process gas.

STEEL PROCESSING

After steel has been refined in a furnace, it is tapped into ladles, where the necessary alloying elements and deoxidizers may be added. The molten steel may be teemed into ingots for later use, or be formed directly into slabs by the continuous-casting process.

Steel teemed into ingot molds is allowed to solidify. The ingot molds are then removed or stripped, and the hot ingots are placed into soak-

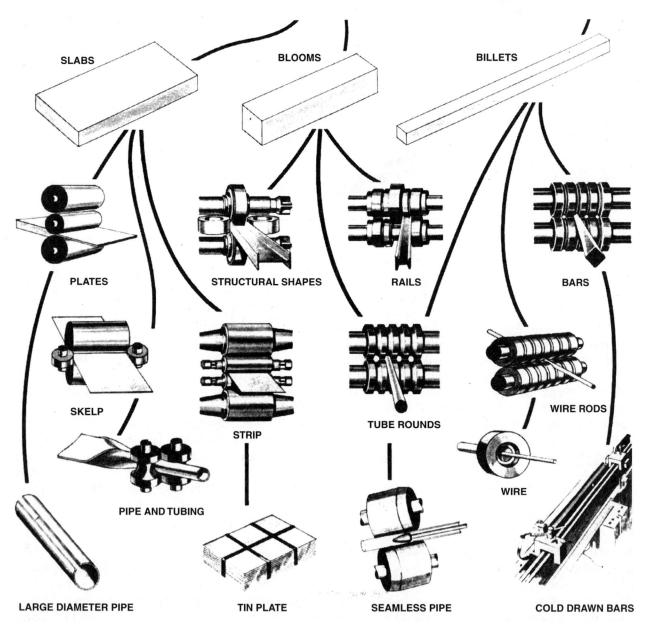

Fig. 11-10 A wide variety of steel shapes and sizes are produced by rolling mills. (Courtesy of American Iron and Steel Institute)

ing pits at 2200°F (1204°C) to bring them to a uniform temperature. The reheated ingots are then sent to rolling mills, where they are rolled into various shapes such as blooms, billets, and slabs, Fig. 11-10.

- **Blooms** are generally rectangular or square and are larger than 36 sq. in. (232 cm²) in cross-sectional area. They are used to manufacture structural steel and rails.

- **Billets** may be rectangular or square, but are less than 36 sq. in. (232 cm²) in cross-

sectional area. They are used to manufacture steel rods, bars, and pipes.

- **Slabs** are usually thinner and wider than billets. They are used to manufacture plate, sheet, and strip steel.

Strand or Continuous Casting

Strand or **continuous casting**, Fig. 11-11, is the most modern and efficient method of converting molten steel into semi-finished shapes such as blooms, billets, and slabs. Continuous casting has

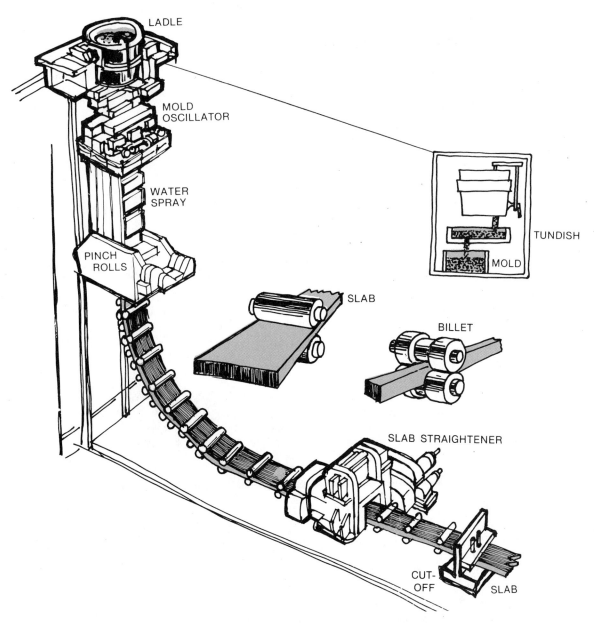

Fig. 11-11 The continuous-casting process is used to manufacture steel blooms and slabs. (Courtesy of American Iron and Steel Institute)

become the main method of producing blooms or slab steel. About 95% of the semi-finished steel produced in the United States, Europe, and Japan is made by continuous casting.

Molten steel from the furnace is transported in a ladle to the top of the strand or continuous caster and poured into the tundish. The tundish acts as a reservoir, permitting the empty ladle to be removed and the full ladle to be poured without interrupting the flow of molten metal to the caster. The steel is stirred continuously by a nitrogen lance or by electromagnetic devices.

The molten steel drops in a controlled flow from the tundish into the mold section. Cooling water in the mold wall quickly chills the outside of the metal to form a solid skin, which becomes thicker as the steel strand descends through the cooling system. As the strand reaches the bottom of the machine, it becomes solid throughout. The solidified steel is moved in a gentle curve by bending rolls until it reaches a horizontal position. The strand is then cut into required lengths by a traveling cutting torch. In some strand casting machines, the solidified steel is cut when it is in the vertical position. The slab or billet then topples to the horizontal position and is taken away.

● MINIMILLS

The integrated coke making, ironmaking, and steelmaking facilities that provided steel companies with good service over many years are now old, inefficient, and impractical to repair and maintain. Since the 1960s, the integrated (big steel) mills, have been facing fierce competition from offshore steel manufacturers. Companies with integrated mills have been looking for technology which would enable them to produce a better quality product at lower cost. **Minimills** have provided manufacturers with a less expensive, more flexible, electric-furnace steelmaking process.

Minimills are smaller, faster, and more efficient than the large-size integrated mills. They use DC electric-arc furnaces and avoid the coke and ironmaking steps used in larger steel mills. The furnaces are fast, producing 180 to 200 tons of steel per hour at 20% lower capital and operating costs. They have become so popular that by the early 1990s, 40% of the steel made was produced in minimills.

The Minimill Process

The minimill process of steelmaking is basically the same as that of integrated mills; it uses raw materials, furnacing, casting tower, soaking furnace, and the finishing mill.

1. **RAW MATERIALS**—Most of the steel produced in minimills is made of scrap steel. However, because the supply and the cost vary so greatly, many steelmakers are using pig iron, hot-briquetted iron, direct-reduced iron, and iron carbide to provide themselves with a steady supply of raw materials at fixed prices.

2. **FURNACES**—In most cases, DC, electric-arc furnaces are used in minimills, Fig. 11-12. They are 22 ft. (6.7 m) diameter, bottom-tap furnaces which use 24 electrodes and have water-cooled roofs and sidewalls. These furnaces, which can produce from 50 to 200 tons per hour, are generally smaller than the furnaces used by integrated mills, cost less, are more flexible, and create less pollution.

3. **THE MELTING PROCESS**—The charge of scrap iron and iron supplements is placed into the DC electric-arc furnace, Fig. 11-12, and brought to a temperature of approximately 2800° to 2900°F (1538° to 1593°C) and held there for about one hour.

The molten metal is moved from the furnace to a **ladle metallurgical station** which has a lid-type vacuum degassing unit. This unit allows for stirring, removing impurities, adding alloys,

Fig. 11-12 The DC electric-arc furnaces are generally used in minimills because they are fast and cost-efficient. (Courtesy of Nucor Corp.)

Fig. 11-13 The finishing mill is used to roll hot steel slabs to shape and thickness. (Courtesy of Nucor Corp.)

and controlling the temperature. The vacuum degassing unit is only used when steel with very low carbon and/or nitrogen is required.

4. **CASTING TOWER**—The metal goes to the casting tower from the ladle station and is fed into the tundish. As the liquid steel flows out of the tundish, it passes through a water-cooled mold where it begins to solidify and take shape. The temperature of the slab coming out of the containment section is about 1800°F (980°C); the slab is traveling at approximately 13 ft./min. (4 m/min).

5. **SOAKING FURNACE**—As the slabs go into the soaking furnace, where they are brought up to and held at the required temperature, they are sheared to lengths of 138 to 150 ft. (42 to 46 m). After leaving the soaking furnace, the slabs pass a water spray to remove scale.

6. **FINISHING MILL**—The finishing mill reduces the slab thickness from 2 to .100 in. (50 to 2.5 mm), Fig 11-13. The strip then passes through a cooling section where it is cooled from the top and bottom by water sprays. The strip, which is at a temperature of 986 to 1290°F (530 to 700°C), is rolled into 76 in. (1930 mm) diameter coils.

SUMMARY

*   Steel can be manufactured with a wide variety of properties; this makes it a versatile material with many uses.

*   Pig iron is the primary ingredient in steel and cast iron.

*   Most steel is made in basic oxygen furnaces. Some specialty steels are made in electric-arc furnaces.

*   Small amounts of alloying metals are added during steelmaking to give different types of steels their desirable properties.

*   Direct steelmaking is a relatively new process that produces steel directly from iron ore. It is designed to reduce pollution, improve efficiency, and cut costs.

*   Continuous casting is the primary method used to convert molten steel into semi-finished shapes.

*   Minimills which use electric-arc furnaces, are gaining greater use because they are smaller, more flexible, and more efficient then large, integrated mills.

KNOWLEDGE REVIEW

1. Why is steel called our most versatile metal?

2. Name two factors that have caused changes in the iron and steel industries during the past twenty years.

Raw Materials

3. Where are the chief sources of iron ore in North America?

4. Name two methods of mining iron ore.

5. Name and describe three types of iron ore.

6. Briefly describe the pelletizing process for iron ore.

7. How is coal converted into coke?

8. What purpose does limestone serve in the steelmaking process?

Manufacture of Pig Iron

9. What type of furnace is used to manufacture pig iron?

10. List the raw materials used to manufacture pig iron.

11. In point form, describe the operation of a blast furnace.

Direct Ironmaking

12. What is the aim of the direct ironmaking process?

13. Name three main advantages of coal-based ironmaking or bath smelting.

Manufacture of Cast Iron

14. In what type of furnace is cast iron manufactured?

15. Name four types of cast iron and give one use for each.

Manufacture of Steel

16. Name two furnaces that are used to convert pig iron into steel.

17. Describe the basic oxygen furnace.

18. List the main steps in the operation of the basic oxygen furnace.

19. Explain why the electric furnace is used to produce fine alloy and tool steels.

20. List the main steps involved in producing steel in an electric furnace.

Direct Steelmaking

21. What is the purpose of the direct steelmaking process?

22. Name the four main steps in direct steelmaking.

23. Why will direct steelmaking be almost pollution-free?

Steel Processing

24. In point form, describe the continuous casting process used to convert molten steel into blooms, billets, and slabs.

Minimills

25. Name five advantages of minimills over integrated mills.

26. Describe the melting process.

27. What is the purpose of the ladle metallurgical station?

UNIT 12

Metals and Their Properties

Understanding the properties and heat treatment of metals has become increasingly important to machinists during the past two decades. The study of metal properties and the development of new alloys has resulted in metals with increased tensile strength and reduced weight for the manufacture of better-quality products.

The most commonly used metals are ferrous metals, those which contain iron. The composition and properties of ferrous materials may be changed by the addition of various alloying elements during manufacture to impart the desired qualities to the material. Cast iron, machine steel, carbon steel, alloy steel, and high-speed steel are all ferrous metals, each having different properties.

OBJECTIVES

After completing this unit, you should be able to:

- List the five common terms used to describe the properties of metals.
- List the types and uses of ferrous and nonferrous metals.
- Describe the common methods of identifying metals.

KEY TERMS

ferrous metals
spark testing

nonferrous alloys

nonferrous metals

● PROPERTY DEFINITIONS

To better understand the use of the various metals, one should be familiar with the following metallic properties:

- **Brittleness**, Fig. 12-1, is the property of a metal which permits no permanent distortion before breaking. Cast iron is a brittle metal; it will break rather than bend under shock or impact.

- **Ductility**, Fig. 12-2, is the ability of the metal to be permanently deformed without breaking. Metals such as copper and machine steel, which may be drawn into wire, are ductile materials.

Fig. 12-3 Elastic metals return to their original shape after the load has been removed. (Courtesy of Praxair, Inc.)

Fig. 12-1 Brittle metals are not flexible and break easily. (Courtesy of Praxair, Inc.)

- **Elasticity**, Fig. 12-3, is the ability of a metal to return to its original shape after any force acting upon it has been removed. Properly heat-treated springs are good examples of elastic materials.

- **Hardness**, Fig. 12-4, may be defined as the resistance to forcible penetration or plastic deformation.

- **Malleability**, Fig. 12-5, is that property of a metal which permits it to be hammered or rolled into other sizes and shapes.

- **Toughness** is the property of a metal to withstand shock or impact. Toughness is the opposite of brittleness.

- **Fatigue-Failure** is the point at which a metal breaks, cracks, or fails as a result of repeated stress.

Fig. 12-2 Ductile metals can easily be formed into various shapes. (Courtesy of Praxair, Inc.)

Fig. 12-4 Hard metals resist penetration and plastic deformation. (Courtesy of Praxair, Inc.)

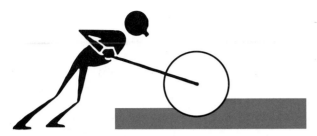

Fig. 12-5 Malleable metals can easily be formed or rolled to shape. (Courtesy of Praxair, Inc.)

● FERROUS METALS

The three general classes of ferrous metals are steel, cast iron, and wrought iron. **Ferrous metals** are made up principally of iron, which is magnetic. Steels are the most important ferrous metals used in machine shop work and they are generally classified by their carbon content, Fig. 12-6.

Types of Steel

As described in unit 11, steel can be custom-made to fit a wide range of requirements. By using various chemical and alloying elements, steels with many different properties can be produced.

Low-Carbon Steel

Low-carbon steel, commonly called machine steel, contains from 0.10 to 0.30% carbon. This steel, which is easily forged, welded, and machined, is used for making such things as chains, rivets, bolts, and shafting.

Medium-Carbon Steel

Medium-carbon steel contains from 0.30 to 0.60% carbon and is used for heavy forgings, car axles, rails, etc.

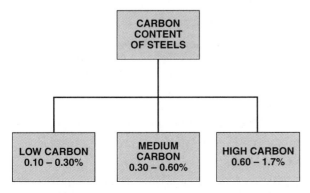

Fig. 12-6 The carbon content of the most commonly used steels. (Courtesy of Kelmar Associates)

High-Carbon Steel

High-carbon steel, commonly called tool steel, contains from 0.60 to 1.7% carbon and can be hardened and tempered. Hammers, crowbars, etc., are made from steel having 0.75% carbon. Cutting tools such as drills, taps, reamers, etc., are made from steel having 0.90 to 1.00% carbon.

Alloy Steels

Alloy steels are steels which have certain metals, such as chromium, nickel, tungsten, vanadium, etc., added to them to give the steel additional characteristics not found in plain carbon steels. By the addition of various alloys, steel can be made resistant to rust, corrosion, heat, abrasion, shock, and fatigue.

High-Speed Steels

High-speed steels contain various amounts and combinations of tungsten, chromium, vanadium, cobalt, and molybdenum. Cutting tools made of such steels are used for machining hard materials at high speeds and for taking heavy cuts. High-speed steel cutting tools are noted for maintaining a cutting edge at temperatures where most steels would break down.

High-Strength, Low-Alloy Steels

High-strength, low-alloy steels contain a maximum carbon content of 0.28% and small amounts of vanadium, columbium, copper, and other alloying elements. They have higher strength than medium-carbon steels and are less expensive than other alloy steels. These steels develop a protective coating when exposed to the atmosphere and therefore do not require painting.

Chemical Elements in Steel

Small amounts of chemical elements present in steel can enhance steel properties. However, some elements can also have a detrimental effect.

- **Carbon** in steel may vary from 0.01 to 1.7%. The amount of carbon will determine the steel's brittleness, hardness, and strength.

- **Manganese** in low-carbon steel makes the metal ductile and of good bending quality. In high-speed steel, it toughens the metal and raises its critical temperature. Manganese content usually varies from 0.39 to 0.80%, but may run higher in special steels.

- **Phosphorus** is an undesirable element which makes steel brittle and reduces its ductility. In satisfactory steels, the phosphorus content should not exceed 0.05%.

- **Silicon** is added to steel in order to remove gases and oxides, thus preventing the steel from becoming porous and oxidizing. It makes the steel harder and tougher. Low-carbon steel contains about 0.20% silicon.

- **Sulfur**, an undesirable element, causes crystallization of steel (hot shortness) when the metal is heated to a red color. A content of 0.04% sulfur is used in some steels to improve machinability, tool life, and produce higher surface finishes.

Alloying Elements in Steel

Alloying elements may be added during the steelmaking process to produce certain qualities in the steel. Some of the more common alloying elements are chromium, molybdenum, nickel. tungsten, and vanadium.

- **Chromium** in steel imparts hardness and wear resistance. It gives steel a deeper hardness penetration than other alloying metals. It also increases resistance to corrosion.

- **Molybdenum** allows cutting tools to retain their hardness when hot. Because molybdenum improves steel's physical structure, it gives steel a greater ability to harden.

- **Nickel** in steel improves its toughness, impact properties and resistance to fatigue-failure. and corrosion.

- **Tungsten** increases the strength and toughness of steel and its ability to harden. It also gives cutting tools the ability to maintain a cutting edge even at a red heat.

- **Vanadium** in amounts up to 0.20% will increase steel's tensile strength and produce a finer grain structure in the steel. Vanadium steel is usually alloyed with chromium to make springs, gears, wrenches, car axles, and many drop-forged parts.

● NONFERROUS METALS

Nonferrous metals are metals that contain little or no iron. They are resistant to corrosion and are non-magnetic. In machine shop work, nonferrous metals are used where ferrous metals would be unsuitable. The most commonly used nonferrous metals are aluminum, copper, lead, nickel, tin and zinc.

Aluminum

Aluminum is made from an ore called bauxite. It is a white, soft metal used where a lightweight, non-corrosive metal is required. Aluminum is usually alloyed with other metals to increase its strength and stiffness. It is used extensively in aircraft manufacture because it is only one-third as heavy as steel.

Copper

Copper is a soft, ductile, malleable metal which is very tough and strong. It is reddish in color and second only to silver as an electrical conductor. Copper is the primary metal in brasses and bronzes.

Lead

Lead is a soft, malleable, heavy metal which has a melting point of about 620°F (327°C). It is corrosion-resistant and used for lining vats and tanks, and for covering cables. It is also used for making alloys such as babbitt and solder.

Nickel

Nickel is a very hard, corrosion-resistant metal. It is used as a plating agent on steel and brass, and is added to steel to increase its strength and toughness.

Tin

Tin is a soft, white metal having a melting point of about 450°F (232°C). It is very malleable and corrosion-resistant. Used in the manufacture of tin-plate and tin-foil, it is also used for making such alloys as babbitt, bronze, and solder.

Zinc

Zinc is a bluish-white element which is fairly hard and brittle. It has a melting point of about 790°F (420°C) and is used mainly to galvanize iron and steel.

Nonferrous Alloys

A **nonferrous alloy** is a combination of two or more nonferrous metals completely dissolved in each other. Nonferrous alloys are made when certain qualities of both original metals are desired. Some common nonferrous alloys are:

Brass

Brass is an alloy of approximately two-thirds copper and one-third zinc. Sometimes 3% lead is added to make it easy to machine. Its color is normally a bright yellow, but this varies slightly according to the amounts of alloys it contains. Brass is widely used for small bushings, plumbing and radiator parts, fittings for water cooling systems, and miscellaneous castings.

Bronze

Bronze is an alloy composed mainly of copper and tin. Some types of bronze contain such additions as lead, phosphorus, manganese, and aluminum to give them special qualities. Bronze is harder than brass and resists surface wear. It is used for machine bearings, gears, propellers, and miscellaneous castings.

Babbitt

Babbitt is a soft, grayish-white alloy of tin and copper. Antimony may be added to make it harder, while lead is usually added if a softer alloy is required. Babbitt is used in bearings of many reciprocating engines.

● IDENTIFICATION OF METALS

Because the machinist's work consists of machining metals, it is advantageous to learn as much as possible about the various metals used in the trade. It is often necessary to determine the type of metal being used by observing its physical appearance. Some of the more common

machine shop metals and their appearance, use, etc., are found in Table 12-1. Metals are usually identified by one of four methods:

1. by their appearance,
2. by spark testing,
3. by a manufacturer's stamp,
4. by a code color painted on the bar,

The latter two methods are most commonly used and are probably the most reliable. Each manufacturer, however, may use its own system of stamps or code colors.

Spark Testing

Any ferrous metal, when held in contact with a grinding wheel, will give off characteristic sparks. Small particles of metal, heated to red or yellow heat, are hurled into the air, where they come in contact with oxygen and oxidize or burn. An element such as carbon burns rapidly, resulting in a bursting of the particles. Depending on the composition of the metal that is being ground, spark bursts will vary in color, intensity, size, shape, and the distance they fly. **Spark testing** may be used to identify a number of metals.

Low-Carbon or Machine Steel

Fig. 12-7A, produces sparks in long, light-yellow streaks with little tendency to burst.

Table 12-1   Identification of metals

Metal	Carbon Content	Appearance	Method of Processing	Uses
Cast iron (C.I.)	2.5 to 3.5%	Grey, rough sandy surface	Molten metal poured into sand molds	Parts of machines, such as lathe beds, etc.
Machine steel (M.S.)	0.10 to 0.30%	Black, scaly surface	Put through rollers while hot	Bolts, rivets, nuts machine parts
Cold rolled or cold drawn (C.R.S) (C.D.S)	0.10 to 0.30%	Dull silver, smooth surface	Put through rollers or drawn through dies while cold	Shafting, bolts, screws, nuts
Tool steel (T.S.)	0.60 to 1.5%	Black, glossy	Same as machine steel	Drills, taps, dies, tools
High speed steel (H.S.S.)	Alloy steel	Black, glossy	Same as machine steel	Dies, tools, taps, drills, toolbits
Brass	…	Yellow (various shades), rough if cast, smooth if rolled	Same as cast iron, or rolled to shape	Bushings pump parts, ornamental work
Copper	…	Red-brown, rough if cast, smooth if rolled	Same as cast iron, or rolled to shape	Soldering irons, electric wire, water pipes

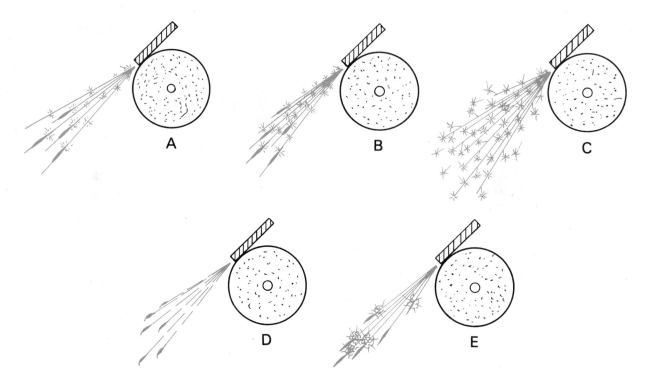

Fig. 12-7 Spark testing is one method which may be used to identify ferrous metals. (Courtesy of Norton Co.)

Medium-Carbon Steel

Fig. 12-7B, is similar to low-carbon steel, but has more sparks which burst with a sparkler effect because of the greater percentage of carbon in the steel.

High-Carbon or Tool Steel

Fig. 12-7C, shows numerous little yellow stars bursting very close to the grinding wheel.

High-Speed Steel

Fig. 12-7D, produces several interrupted spark lines with a dark red, ball-shaped spark at the end.

Cast Iron

Fig. 12-7E, shows a definite torpedo-shaped spark with a feather-like effect near the end. It changes from dark red to a gold color.

● SHAPES AND SIZES OF METALS

Due to the wide variety of work performed in a machine shop and the necessity of conserving machining time, as well as reducing the amount of metal cut into steel chips, metals are manufactured in a wide variety of shapes and sizes. When ordering steel for work that must be

machined, it is recommended that it be purchased a little larger than the finished size to allow for the machining operation. Although many factors determine exactly how much larger the piece should be, generally 1/16 in. (1.5 mm) oversize on each surface to be machined is adequate. For example, if a piece of round work must be finished to 3/4 in. (19 mm) diameter, it is wise to purchase 7/8 in. (22 mm) diameter material.

Specifications for Purchasing

There is a proper method for specifying the sizes and dimensions of metal when ordering (see Fig. 12-8).

1. **Round material** has only two dimensions; therefore, when ordering, specify the diameter first and then the length.

2. **Flat or rectangular material** has three dimensions: thickness, width, and length, and should be ordered in that sequence.

3. **Square material** has three dimensions; however, the thickness and width are the same. When ordering, specify the thickness (or width) and then the length.

4. **Hexagonal material** has only two dimensions: the distance across flats and the length, and should be ordered in that sequence.

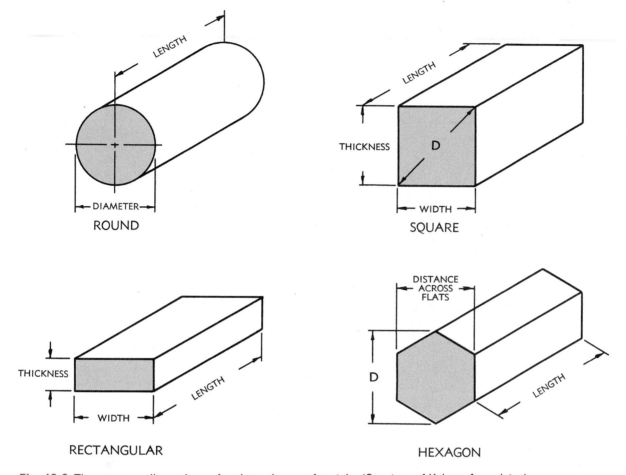

ROUND

SQUARE

RECTANGULAR

HEXAGON

Fig. 12-8 The common dimensions of various shapes of metals. (Courtesy of Kelmar Associates)

SUMMARY

- Ferrous metals are made primarily of iron and include steel, cast iron, and wrought iron.

- Nonferrous metals, such as aluminum and copper, do not contain iron. They are corrosion-resistant and are often alloyed with other metals.

- Steels are generally classified by their carbon content. They can be produced with specific properties by alloying with small amounts of other metals.

- Brass and bronze are nonferrous copper alloys that can be processed in machine shops.

- Metals are usually identified by their appearance, by spark testing, by a manufacturer's stamp, or by a code color painted on the material.

- In spark testing, ferrous metals can be identified by observing the characteristic sparks that result when the metals are ground with a grinding wheel.

KNOWLEDGE REVIEW

1. Briefly define the following metal properties:
   a. brittleness
   b. ductility
   c. hardness
   d. malleability

Ferrous Metals

2. Define a ferrous metal.

3. Give the carbon content and one use for each of the following:

   a. low-carbon steel

   b. medium-carbon steel

   c. high-carbon steel

4. What are the advantages of alloy steels?

5. Describe the composition of high-speed steels.

6. Why is high-speed steel especially valuable for the manufacture of cutting tools?

7. Explain how carbon affects steel.

8. What two alloys help cutting tools maintain their hardness and cutting edge when hot?

Nonferrous Metals

9. Define nonferrous metals and state why they are used in machine shop work.

10. Briefly describe and give one use for:

    a. aluminum

    b. copper

    c. nickel

    d. tin

11. What is a nonferrous alloy?

Identification of Metals

12. Name four methods of identifying metals.

13. Describe the appearance of the following metals and give one use for each.

    a. cast iron

    b. machine steel

    c. tool steel

    d. brass

14. Why do sparks from different materials vary?

15. What are the spark characteristics of:

    a. high-carbon steel?

    b. high-speed steel?

Shapes and Sizes of Metals

16. Why are metals manufactured in a wide variety of shapes and sizes?

17. How much larger than the finished size should work that requires machining be ordered?

18. Explain how round and flat material should be ordered.

SECTION

8

Principles of Metal Cutting

For any country to survive in the world marketplace, it must adopt new manufacturing technologies and make them as efficient and trouble-free as possible. Although computers and numerical control make a machine tool more productive and produce quality goods, this productivity is limited by the life and accuracy of the cutting tool. If the cutting life is short, the machine tool must be shut down repeatedly to replace the worn tool, reducing productivity and increasing the cost of manufacturing.

Since the amount of material that is cut into chips each year amounts to billions of dollars, any improvement in machining procedures or cutting tool materials could result in savings of both time and money. Therefore, it is important to have a good knowledge of the types and uses of cutting tools used for lathes (turning centers) and milling machines (machining centers), since most machining is done on these two types of machines.

Figures courtesy of (clockwise from far left): Cleveland Twist Drill Co., Carboloy Inc., Cincinnati Milacron Inc., The Association for Manufacturing Technology, The Association for Manufacturing Technology, The Association for Manufacturing Technology.

UNIT 13

Metal Cutting

The continual development of new metals and alloys makes it important to know what cutting tools and conditions are required for proper machining. These new materials generally have special properties which make them harder, more wear-resistant, and as a result, harder to machine. Some of the newer cutting tools such as ceramic, diamond, and cubic boron nitride, have proven to be indispensable in machining space-age alloys.

OBJECTIVES

After completing this unit, you should be able to:

- Describe what causes each chip type to form.
- Recognize and describe the application of high-speed, cast alloy, and carbide tools.
- Recognize and describe the application of ceramic, cermet, and polycrystalline tools.

KEY TERMS

cemented carbide
continuous chip with
  built-up edge
machinability

cermet
discontinuous
  (segmented) chip
rake angle

continuous chip
cubic boron nitride (CBN)
superabrasive

COMPETITIVE REQUIREMENTS

A large portion of manufacturing operations in the world consists of machining metal to size and shape. To be competitive, it is important that machining operations be as cost-efficient as possible. This requires a good knowledge of metals, cutting tools, and machining conditions and processes. There must be an economical balance between cutting tool life, productivity, part accuracy, and surface finish to produce the highest-quality parts in the shortest period of time. This can only be achieved if we use competitive tools and machining processes. A company cannot compete using 50- to 75-year old high-speed steel and carbide tools and machining processes when other manufacturers are using cutting tools and processes which are up to 700 times more efficient and productive.

CHIP FORMATION

Although the use of tools to cut metal has been around for hundreds of years, the details of the cutting process were not very well understood. It was formerly thought that the metal in front of the cutting tool split, similar to what occurs when an axe splits wood, Fig. 13-1. Research revealed that as the cutting tool enters the

Fig. 13-1 An axe splitting wood was used incorrectly to illustrate the action of producing a chip. (Courtesy of Cincinnati Milacron Inc.)

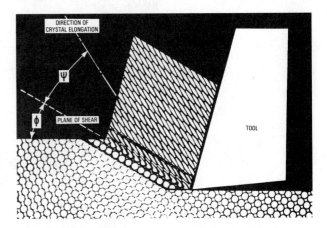

Fig. 13-2 This schematic drawing illustrates the formation of a continuous chip and shows the crystal deformation which occurs in the shear zone. (Courtesy of Cincinnati Milacron Inc.)

metal, it causes the metal in front of it to become plastic, compress, and deform. Under constant tool pressure, the softened metal starts to form a chip and separate from the workpiece along the shear plane, Fig. 13-2. The heat caused by the plastic deformation of the metal and the chip sliding up the cutting tool face can shorten the life of the cutting tool. Therefore, in most machining operations, cutting fluid is applied to the chip-tool interface to reduce friction and cool the tool and the workpiece.

In order to understand what occurs when a chip is produced, it is important to know the more important terms associated with metal cutting.

Built-Up Edge

Built-up edge is a layer of compressed metal formed when the material being cut sticks to and builds up on the cutting tool, Fig. 13-3.

Chip-Tool Interface

Chip-tool interface is the portion of the face of the cutting tool along which the chip slides as it separates from the metal, Fig. 13-3.

Plastic Deformation and Crystal Elongation

Plastic deformation and crystal elongation is the distortion of the crystal structure of the metal which occurs during a machining operation, Fig. 13-4.

Shear Plane and Shear Zone

Shear plane and shear zone is the area where the plastic deformation of the metal occurs, as seen in Fig. 13-2.

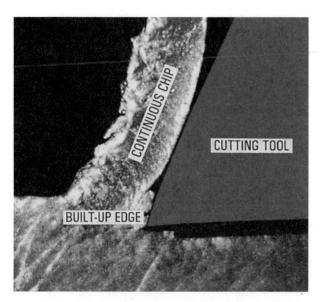

Fig. 13-3 A built-up edge being formed on the cutting tool at the chip-tool interface. (Courtesy of Cincinnati Milacron Inc.)

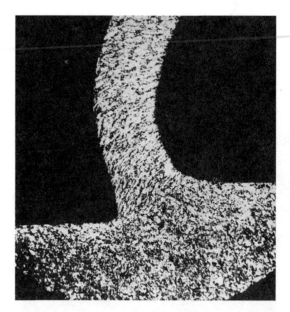

Fig. 13-5 A continuous chip is considered ideal for an efficient cutting action. (Courtesy of Cincinnati Milacron Inc.)

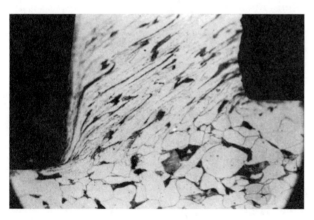

Fig. 13-4 A photomicrograph of a chip forming that shows the plastic deformation and crystal elongation. (Courtesy of Cincinnati Milacron Inc.)

The type of chip produced during any machining operation will depend on whether there is a smooth plastic deformation of the metal, or whether it ruptures and breaks. Generally, a plastic flow occurs in machining ductile metals, while brittle materials, such as cast iron, produce chips which rupture and break.

Chip Types

Three distinct types of chips are produced during a machining operation: continuous, continuous with a built-up edge, and discontinuous.

The **continuous chip**, Fig. 13-5, produces a ribbon-like flow of metal and is considered ideal for efficient cutting action because of the better surface finish it produces. To reduce the amount of friction which occurs when the chip slides up the cutting tool face, the tool should have a good rake angle and cutting fluid should be applied to the chip-tool interface. A shiny surface on the underside of the continuous chip usually indicates an ideal cutting condition because there is little resistance to chip flow. The continuous chip is generally produced when ductile materials such as aluminum, machine steel, or other soft steels are machined.

The **continuous chip with a built-up edge**, Fig. 13-6, occurs when low-carbon machine steel is cut with a high-speed steel tool without the use of cutting fluids. Small particles of the metal tend to weld themselves to the cutting tool face and continue to build up. During the cutting action, some of these particles are torn free as the chip slides along the cutting tool face. These particles imbed into both the chip and the workpiece. This cycle is repeated as the cutting action continues, resulting in the machined surface being embedded with built-up fragments that appear coarse. This type of chip produces a poor surface finish and shortens tool life due to two factors:

1. The fragments of the built-up edge have an *abrasive effect* on the underside of the tool's cutting edge (flank).

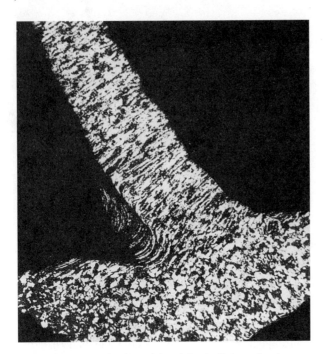

Fig. 13-6 A continuous chip with a built-up edge is produced when metal sticks to the cutting-tool edge. (Courtesy of Cincinnati Milacron Inc.)

2. A *depression (cratering effect)* is caused a short distance away from the cutting edge where the chip contacts the tool face. As this crater continues, it spreads closer to the cutting edge until the edge fractures or breaks down.

The **discontinuous (segmented) chip**, Fig. 13-7, is produced by brittle materials such as cast iron, hard bronze, etc., which are not ductile enough to deform continuously. As the cutting tool contacts the metal, a little compression occurs until it reaches a point of rupturing or segmenting along the shear angle or plane. The cycle of material fracturing ahead of the cutting tool is repeated continuously, and as a result of this rupturing and segmenting, a poor surface finish is produced.

● MACHINABILITY OF METALS

Machinability is a term that describes the ease or difficulty with which a metal can be machined. Machinability can be measured by the life of the cutting tool or the material-removal rate (MRR) in relation to the cutting speed used.

The material-removal rate of any metal is affected by the microstructure of the metal and the shape of the cutting tool. The microstructure of a metal can be modified by annealing and nor-

Fig. 13-7 A discontinuous chip is produced when brittle metal is cut under poor cutting conditions. (Courtesy of Cincinnati Milacron Inc.)

malizing which will change its ductility and shear strength. The addition of certain chemical elements and cold working can also improve the machinability of a metal. The most common metals and their properties are described in the following list:

Low-Carbon (Machine) Steel

Low-carbon (machine) steel contains large areas of ferrite (soft and ductile), mixed with small areas of pearlite (low ductility, high strength). This type of steel has a desirable microstructure and is easy to machine.

High-Carbon Steel

High-carbon steel contains more pearlite than machine steel and therefore is more difficult to cut.

Alloy Steel

A combination of two or more metals give alloy steel special qualities. It is more difficult to cut than low- or high-carbon steel.

Cast Iron

Generally consists of compound pearlite, fine ferrite and iron carbide, and flakes of graphite. It is usually easy to machine; however, if there is sand on the harder outer surface, cast iron can be difficult to cut.

● CUTTING TOOL DESIGN

The shape of the cutting tool will generally determine the material-removal rate (MRR). One of the keys to long cutting tool life is reducing the friction between the chip and the tool. The proper **rake angle** on the cutting-tool face allows the chips to escape freely and reduces the amount of cutting pressure (power) required for the machining operation. Honing the cutting tool face will reduce the friction on the chip-tool interface, produce a better surface finish, and reduce the size of the built-up edge.

The cutting tool rake angle affects the shear angle—the plane in which the softened material separates from the work material. As the shear angle becomes smaller, the chip becomes thicker, requiring more power for machining.

If a large rake angle is ground on the cutting tool, a large shear angle is created with the following results:

• A thin chip is produced.

• The shear zone is fairly short.

• Less heat is generated in the shear zone.

• A good surface finish is produced.

• Less power is required for machining.

There are two types of rake angles found on cutting tools—positive and negative rake.

A *positive* rake angle, Fig. 13-8, causes the chip to flow easily down the face of the toolbit. Positive rake-angle cutting tools are widely used on low-yield strength or ductile materials which are not very hard or abrasive. While positive rake-

angle tools remove metal efficiently, they are not always recommended for all work materials or cutting applications. The following factors must be considered when the type and amount of rake angle for a cutting tool are being determined:

1. cutting tool material and shape,

2. metal type and hardness,

3. continuous or interrupted machining,

4. strength of the cutting edge.

A *negative* rake angle, Fig. 13-9, which protects the cutting tool edge, is recommended for work with interrupted surfaces, and tough or abrasive characteristics. A negative rake angle creates a small shear angle and causes more friction and heat to be generated. An increase in heat may be considered a disadvantage, however, it is desirable when machining tough metals with carbide cutting tools. Carbide tool face milling cutters are a good example of using negative rake for interrupted and high-speed cutting.

Other advantages of negative rake on cutting tools are:

• Cutting-tool life is extended because the shock of the work meeting the cutting tool is absorbed on the face and not the tool point.

• The metal's hard outer scale does not contact the cutting edge.

• Higher cutting speeds can be used.

• Tool strength is increased.

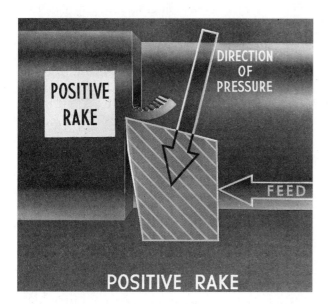

Fig. 13-8 A positive rake angle allows the chip to flow freely along the chip-tool interface. (Courtesy of A.C. Wickman Ltd.)

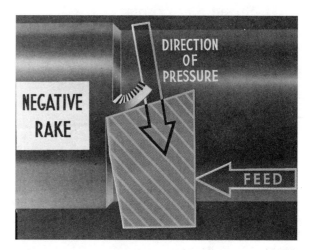

Fig. 13-9 A negative rake protects the cutting point and edge, especially when cutting hard materials. (Courtesy of A.C. Wickman Ltd.)

- Surface with interrupted cuts can easily be machined.

The shape of a chip can be changed to improve the cutting action and reduce the power required. A toolbit producing a continuous-straight ribbon chip on a lathe can be altered to produce a continuous-curled ribbon by:

- Changing the angle of keenness (the included angle formed by the side rake and side clearance) on the toolbit.
- Grinding a chipbreaker behind the cutting edge.

A large helix angle on the milling cutter will improve the cutting performance by providing a good shearing action.

● CUTTING TOOL MATERIALS

The type of cutting tool and the way it is used in a machining process determines the efficiency of a metal removal operation. It is very important to select the proper cutting tool to suit the work material and the type of machining operation. The most common cutting tool materials used in metalworking operations are high-speed steel, cast alloys, cemented carbides, ceramics, and cermets. Polycrystalline diamond and cubic boron nitride (CBN) tools are revolutionizing metal-removal operations throughout the world.

In order for cutting tools to remove metal efficiently, they must be hard, wear-resistant, shock resistant, and capable of maintaining their cutting edge during the machining operation.

High-Speed Steel Tools

High-speed steel tools are commonly used in training programs because they are general-purpose tools and are relatively low in cost. They contain combinations of tungsten, chromium, vanadium, molybdenum, and cobalt. Two types of high-speed steel tools are generally used:

- **Tungsten-base tools**, known as T1 or 18-4-1, contain about 18% tungsten, 4% chromium, and 1% vanadium.
- **Molybdenum-base tools**, known as M1 or 8-2-1, contain about 8% molybdenum, 2% tungsten, 1% vanadium, and 4% chromium.

Cast Alloy Tools

Cast alloy (stellite) tools have high hardness, high wear-resistance, and excellent red-hardness qualities. They contain 25 to 35% chromium, 4 to 25% tungsten, 1 to 3% carbon, with the remainder being cobalt. They are weaker than high-speed steel tools, but can be operated at 2 to 2½ times as fast.

Cemented-Carbide Tools

Cemented carbides are produced by a powder metallurgy process. These cutting tools are largely composed of tiny powder particles of tungsten carbides (carburized tungsten), carbon powder, and cobalt (the binder) which are sintered together at temperatures between 2550 and 2730°F (1400–1500°C).

The process for manufacturing cemented-carbide tools involves the steps shown in Fig. 13-10.

1. **Blending:** mixing the right amount of carbide powders and cobalt together for the type of cemented carbide required.
2. **Compacting:** molding the green powder into size and shape in a press.
3. **Presintering:** heating the green compacts to approximately 1500°F (815°C) in a furnace to hold their size and shape.
4. **Sintering:** the final heating process, 2550 to 2730°F (1400 to 1500°C), to cement the carbide powders into a dense structure of extremely hard crystals.

Straight tungsten carbide, used for machining a variety of metals ranging from gray cast iron to stainless steel, is composed of 94 to 97% tungsten carbide, with the rest being cobalt. Powdered metals such as titanium, tantalum, molybdenum, and niobium, are also used in the manufacture of various cemented carbide types. Four distinct types of carbides are produced:

1. straight tungsten carbide,
2. crater-resistant carbide,
3. titanium carbide,
4. coated carbide.

Carbide tools are available as throwaway inserts, Fig. 13-11A, or brazed-tip tools, Fig. 13-11B.

Machining with Carbide Tools

Cemented-carbide tools provide many advantages in machining operations. They are much harder and more wear-resistant than high-speed steel. They can be operated at cutting speeds two to three times faster than high-speed steel tools. At temperatures around 1400°F

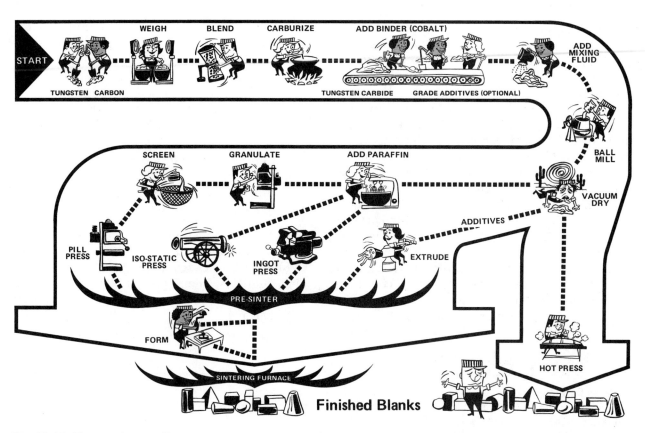

Fig. 13-10 The powder metallurgy process of manufacturing cemented-carbide tools. (Courtesy of Carboloy Inc.)

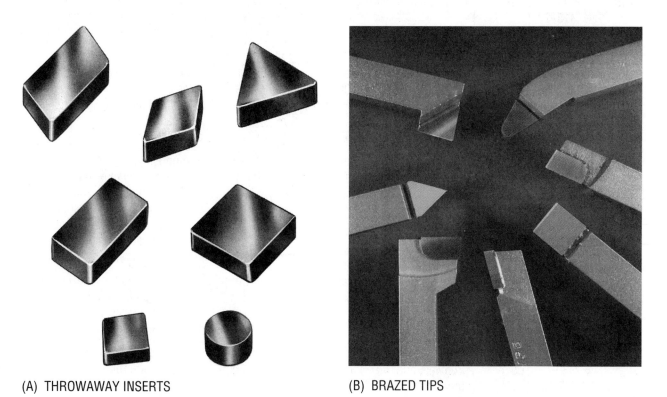

(A) THROWAWAY INSERTS   (B) BRAZED TIPS

Fig. 13-11 A variety of throwaway and brazed-tip carbide tools. (Courtesy of Hertel Carbide Ltd.)

(760°C), carbide tools keep their hardness and produce good surface finishes. They have excellent wear resistance but have a tendency to be somewhat brittle and require care in their use. The harder-grade carbides are more brittle, while the softer grade carbides are tougher but less wear-resistant. The very hard carbides are widely used for finishing cuts. For general-purpose cutting, use carbides which have the highest hardness and are strong enough to avoid chipping and breaking under the machining condition.

The application of hard, wear-resistant coatings of carbides, nitrides, and oxides to carbide inserts have improved the cutting performance of these tools. The major benefits they offer are longer tool life, increased productivity, and reduced machining costs. The most common coatings for cemented-carbide tools, their characteristics, and their uses, are as follows:

- **Titanium carbide:** High wear and abrasion resistance at moderate speeds. Used in roughing and finishing applications.
- **Titanium nitride:** Extremely hard (Rc 80) with excellent lubricating properties. Provides good crater resistance and minimizes edge buildup. Used for heavy roughing cuts at higher speeds.
- **Ceramic oxide:** Provides chemical stability and maintains hardness at high temperatures. Used for roughing and finishing at high speeds

There are also double- and triple- coated inserts available which combine the features of one or more coatings to give the insert special qualities. See Table 13–1 for a list of coated and uncoated carbide inserts and their applications.

Table 13–1 Carbide grades for metal cutting

Coated Carbides					
Valenite Grade	**ISO Class**	**Industry Class**	**Application**	**Materials**	**Working Methods and Conditions**
VN5	P10–25 M15–20	C5–C7	Turning and boring	Steel, cast steel, malleable cast iron, stainless steel	A TiN coated grade for roughing and finishing. Has excellent crater and deformation resistance.
VN8	P10–30 M20–30 K10–30	C2 C7	Turning, boring, milling, threading, and grooving	Cast iron, steels, 300 and 400 series stainless steels, PH stainless steels	A very well balanced TiN coated grade suitable for a broad range of applications. Has outstanding crater and impact resistance at low to high speeds.
VO1	P01–30 M10–30 K01–20	C2 C8	Turning, boring, and milling	Cast iron, stainless steels, alloyed steels, carbon steels	A composite ceramic coated grade providing maximum resistance to built-up edge. Suitable for operations ranging from roughing to finishing at medium to high speeds.
VO5	P01–30 M10–30 K01–20	C2, C8	Turning, boring, and milling	Cast iron, alloy steel, stainless steel, and carbon steel	A composite ceramic coated grade optimized for wear resistance with good impact and built-up edge resistance. A good choice for machining difficult materials.
V1N	P30-45 M30-40 K25-45	C1, C2, C5	Milling, turning, grooving, and threading	Cast iron, steels, high temperature exotics, 300 and 400 series stainless steels, and PH stainless steels	A TiN coated heavy duty grade used in severe roughing and interrupted cuts at slow speeds.

Table 13–1   *(Continued)*

Valenite Grade	ISO Class	Industry Class	Application	Materials	Working Methods and Conditions
V88	P05–30 M10–30 K05–30	C2 C7	Turning, boring, and milling	Cast iron, steel, and alloy steel	A TiC coated grade with excellent flank wear resistance for use in applications where abrasive wear is the primary failure mode.
VX8	P15–30 M15–30 K15–30	C2, C5	Turning, boring, milling, and threading	Cast iron, steels, high temperature exotics, and stainless steels	A TiC and TiN coated grade for moderate to heavy cuts with medium to heavy feeds. Optimized for flank wear resistance.
Uncoated Carbide					
VC2	M10–20 K10–20	C2	Turning, boring and milling	Cast iron, copper, brass, non-ferrous alloys, high temperature exotics, stone, and plastics	General purpose grade of high toughness and resistance against flank wear at low to medium cutting speeds.
VC3	K01–05	C3, C4	Precision turning, boring and milling	Cast iron, aluminum, high temperature exotics, and non-ferrous materials	Wear-resistance grade for finishing cuts, low to medium feed rates under rigid conditions.
VC5	P20–30 M20–40	C5	Turning, boring and milling	Steel, cast steel, malleable cast iron, 400/500 series stainless steels	General purpose grade covering a wide range of applications, low to medium cutting speeds, high feeds and depths of cut. Has good deformation resistance.
VC7	P05–15	C7	Turning, boring, grooving, and threading	Steel, cast steel, malleable cast iron, 400/500 series stainless steels	Light roughing to finishing at low to moderate feeds. Good crater and deformation resistance.
VC8	P01–10	C8	Precision turning and boring	Steel, cast steel, malleable cast iron, 400/500 series stainless steels	High-speed finishing grade with best thermal and deformation resistance.
VC27	P15–30 M15–30 K20–30	C2	Turning and milling	Steel, cast steel, alloyed cast irons, cast alloys, exotics	General purpose fine grain grade with improved toughness and wear resistance for turning and milling.
VC28	M20–30 K15–30	C2	Milling	Cast and alloy irons	General purpose grade for roughing to finishing in cast irons.
VC29	M10–20 K10–20	C2, C3	Turning, boring, and milling	Stainless steels, irons, exotics, and non-ferrous metals	Fine grain grade for finishing of exotic irons and non-ferrous metals.
VC35M	P20–35	C5	Milling	Carbon, alloy steel, and stainless steel	General purpose steel milling grade for moderate roughing to finishing.
VC101	M30–40 K30–40	C1	Turning, boring, and milling	Iron, 200/300 stainless steel and exotics	Fine grain heavy duty grade for roughing at low to moderate speeds.

Courtesy of Valinite Inc.

Ceramic Cutting Tools

Ceramic (cemented oxide) cutting tools were introduced in the mid-1950s. Since that time, ceramic cutting tools have been greatly improved with increased uniformity, quality, and double the strength. They are used for machining hard ferrous metals and cast iron, resulting in lower costs and increased productivity (three to four times the speed of carbides).

Ceramic tools are brittle in comparison to other tools and require machines and setups which are rigid and free of vibration to prevent tool breakage. They also require a machine which is capable of high speeds without slowing down under pressure of the cut.

Manufacture of Ceramic Tools

Most ceramic (cemented oxide) cutting tools are manufactured primarily from aluminum oxide. Micron size (fine) grains of alpha alumina (a hydrated form of aluminum oxide) are used in cold or hot pressing operations to produce ceramic tool inserts. In cold pressing, micron size alumina powder is compressed into required shape and then sintered in a furnace at 2912° to 3092°F (1600° to 1700°C). Hot pressing combines the forming and heating process, and sometimes titanium, magnesium, and chromium oxides are added to aid in sintering and to retard growth. After forming, the inserts are finished with a diamond-impregnated grinding wheel.

Throwaway inserts (round, square, triangular, and rectangular), Fig. 13-12, can be fastened in a mechanical holder and are the most economical type of ceramic cutting tool. Designed to be indexable, dull inserts can be quickly replaced by turning the insert to present a new sharp edge. When all edges on the tool are dull, the cutting tool is thrown away, reducing downtime and eliminating resharpening costs. On cemented ceramic tools, the ceramic insert is bonded to a steel shank holder by means of an epoxy glue. This method of fastening the ceramic insert eliminates the strains caused by clamping in mechanical holders.

Cermets

Cermets are cutting tools made of various ceramic (aluminum oxide) and metallic (titanium carbide and titanium nitride) combinations. Other materials, such as vanadium carbide, molybdenum carbide, and zirconium carbide, may be added to produce various tool

Fig. 13-12 Ceramic throwaway inserts are used for high-speed finishing operations on difficult-to-cut metals. (Courtesy of Hertel Carbide Ltd.)

characteristics. The characteristics of cermet cutting tools are illustrated in Fig. 13–13.

Fine powders of the materials are mixed, compacted, and squeezed together under intense heat and pressure that bond the ceramic particles with the metal. Cermet inserts are available in a variety of sizes, in coated and uncoated grades. Cermet inserts are capable of high-speed finishing and semi-finishing of steel, stainless steel, and cast iron. They can cut wet or dry, but coolant is preferred when they are used for finish cuts to provide a better surface finish. In many cases, industry uses high speeds, light feed rates and finish turning to eliminate the finish grinding operation. Cermets have also proved successful in ordinary turning and boring, threading and grooving, and milling operations. In threading and grooving, they outperform carbide tools in wear resistance,

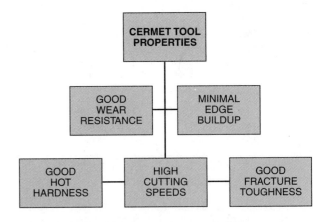

Fig. 13-13 The characteristics of cermet tools give them high productivity in the correct application. (Courtesy of Kelmar Associates)

chemical stability, and edge strength. Depending on the material to be cut, cermets can be used at speeds up to 2000 sf/min. (surface feet per minute).

Polycrystalline Tools (Superabrasives)

The development of cutting tool materials was relatively slow from the early 1900s progressing from carbon tool steel, high-speed steel, cemented carbides, and ceramic tools until the mid 1950s.

Fig. 13–14 traces the development of metal cutting tools to suit new metals and to increase productivity. In 1957, the General Electric Company announced a major breakthrough with the manufactured diamond. This was followed in 1969 with the introduction of an entirely new material—**Cubic boron nitride (CBN).** These two materials were classified as **superabrasives** since they both have the qualities of super hardness and super wear-resistance. Superabrasive cutting tools and grinding wheels have been revolutionizing metal-removal processes throughout the world by increasing productivity, producing better-quality products, and reducing manufacturing costs. They are having a greater impact on some manufacturing operations than cemented-carbide tools had on high-speed cutting tools. Cutting speeds of 1000 surface feet per minute (sf/min.) or higher are common with these high-efficiency superabrasive cutting tools.

Superabrasive Properties

Manufactured diamond is used to machine or grind hard, abrasive composite, nonferrous, and nonmetallic materials. CBN is used to machine and grind ferrous materials such as hardened steels, tool steels, and superalloys that have proved to be very difficult, if not impossible, to cut with conventional tool materials.

The superabrasives diamond and CBN possess properties unmatched by conventional abrasives such as aluminum oxide and silicon carbide. The hardness, abrasion resistance, compressive strength, and thermal conductivity of superabrasives make them logical choices for many difficult grinding and machining applications. Superabrasives can penetrate the hardest materials known, making difficult material-removal applications routine operations. Fig. 13–15 compares four important properties of the superabrasives with those of the two most common conventional abrasives—aluminum oxide and silicon carbide.

- **Hardness**—The hardness property is very important for a cutting tool. The harder the cutting tool is with respect to the workpiece, the more easily it can cut. Fig. 13-15A compares the hardness of diamond and CBN with the same properties of the two major conventional abrasives. Diamond between 7000 and 10000 on the Knoop hardness scale is the hardest known substance, while CBN at 4700 on the scale is second only to diamond in hardness.

- **Abrasion resistance**—Fig. 13-15B shows the relative abrasion resistance of diamond and CBN in relation to the conventional abrasives. Diamond has about three times the abrasion resistance of silicon carbide, while CBN has about four times the abrasion resistance of aluminum oxide. This high-abrasion resistance makes superabrasive cutting tools ideal for machining hard, tough materials at high cutting speeds because they maintain their sharp cutting

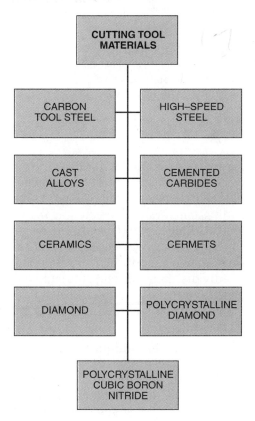

Fig. 13-14 A history of the development of cutting tool materials during the 1900s. (Courtesy of Kelmar Associates)

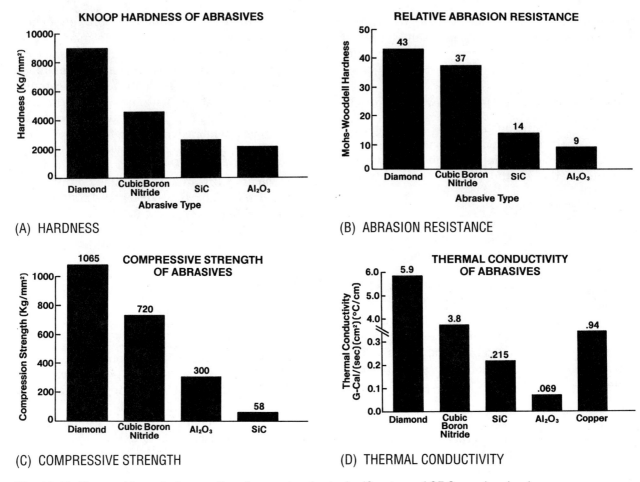

Fig. 13-15 The most important properties of superabrasive tools. (Courtesy of GE Superabrasives)

edges much longer than conventional tools, thereby increasing productivity, while at the same time producing parts which are dimensionally accurate.

- **Compressive strength**—Fig. 13–15C illustrates the compressive strength of superabrasives and conventional abrasives. Compressive strength is defined as the maximum stress in compression that a material will take before it ruptures and breaks. Diamond and CBN crystals have nearly the same density and as a result, have excellent qualities to withstand the forces created during the high metal-removal rates and the shock of severe interrupted cuts.

- **Thermal conductivity**—Fig. 13–15D shows that diamond and CBN superabrasives have excellent thermal conductivity which allows

greater heat dissipation (transfer), especially when cutting hard, abrasive, or tough materials at high material-removal rates. The high cutting temperatures created at the cutting tool-workpiece interface would weaken or soften conventional cutting tool materials. Heat generated during the operation is rapidly dissipated through the superabrasive material into the grinding wheel or tool, thus reducing the risk of thermally damaging the workpiece material.

Because of the unique combination of properties—hardness, abrasion resistance, compressive strength, and thermal conductivity— the superabrasives have achieved a position of major importance in many industries. Four major factors account for the broad acceptance and continuing growth in the use of these truly "super" materials.

1. The increasing precision of modern machining operations has increased the use of superabrasive tools. A superabrasive tool keeps its cutting edge almost unchanged throughout most of its useful life, holding the tolerances set, and in many instances reducing scrap to zero or near zero.

2. Automation has placed great importance on continuity of production. As an operation becomes more automatic, machine downtime becomes more costly. It is not unusual for a properly designed superabrasive tool to produce 10 to 100 times the number of pieces formerly produced by conventional tooling.

3. The new workpiece materials, many of them very hard and/or abrasive, have created machining problems that often can only be solved with superabrasive tools.

4. More people are learning to use superabrasive tools on the same jobs formerly done by conventional abrasives or tool materials, but they are doing it better, faster, and at a lower cost.

Machining with Polycrystalline Tools

Polycrystalline diamond (PCD) tools are used for cutting hard, abrasive, nonferrous and nonmetallic materials, while polycrystalline cubic boron nitride (PCBN) tools are used for cutting ferrous materials. Since a great percentage of the material used in manufacturing is ferrous, this section will deal only with PCBN tools.

The superhardness of CBN, especially at high temperature, combined with the toughness and strength of the tungsten carbide base, allows the machining of hardened steels, abrasive cast irons, and tough high-temperature alloys at high removal rates with long tool life even when heavy or interrupted cuts are taken. The polycrystalline form of CBN features non-directional, consistent properties that resist chipping and cracking, and provide uniform hardness and abrasion resistance in all directions. The properties of polycrystalline CBN that make it an ideal cutting tool material are super hardness, super wear-resistance, high compressive strength, and excellent thermal conductivity.

PCBN Tools

Polycrystalline CBN cutting tool blanks and inserts are a combination of a layer of CBN, the hardest material known next to diamond, bonded (fused) to a cemented tungsten carbide substrate (base). The individual CBN grains are grown together, as a layer approximately .020 in. (0.5 mm) thick, on top of a cemented carbide substrate (base) using an advanced high-pressure, high-temperature process, Fig. 13–16. This produces a compact cutting tool blank or insert which has the high hardness of CBN and good impact resistance because of the random grain orientation (placement), and the strength of the tungsten carbide base which provides excellent mechanical support, and has high thermal (heat) conductivity and a fairly low coefficient thermal expansion.

PCBN tools are available in most regular carbide shapes and sizes to suit various machining applications:

- **Tipped inserts** are available in most regular carbide shapes and are generally the most economical to purchase. These inserts have the same wear life per cutting edge as a full-faced insert, however, they have only one cutting edge.

- **Full-faced inserts**, Fig. 13-17, are available in a wide range of shapes such as squares, rounds, and triangles. Full-faced inserts are generally very cost-effective, since for a very low cost the tool manufacturer can downsize them repeatedly, providing new cutting edges.

- **Brazed-shank insert** tools are available in most common cutting tool forms. They can be specially ordered from the manufacturer to suit a wide variety of machining applications.

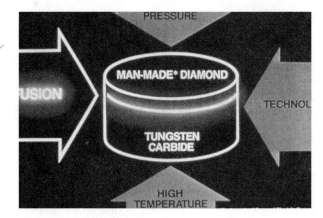

Fig. 13-16 Polycrystalline tool blanks and inserts consist of a thin layer of diamond or CBN bonded to a cemented carbide substrate (base). (Courtesy of GE Superabrasives)

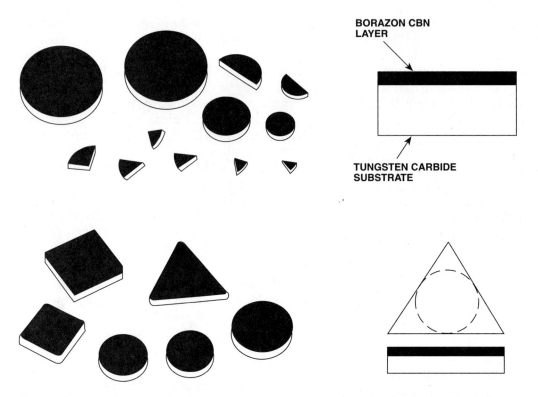

Fig. 13-17 Polycrystalline tool blanks are available in a wide variety of shapes and sizes. (Courtesy of GE Superabrasives)

Applications of PCBN Tools

PCBN cutting tools are used on lathes and turning centers for machining round surfaces, and on milling machines and machining centers for machining flat surfaces. These cutting tools have been used successfully for straight turning, facing, boring, grooving, profiling, and milling operations. PCBN blank tools and inserts are capable of removing material at much higher rates than conventional cutting tools, with far longer tool life. This results in a greater increase in productivity and lower cost per piece machined.

PCBN tool blanks and inserts were designed for hard, difficult-to-cut materials and hard abrasive materials that were previously ground because they were thought to be unmachinable. Wherever PCBN cutting tools were used to replace a grinding operation, machining time was reduced by one to ten times because of the higher metal-removal rate.

The four general classes of metals where PCBN cutting tools have found the best applications and proven to be cost-effective are:

1. **Hardened ferrous metals** over 45 Rc hardness. These include hardened steels such as 4340, 8620, M-2, T-15, etc., and cast irons such as chilled iron, Ni-Hard, etc.

2. **Abrasive ferrous metals**, such as cast irons ranging from 180 to 240 Brinell hardness and pearlistic gray cast iron, Ni-Resist, etc.

3. **Heat-resistant alloys**—most frequently high-cobalt ferrous alloys used as flame-applied hard surfacing.

4. **Superalloys**, such as high-nickel alloys used in the aerospace industry for jet engine parts, etc., are cut efficiently with PCBN cutting tools.

The best applications for PCBN cutting tools are on materials where conventional cutting tool edges of cemented carbides and ceramics break down too quickly. PCBN tools are especially important on expensive machining systems such as numerical control machine tools and Flexible Manufacturing Systems. Their long-lasting cutting edges are capable of transferring the accuracy of computer-controlled machine tools or systems to the workpiece, thereby producing accurate parts, increasing productivity, and reducing expensive machine downtime on high investment equipment.

● CUTTING TOOL AND MACHINE SETUP

To obtain the best performance with any type of cutting tool, certain precautions should be observed in both the machine setup and cutting operation. The machine tool used should be equipped with heat-treated gears, have enough power to maintain a steady cutting speed, and remain rigid in both tool and machine setup throughout the machining operation. It is also important that the machine tool have good bearings and close-fitting slides to avoid chatter and vibrations which cause premature tool failure, Fig. 13-18.

Cutting Tool Guidelines

The use of a cutting tool made of the proper material, with the proper rake and clearances, is very important to the efficiency of the machining operation. Although it is almost impossible to give guidelines for every work material, cutting tool, and machining operation, the following usually apply to most operations:

1. Use a cutting tool with the proper rake and clearances for the material being cut.

2. Hone the cutting edge for better surface finishes, increased keenness, and a longer tool life.

3. Always use a side cutting edge angle large enough to protect the nose of the cutting tool (weakest point) from shock and wear as it contacts the work.

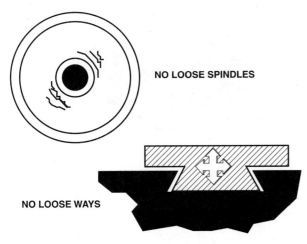

Fig. 13-18 The most efficient cutting action occurs on machines with no vibration in the spindle and no looseness in the slides. (Courtesy of GE Superabrasives)

4. Use the largest nose radius possible. Too large a radius will cause chattering; too small a radius will cause the tool point to break down quickly.

5. Use positive rake tools for soft ductile materials; negative rake tools for hard, abrasive materials.

6. Hold the tool short and securely as possible to prevent vibration.

7. Set the cutting tool point as close to center as possible.

Machine Tool Guidelines

It is much easier to produce accurate work and reduce the amount of scrap on a machine that is in good condition than on one in poor condition. For the most efficient metal removal, the machine tool should be rigid, have sufficient power to maintain a constant cutting speed, have good spindle bearings, and no looseness in the ways or slides, Fig. 13-18. It is important to observe the following guidelines:

1. Always use the speeds, feeds, and depth of cuts recommended by the manufacturer for the material being cut and the cutting tool used.

   - Too high speed causes rapid tool failure.
   - Too low speed reduces productivity.
   - Too high feed produces a rough surface finish and may cause the tool to break down quickly.
   - Too low feed causes rubbing, may work-harden the material, and reduces productivity.

2. Be sure that the workpiece is held securely and the machine is as rigid as possible.

3. Never touch the cutting tool against stationary work, or stop the machine while the feed is engaged; this will break the cutting edge.

4. Always use sharp cutting tools to ensure an efficient cutting action and accurate work. Dull tools can be recognized when the following occur:

   - poor surface finishes,
   - inaccurate work being produced,
   - changes in the color or shape of the chip during machining.

TOOL LIFE

The life of any cutting tool generally depends on the wear and abrasion which occur at the cutting edge. Three types of wear are common with cutting tools: flank, nose, and crater.

- **Flank wear**, Fig. 13-19A, is the wear that takes place along the side of the cutting edge because of the friction between the side of the cutting tool and the metal being machined. Too much flank wear increases friction, requiring more power for machining. When the flank wear becomes .015 to .030 in. (0.38 to 0.76 mm) long, the tool should be resharpened.

- **Nose wear**, Fig. 13-19B, is the wear that takes place on the nose or point of the cutting tool due to the friction between the nose and the metal being machined. Nose wear on a cutting tool affects the quality of the surface finish produced on the workpiece.

- **Crater wear**, Fig. 13-19C, is the depression or groove which occurs a short distance away from the cutting edge due to the chips sliding along the chip-tool interface. Too much crater wear (depression) will eventually break the cutting edge down.

The following factors affect the life of a cutting tool:

1. the type, hardness, and microstructure of the material being cut,
2. the surface characteristics of the work material (smooth or scaly),
3. the cutting-tool material and shape,
4. the type of machining operation being performed,
5. speed, feed, and depth of cut.

EFFECTS OF TEMPERATURE AND FRICTION

Excessive temperatures and friction during machining can cause serious problems. A machinist must understand cutting processes and material properties in order to minimize these problems.

Temperature

In the process of cutting metal, heat is generated in the cutting zone by the plastic deformation of the material and the chip sliding up the cutting tool face. The amount of heat gener-

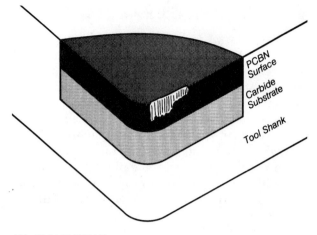

(A) FLANK WEAR

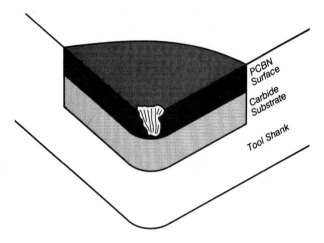

(B) NOSE WEAR

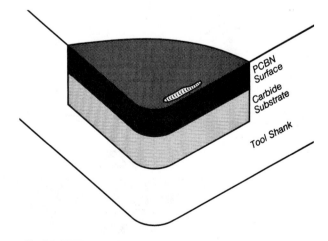

(C) CRATER WEAR

Fig. 13-19 Types of cutting tool wear (Courtesy of GE Superabrasives)

ated varies with each type of metal and increases with higher speeds and an increase in the metal-removal rate. The lowest heat is generated when cutting materials of low tensile strength, such as brass and aluminum. Harder materials such as stainless steel and titanium generate the most heat due to their resistance to deformation. The temperature of the machining operation affects the cutting tool life, quality of surface finish, rate of production, and workpiece accuracy.

The heat generated in the cutting zone can come close to the melting temperature of the metal which can affect the life of the cutting tool. High-speed steel tools cannot stand the same high temperatures as cemented-carbide tools without the cutting edge breaking down. However, high-speed cutting tools can still maintain their cutting edge at "red hardness", which occurs at temperatures above 900°F (482°C). At temperatures which exceed 1000°F (538°C), the edge of the cutting tool begins to break down.

Cemented-carbide tools maintain their cutting edge and can be used effectively at temperatures as high as 1600°F (871°C). These tools, that are harder than high-speed steel tools, have greater wear resistance and can be used at higher cutting speeds.

Friction

Friction, and the heat it causes, is one of the main reasons for tool failure. In order to maintain good cutting action, it is important that friction between the chip and the tool face be kept to a minimum. As friction increases, the possibility of a built-up edge forming increases. As the size of the built-up edge increases, more friction is generated, causing a breakdown of the cutting edge and poor surface finish. Each time the machine is stopped to regrind or replace the cutting tool, productivity drops.

The temperature caused by the friction can also affect the accuracy of the product. While the workpiece may not reach the same temperature as the cutting tool point, it may become hot enough to cause the material to expand. A part machined to size in this condition will be smaller than the machined size when it cools and returns to room temperature. A good grade of cutting fluid applied to the chip-tool interface can reduce friction and also cool the tool and the workpiece.

● SURFACE FINISH

A number of factors affect the surface finish produced during a machining operation. These include the feed rate, the nose radius of the tool, the cutting speed, and the temperature generated during machining (Table 13–2). To calculate the approximate micro finish for any given nose radius use the following formula

$$AA = \frac{Feed^2}{24 \times radius}$$

NOTE: $(Rq = 1.11 \times AA)$
AA   = Arithmetical Average
Rq   = Root Mean Square

Table 13-2   Approximate surface finishes

Feed In./Rev	1/16 in. Tool Nose Radius							
	Carbide				Coated Carbide and Ceramic			
	Approximate Surface Finish		Finish with 20% Safety Factor		Approximate Surface Finish		Finish with 20% Safety Factor	
	Rq	Ra	Rq	Ra	Rq	Ra	Rq	Ra
.003	7	6	8	7	4	4	5	4
.004	12	10	14	12	7	6	8	8
.005	18	16	22	19	11	10	13	12
.006	26	23	31	28	16	14	19	17
.007	36	32	43	38	23	20	27	24
.008	47	41	56	50	29	26	35	31
.010	74	65	88	78	45	40	54	48
.012	125	111	149	133	65	58	79	70

There is a relationship between high temperature and a rough surface finish. Metal particles have a tendency to stick to the cutting tool at high temperatures and form a built-up edge.

- The results of machining a piece of aluminum without cutting fluid at 200°F (93°C) is illustrated in Fig. 13-20A. The rough surface finish shows the presence of a built-up edge on the cutting tool.

- The same piece of aluminum was machined under the same conditions, without cutting fluid but at a room temperature of 75°F

(A) 200° F (93° C)

(B) 75° F (24° C)

(C) −60° F (−50° C)

Fig. 13-20 A comparison of surface finishes when machining aluminum at different temperatures. (Courtesy of Cincinnati Milacron Inc.)

(24°C), Fig. 13-20B. The differences in surface finishes between the samples in Fig. 13-20A and 13-20B are very noticeable.

- When the piece of aluminum was cooled and machined at temperatures of −60°F (−50°C), a further improvement was noted, Fig. 13-20C. By cooling the work material to −60°F (−50°C), the temperature of the cutting tool edge was reduced, which resulted in better surface finish than that produced at 200°F (93°C).

SUMMARY

- Chip formation involves compression and deformation of the metal workpiece. The type of chip produced depends on the type of metal being machined.

- A proper rake angle on a cutting-tool face can reduce friction and prolong tool life.

- Common cutting-tool materials include high-speed steels, cast alloys, cemented carbides, ceramics, and cermets.

- Superabrasives such as manufactured diamond and cubic boron nitride (CBN) possess super hardness and wear resistance. They are used in difficult machining and grinding operations.

- Proper setup of cutting tools and machines is necessary to maximize cutting performance.

- Tool life depends on wear and abrasion occurring at the cutting edge.

- High temperatures and friction can adversely affect the efficiency of the machining operations and the accuracy of the finished parts.

KNOWLEDGE REVIEW

1. Why is it important that every manufacturing operation be as efficient and trouble-free as possible?

Metal Cutting

2. What knowledge is required to make manufacturing operations competitive with off-shore manufacturers?

3. Briefly explain what occurs as the cutting tool enters the metal.

4. Why is cutting fluid used during machining operations?

Chip Formation

5. Explain the following terms:
   a. built-up edge
   b. chip-tool interface
   c. shear zone

6. List the conditions that cause the following chips:
   a. continuous
   b. continuous with built-up edge
   c. discontinuous

Machinability of Metals

7. Name two factors that affect the material-removal rate of any metal.

8. Why are the following more difficult to machine than low-carbon steel?
   a. high-carbon steel
   b. alloy steel

Cutting Tool Design

9. List five effects of a large rake angle on a cutting tool.

10. What factors must be considered to determine the type and amount of rake angle?

11. State five advantages of a negative rake angle on a cutting tool.

Cutting Tool Materials

12. Name five qualities a cutting tool should have to remove metal efficiently.

13. Name and list the composition of two types of high-speed steel tools.

14. From what materials are cemented carbide tools made?

15. List the four main steps in the manufacture of cemented carbides.

16. Name the three most common cemented carbide coatings and the characteristics of each.

17. For what purpose are ceramic tools used?

18. What are cermets and where are they used?

19. What effect did superabrasives have on metal-removal processes?

20. What types of materials are machined or ground with
    a. diamond?
    b. cubic boron nitride?

21. Name the four main properties of super-abrasives.

22. How are PCBN tools made?

23. Name the four general areas where PCBN tools have proven to be cost effective.

Cutting Tool and Machine Setup

24. List four requirements a machine tool should have to obtain the best performance with a cutting tool.

25. What are four of the most important guidelines for an efficient machining operation?

26. What results may be expected from the following?
    a. too high speed
    b. too low speed
    c. too high feed
    d. too low feed

Tool Life

27. List the factors that affect the life of a cutting tool.

Effects of Temperature and Friction

28. How is heat generated in the cutting zone?

29. Name the things that can be affected by the temperature of a machining operation.

30. What factors affect the surface finish produced during a machining operation?

UNIT 14

Cutting Fluids

Cutting fluids are very important to minimize or reduce the effects of friction and heat in a machining operation. They affect the performance of the cutting tool by increasing the material-removal rate and improving part quality and dimensional accuracy, which in turn, reduces manufacturing costs. Excessive heat causes thermal damage to the microstructure of the workpiece, which can result in rapid part wear and premature failure. When selecting a cutting fluid, both the cutting and noncutting functions must be considered to get the best results.

OBJECTIVES

After completing this unit, you should be able to:

- Explain the function and importance of cutting fluids in metal-removal operations.
- Describe the characteristics and advantages of the most common types of cutting fluids.
- Select the proper cutting fluid to suit various metals and machining operations.

KEY TERMS

active cutting oils   chip-tool interface   emulsifiers
extreme-pressure oils   inactive cutting oils   microstructure

● CUTTING FLUID FUNCTIONS

A good cutting fluid must serve two important functions. First, it provides lubrication—this reduces the amount of friction at the interface. Second, a cutting fluid provides cooling—it removes heat produced during machining operations.

Lubrication

In a machining operation, about one-third of the heat generated is caused by the external friction of the chip sliding over the cutting tool face, and two-thirds of the heat is caused by the resistance of the metal atoms of the workpiece being removed. Introducing an effective lubricant into the tool/work interface will produce the following results:

- The heat generated by friction is reduced because the chip slides easily over the tool face or abrasive grain.

- The shear angle increases and the path of shear decreases, producing a thinner chip.

- There is less molecular disturbance, which reduces internal friction and heat.

Cooling

Cooling is the process of removing heat from the cutting tool, the chip, and the workpiece. In order for a cutting fluid to do this effectively, it must be able to transfer heat quickly and be able to absorb a large quantity of heat without increasing its own temperature very much.

● CUTTING FLUID BENEFITS

The selection and proper application of a cutting fluid can produce the following advantages, Fig. 14-1:

- **Reduction of tool costs:** Cutting fluids reduce tool wear and the need for regrinding or truing and dressing, which means that the tools will cut longer and reduce machine downtime.

- **Increased productivity:** Because cutting fluids reduce the friction and heat of a machining or grinding operation, higher speeds and feeds can be used which result in higher material-removal rates.

- **Reduction of labor costs:** Since cutting tools last longer and require less regrinding when

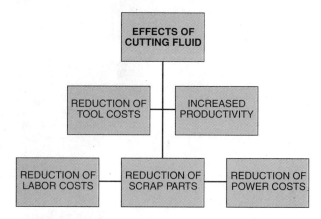

Fig. 14-1 The advantages of cutting fluids in a machining or grinding operation. (Courtesy of Kelmar Associates)

cutting fluids are used, there is less machine downtime, which reduces the labor cost per part.

- **Reduction of power costs:** Cutting fluid reduces the friction created during a machining or grinding operation: therefore, the amount of power required to remove material is reduced with a saving in power costs.

- **High surface finishes:** When the correct fluid is properly directed into the work/tool interface in sufficient volume, it washes away the chip. This prevents it from being caught between the tool and work where it would cause minute scratches and result in a poorer surface finish.

- **Accurate part geometry:** The effective use of cutting fluids reduces the friction and heat created during the machining operation. The rapid removal of heat stops the work from expanding and prevents its **microstructure** from being damaged. This results in the production of parts with accurate geometry.

● NONCUTTING FUNCTIONS

The noncutting functions that a good cutting fluid should have can be important in a productive machining operation. These functions include:

- **Rust and corrosion prevention:** Cutting fluids should protect the machine and the parts being ground against rust and corrosion by leaving a thin protective residual

Fig. 14-2 A good cutting fluid should protect the machine and workpieces against rust and corrosion. (Courtesy of *Modern Machine Shop Magazine*, Copyright 1994, Gardner Publications Inc.)

film on these surfaces after the water has evaporated, Fig. 14-2.

- **Bacterial control:** The effects of bacterial growth in cutting fluids can radically affect, and eventually destroy, the stability of almost all water-based cutting fluids and cause them to turn rancid. To stop or slow the growth of bacteria, it is very important that good housekeeping procedures be followed and pure water be used for mixing with the fluid concentrate.

- **Nonflammability:** High production rates generally produce considerable heat which can be damaging to the cutting tool and a potential fire hazard. Cutting fluids should not burn easily and should preferably be noncombustible. They should not smoke excessively, form gummy deposits which may cause the machine slides to become sticky, or clog the circulating system, Fig. 14-3.

- **Stability:** Good cutting fluids contain quality raw products which are durable and re-

Fig. 14-3 A cutting fluid that smokes excessively causes an unpleasant work environment. (Courtesy of *Modern Machine Shop Magazine*, Copyright 1994, Gardner Publications Inc.)

sist breakdown in storage or use. If these products are properly installed and maintained, they can often last four to six times longer than other cutting fluids, with a corresponding saving in costs.

- **Relatively low viscosity:** Good surface finish depends on clean cutting fluids flushing away the chips and a filtration system to remove these particles from the fluid before they are recycled.

● TYPES OF CUTTING FLUIDS

There are four main types of cutting fluids—chemical fluids, emulsions, semichemical fluids, and straight cutting oils, each with specific advantages and applications.

Three of the four types of cutting fluids are mostly water, and the hardness of the water has a major effect on the coolant quality. Very hard water and the minerals that it contains can cause rusting, staining, or corrosion of both machine and workpiece. For the best coolant life and performance, it is advisable that the water used in mixing fluid concentrates be as pure as possible. In most cases, this means that water available locally should be deionized (passed through an ion exchange resin to remove all impurities and minerals).

Water provides the best cooling because it can absorb and carry away more heat than most other fluids. However, water alone causes rust and is a very poor lubricant. Oil, on the other hand, is an excellent lubricant but a very poor coolant, and is flammable. Since neither water nor oil is a good cutting fluid by itself, water-soluble fluids have been manufactured which provide cooling, lubrication, corrosion resistance, etc. It is *very important* that the mixing procedures and the coolant concentrations specified by the manufacturer be closely followed to achieve the best results.

Chemical Fluids (Synthetic Fluids)

Chemical cutting fluids are stable, pre-formed emulsions that contain very little oil and mix very easily with water. These fluids depend upon chemical agents for lubrication, rust prevention, bacterial control, water softening, etc. These clean, transparent fluids possess excellent cooling and fair lubricating qualities, and are generally used for moderate to high stock-removal grinding and machining operations where a high surface finish is required.

Chemical cutting fluids are available in three types:

> *true solution fluids, wetting-agent types, and wetting-agent types with EP (extreme pressure) lubricants.*

True Solution Fluids

True solution fluids are generally clear, transparent solutions which contain mostly rust inhibitors to prevent rust, and provide the rapid removal of heat during machining operations. Sometimes a dye is added to color the water and this concentrate is then mixed with one part to 50 to 250 parts water, depending on the machining application.

Wetting Agent Fluids

Wetting-agent fluids contain chemical agents which improve the wetting action of water, providing more uniform heat removal and lubricating action. Wetting-type fluids have small particle size which allows the fluid to thoroughly penetrate and "wet" all contacting surfaces during machining operations.

The third category, EP (extreme pressure) lubricants, is described in the next section, which deals with emulsions.

Fig. 14-4 Cutting oils are usually mineral oils containing additives to make them soluble in water. (Courtesy of *Modern Machine Shop Magazine*, Copyright 1994, Gardner Publications Inc.)

Emulsions (Soluble Oils)

Emulsions, or soluble oils, are mineral oils that contain a soap-like material (emulsifier) that makes them mix in water into a milky-white solution, Fig. 14-4. These **emulsifiers** break the soluble oil into minute (very small) particles and keep them separated in the water for a long period of time.

There are three types of emulsifiable or soluble oils available: emulsifiable mineral oils, superfatted mineral oils, and extreme pressure (EP) emulsifiable oils.

Emulsifiable Mineral Oils

Emulsifiable mineral oils are mineral oils that contain various additives to make the oil mix with the water. These oils, which have good cooling and lubricating qualities, are low in cost and are used for general machining applications.

Superfatted Mineral Oils

Superfatted emulsifiable oils are emulsifiable mineral oils to which some fatty oils have been added. They are used for heavy machining operations because of the added lubricant they provide.

Extreme-Pressure (EP) Emulsifiable Oils

Extreme-pressure (EP) emulsifiable oils are superfatted emulsifiable mineral oils to which chlorine, sulfur, and phosphorus have been added to provide extra lubrication qualities. **Extreme-pressure oils** are primarily used in extreme machining conditions where it is necessary to reduce friction at the tool/workpiece interface.

Semichemical Fluids

Semichemical fluids are made by combining various percentages of chemical fluids and grinding oils to suit various types of metals, their microstructure, and the type of machining operation. Semichemical fluids can be mixed to suit the specific qualities required for any of the variables associated with fluid selection. There are two main types of semichemical cutting fluids which can run rich and not foam:

- The *plain type* that contains polymers, rust inhibitors, and from 5 to 18% mineral oils.
- The *extreme-pressure type* that is the plain type with chlorine and sulfur added to provide better lubrication.

It is wise to check with the manufacturer when selecting a cutting fluid for special conditions.

Straight Cutting Oils

Straight cutting oils, not mixed with water, vary considerably in their chemical composition and viscosity; this affects their effectiveness and use. These cutting oils are classified under two types, inactive and active. These terms relate to the oil's chemical ability or activity to react with a metal surface at high temperatures to protect it and improve the grinding action.

Inactive Cutting Oils

Inactive cutting oils contain sulfur that is so firmly attached to the oil that very little is released during the machining operation to react with the work surface. Inactive oils fall into four general categories:

- *Straight mineral oils* provide excellent lubrication but are not as effective in dissipating heat. Because of their low viscosity, they have good wetting and penetrating qualities. Straight mineral oils are commonly used for machining nonferrous materials such as aluminum, brass, and magnesium.

- *Fatty oils*, such as lard and sperm oils, are finding limited use as cutting fluids. They are generally used for severe machining operations on tough nonferrous metals where a sulfurized oil may cause discoloration.
- *Fatty-mineral oil* blend combinations provide better wetting and penetrating qualities than straight mineral oils. These blends provide better surface finishes on ferrous and nonferrous metals.
- *Sulfurized fatty-mineral oil blends* are made by adding sulfur to fatty-mineral oil blends. This oil provides excellent lubricity and anti-weld properties, especially where grinding or machining pressures are high. These blends, that can be used on ferrous and nonferrous materials, produce high surface finishes.

Active Cutting Oils

Active cutting oils contain sulfur that is not firmly attached to the oil; therefore, it is released during the machining operation and reacts with the work surface. Active oils fall into three categories:

- *Sulfurized mineral oils*, containing 0.5 to 0.8% sulfur, are generally light-colored, transparent, and have good cooling and lubricating qualities. They are useful for machining carbon steels and tough ductile metals. These oils will react with copper and its alloys, causing stains and discoloration.
- *Sulfochlorinated mineral oils* contain up to 3% sulfur and 1% chlorine to prevent grinding wheel loading and prolong wheel life. They are recommended for tough low-carbon and nickel-chrome alloy steels. They are widely used in thread-cutting operations.
- *Sulfochlorinated fatty-oil blends* contain more sulfur than other types, to provide added lubricity for heavy-duty machining operations.

● CUTTING FLUID APPLICATION

The life of the cutting tool or grinding wheel and the efficiency of the metal-removal operation are affected by how the cutting fluid is applied. To be effective, a continuous supply of cutting fluid must be supplied at the **chip-tool interface** to allow the chip to slide freely up the tool face and to cool the workpiece and cutting tool. A good general rule of thumb is that the inside diameter of the fluid supply nozzle should be at least three-quarters of the width of the cutting tool face being used for the cut. Since the metal-removal operations vary from machine to machine, the methods of supplying cutting fluid also vary.

Turning Operations

In operations on lathes, boring machines, turning and chucking centers, cutting fluid must be applied to the portion of the cutting tool producing the chip. For general facing and turning operations, one nozzle supplying fluid directly over the tool is an acceptable application, Fig. 14-5. Usually two nozzles are used for heavy-duty machining operations, one above and one below the cutting tool, to keep the tool and work cool, Fig. 14-6.

Drill Press Operations

On drill press tools such as drills, reamers, taps, etc., it is difficult to get cutting fluid to the area producing the chip. The most effective way of applying fluids is through the use of "oil-feed" drills and hollow reamers to lubricate the cutting edges and flush out the chips, Fig. 14-7. If this type of tool is not available, use a generous supply of cutting fluid on top of the work so that some reaches the cutting edges of the tool. Also, withdraw the tool from the hole occasionally to get rid of the chips.

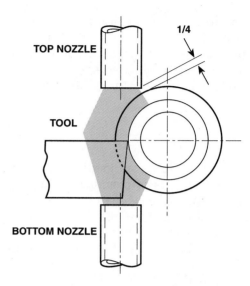

Fig. 14-6 Top and bottom nozzles should be used to supply cutting fluid for heavy-duty turning operations. (Courtesy of Cincinnati Milacron Inc.)

Milling Operations

In any type of milling operation, it is desirable to use two nozzles, one on each side of the cutter, to cool the tool and work, and to wash away chips. The nozzles should be at least three-quarters the width of the cutter to be effective, Fig. 14-8. When face milling, a ring-type fluid distributor, Fig. 14-9, is preferred because it can flood the cutter completely, thereby increasing its life almost 100 percent.

Grinding Operations

Grinding wheels have hundreds of thousands of cutting edges (abrasive grains) which produce minute (very small) chips. As a result, a grinding operation generally produces more heat than a machining operation and cutting fluid, supplied in large quantities, is essential. The three most common grinding operations are surface, cylindrical, and internal.

Surface Grinding

Several methods are used to apply fluid for surface grinding operations some being better than others for a specific operation.

- The *flood* method is most commonly used because it supplies a steady supply of cutting fluid to the wheel and workpiece. Wherever possible, use two coolant nozzles, one on either side of the wheel, as close as possi-

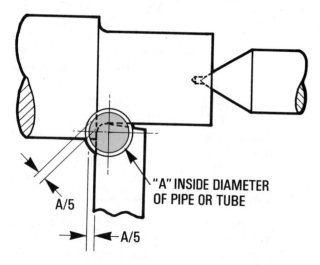

Fig. 14-5 Cutting fluid supplied through one nozzle is generally enough for most turning and facing operations. (Courtesy of Cincinnati Milacron Inc.)

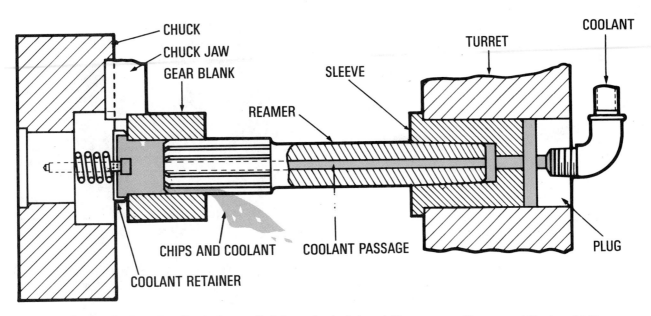

Fig. 14-7 Cutting fluid can be effectively supplied through a hole in a drill or reamer. (Courtesy of Cincinnati Milacron Inc.)

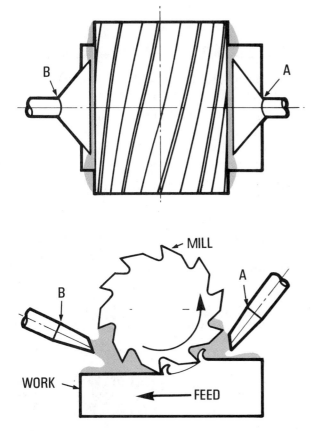

Fig. 14-8 Cutting fluid is being supplied to both sides of a slab milling cutter. (Courtesy of Cincinnati Milacron Inc.)

Fig. 14-9 A ring-type distributor is a good means of supplying cutting fluid when face milling. (Courtesy of Cincinnati Milacron Inc.)

ble to the wheel, Fig. 14-10. It is good practice to use a dummy block, slightly lower than the surface to be ground, to ensure that the work surface always has a good supply of coolant, Fig. 14-11.

- *Mist spray* method uses coolant that is siphoned from a reservoir by a stream of air, and a fine spray is directed to the wheel/work interface.

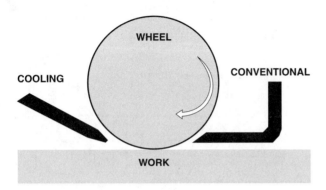

Fig. 14-10 Two nozzles used in flood cooling to cool the work and wash away the grinding grit and swarf. (Courtesy of GE Superabrasives)

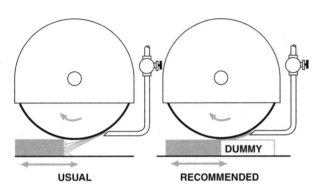

Fig. 14-11 A dummy block at the right end of the work-piece ensures that the work surface always receives a good supply of coolant. (Courtesy of GE Superabrasives)

- *Through-the-wheel* method has the coolant fed into a special wheel flange. Centrifugal force carries the coolant through the voids to the periphery of the grinding wheel. This cooling system works best on open-structure wheels where the voids are larger.

Cylindrical Grinding

The entire wheel/work contact area should be flooded with a steady stream of cutting fluid for cylindrical grinding operations. A fan-shaped nozzle, Fig. 14-12, a little wider than the grinding wheel face is recommended for the most efficient grinding operation.

Internal Grinding

Internal grinding practices recommend that the grinding wheel used be as large as possible, approximately 75% of the bore diameter. Since this does not leave much room for the coolant to

Fig. 14-12 A fan-shaped nozzle floods the grinding area with cutting fluid. (Courtesy of Cincinnati Milacron Inc.)

flush the chips and abrasive particles out of the hole, a good supply of cutting fluid should be used when internal grinding.

SUMMARY

- Cutting fluids have two primary functions—lubricating and cooling.

- The benefits of cutting fluids include reductions in friction, corrosion, tool wear, poor surface finishes, scrap, machine downtime, and production costs.

- The main types of cutting fluids are chemical fluids, emulsions, semichemical fluids, and straight cutting oils.

- Inactive cutting oils contain sulfur that does not react with the work surface. Active cutting oils contain sulfur that does react with the work surface.

- Methods of supplying cutting fluids to the workpiece vary according to the type of metal-removal operations.

KNOWLEDGE REVIEW

1. Name two important functions of a cutting fluid.

2. What advantages may be expected from the selection and proper application of a cutting fluid?

3. List the noncutting functions of a good cutting fluid.

4. Name the four main types of cutting fluids.

5. For what purpose are the following used?
   a. wetting-agent fluids with EP lubricants
   b. emulsifiable mineral oils
   c. extreme-pressure (EP) emulsifiable oils
   d. straight mineral oils
   e. sulfurized mineral oils

SECTION

9

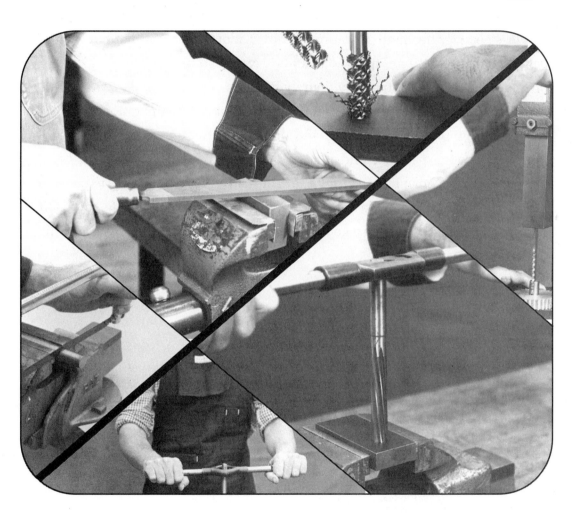

Bench and Hand Tools

At one time it was very important for a machinist to be highly skilled in the use of hand tools. The master tradespeople of the early twentieth century were known for the high skill and craftsmanship they had developed with hand tools. As newer and more accurate machine tools were developed, there was less need for the hand operations of old. Today it is important that we recognize the fact that if an operation can be performed on a machine, it can be done faster and more accurately than by hand.

Hand tools are still necessary for some machine shop operations, such as sawing, filing, polishing, tapping, and threading. It is important that the apprentice, through patience and practice, become skilled in the use of hand tools. Hand tools must be used with the same care given to the more expensive machine tools. A reasonable amount of care will keep tools in safe and good working condition. A sign of a good machinist is the excellent condition of his or her tools.

Figures Courtesy of (clockwise from far left): Kelmar Associates, Kelmar Associates, DoALL Co., Kelmar Associates, Kelmar Associates, Kelmar Associates.

UNIT 15

Hand Tools

Hand tools used in machine shop work are those that use hand or muscle power as their driving or turning force. These tools fall into two categories: cutting tools and noncutting tools. The cutting tools are generally used to remove metal or cut special forms in metal. The noncutting tools are used to hold or turn the workpiece. It is important that all hand tools be used properly and with care in order to get the best results and prevent accidents.

OBJECTIVES

After completing this unit, you should be able to:

- Recognize and properly use tools such as hammers and metal stamps.
- Select and properly use hacksaws for hand-cutting metals.
- Select and properly use files for removing surplus metal and smoothing surfaces.

KEY TERMS

file	hacksaw	hammer
metal stamp	pitch	vise

● THE MACHINIST'S VISE

The machinist's **vise** is a work-holding device used to hold work for such operations as sawing, filing, chipping, tapping, threading, etc.

Vises are made in a variety of sizes to hold work of many sizes and shapes. Special vise jaw inserts are available for some vises to hold round, square, and odd-shaped workpieces. Some vises are equipped with a swivel base that allows the vise to be turned to any position, Fig. 15-1. To hold work with finished surfaces, it is wise to place jaws made of aluminum, brass, or copper over the regular jaws to protect the work surface.

● HAMMERS

The *ball-peen hammer*, or machinist's hammer as it is more commonly called, is the **hammer** generally used in machine shop work. The rounded top, Fig. 15-2, is called the peen, and the bottom is known as the face. Both are hardened and tempered to prevent wear and resist deformation. Ball-peen hammers are made in a variety of sizes, with the weight of the head ranging from approximately 4 to 28 oz. (110 to 790 g). The smaller sizes are used for layout work, while the larger ones are used for general bench work.

The *toolmaker's or layout hammer*, Fig. 15-3, a handy tool for tool and diemakers, is used for accurately spotting center and layout lines. It is lightweight and has a magnifying glass built into

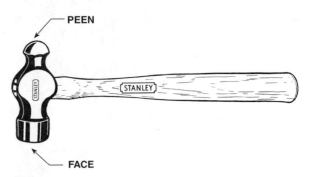

Fig. 15-2 A ball-peen hammer is the most commonly used hammer in machine shop work. (Courtesy of Stanley Tools, Division of The Stanley Works.)

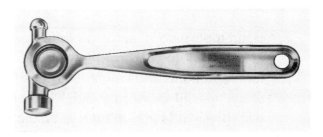

Fig. 15-3 A toolmaker's hammer should be used to make accurate layouts. (Courtesy of The L.S. Starrett Co.)

the head which makes it easy to locate and strike the punch in the proper location.

Soft-faced hammers, Fig. 15-4, are used in assembly and setup work because they do not mar the finished surface of the work. These hammers have pounding surfaces made of brass, plastic, lead, rawhide, or hard rubber.

Hammer Use and Safety

When using a hammer, hold it at the end of the handle. This position provides greater striking force and balance for the hammer than if it is gripped near the head. It also helps to keep the hammer face flat on the work being struck, while

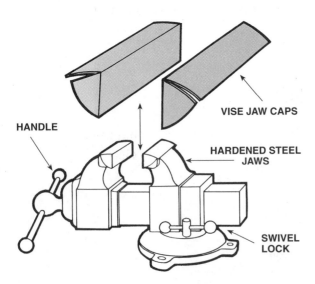

Fig. 15-1 A swivel-base vise allows work to be held and swiveled to any position in a 360° arc. (Courtesy of Kelmar Associates)

Fig. 15-4 A soft-faced hammer will not mar the finished surface of a workpiece. (Courtesy of Stanley Tools, Division of The Stanley Works)

minimizing the chance of damage to the face of the work.

The following safety precautions should be observed when using a hammer:

1. **A hammer with a loose head is dangerous. Always keep the hammer handle tightly secured in the head with a suitable wedge.**

2. **Replace the handle on any hammer if it is cracked or does not appear to be sound— do not wait for a serious accident to happen.**

3. **Never use a hammer with a greasy handle or when hands are oily.**

4. **Never strike the face of a hammer against hardened steel. A metal chip may fly off and cause a serious injury.**

● METAL STAMPS

A **metal stamp** or stencil, Fig. 15-5, is used to mark or identify workpieces. They are made in a variety of sizes, with letters or numbers from 1/32 to 1/2 in. (0.8 to 12.7 mm) high. Metal stamps should never be used on hardened metal, and if used on cast iron or hot rolled steel, the hard outer scale of the metal should first be removed by grinding, chipping, or machining to prevent damage to the stamp characters.

To Use Metal Stamps

Proper use of metal stamps is an important part of machine shop operations. Errors and confusion will result if materials are not identified legibly and correctly. The following steps describe the proper procedure for using metal stamps.

1. Place the work in a vise or on a flat steel bench plate.

2. Lay out a baseline to indicate where the stamping is to be located.

3. Lay out a line which should be the center of the lettering.

4. Select the letter or number which would be the center of the name to be stamped.

5. Hold the metal stamp so that the letter or trademark on its side is facing you. This will ensure that the letter will be imprinted "right-side up." Before stamping a workpiece, it is good practice to test the stamp on a piece of wood to be sure that the letter is right side up.

6. Place the edge of the metal stamp on the layout line at the center of the layout.

7. Stamp the middle letter on the center line.

8. Stamp all the letters to the right of the center line and then work from the center to the left, thus balancing the stamping about the center line.

To Use the Three-Step Method for Large Metal Stamps

When using metal stamps larger than 1/4 in. (6.35 mm), the three-step method is advisable in order to produce clear, sharp impressions.

1. Place the stamp on the baseline inclined toward you and strike it sharply with a medium-sized ball-peen hammer, Fig. 15-6A.

2. Replace the stamp in the impression, inclined only slightly, and strike it sharply, Fig. 15-6B.

3. Replace the stamp in the impression, hold it vertically, and strike it sharply to complete the letter, Fig. 15-6C.

● HAND HACKSAWS

The **hacksaw** is a hand tool used to cut metal. The pistol-grip hacksaw, Fig. 15-7, consists of four main parts: handle, frame, blade, and adjusting wing nut. The frame on most hacksaws can be flat or tubular. Most hacksaws have adjustable frames to accommodate various hacksaw blade lengths.

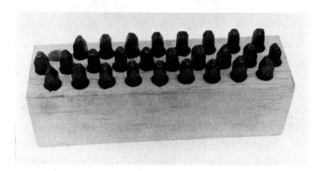

Fig. 15-5 Metal stamps are available in various sizes to identify parts or workpieces. (Courtesy of Kelmar Associates)

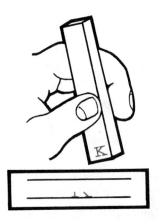
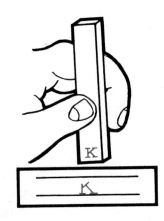
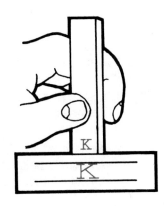

Fig. 15-6 The three-step method is effective when impressing large-size metal stamps. (Courtesy of Kelmar Associates)

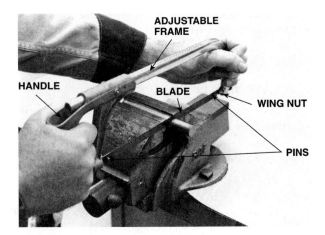

Fig. 15-7 The main parts of a hand hacksaw. (Courtesy of Kelmar Associates)

Hacksaw Blades

Hacksaw blades are made of high-speed, molybdenum or tungsten alloy steel that has been hardened and tempered. The saw blades generally used are 1/2 in. (12.7 mm) wide, and .025 in. (0.63 mm) thick. Common lengths of hacksaw blades are 8, 10, and 12 in. (200, 250, and 300 mm). There is a hole at each end of the blade for mounting it on the hacksaw frame.

The distance between each tooth on a blade is called the **pitch**. A pitch of 1/18 represents 18 teeth per inch. The most common blades have 14, 18, 24, or 32 teeth per inch. An 18-tooth blade (18 teeth per inch) is recommended for general use. It is important to use the right pitch for the work being cut. Select a blade as coarse as possible in order to provide plenty of chip clearance and cut through the work quickly. The blade selected

should have *at least two teeth in contact with the work* so that the work cannot jam between the teeth and strip the teeth from the saw blade.

Mounting a Blade on a Hand Hacksaw

1. Select the proper blade for the job, Fig. 15-8.
2. Adjust the frame to the length of the blade.
3. Place one end of the blade on the back pin (near the wing nut).

> **Be sure that the teeth of the blade point away from the handle.**

4. Place the other end of the blade on the front pin.
5. Tighten the wing nut until the blade is just snugged up, Fig. 15-9.

> Do not tighten the blade too much, as the blade may be broken or the hacksaw frame bent.

To Use a Hand Hacksaw

Care must be given to the proper method for using hacksaws in order to use the tool properly and avoid accidents.

1. Check that the pitch is proper for the job and be sure the teeth point away from the handle.
2. Adjust the tension on the saw blade as tight as two-finger pressure will permit.

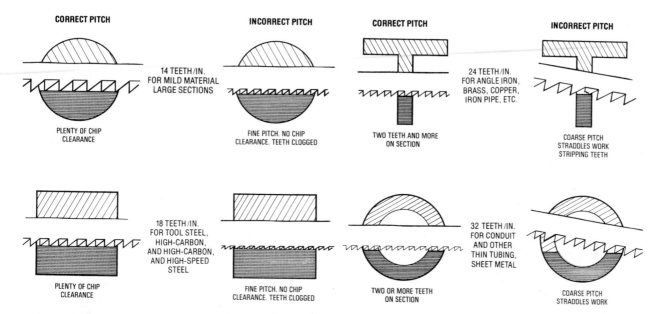

Fig. 15-8 Recommended saw pitches for various types and sizes of workpieces. (Courtesy of Kelmar Associates)

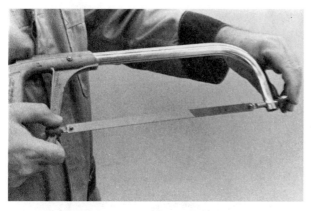

Fig. 15-9 Be sure that the saw teeth point away from the handle and tighten the wing nut securely. (Courtesy of Kelmar Associates)

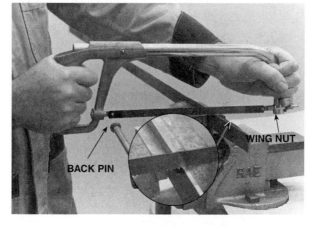

Fig. 15-10 The correct method of holding a hand hacksaw. (Courtesy of Kelmar Associates)

Too much pressure will bend the frame and damage the blade.

Too little pressure will allow the blade to bend, produce an inaccurate cut, and may break the blade.

3. Mark the position of the cut on the workpiece with a layout line or a nick with a file.

4. Mount the stock in the vise so that the cut will be made about 1/4 in. (6 mm) from the vise jaws.

5. Hold the hacksaw as shown in Fig. 15-10 and assume a comfortable stance.

6. Position the blade on the work just outside the layout line or in the file nick.

7. Apply down-pressure on the forward stroke and release it on the return stroke. Use a speed of about 50 strokes/min.

If the cut does not start in the proper place, file a V-shaped nick at the cut-off mark to guide the saw blade.

8. When nearing the end of the cut, slow down and ease up on the down-pressure to control the saw as it breaks through the material.

Fig. 15-11 Use a piece of wood on each side of thin metal when cutting it with a hacksaw. (Courtesy of Kelmar Associates)

9. When cutting thin material, hold the saw at an angle so that at least two teeth will bear on the work at all times as shown in Fig. 15-10. Sheet metal and other thin material may be clamped between two thin pieces of wood. The cut is then made through all three pieces as shown in Fig. 15-11.

Starting a new blade in an old cut will bind and ruin the "set" of the new blade. It is advisable to turn the work and start a cut in another place.

● **FILES**

A **file**, Fig. 15-12, is a hand cutting tool with many teeth, used to remove burrs, sharp edges, and surplus metal, and produce finished surfaces. Files are made of high-carbon steel, hardened and tempered, and are used in machine shop work when it is impractical to use machine tools. The file is a useful tool for removing ma-

chine marks, for tool and die making, and for the fitting of machine parts. Files are manufactured in a variety of shapes and sizes, each having a specific purpose. They are divided into two classes, single- and double-cut files.

Single-Cut Files

These files have a single row of parallel teeth across the face at an angle from 65 to 85°, Fig. 15-13. Single-cut files are used when a smooth surface is desired, and when harder metals are to be finished.

Double-Cut Files

These files have two rows of teeth crossing each other, one row being finer than the other, Fig. 15-13. The two rows crossing each other produce hundreds of sharp cutting teeth which remove metal quickly and make for easy clearing of chips.

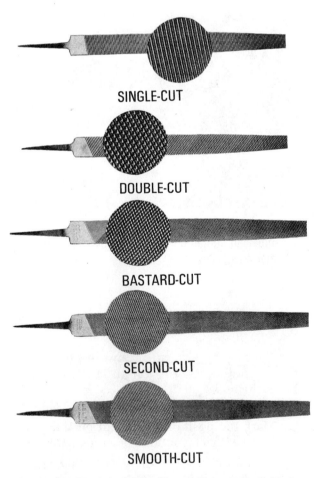

SINGLE-CUT

DOUBLE-CUT

BASTARD-CUT

SECOND-CUT

SMOOTH-CUT

Fig. 15-13 Single- and double-cut files are available in several degrees of coarseness. (Courtesy of Delta File Works)

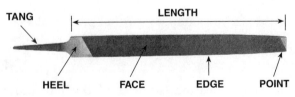

TANG LENGTH

HEEL FACE EDGE POINT

Fig. 15-12 The main parts of a hand file. (Courtesy of Kelmar Associates)

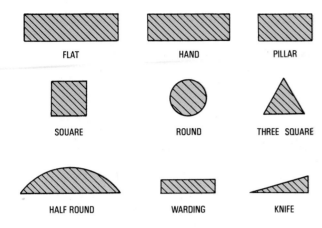

Fig. 15-14 Cross-sections of the most common files used in a machine shop. (Courtesy of Kelmar Associates)

Degrees of Coarseness

Both single- and double-cut files are manufactured in various degrees of coarseness. On larger files this is indicated by the terms rough, coarse, bastard, second-cut, smooth, and dead smooth. The bastard, second-cut, and smooth files are the ones most commonly used in machine shops. On smaller files, the degree of coarseness is indicated by numbers from 00 to 8, number 00 being the coarsest.

Shapes of Files

Files are manufactured in many shapes, Fig. 15-14, and may be identified by their cross-section, shape, or special use. The types of files most commonly used in a machine shop are the mill, hand, flat, round, half-round, square, three-square (triangular), pillar, warding, and knife.

Care of Files

Proper care, selection, and use of files are important factors if good results are to be obtained during the filing operation. In order to preserve the life of a file, the following points should be observed:

1. Use a *file card* to keep the file clean and free of chips.

2. Do not knock a file on a vise or hard metallic surface to clean it.

3. Do not apply too much down-pressure on a new file.

This tends to break down the cutting edges quickly. Too much pressure also causes "pinning"(small particles of metal become wedged between the file teeth), which produces scratches on the work surface.

4. Never use a file as a pry or hammer.

A file is a hardened tool that can break easily. This may cause small metal pieces to fly and cause a serious eye injury.

5. Always store files where they will not rub against other files. Hang or store them separately to preserve their sharp cutting edges.

Hints on Using Files

The following general points should be observed when filing:

1. Never use a file without a handle.

The tang has a sharp point, and if the file should slip, it could easily puncture a hand or arm.

2. Always be sure that the handle is tight on the file; loose handles are dangerous.

3. To produce a flat surface when cross-filing, the right hand, forearm, and the left hand should be held in a horizontal position, Fig. 15-15. Do not rock the file, but push it across the surface in a straight line.

Fig. 15-15 The proper method of holding a file for general filing. (Courtesy of Kelmar Associates)

4. A file cuts only on the forward stroke. When filing, apply down-pressure on the forward stroke only, and relieve the pressure on the return stroke.

Applying pressure on the return stroke will dull a file quickly.

5. Work to be filed should be held in a vise at about elbow height.

Small, fine work may be held higher, while heavier work, requiring much removal of metal, should be held lower.

6. Never rub the hand or fingers over a surface being filed.

Grease or oil from a hand touching a work surface will cause the file to slide over, instead of cutting the work. Oil will also cause the filings to stick between the teeth, causing the file to clog. To prevent a file from clogging, clean the file teeth with a file card, Fig. 15-16, and then rub chalk over the file surface.

Filing Practice

At one time, filing was an important skill that every good machinist had to master. During the twentieth century, the need for filing has decreased because new machine tools are able to perform the operations usually done by hand, faster and more accurately. They are also able to produce very flat surfaces and high surface finishes in much less time than is possible by hand. In the latter part of the twentieth century, it would be frowned upon if a person in the trade used a file to perform an operation which could be performed on a machine.

However, there are times when it may be necessary or more convenient to use a file. Therefore, it is wise for a machinist to know the correct methods of filing.

A clean file cuts easily and produces a better surface finish. Use a file card occasionally to remove the cuttings from the file teeth.

Filing Techniques

Depending on the task to be accomplished, different files and techniques are used. Stroke length, stroke direction, and pressure can all be modified.

Cross-Filing

Cross-filing is used when metal is to be removed rapidly or the surface made flat before finishing by draw filing.

For rough filing, use a double-cut file and cross the stroke at regular intervals to keep the surface flat and straight, Fig. 15-17. When finishing, use a single-cut file and take shorter strokes to keep the file flat on the work surface. The down-pressure on the file should be applied with the fingers of the opposite hand which should be kept over the workpiece, Fig. 15-18, during the finishing operation.

The work surface should be tested occasionally for flatness with the edge of a steel rule, Fig. 15-19A. A steel square should be used to test the squareness of one surface to another, Fig. 15-19B.

Draw Filing

Draw filing is used to produce a straight, square surface with a finer surface finish than is produced by straight filing. A single-cut is used and pressure is applied to the file just above the edge of the workpiece, Fig. 15-20. The file is alternately pulled and pushed lengthwise along the surface of the work. The file should be held

Fig. 15-16 Clean the file with a file card at regular intervals to remove the filings from the file teeth. (Courtesy of Kelmar Associates)

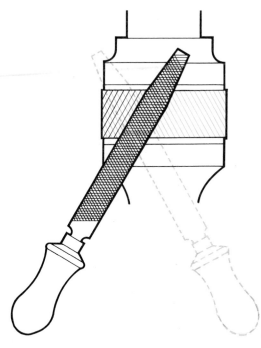

Fig. 15-17 Crossing the file at regular intervals helps to produce a flat surface. (Courtesy of Kelmar Associates)

flat and pressure applied only on the forward stroke. On the return stroke, the file should be slid back without applying pressure and without lifting the file from the work.

 After draw filing, remove the burrs from the edges of the workpiece to avoid painful cuts to the fingers and hand.

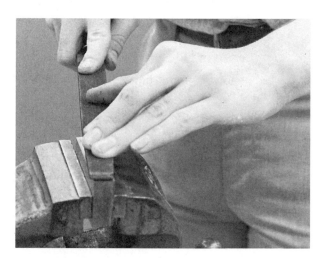

Fig. 15-18 Use only short strokes and keep down-pressure on the file that is over the work to produce a flat surface. (Courtesy of Kelmar Associates)

Fig. 15-19A Testing the flatness of a work surface with a rule. (Courtesy of Kelmar Associates)

Fig. 15-19B Testing a workpiece for squareness with an adjustable square. (Courtesy of Kelmar Associates)

Fig. 15-20 Finger pressure should not extend beyond the edge of the workpiece when draw filing. (Courtesy of Kelmar Associates)

SUMMARY

- Hand tools must be properly used and cared for in order to produce parts efficiently, accurately, and safely.
- The ball-peen hammer is the type of hammer used most often in machine shops.
- Metal stamps are used to identify workpieces.
- Hand hacksaws are used to cut metals; blades should have the proper pitch for each job.
- Files are hand tools that are used to perform a variety of metal operations.

KNOWLEDGE REVIEW

The Machinist's Vise

1. How should the finished surface of a workpiece be protected when it is held in a vise?

Hammers

2. Describe the machinist's hammer.
3. Why are soft-faced hammers used?
4. Of what materials are soft-faced hammers made?
5. Why should a hammer be gripped at the end of the handle?
6. State three safety precautions that should be observed when using a hammer.

Metal Stamps

7. What should be done to cast iron or hot rolled steel before using metal stamps on it?
8. List the procedure for balancing stamping on a centerline.

Hand Hacksaw

9. In what direction should the teeth of a hacksaw blade point?

10. Of what materials are hacksaw blades manufactured?
11. Explain why two teeth of a saw blade should be in contact with the work at all times.
12. What pitch blade is recommended for:

    a. general work?

    b. tool steel?

    c. angle iron and brass?

13. On what stroke is down-pressure applied when hacksawing?
14. How can thin pieces of metal be sawn?
15. Describe the procedure for starting a new blade in an old cut.

Files

16. Name five parts of a file.
17. Of what material are files made?
18. Describe and state the purpose of:

    a. single-cut files

    b. double-cut files

19. Name the degrees of coarseness for larger files.
20. How is the degree of coarseness indicated on small files?

Using Files

21. Why should a file never be used without a handle?
22. On which stroke should down-pressure be applied to the file? Explain why.
23. Why should a hand never be rubbed across a surface being filed?
24. How can a file be prevented from clogging?

Filing Practice

25. Describe the procedure for filing a flat surface.
26. How can work be tested for flatness?

U N I T
16

Threading and Reaming

Whenever possible, threads should be cut on machine tools where they can be accurately controlled and the thread cut will be of high quality. Sometimes it may be necessary, due to the size and shape of the workpiece, or because only a few parts are required, to cut the thread with hand tools. By using a reasonable amount of care, fairly accurate internal threads can be cut with a tap; external threads can be cut with a die.

Hand reamers are cutting tools used to bring a drilled or bored hole to accurate size and produce a good surface finish. Hand reamers are available in inch and metric sizes, and consist of the solid, expansion, adjustable, and taper varieties. All hand reamers have a square cut on the shank end so that a tap wrench can be used to turn the reamer into the work.

OBJECTIVES

After completing this unit, you should be able to:

- Identify and know the purpose of the three taps in a set.
- Calculate the tap drill size for inch and metric threads.
- Cut internal and external threads with taps and dies.
- Ream an accurate hole with a hand reamer.

KEY TERMS

blind hole   tap   threading
through hole

HAND TAPS

A **tap** is a cutting tool used to cut internal threads. It may be made of high-quality tool steel, high-speed steel, and various alloy steels. Special treatments, such as the titanium nitride coating of drills and taps, allow taps to be run at higher speeds and feeds, thereby increasing productivity as much as tenfold. The coating improves cutting-tool life and, at the same time, produces superior finishes on holes and threads.

The most common taps have two or three flutes cut lengthwise across the threads in order to form cutting edges, provide clearance for the chips, and admit cutting fluid to lubricate the tap. Four flutes are used less often because they tend to chip more easily due to the thin cutting web. The end of the tap shank is square so that a tap wrench can be used to turn it into a hole.

Hand taps are usually made in sets of three, called taper, plug, and bottoming, Fig. 16-1.

A *taper tap* is tapered approximately six threads from the end and is used to start a thread easily. It can be used for tapping a **hole** which goes all the way through the work, as well as for starting a **blind hole**, which does not go all the way through the work.

A *plug tap* is tapered for approximately three threads from the end. Sometimes the plug tap is the only tap used to thread a hole going through a piece of work.

A *bottoming tap* is not tapered, but chamfered at the end for one thread. It is used for threading to the bottom of a blind hole. When tapping a blind hole, first use the taper tap, then the plug tap, and complete the hole with a bottoming tap.

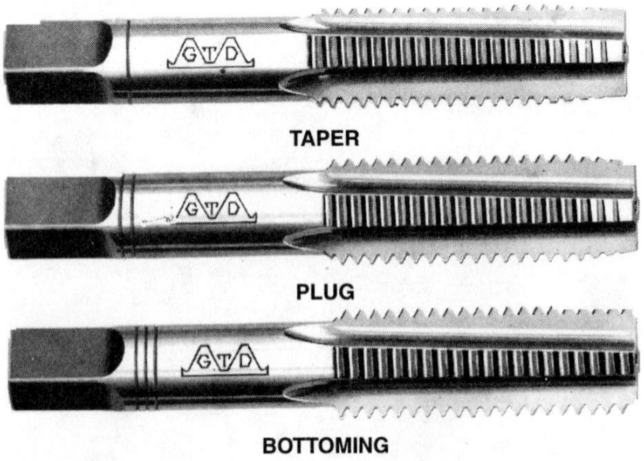

Inch Taps

Inch taps are available in a large variety of sizes, thread pitches, and thread forms. (See Table 5 in the Appendix.) The major diameter, number of threads per inch, and type of thread are usually found stamped on the shank of a tap.

For example: 1/2 in.—13 N.C. represents:

a. 1/2 in. = major diameter of the tap,

b. 13 = number of threads per inch,

c. N.C. = National Coarse (a type of thread).

Metric Taps

The International Standards Organization (ISO) has developed a standard metric thread which is used in the United States, Canada, and many other countries throughout the world. This series has only 25 thread sizes, ranging from 1.6 to 100 mm diameter. (See Table 6 in the Appendix.)

Metric taps are identified with the letter M followed by the nominal diameter of the thread in millimeters times the pitch in millimeters. A tap with the markings M 2.5 × 0.45 indicates:

M = metric thread,

2.5 = nominal diameter of the thread in millimeters,

0.45 = pitch of the thread in millimeters.

TAP WRENCHES

Tap wrenches are available in two types and in various sizes to suit the size of the tap being used.

The double-end adjustable tap wrench, Fig. 16-2A, is available in several sizes, but is generally used for larger taps and in open places where there is room to turn the tap wrench. Because of the greater leverage obtained when using this wrench, it is important not to use a large wrench to turn a small tap, because small taps are easily broken.

The adjustable T-handle tap wrench, Fig. 16-2B, is generally used for small taps or in confined areas where it is not possible to produce 75% of a full thread.

When used with very small number taps, the body of the wrench should be turned with

TAPER

PLUG

BOTTOMING

Fig. 16-1 A standard set of hand taps consists of a taper, plug, and bottoming tap. (Courtesy of Greenfield Industries, Inc.)

Fig. 16-2A Double-end adjustable tap wrenches are used for larger taps. (Courtesy of Kelmar Associates)

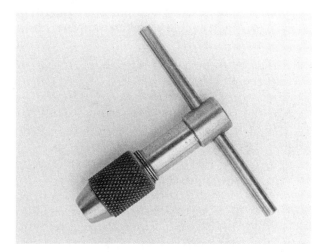

Fig. 16-2B A T-handle tap wrench is generally used for smaller taps. (Courtesy of Kelmar Associates)

the thumb and forefinger to advance the tap into the hole. The handle is usually used when threading with larger taps, which will not break as easily as small ones.

NOTE
Always use the proper size tap wrench for the size of tap being used.

● **TAP DRILL SIZE**

Before a tap is used, a hole must be drilled in the workpiece to the correct tap drill size, Fig. 16-3. The tap drill size (T.D.S.) is the size of the drill that should be used to leave the proper amount of material in the hole for a tap to cut a thread.

The tap drill, which is always smaller than the tap, leaves enough material in the hole for the tap to produce 75% of a full thread.

Tap Drill Sizes for Inch Threads

When a chart is not available, the tap drill size for any American National or Unified thread

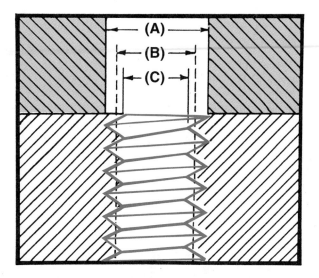

Fig. 16-3 The cross-section of a tapped hole. (A) Major diameter (B) Tap drill size (C) Minor diameter. (Courtesy of Kelmar Associates)

can be easily found by applying this simple formula:

$$T.D.S. = D - \frac{1}{N}$$

$T.D.S.$ = tap drill size
D = major diameter of tap
N = number of threads per inch

EXAMPLE: Find the tap drill size for a 5/8 in. – 11 UNC tap.

$$T.D.S. = \frac{5}{8} - \frac{1}{11}$$
$$= .625 - .091$$
$$= .534 \text{ in.}$$

The nearest drill size to .534 in. is .531 in. (17/32 in.). Therefore 17/32 in. is the tap drill size for a 5/8 in.—11 UNC tap. See Table 5 in the Appendix for a complete tapping drill chart for American National and Unified threads.

Tap Drill Sizes for Metric Threads

The tap drill sizes for metric threads may be calculated by subtracting the pitch from the nominal diameter.

$$T.D.S. = D - P$$

EXAMPLE: Calculate the tap drill size for a M12 × 1.75 thread.

$$T.D.S. = 12 - 1.75$$
$$= 10.25 \text{ mm}$$

Refer to the Appendix for tap drill sizes for metric threads.

● TAPPING A HOLE

Tapping is the operation of cutting an internal thread using a tap and tap wrench. Because taps are hard and brittle, they are easily broken. Before any hole is tapped, both ends should be countersunk slightly larger than the tap diameter. This eliminates the burr at the start of the thread, and reduces the possibility of the end of the thread being damaged.

> Extreme care must be used to prevent breakage when tapping a hole. A broken tap in a hole is very difficult to remove and often results in scrapping the work.

Some of the most common causes of tap breakage and the steps to correct them are as follows:

1. The tap drill hole is too small.

> Be sure to drill the correct-size tap drill hole.

2. The tap's cutting edges are dull.

> Check each tap before use to be sure the cutting edges are sharp.

3. The tap was started out of alignment.

> Check the tap for squareness after it has entered the hole two full turns, and correct the alignment.

4. Too much pressure applied on one side of the tap wrench while trying to align the tap.

> Remove the tap from the hole and apply only slight pressure on the wrench while aligning the tap. Repeat as often as necessary until the tap is square to the work surface.

5. Chips clog in the flutes while tapping.

> Clean the hole and tap of chips occasionally, and use cutting fluid while tapping.

To Tap a Hole by Hand

1. Select the correct size and type of tap for the job.
   - Use a plug tap for through holes.
   - Use the taper, plug, and bottoming tap, in that sequence, when tapping blind holes.
2. Select the correct tap wrench for the size of tap being used.

> Too large a tap wrench can result in a broken tap.

3. Apply a suitable cutting fluid.

> No cutting fluid is required when tapping brass or cast iron.

4. Place the tap in the hole as near to vertical as possible.
5. Apply equal down-pressure on both handles, and turn the tap clockwise (for right-hand thread) for about two turns.
6. Remove the tap wrench and check the tap for squareness. Check at two positions 90° to each other, Fig. 16-4.
7. If the tap has not entered squarely, remove it from the hole and restart it by applying *slight* pressure in the direction from which the tap leans, Fig. 16-5.

Fig. 16-4 Check a tap for squareness at two positions 90° apart. (Courtesy of Kelmar Associates)

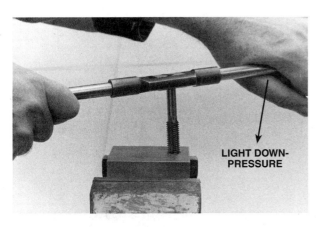

Fig. 16-5 To correct the alignment of a tap, apply slight down-pressure in the direction from which the tap leans while the tap is being turned. (Courtesy of Kelmar Associates)

> Be careful not to exert too much pressure in the straightening process, otherwise the tap may be broken.

8. When a tap has been properly started, feed it into the hole by turning the tap wrench. Down-pressure is no longer required, since the tap will thread itself into the hole.

9. Turn the tap clockwise one-quarter of a turn and then turn it backward about one-half turn to break the chip. This must be done with a steady motion to avoid breaking the tap.

NOTE

When tapping blind holes, use all three taps in order: taper, plug, and then the bottoming tap. Before using the bottoming tap, remove all the chips from the hole and be careful not to hit the bottom of the hole with the tap.

Tapping Lubricants

The use of a suitable cutting lubricant when tapping results in longer tap life, better finish, and greater production. The recommended tapping lubricants for the more common metals are listed below. A more complete list will be found in Table 8 of the Appendix.

Material	*Lubricant*
Machine steel (hot and cold rolled)	soluble oil, lard oil
Tool steel (carbon and high-speed)	mineral lard oil sulfur-base oil
Malleable iron	soluble oil
Cast iron	dry
Brass and bronze	dry

THREADING DIES

A **threading** die is used to cut external threads on round work. The most common threading dies are the adjustable and solid types. The round adjustable die, Fig. 16-6A, is split on one side and can be adjusted to cut slightly over or under-sized threads. It is mounted in a die stock, Fig. 16-7, which provides the means of turning the dies onto the work.

The solid die, Fig. 16-6B, cannot be adjusted and is generally used for recutting damaged or oversized threads. Solid dies are turned onto the thread with a special die stock, or adjustable wrench.

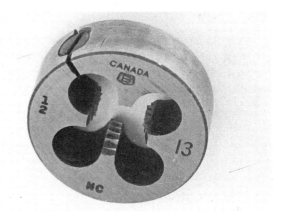

(A) ROUND ADJUSTABLE DIE

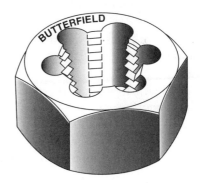

(B) SOLID DIE

Fig. 16-6 Dies are used to cut external threads on a workpiece. (Courtesy of Union Butterfield Corp.)

Fig. 16-7 Start the tapered end of the die on the chamfered end of the workpiece. (Courtesy of Kelmar Associates)

To Thread with a Hand Die

The threading process requires the machinist to work carefully to produce usable parts and avoid damage. The following describes the procedure to be used.

1. Chamfer the end of the workpiece with a file or on the grinder.

2. Fasten the work securely in a vise. Hold small diameter work short to prevent it from bending.

3. Select the proper die and die stock.

4. Lubricate the tapered end of the die with a suitable cutting lubricant, Fig. 16-7.

5. Place the tapered end of the die squarely on the work.

6. Apply down-pressure on both die stock handles and turn clockwise several turns, Fig. 16-8.

7. Check the die to see if it has started squarely with the work.

8. If it is not square, remove the die from the work and restart it squarely, applying slight pressure while the die is being turned.

9. Turn the die forward one turn, and then reverse it approximately one-half of a turn to break the chip.

10. Apply cutting fluid frequently during the threading process.

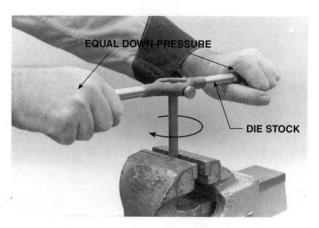

Fig. 16-8 Apply equal down-pressure on diestock handles and turn clockwise two or three turns. (Courtesy of Kelmar Associates)

 When cutting a long thread, keep the arms and hands clear of the sharp threads coming through the die.

If the thread must be cut to a shoulder, remove the die and restart it with the tapered side of the die facing up. Complete the thread, being careful not to hit the shoulder, otherwise the work may be bent and the die broken.

● HAND REAMING

A *hand reamer* is a cutting tool used for finishing drilled or bored holes to an accurate size and shape. Three of the more common reamers, the hand, adjustable, and taper, are shown in Fig. 16-9A. The hand reamer is straight for nearly the full length of its flutes, but is tapered slightly on the end for a distance equal to its diameter to enable it to start in a hole easily. The *adjustable reamer*, Fig. 16-9B, can be adjusted about 1/32 in. over or under the nominal size for finishing holes. The *taper reamer*, Fig. 16-9C, is available in standard taper sizes to finish tapered holes.

Hand reamers are provided with a square at one end to fit a tap wrench and should never be used under mechanical power. A suitable cutting lubricant should be used and not more than .005 in. (0.12 mm) should be removed with a hand reamer.

Never turn a reamer backwards or the cutting edges will be damaged.

(A) HAND REAMER

(B) ADJUSTABLE REAMER

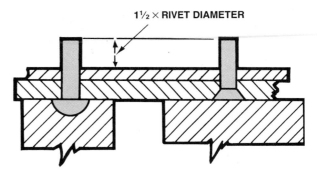

(C) TAPER REAMER

Fig. 16-9 Common types of hand reamers. (Courtesy of Cleveland Twist Drill Co.)

Inch solid hand reamers are available in sizes from 1/8 to 1½ in. in diameter in steps of 1/64 in.

Metric solid hand reamers are available in sizes from 1 to 13 mm in steps of 0.5 mm and from 13 to 26 mm in steps of 1 mm.

● **METAL FASTENERS**

In machine shop work, many methods are used to fasten work together. Some of the common methods include riveting and fastening with screws and dowel pins.

Rivets

A rivet is a metal pin made of soft steel, brass, copper, or aluminum, with a head at one end. The two most common types are the round head and the countersunk head, Fig. 16-10. Riveting consists of placing a rivet through holes in two or more pieces of metal and then forming a head on the other end of the rivet with a ball-peen hammer. To rivet, use the following procedure.

1. Drill holes in the metal pieces 1/64 to 1/32 in. (0.4 to 0.8 mm) larger than the body of the rivet. If a countersunk rivet is being used, countersink for the head.

2. Insert a rivet through the holes, having it extend past the work 1½ times the diameter of the rivet.

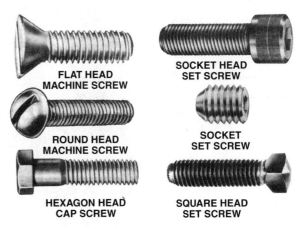

Fig. 16-10 Two workpieces set up for assembly by riveting. (Courtesy of Kelmar Associates)

3. Place round head rivets in a metal block recessed for the shape of the head, Fig. 16-10. Countersunk rivets may be placed on a flat block.

4. Use a ball-peen hammer to form a head on the body of each rivet.

Machine Screws

Machine screws are widely used by the machinist for assembly work. A hole slightly larger than the body size of the screw may be drilled through the workpieces, the screw inserted, and a nut placed on the end of it. Another method of using machine screws is to drill a clearance hole through one piece and drill and tap the other piece to fit the thread of the screw. Some of the more common screws are shown in Fig. 16-11.

Work being fastened by flathead and socket-head screws must be recessed so that the head of the screw is flush with the surface of the work. Flathead screws are recessed with an 82° coun-

FLAT HEAD MACHINE SCREW

SOCKET HEAD SET SCREW

ROUND HEAD MACHINE SCREW

SOCKET SET SCREW

HEXAGON HEAD CAP SCREW

SQUARE HEAD SET SCREW

Fig. 16-11 Threaded fasteners are available in a wide variety of types and sizes. (Courtesy of Kelmar Associates)

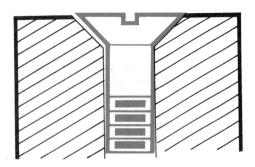

Fig. 16-12 The workpiece is countersunk to the correct depth when the head of the screw is flush with the work surface. (Courtesy of Kelmar Associates)

tersink, Fig. 16-12. To recess work for socket-head screws, Fig. 16-13, a counterbore or flat bottom drill is used.

Self-Tapping Screws

Self-tapping screws, Fig. 16-14, are designed to cut a thread and therefore eliminate the

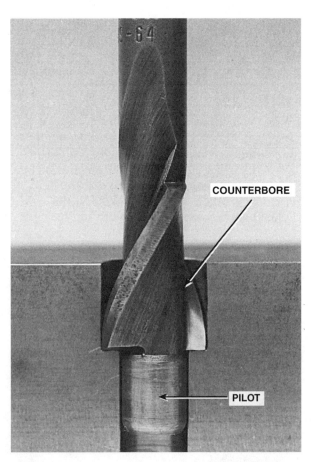

Fig. 16-13 A hole must be counterbored when socket-head cap screws are used. (Courtesy of Kelmar Associates)

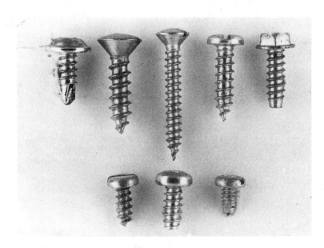

Fig. 16-14 A variety of self-tapping screws. (Courtesy of Kelmar Associates)

need for tapping a hole. Self-tapping screws cut a mating thread in the work material which fits snugly to the body of the screw. This close-fitting thread keeps the screw from backing off or coming loose, even under vibrating conditions. Self-tapping screws are used when assembling thin metal parts, nonferrous materials, and plastics.

Dowel and Taper Pins

Dowel pins, Fig. 16-15A, are used to locate one piece of work accurately with another. A hole is drilled and reamed through both workpieces simultaneously, and then the dowel pin is forced into the hole to keep both pieces in alignment. A tapered hole must be reamed to accommodate taper pins, Fig. 16-15B.

(A) DOWEL PIN

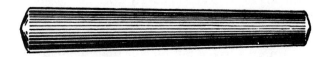

(B) TAPER PIN

Fig. 16-15 Dowel and taper pins are used to locate parts correctly. (A) (Courtesy of H. Paulin & Co. Ltd.), (B) (Courtesy of Kelmar Associates)

● WRENCHES

Many types of wrenches are used in machine shop work, each being suited to a specific purpose. The name of a wrench is derived from either its shape, its use, or its construction. Various types of wrenches used in a machine shop are illustrated in Fig. 16-16.

A *single-end wrench* is one that fits only one size of bolt, head, or nut. The opening is generally offset at a 15° angle to permit complete rotation of a hexagonal nut in only 30° by "flopping" the wrench.

A *double-end wrench* has a different size opening at each end. It is used in the same manner as a single-end wrench.

The *adjustable wrench* is adjustable to various size nuts and is particularly useful for odd size nuts. Unfortunately, this type of wrench, when not properly adjusted to the flats of the nut, will damage the corners of the nut. Adjustable wrenches should not be used on nuts which are very tight on the bolt. When excess pressure is applied to the handle, the jaws tend to spring, causing damage to the wrench and the corners of the nut. When using an adjustable wrench, the solid jaw should point in the direction of the force being applied, Fig. 16-17.

A *toolpost wrench* is a combination open-end wrench and a box-end wrench. The box end is used on toolpost screws and often on lathe carriage locking screws. In order for the toolpost screw head not to be damaged, it is important that only this type of wrench be used.

Box-end or *twelve point wrenches* are capable of operating in close quarters. The box-end wrench has twelve notches cut around the inside of the face. This type of wrench completely surrounds the nut and will not slip.

The *pin spanner wrench* fits around the circumference of a round nut. The pin on the wrench fits into a hole in the periphery of the nut.

(A) SINGLE-END WRENCH

(B) DOUBLE-END WRENCH

(C) ADJUSTABLE WRENCH

(D) TOOLPOST WRENCH

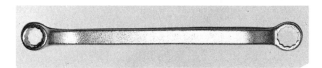

(E) BOX-END WRENCH

(F) PIN SPANNER WRENCH

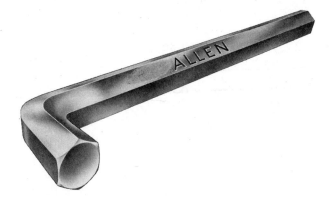

(G) ALLEN WRENCH

Fig. 16-16 Common wrenches used in machine shop work. (Courtesy of Acme Screw & Gear, J.H. Williams Co., and Kelmar Associates)

Fig. 16-17 The correct method of using an adjustable wrench. (Courtesy of Kelmar Associates)

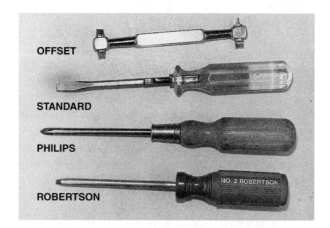

Fig. 16-18 Common types of screwdrivers used in mechanical work. (Courtesy of Kelmar Associates)

The *socket setscrew wrench*, commonly called the *Allen wrench*, is hexagonal and fits into the holes in safety setscrews or socket-head setscrews. It is made of tool steel in various sizes to suit the wide range of screw sizes. It is identified by the distance across the flats; this distance is usually one-half the outside diameter of the set screw thread in which it is used.

Using Wrenches

Observe the following guidelines when using wrenches:

1. Always select a wrench that fits the nut or bolt properly.

 A wrench that is too large may slip off the nut and cause an accident.

2. Whenever possible, *pull* rather than push on a wrench in order to avoid injury if the wrench should slip.

3. Always be sure that the nut is fully seated in the wrench jaw.

4. Use a wrench in the same plane as the nut or bolt head.

5. When tightening or loosening a nut, a sharp, quick jerk is more effective than a steady pull.

6. A drop of oil on the threads when assembling a bolt and nut will ensure easier removal later.

● SCREWDRIVERS

Screwdrivers, Fig. 16-18, are made in a variety of shapes, types, and sizes. The standard or common screwdriver is used on slotted-head screws. It consists of three parts: the blade, the shank, and the handle. Although most shanks are round, those on heavy-duty screwdrivers are generally square. This permits the use of a wrench to turn the screwdriver when extra torque is required.

The *offset* screwdriver is designed for use in confined areas where it is impossible to use a standard screwdriver. The blades on the ends are at right angles to each other. The screw is turned one-quarter of a turn with one end and then one-quarter of a turn with the other end.

Other commonly used screwdrivers are the *Robertson*, which has a square tip or blade, and the *Phillips*, which has a cross- or x-shaped point. Both types are made in different sizes to suit the wide range of screw sizes.

SUMMARY

- Taps are used to cut internal threads. They are usually made in sets of three: taper, plug, and bottoming.

- Tap drill size must be carefully selected in order to leave the proper amount of material in the hole for cutting threads.

- Threading dies are used to cut external threads on round workpieces. The adjustable and solid types are the most common.

- A hand reamer is used to finish drilled or bored holes to an accurate size and shape.

KNOWLEDGE REVIEW

Hand Taps

1. Define a tap.
2. What is the purpose of the flutes on a tap?
3. Explain the following information found on the shank of a tap:

   a. 1/2 in. – 10 N.F.

   b. M 30 × 3.5

4. Describe and state the purpose of:

   a. taper tap

   b. plug tap

   c. bottoming tap

Tap Wrenches

5. Name two types of tap wrenches and state where each is used.

Tap Drill Size

6. Define a tap drill.
7. What percentage of a thread will a tap cut after the hole has been drilled to tap drill size?
8. Using the tap drill formula, calculate the tap drill size for:

   a. 7/16 in. – 14 N.C.

   b. 3/4 in. – 10 N.C.

   c. M 8 × 1.25

   d. M 30 × 3.5

Tapping a Hole

9. Define the operation of tapping.
10. What care should be used while tapping a hole?
11. How should a tap be tested for squareness?
12. Explain the procedure for correcting a tap which has not started squarely.
13. Why should the tap be turned backwards after every quarter turn?

14. Describe the procedure for tapping a blind hole.

Tapping Lubricants

15. Why is a cutting lubricant used while tapping?
16. What lubricant is recommend for:

   a. machine steel?

   b. cast iron?

Threading Dies

17. For what purpose are threading dies used?
18. State the purpose of the adjustable die and the solid type die.
19. Explain the procedure for starting a die on work.
20. What procedure should be followed when it is necessary to thread up to a shoulder?

Hand Reaming

21. Describe and state the purpose of a hand reamer.
22. How much material should be removed with a hand reamer?

Metal Fasteners

23. Describe the procedure for riveting two pieces of metal together.
24. What tools are used to recess the work when using:

   a. flat head screws?

   b. socket head cap screws?

25. What is the purpose of dowel pins?

Wrenches

26. Name six types of wrenches used in a machine shop.
27. Why should an adjustable wrench not be used on an extremely tight nut?
28. Why is it advisable to pull rather than push a wrench?

Screwdrivers

29. Name three parts of a standard screwdriver.
30. What is the purpose of a square shank on a heavy-duty screwdriver?

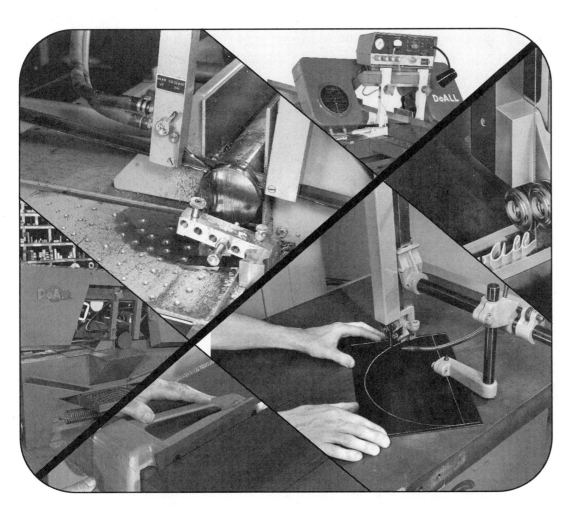

Power Saws

Archaeological discoveries have shown that the development of the first crude saw closely followed the origin of the stone axe and knife. The sharp edge of a stone was serrated or toothed to form a saw. This instrument cut by scraping away particles of objects softer than itself. A great improvement in the quality of saws followed the appearance of copper, bronze, and ferrous metals. With modern steels and hardening methods, a wide variety of saw blades is available to suit hand hacksaws and machine power saws.

The two most common types of power saws used in machine shops and manufacturing industries are the cut-off saws and the contour bandsaws. *Cut-off saws* are used to cut a workpiece to a rough length from bar stock before it is finished to size by machining. *Contour bandsaws* are used to rough-cut a workpiece close to size and shape by slotting, notching, and angular and radius cutting before it is finished by machining.

Figures courtesy of DoALL Co.

U N I T
17

Cut-Off Saws

Cut-off saws are used to rough-cut work to the required length from a longer piece of metal. The most common types of cut-off saws used in machine shop work are the power hacksaw, the horizontal bandsaw, the abrasive cut-off saw, and the circular cut-off saw.

OBJECTIVES

After completing this unit, you should be able to:

- Describe the function of the main parts of a horizontal bandsaw.
- Select the proper saw blade pitch for cutting various workpiece materials and shapes.
- Install a saw band properly and be able to cut off work to accurate length.

KEY TERMS

coarseness   horizontal bandsaw   run-out

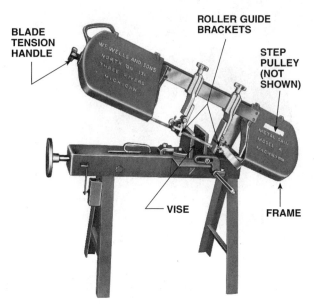

BLADE TENSION HANDLE

ROLLER GUIDE BRACKETS

STEP PULLEY (NOT SHOWN)

VISE

FRAME

Fig. 17-1 The main operative parts of a horizontal bandsaw. (Courtesy of KTS Industries)

● TYPES OF CUT-OFF SAWS

The *power hacksaw*, a reciprocating type of saw, is usually fastened to the floor. The saw frame and blade travel back and forth, with down-pressure being applied automatically only on the forward stroke. The power hacksaw finds limited use in machine shop work because it cuts only on the forward stroke, which results in considerable wasted motion.

The *horizontal bandsaw*, Fig. 17-1, has a flexible, belt-like "endless" blade that cuts continuously in one direction. The thin, continuous blade travels over the rims of two pulley wheels and passes through roller guide brackets that support the blade and keep it running true.

Horizontal bandsaws are available in a wide variety of types and sizes, and are becoming increasingly popular because of their high productivity and versatility. On some models, casters are supplied with the machine so that it can be moved easily from place to place.

The *abrasive cut-off saw*, Fig. 17-2, uses a thin, abrasive grinding wheel, revolving at high speed, to cut off material. It can cut metals, glass, ceramics, etc. quickly and accurately, whether the material is hardened or not. This cut-off operation can be performed dry; however, cutting fluid is generally used to keep the work and saw cooler and to produce a better surface finish.

The *cold circular cut-off saw*, Fig. 17-3, uses a circular saw blade similar to the one

Fig. 17-2 The abrasive cut-off saw can cut hardened metals, glass, and ceramics. (Courtesy of Everett Industries, Inc.)

Fig. 17-3 A cold circular cut-off saw is used for cutting soft or unhardened metals. (Courtesy of Everett Industries, Inc.)

used on a wood-cutting table saw. The saw blade is generally made of chrome-vanadium steel; however, carbide-tipped blades are used for some cutting applications. These saws produce very accurate cuts on materials such as aluminum, brass, copper, machine steel, and stainless steel.

● HORIZONTAL BANDSAW

The most common type of cut-off saw used in a school shop is the **horizontal bandsaw**, because it is easy to operate and cuts material quickly and accurately. The main operative parts of the horizontal bandsaw are shown in Fig. 17-1.

The *saw frame*, hinged at the motor end, has two pulley wheels mounted on it over which the continuous blade passes.

The *step pulleys* at the motor end are used to vary the speed of the continuous blade to suit the type of material being cut.

The *roller guide brackets* provide rigidity for a section of the blade and can be adjusted to accommodate various widths of material. These brackets should be adjusted to just clear the width of the work being cut.

The *blade tension handle* is used to adjust the tension on the saw blade. The blade should be adjusted to prevent it from wandering and twisting.

The *vise*, mounted on the table, can be adjusted to hold various sizes of workpieces. It can also be swiveled for making angular cuts on the end of a piece of material.

● SAW BLADES

High-speed tungsten and high-speed molybdenum steel are commonly used in the manufacture of saw blades, and for the power hacksaw they are usually hardened completely. Flexible blades used on bandsaws have only the saw teeth hardened.

Saw blades are manufactured in various degrees of **coarseness** ranging from 4 (coarse) to 14 (fine) pitch. When cutting large sections, use a coarse or 4-pitch blade, which provides the greatest chip clearance and helps to increase tooth penetration. For cutting tool steel and thin material, a 14-pitch blade is recommended. A 10-pitch blade is recommended for general purpose sawing. *Metric* saw blades are now available in similar sizes, but in teeth per 25 millimeters of length rather than teeth per inch. Therefore, the pitch of a blade having 10 teeth per 25 mm would be 10 ÷ 25 mm or 0.4 mm. Always select a saw blade as coarse as possible, but make sure that two teeth of the blade will be in contact with the work at all times.

If less than two teeth are in contact with the work, the work can be caught in the tooth space (gullet), which would cause the teeth of the blade to strip or break.

Installing a Saw Blade

When replacing a blade, always make sure that the teeth are pointing in the direction of saw travel or toward the motor end of the machine. The blade tension should be adjusted to prevent the blade from twisting or wandering during a cut. If it is necessary to replace a dull blade before a cut is finished, rotate the work one-quarter of a turn in the vise. This will prevent the new blade from jamming or breaking in the cut made by the worn saw.

 Always be sure that the machine is stopped when setting up the workpiece or installing a saw blade.

To install a saw blade, use the following procedure:

1. Loosen the blade tension handle.

2. Move the adjustable pulley wheel forward slightly.

3. Mount the new saw band over the two pulleys.

 Be sure that the saw teeth are pointing toward the motor end of the machine.

4. Place the saw blade between the rollers of the guide brackets, Fig. 17-4.

5. Tighten the blade tension handle only enough to hold the blade on the pulleys.

6. Start and quickly stop the machine in order to make the saw blade revolve a turn or two. This will seat the blade on the pulleys.

7. Tighten the blade tension handle as tightly as possible with *one hand*.

● SAWING

For the most efficient sawing, it is important that the correct type and pitch of saw blade be selected and that the saw be run at the proper speed for the material being cut. Use finer-tooth blades when cutting thin cross-sections and

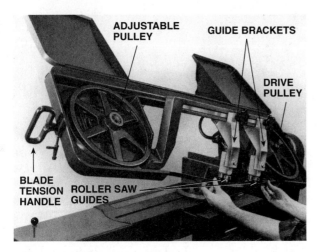

Fig. 17-4 Installing a blade on a horizontal bandsaw. (Courtesy of KTS Industries)

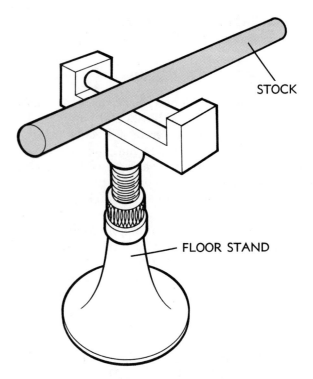

Fig. 17-5 A floor stand is used to support long workpieces held in a cut-off saw. (Courtesy of Kelmar Associates)

extra-hard materials. Coarser-tooth blades should be used for thick cross-sections and material which is soft and stringy. The blade speed should suit the type and thickness of the material being cut. Too fast a blade speed or excessive feeding pressure will dull the saw teeth quickly and cause an inaccurate cut. Too slow a blade speed does not use the saw efficiently, wastes time, and results in lower production.

To Saw Work to Length

1. Check the solid vise jaw with a square to make sure it is at right angles to the saw blade; correct if necessary.

2. Place the material in the vise, supporting long pieces with a floor stand, Fig. 17-5.

3. Lower the saw blade until it just clears the top of the work. Keep the blade in this position by engaging the ratchet lever or by closing the hydraulic valve.

4. Adjust the roller guide brackets until they *just clear* both sides of the material to be cut.

 Never attempt to mount, measure, or remove work unless the saw is stopped.

5. Hold a steel rule against the edge of the saw blade and move the material until the correct length is obtained.

6. Always allow 1/16 in. (1.5 mm) longer than required for each 1 in. (25 mm) of thickness to compensate for any saw **run-out** (slightly angular cut caused by hard spots in steel or a dull saw blade).

7. Tighten the vise and recheck the length from the blade to the end of the material to make sure the work has not moved.

8. Raise the saw frame slightly, release the ratchet lever or open the hydraulic valve, and then start the machine.

9. Lower the blade slowly until it just touches the work.

10. When the cut has been completed, the machine will shut off automatically.

Sawing Hints

Cutting materials to length is an important part of machine shop processing. Properly done, it increases productivity and minimizes waste and damage. The following tips are provided to aid in sawing operations:

1. Guard long material at both ends to prevent anyone from coming in contact with it.

2. Use cutting fluid whenever possible to help prolong the life of the saw blade and to help the blade cut faster.

Fig. 17-6 The stop gage should be set when cutting many pieces of the same length. (Courtesy of KTS Industries)

Fig. 17-7 A spacer block should be used when clamping short work to prevent the vise jaw from twisting. (Courtesy of Kelmar Associates)

3. When sawing thin pieces, hold the material flat in the vise to prevent the saw teeth from jamming and breaking.

4. Do not apply extra force to the saw frame, as this will generally twist the blade and cause work to be cut out of square.

5. When several pieces of the same length are required, set the stop gage which is supplied with most cut-off saws, Fig. 17-6.

6. When holding short work in a vise, be sure to place a short piece of steel of the same width close to the opposite end of the vise. This will prevent the vise from twisting when it is tightened, Fig. 17-7.

SUMMARY

- Horizontal bandsaws are widely used in machine shops because of their productivity and versatility.

- The required saw blade speed and coarseness are determined by the type and thickness of the material being cut.

- Saw blades are typically made of high-speed steels.

- Saw blade coarseness is designated by pitch, which ranges from 4 (coarse) to 14 (fine).

- When a saw blade is being changed, the teeth must be pointing toward the motor end of the machine.

KNOWLEDGE REVIEW

Cut-Off Saws

1. Name and describe the cutting action of two types of cut-off saws.

Horizontal Bandsaw Parts

2. What is the purpose of the roller guide brackets?

3. How tight should the blade tension handle be adjusted?

Saw Blades

4. Of what materials are saw blades made?

5. What pitch blade is recommended for:

   a. large sections?

b. thin sections?

c. general purpose sawing?

6. Why should two teeth of a saw blade be in contact with the work at all times?

Installing a Blade

7. In what direction should the teeth of a saw blade point?

8. How should the workpiece be set when sawing partially cut work with a new saw blade?

Sawing

9. What may happen if too fast a blade speed is used?

10. How can a vise be checked for squareness?

11. How should the roller guide brackets be set when cutting work?

12. Explain how thin work should be held in the vise.

13. What precautions should be used when holding short work in a vise?

UNIT 18

Contour-Cutting Bandsaw

The contour-cutting (vertical) bandsaw has been widely accepted as a fast and economical method of cutting metal and other materials since its development in the early 1930s. It has provided industry with a fast and accurate method of sawing, filing, and polishing straight and contour shapes.

OBJECTIVES

After completing this unit, you should be able to:

- State the purpose of the main operative parts of a contour bandsaw.
- Recognize three common tooth forms and sets, and know the application of each.
- Calculate the length of a saw band required for a two-pulley bandsaw.

KEY TERMS

contour bandsaw

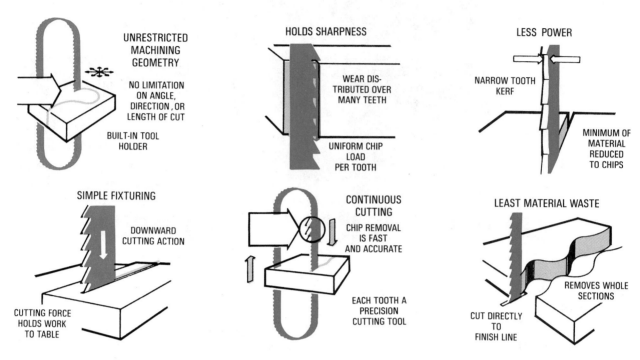

Fig. 18-1 The main advantages of a contour-cutting bandsaw. (Courtesy of DoALL Co.)

CONTOUR BANDSAW PARTS

The **contour bandsaw** has a flexible "end-less" blade, which cuts continuously in one direction. This blade is fitted around two or three vertically mounted pulleys with flat rubber-tired rims on which the blades ride. Saw guides immediately above and below the table of the saw help to guide and support the vertical saw blade.

The contour-cutting bandsaw has many advantages in the metal-cutting trade and some of the more common ones are illustrated in Fig. 18-1. Although contour bandsaws are available in a wide variety of sizes and shapes, each machine contains parts which are basic to all machines, Fig. 18-2.

- The job selector dial lists the recommended cutting speeds for various materials and pitches of blade.

- The upper pulley is used to support, tension, and adjust the tracking of the saw band. The tension is adjusted by the tension handwheel.

- The lower pulley is driven by a variable-speed pulley and can be adjusted to various speeds by the speed change handwheel.

- The upper and lower saw guides help support the blade and prevent it from wandering.

- The table can be tilted up to 45° for the cutting of angular surfaces.

- The butt welder, supplied with most machines, is used to weld and anneal saw bands.

TYPES OF SAW BLADES

Saw blades used on contour bandsaws are made of carbon-alloy steel, high-speed steel, and tungsten-carbide tipped. They are available in three types of tooth forms, Fig. 18-3, and each is available in three types of set, Fig. 18-4.

- The *precision* or *regular tooth* blade is a general-purpose blade, used where a fine finish and accurate cut are required.

- The *claw* or *hook tooth* blade allows rapid removal of chips without increasing friction heat. It is used for sawing wood, plastics, nonferrous metals, and large sections of ferrous castings and machine steel.

- The *buttress* or *skip tooth* blade has wide tooth spacing for good chip clearance and

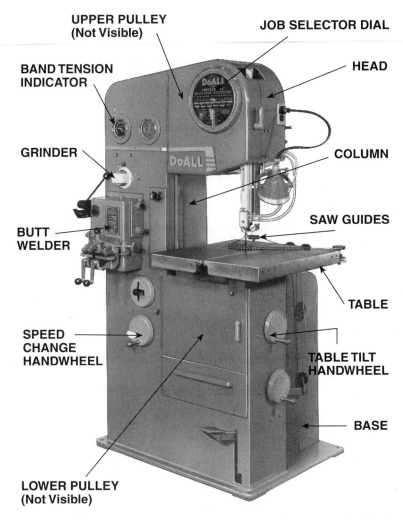

Fig. 18-2 The main parts of a contour bandsaw. (Courtesy of DoALL Co.)

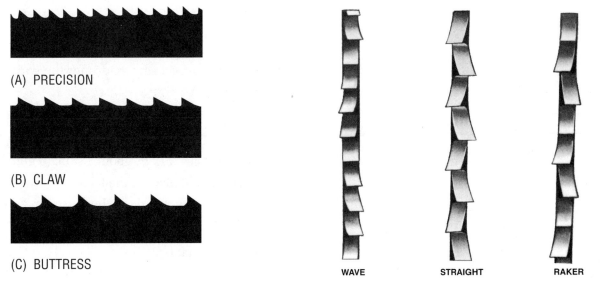

(A) PRECISION

(B) CLAW

(C) BUTTRESS

Fig. 18-3 The most commonly used saw-tooth forms. (Courtesy of DoALL Co.)

WAVE   STRAIGHT   RAKER

Fig. 18-4 The common tooth sets found on most saw blades. (Courtesy of DoALL Co.)

high-speed cutting of nonferrous metals wood, plastics, etc.

On large work sections, a coarse pitch blade should be used. For thin sections, select a fine pitch blade. As in all other methods of sawing, a general rule to follow is that not less than two teeth should be in contact with the work at all times. Use the thickest blade possible for strength.

BLADE LENGTH

Metal-cutting saw bands are usually packaged in coils about 100 to 500 ft. (30 to 150 m) in length. The length required for a machine is cut from the coil and its two ends are then welded together.

To calculate the length required for a two-wheel bandsaw, take twice the center distance (CD) between the pulleys and add to it the circumference of one pulley (PC). This is the total length of the saw band, Fig. 18-5.

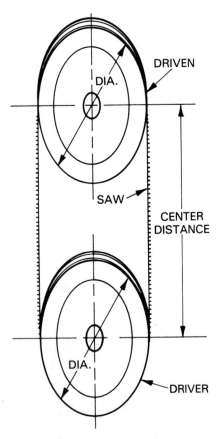

Fig. 18-5 The dimensions necessary for calculating the saw blade length for a two-pulley bandsaw. (Courtesy of DoALL Co.)

When measuring the center distance between wheels, make sure the top wheel is lowered approximately 1 in. (25 mm) from its top position. This will allow for stretching when the blade is tensioned.

EXAMPLE: Calculate the length of a saw blade required to fit a two-pulley contour bandsaw with 28 in. diameter pulleys having a center-to-center distance of 48 in.

SOLUTION:

$$\begin{aligned} \text{Blade length} &= 2\,(CD) + (PC) \\ &= 2\,(48) + (28 \times 3.1416) \\ &= 96 + 87.9 \\ &= 183.9 \text{ in.} \end{aligned}$$

CONTOUR BANDSAW OPERATIONS

With the proper attachments, a wide variety of operations can be performed on a contour bandsaw. The procedures for performing some of the more common operations are listed in the following sections.

To Weld a Saw Band

Most contour bandsaws contain an attachment or accessory for welding a saw band to make a continuous loop. Since the operation of these welding units varies from machine to machine, follow the instructions in the operator's manual when welding a saw band.

To Mount a Saw Blade

The following describes the steps to be used in mounting the saw blade:

1. Select the correct saw guides for the width of blade being used.

2. Mount both upper and lower saw guides on the machine, Fig. 18-6.

3. Adjust the upper saw guide until it clears the top of the work by about 1/4 in. (6 mm).

4. Insert the saw plate in the table.

5. Place the saw band on both upper and lower pulleys, making sure that the teeth are pointed in the direction of band travel (toward the table).

6. Adjust the upper pulley with the tension handwheel until some tension is registered on the tension gage.

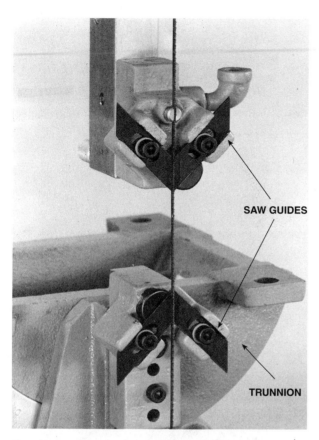

SAW GUIDES

TRUNNION

Fig. 18-6 Upper and lower saw guides are used to support the saw band and prevent it from wandering. (Courtesy of DoALL Co.)

7. Set the gearshift lever to neutral and turn the upper pulley by hand to be sure that the saw blade is tracking properly on the pulley.

8. Re-engage the gearshift lever and start the machine.

9. Adjust the blade to the recommended tension as indicated on the tension gage by using the tension handwheel.

10. Check that the back of the blade just touches the back of the saw guides.

> If the blade is not tracking properly, use the handwheel at the rear of the frame (at the opposite side to the speed change handwheel) to tilt the upper pulley.

To Saw to a Layout

The following procedure should be used to saw to a layout in order to produce a workpiece that is ready for finishing operations:

1. Set the machine to the proper speed for the type of blade and the material being cut. Consult the job selector, Fig. 18-2.

2. Use the work-holding jaw, Fig. 18-7, or a piece of wood to feed the work into the saw.

 Keep the fingers well clear of the moving saw blade.

3. Feed the work into the saw blade at a steady rate. Do not apply too much pressure and crowd the blade.

4. Cut to within approximately 1/32 in. (0.8 mm) of the layout lines. This leaves material for the finishing operation.

5. Never attempt to cut too small a radius with a wide saw. This will damage the blade and the drive wheel.

6. The table may be tilted up to 45° for cutting angular surfaces, Fig. 18-8.

To Mount a File Band

The following procedure should be used when mounting a file band:

1. Remove the saw blade, upper and lower saw guides, and the saw plate.

Always shut off the machine power switch before making any adjustments or installing accessories.

Fig. 18-7 The workholding jaw or a piece of wood should be used to feed work into a revolving saw band. (Courtesy of DoALL Co.)

Fig. 18-8 The table in a horizontal position and tilted to 45° are shown in this composite photograph. (Courtesy of DoALL Co.)

2. Mount the correct file guide support on the lower post block, Fig. 18-9.

3. Insert the file plate into the table.

4. Mount the file guide to the upper post.

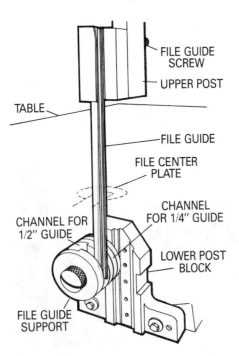

Fig. 18-9 Mount the correct file guide and file guide support for the size of file band to be used. (Courtesy of DoALL Co.)

5. Lower the upper post until the file guide is below the center of the file guide support.

6. Thread the file band upward through the hole in the table center disk.

Make sure the unhinged end of the file segment is pointing upward.

7. Join the ends of the file band, Fig. 18-10, and then apply the proper amount of tension to the file.

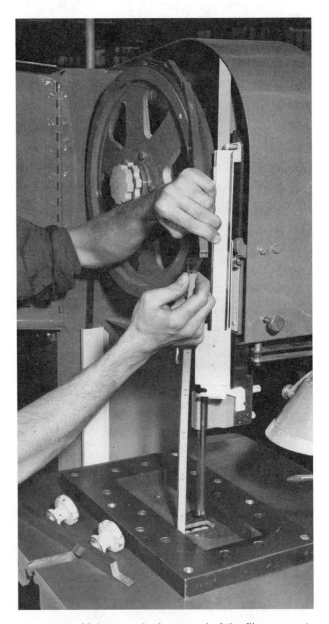

Fig. 18-10 Make sure the loose end of the file segment is pointing up when mounting a file band. (Courtesy of DoALL Co.)

8. If necessary, adjust the upper pulley to make the file band track freely in the file guide channel.

To File on a Contour Bandsaw

1. Consult the job selector, Fig. 18-2, and set the machine to the proper speed.

> The best filing speeds are between 50 and 100 ft. (15 and 30 m) per minute.

2. Apply light work pressure to the file band.

> It will produce a better finish and prevent the teeth from becoming clogged.

3. Keep moving the work sideways (back and forth) against the file to prevent filing grooves in the work.

4. Use a file card to keep the file clean. Loaded files cause bumpy filing and scratches in the work.

 Stop the machine before attempting to clean the file.

SUMMARY

- The contour bandsaw has been widely used for sawing, filing, and polishing operations.

- When sawing large work sections, a coarse-pitch blade should be used; for thin sections, use a fine-pitch blade.

- The length of saw band required for a two-wheel bandsaw is equal to twice the center distance between the pulleys plus the circumference of one pulley.

- When filing on a contour bandsaw, use a file card to keep the file clean and prevent file loading.

- When sawing to a layout, cut to within about 1/32 in. of the layout lines to leave material for the finishing operation.

KNOWLEDGE REVIEW

Contour-Cutting Bandsaw

1. List six advantages of the contour bandsaw.
2. Explain the purpose of the following:
   a. upper pulley
   b. lower pulley
   c. saw guides

Types of Saw Blades

3. Name and state the purpose of three types of saw blades.
4. What general rule should be followed when selecting the pitch of a saw blade?

Blade Length

5. Calculate the length of a saw blade for a two-pulley bandsaw with 26 in. diameter pulleys and a center-to-center distance of 48 in.

Mounting a Saw Blade

6. In what position should the upper saw guide be set?
7. In what direction should the teeth of a saw blade point?
8. How can the upper pulley be adjusted if the saw is not tracking properly?

Sawing to a Layout

9. How should work be fed into a saw blade?
10. What would be the result of trying to cut a small radius with a wide blade?
11. How can angular surfaces be cut on a contour bandsaw?

Filing

12. List the procedure for mounting a file band on the bandsaw.
13. What speeds are recommended for filing?
14. Why should only light work pressure be applied to the file band?
15. How can a loaded or clogged file be cleaned?

SECTION

11

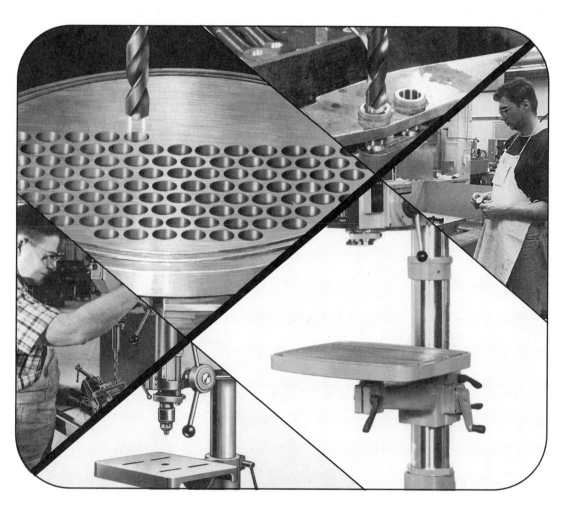

Drill Presses

SECTION 11

The drill press, probably the first mechanical device developed by man, is one of the most commonly used machines in a machine shop. The main purpose of a drill press is to grip, revolve, and feed a twist drill in order to produce a hole in a piece of metal or other material. Main parts of any drill press include the spindle, which holds and revolves the cutting tool, and the table upon which the work is held or fastened. The revolving drill or cutting tool is generally fed into the workpiece manually on bench type drill presses, and either manually or automatically on floor type drill presses. The variety of cutting tools and attachments that are available also allows operations such as drilling, reaming, countersinking, counterboring, tapping, spot facing, and boring to be performed Fig. 19-1.

Figures courtesy of (clockwise from far left): Delta International Machinery Corp., Cleveland Twist Drill Co., Cleveland Twist Drill Co., DoALL Co., DoALL Co., Clausing Industrial, Inc.

UNIT 19

Drill Press Types

A **drill press** is a machine used for drilling operations available in a wide variety of types and sizes to suit different types and sizes of workpieces. These range from the small hobby-type drill press to the larger, more complex, and computer numerical control machines used by industry for precision machining with high productivity. The most common machines found in a machine shop are the bench-type sensitive drill press and the floor-type drill press. Other drill presses, such as the upright, post, radial, horizontal, gang, portable, multiple spindle, and computer numerical control types, are variations of the standard machine and are generally designed for specific purposes.

OBJECTIVES

After completing this unit, you should be able to:

- Recognize and know the purpose of two types of drill presses.
- Know the function and purpose of the main operative parts of a drill press.
- Observe proper safety practices when operating a drill press.

KEY TERMS

drill chuck	drill press	drill sleeve
drill socket	keyless drill chuck	key-type chuck

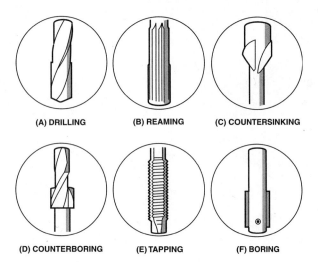

(A) DRILLING (B) REAMING (C) COUNTERSINKING

(D) COUNTERBORING (E) TAPPING (F) BORING

Fig. 19-1 Common operations performed on a drill press. (Courtesy of Kelmar Associates)

DRILL PRESS SIZES

Drill presses are available in a wide variety of sizes to suit the purpose for which they will be used. The size of any drill press is generally given as the distance from the edge of the column to the center of the drill press spindle. For example, on a 10 in. drill press a hole could be drilled 10 in. from the edge of the workpiece; on a 250 mm drill press it would be possible to drill a hole in a workpiece 250 mm from its edge. The vertical capacity of a drill press is measured when the head is in the highest position and the table is in its lowest position

DRILL PRESS PARTS

Although drill presses are manufactured in a wide variety of types and sizes, all drilling machines contain certain basic parts. The main parts on the bench- and floor-type models are base, column, table, and drilling head, Fig. 19-2A. The floor type model is larger, has a longer column, and usually has a table-raising mechanism, Fig. 19-2B.

Base

The base, usually made of cast iron, provides stability for the machine and rigid mounting for the column. The base is usually provided with holes so that it may be bolted to a table or bench to keep it rigid. The slots or ribs in the base allow the workholding device or the workpiece to be clamped to the base.

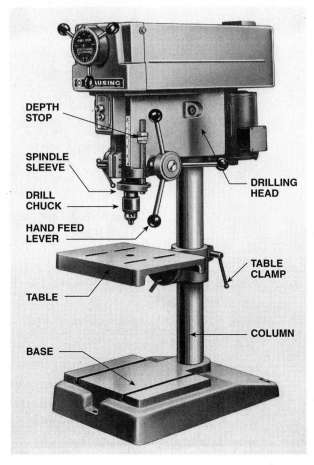

DEPTH STOP

SPINDLE SLEEVE

DRILL CHUCK

HAND FEED LEVER

TABLE

BASE

DRILLING HEAD

TABLE CLAMP

COLUMN

Fig. 19-2A The main operative parts of a sensitive drill press. (Courtesy of Clausing Industrial, Inc.)

Column

The column is an accurate, vertical, cylindrical post which fits into the base. The table, which is fitted on the column, may be adjusted to any point between the base and head. The head of the drill press is mounted near the top of the column.

Table

The table, either round or rectangular in shape, is used to support the workpiece to be machined. The table, whose surface is at 90° to the column, may be raised, lowered, and swiveled around the column. On some models it is possible to tilt and lock the table in either direction for drilling holes on an angle. Slots are provided in most tables to allow jigs, fixtures, or large workpieces to be clamped directly to the table.

Drilling Head

The head, mounted close to the top of the column, contains the mechanism that is used to

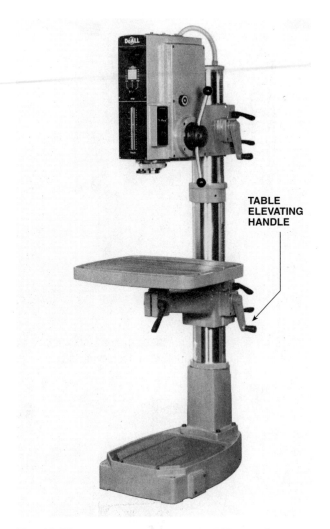

Fig. 19-2B The floor type or upright drill press has an elevating mechanism to raise and lower the table. (Courtesy of DoALL Co.)

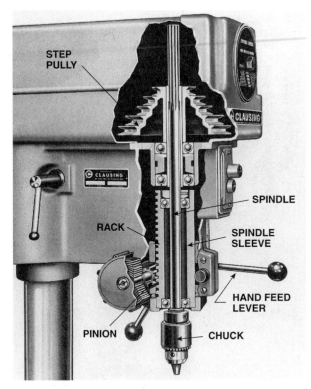

Fig. 19-3 A cutaway showing the construction of a sensitive drill press drilling head. (Courtesy of Clausing Industrial, Inc.)

revolve the cutting tool and advance it into the workpiece. The spindle, which is a round shaft that holds and drives the cutting tool, is housed in the spindle sleeve or quill. The spindle sleeve does not revolve, but is moved up and down by the hand feed lever that is connected to the pinion on the rack of the spindle sleeve, Fig. 19-3. The end of the spindle may have a tapered hole to hold taper shank tools, or it may be threaded or tapered for attaching a drill chuck, (as in Fig 19-2A).

The hand feed lever is used to control the vertical movement of the spindle sleeve and the cutting tool. A depth stop, attached to the spindle sleeve, can be set to control the depth that a cutting tool enters the workpiece.

● TOOL-HOLDING DEVICES

The drill press spindle provides a method of holding and driving the cutting tool. It may have a tapered hole to accommodate taper shank tools, or its end may be tapered or threaded for mounting a drill chuck. Although there is a variety of drill press toolholding devices and accessories, the most common in a machine shop are drill chucks, drill sleeves, and drill sockets.

Drill Chucks

Drill chucks are the most common devices used on a drill press for holding straight-shank cutting tools. Most drill chucks contain three jaws that move simultaneously (all at the same time) when the outer collar is turned or, on some types of chucks, when the outer collar is raised. The three jaws hold the straight shank of a cutting tool securely and cause it to run accurately. There are two common types of drill chucks: the key type and the keyless type.

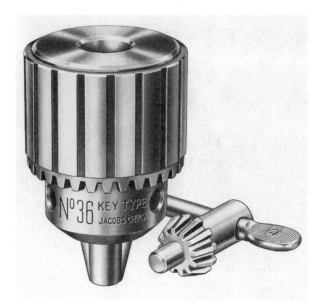

Fig. 19-4A The standard key-type chuck used on most drill presses. (Courtesy of Jacobs© Chuck Manufacturing Co.)

Fig. 19-4B The internal mechanism of a key chuck. (Courtesy of Jacobs© Chuck Manufacturing Co.)

The **key-type chuck**, Fig. 19-4A, can be provided with a tapered arbor, which fits into the tapered hole of the spindle, or it may have a tapered or threaded hole for fastening to the end of the drill press spindle. A key is used to turn the outer collar, which causes the three jaws to move simultaneously to grip a straight shank cutting tool, Fig. 19-4B.

The **keyless drill chuck**, Fig. 19-5A, is generally used in production work because some models allow cutting tools to be inserted and removed while the machine is in operation. The outer collar is turned by hand, which moves the three jaws at the same time to tighten on the tool shank, Fig. 19-5B.

Drill Sleeves and Sockets

The size of the tapered hole in the drill press spindle is generally in proportion to the size of the machine: the larger the machine, the larger the spindle hole. The size of the tapered shank on cutting tools is also manufactured in proportion to the diameter of the tool. A **drill sleeve**, Fig. 19-6A, is used to adapt the cutting tool shank to the machine spindle if the taper on the cutting tool is smaller than the tapered hole in the spindle.

A **drill socket**, Fig. 19-6B, is used when the hole in the drill press spindle is too small for the taper shank of the drill. The drill is first mounted

Fig. 19-5A Keyless chucks are mainly used in production work. (Courtesy of Jacobs© Chuck Manufacturing Co.)

in the socket, and then the socket is inserted into the drill press spindle. Drill sockets may also be used as extension sockets to provide extra length necessary for some drilling operations.

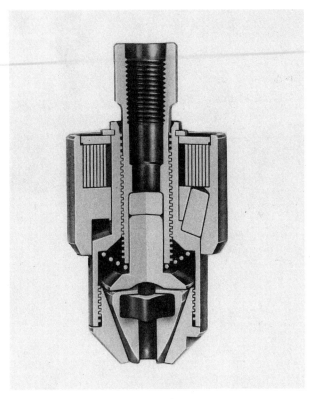

Fig. 19-5B The internal mechanism of a keyless chuck. (Courtesy of Jacobs© Chuck Manufacturing Co.)

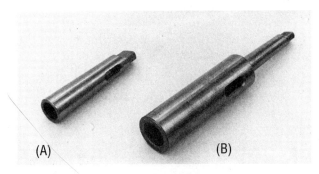

Fig. 19-6 Drill sleeves and sockets are used to fit the tapered shank of a cutting tool to the drill press spindle. (A) Drill sleeve (B) Drill socket. (Courtesy of Kelmar Associates)

● MOUNTING AND REMOVING TAPER SHANK TOOLS

Before a taper shank tool is mounted in a drill press spindle, be sure that the external taper of the tool shank and the internal taper of the spindle are thoroughly cleaned. Align the tang of the tool with the slot in the spindle hole and,

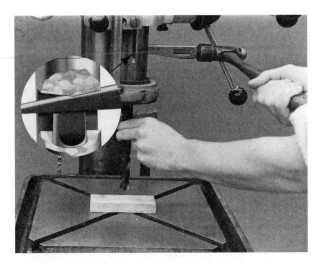

Fig. 19-7 Place a wooden block on the table to prevent damage when removing a drill with a drill drift. (Courtesy of Kelmar Associates)

with a sharp upward snap, force the tool into the spindle.

A drift, a wedge-shaped tool, is used to remove a taper-shank tool from the drill press spindle, Fig. 19-7. The drift should be inserted in the spindle slot with its *rounded edge up*. Place a wooden block on the drill press table to prevent the drill from marring the table when it is removed. Lower the spindle until the end of the cutting tool is within 1 in. (25 mm) of the block of wood and lock the spindle in this position. Sharply strike the end of the drift with a hammer to remove the tool from the drill press spindle.

● DRILL PRESS SAFETY

The drill press is probably the most common machine tool used in industry, school shops, and workshops. Because it is so common, good safety practices, which can prevent accidents, are often ignored. Before operating a drill press, a person should be familiar with some common sense safety rules to avoid an accident and personal injury.

Safety Rules

The following procedure practices and precautions should be observed when operating drill presses:

1. Never wear ties or loose clothing around machinery, Fig. 19-8.

Fig. 19-8 Loose clothing should not be worn around machinery because it can easily be caught in revolving cutting tools or machine parts. (Courtesy of Kelmar Associates)

 Roll up sleeves to above elbow height to prevent them from getting caught in the machine.

2. Long hair should be protected by a hair net or shop cap to prevent it from becoming caught in the revolving parts of the drill press.

3. Never wear rings, watches, bracelets, or necklaces while working in a machine shop, Fig. 19-9.

Fig. 19-9 It is dangerous practice to wear rings and watches in a machine shop. (Courtesy of Kelmar Associates)

Fig. 19-10 Poor housekeeping generally leads to accidents. (Courtesy of Kelmar Associates)

4. Always wear safety glasses when operating any machine.

5. Never attempt to set the speeds, adjust or measure the work until the machine is completely stopped.

6. Keep the work area and floor clean and free of oil and grease, Fig. 19-10.

 Clean floors can prevent most slipping and falling accidents.

7. Never clamp taper shank drills, end mills, or non-standard tools in a drill chuck.

8. Never leave a chuck key in a drill chuck *at any time*, Fig. 19-11.

Fig. 19-11 Never leave chuck wrenches in a drill chuck; this is a potential accident waiting to happen. (Courtesy of Kelmar Associates)

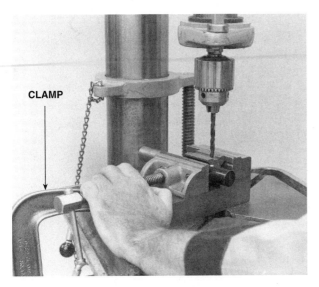

CLAMP

Fig. 19-12 It is good practice to use a clamp or table stop on the left side of the table when drilling. (Courtesy of Kelmar Associates)

If the drill press is turned on accidentally, the wrench will fly out and may cause a serious injury.

9. Always use a brush to remove chips.

10. Never attempt to hold work by hand when drilling holes larger than 1/2 in. (12.7 mm) in diameter.

Use a clamp or table stop to prevent the work from spinning, Fig. 19-12.

11. Ease up on the drilling pressure as the drill breaks through the workpiece.

This will prevent the drill from pulling into the work and breaking.

12. Always remove the burrs from a hole that has been drilled.

SUMMARY

- The purpose of a drill press is to hold, revolve, and feed a twist drill to produce a hole in a workpiece.

- The two most common types of drill presses are the bench-type and the floor-type.

- Drill chucks are used to hold straight-shank cutting tools in a drill press.

- Drill sleeves and drill sockets are toolholding devices used to accommodate a variety of spindle/cutting-tool size combinations.

- Personal grooming, including proper care of clothing, hair, and jewelry, plays an important part in maintaining drill-press safety.

KNOWLEDGE REVIEW

1. Name six operations which can be performed on a drill press.

Drill Press Types and Construction

2. Name two common types of drill presses.

3. List four main parts of a drill press and give one use for each.

4. State the purpose of the following parts:
    a. spindle   c. hand feed lever
    b. spindle sleeve   d. depth stop

Tool-Holding Devices

5. For what purpose are drill chucks used?

6. How do the jaws on most drill chucks operate ?

7. State the purpose of:
    a. a drill sleeve   b. a drill socket

8. How should a taper shank tool be mounted in a drill press spindle?

9. How should a taper shank tool be removed from a drill press spindle?

Drill Press Safety

10. Why is loose clothing and long hair considered dangerous around machinery?

11. Name three other safety rules you consider most important.

U N I T 20

Twist Drills

A **twist drill** is an end-cutting tool used to produce a hole in a piece of metal or other material. The most common drill manufactured has two cutting edges (lips) and two straight or helical flutes, that provide the cutting edges, admit cutting fluid, and allow the chips to escape during the drilling operation. Two cutting actions occur on a drill, one at the chisel edge and the other at the lips or cutting edges. The chisel edge has a negative rake and does not produce a good cutting action. It produces a very small chip, but creates forces to deform the metal similar to a center-punch indent, Fig. 20-1. The most efficient cutting action occurs as the lips or cutting edges of the drill contact the metal.

OBJECTIVES

After completing this unit, you should be able to:

- Recognize and state the purpose of four different twist drills, and know in which drill sizes they are available.
- Identify and state the purpose of four main parts of a drill.
- Calculate the speeds and feeds required for various materials and drill sizes.

KEY TERMS

chisel edge   cutting edge   feed
spindle speed

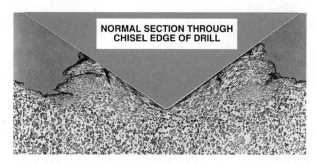

Fig. 20-1 A photomicrograph showing the metal deformation and flow which occur under the chisel edge of a twist drill. (Courtesy of National Twist Drill & Tool Co.)

● TWIST-DRILL MATERIALS

The most common twist drills used in a machine shop are made of high-speed steel and cemented carbides. A recent development is the coating of regular drills with titanium carbide or other wear-resistant materials to improve their performance.

High-speed steel drills are the most commonly used drills, since they can be operated at good speeds and the cutting edges can withstand heat and wear.

Cemented-carbide drills, which can be operated much faster than high-speed steel drills, are used to drill hard materials. Cemented-carbide drills have found wide use in production work because they can be operated at high speeds, the cutting edges do not wear rapidly, and they are capable of withstanding higher heat. Carbide drills are finding more and more applications in general drilling work.

The application of ultra-thin, hard, wear-resistant coatings of titanium carbide, aluminum oxide, and titanium nitride to steel and carbide cutting tools has resulted in the following advantages:

- Longer tool life: As much as two to ten times.

- Increased productivity: Cutting speeds have increased by 25 to 90%.

- Improved workpiece quality: Tools operate at lower temperatures and produce better surface finishes.

- Reduced manufacturing costs: fewer tool changes and less machine downtime mean lower costs.

- Reduced machining forces: Power requirements can be lowered by as much as 15%.

● TWIST DRILL PARTS

A twist drill, Fig. 20-2, may be divided into three main sections: the shank, body, and point.

Shank

The shank is the part of the drill that fits into a holding device that revolves the drill. The shank of a twist drill may be either straight or tapered, Fig. 20-2. Straight shanks are generally provided on drills up to 1/2 in. (12.7 mm) in diameter, while drills over 1/2 in. diameter usually have tapered shanks. Straight-shank drills are held in some type of drill chuck, while taper-shank drills fit into the internal taper of the drill press spindle.

The *tang*, at the small end of the tapered shank, is machined flat to fit the slot in the drill press spindle. Its main purpose is to allow the drill to be removed from the spindle with a drift without damaging the shank. The tang may also prevent the shank from turning in the drill press spindle because of a poor fit on the taper or too much drilling pressure.

Body

The body of a twist drill consists of the portion between the shank and the point. The body contains the flutes, margin, body clearance, and web of the drill.

- The *flutes* on most drills consist of two or more helical grooves cut along the body of the drill from the point to the start of the shank.

> The flutes form the cutting edges of the drill, provide them with rake, admit cutting fluid, and allow chips to escape during the drilling operation.

- The *margin* is the narrow, raised section on the body next to the flutes.

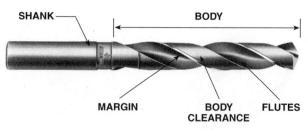

Fig. 20-2 The main parts of a twist drill. (Courtesy of Cleveland Twist Drill Co.)

The diameter of the drill is measured across the margin, which extends the full length of the flutes.

- The *body clearance* is the undercut portion of the body between the margin and the flute.

This reduces the amount of friction between the drill and the hole.

- The *web* is the thin metal partition in the center of the drill which extends the full length of the flutes. This part forms the chisel edge at the cutting end of the drill.

The web gradually increases in thickness toward the shank to give the drill strength, Fig. 20-3.

Point

The point of a twist drill consists of the entire cone-shaped cutting end of the drill. The drill point consists of the chisel edge, cutting edges or lips, lip clearance, and heel, Fig. 20-4.

The shape and condition of the point are very important to the cutting action of the drill.

- The *chisel edge* is the portion which connects the two cutting edges. It is formed by the intersection of the cone-shaped surface of the point. If a drill does not have a chisel point, it is wise to first drill a lead or pilot hole in the workpiece, slightly larger than the thickness of the drill web, to relieve some of the pressure on the drill point.

The cutting action of the chisel edge is not very good.

- The *cutting edges or lips* are formed by the intersection of the flutes and the cone-shaped point.

Both lips must be the same length and have the same angle, so that the drill will run true and not cut a hole larger than the size of the drill.

The effects of unequal lip lengths are shown in Fig. 20-5.

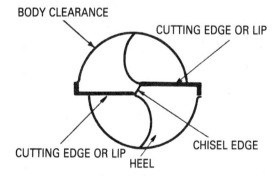

Fig. 20-3 The web thickness increases towards the shank to strengthen a drill. (Courtesy of Cleveland Twist Drill Co.)

Fig. 20-4 The parts of a twist drill point. (Courtesy of Cleveland Twist Drill Co.)

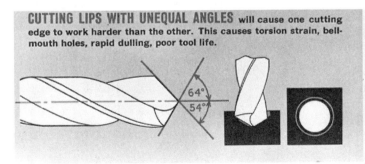

CUTTING LIPS WITH UNEQUAL ANGLES will cause one cutting edge to work harder than the other. This causes torsion strain, bell-mouth holes, rapid dulling, poor tool life.

64°
54°

Fig. 20-5 Unequal cutting lip lengths produce oversize holes. (Courtesy of Greenfield Industries, Inc.)

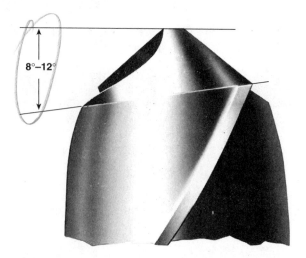

Fig. 20-6 The general purpose drill has a lip clearance of 8 to 12°. (Courtesy of Cleveland Twist Drill Co.)

- The *lip clearance* is the relief which is ground on the point of the drill extending from the cutting lips back to the heel, Fig. 20-6. Lip clearance allows the lips of the drill to cut into the metal without the heel rubbing.

The average lip clearance is from 8 to 12°, depending on the type of the material to be drilled.

Drill Point Angles and Clearances

For general-purpose drilling, the drill point should be ground to an included angle of 118°, Fig. 20-7A, and the lip clearance should range from 8 to 12°. The drill point for hard materials should be ground to an included angle from 135 to 150°, Fig. 20-7B, and the lip clearance should be from 8 to 10°. For drilling soft materials, the drill point should be ground to an included angle of 90°, Fig. 20-7C, with the lip clearance ranging from 15 to 18°.

● SYSTEMS OF DRILL SIZES

Twist drills are available in both inch and metric sizes. Inch drills are designated by fractional, number, and letter systems. Metric drills are available in various set ranges. The size of straight-shank drills is marked on the shank, while taper shank drills are generally stamped on the neck between the body and the shank.

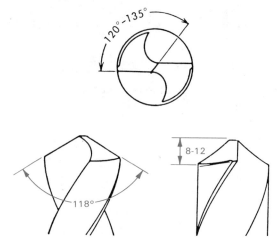

(A) GENERAL PURPOSE

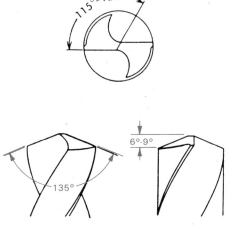

(B) HARD MATERIAL

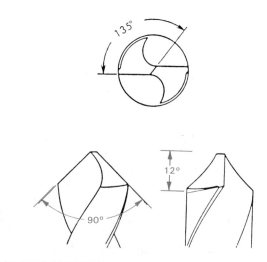

(C) SOFT MATERIAL

Fig. 20-7 Drill point angles and clearances for drilling various materials. (Courtesy of Kelmar Associates)

Inch Drills

The three types of designations for inch drills are defined as follows:

- **Fractional** inch drills are available in sizes from 1/64 to 3½ in. in diameter, varying in steps of 1/64 in. from one size to the next. Drills larger than 3½ in. in diameter must be ordered specially from the manufacturer.

- **Number** size drills range from the #1 drill (.228 in.) to the #97 drill (.0059 in.). The most common number drill set contains drills from #1 to #60. The large range of sizes enables almost any hole between .0059 to .228 in. to be drilled.

- **Letter** size drills range from A to Z. The letter A drill is the smallest in the set (.234 in.) and the letter Z is the largest (.413 in.).

Metric Drills

Metric size drills are available in various sets, but are not designated by various systems. The miniature metric drill set ranges from 0.04 to 0.99 mm in steps of 0.01 mm. Straight-shank metric drills are available in sizes from 0.5 to 20 mm, ranging in steps of 0.02 to 1 mm, depending on the size. Taper-shank metric drills are available in sizes from 8 to 80 mm.

See the tables in the Appendix for the number, letter, and metric drill sizes.

Measuring the Size of a Drill

In order to produce a hole to the required size, it is important that the correct size drill is used to drill the hole. It is good practice to always check a drill for size before drilling a hole. Drills may be checked for size by two methods: with a drill gage, Fig. 20-8A, and with a micrometer, Fig. 20-8B. When a drill is being checked for size with a micrometer,

NOTE

Always be sure that the measurement is taken across the margin of the drill.

CUTTING SPEEDS AND FEEDS

The selection of the proper speeds and feeds for the cutting tool and the type of material

Fig. 20-8A Checking the size of a drill with a drill gage. (Courtesy of Kelmar Associates)

Fig. 20-8B Checking the size of a drill by measuring across the margins with a micrometer. (Courtesy of Kelmar Associates)

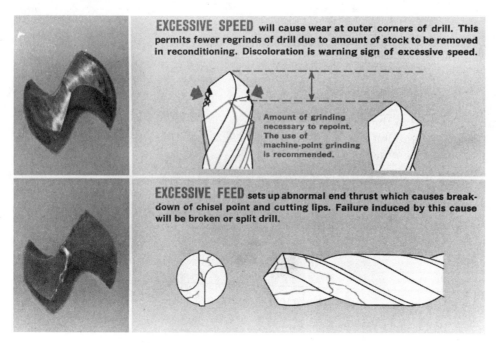

Fig. 20-9 Excessive speeds and feeds will damage a drill and shorten its life. (Courtesy of Greenfield Industries, Inc.)

being drilled are important factors which the operator must consider. These two factors affect the amount of time required to complete an operation (production rate) and how long a cutting tool will perform satisfactorily (tool life). Time will be wasted if the speed and feed are too low, while the cutting tool will wear too fast if the speed and feed are set too high. The importance of using the proper speeds and feeds is shown in Fig. 20-9.

Cutting Speed

The speed at which a twist drill should be operated is often referred to as cutting speed, surface speed, or peripheral speed. **Cutting speed** may be defined as the distance, in either surface feet or surface meters, that a point on the circumference of the drill travels in one minute. For example, if tool steel has a recommended cutting speed of 100 ft. (30 m) per minute, the drill press speed should be set so that a point on the circumference of the drill will travel 100 ft. (30 m) in one minute. The wide range of drill sizes used for drilling holes in the various kinds of metals requires an equally wide range of speeds at which the drills can be efficiently operated.

NOTE

The size of the drill, the material it is made from, and the type of material to be drilled must all be taken into account when determining a safe and efficient speed at which to operate a drill press.

As a result of many years of research, cutting tool and steel manufacturers recommend that various types of metal be machined at certain cutting speeds for the best production rates and long tool life. The recommended cutting speeds for various materials are listed in Table 20-1.

Whenever reference is made to the speed of a drill, the cutting speed in *surface feet per minute or in surface meters per minute is implied*, and not in revolutions per minute (r/min), unless specifically stated.

Revolutions per Minute

The number of revolutions necessary to produce the desired cutting speed is called r/min (revolutions per minute). A small drill operating at the same r/min as a larger drill will travel fewer

Table 20-1 Cutting speeds for high-speed steel drills

Size		Steel Casting		Tool Steel		Cast Iron		Machine Steel		Brass and Aluminum	
		Cutting Speeds									
Inch	Milli-meter	Ft./Min. 40	M/Min. 12	Ft./Min. 60	M/Min. 18	Ft./Min. 80	M/Min. 24	Ft./Min. 100	M/Min. 30	Ft./Min. 200	M/Min. 60
		Revolutions per Minute									
1/16	2	2445	1910	3665	2865	4890	3820	6110	4775	12225	9550
1/8	3	1220	1275	1835	1910	2445	2545	3055	3185	6110	6365
3/16	4	815	955	1220	1430	1630	1910	2035	2385	4075	4775
1/4	5	610	765	915	1145	1220	1530	1530	1910	3055	3820
5/16	6	490	635	735	955	980	1275	1220	1590	2445	3180
3/8	7	405	545	610	820	815	1090	1020	1305	2035	2730
7/16	8	350	475	525	715	700	955	875	1195	1745	2390
1/2	9	305	425	460	635	610	850	765	1060	1530	2120
5/8	10	245	350	365	520	490	695	610	870	1220	1735
3/4	15	205	255	305	380	405	510	510	635	1020	1275
7/8	20	175	190	260	285	350	380	435	475	875	955
1	25	155	150	230	230	380	305	380	380	765	765

feet or meters per minute and naturally will cut more efficiently at a higher number of r/min.

For example, in one revolution the circumference of the 1/2 in. diameter drill in Fig. 20-10A would travel 1/2 × 3.1416 (π) or 1.57 in. The 1 in. diameter drill in Fig. 20-10B would travel 1 × 3.1416 or 3.14 in. in the same revolution since it has a larger circumference.

It is necessary to calculate the proper r/min. for each size drill to be able to set it to the correct cutting speed.

(A) 1/2" DIAMETER

(B) 1" DIAMETER

Fig. 20-10 Smaller drills must be run faster to have the same cutting speed as large drills. (Courtesy of Cleveland Twist Drill Co.)

To find the number of revolutions per minute at which a drill press spindle must be set to obtain a certain cutting speed, the following information must be known:

1. The recommended cutting speed of the material to be drilled,

2. The type of material from which the drill is made,

3. The diameter of the drill.

To calculate the **spindle speed** for any machine, divide the cutting speed of the metal by the circumference of the rotating part which may be a drill, a milling cutter, or the workpiece in a lathe.

Apply one of the following formulas to calculate the spindle speed (r/min) at which the drill press should be set. (Table 20-1 lists the cutting speeds for various materials.)

Inch Drills

$$r/min = \frac{CS \times 12}{\pi \times D}$$

CS = cutting speed of the material in feet per minute

D = diameter of the drill in inches.

π = 3.416

Since only a few machines are equipped with variable speed drives, which allow them to be set to the exact calculated speed, a simplified formula can be used to calculate r/min. The π (3.1416) on the bottom line of the formula will divide into 12 of the top line approximately four times. This results in a simplified formula which is close enough for most drill presses.

$$r/min = \frac{CS \times 4}{D}$$

EXAMPLE: Calculate the r/min at which the drill press should be set to drill a 1/2 in. diameter hole in a piece of machine steel. (See Table 20-1 for the cutting speed of machine steel.)

$$r/min = \frac{CS \times 4}{D}$$
$$= \frac{100 \times 4}{D}$$
$$= 800$$

When it is not possible to set the drill press to the exact speed, always set it to the closest speed *under* the calculated speed.

Metric Drills

$$r/min = \frac{CS \text{ (meters)}}{\pi \times D \text{ (millimeters)}}$$

Since the cutting speed is usually given in meters and the diameter of a workpiece is expressed in millimeters, it is necessary to convert the meters in the numerator to millimeters so that both parts of the equation are in the same unit. Therefore multiply the CS in meters by 1000 to bring it to millimeters.

$$r/min = \frac{CS \times 100}{\pi \times D}$$

EXAMPLE: Calculate the r/min at which a drill press should be set to drill a 12 mm hole in a piece of machine steel.

$$r/min = \frac{30 \times 1000}{3.1416 \times 12}$$
$$= \frac{30000}{37.699}$$
$$= 796$$

Since only a few machines are equipped with a variable speed drive, a simplified formula,

which is suitable for most drilling operations, can be derived by dividing π (3.1416) into 1000.

$$r/min = \frac{CS \times 1000}{\pi \times D}$$
$$= \frac{CS \times 320}{D}$$

The same drilling problem, solved by using the simplified formula, is as follows:

$$r/min = \frac{30 \times 320}{12}$$
$$= \frac{9600}{12}$$
$$= 800$$

Factors Affecting Drill Speed (r/min)

The calculated drill speed may have to be varied slightly to suit the following factors:

- the type and condition of the machine,
- the accuracy and finish of the hole required,
- the rigidity of the work setup,
- the use of cutting fluid.

Feed

Feed is the distance that a drill advances into the work for each complete revolution. The feed rate is important because it affects both the life of the drill and the rate of production. Too coarse a feed may cause the cutting edges to break or chip, while too fine a feed causes a drill to chatter, which dulls the cutting edges, as shown in Fig. 20-9. The recommended feeds per revolution for millimeters and fractional inch size drills are listed in Table 20-2.

Table 20-2 Recommended drill feeds

Drill Size		Feed per Revolution	
Inch	Millimeters	Inches	Millimeters
1/8 or less	3 or less	.001–.002	0.02–0.05
1/8–1/4	3–6	.002–.004	0.05–0.10
1/4–1/2	6–12	.004–.007	0.10–0.17
1/2–1	12–15	.007–.015	0.17–0.37
1–1½	25–38	.015–.25	0.37–0.63

SUMMARY

- Two cutting actions of a twist drill occur, one at the chisel edge and the other at the lips or cutting edges.
- Cemented-carbide drills operate at higher speeds and withstand higher heat levels than high-speed steel drills.
- Tool life and productivity are affected by the type of material being drilled and the operator's choice of drill speed.
- In selecting a drill speed, the drill size, drill material, and material to be drilled must all be considered.
- Drill speed usually refers to cutting speed in surface feet (or meters) per minute.
- The rate of feed affects both the life and the productivity of a drill.

KNOWLEDGE REVIEW

Twist Drills

1. Describe the most common twist drill manufactured.

2. State where the two cutting actions occur on a drill.

3. Of what two materials are twist drills manufactured?

4. Name four of the most important advantages of wear-resistant coatings on drills.

5. State the purpose of each part:

   a. tang   d. body clearance

   b. flutes   e. web

   c. margin   f. lip clearance

6. What is the recommended point angle and lip clearance for general purpose drilling?

Systems of Drill Sizes

7. Name the three systems used to designate inch drill sizes. In what ranges are metric drills available?

8. What size are the following drills: #1, #29, #60, Letters A, F, K, Z?

9. Name two methods of measuring the size of a drill.

Cutting Speeds and Feeds

10. Why are speeds and feeds so important to the life of a cutting tool and the production rate?

11. Define:

    a. cutting speed

    b r/min

12. Calculate the r/min required to drill machine steel (100 CS) with the following drills: 5/8 in., 3/4 in., 1 in., 1¼ in.

13. Name the two factors which may affect the calculated drill speed.

14. Define drilling feed.

15. Explain the effects of:

    a. too coarse, and

    b. too fine, a drilling feed.

U N I T 21

Producing and Finishing Holes

Holes can be produced and finished in a drill press on a wide variety of workpiece sizes and materials. Holes may be required on a workpiece as clearance for screws, bolts, or other components that may not require the highest degree of accuracy. Holes may have to be finished to an accurate size to suit dowel pins and mating parts, and they would require another operation such as reaming to finish. Although the size or shape of the drill may vary to suit the workpiece material, the operation of producing a hole is basically the same and the proper procedure should be followed. For best results, always use the proper speeds and feeds to suit the workpiece material and the size of the drill. Wherever possible, use cutting fluids to prolong tool life and increase productivity.

OBJECTIVES

After completing this unit, you should be able to:

- Select and properly apply cutting fluid for various drill press operations.
- Drill holes, up to and over 1/2 in. diameter, in various types of materials.
- Perform drill press operations such as counterboring, reaming, tapping, etc.

KEY TERMS

center drill   cutting fluid   pilot hole

CUTTING FLUIDS

The main purpose of a **cutting fluid** is to reduce the amount of heat created during the drilling operation by the friction of the revolving drill against the workpiece. The heat and friction dull the drill quickly and produce a poor finish in the hole. The purposes of cutting fluids are to:

- keep the cutting edges cool,
- prolong the life of the cutting tool,
- permit the use of faster cutting speeds,
- aid in the chip removal,
- produce a better surface finish in the hole,
- increase productivity.

Types of Cutting Fluids

There are many types of cutting fluids available as shown in Table 21-1, and each is suitable for a specific machining operation or work material. They fall into three general categories: cutting oils, emulsifiable oils, and chemical cutting fluids. Their main purposes are to prolong tool life and increase manufacturing productivity, Fig. 21-1.

Cutting oils

Cutting oils are generally mineral oils which contain certain additives to improve the drill's cutting action, Fig. 21-2. Cutting oils fall into two general categories: active *cutting oils*, used for drilling ferrous metals; and inactive *cutting oils*, used for drilling nonferrous metals.

Emulsifiable or Soluble Oils

These are oils which are first mixed with an emulsifier that breaks the oil up into minute particles. This concentrate is then mixed with water at a ratio as high as fifty parts of water to one part

Table 21-1   Cutting lubricants for drill press work

Material	Drilling	Reaming	Tapping
Machine steel (hot and cold rolled)	Soluble oil Mineral lard oil Sulfurized oil	Mineral lard oil Sulfurized oil Soluble oil	Soluble oil Mineral lard oil
Tool steel (carbon and high-speed)	Soluble oil Mineral lard oil Sulfurized oil	Lard oil	Sulfurized oil Lard oil
Alloy steel	Soluble oil Mineral lard oil Sulfurized oil	Soluble oil Sulfurized oil Mineral lard oil	Sulfurized oil Lard oil
Brass and bronze	Dry Lard oil Kerosene mixture	Dry Soluble oil Lard oil	Soluble oil Lard oil
Copper	Dry Mineral lard oil Kerosene Soluble oil	Soluble oil Lard oil	Soluble oil Lard oil
Aluminum	Soluble oil Kerosene Lard oil	Soluble oil Kerosene Mineral oil	Soluble oil Kerosene and lard oil
Monel metal	Lard oil Soluble oil	Lard oil Soluble oil	Lard oil
Malleable iron	Dry Soda water	Dry Soda water	Lard oil Soda water
Cast iron	Dry Air jet Soluble oil	Dry Soluble oil Mineral lard oil	Dry Sulfurized oil Mineral lard oil

Chemical cutting fluids can be used for most of the above operations. Follow the manufacturer's recommendations for use and mixture.

Fig. 21-1 Cutting fluids improve the cutting action and result in higher production rates. (Courtesy of *Modern Machine Shop Magazine*, copyright 1994, Gardner Publications Inc.)

Fig. 21-2 Cutting oils are usually mineral oils with additives to provide better cutting action on certain materials. (Courtesy of *Modern Machine Shop Magazine*, copyright 1994, Gardner Publications Inc.)

of concentrate. Because of the good cooling qualities of soluble oils, they are suitable for high cutting speeds and where considerable heat is generated.

Chemical Cutting Fluids

These fluids contain water-soluble chemicals which give the fluid cooling, lubricating, and anti-weld properties. These synthetic cutting fluids are efficient and cleaner than cutting oils and soluble oils, and they are finding wide use. See Table 21-1 for the cutting fluids recommended for various materials and drill press operations.

● DRILLING CENTER HOLES

Work that is to be turned between the centers on a lathe must have a hole drilled in each end so that the work may be supported by the lathe centers. Although center holes in the work are more easily and accurately drilled on a lathe, they are often machined on a drill press because of the shape of the workpiece or the equipment available.

After the center locations have been laid out on both ends of the workpiece, a combined drill and countersink, commonly called a **center drill**, is used to drill the center holes.

Two types of center drills are available: the plain or regular type, Fig. 21-3A, and the bell type, Fig. 21-3B. The *regular* type has a 60° angle with a small clearance drill located on the end. The *bell* type has a secondary bevel near the large

(A) REGULAR

(B) BELL

Fig. 21-3 The two common types of center drills. (Courtesy of Cleveland Twist Drill Co.)

diameter, which produces a clearance angle near the top of the hole. This clearance angle prevents the top edge of the 60° bearing surface from becoming burred or damaged. Center drills are available in a wide variety of sizes to suit different diameters of work. Table 21-2 lists information regarding regular and bell-type center drills and the diameters for which each should be used.

In order to provide a good bearing surface for the work on the lathe centers, it is important that the center holes be drilled to the correct size (see Table 21-2). The center hole illustrated in Fig. 21-4A is too shallow and will not provide an adequate bearing surface. The center hole in Fig. 21-4B is too deep and will not allow the taper of

Table 21-2 Center drill sizes

Size		Work Diameter		Diameter of Countersink C	Drill Point Diameter	Body Size
Regular Type	Bell Type	Inches	Millimeters			
1	11	3/16–5/16	3–8	3/32	3/64	1/8
2	12	3/8–1/2	9.5–12.5	9/64	5/64	3/16
3	13	5/8–3/4	15–20	3/16	7/64	1/4
4	14	1–1½	25–40	15/64	1/8	5/16
5	15	2–3	50–75	21/64	3/16	7/16
6	16	3–4	75–100	3/8	7/32	1/2
7	17	4–5	100–125	15/32	1/4	5/8
8	18	6 and over	150 and over	9/16	5/16	3/4

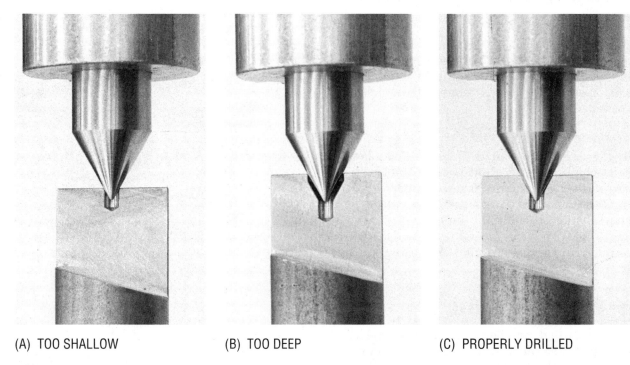

(A) TOO SHALLOW   (B) TOO DEEP   (C) PROPERLY DRILLED

Fig. 21-4 For center holes to support the workpiece properly, they must be drilled accurately. (Courtesy of Kelmar Associates)

the lathe center to contact the taper of the center hole. The center hole illustrated in Fig. 21-4C is drilled to the proper depth, which provides a good bearing surface for the lathe centers.

To Drill a Lathe Center Hole

Center holes should be as smooth and accurate as possible to provide a good bearing surface

and reduce the friction and wear between the work and lathe centers.

1. Check the center hole layout to be sure that it is in the center of the work.

2. Obtain the correct size center drill to suit the diameter of the work to be drilled from Table 21-2.

3. Fasten the center drill in the drill chuck.

To prevent breakage, do not have more than 1/2 in. (12.7 mm) of the center drill extending beyond the drill chuck.

4. Set the drill press to the proper speed for the size of the center drill being used.

 NOTE

A speed of 1200 to 1500 r/min is suitable for most center drills.

5. Fasten a clamp or table stop on the left side of the table to prevent the vise from swinging during the drilling operation, Fig. 21-5.

6. Set the vise on its side on a clean drill press table.

7. Press the work firmly against the bottom of the vise, and then tighten it securely.

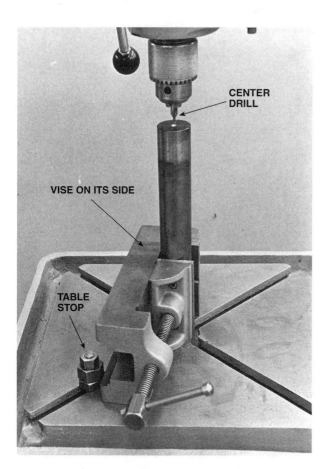

Fig. 21-5 The work is held in a vise set on its side to keep it square with the table for drilling center holes. (Courtesy of Kelmar Associates)

8. With the vise against the table clamp or stop, locate the center-punch mark of the work under the drill point.

9. Start the drill press and use the hand feed lever to carefully feed the center drill into the work.

Too heavy a feed pressure may break the drill point.

10. Frequently raise the drill from the work and apply a few drops of cutting fluid.

11. Continue drilling until the top of the countersunk hole (C) is the correct size, (see Table 21-2).

12. If the center hole is not smooth, apply a little cutting fluid and lightly bring the center drill into the hole again.

To Spot a Hole with a Center Drill

The chisel edge at the end of the web of a drill is generally larger than the center-punch mark on the work, and therefore it is difficult to start a drill at the exact marked location. To prevent a drill from wandering off center, it is considered good practice to first spot every center-punch mark with a center drill. The small point on the center drill will accurately follow the center-punch mark and provide a guide for the larger drill which will be used. To spot a hole, use the following procedure:

1. Mount a small size center drill in the drill chuck.

2. Mount the work in a vise or set it on the drill press table.

Do not clamp the work or the vise.

3. Set the drill press speed to about 1500 r/min.

4. Bring the point of the center drill into the center-punch mark and allow the work to center itself with the drill point, Fig. 21-6.

5. Continue drilling until about one-third of the tapered section of the center drill has entered the work.

6. Spot the location of all the holes which are to be drilled.

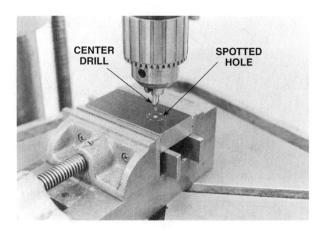

Fig. 21-6 A center drill can be used to accurately spot the location of holes. (Courtesy of Kelmar Associates)

● **DRILLING A HOLE**

To drill a hole accurately and safely, it is wise to always observe the following points:

- Measure the drill with a micrometer or gage to be sure that it is the correct size, especially if the hole is to be reamed or tapped later.

- Always use a sharp drill which is correctly ground for the material being drilled.

- Clamp the work properly to prevent inaccurate work and accidents.

- Set the drill press to the proper speed and feed to avoid damaging the twist drill or the machine and to prevent an accident.

- Withdraw the drill frequently to remove chips and add cutting oil.

> This can prevent breakage, especially when using small diameter drills.

- Ease up on drilling pressure as the drill starts to break through the work.

- Use cutting fluids whenever possible to prolong tool life and produce a better hole.

To Drill a Hole in Work Held in a Vise

The most common method of holding small workpieces is by means of a vise, which may be held by hand against a table stop or clamped to the table. When drilling holes larger than 1/2 in. (12.7 mm) in diameter, the vise should be clamped

to the table. Use the following procedure to drill the hole:

1. Spot the hole location with a center drill.

2. Mount the correct size drill in the drill chuck.

3. Set the drill press to the proper speed for the size of drill and the type of material to be drilled.

4. Fasten a clamp or stop on the left side of the table, Fig. 21-7.

5. Mount the work on parallels in a drill vise, and tighten it securely.

> Be sure to position the parallels so that they are away from the hole to be drilled.

6. With the vise against the table stop, locate the spotted hole under the center of the drill.

Fig. 21-7 The vise should be held against a table stop or clamp by hand when drilling holes over 1/2 in. (12.7 mm) in diameter. (Courtesy of Kelmar Associates)

7. Start the drill press spindle and begin to drill the hole.

   a. For holes up to 1/2 in. (12.7 mm) in diameter, hold the vise against the table stop by hand, Fig. 21-7.

OR

   b. For holes over 1/2 in. (12.7 mm) in diameter

     i. Lightly clamp the vise to the table with another clamp, Fig. 21-8.

     ii. Drill until the full drill point is into the work.

     iii. With the drill revolving, keep the drill point in the work and tighten the clamp, holding the vise securely.

8. Raise the drill occasionally and apply cutting fluid during the drilling operation.

9. Ease up on the drilling pressure as the drill starts to break through the workpiece.

To Drill Work Fastened to a Drill Table

Sometimes, because of the size or shape of the workpiece, it is necessary to clamp the work directly to the table. When work is clamped to the table, always be sure to use parallels between the table and work, so that the drill will not cut into the table.

1. Spot the hole location with a center drill.

2. Mount the correct size drill and set the drill press speed and feed.

3. Set the work on a suitable set of parallels.

Fig. 21-8 The vise should be clamped to the table when drilling holes over 1/2 in. (12.7 mm) in diameter. (Courtesy of Kelmar Associates)

Fig. 21-9 The clamps correctly set for clamping a workpiece to the drill press table. (Courtesy of Kelmar Associates)

> Be sure the parallels are away from the position where the hole is to be drilled.

4. Locate the spotted hole under the point of the twist drill.

5. Select suitable clamps, step blocks, and bolts and position them on the work as shown in Fig. 21-9.

6. Lightly tighten each clamp.

7. Start the drill revolving and feed the drill until about one-half the drill point has entered the work.

> If the work is lightly clamped, the spotted hole will align itself with the drill point.

8. While the revolving drill is still in contact with the work, *tighten both clamps securely.*

9. Occasionally apply cutting fluid during the drilling operation.

10. Ease up on the drilling pressure as the drill begins to break through the work.

To Drill a Pilot Hole for Large Drills

As the size of a drill increases, the thickness of the web of the drill increases to strengthen the drill. The thicker web results in a longer chisel point as the drill gets shorter. Some larger drills without a chisel point require more pressure to feed them into the work because the point does

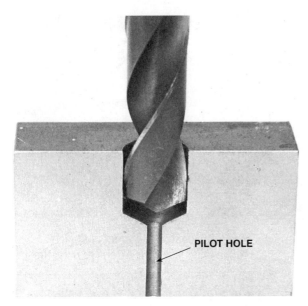

Fig. 21-10 A pilot hole reduces drilling pressure and prevents the drill from wandering. (Courtesy of Kelmar Associates)

not cut. To relieve some drilling pressure and provide a guide for the larger drill to follow, a **pilot hole** slightly larger than the web thickness, is first drilled in the work at the hole location. The size of pilot hole should be only slightly larger than the thickness of the web of the drill to be used, Fig. 21-10. If the pilot hole is drilled too large, the following drill may chatter, drill the hole out of round, or damage the top of the hole. Care must be used to drill the pilot hole on center, because the larger drill *will follow the pilot hole*. This method may also be used to drill average-size holes when the drill press is small and does not have sufficient power to drive the drill through the solid metal.

To Drill Round Work in a V-Block

V-blocks are used to hold round work accurately for drilling. The round material is seated in the accurately machined groove and small diameters are held securely with a U-shaped clamp. Larger work may be supported by a V-block in a drill vise or it may be mounted on V-blocks and clamped to the table.

1. Select a V-block to suit the diameter of round work to be drilled.

If the work is long, use a pair of V-blocks.

Fig. 21-11 Using a square and rule to align the center punch mark on round work. (Courtesy of Kelmar Associates)

2. Mount the work in the V-block, and then rotate it until the center-punch mark is in the center of the workpiece.

Check that the distance from both sides is equal with a rule and square, Fig. 21-11.

3. Tighten the U-clamp securely on the work in the V-block.

OR

Hold the work and V-block in a vise as shown in Fig. 21-12.

4. Spot the hole location with a center drill.

5. Mount the proper size drill and set the machine to the correct speed.

6. Drill the hole, being sure that the drill does not hit the V-block or vise when it breaks through the work.

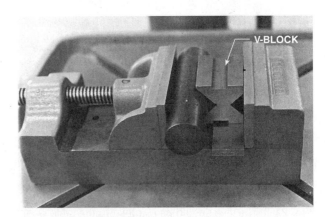

Fig. 21-12 The round work and V-block held in a vise ready for drilling. (Courtesy of Kelmar Associates)

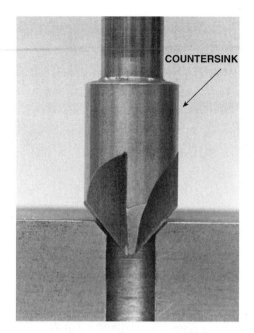

Fig. 21-13A A countersink is used to enlarge the top of a hole in the shape of a cone. (Courtesy of Kelmar Associates)

Fig. 21-13B After countersinking, the head of the machine screw should be flush with the work surface. (Courtesy of Kelmar Associates)

● DRILL PRESS OPERATIONS

With the use of various cutting tools, a variety of operations can be performed on a drill press. Some of the more common drill press operations are briefly described as follows:

Countersinking

Countersinking is the process of enlarging the top of a hole to the shape of a cone. Countersinks, Fig. 21-13A, are available with included angles of 60° and 82°. The 60° countersink is used for producing lathe center holes, while the 82° countersink is used to produce the tapered hole which accommodates a flat head bolt or machine screw, Fig. 21-13B. Countersinks may also be used to remove burrs from the top of a drilled hole by slightly chamfering the edge.

When countersinking for a flat-head machine screw, the hole should be countersunk so that the head of the screw is flush with the top of the work surface, Fig. 21-13B. The speed for countersinking is generally about one-quarter of the recommended drilling speed.

Counterboring

Counterboring is the operation of enlarging the top of a previously drilled hole, Fig. 21-14. Counterbores are available in a variety of types

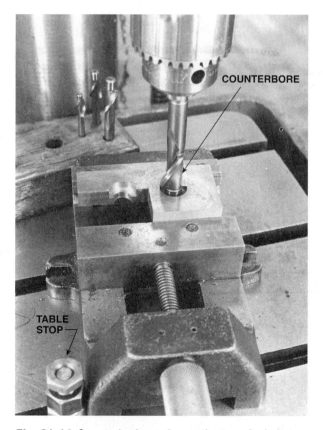

Fig. 21-14 Counterboring enlarges the top of a hole to accommodate the head of a screw. (Courtesy of Kelmar Associates)

and sizes, with pilots which may be either solid or interchangeable. Holes are counterbored to create an enlarged hole with a square shoulder to accommodate the head of a bolt or cap screw or the shoulder on a pin. The speed for counterboring is usually about one-quarter of the drilling speed.

Reaming

The purpose of reaming is to bring a drilled or bored hole to size and shape, and to produce a good surface finish in the hole, Fig. 21-15. Speed, feed, and reaming allowance are three factors that can affect the accuracy of a reamed hole. Approximately 1/64 in. (0.4 mm) is left for reaming holes up to 1/2 in. (12.7 mm) in diameter; 1/32 in. (0.8 mm) is recommended for holes over 1/2 in. (12.7 mm) in diameter. The speed for reaming is generally about one-half of the drilling speed.

There are two types of reamers used in machine shop work: hand reamers and machine reamers. Hand reamers have a square on one end and are used to remove no more than .005 in. (0.12 mm) from a hole. Machine reamers have a straight or tapered shank and are used under power.

Boring

Boring is the process of enlarging a previously drilled or cored hole to produce a straight hole and bring it to an accurate size. This operation is rarely performed on a drill press because the machine was not designed for this purpose. However, if absolutely necessary, it may be performed with a single-point cutting tool mounted

Fig. 21-16 The operation of boring produces a true, round hole that is accurate to size. (Courtesy of Kelmar Associates)

in a boring bar, which is held in the drill press spindle, Fig. 21-16.

In order to bore effectively in a drill press, the table must be equipped with a hole in the center, which serves as a guide or support for the end of the boring bar and the workpiece must be securely clamped to the table.

Spot-Facing

Spot-facing is the operation of smoothing and squaring the surface around the top of a hole to provide a flat seat for the head of a cap screw or nut. A boring bar, with a pilot on its end to fit the hole, is fitted with a double-edged cutting tool, Fig. 21-17. When spot-facing, it is important that the work be securely clamped and the drill press set to about one-quarter of the drilling speed.

Tapping

Tapping in a drill press can be performed either by hand or with the use of a tapping attachment. The advantage of tapping a hole in a drill press is that the tap can be started squarely and kept that way throughout the entire length of the hole being threaded. The following steps describe tapping in a drill press:

1. Mount the work on suitable parallels, and lightly clamp the work to the drill press table. Work may be held in a vise which is lightly clamped to the table.

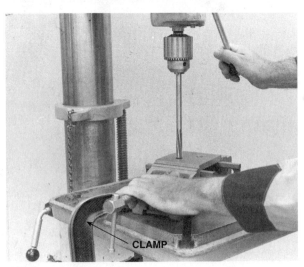

CLAMP

Fig. 21-15 Reaming brings a hole to accurate size and produces a smooth finish. (Courtesy of Kelmar Associates)

Fig. 21-17 Spot-facing produces a flat seat around the top of a hole. (Courtesy of Kelmar Associates)

2. Mount a center drill in the drill chuck, and adjust the drill press table or the work until the center-punch mark aligns with the point of the center drill.

3. Spot the hole with a center drill and then tighten the clamps securely.

4. Drill the hole to the correct tap drill size for the size of tap to be used, then countersink the top of the hole slightly larger than the tap diameter.

> The work or table must not be moved after drilling the hole; otherwise the alignment will be disturbed and the tap will not enter squarely.

5. Mount a stub center in the drill chuck, Fig. 21-18.

OR

Remove the drill chuck and mount a special center in the drill press spindle.

6. Fasten a tap wrench on the correct size tap and place it into the hole.

7. Lower the drill press spindle until the stub center point fits into the center hole in the end of the tap shank.

8. Turn the tap wrench clockwise two full turns to start the tap into the hole, and at the same time keep the center in light contact with the tap.

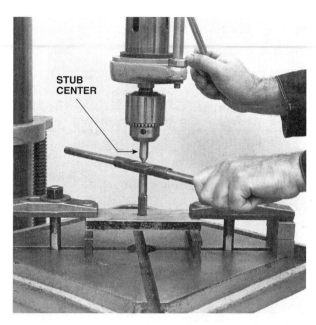

Fig. 21-18 A stub center keeps the tap aligned when tapping holes in a drill press. (Courtesy of Kelmar Associates)

9. Continue to tap the hole in the usual manner while keeping the tap aligned by applying light pressure on the drill press downfeed lever.

A tapping attachment may be mounted in the drill press spindle to rotate the tap by power. Special two- or three-fluted gun taps are used for tapping under power because of their ability to clear chips. The speed for tapping under power generally ranges from 60 to 100 r/min.

SUMMARY

- Drilling speeds and feeds must suit the workpiece material and the drill size.

- The main purposes of cutting fluids are to prolong tool life and increase productivity.

- The three main types of cutting fluids are cutting oils, emulsifiable oils, and chemical cutting fluids.

- Center drills are used to drill center holes in work that is to be turned between centers.

- In addition to producing holes, a drill press can be used for countersinking, counterboring, reaming, boring, spot-facing, and tapping.

KNOWLEDGE REVIEW

Cutting Fluids

1. List four purposes of a cutting fluid.
2. State where the following should be used:
   a. cutting oils
   b. emulsifiable or soluble oils
   c. chemical cutting fluids

Drilling Center Holes

3. State the difference between a regular type and a bell type center drill.
4. Why should center holes not be drilled
   a. too shallow? b. too deep?
5. Why is it important that center holes be drilled as smoothly and accurately as possible?
6. How much of the center drill should extend beyond the drill chuck?
7. At what speed should the drill press be set for center drilling?
8. Why is a center drill suitable for spotting a hole?
9. How deep should holes be spotted?

Drilling a Hole

10. List four important points that should be observed in order to drill a hole accurately and safely.

11. When should a vise be clamped to the table?
12. What is the purpose of fastening a stop to the left side of the table?
13. When should the drilling pressure be eased?
14. Why should parallels be used between the table and the workpiece?
15. What size pilot hole should be drilled in relation to the larger drill which will be used?

To Drill Round Work

16. Explain the procedure for locating the center-punch mark in the center of a V-block.
17. How can work be held in a V-block for drilling a hole?

Drill Press Operations

18. What countersink should be used to provide a seat for a flat-head machine screw ?
19. State two purposes of reaming.
20. When tapping in a drill press, why should the work or table not be moved after the hole is drilled?
21. How is the tap kept in alignment during the tapping process?

The Engine Lathe

The lathe, probably one of the earliest machine tools, is one of the most versatile and widely used machines. Because a large percentage of the metal cut in a machine shop is cylindrical, the basic lathe has led to the development of turret lathes, screw machines, boring mills, computer numerical controlled lathes, and turning centers. The progress in the design of the basic engine lathe and its related machines has been responsible for the development and production of thousands of products we use and enjoy every day.

Figures courtesy of (clockwise from far left): Valenite Inc., Emco Maier Corp., Emco Maier Corp., Emco Maier Corp., MTI Corp., Colchester Lathe Co. Ltd.

UNIT 22

Lathe Types and Construction

The main function of the engine lathe is to turn cylindrical shapes on work-pieces, Fig. 22-1. This is done by rotating the metal held in a work-holding device while a stationary cutting tool is forced against its circumference. Fig. 22-2 shows the cutting action of a cutting tool on work being machined in a lathe. Some of the common operations performed on a lathe are facing, taper turning, parallel turning, thread cutting, knurling, boring, drilling, and reaming. The engine lathe is the backbone of a machine shop, and a thorough knowledge of it is essential for any machinist.

OBJECTIVES

After completing this unit, you should be able to:

- Name and state the purpose of three types of engine lathes.
- Identify and state the purpose of the main operative parts.
- Operate the lathe in a safe manner by observing the safety precautions.

KEY TERMS

digital readout system
engine lathe with digital readout
micrometer collar

engine lathe
numerically controlled turning center

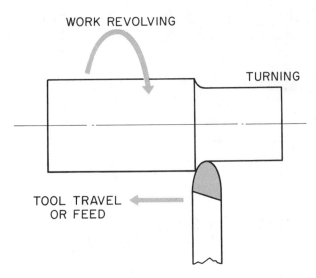

Fig. 22-1 The cutting action of an engine lathe. (Courtesy of Kelmar Associates)

Fig. 22-2 A roughing cut being taken on a workpiece. (Courtesy of Valenite Inc.)

● TYPES OF LATHES

The engine lathe has kept pace with technological changes in order to increase productivity and improve part quality. The addition of a digital readout systems and the incorporation of

limited CNC technology have closed the gap between the engine lathe and CNC chucking and turning centers. The most common engine lathes available are:

The **engine lathe**, Fig. 22-3, has long been one of the basic machine tools. Because it was one of the earlier machine tools developed, it was responsible for producing many parts that went into the development of other machine tools. Its prime function is to produce cylindrical forms using a stationary tool that is brought into contact with a revolving workpiece. The cutting tool's length movement is controlled by a carriage handwheel whose pinion is attached to a rack on the bed of the lathe. The cross movement is controlled by a crossfeed screw to which a micrometer collar, graduated in thousandths of an inch or hundredths of a millimeter, is attached. The accuracy of both movements, and therefore the accuracy of the work produced, is controlled by the machine operator. Both movements are subject to errors in calculation, measurement, or operator judgment. It takes an operator many years of practice to develop skills required to produce work to within an accuracy of .001 in. (0.02 mm).

The **engine lathe with digital readout**, Fig. 22-4, improves both accuracy and performance of any lathe. This system helps to bypass mechanical and operator errors by making independent measurements of the cutting tool's location and workpiece measurements to within .0001 in. (0.002 mm) or less. The advantages of machining with a digital readout (DRO) system are:

- Very accurate positioning of the cutting tool, resulting in accurate workpiece dimensions.

- The time required for positioning the tool is reduced by 50%.

- There is no need for the operator to make pencil-and-paper calculations where errors could arise.

- Scrap parts are almost eliminated because the operator can visually see the dimensions of the part on the DRO.

- Operator efficiency is improved, resulting in increased productivity and lower manufacturing costs.

The **digital readout system** consists of two linear spars and optical reading heads which are attached to the saddle and cross-slide, and to the bed and carriage of the lathe, Fig. 22-4. These in

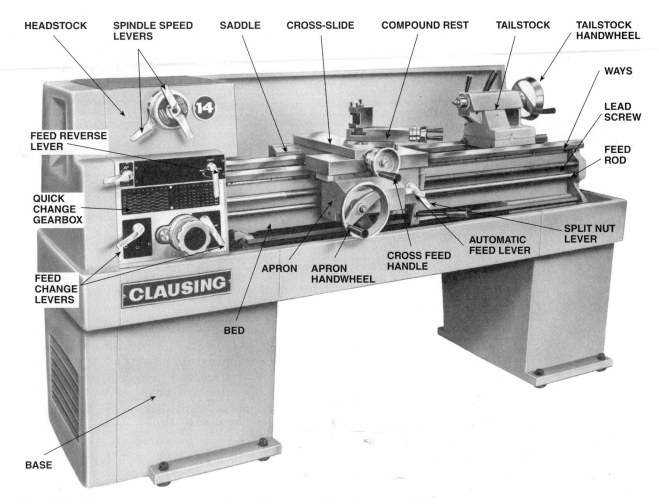

HEADSTOCK   SPINDLE SPEED LEVERS   SADDLE   CROSS-SLIDE   COMPOUND REST   TAILSTOCK   TAILSTOCK HANDWHEEL

WAYS

FEED REVERSE LEVER

LEAD SCREW

FEED ROD

QUICK CHANGE GEARBOX

SPLIT NUT LEVER

AUTOMATIC FEED LEVER

FEED CHANGE LEVERS

CLAUSING

APRON   APRON HANDWHEEL   CROSS FEED HANDLE

BED

BASE

Fig. 22-3 The main parts of an engine lathe. (Courtesy of Clausing Industrial, Inc.)

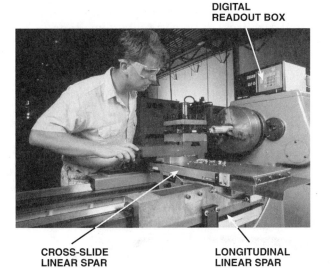

DIGITAL READOUT BOX

CROSS-SLIDE LINEAR SPAR

LONGITUDINAL LINEAR SPAR

Fig. 22-4 A lathe equipped with a digital system can produce very accurate work quickly. (Courtesy of MTI Corp.)

turn, are connected to the digital readout box (display) which indicates the cross movement of the tool in the X axis and the longitudinal movement in the Z axis in inches or millimeters.

The procedure for turning diameters to size is as follows:

1. Make a trial cut at the end of a workpiece.

2. Use a micrometer to measure the diameter of the turned section.

3. Push a button on the DRO to record the diameter.

 NOTE

The same procedure should be used to set the DRO to length.

Fig. 22-5 Some CNC programming and digital readout make a lathe more productive. (Courtesy of Emco Maier Corp).

Fig. 22-6 CNC turning centers are used for high production where very accurate parts are required. (Courtesy of Emco Maier Corp.)

4. Continue to cut the part until the desired dimensions show on the display.

The engine lathe with digital readout and limited programming features, Fig. 22-5, is a further improvement to make the engine lathe more productive. This dual feature allows the machine to be used as a standard lathe for conventional machining and also to use the built-in programming feature for repetitive steps or operations. The digital readout on the display gives the exact location of the cutting tool, and at the same time, the workpiece dimensions in the X and Z axes in inches and millimeters for accurate machining.

The digital readout control can retain information about an operation in its memory for later use. The memory can be edited if there is a need to change any steps or information held in the memory. The constant-speed feature is ideal for facing operations where the cutting speed is constantly changing as the surface is being machined. The programmable thread feature bypasses the gears in the quick-change gearbox for cutting inch or metric threads.

The advantages this lathe offers are as follows:

- Both conventional machining and programming features are available to suit the machining operation.

- Tool setup time is greatly reduced as a result of tool presetting and programming features.

- It provides the best cutting performance and longest tool life.

- There is a large increase in productivity and an improvement in part accuracy and quality.

The **numerical controlled turning center**, Fig. 22-6, has the same basic parts as an engine lathe, but all its movements are CNC-controlled. This type of machine is specially designed for production work that requires great precision and high productivity. Many of the turning centers have slant beds for easy removal of the large amount of chips produced.

The programmable rotating tool turrets on CNC turning and chucking centers may contain as many as six or more different tools. Some of the newer tool turrets contain motorized pockets which provide a drive for tools to perform drilling, tapping, and milling operations on the round part being machined. Larger CNC turning and machining centers usually have two turrets and can perform machining operations on the front and back of the part at the same time.

All these features produce more finished parts per hour with faster cycles and fewer part changes. The accuracy of the part is excellent because the repeatability of tool positioning in the X axis is accurate to within 60 millionths of an inch.

● **LATHE SIZE**

The size of an engine lathe is determined by the largest diameter of work that may be revolved or swung over the bed and the longest work that can be held between centers, Fig. 22-7.

Lathes used in training programs may have a swing of 9 to 13 in. (230 to 330 mm) and a bed length from 20 to 60 in. (500 to 1500 mm). Lathes used in industry may have swings of 9 to 30 in.

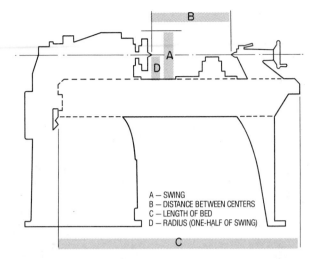

Fig. 22-7 The engine lathe size is determined by the size of work swung over the bed and length held between centers. (Courtesy of South Bend Lathe Corp.)

(230 to 760 mm) and a capacity of 16 in. to 12 ft. (400 mm to 3½ m). A typical lathe may have a 13 in. (330 mm) swing, a 6 ft.(2 m) long bed, and a capacity to turn work 36 in. (1 m) between centers.

● LATHE PARTS

The main function of a lathe is to provide a means of rotating a workpiece against a cutting tool, thereby removing metal. All lathes, regardless of design or size, are basically the same, refer to Fig. 22-3 on page 215, and serve three functions. They provide:

- a support for the lathe accessories or the workpiece,
- a way of holding and revolving the workpiece,
- a means of holding and moving the cutting tool.

Bed

The bed is a heavy, rugged casting made to support the working parts of the lathe. On its top section are machined ways that guide and align the major parts of the lathe. Many lathes are made with flame-hardened and ground ways to reduce wear and maintain accuracy.

Headstock

The headstock is fastened on the left side of the bed. The *headstock spindle*, a hollow cylin-

drical shaft supported by bearings, provides a drive from the motor to work-holding devices. A live center and sleeve, a faceplate, or a chuck can be fitted to the spindle nose to hold and drive the work. The live or driving center has a 60° point which provides a bearing surface for the work to turn between centers.

Most modern lathes are geared-head, Fig. 22-8, and the spindle is driven by a series of gears in the headstock. This allows the setting of a number of different spindle speeds to accommodate different types and sizes of work.

The *feed reverse lever* can be placed in three positions. The right-hand position provides a *forward* direction to the feed rod and lead screw, the center position is *neutral*, and the left-hand position *reverses* the direction of the feed rod and lead screw.

Quick-Change Gearbox

The quick-change gearbox, which contains a number of different sized gears, provides the feed rod and lead screw with various speeds for turning and thread-cutting operations. The feed rod and lead screw provide a drive for the carriage when either the *automatic feed lever* or the *split-nut lever* is engaged.

Carriage

The carriage supports the cutting tool and is used to move it along the bed of the lathe for turning operations. The carriage consists of three

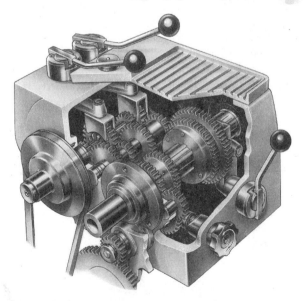

Fig. 22-8 A gear-drive headstock provides a positive drive for a workpiece. (Courtesy of Clausing Industrial, Inc.)

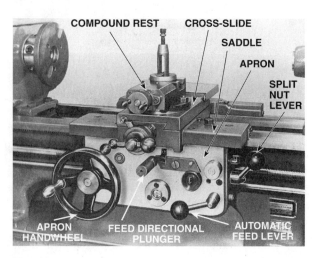

Fig. 22-9 The main parts of a lathe carriage. (Courtesy of Kelmar Associates)

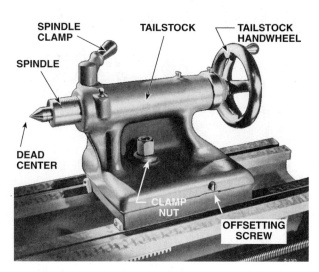

Fig. 22-10 The main parts of a tailstock assembly. (Courtesy of South Bend Lathe Corp.)

main parts, the saddle, the cross-slide, and the apron, Fig. 22-9.

The *saddle*, an H-shaped casting mounted on the top of the lathe ways, supports the cross-slide which provides a cross movement for the cutting tool. The *compound rest* is used to support the cutting tool and can be swiveled to any angle for taper-turning operations. The cross-slide and the compound rest are moved by feed screws. Each of these is provided with a graduated collar to make accurate settings possible for the cutting tools.

The *apron* is fastened to the saddle and contains the feeding mechanisms that provide automatic feed to the carriage. The automatic feed lever is used to engage power feeds to the carriage and the cross-slide. The *apron handwheel* can be turned by hand to move the carriage along the bed of the lathe. This handwheel is connected to a gear that meshes in a rack fastened to the bed of the lathe. The *feed directional plunger* can be shifted into three positions. The "in" position engages the longitudinal feed for the carriage. The "center or neutral" position is used in thread-cutting to allow the split-nut lever to be engaged. The "out" position is used when an automatic crossfeed is required.

Tailstock

The tailstock, Fig. 22-10, is made up of two units. The top half can be adjusted on the base by two adjusting screws for aligning the tailstock and headstock centers for parallel turning. These screws can also be used for offsetting the tailstock for taper-turning between centers. The tail-

stock can be locked in any position along the bed of the lathe by tightening the *clamp lever* or *nut*. One end of the dead center is tapered to fit into the tapered hole in the tailstock spindle, while the other end has a 60° point to provide a bearing support for the work turned between centers. Other standard tapered tools, such as reamers and drills, can be held in the *tailstock spindle*. A spindle binding lever or lock handle is used to hold the tailstock spindle in a fixed position. The *tailstock handwheel* moves the tailstock spindle in or out of the tailstock casting. It can also be used to provide a hand feed for drilling and reaming operations.

Graduated Micrometer Collars

Graduated **micrometer collars** are sleeves or bushings that are mounted on the compound rest and crossfeed screws, Fig. 22-11. They assist the lathe operator in setting the cutting tool accurately to remove the required amount of material from the workpiece. The micrometer collars on lathes using the inch system of measurement are usually graduated in thousandths of an inch (.001 in.). The collars on lathes using the metric system of measurement are usually graduated in fiftieths of a millimeter (0.02 mm).

The crossfeed graduated collars may indicate the *distance that the cutting tool has moved* toward the work, or the *amount that will be removed* from the work diameter. Two types of graduated collars are provided with lathes: standard collars where the amount of material

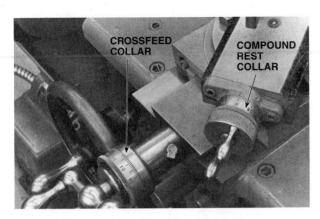

Fig. 22-11 Micrometer collars help the operator to set an accurate depth of cut. (Courtesy of Kelmar Associates)

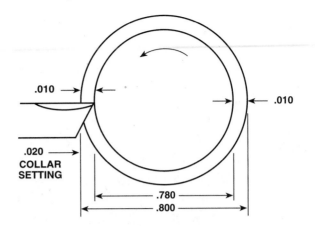

Fig. 22-12B On direct-reading graduated collars, a .020 in. setting will remove .020 in. from the diameter. (Courtesy of Kelmar Associates)

removed is twice the collar setting, and direct-reading collars where the amount removed is equal to the collar setting. Therefore, it is important for the operator to take a light *trial cut* from the workpiece to test the collar graduations before setting a depth of cut.

Inch Lathes

The circumference of the crossfeed and compound rest screw collars on lathes using the *inch system* of measurement is usually divided into 100 or 125 equal divisions, each having a value of .001 in. Therefore, if the crossfeed screw standard collar is turned *clockwise* 20 graduations, the tool will be moved .020 in. towards the work. Because the work revolves, a .020 in. depth of cut will be taken from the entire work circumference, reducing the diameter .040 in. (2 × .020 in.), Fig. 22-12A. On lathes with direct-reading collars, a .020 in. collar setting will remove .020 in. from the diameter, Fig. 22-12B.

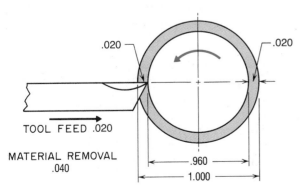

Fig. 22-12A On standard graduated collars, a .020 in. setting will remove .040 in. from the diameter. (Courtesy of Kelmar Associates)

Metric Lathes

The circumference of the crossfeed and compound rest screw collars on lathes using the *metric system* of measurement is usually divided into 200 or 250 equal divisions, each having a value of 0.02 mm. Therefore, if the crossfeed screw is turned *clockwise* 30 graduations, the cutting tool will be moved 30 × 0.02 mm or 0.6 mm toward the work. Because the work in a lathe revolves, a 0.6 mm depth of cut will remove 1.2 mm from the diameter of a workpiece.

● SAFETY PRECAUTIONS

A lathe, as well as all machine tools, can be very dangerous if not operated properly, even though it is equipped with various safety guards and features. It is the duty of the operator to observe various safety precautions and prevent accidents. Everyone must realize that a clean and orderly area around a machine will go a long way in preventing accidents.

The following are some of the more important safety regulations which should be observed when operating a lathe:

1. Always wear safety glasses when operating any machine, Fig. 22-13.

2. Never attempt to run a lathe until you are familiar with its operation.

3. Never wear loose clothing, rings, bracelets, necklaces, or watches when operating a lathe.

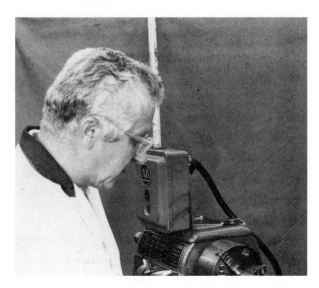

Fig. 22-13 Safety glasses protect your eyes in a machine shop. (Courtesy of Kelmar Associates)

Fig. 22-15 Leaving a wrench in a chuck is a dangerous practice which could result in serious injury. (Courtesy of Clausing Industrial, Inc.)

 These could be caught by the revolving parts of a lathe and cause a serious accident.

4. Always stop the lathe before making measurements of any kind.

5. Always use a brush to remove chips.

 DO NOT handle them by hand; they are very sharp, Fig. 22-14.

6. Before mounting or removing accessories, always shut off the power supply to the motor.

Fig. 22-14 Always use a brush to remove chips; using your hand is a dangerous practice. (Courtesy of Kelmar Associates)

7. Do not take heavy cuts on long slender pieces.

 This could cause the work to bend and fly out of the machine.

8. Do not lean on the machine. Stand erect, keeping your face and eyes away from flying chips.

9. Keep the floor around a machine clean and free of grease, oil, and other materials that could cause dangerous falls.

10. Never leave a chuck wrench in a chuck.

 If the machine is started, the wrench will fly out and possibly injure someone, Fig. 22-15.

SUMMARY

- The primary function of an engine lathe is to turn cylindrical shapes by bringing a stationary tool into contact with a revolving workpiece.

- Digital readouts and programming features improve accuracy, reduce errors, and increase efficiency and productivity.

- Engine lathe size is designated by the diameter of the largest work that can be revolved and the longest work that can be held between centers.

- On standard micrometer collars, the amount of material removed is equal to twice the collar setting. On direct-reading collars, the amount of material removed is equal to the collar setting.

KNOWLEDGE REVIEW

1. Explain the cutting action of a lathe.

2. Name six operations which may be performed on a lathe.

3. Name three types of engine lathes.

4. List four advantages of machining with a digital readout system.

5. What are the main benefits of a CNC turning center?

6. How is the size of a lathe determined?

Lathe Parts

7. Name three parts of the headstock.

8. What is the purpose of the quick-change gearbox?

9. Name the three main parts of the carriage, and state one purpose for each.

10. For what purpose are the following used?

    a. compound rest

    b. apron handwheel

    c. feed directional plunger

11. What purpose does the tailstock serve?

Graduated Micrometer Collars

12. What is the purpose of graduated micrometer collars?

13. What is the value of each graduation on the micrometer collar of lathes using:

    a. the inch system of measurement?

    b. the metric system of measurement?

14. If a .050 in. depth of cut is set on a standard crossfeed graduated collar, how much material would be removed from the diameter of a workpiece?

15. What rule should be followed when setting a depth of cut on standard graduated collars where the work revolves?

Safety Precautions

16. Why is loose clothing dangerous around machines?

17. Why should metal chips not be handled by hand?

18. Explain why heavy cuts should not be taken on long, slender pieces.

UNIT
23

Lathe Accessories and Tooling

Many accessories and tooling are available for a conventional lathe to increase its flexibility and enable an operator to produce parts more easily and accurately. These accessories and tooling are available for the following lathe parts:

1. Spindle accessories that are generally used to hold and drive the work.

2. Carriage accessories that are used to support the workpiece.

3. Compound rest tooling that is generally used to hold the cutting tools.

4. Tailstock accessories used to support the work and hold cutting tools such as drills, reamers, taps, etc.

OBJECTIVES

After completing this unit, you should be able to:

- Select the proper spindle accessories for the size and shape of the workpiece.
- Understand the purpose of carriage accessories.
- Select the proper cutting tools for each machining operation.

KEY TERMS

chuck   lathe center   toolbit
toolpost

● WORK-HOLDING DEVICES

Work-holding devices are used to hold and drive the work in a lathe while cutting operations are performed. Lathe centers, drive plates, chucks, mandrels, and steady rests are some of the common work-holding devices used for lathe work, Fig. 23-1.

Types of Lathe Spindle Noses

There are two common types of lathe spindle noses upon which accessories such as drive plates and chucks are mounted. They are the tapered spindle nose and the cam-lock spindle nose. The threaded spindle nose may still be found on older lathes.

The *tapered spindle nose*, Fig. 23-2, has a 3-1/2 in. taper per foot. The chuck or drive plate is fitted on the taper and aligned by the drive key. The threaded lock ring, fitted at the left end of the spindle nose, threads onto the chuck or drive plate and locks it in position. The *cam-lock spindle nose*, Fig. 23-3, has a very short taper (3 in. taper per foot). The chuck or drive plate is located by the taper on the spindle and held in position by three cam-lock devices.

Lathe Centers

Most turning operations can be performed between the centers on a lathe. Centers are provided in a variety of types to suit the job required. Probably the most common **lathe center** used in school shops is the solid 60° center with

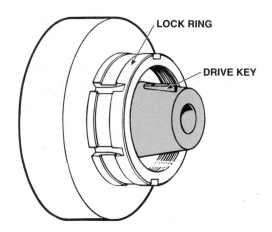

Fig. 23-2 The type-L spindle nose holds accessories on by the lock ring. (Courtesy of Kelmar Associates)

Fig. 23-3 The type D-1 cam-lock spindle nose holds accessories on by three cam locks. (Courtesy of Kelmar Associates)

a Morse taper shank, Fig. 23-4A. This type of center is usually made from a high-speed steel or from machine steel with hard carbide inserts or tips.

The *revolving or live tailstock center*, Fig. 23-4B, is used to replace the standard solid dead center for many applications. This type of center contains precision anti-friction bearings to take both radial and axial thrusts. These centers are required when machining work at high speeds with carbide cutting tools, where the heat of friction causes the work to expand rapidly. The end or axial thrust created by the heated work is taken by the revolving center and no adjustment

Fig. 23-1 Work set up for machining between lathe centers. (Courtesy of Kelmar Associates)

Fig. 23-4A Lathe centers can be made of hardened high-speed steel, or supplied with a carbide insert. (Courtesy of DoALL Co.)

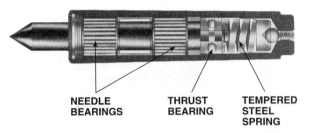

Fig. 23-4B Revolving tailstock centers are used when machining work at high speeds. (Courtesy of Concentric Tool Corp.)

Fig. 23-5 A workpiece mounted on a mandrel for machining. (Courtesy of South Bend Lathe Corp.)

Fig. 23-6 The 3-jaw universal chuck is used to hold finished round and hexagonal workpieces. (Courtesy of Clausing Industrial, Corp.)

is normally required. Because these centers require no center lubricant, they are particularly useful when turning long shafts, where adjusting (and lubricating) the dead center during the cutting operation would affect the finish of the workpiece.

Mandrels

A mandrel is used to hold an internally machined workpiece between centers so that further machining operations will be concentric with the bore. The mandrel, when pressed into a hole, allows the work to be mounted between centers and also provides a means of driving the workpiece, Fig. 23-5.

Chucks

Chucks are work-holding devices that can securely hold a variety of workpiece shapes. The most commonly used chucks for lathe work are 3-jaw universal, 4-jaw independent, combination, and collet.

The *3-jaw universal chuck*, Fig. 23-6, is used to hold round and hexagonal work quickly

to within thousandths of an inch or few hundredths of a millimeter of accuracy. The three jaws move simultaneously (at the same time) when adjusted by the chuck wrench because of the scroll plate into which all three jaws fit. Three-jaw chucks, which should *only* be used to hold round or hexagonal workpieces, range in sizes from approximately 4 to 16 in. (100 to 400 mm) in diameter. They are usually provided

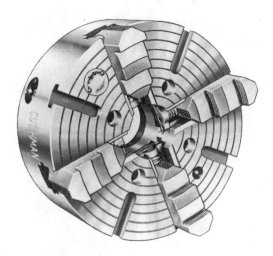

Fig. 23-7 The 4-jaw independent chuck can be adjusted to hold work of any shape. (Courtesy of TSD of Cushman Industries; Davliey-Bulland, Inc.)

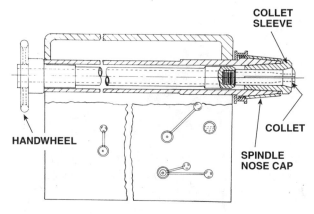

Fig. 23-8 A cross-section of a collet chuck being held in a headstock spindle with a draw bar. (Courtesy of Kelmar Associates)

Fig. 23-9 The Rubber-Flex chuck can hold workpieces with a wider range of diameters. (Courtesy of Jacobs© Chuck Manufacturing Co.)

with two sets of jaws, one for outside chucking and the other for inside chucking.

The *4-jaw independent chuck*, Fig. 23-7, has four jaws, each of which can be adjusted independently by a square-end chuck wrench. It can be used to hold round, square, hexagonal, and irregular-shaped workpieces securely because of its four jaws. The jaws can be reversed to hold work by the inside diameter. Since each jaw can be adjusted independently, any workpiece can be adjusted to run absolutely true in a 4-jaw chuck.

A *combination chuck* has the mechanical features of both independent and universal chucks. The jaws can be operated individually by a screw, or universally by turning the adjusting socket which operates the bevel gear-driven scroll. This chuck is used for the same purposes as are the universal and independent chucks.

The *collet chuck*, Fig. 23-8, is a draw-in chuck or collet that fits into the headstock spindle. The handwheel is used to tighten the collet on the work. Only true, round, and sized work can be held in a collet because of its spring limitations. There are collets for round, square, and hexagonal work.

The *Jacobs Rubber-Flex collet chuck*, Fig. 23-9, has a wider range than the draw-in collet chuck. Instead of a draw-in bar, an impact-tightening handwheel is used to close and release the collets on the workpiece. A set of 11 Rubber Flex collets, each having a range of almost 1/8 in. (3 mm), makes it possible to hold a wide range of work diameters.

CARRIAGE ACCESSORIES

The most common accessories mounted on the lathe carriage or bed, are the steady rest and the follower rest. Both are used to provide a support for long, slender workpieces to prevent them from springing or bending while being machined.

Steady Rests

A steady rest, Fig. 23-10, is used to support long work held in a chuck or between lathe centers. It is fastened to the lathe bed and its three jaws are adjusted to lightly contact the work diameter and prevent the work from springing during a machining operation.

The follower rest, Fig. 23-11, is mounted on

Fig. 23-10 A steady rest, mounted to the lathe bed, is used to support long, slender work for machining. (Courtesy of Kelmar Associates)

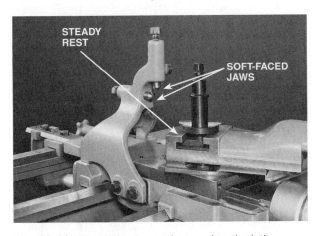

Fig. 23-11 The follower rest, fastened to the lathe carriage, travels along the work to prevent it from springing during machining. (Courtesy of Cincinnati Milacron Inc.)

the saddle and supports the top and back of the work being turned. It prevents the work from springing up and away from the cutting tool when a cut is being taken on long work.

● COMPOUND REST TOOLING

The tooling systems used on conventional lathes consist of a cutting tool and a means of holding the tool for machining operations. Many different types of tooling systems have been developed for conventional lathes to improve their accuracy and make them more productive.

Cutting Tools

To machine metal in a lathe, a cutting tool called a **toolbit** is used. Toolbits used in training programs are either high-speed steel or carbide-tipped tools. Carbide toolbits are commonly used in industry because they can withstand the heat and friction created by high cutting speeds. They are not used extensively in school shops because of their cost and the higher-horsepower equipment required to machine work with carbide cutting tools. Toolbits are made in a variety of sizes and shapes for use with different size machines and different applications.

In order to cut the various shapes desired, specially shaped toolbits are used, Fig. 23-12. Left-hand toolbits have their cutting edge on the right-hand side and are used for turning work toward the tailstock. Right-hand cutting tools have the cutting edge on the left-hand side and are used for cutting toward the headstock. A lathe cutting tool is generally known by the operation it performs. For example, a roughing tool is used

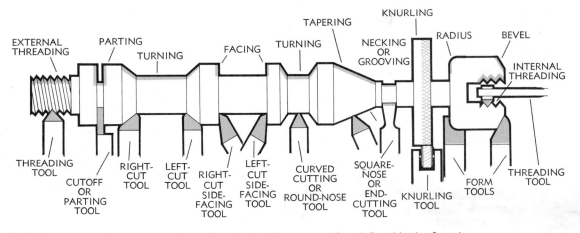

Fig. 23-12 Common cuts made by lathe cutting tools. (Courtesy of South Bend Lathe Corp.)

to rough-turn work, a threading tool is used for thread cutting, etc.

To produce various surfaces, faces, and forms, the cutting edge of the toolbit must be precisely ground. The cutting edge must be shaped to the proper form and then relieved with clearance angles to allow the edge to cut into the metal. The *end-relief* (clearance) angle is ground on the end of the toolbit to allow the point to be fed into the work. The *side relief* (clearance) angle is ground below the cutting edge to allow the toolbit to be fed lengthwise along the work. (See "Grinding a Lathe Toolbit" in Section 14, Grinders.)

Toolholders and Toolposts

A variety of lathe toolholders and toolposts are available to suit various operations, manufacturing sequences, or the types of cutting tools being used.

Toolbits can be used in three different types of toolholders. For general turning, a straight toolholder, Fig. 23-13A, is used. When machining work to the right or squaring a right-hand surface, a right-hand toolholder, Fig. 23-13B, is used. When turning work toward the headstock or squaring a left-hand surface, a left-hand toolholder, Fig. 23-13C, is used.

Toolholders for carbide-tipped tools are similar to those for high-speed steel toolbits. The major difference between the two types is that the carbide toolholder has the hole for the toolbit parallel to the base, Fig. 23-13D. This is because carbide cutting tools require little or no back rake, which is provided by the slanted hole in regular toolholders.

Standard (Round) Toolpost

The standard or round **toolpost**, Fig. 23-14, generally supplied with a lathe, consists of a round toolpost which fits into the T-slot of the compound rest. A concave ring, rocker, and toolpost screw provide a means of adjusting and holding the tool to the proper height. Standard toolposts are seldom used in machine shops because of the more accurate and efficient turret and quick-change indexable toolposts.

Turret-Type Toolpost

The turret-type toolpost, Fig. 23-15, is designed to hold four cutting tools, which can be easily indexed for use as required. Several operations, such as turning, grooving, threading, and

(A) STRAIGHT

(B) RIGHT-HAND

(C) LEFT-HAND

(D) CARBIDE

Fig. 23-13 Common lathe toolholders. (Courtesy of J.H. Williams Co., Division of Snap-On Inc.)

parting, may be performed on a workpiece by loosening the locking handle and rotating the holder until the desired toolbit is in the cutting position. This reduces the setup time for various toolbits, thereby increasing production.

Quick-Change Toolpost and Holders

Quick-change toolholders are made in different styles to accommodate different types of cutting tools. Each holder is dovetailed and fits on a dovetailed toolpost, Fig. 23-16A, which is mounted on the compound rest.

Fig. 23-14 The standard round toolpost is supplied with all lathes. (Courtesy of Kelmar Associates)

Fig. 23-16A The dovetailed toolpost allows tools to be changed quickly and accurately. (Courtesy of Dorian Tool International)

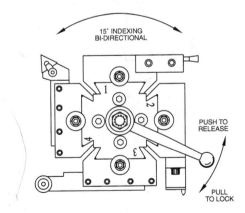

Fig. 23-15 The 4-station turret toolpost can be indexed quickly for different operations on a workpiece. (Courtesy of Dorian Tool International)

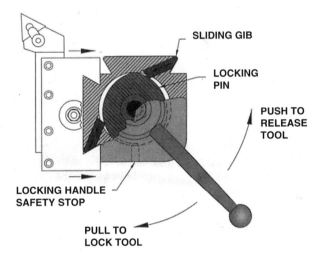

Fig. 23-16B Preset tools can be changed quickly in two positions—at right angles, or parallel to the work. (Courtesy of Dorian Tool International)

The tool is held in position by setscrews and is generally sharpened in the holder, after which it is preset to a gage. When a tool becomes dull, the unit (the holder and tool) may be replaced with another preset unit. Each toolholder,

Fig. 23-16B, fits onto the dovetail on the toolpost and is locked in position by means of a clamp. A knurled nut on each holder provides for vertical adjustment of the unit.

The *super-six index turret*, Fig. 23-17, can increase the productivity of the lathe operator, especially when multiple operations requiring more than one tool are required. This rotary indexing turret, which can be fitted to most compound rests, may contain as many as six cutting tools for external and internal machining operations. The tools can be indexed from one to an-

Fig. 23-17 The super-six index turret can increase the productivity and accuracy of a lathe operator. (Courtesy of Dorian Tool International)

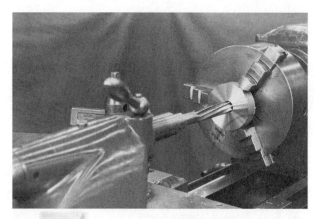

(B) REAMING

(C) TAPPING

Fig. 23-18 Tailstock tooling can be used to perform internal operations. (Courtesy of Kelmar Associates)

other in less than one second with high-precision accuracy.

● TAILSTOCK TOOLING

The prime purpose of tailstock tooling is to hold the dead or revolving center to provide support for a workpiece being machined between centers. The Morse taper hole in the spindle also allows it to hold cutting tools and accessories for internal operations such as drilling, reaming, and tapping, Fig. 23-18.

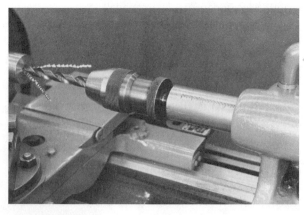

(A) DRILLING

Fig. 23-18 (Continues)

SUMMARY

- Two types of lathe centers are used for supporting work between centers. The solid tailstock center must be lubricated; revolving tailstock centers do not require lubrication.
- Some chucks have jaws that move simultaneously when being adjusted, others have jaws that move independently
- Steady rests and follower rests provide support for long, slender workpieces.
- Toolbits are cutting tools used to machine metal in a lathe.
- Toolposts hold a variety of cutting tools that can quickly be indexed and positioned when changing cutting operations.

KNOWLEDGE REVIEW

1. Name the four areas of a lathe where accessories and tooling are available to increase productivity.

Work-Holding Devices

2. Name and describe two types of lathe spindle noses.

3. What purpose do lathe centers serve?

4. For what operations are revolving tailstock centers particularly useful?

5. Name two types of chucks and explain how each operates.

6. For what purpose are the following used?

    a. collet chucks

    b. mandrels

    c. steady rests

    d. follower rests

7. What is the advantage of a Jacobs chuck over a draw-in collet chuck?

Cutting Tools and Toolholders

8. Explain the difference between left-hand and right-hand toolbits.

9. What is the purpose of end relief and side relief on a toolbit?

10. Name three lathe toolholders and state the purpose of each.

11. How does the carbide toolholder differ from other toolholders?

12. What is the advantage of using a turret type toolpost?

13. What is one advantage of the quick-change toolpost?

14. State two purposes of tailstock tooling.

U N I T
24

Cutting Speeds and Feeds

The rate or speed that work revolves in a lathe can have a major effect on the life of the cutting tool and the rate at which work is produced. *Too high a speed* will quickly dull the cutting tool and result in much time spent on resharpening or resetting a tool. *Too low a speed* results in a poor cutting action, wasted time, and low production. For the conditions that will produce the longest tool life and the best productivity, it is important that the speeds and feeds be set correctly for each different diameter, work material, and cutting tool.

OBJECTIVES

After completing this unit, you should be able to:

- Calculate and set the proper speeds to suit various types of work materials.
- Set the feed to suit the work material, diameter, and cutting tool used.
- Calculate the time required to make a cut or machine a part.

KEY TERMS

cutting speed   feed   spindle speed

● CUTTING SPEED

The **cutting speed** for lathe work may be defined as the rate at which a point on the circumference of the work passes the cutting tool in one minute. Cutting speed is expressed in feet per minute (ft./min) or in meters per minute (m/min). For example, if tool steel has a cutting speed of 60 ft./min (18 m/min), the lathe speed (r/min) should be set so that 60 ft. (18 m) of the work circumference passes the cutting tool in one minute. Setting the correct speed and feed for each cut or operation will provide the longest tool life and the best productivity.

The recommended cutting speeds (CS) for various materials are listed in Table 24-1. These cutting speeds have been determined by cutting tool and metal manufacturers as being the best for increasing production rates and cutting tool life.

Spindle Speed Calculations

To be able to calculate the number of revolutions per minute or **spindle speed,** at which to set a lathe, the diameter of the work and the cutting speed of the material must be known. The revolutions per minute at which the lathe should be set can be found by applying one of the following simplified formulas for inch or metric calculation. The standardized spindle speed formulas (and their development) may be found in Section 11, Drill Presses.

Inch Calculations

The most accurate formula for calculating the r/min is derived from

$$\frac{CS \times 12}{\pi \times D} \quad \text{which results in} \quad \frac{CS \times 3.82.}{D}$$

Since most machines cannot be set to the exact calculated speed and for easy calculating, the 3.82 has been rounded off to 4.

$$r/min = \frac{CS \times 4}{D}$$

where:

CS = cutting speed of the metal in ft./min.

D = diameter of the workpiece in inches

EXAMPLE: Calculate the r/min required to finish turn a 2 in. diameter piece of machine steel. Table 24-1 lists the cutting speed for machine steel as 100.

$$r/min = \frac{CS \times 4}{D}$$
$$= \frac{100 \times 4}{2}$$
$$= 200$$

Metric Calculations

The simplified formula for determining the spindle speed when the cutting speed is given in meters is:

$$r/min = \frac{CS \times 320}{D}$$

where:

CS = cutting speed in m/min

D = diameter of work in millimeters

EXAMPLE: Calculate the r/min required to finish turn a 60 mm diameter piece of machine steel.

$$r/min = \frac{30 \times 320}{60}$$
$$= 160$$

Table 24-1   Lathe cutting speeds in feet per minute and meters per minute using a high-speed toolbit

| Material | Turning and Boring | | | | Threading | |
| | Rough Cut | | Finish Cut | | | |
	Ft./Min	M/Min	Ft./Min	M/Min	Ft./Min	M/Min
Machine steel	90	27	100	30	35	11
Tool steel	70	21	90	27	30	9
Cast iron	60	18	80	24	25	8
Bronze	90	27	100	30	25	8
Aluminum	200	61	300	93	60	18

Since most lathes are provided with a limited number of speed settings, the simplified inch and metric formulas are acceptable for most calculations. When it is not possible to set the lathe to the exact spindle speed, always set it to the next *lower* speed.

Setting Lathe Speeds

Engine lathes are designed to operate at various spindle speeds for machining different-sized diameters and types of work material. These speeds are measured in revolutions per minute and are changed by means of gear levers or a variable-speed adjustment. When setting the spindle speed, *always set the machine as close as possible to the calculated speed, but never higher.* If the cutting action is satisfactory, the speed may be increased slightly; however, if the cutting action is not satisfactory or the work vibrates or chatters, reduce the speed and increase the feed.

On the geared-head lathe, Fig. 24-1, speeds are changed by moving the speed levers into their proper positions according to the r/min chart that is fastened to the headstock. While shifting the lever positions, place one hand on the drive plate or chuck, and turn it slowly by hand. This will enable the levers to engage the gear teeth without clashing.

 Never change speeds when the lathe is running.

Some lathes are equipped with a variable-speed headstock, and any speed within the range of the machine can be set. The spindle speed is set while the lathe is running by turning a speed control knob until the desired speed is indicated on the speed dial.

● LATHE FEED

The **feed** of a lathe is defined as the distance in inches or millimeters the cutting tool advances along the length of the work for every revolution of the spindle. For example, if the lathe is set for a .008 in. (0.2 mm) feed, the cutting tool will travel along the length of the work .008 in. (0.2 mm) for every complete turn that the work makes. The feed of an engine lathe is dependent upon the rate at which the lead screw or feed rod revolves. This is controlled by the change gears in the quick-change gearbox, Fig. 24-2A. The chart in Fig. 24-2B shows a range of gearbox settings.

Fig. 24-1 The speed (r/min) of a geared-head lathe is controlled by the position of the gear-change levers. (Courtesy of Clausing Industrial Inc.)

Whenever possible, only two cuts should be taken to bring a diameter to size: a roughing cut and a finishing cut. Since the purpose of a roughing cut is to remove excess material quickly and surface finish is not too important, a coarse feed should be used. The finishing cut is used to bring the diameter to size and produce a good surface finish; therefore a finer feed should be used. For general purpose machining, a .010 to .015 in. (0.25 to 0.4 mm) feed for roughing and a .003 in. to .005 in. (0.07 to 0.12 mm) feed for finishing is recommended. Table 24-2 lists the recommended feeds for cutting various materials when using a high-speed steel cutting tool.

To Set the Lathe Feed

Use the following procedure to set the lathe feed.

1. From the quick-change gearbox chart, Fig. 24-2B, select the amount of feed required.

Fig. 24-2A The quick-change gearbox controls the feed rate of the lathe. (Courtesy of Clausing Industrial Inc.)

LEVERS		ENGLISH—THREADS PER INCH					METRIC—PITCH IN M/M			
		SLIDING FEEDS IN THOUSANDS					SURFACING $\frac{1}{4}$ SLIDING			
D	B	60	56	52	48	44	40	38	36	32
		.5 mm	.005	.005	.006	.006	0.75 mm	0.007	0.008	0.009
C	B	30	28	26	24	22	20	19	18	16
		1 mm	.010	.011	1.25 mm	.013	1.5 mm	0.015	0.016	0.017
D	A	15	14	13	12	11	10	$9\frac{1}{2}$	9	8
		2 mm	.020	.021	2.5 mm	.025	3 mm	0.029	0.031	0.34
C	A	$7\frac{1}{2}$	7	$6\frac{1}{2}$	6	$5\frac{1}{2}$	5	$4\frac{3}{4}$	$4\frac{1}{2}$	4
		4 mm	.039	.042	5 mm	.050	6 mm	0.058	0.061	0.068

Fig. 24-2B The quick-change feed and thread chart indicates what the gearbox can be set to. (Courtesy of Kelmar Associates)

2. Disengage the tumbler level #4, Fig. 24-2A, by pulling out the knob and moving the lever assembly down.

3. Slide it over until it aligns vertically with the row the desired feed is in.

4. Move the lever assembly up, until its pin engages into the correct hole in the gearbox housing.

5. Follow the row in which the selected feed is found to the left and set the feed change levers (#1 and #2) to the letters indicated on the feed chart.

6. Set lever #3 (lead screw engaging lever) to the down position.

NOTE

Before turning on the lathe, be sure all levers are fully engaged by turning the headstock spindle by hand. If all the levers are correctly engaged, the feed rod should turn as the spindle is revolved.

DEPTH OF CUT

The depth of cut in lathe work may be defined as the depth of the chip that is taken by the cutting tool. Figure 24-3 shows a standard (radius) collar setting of .125 in. cut being taken from a piece of work 2.000 in. in diameter, reducing the work by .250 in. to 1.750 in. in diameter. If a 1 mm depth of cut was taken from a 25 mm diameter, the work would be reduced to 23 mm. In rough turning, the maximum depth of cut depends upon the condition of the machine, the type of cutting tool used, and the rigidity of the workpiece. In finish turning, the depth of cut, which should never be less than .004 in. (0.1 mm), is dependent on the type of work, the type of cutting tool, and the surface finish required.

MACHINING TIME

The time required to machine a part is very important to the efficient operation of any manufacturing company. The management must know how long it will take to produce a part in order to

Table 24-2 Feeds for various materials (using a high-speed steel cutting tool)

Material	Rough Cuts		Finish Cuts	
	Inches	Millimeters	Inches	Millimeters
Machine Steel	.010–.020	0.25–.5	.003–.010	0.07–0.25
Tool Steel	.010–.020	0.25–.5	.003–.010	0.07–0.25
Cast Iron	.015–.025	0.40–.65	.005–.012	0.13–0.3
Bronze	.015–.025	0.40–.65	.003–.010	0.07–0.25
Aluminum	.015–.03	0.40–.75	.005–.010	0.13–0.25

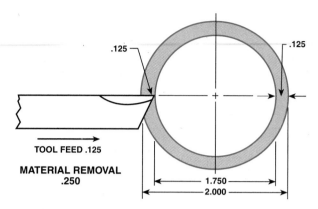

Fig. 24-3 A standard (radius) graduated collar setting of .125 in. will reduce the diameter by .250 in. (Courtesy of Kelmar Associates)

TOOL FEED .125

MATERIAL REMOVAL .250

accurately estimate its cost and schedule the machines, tools, and operator time required. All these factors play an important part of manufacturing; without them, confusion can result and delivery time to the customer becomes uncertain.

Calculating Machining Time

To calculate the time required to machine any workpiece, factors such as speed, feed, and length of cut must be considered. By applying the following formula, the time required to take a cut can be readily calculated.

$$\text{Machining time} = \frac{\text{length of cut}}{\text{feed} \times \text{r/min}}$$

EXAMPLE: Calculate the time required to take a roughing cut (.015 feed) from a 20 in. long piece of 2 in. diameter machine steel.

$$\text{r/min} = \frac{CS \times 4}{D}$$

$$= \frac{90 \times 4}{2}$$

long Piece of 2 IN

$$= 180$$

$$\text{Machining time} = \frac{\text{length of cut}}{\text{feed} \times \text{r/min}}$$

$$= \frac{20}{.015 \times 180}$$

form/inch

feed

$$= 7.4 \text{ min}$$

Example: Calculate the time required to take a finish cut (0.10 mm feed) from a 320 mm long piece of 30 mm diameter machine steel.

$$\text{r/min} = \frac{CS \times 320}{D}$$

$$= \frac{30 \times 320}{30}$$

$$= 320$$

$$\text{Machining time} = \frac{\text{length of cut}}{\text{feed} \times \text{r/min}}$$

$$= \frac{320}{0.10 \times 320}$$

$$= 10 \text{ min}$$

SUMMARY

- Cutting speed is an important factor in determining tool life and productivity.

- Spindle speeds are calculated using standard formulas. They should be set as close as possible to calculated speeds, but never higher.

- Lathe feed is the distance the cutting tool advances along the workpiece length for each revolution of the spindle.

- In rough turning, the maximum depth of cut depends on machine condition, type of cutting tool, and workpiece rigidity.

- Machining time is calculated from standard formulas, and depends on length of cut, feed, cutting speed, and r/min.

KNOWLEDGE REVIEW

Cutting Speeds and Feeds

1. How is cutting speed defined?

2. What will result if the lathe speed is
   a. too slow?
   b. too fast ?

3. Calculate the r/min required to take a rough cut from the following: (See Table 24-1 for the cutting speeds of various materials.)
   a. A 3/4 in. diameter piece of machine steel
   b. A 2-1/2 in. diameter piece of cast iron
   c. A 75 mm piece of machine steel
   d. A 44 mm piece of tool steel

4. Define lathe feed.

5. What feed, in both inch and metric units, is recommended for general purpose machining for:

a. rough cuts ?

b. finish cuts ?

Machining Time

6. State three reasons why machining time is very important in manufacturing.

7. Calculate the time required to take a finish cut from:

a. 1 in. diameter piece of machine steel 12 in. long (CS 100, Feed .007 in.)

b. 30 mm diameter piece of tool steel 250 mm long (CS 27, Feed 0.15 mm)

U N I T
25

Mount and Remove Accessories

Lathe accessories are often removed from the **headstock** spindle and the **tail-stock** to allow other accessories to be mounted. This is often necessary because of the size or shape of the workpiece, or the machining operation that must be performed. It is very important that this operation be done safely and in the proper sequence to have the accessories seated properly and to preserve the accuracy of the lathe parts.

OBJECTIVES

After completing this unit, you should be able to:

- Mount and remove headstock and tailstock centers accurately.
- Align the lathe centers by three different methods.
- Mount and remove headstock accessories.
- Drill accurate center holes to support work between lathe centers.

KEY TERMS

cam-lock spindle nose   headstock   tailstock
tapered spindle nose

● MOUNTING AND REMOVING LATHE CENTERS

Lathe centers are removed to allow other accessories to be mounted on a lathe, to clean the centers, to obtain center trueness, and to replace worn centers.

In order to machine parallel work that is concentric with the center holes in the work, it is very important that the centers be correctly mounted. Dirt, burrs, or chips on the tapered shanks of the lathe centers or in the lathe spindle holes, can prevent the center from seating properly and result in spoiled work.

Mounting a Center in the Headstock

Use the following procedure to mount a center in the headstock.

1. Clean the tapered hole in the spindle with a cloth wrapped around a stick, Fig. 25-1.

● **Never run the lathe when mounting and removing centers. SHUT OFF THE ELECTRICAL POWER.**

2. Remove all burrs from the center sleeve and lathe center with a file or hone, Fig. 25-2.

3. Clean the internal taper of the center sleeve and the taper on the lathe center.

4. Seat the center into the center sleeve with a sharp snap.

5. Insert the center sleeve into the lathe spindle for a short distance, Fig. 25-3.

6. With a sharp snap, seat it into the headstock spindle.

Fig. 25-1 The lathe spindle hole should be thoroughly cleaned before a center is inserted. (Courtesy of Kelmar Associates)

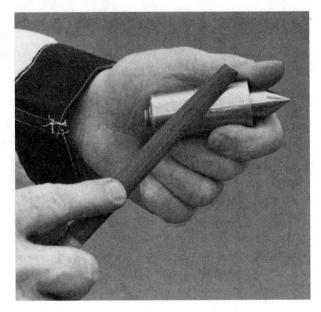

Fig. 25-2 Remove burrs from the lathe center taper to ensure that it runs true. (Courtesy of Kelmar Associates)

Fig. 25-3 Insert the center assembly a short distance into the lathe spindle and then give it a sharp snap to seat it properly. (Courtesy of Kelmar Associates)

If the center is not inserted with a sharp snap, the pressure of the cut will force the center into the headstock and the work will become loose between centers.

Removing a Headstock Center

Use the following procedure to remove a headstock center.

1. Be sure that the lathe is stopped.

2. Wrap a cloth around the headstock center and hold it with one hand, Fig. 25-4.

Fig. 25-4 Hold the lathe center in a cloth when removing it from the lathe spindle to prevent hand injury. (Courtesy of Kelmar Associates)

 Holding the lathe center with a cloth will prevent hand injury from the sharp point of the center.

3. Insert a knockout bar in the headstock spindle.

4. Tap the back of the center with the knockout bar until it is removed from the headstock spindle.

5. Clean and store the center where it will not be damaged.

Mounting a Center in the Tailstock

Use the following procedure to mount a center in the tailstock.

1. Clean the tapered hole in the tailstock spindle and the taper on the lathe center.

2. Extend the tailstock spindle about 2 in. (50.8 mm) beyond the tailstock.

> This will ensure that the center seats itself in the spindle and does not hit the knockout screw, Fig. 25-5.

3. Insert the center into the tailstock for a short distance.

4. With a sharp snap, seat it into the tailstock spindle.

Removing a Tailstock Center

Use the following procedures to remove a tailstock center.

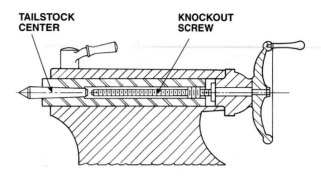

Fig. 25-5 Turn the tailstock handwheel counterclockwise until the knockout screw ejects the lathe center. (Courtesy of Kelmar Associates)

1. Wrap a cloth around the tailstock center and hold it with one hand.

2. Turn the tailstock handwheel *counterclockwise* to draw the spindle into the tailstock.

3. When the end of the center contacts the knockout screw, it will be removed from the tailstock spindle, Fig. 25-5.

4. Clean and store the center where it will not be damaged.

● ALIGNMENT OF LATHE CENTERS

To produce a parallel diameter on work machined between centers, it is important that the headstock and tailstock centers be *in line*. If the lathe centers are not aligned with each other, the diameter that is cut will be tapered (larger at one end than the other end). There are three common methods used to check the alignment of lathe centers:

1. By having the lines on the back of the tailstock in line. (This should always be done as a first step in aligning centers.)

2. By taking a light trial cut to the same depth setting near each end of the work and measuring the diameters with a micrometer.

3. By using a dial indicator and a parallel test bar. (This method is the most accurate.)

The lathe tailstock consists of two halves, the baseplate and the tailstock body, Fig. 25-6. The tailstock body can be adjusted either toward or away from the cutting tool by means of the two adjusting screws (G and F). This allows the tailstock center to be aligned with the headstock center.

Fig. 25-6 The tailstock can be adjusted for parallel or taper turning. (Courtesy of Kelmar Associates)

To Align Centers by the Tailstock Graduations

Use the following procedure to align centers by the tailstock graduations.

1. Loosen the tailstock clamp nut or lever.

2. Loosen one adjusting screw and tighten the other one, depending upon the direction the tailstock must be moved.

3. Continue adjusting the screws until the line on the tailstock body matches the line on the baseplate, Fig. 25-6.

4. Tighten the loose adjusting screw to hold the tailstock body in position.

5. Recheck the tailstock graduations and then tighten the tailstock clamp nut or lever.

6. Mount the cutting tool on the left-hand side of the compound rest and on center, Fig. 25-7.

7. Mount the workpiece between centers.

8. Set the lathe for the correct speed.

9. Take a light trial cut about 4 in. (200 mm) long.

10. Stop the machine, and check the diameter at both ends for taper.

11. If there is too much taper, follow steps 1 to 4 and readjust the tailstock in the proper direction.

To Align Centers with the Trial Cut Method

Before using the trial cut method of aligning centers, first check that the graduations on the

Fig. 25-7 Set the toolbit on center and to the left side of the compound rest for facing operations. (Courtesy of Kelmar Associates)

tailstock are in line. Align them visually, if necessary, and then proceed as follows:

1. Take a trial cut from the work at the tailstock end (section A) deep enough to produce a true diameter, Fig. 25-8. This cut should be about 1/4 in. (6 mm) long.

2. Disengage the automatic feed and note the reading on the graduated collar of the crossfeed screw.

3. Turn the crossfeed handle *counterclockwise* to bring the cutting tool away from the work.

Fig. 25-8 A light trial cut, 1/4 in. long, should be taken from the work at the tailstock end. (Courtesy of Kelmar Associates)

Fig. 25-9 A trial cut, at the same graduated collar setting, should be made at both the tailstock and headstock end. (Courtesy of Kelmar Associates)

4. Move the carriage until the cutting tool is about 1 in. (25 mm) from the lathe dog.

5. Start the lathe.

6. Turn the crossfeed handle *clockwise* until it is at the same graduated collar setting as it was at section A.

7. Cut section B about 1/2 in. (12.7 mm) long, Fig. 25-9.

8. Stop the lathe and measure both diameters with a micrometer, Fig. 25-10.

9. If both diameters are the same, the lathe centers are in line.

If the diameters are different, move the tailstock one-half the difference between the two diameters:
a. *Away from the cutting tool if the diameter at the tailstock end is smaller.*
b. *Toward the cutting tool if the diameter at the tailstock end is larger.*

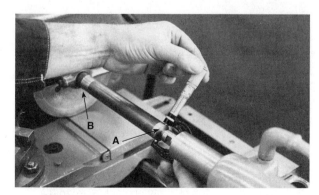

Fig. 25-10 Measure the diameter at section A and B to see if there is any taper on the work. (Courtesy of Kelmar Associates)

10. If necessary, readjust the tailstock and take trial cuts until both diameters are the same.

To Align Centers with a Test Bar and Dial Indicator

Aligning lathe centers with a parallel test bar and dial indicator is usually the fastest and most accurate method of aligning centers. Once the indicator readings at both ends of the bar are the same, the centers are in line and parallel work will be produced. Use the following procedure to align the centers:

1. Clean the lathe centers and the center holes in the test bar.

2. Mount a dial indicator, with the indicator plunger on center and in a horizontal position in the toolpost or on the lathe carriage, Fig. 25-11.

3. Adjust the test bar snugly between centers and then tighten the tailstock spindle clamp.

4. Turn the crossfeed handle until the indicator registers approximately one-quarter of a revolution on section A, then set the indicator bezel to zero, Fig. 25-12.

5. Turn the carriage handwheel to bring the indicator into contact with section B, Fig. 25-13.

6. Compare the two indicator readings. If they are not the same, return the carriage until the indicator again registers on section A.

7. Loosen the tailstock clamp nut and, by means of the tailstock adjusting screws, move the tailstock in the proper direction

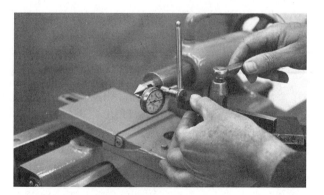

Fig. 25-11 For accurate setup, the indicator plunger should always be set on center. (Courtesy of Kelmar Associates)

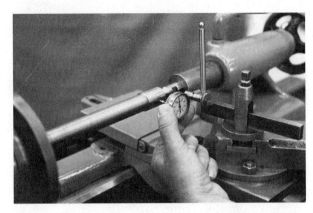

Fig. 25-12 Set the indicator bezel to zero when aligning centers by test bar and dial indicator method. (Courtesy of Kelmar Associates)

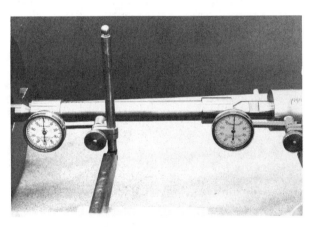

Fig. 25-13 If both indicator readings are the same, the lathe centers are aligned and parallel work will be produced. (Courtesy of Kelmar Associates)

by the amount of the difference between the readings at sections A and B.

8. Tighten the tailstock clamp nut, and repeat steps 4 to 7 to recheck the center alignment.

● REMOVING AND MOUNTING CHUCKS

Lathe accessories such as chucks and drive plates are fitted to the headstock of a lathe to hold various shaped workpieces for machining. The proper procedure for removing and mounting these accessories must be followed in order not to damage the lathe spindle and/or accessories and to preserve the accuracy of the machine. There are three types of lathe spindle noses: the threaded spindle nose, the **tapered spindle nose**, and the **cam-lock spindle nose**.

To Remove a Chuck

The following procedures apply to any accessories which must be removed from a lathe spindle nose:

1. Set the lathe in the slowest speed.

 SHUT OFF THE ELECTRICAL SWITCH.

2. Place a chuck cradle under the chuck, Fig. 25-14.

3. Remove the chuck or accessory by following the steps, outlined, depending on the type of lathe spindle nose.

Threaded Spindle Nose

1. Turn the lathe spindle until a chuck-wrench socket is in the top position.

2. Insert the chuck wrench into the hole and pull it *sharply counterclockwise* (toward you).

OR

1. Place a block or short stick on the lathe bed, under the chuck jaw at the back of the lathe.

2. Revolve the lathe spindle by hand in a *clockwise* direction until the chuck is loosened on the spindle.

3. Remove the chuck from the spindle and store it where it will not be damaged.

Fig. 25-14 A chuck cradle will prevent injury to the hands and damage to the lathe bed when accessories are mounted or removed. (Courtesy of Kelmar Associates)

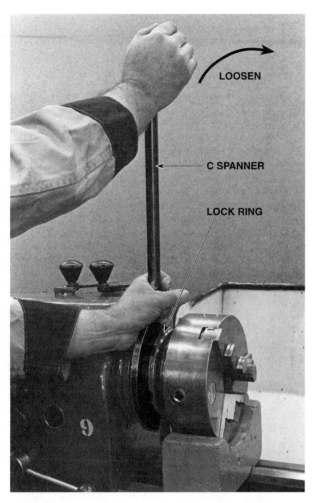

Fig. 25-15 A C-spanner wrench is used to loosen the lock ring on a taper spindle nose. (Courtesy of Kelmar Associates)

Taper Spindle Nose

1. Secure the proper C-spanner wrench.

2. Place it around the front of the lock ring of the spindle with the handle in an upright position, Fig. 25-15.

3. Place one hand on the curve of the spanner wrench to prevent it from slipping off the lock ring.

4. With the palm of the other hand or a soft-faced hammer, *sharply* strike the handle of the wrench in a *clockwise* direction.

5. Hold the chuck with one hand, while turning the lock ring clockwise with the other hand.

6. If the lock ring becomes tight, use the spanner wrench to break the taper contact between the spindle and chuck.

Fig. 25-16 Before storing a chuck, place a cloth in its taper hole to keep out dirt and chips. (Courtesy of Kelmar Associates)

7. Plug the hole in the chuck with a cloth to keep out chips and dirt, Fig. 25-16.

8. Store the chuck with the jaws in the up position.

Cam-Lock Spindle Nose

1. With the proper size wrench, turn each cam-lock *counterclockwise* until its registration line matches the registration line on the spindle nose, or is at the 12 o'clock position.

2. Hold the chuck and *sharply* strike the top of the chuck with the other hand or a soft-faced hammer to remove it from the spindle, Fig. 25-17.

3. Remove and store the chuck properly.

Fig. 25-17 A sharp, downward blow with the hand will break the taper contact between the lathe spindle and the chuck. (Courtesy of Kelmar Associates)

To Mount a Chuck

The following procedures apply to all accessories which are mounted on a lathe spindle nose.

1. Set the lathe to the slowest speed.

 SHUT OFF THE ELECTRICAL SWITCH.

2. Clean all surfaces of the spindle nose and the mating parts of the chuck.

3. Place a cradle block on the lathe bed in front of the spindle and place the chuck on the cradle. (Refer to Fig. 25-14.)

4. Slide the cradle close to the lathe spindle nose and mount the chuck.

5. Mount the chuck or accessory by following the steps outlined below, depending on the type of lathe spindle nose.

Threaded Spindle Nose

1. Revolve the lathe spindle by hand in a *counterclockwise* direction and bring the chuck up to the spindle.

 NEVER USE POWER.

2. If the chuck and spindle are clean and correctly aligned, the chuck should easily thread onto the lathe spindle.

3. When the chuck adapter plate is within 1/16 in. (1.5 mm) of the spindle shoulder, give the chuck a quick turn to seat it against the spindle shoulder.

4. Do not jam a chuck against the shoulder too tightly. It may damage the threads and make the chuck difficult to remove.

Taper Spindle Nose

1. Revolve the lathe spindle by hand until the key on the spindle nose aligns with the keyway in the tapered hole of the chuck, Fig. 25-18.

2. Slide the chuck onto the lathe spindle, and at the same time turn the lock ring in a *counterclockwise* direction, Fig. 25-19.

3. Tighten the lock ring securely with a spanner wrench by striking it sharply downward, Fig. 25-20, when standing at the front of the machine.

Cam-Lock Spindle Nose

1. Align the registration of each cam lock with the registration line on the lathe spindle nose, Fig. 25-21.

2. Revolve the lathe spindle by hand until the clearance holes in the spindle align with the cam-lock studs of the chuck.

Fig. 25-18   The spindle key and the chuck keyway must be aligned when mounting a taper spindle nose accessory. (Courtesy of Kelmar Associates)

Fig. 25-19   Turn the lock ring counterclockwise to draw the chuck onto the lathe spindle. (Courtesy of Kelmar Associates)

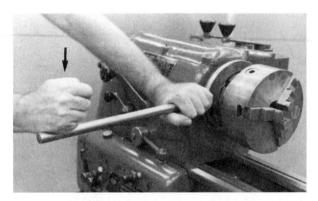

Fig. 25-20 Tighten the lock ring securely with a C-spanner wrench by striking it sharply downward. (Courtesy of Kelmar Associates)

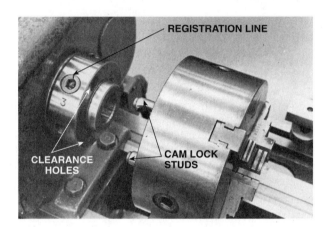

Fig. 25-21 Align the lathe spindle holes with the cam-lock studs before attempting to mount a chuck. (Courtesy of Kelmar Associates)

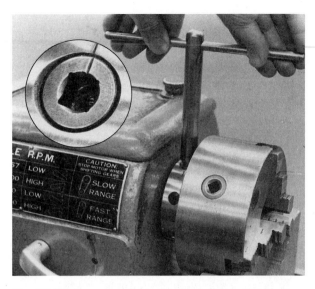

Fig. 25-22 Turn the cam locks clockwise to tighten the chuck on the spindle. (Courtesy of Kelmar Associates)

Fig. 25-23 For accurate work, it is wise to face the end of the work before drilling center holes. (Courtesy of Kelmar Associates)

3. Slide the chuck onto the spindle.

4. Securely tighten each cam lock in a clockwise direction, Fig. 25-22.

To Drill Center Holes

Center holes can be drilled on a lathe in round or hexagonal work, held in a three-jaw chuck, without having to lay out the location of the center.

1. Grip the work short in a three-jaw chuck.

> No more than three times the diameter should extend beyond the chuck jaws.

2. Square the end of the work by facing, Fig. 25-23.

3. Mount a drill chuck in the tailstock spindle.

4. Select the proper center drill to suit the work diameter and fasten it in the drill chuck, Fig. 25-24.

See Table 21-2 in Unit 21, *Drill Presses*, page 202.

5. Check the lines on the back of the tailstock to see that they are aligned, Fig. 25-25. Correct if necessary.

Fig. 25-24 The center drill is held short in a drill chuck in the tailstock. (Courtesy of Kelmar Associates)

Fig. 25-25 The upper and lower lines on the tailstock should be aligned before drilling center holes. (Courtesy of Kelmar Associates)

6. Set the lathe speed to approximately 1200 to 1500 r/min.

7. Move the tailstock until the center drill is close to the work and then lock the tailstock clamp nut.

8. Start the lathe spindle and turn the tailstock handwheel to feed the center drill into the work, Fig. 25-26.

9. Frequently apply cutting fluid and drill the center hole until the top of the hole is to the correct diameter [about 3/16 in. (5 mm) for 3/4 in. (19 mm) diameter work].

Fig. 25-26 Feed the center drill slowly with the tailstock handwheel until the diameter of the holes is correct. (Courtesy of Kelmar Associates)

SUMMARY

- Headstock and tailstock centers are often removed from lathes to permit other accessories to be mounted.

- Lathe centers must be correctly mounted in order to avoid producing inaccurate work.

- Alignment of lathe centers can be checked by adjusting tailstock graduations, by taking a trial cut, or by using a dial indicator and a parallel test bar.

- Procedures for mounting and removing chucks depend on the type of lathe spindle nose—threaded, tapered, or cam-lock.

KNOWLEDGE REVIEW

Removing and Mounting Centers

1. Explain how the center is removed from the headstock and the tailstock.

2. What should be done before replacing a center in the headstock or tailstock?

Alignment of Lathe Centers

3. Why is it important that the headstock and tailstock centers be in line?

4. Name three methods of aligning lathe centers.

5. Briefly describe how to align the centers using the trial cut method.

Mounting and Removing Chucks

6. Why should a chuck cradle be used when mounting or removing a chuck?

7. Briefly explain how to remove a chuck from:

   a. a taper spindle nose

   b. a cam-lock spindle nose

8. When mounting a chuck on a threaded spindle nose, why should the chuck not be jammed against the shoulder?

9. Briefly explain how to mount a chuck on:

   a. a taper spindle nose

   b. a cam-lock spindle nose

Drilling and Reaming Holes

10. At what speed should the lathe be set for drilling center holes?

11. How deep should center holes be drilled for 3/4 in. (19 mm) diameter work?

Mount Work between Centers

Much of the work machined on a lathe in training programs is held between the headstock and tailstock centers. Since both centers can be in a fixed position, it is possible to machine work, remove it from the lathe, and replace it for additional machining with the assurance that all machined diameters will be concentric (true) with each other. In training programs, it is often necessary to remove and replace the work numerous times before it is finally completed. Having the workpiece between centers allows it to be replaced more quickly and accurately than if it is held by other work-holding methods.

OBJECTIVES

After completing this unit, you should be able to:

- Set up various lengths of workpieces accurately between centers.
- Describe the proper machining sequence to follow for machining between centers.
- Set up the cutting tool and toolholder for facing and turning operations.

KEY TERM

facing

● MOUNTING WORK

Work that is mounted between centers can be machined, removed, and set up for additional machining, and still maintain the same degree of accuracy. The cutting tool and the work must be properly set up or damage to the machine, work, and lathe centers will result.

To Set up the Cutting Tool

1. Move the toolpost to the left-hand side of the compound rest.

2. Mount a toolholder so that its setscrew is close to the toolpost.

> For the best rigidity, have about the width of a thumb between the toolholder screw and the toolpost, Fig. 26-1.

3. Set the toolholder so that it is at right angles to the work or pointing slightly toward the tailstock.

4. Insert the desired toolbit in the toolholder so that it extends no more than 1/2 in. (12.7 mm).

5. Use two-finger pressure on the wrench to tighten the toolholder setscrew.

> Tightening the setscrew too tightly may break the toolbit because it is hard and brittle.

6. Adjust the toolholder until the point of the cutting tool is even with the lathe center point.

Fig. 26-1 The toolholder should be held short and the cutting tool tip set to center. (Courtesy of Kelmar Associates)

7. *Tighten the* toolpost screw securely to prevent the toolholder from moving under the pressure of the cut.

To Mount Work between Centers Using a Solid Dead Center

1. Check the live center by holding a piece of chalk close to it while it is revolving.

> If the live center is not running true, the chalk will mark only the high spot.

2. If the chalk marks the high spot, remove the live center from the headstock and clean the tapers on the center and the headstock spindle.

3. Replace the live center and check again for trueness.

4. Adjust the tailstock spindle until it extends about 2½ to 3 in. (65 to 75 mm) beyond the tailstock.

5. Loosen the tailstock clamp nut or lever.

6. Place the lathe dog on the end of the work with the bent tail pointing to the left, Fig. 26-2.

7. Lubricate the dead center with a suitable center lubricant such as center lube, white lead, oil, etc.

> The lubricant is used to reduce the friction and heat caused by the revolving work on the stationary center.

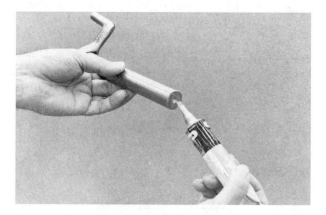

Fig. 26-2 The lathe dog is placed loosely on the left end and center lubricant is applied to the center hole at the other end. (Courtesy of Kelmar Associates)

Fig. 26-3 Tighten the lathe dog on the work and be sure that its tail does not bind in the drive plate slot. (Courtesy of Kelmar Associates)

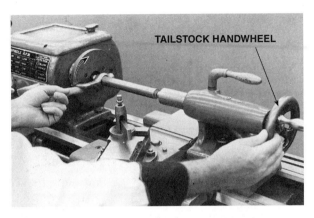

TAILSTOCK HANDWHEEL

Fig. 26-4 Adjust the tailstock handwheel so that the lathe dog drops of its own weight and there is no end play between centers. (Courtesy of Kelmar Associates)

8. Place the end of the work with the lathe dog on the live center.

9. Slide the tailstock toward the headstock until the dead center supports the other end of the work.

10. Tighten the tailstock clamp nut or lever.

11. Adjust the tail of the dog in the slot of the drive plate and tighten the lathe dog screw, Fig. 26-3.

> Be sure that the tail of the dog does not bind in the drive plate slot.

12. Turn the drive plate by hand until the slot and lathe dog are parallel with the bed of the lathe.

13. Hold the tail of the dog up in the slot and tighten the tailstock handwheel only enough to hold the lathe dog in the *up position*, Fig. 26-4.

14. Turn the tailstock handwheel backwards only until the lathe dog drops in the drive plate slot.

15. Hold the tailstock handwheel in this position and tighten the tailstock spindle clamp with the other hand.

16. Check the center adjustment by attempting to move the work endwise.

> There should be no end play.

17. Check the center tension by raising the tail of the dog to the top of the faceplate slot, Fig. 26-4, and remove your hand.

> The tail of the dog should drop of its own weight.

18. Move the carriage to the furthest position (left-hand end) of cut, and revolve the lathe spindle by hand to see that the dog does not hit the compound rest.

To Mount Work between Centers Using a Revolving Tailstock Center

The use of a revolving or live dead center eliminates the need for center lubrication when turning a long workpiece. This is because the center revolves with the work, therefore, there is no friction and heat. Revolving centers are also useful for heavy-duty and high-speed turning. When using this type of center, proceed as follows.

Follow steps 1 to 9, omitting step 7, in the previous section, "To Mount Work between Centers Using a Solid Dead Center", page 252.

Then add the following steps:

10. Tighten the tailstock handwheel until the workpiece is held snugly between centers.

11. Tighten the tailstock spindle clamp.

12. Re-check the setup to be sure that the dog does not bind in the slot of the drive plate.

When facing work using this type of center, always recess the outer end of the center hole to prevent damaging the revolving center with the cutting tool.

● MACHINING SEQUENCE

Proper procedure should be followed when machining any part so that it can be machined accurately and in the shortest time possible. It would be impossible to list the exact sequence of operations that should be followed for every type of workpiece machined on a lathe. The following guidelines, with minor exceptions, apply to work being machined between centers or in a chuck:

1. Rough-turn all diameters and lengths to within 1/32 in. (0.8 mm) of the size required.

2. Machine the largest diameter first and progress to the smallest.

3. Special operations, such as knurling or grooving, should be done next.

> It is not advisable to use a ball or roller bearing-type center for knurling operations because of the possibility of damaging the bearing.

4. Cool the workpiece before starting any finishing operations.

5. Finish turn all diameters and lengths.

> Always start with the largest diameter and work to the smallest.

● MACHINING BETWEEN CENTERS

There are times when it is necessary to machine the entire length of a round workpiece. The workpiece shown in Fig. 26-5 is a typical part which can be machined between the centers on a lathe.

Machining Sequence

1. Cut off a piece of steel 1/8 in. (3.18 mm) longer and 1/8 in. (3.18 mm) larger in diameter than required.

> In this case the diameter of the steel cut off would be 1-5/8 in. (41.3 mm) and its length would be 16-1/8 in. (409.6 mm).

2. Hold the work in a three-jaw chuck, face one end square, and then drill the center hole.

3. Face the other end to length, and then drill the center hole.

4. Mount the workpiece between the centers on a lathe.

5. Rough-turn the largest diameter to within 1/32 in. (0.8 mm) of finish size or 1-17/32 in. (38.9 mm).

> The purpose of the rough cut is to remove excess metal as quickly as possible.

6. Finish-turn the diameter to be knurled.

> The purpose of the finish cut is to cut work to the required size and produce a good surface finish.

7. Knurl the 1-1/2 in. (38.1 mm) diameter.

8. Machine the 45° chamfer on the end.

9. Reverse the work in the lathe, being sure to protect the knurl from the lathe dog with a piece of soft metal.

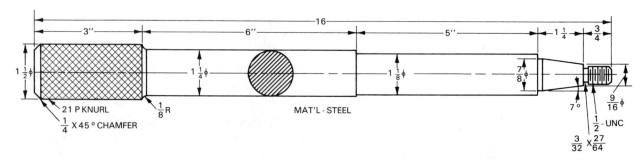

Fig. 26-5 A working print (drawing) of a round workpiece to be machined. (Courtesy of Kelmar Associates)

10. Rough turn the 1-1/4 in. (31.8 mm) diameter to 1-9/32 in. (32.5 mm).

> Be sure to leave the length of this section 1/8 in. (3.18 mm) short [12-7/8 in. (327 mm) from the end] to allow for finishing the 1/8 in. (3.18 mm) radius.

11. Rough-turn the 1-1/8 in. (28.6 mm) diameter to 1-5/32 in. (29.4 mm), Fig. 26-6.

> Leave the length of this section 1/32 in. (0.8 mm) short [6-31/32 in. (177 mm) from the end] to allow for finishing the shoulder.

12. Rough-turn the 7/8 in. (22.2 mm) diameter to 29/32 in. (23 mm) for a distance of 1-31/32 in. (50 mm) from the end.

13. Rough-turn the 1/2 in. (12.7 mm) diameter to 17/32 in. (13.5 mm) for a distance of 23/32 in. (18.2 mm).

14. Cool the work to room temperature before starting the finishing operations.

15. Finish-turn the 1-1/4 in. (31.8 mm) diameter to 12-7/8 in. (327 mm) from the end.

16. Mount a 1/8 in. (3.18 mm) radius tool and finish the corner to the correct length, Fig. 26-7.

17. Finish-turn the 1-1/8 in. (28.6 mm) diameter to 7 in. (178 mm) from the end.

18. Finish-turn the 7/8 in. (22.2 mm) diameter to 2 in. (50.8 mm) from the end.

19. Set the compound rest to 7° and machine the taper to size.

20. Finish-turn the 1/2 in. (12.7 mm) diameter to 3/4 in. (19 mm) from the end.

21. With a cut-off tool, cut the groove at the end of the 1/2 in. (12.7 mm) diameter, Fig. 26-8.

22. Chamfer the end of the section to be threaded.

23. Set the lathe for threading and cut the thread to size.

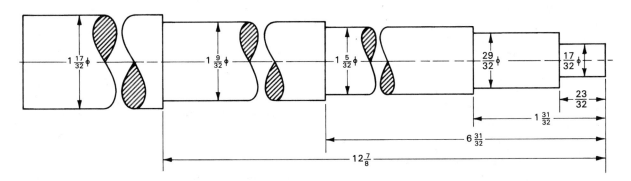

Fig. 26-6 The rough-turn diameters and lengths of the part required. (Courtesy of Kelmar Associates)

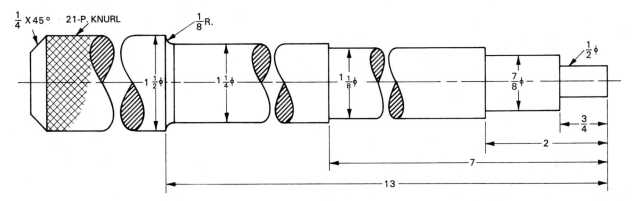

Fig. 26-7 The finish-turn diameters and lengths of the workpiece. (Courtesy of Kelmar Associates)

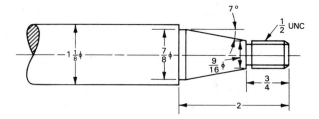

Fig. 26-8 Machining the taper and the thread are the last few operations on the part. (Courtesy of Kelmar Associates)

● FACING WORK BETWEEN CENTERS

Facing is a squaring operation performed on the ends of work after it has been cut off by a saw. In order to produce a flat surface when facing between centers, the lathe centers must be in line. The purposes of facing are:

- To provide a true flat surface, square with the axis of the work:

- To make a smooth surface from which to take measurements;

- To cut work to a required length.

To Face Work between Centers

1. Set the toolholder to the left-hand side of the compound rest.

2. Mount a facing tool in the toolholder, having it extend only about 1/2 in. (12.7 mm), and tighten the toolholder setscrew using only two-finger pull.

3. Adjust the toolholder until the point of the facing tool is at the same height as the lathe center point, then tighten the toolpost lightly, Fig. 26-9.

4. Tap the toolholder lightly until the point of the cutting tool is closest to the work and there is a space along the side, Fig. 26-10.

5. Tighten the toolpost securely to keep the toolholder in position.

6. If the entire end must be faced, insert a half center in the tailstock spindle, Fig. 26-11.

7. With a center punch, mark the length to be faced.

When facing, start at the center and feed outward; the length is correct when the punch mark is cut in half.

Fig. 26-9 The toolholder is gripped short and the cutting tool point is set to center height. (Courtesy of Kelmar Associates)

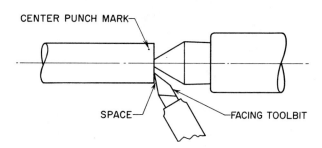

Fig. 26-10 The cutting tool is set properly for facing and the center punch mark indicates the material to be removed. (Courtesy of Kelmar Associates)

8. Bring the toolbit to the end of the work as in Fig. 26-11 by moving the carriage with the apron handwheel.

9. Set the lathe to the correct speed for the material being cut.

10. Move the toolbit in as close to the center of the work as possible and set the depth of cut by using the apron hand wheel.

11. Tighten the carriage lock, using only a two-finger pull on the wrench, Fig. 26-12.

12. Feed the tool out by turning the crossfeed screw handle slowly and steadily with two hands, cutting from the center outward, Fig. 26-13.

13. Repeat operations 9 to 12 until the center-punch mark is cut in half (see Fig. 26-10).

Fig. 26-11 The half center allows the entire end of the part to be faced. (Courtesy of Kelmar Associates)

When facing, all cuts must begin at the center and feed toward the outside.

14. Check the surface for flatness using the edge of a rule, Fig. 26-14.

Fig. 26-12 Lock the carriage, two-finger tight, to prevent it from moving during the facing operation. (Courtesy of Kelmar Associates)

Fig. 26-13 Use two hands to provide a steady feed rate and face from the center outward. (Courtesy of Kelmar Associates)

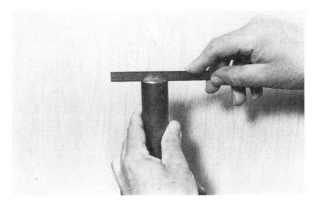

Fig. 26-14 If no light shows between the rule and the work, the surface has been faced flat. (Courtesy of Kelmar Associates)

If the center is high or low, recheck the alignment of the lathe centers and reface the work.

SUMMARY

- Cutting tools and work must be properly set up so accuracy can be maintained when work is mounted, removed, and mounted again for more machining.

- When turning a workpiece between centers, the use of a revolving or live dead center eliminates the need for center lubrication.

- Proper machining sequence calls for starting with the largest diameter and working to the smallest.

- Facing is a squaring operation that produces a flat surface on a workpiece.

KNOWLEDGE REVIEW

Mounting Work between Centers

1. How far should the toolholder extend beyond the toolpost?

2. How tight should the toolholder setscrew be tightened?

3. List six important steps involved in mounting work between centers.

4. How tight should the work be adjusted when using a solid tailstock center?

5. Why is no lubricant required when using a revolving tailstock center?

6. List the general rules for machining round work in a lathe.

Facing Work between Centers

7. State three purposes for facing work.

8. Explain how the cutting tool must be set up for facing.

9. How should the cutting tool be fed in order to face a surface?

10. How can a faced surface be checked for flatness?

U N I T 27

Machining between Centers

In many training programs, work is machined between centers because of the need to take work out of the lathe, and replace it a number of times before the part is actually completed. Much of the work in industry is held in chucks or other work-holding devices, and the part is generally machined in one setup. Industry, however, also uses center turning for machining shafts where it is very useful since parts may have to be removed and replaced more than once.

OBJECTIVES

After completing this unit, you should be able to:

- Set an accurate depth of cut for machining.
- Machine parallel diameters to an accuracy of ±.001 in. (0.02 mm).
- File and polish diameters to within .001 in. (0.02 mm) accuracy should these operations be required.

KEY TERMS

abrasive cloth   finish turning   polishing
rough turning

● PARALLEL TURNING

Work is generally machined on a lathe for two reasons: to produce a true diameter and to cut it to size. Work that must be cut to size and also be the same diameter along the entire length of the workpiece involves the operation of parallel turning. In order to produce a parallel diameter, the headstock and tailstock centers must be in line. See Unit 25 for the different methods that can be used to align lathe centers.

Many factors determine the amount of material that can be removed on a lathe at one time. However, whenever possible, work should be cut to size in two cuts: a roughing cut and a finishing cut.

To Set an Accurate Depth of Cut

The graduated micrometer collar on the crossfeed handle is a very useful tool for accurately setting the depth of a cut. On some lathes, the standard graduations, generally in .001 in. (0.02 mm), indicate the distance the cutting tool moves towards the workpiece. On other lathes, the direct-reading graduations indicate the amount of material which will be removed from a diameter. See Unit 22 for a more detailed explanation of graduated micrometer collars and their use on inch and metric lathes. Use the following procedure to set the depth of cut:

1. Move the toolpost to the left-hand side of the compound rest.

2. Set the cutting tool to center and tighten the toolpost screw *securely*.

3. Start the lathe and move the carriage until the toolbit overlaps the right-hand end of the workpiece by approximately 1/16 in. (1.5 mm).

4. Feed the toolbit in with the crossfeed handle until a light cut is made around the entire circumference of the work.

5. Turn the carriage handwheel until the toolbit just clears the right-hand end of the work.

6. Turn the crossfeed handle clockwise .005 in. (0.12 mm for metric lathes) and set the graduated collar to zero without moving the crossfeed handle.

7. Take a trial cut about 1/4 in. (6 mm) long, Fig. 27-1.

Fig. 27-1 A light trial cut should be taken at the tailstock end before turning any diameter. (Courtesy of Kelmar Associates)

The purpose of this trial cut is to:
- Produce a true diameter on the work.
- Set the cutting tool to the diameter.
- Set the crossfeed graduated collar to the diameter.

8. Stop the lathe and *be sure that the crossfeed handle setting is not moved.*

9. Turn the carriage handwheel until the toolbit clears the right-hand end of the work.

10. Measure the diameter of the trial cut with a micrometer, Fig. 27-2, and then calculate the amount of metal yet to be removed.

Fig. 27-2 Measuring the diameter of the trial cut with a micrometer. (Courtesy of Kelmar Associates)

11. Turn the crossfeed handle clockwise until the standard graduated collar moves one-half the amount of material to be removed.

   For example, if .020 in. (0.5 mm) must be removed, the crossfeed handle should be set in .010 in. (0.25 mm).

12. Take another trial cut 1/4 in. (6 mm) long and measure the diameter with a micrometer.

13. Repeat steps 11 and 12 until the diameter is to the correct size.

● ROUGH TURNING

Rough turning is used to remove most of the excess material as quickly as possible and to true the work diameter. The roughing cut should be taken to within 1/32 in. (0.8 mm) of the finished size of the workpiece. Generally one roughing cut should be taken if up to 1/2 in. (12.7 mm) is to be removed from the diameter. If more than 1/2 in. (12.7 mm) must be removed, take two roughing cuts.

To Rough Turn a Diameter

Use the following procedure to rough turn a diameter:

1. Check the lathe center alignment. See Unit 25.

2. Mount a general-purpose toolholder in the toolpost.

3. Set the toolpost on the left side of the compound rest.

4. Have the toolholder extend as little as possible beyond the toolpost, and set the point of the cutting tool even with the lathe center point, Fig. 27-3.

5. Adjust the toolholder so that it is pointing slightly toward the tailstock, Fig. 27-4A. An incorrectly set toolholder is shown in Fig. 27-4B.

6. Tighten the toolpost screw *securely.*

7. Set the lathe speed for the material being cut. (See Table 24-1.)

8. Set the quick-change gearbox, Fig. 27-5, for the rough-cut feed, generally about .010 to .020 in. (0.25 to 0.5 mm for metric lathes).

9. Take a light trial cut about 1/4 in. (6 mm) long at the right-hand end of the work.

10. Stop the lathe, but *do not move the cross-feed handle setting or the graduated collar.*

11. Turn the carriage handwheel until the cutting tool clears the right-hand end of the work.

12. Measure the diameter of the trial cut, and calculate how much material must be removed.

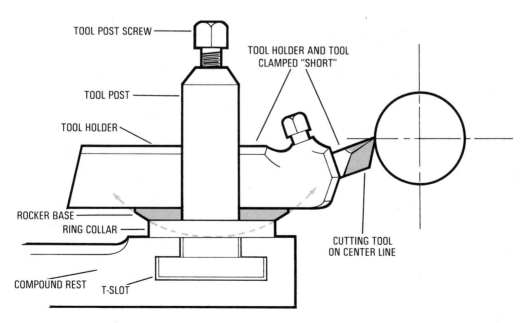

Fig. 27-3 The toolholder should be set as close as possible (about the width of a thumb) to the toolpost, and the cutting tool point set to center height. (Courtesy of Kelmar Associates)

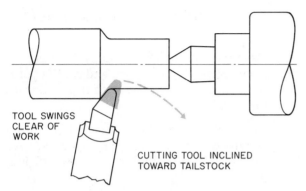

Fig. 27-4A If a correctly set toolholder moves under the pressure of a cut, the cutting tool will swing away from the work. (Courtesy of Kelmar Associates)

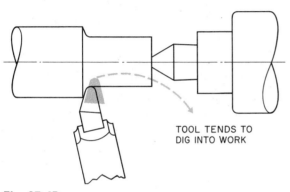

Fig. 27-4B If an incorrectly set toolholder moves under the pressure of a cut, the work will be machined undersize and possibly scrapped. (Courtesy of Kelmar Associates)

> Always leave the rough cut diameter from .030 to .050 in. (0.8 to 1.3 mm) over the finished size required. This will leave enough material for the finish cut, Fig. 27-6.

13. Turn the crossfeed handle clockwise one-half the calculated amount and take a trial cut 1/4 in. (6 mm) long.

14. Measure the diameter with a micrometer and reset the depth of cut if necessary.

15. Turn the rough cut to the required length.

● FINISH TURNING

The purpose of **finish turning** is to bring the workpiece to the required size and to produce a good surface finish. Generally only one finish cut is required since no more than .030 to .050 in. (0.8 to 1.3 mm) should be left on the diameter for the

Fig. 27-5 The quick-change gearbox levers can be set to various positions to set the feed rate for rough and finish cuts. (Courtesy of Kelmar Associates)

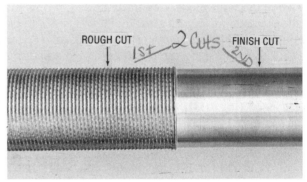

Fig. 27-6 The types of surface finishes produced by rough and finish turning. (Courtesy of Kelmar Associates)

finish cut. The toolbit should have a slight radius on the point, and the lathe should be set for a .003 to .005 in. (0.07 to 0.12 mm) feed. Be sure that the lathe centers are aligned exactly (Unit 25); otherwise a taper will be cut on the workpiece.

To Finish Turn a Diameter

Use the following procedure to finish turn a diameter:

1. Set the lathe speed for finish turning (Table 24-1, page 234).

2. Set the quick-change gearbox for the finish feed at approximately .003 to .005 in. (0.07 to 0.12 mm for metric lathes).

3. Mount and set the finishing toolbit to center.

4. Take a *light trial cut* 1/4 in. (6 mm) long from the diameter at the tailstock end.

5. Disengage the feed and stop the lathe, but *do not move the crossfeed handle setting*.

6. Turn the carriage handwheel until the cutting tool clears the right-hand end of the work.

7. Measure the diameter with a micrometer and calculate the amount of material that must still be removed.

8. Turn the crossfeed handle *clockwise* one-half the calculated amount (the difference between the trial-cut size and the finished diameter) and take a trial cut 1/4 in. (6 mm) long.

9. Stop the lathe and measure the diameter.

10. Reset the crossfeed handle setting if necessary and turn the finish cut to the required length.

To Check a Diameter

Whenever possible, a narrow-groove diameter should be measured with a blade micrometer or a blade vernier caliper. However, if these are not available, the diameter may be measured within a reasonable degree of accuracy with outside calipers that are properly set and used.

1. Stop the machine.

 Never measure work that is revolving—it is dangerous and the measurement taken will be inaccurate.

2. Turn the adjusting screw until the legs of the caliper lightly contact the center of the diameter.

3. Check the setting by holding the spring of the caliper between the thumb and index finger and allowing the legs to drop over the diameter, Fig. 27-7.

If the setting is correct, the caliper should just drop over the diameter by its own weight. Do not force a caliper over a diameter.

4. Adjust the caliper until just a *slight drag* is felt as the caliper legs pass over the diameter.

5. Use a rule to measure the distance between the two caliper legs.

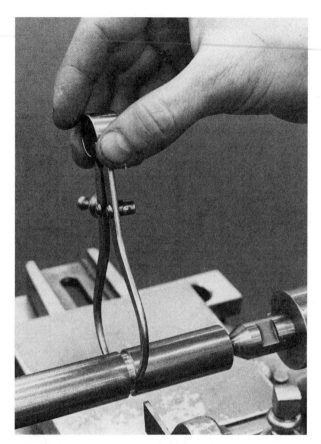

Fig. 27-7 The diameter of a narrow groove can be checked to approximate size with an outside caliper. (Courtesy of Kelmar Associates)

This method of measurement is only as accurate as a rule can be read. If greater accuracy is required when measuring a groove, a knife-edge vernier caliper should be used.

● FILING IN A LATHE

Filing on a lathe is used to remove burrs, tool marks, and sharp corners. It is not considered good practice to file a diameter to size because too much filing tends to produce a diameter which is out of round. The National Safety Council recommends grasping the file handle with the left hand, so that arms and hands can be kept clear of the headstock.

When filing or polishing in a lathe, it is good practice to cover the lathe bed with a piece of paper to prevent filings from getting into the slides and causing excessive wear and damage to the lathe. A cloth is not advisable for this purpose because of the danger of it being caught in the revolving work or the lathe.

NOTE

Before starting to file, it is wise to observe the following:

- *Always remove watches, rings, and bracelets.*
- *Roll coat and shirt sleeves up above the elbows.*
- *Never use a file without a properly fitted handle.*

To File Work in a Lathe

Use the following procedure to file work in a lathe:

1. Cover the lathe bed with paper, Fig. 27-8.

2. Set the lathe at twice the speed used for turning.

3. Adjust the work freely between centers. (Use a rotating dead center if it is available.)

4. Disengage the lead screw by placing the reverse lever in a neutral position.

5. Select a suitable long-angle lathe or mill file.

6. Grip the lathe file handle in the left hand, using the fingers of the right hand to balance and guide the file at the point, Fig. 27-9.

7. Move the file along the work after each stroke, so that each cut overlaps approximately one-half the width of the file.

Fig. 27-8 Cover the lathe bed with paper to prevent small filings from getting into lathe slides and causing damage. (Courtesy of Kelmar Associates)

Fig. 27-9 The left-hand method of filing on a lathe is safer. (Courtesy of Kelmar Associates)

8. Use long strokes and apply pressure only on the forward stroke.

9. Use approximately 35 to 40 strokes/min.

10. If the file loads up with cuttings, clean it with a file brush and rub a little chalk on the file teeth to reduce pinning.

POLISHING

Polishing is a finishing operation that generally follows filing to improve the surface finish on the work. The finish obtained on the diameter is directly related to the coarseness of the **abrasive cloth** used. A fine-grit abrasive cloth produces the best surface finish. Aluminum oxide abrasive cloth should be used for polishing most ferrous metals, while silicon carbide abrasive cloth is used on nonferrous metals.

To Polish Work in a Lathe

Use the following procedure to polish work in a lathe.

1. Be sure that all loose clothing is tucked in to prevent it from becoming caught by the revolving work.

2. Cover the lathe bed with paper, as shown in Fig. 27-8.

3. Set the lathe at twice the filing speed or four times the turning speed and disengage the lead screw and feed rod.

4. Mount work between centers freely with very little end play, or use a revolving dead center.

Fig. 27-10 A high surface finish can be produced with abrasive cloth. (Courtesy of Kelmar Associates)

5. Use a piece of 80 to 100 grit abrasive cloth about 1 in. (25.4 mm) wide for rough polishing.

6. Hold the abrasive cloth as shown in Fig. 27-10 to prevent the top end of the abrasive cloth from wrapping around the work and injuring the fingers.

7. Hold the long end of the abrasive cloth securely with one hand while the fingers of the other hand press the cloth against the diameter, Fig. 27-10.

8. Move the abrasive cloth back and forth at a steady rate along the diameter to be polished.

9. Use a piece of 120 to 180 grit abrasive cloth for finish polishing.

10. Apply a few drops of oil to the abrasive cloth for the final passes along the diameter.

SUMMARY

- Work should generally be cut to size on a lathe in two cuts—a roughing cut and a finishing cut.
- Use one roughing cut to remove up to a half-inch from the diameter; use two roughing cuts if more than a half-inch is to be removed.
- Use a blade micrometer or blade vernier caliper to measure a narrow-groove diameter.
- Work should never be measured while it is still revolving.
- In polishing operations, the surface finish depends on the coarseness of the abrasive cloth.

KNOWLEDGE REVIEW

Parallel Turning

1. When can a parallel diameter be produced on a lathe?

2. State three purposes for a light trial cut before taking a rough or finish cut from a diameter.

3. How close to finished size should the rough cut be taken?

4. What feed should be used for rough turning?

5. What is the purpose of finish turning?

6. What feed should be used for finish turning?

7. Name two methods of measuring the diameter of narrow grooves.

8. How can you tell when the outside caliper setting is correct?

Filing in a Lathe

9. Why is too much filing on a diameter not recommended?

10. How can damage to the lathe be prevented during filing and polishing?

11. List three safety precautions which should be observed when filing.

Polishing

12. What purpose does polishing serve?

13. Explain how the abrasive cloth should be held for polishing in a lathe.

UNIT 28

Knurling, Grooving, Shoulder Turning

Many operations can be performed on a workpiece in a lathe to improve the appearance of the part, increase its strength, or provide clearance for mating parts. Operations such as knurling, grooving, and shoulder turning are a few of the special operations which can be performed on work between centers and work held in a chuck.

OBJECTIVES

After completing this unit, you should be able to:

- Set up and use various knurling tools to produce a good pattern on the work.
- Cut various types of grooves in work held between centers or in a chuck.
- Machine square, filleted, and beveled shoulders to within 1/64 in. (0.39 mm) accuracy.

KEY TERMS

grooving   knurling   shoulder turning

● KNURLING

Knurling is a process of impressing diamond-shaped or straight indentations on the surface of work. The purposes of knurling are to improve the appearance of the work, provide a better grip, and serrate surfaces where parts are locked or keyed together. The operation is done by forcing a knurling tool containing a set of hardened, cylindrical, patterned rolls against the surface of revolving work. Diamond- and straight-pattern rolls in three styles (fine, medium, and coarse) are illustrated in Fig. 28-1.

Knurling tools have a heat-treated body which is held in the toolpost and a set of hardened rolls mounted in a movable head. The knurling tool shown in Fig. 28-2A contains one set of rolls mounted in a self-centering head. The knurling tool in Fig. 28-2B contains three sets of rolls (fine, medium, and coarse) mounted in a revolving head which pivots on a hardened steel pin.

The knurling tools for conventional lathes, Fig. 28-3A, and for CNC machines, Fig. 28-3B, have interchangeable knurling rolls which can be changed quickly from one pattern to another. The knurling rolls in these holders are set at a 30° angle to reduce the knurling pressure and create minimum stress to the lathe spindle and the workpiece.

Fig. 28-2A A knurling tool with one set of rolls in a self-centering head. (Courtesy of J.H. Williams Co., Division of Snap-On Inc.)

Fig. 28-2B A knurling tool with three sets of rolls (coarse, medium, and fine) in a revolving head. (Courtesy of J.H. Williams Co., Division of Snap-On Inc.)

Fig. 28-3A A knurling tool with interchangeable rolls for use in a conventional lathe. (Courtesy of Dorian Tool International)

Fig. 28-1 Coarse, medium, and fine diamond (top) and straight pattern knurling rolls(bottom). (Courtesy of J.H. Williams Co., Division of Snap-On Inc.)

Fig. 28-3B A knurling tool with interchangeable rolls for use with quick-change toolposts. (Courtesy of Dorian Tool International)

To Knurl a Diamond Pattern

Use the following procedures to knurl a diamond pattern.

1. Mount the work between centers with the required length of knurled section marked on the work.

> Use a revolving center in the tailstock only if it has been designed for this purpose.

2. Set the lathe at one-quarter the speed used for turning.

3. Set the quick-change gearbox for a feed of .010 to .020 in. (0.25 to 0.5 mm).

4. Set the center of the floating head of the knurling tool even with the dead center of the lathe, Fig. 28-4.

5. Adjust the knurling tool so that it is at right angles to the work, Fig. 28-5.

6. Tighten the toolpost screw *securely* so that the knurling tool will not move during the knurling operation.

7. Set the knurling tool near the end of the work so that only one-half to three-quarters of the width of the knurling roll is on the work, Fig. 28-6.

> This generally results in easier starting and produces a better knurling pattern.

Fig. 28-5 The knurling tool should be set at 90° to the surface to be knurled. (Courtesy of Kelmar Associates)

Fig. 28-6 For starting a knurling pattern, set the knurling tool so that about 1/2 to 3/4 of the roll width is over the work. (Courtesy of Kelmar Associates)

8. Force the knurling tool into the work approximately .025 in. (0.65 mm) and start the lathe.

OR

Start the lathe and then force the knurling tool into the work until the diamond pattern comes to a point.

9. Stop the lathe and examine the pattern. If necessary, reset the knurling tool.

   a. If the pattern is incorrect, Fig. 28-7, it is usually because the knurling tool is not set on center.

   b. If the knurling tool is on center and the pattern is not correct, it is generally due to worn knurling rolls. In this case, it

Fig. 28-4 The knurling tool should be set to the center of the floating head. (Courtesy of Kelmar Associates)

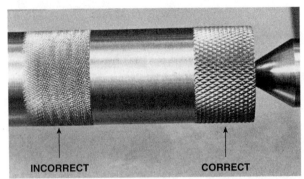

Fig. 28-7 Correct and incorrect knurling patterns. (Courtesy of Kelmar Associates)

Fig. 28-8 Disengaging the automatic feed during a knurling operation will damage the knurling pattern. (Courtesy of Kelmar Associates)

will be necessary to set the knurling tool slightly off square so that the corner of the knurling rolls can start the pattern.

10. Once the pattern is correct, engage the automatic carriage feed and apply cutting fluid to the knurling rolls.

11. Knurl to the proper length.

Do not disengage the feed until the full length has been knurled, otherwise rings will form on the knurled pattern, Fig. 28-8.

12. If the knurling pattern is not to a point, reverse the lathe feed and take another pass across the work.

● GROOVING

Grooving is an operation often referred to as recessing, undercutting, or necking. It is often done at the end of a thread, at the side of a shoulder, or for appearance. Grooves may be any desired shape, but they are generally square, round, or V-shaped, Fig. 28-9.

To Cut a Groove

Use the following procedure to cut a groove:

1. Lay out the location of the groove, using a center punch and layout tools.

2. Set the lathe to one-half the turning speed.

3. Mount the proper-shaped toolbit in the toolholder.

4. Set the cutting tool to center and at 90° to the work.

5. Locate the toolbit on the work at the position where the groove is to be cut.

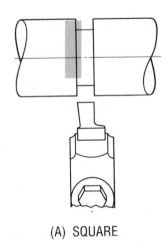
(A) SQUARE

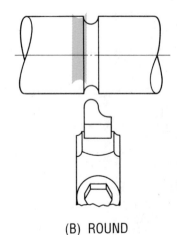
(B) ROUND

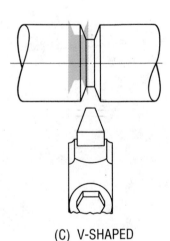
(C) V-SHAPED

Fig. 28-9 Three types of grooves used in machine shop work. (Courtesy of Kelmar Associates)

6. Start the lathe and use the crossfeed handle to feed the cutting tool toward the work until the toolbit lightly marks the work.

7. Hold the crossfeed handle in position and then set the graduated collar to zero.

8. Calculate how far the crossfeed screw must be turned to cut the groove to the proper depth.

9. Apply cutting fluid frequently and groove the work to the proper depth at a steady feed rate.

10. Stop the lathe and check the depth of the groove with outside calipers or a knife-edge vernier caliper.

11. It is desirable to move the carriage by hand a little to the right and left while grooving to overcome chatter.

 Wear safety goggles when grooving work in a lathe.

● SHOULDER TURNING

Whenever more than one diameter is machined on a shaft, the section joining each diameter is called a shoulder or step. The square, filleted, and chamfered shoulders are most commonly used in machine shop work, Fig. 28-10. **Shoulder turning** is the process of machining shoulders to the proper size and shape.

To Machine a Square Shoulder

Use the following procedure to machine a square shoulder:

1. Lay out the length of the shoulder with a center-punch mark or cut a light groove at this point with a sharp toolbit, Fig. 28-11.

2. Rough and finish turn the small diameter to within 1/32 in. (0.8 mm) of the required length.

3. Mount a facing tool and set it for the facing operation.

4. Start the lathe and feed the cutting tool in until it lightly marks the small diameter near the shoulder.

5. Note the reading on the crossfeed graduated collar, Fig. 28-12.

6. Turn the carriage handwheel to start a light cut.

7. Face the shoulder by turning the crossfeed handle *counterclockwise*, cutting from the center outward.

8. Return the crossfeed handle to the original graduated collar setting.

9. Repeat steps 6 and 8 until the shoulder is to the correct length.

To Machine a Filleted Shoulder

Use the following procedure to machine a filleted shoulder:

1. Lay out the length of the shoulder with a center punch mark or by cutting a light groove at this point, as shown in Fig. 28-11.

2. Rough and finish turn the small diameter to the correct length *minus the radius to be cut.*

For example, a 3 in. (76 mm) length with a 1/8 in. (3 mm) radius should be turned 2-7/8 in. (72 mm) long.

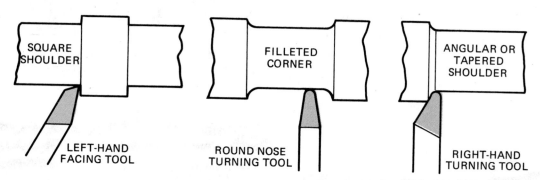

Fig. 28-10 Three common types of shoulders and the turning tools used for each. (Courtesy of Kelmar Associates)

Fig. 28-11A Marking the length of a shoulder with a center punch mark. (Courtesy of Kelmar Associates)

Fig. 28-11B Marking the length of a shoulder by cutting a light groove on the workpiece. (Courtesy of Kelmar Associates)

Fig. 28-12 The cutting tool can be returned to the same position while facing a shoulder if the graduated collar setting is used. (Courtesy of Kelmar Associates)

Fig. 28-13 A radius tool is used to produce a filleted shoulder. (Courtesy of Kelmar Associates)

3. Mount the correct radius toolbit and set it to center, Fig. 28-13.

4. Set the lathe for one-half the turning speed.

5. Start the lathe and feed the cutting tool in until it *lightly marks* the small diameter near the shoulder.

6. Slowly feed the cutting tool sideways with the carriage handwheel until the shoulder is cut to the correct length.

To Machine Angular Shoulders

Long angular shoulders are generally produced by swiveling the compound rest to the required angle and then feeding the compound rest screw. Short chamfers or bevels can be cut by setting the side of the toolbit to the required angle, Fig. 28-14, and feeding it against the workpiece, Fig. 28-15.

Fig. 28-14 Using a bevel protractor to set the side of a cutting tool to an angle. (Courtesy of Kelmar Associates)

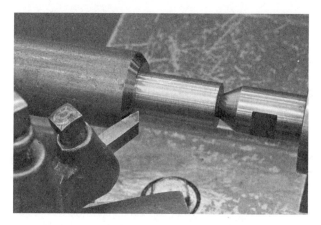

Fig. 28-15 Machining a beveled shoulder with the side of the toolbit. (Courtesy of Kelmar Associates)

SUMMARY

- Knurling, grooving, and shoulder turning can be performed on work between centers or work held in a chuck.

- In knurling, indentations are impressed on the surface of the work.

- Grooves cut into metal workpieces are usually square, round, or V-shaped.

- Shoulders need to be turned when more than one diameter is machined on a shaft. The square, filleted, and chamfered shapes are the most common.

KNOWLEDGE REVIEW

Knurling

1. What are three purposes of knurling?
2. Define the process of knurling.
3. Explain the process of knurling.
4. Explain how the knurling tool should be set up.
5. If the knurling pattern is not correct, how can it be corrected?
6. Why should the lathe feed not be disengaged during the knurling operation?

Grooving

7. For what purpose are grooves used?
8. How can the depth of cut be gaged when grooving?

Shoulder Turning

9. Name three types of shoulders used in machine shop work.
10. To what length should the small diameter be turned when cutting a filleted shoulder?
11. How can short, angular shoulders be cut?

U N I T
29

Taper Turning

Tapers are widely used in the mechanical trades to quickly and accurately align mating parts. Common examples of tapers in use are the wheel axles on automobiles, the shank on lathe centers, and spindles on machine tools. Most revolving spindles on machine tools have a taper hole to hold accessories or cutting tools. This equipment can be quickly positioned and its accuracy maintained, yet all it takes is a slight tap to remove it. Every machinist should know how to machine and fit tapers to suit the variety of uses they have in machine shop work.

OBJECTIVES

After completing this unit, you should be able to:
- Describe the purpose of self-holding and self-releasing tapers.
- Calculate and cut short tapers using the compound rest.
- Calculate and machine tapers by offsetting the tailstock.
- Cut tapers using a taper attachment.

KEY TERMS

inch taper   metric taper   taper

TAPERS

A **taper** may be defined as a uniform increase or decrease in the diameter of a piece of work measured along its length. **Inch tapers** are expressed in taper per foot or taper per inch. **Metric tapers** are expressed as a ratio of 1 mm per unit of length; for example 1:20 taper would have a 1 mm change in diameter in 20 mm of length. A taper provides a rapid and accurate method of aligning machine parts and an easy method of holding tools such as twist drills, lathe centers, and reamers. The American Standards Association classifies tapers used on machines and tools as self-holding tapers and self-releasing or steep tapers.

Self-holding tapers are those that remain in position due to the wedging action of the taper. The inch tapers of this series are composed of the Morse, Brown and Sharpe, and 3/4 in./ft., see machine tapers, Table 29-1. *Steep* or *self-releasing tapers*, such as those used on milling machine arbors and accessories, are held in the machine by a drawbolt and are driven by lugs or keys.

Tapers are designed to accurately align parts when assembled properly. In order to preserve the accuracy and efficiency of tapers (holes and shanks), it is very important that tapers be free of chips, dirt, burrs, or nicks. Tapers can be permanently damaged if they are not kept clean. It is very important to wipe tapers dry before assembly because an oily taper will not hold securely.

Inch Tapers

Some of the tapers included in Table 29-1 are taken from the Morse, and Brown and Sharpe

Table 29-1 Basic dimensions of self-holding tapers

Number of Taper	Taper per Foot	Diameter at Gauge Line (A)	Diameter at Small End (D)	Length (P)	Series Origin
1	.502	.2392	.200	$^{15}/_{16}$	Brown and Sharpe taper series
2	.502	.2997	.250	$1^{3}/_{16}$	
3	.502	.3752	.3125	$1^{1}/_{2}$	
*0	.624	.3561	.252	2	Morse taper series
1	.5986	.475	.369	$2^{1}/_{8}$	
2	.5994	.700	.572	$2^{9}/_{16}$	
3	.6023	.938	.778	$3^{3}/_{16}$	
4	.6233	1.231	1.020	$4^{1}/_{16}$	
$4^{1}/_{2}$	.624	1.500	1.266	$4^{1}/_{2}$	
5	.6315	1.748	1.475	$5^{3}/_{16}$	
6	.6256	2.494	2.116	$7^{1}/_{4}$	
7	.624	3.270	2.750	10	
200	.750	2.000	1.703	$4^{3}/_{4}$	3/4 in. taper per foot series
250	.750	2.500	2.156	$5^{1}/_{2}$	
300	.750	3.000	2.609	$6^{1}/_{4}$	
350	.750	3.500	3.063	7	
400	.750	4.000	3.516	$7^{3}/_{4}$	
450	.750	4.500	3.969	$8^{1}/_{2}$	
500	.750	5.000	4.422	$9^{1}/_{4}$	
600	.750	6.000	5.328	$10^{3}/_{4}$	
800	.750	8.000	7.141	$13^{3}/_{4}$	
1000	.750	10.000	8.953	$16^{3}/_{4}$	
1200	.750	12.000	10.766	$19^{3}/_{4}$	

*Taper #0 is not a part of the self-holding taper series. It has been added to complete the Morse taper series.

series. The following describes these and others that are used in machine shop work.

1. *Morse taper*, approximately 5/8 in. tpf (taper/ft.), is a standard taper used for twist drills, reamers, end mills, and lathe center shanks. The Morse taper has eight standard sizes from 0 to 7.

2. *Brown and Sharpe taper*, 1/2 in. tpf (taper/ft.), is a standard taper used in all Brown and Sharpe machines, cutters, and drive shanks.

3. *Jarno taper*, .600 in. tpf, is used for some machine spindles.

4. *Standard taper pin*, 1/4 in. tpf, is a standard for all tapered pins used in the fabrication of machinery. They are listed by numbers from 0 to 10.

5. *Standard milling machine taper*, 3-1/2 in. tpf, is a self-releasing taper used exclusively on milling machine spindles and equipment.

Metric Tapers

Metric tapers are expressed as a ratio of one millimeter per unit of length. In Fig. 29-1, the work tapers one millimeter in a distance of twenty millimeters. This taper would then be expressed as a ratio of 1:20 and would be indicated on a drawing as Taper = 1:20.

Since the work tapers 1 mm in 20 mm of length, the diameter at a point 20 millimeters from the small diameter, d, will be 1 mm larger $(d + 1)$.

Some common metric tapers are:

Milling machine spindle—1:3.43,

Morse taper shank—approximately 1:20,

Tapered pins and pipe threads—1:50.

● INCH TAPER CALCULATIONS

Most inch tapers cut on workpieces are expressed in taper per foot, taper per inch, or de-

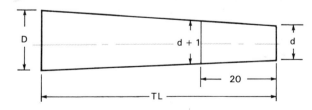

Fig. 29-1 The characteristics of a metric taper. (Courtesy of Kelmar Associates)

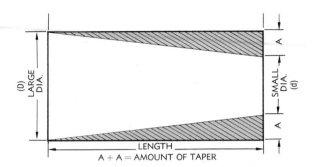

Fig. 29-2 The main parts of an inch taper. (Courtesy of Kelmar Associates)

grees. If this information is not supplied, it is generally necessary to calculate the taper per foot of the workpiece. Taper per foot is the amount of difference between the large diameter and the small diameter of the taper in 12 in. of length. For example, if the tapered section on a piece of work is 12 in. long and the large diameter is 1 in. and the small diameter is 1/2 in., the taper per foot would be the difference between the large and small diameters, or 1/2 in. The main parts of an inch taper are: the amount of taper, the length of the tapered part, the large diameter, and the small diameter, Fig. 29-2.

Since not all tapers are 12 in. long, if the small diameter, large diameter, and length of the tapered section are known, then the taper per foot can be calculated by applying the following formula;

$$tpf = \frac{(D - d) \times 12}{T.L.}$$

where:

D = diameter at the large end of the taper

d = diameter at the small end of the taper

$T.L.$ = total length of the tapered section

EXAMPLE: To calculate the taper per foot of the workpiece shown in Fig. 29-3:

$$tpf = \frac{\left(1\frac{1}{4} - 1\right) \times 12}{3}$$

$$= \frac{\frac{1}{4} \times 12}{3}$$

$$= 1 \text{ in.}$$

If taper per inch is required, divide taper per foot by 12. For example, the 1 in. tpf of the previous problem would have .083 taper/in. (1 in. ÷ 12 in.).

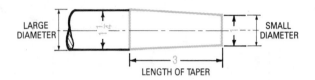

Fig. 29-3 The dimensions of an inch-tapered work-piece. (Courtesy of Kelmar Associates)

After the taper per foot has been calculated, no further calculations are necessary if the taper is to be cut on a lathe with a taper attachment. If the taper is to be cut by the tailstock offset method, the amount to offset the tailstock must be calculated.

METRIC TAPER CALCULATIONS

If the small diameter, d, the unit length of taper, k, and the total length of taper, l, are known, the large diameter, D may be calculated.

In Fig. 29-4, the large diameter, D, will be equal to the small diameter plus the amount of taper. The amount of taper for the unit length, k, is $(d + 1) - d$, or 1 mm.

Therefore the amount of taper per millimeter of unit length = $1/k$

The *total amount of taper* will be the taper per millimeter, $1/k$, multiplied by the total length of taper, (l).

$$\text{Total taper} = 1/k \times l \text{ or } l/k$$
$$D = d + \text{total amount of taper}$$
$$D = d + l/k$$

EXAMPLE: Calculate the large diameter, (D), for a 1:30 taper having a small diameter of 10 mm and a length of 60 mm.

SOLUTION: Since taper is 1:30

$$k = 30$$
$$D = d + l/k$$
$$= 10 + {}^{60}\!/_{30}$$
$$= 10 + 2$$
$$= 12 \text{ mm}$$

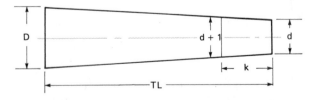

Fig. 29-4 The dimensions of a metric taper. (Courtesy of Kelmar Associates)

TAILSTOCK OFFSET CALCULATIONS

The tailstock offset method is often used to produce tapers in a lathe on work turned between centers when a taper attachment is not available. To produce a taper, the amount to offset the tailstock must first be calculated by applying one of the following formulas:

Inch Tailstock Offset Calculations

$$\text{Tailstock offset} = \frac{tpf \times O.L.}{12 \times 2}$$

where:

tpf = taper per foot
$O.L.$ = Overall length of work
12 = inches per foot
2 = the offset is taken from the center line of the work

To calculate the tailstock offset for a 10 in. long piece of work which has a 3/4 in. taper per foot:

$$\text{Tailstock offset} = \frac{{}^{3}\!/_{4} \times 10}{24}$$
$$= 3/4 \times 1/24 \times 10$$
$$= 5/16 \text{ in.}$$

In cases where it is not necessary to find the taper per foot, a simplified formula can be used to calculate the amount of tailstock offset:

$$\text{Tailstock offset} = \frac{O.L.}{T.L.} \times \frac{(D - d)}{2}$$

where:

$O.L.$ = overall length of work
$T.L.$ = length of the tapered section
D = diameter at the large end of the taper
d = diameter at the small end of the taper

EXAMPLE: Using the simplified formula to find the tailstock offset for the following piece of work: large diameter is 1 in., small diameter is 23/32 in., the length of the taper is 6 in., and the overall length of the work is 18 in.

SOLUTION:

$$\text{Tailstock offset} = \frac{18}{6} \times \frac{(1 - {}^{23}\!/_{32})}{2}$$
$$= \frac{18}{6} \times \frac{9}{64}$$
$$= \frac{27}{64} \text{ in.}$$

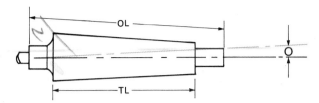

Fig. 29-5 Using the tailstock offset method to turn a metric taper. (Courtesy of Kelmar Associates)

Metric Tailstock Offset Calculations

If the metric taper in Fig. 29-5 is to be turned by off-setting the tailstock, the amount of offset is calculated as follows:

$$\text{Offset, } (o) = \frac{D-d}{2 \times l} \times L$$

where:

D = large diameter
d = small diameter
l = length of taper
L = length of work

EXAMPLE: Calculate the tailstock offset required to turn a 1:30 taper, 60 mm long on a workpiece 300 mm long. The small diameter of the tapered section is 20 mm.

SOLUTION:

Large diameter of taper, $D = d + \frac{1}{k}$

$$= 20 + \frac{60}{30}$$
$$= 20 + 2$$
$$= 22 \text{ mm}$$

$$\text{Tailstock offset} = \frac{D-d}{2 \times l} \times L$$
$$= \frac{22-20}{2 \times 60} \times 300$$
$$= \frac{2}{120} \times 300$$
$$= 5 \text{ mm}$$

● TAPER ATTACHMENTS

Turning a taper using the taper attachment provides many advantages in producing both internal and external tapers. The most important are:

1. Setup is simple. The taper attachment is easy to connect and disconnect.

2. Live and dead centers are not adjusted, so center alignment is not disturbed.

3. Greater accuracy can be achieved, since one end of the guide bar is graduated in degrees or in inches of taper per foot, and the other end in a ratio of 1 mm per unit of length.

4. The taper can be produced between centers or on work held in a chuck or a collet, regardless of the length of the work.

5. Internal tapers can be produced with the same taper setup as for external tapers.

6. A great range of tapers can be produced, and this is of special advantage when production is a factor and various tapers are required on a unit.

There are two common types of taper attachments in use:

a. the plain taper attachment, (see Fig. 29-7)
b. the telescopic taper attachment.

To use the plain taper attachment, the crossfeed screw nut must be disengaged from the crossslide. When a telescopic taper attachment is used, the crossfeed screw is not disengaged, and the depth of cut can be set by the crossfeed handle.

Inch Taper Attachment Offset Calculations

Most tapers cut on a lathe with the taper attachment are expressed in taper per foot. If the taper per foot of the taper on the workpiece is not given, it may be calculated by using the following formula:

$$\text{Taper per foot} = \frac{(D-d) \times 12}{T.L.}$$

EXAMPLE: Calculate the taper per foot for a taper with the following dimensions: large diameter, D is $1\frac{3}{8}$ in., small diameter, d, is 15/16 in., length of tapered section, T.L. is 7 in.

$$tpf = \frac{(1\frac{3}{8} - 15/16) \times 12}{7}$$
$$= \frac{7/16 \times 12}{7}$$
$$= 3/4 \text{ in.}$$

Metric Taper Attachment Offset Calculations

When the taper attachment is used to machine a taper, the amount the guide bar is set over may be determined as follows:

- If the angle of taper is given on the print, set the guide bar to one-half the included angle, Fig. 29-6.
- If the angle of taper is not given on the print, use the following formula to find the amount of guide bar setover:

$$\text{Guide bar setover} = \frac{D-d}{2} \times \frac{L}{l}$$

where:

D = large diameter of taper
d = small diameter of taper
l = length of taper
L = length of taper attachment guide bar

EXAMPLE: Calculate the amount of setover for a 500 mm long guide bar to machine a 1:50 × 250 mm long taper on a workpiece. The small diameter of the taper is 25 mm.

$$\text{Large diameter of taper, } D = d + \frac{1}{k}$$
$$= 25 + \frac{250}{50}$$
$$= 30 \text{ mm}$$

$$\text{Guide bar setover} = \frac{D-d}{2} \times \frac{L}{l}$$
$$= \frac{30-25}{2} \times \frac{500}{250}$$
$$= \frac{5}{2} \times 2$$
$$= 5 \text{ mm}$$

● TAPER TURNING

Tapers can be cut on a lathe by using the taper attachment, offsetting the tailstock, and by setting the compound rest to the angle of the taper. The method used will depend on the length of the taper, the angle of the taper, how the work is held in the lathe, and the number of pieces to be cut. Taper attachments can be used to cut internal and external tapers on workpieces of various lengths.

To Cut a Taper Using a Taper Attachment

Use the following procedure to cut a taper using a taper attachment.

The procedure for machining a taper using either a plain or telescopic taper attachment is basically the same with only minor adjustments required. The procedure for setting the plain taper attachment, Fig. 29-7, and cutting a taper is as follows:

1. Clean and oil the guide bar.

2. Loosen the guide bar locknuts so that it is free to move on the base plate.

3. By adjusting the locking screws, offset the end of the guide bar the required amount, or for inch tapers, set the taper attachment to the required taper per foot.

4. Tighten the guide bar locknuts.

5. Swivel the compound rest so that it is at about 30° to the cross-slide, Fig. 29-8.

6. Set the cutting tool to center and tighten the toolpost securely.

7. Mount the work in the lathe and mark the length to be tapered.

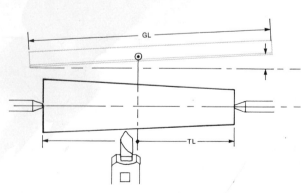

Fig. 29-6 Using the taper attachment to turn a metric taper. (Courtesy of Kelmar Associates)

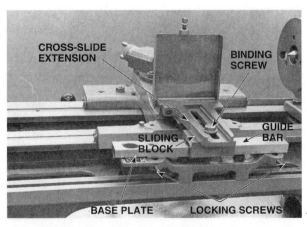

Fig. 29-7 The main parts of a plain taper attachment. (Courtesy of Kelmar Associates)

Fig. 29-8 The compound rest should be set at 30° for taper turning. (Courtesy of Kelmar Associates)

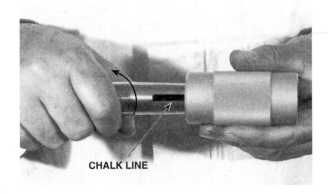

CHALK LINE

Fig. 29-9 Using a taper ring gage to check the accuracy of a taper. (Courtesy of Kelmar Associates)

8. Feed the cutting tool in until it is about 1/4 in. (6 mm) from the diameter of the work.

9. Remove the binding screw that connects the cross-slide and the crossfeed screw nut.

10. Use the binding screw to connect the cross-slide extension to the sliding block using two-finger pressure on the wrench.

11. Insert a plug in the hole where the binding screw was removed to keep chips and dirt from damaging the crossfeed screw.

12. Move the carriage until the cutting tool clears the right-hand end of the work by about 1/2 in. (12.7 mm).

13. Take a light trial cut for about 1/16 in. (1.5 mm) and check the taper for size.

14. Set the depth of the roughing cut about 1/16 in. (1.5 mm) larger than the finish size and rough cut the taper to the required length.

15. Check the taper for fit, Fig. 29-9. See the section "To Check a Taper With a Ring Gage" on page 283 in this unit.

16. Readjust the taper attachment setting if necessary and take a light trial cut from the taper.

17. When the taper fit is correct, cut the taper to size.

Tailstock Offset Method

The tailstock offset method of cutting tapers *should only be used* when a lathe is not equipped with a taper attachment and the work is mounted between centers. The tailstock center must be moved out of line with the headstock center

enough to produce the desired taper, Fig. 29-10. Since the tailstock can only be offset a certain amount, the range of tapers which can be cut by this method is limited. To offset the tailstock, use the following procedure:

To Offset the Tailstock

1. Calculate the amount the tailstock must be offset to cut the desired taper on the work.

2. Loosen the tailstock clamp nut.

3. Loosen one tailstock adjusting screw and tighten the opposite one until the tailstock offset is correct, Fig. 29-11.

4. Tighten the adjusting screw that was loosened and recheck the offset with a rule.

5. Correct the setting if necessary, and then tighten the tailstock clamp nut.

6. Mount the work between centers and cut the taper to size by following operations 12

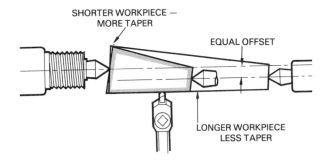

SHORTER WORKPIECE — MORE TAPER

EQUAL OFFSET

LONGER WORKPIECE LESS TAPER

Fig. 29-10 The amount of taper cut using the tailstock offset method varies with the length of each workpiece. (Courtesy of Kelmar Associates)

Fig. 29-11 Checking the offset of the tailstock for taper turning. (Courtesy of Kelmar Associates)

to 15 of the section "To Cut a Taper Using a Taper Attachment" on page 281 in this unit.

Compound Rest Method

The compound rest is used to cut short, steep tapers that are given in degrees. The compound rest must be set to the required angle, and then the cutting tool is advanced along the taper using the compound rest feed handle.

Use the following procedure to cut a taper using the compound rest.

1. Check the print for the angle of the taper in degrees.

2. Loosen the compound-rest locknuts.

3. Swivel the compound rest to the required angle, Fig. 29-12.

> - If the included angle is given as in Fig. 29-12A, set the compound rest to one-half the included angle.
> - If the angle is given on one side only, as in Fig. 29-12B, set the compound rest to that angle.

4. Tighten the compound rest lock nuts, using only a two-finger pull on the wrench to avoid stripping the thread on the compound-rest studs, Fig. 29-13.

5. Set the toolbit on center and then swivel the toolholder so that is it at 90° to the compound rest, Fig. 29-14.

6. Turn the compound-rest handle counterclockwise to bring the compound-rest slide back far enough for the length of the taper to be cut.

(A) INCLUDED ANGLE GIVEN ON DRAWING

(B) ANGLE GIVEN ON ONE SIDE ONLY

Fig. 29-12 The direction for swinging the compound rest to cut various angles. (Courtesy of Kelmar Associates)

Fig. 29-13 When tightening the compound rest locknuts, use only a two-finger pull on the wrench. (Courtesy of Kelmar Associates)

> If the taper length is 2 in., bring the compound rest back $2\frac{1}{2}$ in. to allow for clearance.

7. Bring the toolbit close to the diameter to be cut using the carriage handwheel and crossfeed handle.

8. Tighten the carriage lock to prevent movement during the tapering operation, Fig. 29-15.

9. Cut the taper by turning the compound rest feed screw.

10. Check the taper for size and angle.

Fig. 29-14 Set the toolholder at 90° to the side of the compound rest and the tip of the cutting tool on center. (Courtesy of Kelmar Associates)

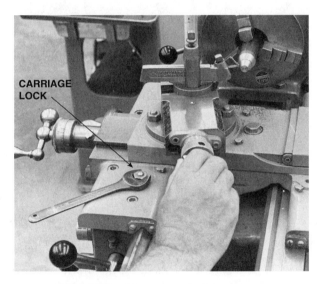

CARRIAGE LOCK

Fig. 29-15 Tighten the carriage lock to prevent movement during tapering with the compound rest. (Courtesy of Kelmar Associates)

● **CHECKING A TAPER**

External tapers can be checked for accuracy of size or fit by using a taper ring gage, a standard micrometer, or a special taper micrometer.

To Check a Taper with a Ring Gage

To check the accuracy of a taper use the ring gage as described in the following procedure:

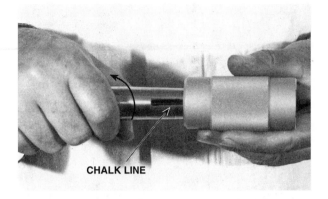

CHALK LINE

Fig. 29-16 Checking a taper for accuracy using a taper ring gage. (Courtesy of Kelmar Associates)

1. Draw three equally spaced light lines with chalk or mechanics' blue along the length of the taper, Fig. 29-16.

2. Insert the taper into the gage and turn *counterclockwise* one-half turn, then remove it for inspection.

3. If the chalk is rubbed from the whole length of the taper, the taper is correct.

4. If the chalk lines are rubbed from only one end, the taper is incorrect.

5. Make slight adjustments to the taper setup and take trial cuts until the taper is correct.

To Check an Inch Taper with a Standard Micrometer

To check the accuracy of a taper use the micrometer as described in the following procedure:

1. Calculate the amount of taper per inch of the taper.

2. Clean the tapered section of the work and apply layout dye as shown in Fig. 29-17.

3. Lay out two lines exactly 1 in. apart, Fig. 29-17.

4. Measure the taper with a micrometer at both lines so that the left edge of the micrometer anvil and spindle just touch the line.

5. Subtract the difference between the two readings, and compare the answer with the required taper per inch.

6. If necessary, adjust the taper attachment setting to correct the taper.

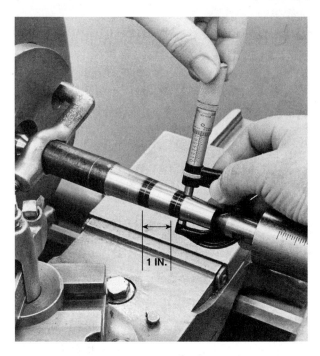

Fig. 29-17 Checking the accuracy of a taper using a micrometer. (Courtesy of Kelmar Associates)

Greater accuracy is possible if the length of the work permits the lines to be laid out 2 or 3 in. apart. Determine the difference in diameters at the lines and divide by the distance between the lines (in inches) to determine the taper per inch.

To Check a Metric Taper with a Metric Micrometer

To check a metric taper use the micrometer as described in the following procedure:

1. Check the drawing for the taper required.
2. Clean the tapered section of the work and apply layout dye.
3. Lay out two lines on the taper which are the same distance apart as the second number in the taper ratio.

EXAMPLE: If the taper is 1:20, the lines would be 20 mm apart.

If the work is long enough, lay out the lines at double or triple the length of the tapered section and increase the difference in diameters by the appropriate amount. For instance, on a 1:20 taper the lines may be laid out 60 mm apart or three times the unit length of the taper. Therefore the difference in diameters would be 3 × 1, or 3 mm. This will give a more accurate check of the taper.

4. Measure the diameters carefully with a metric micrometer at the two lines.

The difference between these two diameters should be 1 mm for each unit of length.

5. If necessary, adjust the taper attachment setting to correct the taper.

SUMMARY

- A taper is a uniform increase or decrease in the workpiece diameter measured along the length of the work.
- Self-holding tapers are held in position by the wedging action of the taper. Self-releasing tapers are held in the machine by a drawbolt.
- Inch tapers are expressed in taper per foot, taper per inch, or degrees. Metric tapers are expressed as a ratio of one millimeter per unit of length.
- On a lathe, tapers can be cut by using the taper attachment, offsetting the tailstock, or setting the compound rest to the taper angle.
- Tapers can be checked for accuracy using a taper ring gage, a standard micrometer, or a special taper micrometer.

KNOWLEDGE REVIEW

Tapers

1. Why are tapers so important to the mechanical trades?
2. Name three places where tapers are used.
3. Define a taper.
4. How are tapers generally expressed?
5. Why is it very important to keep tapers clean?

6. Name four common inch tapers and give the taper per foot for each.

7. A tapered pin has a taper of 3 mm in 120 mm. How would this taper be indicated on a drawing?

8. For three Morse tapers, state the amount of taper for each.

Taper Calculations

9. Name four parts of a taper.

10. Calculate the taper per foot for the following:

   a. Large diameter = 1.625 in.

   Small diameter = 1.425 in.

   Length of taper = 3 in.

   b. Large diameter = 7/8 in.

   Small diameter = 7/16 in.

   Length of taper = 6 in.

11. Calculate the tailstock offset for the problems in question 10.

   a. the length of the work in problem (a) is 10 in.

   b. the length of the work in problem (b) is 9 in.

12. Calculate the tailstock offset for the following using the simplified tailstock offset formula for inch tapers:

   a. L.D. = 3/4 in., S.D. = 17/32 in.,

   Length of taper = 6 in.,

   Overall length of work = 18 in.

   b. L.D. = 7/8 in., S.D. = 25/32 in.,

   Length of taper = 3½ in.,

   Overall length of work = 10½ in.

13. a. A metric taper pin has a taper of 1:50. If the small diameter of the pin is 6 mm and the pin is 75 mm long, what is the large diameter of the pin?

   b. If the taper is 1:30, the small diameter is 9 mm, and the length is 105 mm, what is the large diameter of the pin?

14. Calculate the tailstock offset for the problems in question 13.

   a. the length of the work in problem (a) is 120 mm.

   b. the length of the work in problem (b) is 150 mm.

Taper Turning

15. List four advantages of cutting a taper using a taper attachment.

16. What is the difference between a plain and a telescopic taper attachment?

17. Name four factors that determine how a taper will be cut in a lathe.

18. How should the compound rest be set when cutting a taper on a plain taper attachment?

19. What adjustments must be made to offset a tailstock?

20. What type of tapers are generally cut with the compound rest?

21. Why should the carriage be locked when cutting tapers with the compound rest?

22. How is the cutting tool fed when cutting tapers using the compound rest?

Checking a Taper

23. State two methods of checking a taper.

24. Describe how a taper may be checked by one of these methods.

UNIT 30

Machining in a Chuck

The operations for machining work in a **chuck** are basically the same as for machining between centers. The work machined in a chuck is generally short in length and usually does not require additional support. However, if a length of more than three times the diameter extends beyond the chuck jaws, the work should be supported by a steady rest or tailstock center. Use a revolving tailstock center whenever possible, because the expansion of the work during the machining operation will not affect this center. Although any type of machining operation can be performed on work held in a chuck, the most common operations are facing, turning to size, and cutting off.

OBJECTIVES

After completing this unit, you should be able to:

- Face work to accurate lengths.
- Turn diameter in a three- and four-jaw chuck to within .001 in. (0.02 mm) accuracy.
- Groove and cut off work held in a chuck.
- Drill, ream, and tap holes in work held in a chuck.

KEY TERMS

chuck   cutting off   drilling   facing

● HINTS ON CHUCK WORK

If possible, all diameters should be machined on a workpiece in one setup before it is removed from a three- or four-jaw chuck. If the work is removed and then replaced in the chuck for further machining, it will probably not run true, or it will take some time to bring it back to the original trueness.

1. Remove the headstock center and the spindle sleeve before mounting a chuck.

2. Clean the lathe spindle and the chuck adapter before mounting the chuck.

3. Tighten the chuck jaws around the most rigid part of the work to prevent distortion of the workpiece.

4. If the work projects more than three times the diameter of the stock, it should be supported by a revolving tailstock center or steady rest, Fig. 30-1.

5. Never grip the work on a diameter smaller than the diameter to be machined unless absolutely necessary.

> When gripping on small diameters, take light cuts to prevent the work from being bent.

6. Tighten the chuck securely so that the workpiece is not moved into the chuck by the pressure of the cut.

7. Always set the toolbit point on center. If the toolbit is set too low, the work may be bent.

Fig. 30-1 The work should not extend more than three times its diameter beyond the chuck jaws unless it is supported. (Courtesy of Kelmar Associates)

8. Position the toolpost on the left side of the compound rest.

> This prevents the jaws from striking the compound rest when cutting close to the chuck.

9. Set the toolholder at 90° to the work or slightly to the right to prevent the tool from "digging in" on a heavy cut.

> The toolholder should be pointing to the left only slightly for finishing cuts or facing operations.

10. Move the carriage until the toolbit is at the extreme left end of travel and the toolbit is within 1/8 in. (317 mm) of the work surface.

> Rotate the chuck one turn by hand to see that the jaws do not strike the compound rest.

11. *Never* use an air hose to clean a chuck.

12. Oil the chuck sparingly, being sure to wipe off excess oil to prevent it from flying when the chuck is in motion.

13. Store chucks in suitable compartments when not in use to prevent damage.

14. *Never* leave a chuck wrench in a chuck.

● THREE-JAW CHUCK WORK

The three-jaw universal chuck is used to hold round or hexagonal work for machining. Since all jaws in the chuck move simultaneously when a chuck wrench is turned, with proper care this chuck should be able to hold work to within .002 in. (0.05 mm) of concentricity, even after long use. As the scroll plate which moves all three jaws becomes worn, it may lose some of its accuracy, Fig. 30-2.

Three-jaw chucks are supplied with two sets of jaws, a regular set and a reversed set, Fig. 30-3. The *regular set* is used to grip outside diameters and also inside diameters of large work. The *reversed set* is used to grip the outside of large diameter work. All chuck jaws are stamped with the same serial number as the chuck and also numbered 1, 2, or 3 to match the chuck slot in to which they fit. They have been fitted and ground true for that chuck and must never be used on another chuck.

Fig. 30-2 The three jaws of a universal chuck mesh with the scroll plate and move in or out simultaneously. (Courtesy of Clausing Industrial, Inc.)

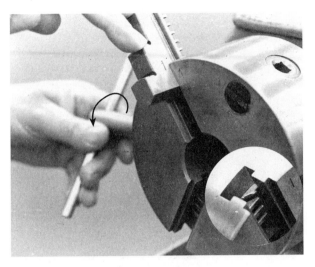

Fig. 30-4 The scroll plate should engage jaw #1 first.

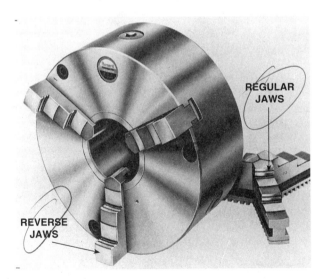

Fig. 30-3 All three jaw chucks are supplied with two sets of jaws; one for external work, the other for internal work. (Courtesy of Clausing Industrial, Inc.)

Assembling the Jaws in a Three-Jaw Universal Chuck

When chuck jaws are changed in a universal chuck, care must be taken to assemble them in the proper order; otherwise the chuck jaws will not run true. All chucks and the jaws supplied with each are marked with the same serial number. It is *very important* to match the serial numbers of the chuck and the jaws being inserted in it.

1. Thoroughly clean the jaws and the jaw slides in the chuck.

2. Turn the chuck wrench clockwise until the start of the scroll thread is almost showing at the back edge of slide 1, Fig. 30-4.

> Be sure that the start of the scroll thread does not protrude into the jaw slide.

3. Insert jaw 1 (in slot 1) and press down with one hand while turning the chuck wrench clockwise with the other, Fig. 30-4.

4. After the scroll thread has engaged in the jaw, continue turning the chuck wrench clockwise until the start of the scroll thread is near the back edge of groove 2.

5. Insert jaw 2 and repeat steps 3 and 4, Fig. 30-5.

6. Insert the third jaw in the same manner, Fig. 30-6.

● FOUR-JAW INDEPENDENT CHUCKS

When it is necessary to hold a workpiece firmly and have it run absolutely true, it is generally mounted in a four-jaw independent chuck. Each jaw can be adjusted independently, and work can be trued to within .0005 in. (0.01 mm) accuracy or less. The jaws of this chuck are re-

Fig. 30-5 Starting jaw #2 into the universal chuck. (Courtesy of Kelmar Associates)

Fig. 30-6 Starting jaw #3 to complete the operation of changing chuck jaws. (Courtesy of Kelmar Associates)

Fig. 30-7A The chuck jaws are set to hold large-diameter work. (Courtesy of Kelmar Associates)

CHUCK RINGS

Fig. 30-7B The jaws in the normal position are holding the inside diameter of a workpiece. (Courtesy of Kelmar Associates)

versible and allow a wide range of work to be gripped either externally or internally, Figs. 30-7A and 30-7B. Most work in industry is held in four-jaw chucks because of their versatility and greater holding power.

Round, square, octagonal, ~~hexagonal~~, and irregularly shaped workpieces can be held in a four-jaw independent chuck. Work can be adjusted to run concentric or off center as required. The face of the chuck has a number of evenly spaced concentric grooves which permit quick and approximate positioning of the chuck jaws.

● TO FACE WORK IN A CHUCK

The purposes of **facing** work in a chuck are to obtain a true flat surface, to have an accurate measuring surface, and to cut the work to length.

1. Set the work in the chuck so that no more than three times its diameter extends beyond the chuck jaws. (Distance X in Fig. 30-8.)

2. Swivel the compound rest 30° to the right if only one surface on the work must be faced.

OR

Fig. 30-8 The work should not extend more than three times its diameter beyond the chuck jaw unless it is supported. (Courtesy of Kelmar Associates)

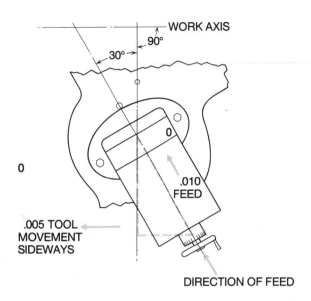

Fig. 30-10 Accurate facing is possible when the compound rest is set at 30°. (Courtesy of Kelmar Associates)

Swivel the compound rest 90° to the cross-slide if a series of steps or shoulders must be faced to accurate length on the same workpiece.

3. Fasten a facing toolbit in the toolholder and set its point to center height.

4. Adjust the toolholder until the point of the facing tool is closest to the work and there is a space left along the side, Fig. 30-9.

5. Move the carriage until the toolbit starts a light cut at the center of the surface to be faced.

6. Lock the carriage in position and set the depth of cut with the compound rest handle.

   • Twice the amount to be removed if the compound rest is set at 30°, Fig. 30-10.

   • The same as the amount to be removed if the compound rest is set at 90° to the cross-slide, Fig. 30-11.

7. Face the work to length.

Fig. 30-9 A facing tool set up for facing the end of a workpiece held in a chuck. (Courtesy of Kelmar Associates)

Fig. 30-11 Set the compound rest at 90° when a series of steps or shoulders must be faced to accurate lengths. (Courtesy of Kelmar Associates)

● TURNING DIAMETERS

Much lathe work is machined between centers because the work setup is quick and simple. However, there are many workpieces which cannot be held between centers, and this type of work is generally held in a three- or four-jaw chuck for machining. Since turning in a chuck is similar to turning between centers, the operations of rough and finish turning only will be briefly reviewed. For a more detailed explanation of these operations refer to Unit 27. Whenever possible, machine the work to size in two cuts, one rough cut and one finish cut.

Rough and Finish Turning

1. Mount the work in the chuck having no more than three times the diameter extending beyond the chuck jaws.

2. Mount a general-purpose toolbit in a straight or left-hand toolholder.

3. Move the toolpost to the left side of the compound rest and grip the toolholder short.

4. Set the cutting tool point to center and tighten the toolpost screw securely, Fig. 30-12.

5. Set the lathe speed and feed for rough turning.

6. Take a trial cut, to a true diameter, 1/4 in. (6.35 mm) long at the end of the work.

> The trial cut produces a true diameter, sets the cutting tool to the diameter, and sets the crossfeed graduated collar to the diameter.

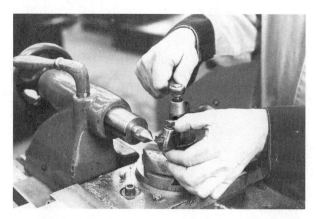

Fig. 30-12 The toolholder should be gripped short and the cutting tool point set to center. (Courtesy of Kelmar Associates)

7. Stop the lathe and measure the diameter of the trial cut, Fig. 30-13.

8. Set the radius crossfeed graduated collar in one-half the amount of metal to be removed for the rough cut.

9. Take a trial cut 1/4 in. (6.35 mm) long and measure the diameter again, Fig 30-13.

> Take the rough cut to within 1/32 in. (0.79 mm) of the finish size.

10. Rough turn the diameter to the required length.

11. Hone the cutting tool or replace it with a finish turning toolbit.

12. Before any finish turning operations, *be sure that the work is at room temperature.*

13. Set the speeds and feeds for finish turning.

14. Take a light trial cut 1/4 in. (6.35 mm) long and calculate the amount to be removed.

15. Set the crossfeed handle in one-half the amount to be removed and take another trial cut.

16. Finish turn the diameter to the required length.

To Cut off Work Held in a Chuck

Cut-off tools, often called parting tools, are used for cutting off work projecting from a chuck, for grooving, and for undercutting. The inserted

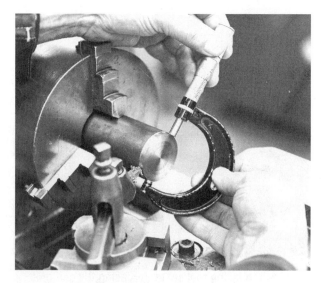

Fig. 30-13 Measuring the diameter of the trial cut. (Courtesy of Kelmar Associates)

STRAIGHT HOLDER

LEFT-HAND OFFSET

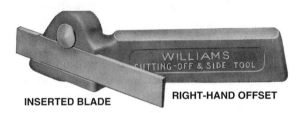

INSERTED BLADE   RIGHT-HAND OFFSET

Fig. 30-14 Three types of inserted-blade cut-off tools. (Courtesy of J.H. Williams Co., Division of Snap-On Tools Inc.)

blade type of parting tool is most commonly used and is provided in three types, Fig. 30-14.

Hints on Cutting Off

Since cutting-off tools are thin and more fragile than most lathe tools, it is important that certain precautions be observed in their use. The following suggestions should help eliminate most of the problems encountered during cutoff operations:

1. Work must be held securely in the chuck or collet.

2. Extend the workpiece only enough to allow the cut to be made as close to the chuck jaws as possible.

3. Work with more than one diameter should be held on the larger diameter when cutting off.

4. Extend the cutoff blade beyond the toolholder only enough to cut off the part; one-half the diameter plus 1/8 in. (3.17 mm) for clearance.

Cut-Off Procedure

Use the following procedure when **cutting off** work held in a chuck:

Fig. 30-15 The cut-off tool should be gripped short to prevent vibration and springing. (Courtesy of Kelmar Associates)

1. Mount the work in the chuck, with the part to be cut off as close to the jaws as possible.

2. Mount the cutting tool on the left-hand side of the compound rest and as close to the toolpost as possible to minimize vibration, Fig. 30-15.

3. Have the cutting blade extend beyond the holder half the diameter of the work to be cut, plus 1/8 in. (3 mm) for clearance, Fig. 30-16.

Fig. 30-16 The cut-off blade should only extend slightly more than half the diameter of the work. (Courtesy of Kelmar Associates)

Fig. 30-17 Using a rule to set the cut-off tool for the proper length. (Courtesy of Kelmar Associates)

4. Set the cutting tool to center and at 90° to the centerline of the work, and tighten the toolpost screw securely.

5. Set the lathe to one-half of the turning speed.

6. Move the cutting tool into position for the proper length of cut, Fig. 30-17.

7. Lock the carriage by tightening the carriage lock screw.

8. Start the lathe, and feed the cut-off tool steadily into the work using the crossfeed handle.

NOTE

Cut brass and cast iron dry, but use cutting fluid for steel.

9. Before the cut is completed, remove the burrs from each side of the groove with a file.

To avoid chatter, keep the tool cutting steadily, and apply cutting fluid during the operation. Feed slowly when the part is almost cut off.

● DRILLING, REAMING, AND TAPPING

Internal operations such as drilling, reaming, and tapping can be performed on work held in a lathe chuck while the cutting tool is held in the tailstock. The cutting process in a lathe is exactly opposite to that in a drill press where the work is stationary and the cutting tool revolves. The stationary cutting tool is held and guided by the tailstock, which *must be in line with the headstock.* The speed at which to set the machine is calculated using the same formula, but the diameter of the drill is used in the calculations.

Drilling Center Holes

Center holes are drilled into the ends of workpieces so they can be held and machined between lathe centers. Round and hexagonal work can be held in a three-jaw chuck and a center drill, held in the tailstock, is brought into contact with the revolving work to drill the center holes. See Unit 25 for a detailed explanation of the procedure for drilling center holes in a lathe.

Drilling a Hole

Drilling a hole may be defined as the process of producing a hole where none existed previously. A hole may be drilled quickly and accurately in work held in a chuck. Straight-shank drills are generally held in a drill chuck mounted in the tailstock, while taper-shank drills are mounted directly in the tailstock spindle. Use the following procedure for drilling holes:

1. Mount the work true in a chuck.

2. Face the end of the workpiece.

3. Set the lathe to the proper speed for the type of material to be drilled.

4. Check the tailstock center and make sure that it is in line.

5. With a center drill, spot the hole until about one-half of the tapered portion of the center drill enters the work, Fig. 30-18.

Fig. 30-18 Drilling the center hole to provide a guide for the following drill. (Courtesy of Kelmar Associates)

Fig. 30-19 A drill mounted in a drill chuck is supported by the back of the toolholder when starting the drill. (Courtesy of Kelmar Associates)

Fig. 30-20 A taper-shank drill mounted in the tailstock spindle can be prevented from turning by clamping a lathe dog on its body. (Courtesy of Kelmar Associates)

6. a. If the hole to be drilled is about 1/2 in. (12.7 mm) or less, mount the correct size of drill in a drill chuck mounted in the tailstock spindle and support it with the back end of a toolholder, Fig. 30-19.

   b. If a hole over 1/2 in. (12.7 mm) is to be drilled, mount the tapered shank of the drill in the tailstock spindle and fasten a lathe dog on the body close to the tailstock spindle. The tail of the lathe dog should rest on the top of the compound rest, Fig. 30-20.

> This setup will prevent the drill from turning and damaging the taper in the tailstock spindle.

7. Bring the spindle back into the tailstock, the depth of the hole to be drilled plus 1 in. (25 mm) for clearance, to make sure there is enough travel.

> It is good practice to bring the spindle in to the 1 in. graduation before starting to drill. This ensures enough travel plus allowing the graduations to be used to measure the depth of the hole.

8. Start the lathe and turn the tailstock handwheel to feed the drill into the work.

9. Apply cutting fluid frequently and drill the hole to depth.

10. Check the depth of the hole using a rule or the graduations on the tailstock spindle.

11. Always *ease up* the drill pressure as a drill starts to break through the work.

Reaming a Hole

Reaming may be performed on a lathe to bring a drilled or bored hole to an accurate size and to produce a good surface finish. Use the following procedure to ream a hole:

1. Check that the tailstock center is in line.

2. Mount the work in a chuck.

3. Face and center drill the work.

4. Select the proper size drill to leave material in a hole for reaming. Use the following guidelines for drill size:
    a. 1/64 in. smaller than the finish size for holes up to 1/2 in. diameter (0.4 mm smaller for holes up to 12 mm diameter).
    b. 1/32 in. smaller than the finish size for holes over 1/2 in. diameter (0.8 mm smaller for holes over 12 mm diameter).

5. Apply cutting fluid and drill the hole to the proper depth.

6. Mount the reamer in the drill chuck or tailstock spindle, Fig. 30-21.

7. Set the lathe to one-half the drilling speed.

8. Apply cutting fluid and turn the tailstock handwheel to feed the reamer into the hole.

9. Remove the reamer and store it where it will not be nicked or damaged.

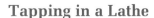

Fig. 30-21 Reaming brings a hole to size and produces a good surface finish. (Courtesy of South Bend Lathe Corp.)

Tapping in a Lathe

Internal threads may be cut on a lathe by using the proper-size tap. A standard tap may be used for this operation; however, a two-flute gun tap is preferred because the chips are cleared ahead of the tap. The tap is aligned by placing the point of the dead center in the shank end of the tap to guide it while it is being turned with a tap wrench. The lathe spindle is locked, and the tap is turned by hand. Use the following procedure for the tapping operation:

1. Mount the work in the chuck and face the end.

2. Center drill a hole so that the top edge of the hole is slightly larger than the tap diameter.

3. Select the proper tap drill size for the tap to be used. See the Appendix, Table 8.

4. Set the lathe to the proper speed for the diameter of the drill being used.

5. Drill the tap drill hole to the required depth, using cutting fluid if required.

6. Stop the lathe and lock the spindle, or put the lathe in its slowest speed.

7. Select the proper size taper tap and mount it in a tap wrench.

8. Place the tapered end of the tap in the hole and support the other end with the dead center of the lathe, Fig. 30-22.

9. Apply cutting fluid and start the tap into the hole by slowly turning the tap wrench clockwise.

Fig. 30-22 The end of the tap is supported by the tailstock center. (Courtesy of Kelmar Associates)

Fig. 30-23 The center should be kept in light contact with the tap during the tapping operation. (Courtesy of Kelmar Associates)

10. Keep the dead center in light contact with the shank of the tap during this operation by turning the tailstock handwheel while turning the tap with the other hand, Fig. 30-23.

11. Back off the tap every half turn to break the chips and apply cutting fluid frequently.

12. Remove the taper tap and complete tapping the hole with a plug or bottoming tap.

SUMMARY

- Chucks are generally used to hold short work that does not require additional support.

- The three-jaw universal chuck, where all the jaws move simultaneously, is used to hold round or hexagonal work.

- In a four-jaw chuck, each jaw can be adjusted independently; this adds to the chuck's versatility and good holding power.

- Facing in a chuck is used to obtain a true flat surface, to produce an accurate measuring surface, and to cut work to length.

- Drill chucks, mounted in the tailstock, are used to hold straight-shank drills.

KNOWLEDGE REVIEW

Hints on Chuck Work

1. Why should chuck jaws grip on the most rigid part of the work?

2. How should work sticking out of the chuck more than three times its diameter be supported?

3. When gripping on small diameters, what precaution should be taken?

Three-Jaw Chuck Work

4. For what purposes are three-jaw chucks used?

5. Name and state the purpose of the two sets of jaws supplied with each three-jaw chuck.

6. Why must chuck jaws be assembled in a chuck in the proper order?

7. What is the purpose of a four-jaw chuck?

Facing in a Chuck

8. How far should the work extend beyond the jaws when facing in a chuck?

9. State two ways that the compound rest can be set to accurately face work to length.

10. How should the facing tool be set?

11. If .020 in. (0.5 mm) must be faced off a surface with the compound rest set at 30°, how far should the compound rest handle be turned in?

Turning Diameters

12. How many cuts should be taken to machine work to size?

13. State three purposes of a light trial cut.

14. How close to finish size should the rough cut be taken?

Cutting Off

15. How should the cutting-off tool be set in a lathe?

16. Name two things that will help avoid chatter during cutting off.

Drilling, Reaming, Tapping

17. Name two methods that can be used to hold drills in the tailstock.

18. Describe how the drill is mounted in a lathe to drill a 3/4 in. (19 mm) hole in a workpiece held in a chuck.

19. How far should the spindle be brought back into the tailstock before starting to drill a hole?

20. How can the depth of a hole be checked while it is being drilled on a lathe?

21. How much material should be left in holes up to 1/2 in. (12.7 mm) in diameter for reaming?

22. At what speed should the lathe be set for reaming?

23. What type of tap should be used for tapping in a lathe?

24. How can the tap be guided to ensure that the thread will be true with the hole?

U N I T 31

Threads and Thread Cutting

Screw threads have been used throughout history as fastening devices to hold parts together. They are also used to transmit power and increase its effect, as in a car jack to convey material, as in the feed screw of a meat grinder, and to control movement, as in a micrometer. The threads for most of these purposes, with the exception of those for conveying material, are cut on a lathe or some form of thread-producing machine. Machinists are often called upon to cut threads, therefore this knowledge is very important to anyone associated with machine shop practice.

OBJECTIVES

After completing this unit, you should be able to:

- Recognize common thread forms, list their characteristics, and describe the purpose of each.
- Calculate various thread dimensions necessary to cut threads.
- Set up a lathe and cut external 60° threads.
- Set up a lathe and cut external metric threads.

KEY TERMS

pitch   quick-change gearbox   thread   thread-chasing dial

● STANDARD THREAD FORMS

A **thread** is a helical ridge of uniform section formed on the inside or outside of a cylinder or a cone. Some of the common types of thread forms are shown in Fig. 31-1.

1. *American National Standard Thread*, Fig. 31-1A, is listed under three main divisions: National Coarse, National Fine, and National Series. This thread is commonly known as a locking thread form in America. The new ISO metric threads will be used for the same purposes as these threads.

2. *Unified Screw Thread*, Fig. 31-1B, was the result of a need for a common system for use in the United States, Canada, and England. This thread incorporates the features of the American National Form and the British Standard Whitworth threads. Threads in the Unified series are interchangeable with American National and Whitworth threads of the same pitch and diameter.

3. *International Metric Thread*, Fig. 31-1C, is a standard thread currently used throughout Europe. It is used in North America mainly on instruments and spark plugs.

4. *American National Acme Thread*, Fig. 31-1D, is generally classified as a power transmission thread.

5. *Square Thread*, Fig. 31-1E, is used for maximum transmission power. Because of its shape, friction between its matching threads is kept to a minimum.

6. *ISO Metric Threads*, Fig. 31-1F. Over the past several decades, one of the world's major industrial problems has been the lack of an international thread standard whereby

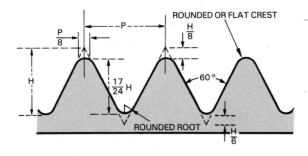

(A) AMERICAN NATIONAL STANDARD THREAD

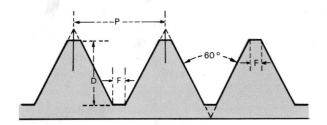

(B) UNIFIED SCREW THREAD

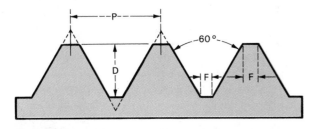

(C) INTERNATIONAL METRIC THREAD

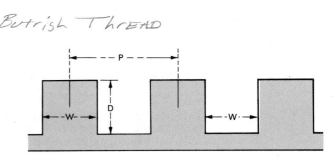

(D) AMERICAN NATIONAL ACME THREAD

Butrish Thread

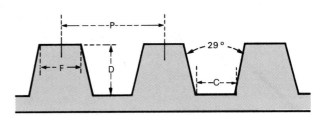

(E) SQUARE THREAD

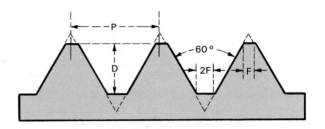

(F) ISO METRIC THREAD

Fig. 31-1 Common thread forms and specifications. (Courtesy of Kelmar Associates)

Table 31-1   ISO metric pitch diameter combinations

Nominal Diameter	Thread Pitch	Nominal Diameter	Thread Pitch
1.6	0.35	20	2.5
2	0.4	24	3
2.5	0.45	30	3.5
3	0.5	36	4
3.5	0.6	42	4.5
4	0.7	48	5
5	0.8	56	5.5
6	1	64	6
8	1.25	72	6
10	1.5	80	6
12	1.75	90	6
14	2	100	6
16	2		

the thread standard used in any country could be interchanged with that of another country. The International Standardization Organization for (ISO) has agreed upon a standard metric thread profile, the sizes and pitches for the various threads, in the new ISO Metric Thread Standard. The new series has only twenty-five thread sizes ranging in diameter from 1.6 mm to 100 mm. See Table 31-1 for this series. These metric threads are identified by the letter M, the nominal diameter and the pitch. For example, a metric thread with an outside diameter of 5 mm and a pitch of 0.8 mm would be identified as follows: M 5 × 0.8.

The new ISO series will simplify thread design, will generally produce stronger threads for a given diameter and pitch, and will reduce the large inventory of fasteners now required by industry.

The new ISO metric thread, Fig. 31-1F, has a 60° included angle and a crest equal to 0.125 times the pitch, which is similar to the National Form thread. The main difference is the depth of thread, which is 0.54127 times the pitch. Because of these dimensions, the root of the thread is larger than that of the National Form thread. The root of the new ISO metric thread is 1/4 of the pitch (0.25P).

● THREAD TERMS AND CALCULATIONS

Screw threads form a very important part of every component made, from a tiny wristwatch to

a large earthmover. To understand thread theory and screw cutting, it is important to know the parts of a thread, Fig. 31-2. All threads have common thread terms. The American National Thread Series and the new ISO metric thread are the only threads fully explained in this book. The following list defines important thread items:

- *ANGLE OF THREAD*—The angle included between the sides of the thread; for example, the thread angle of the new ISO Metric Thread and that of the American National Form is 60°.

- *MAJOR DIAMETER*—The largest diameter of the thread on the screw or nut.

- *MINOR DIAMETER*—The smallest diameter of an external or internal screw thread.

- *NUMBER OF THREADS*—The number of roots or crests per inch of the threaded length. This term does not apply to metric threads.

- *PITCH*—The distance in inches from a point on one thread to the corresponding point on the next thread measured parallel to the axis. It is expressed in millimeters for metric threads.

- *LEAD*—The distance a screw thread advances axially for one complete revolution.

- *CREST*—The top surface joining the two sides of a thread.

- *ROOT*—The bottom surface joining the sides of two adjacent threads.

- *SIDE*—The surface of the thread which connects the crest with the root.

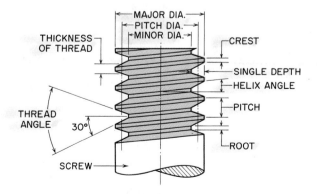

Fig. 31-2 The main parts of a screw thread. (Courtesy of Kelmar Associates)

- *DEPTH OF THREAD*—The distance between the crest and the root of a thread, measured perpendicular to the axis.

Calculations for American National Form Thread

P = Pitch of thread

$$= \frac{1}{\text{number of threads per inch}}$$

D = Depth of thread

$$= .61343 \times \text{ pitch}$$

$$= \frac{.61343}{\text{number of threads per inch}}$$

F = Width of flat on crest or root

$$= \frac{P}{8}$$

$$= \frac{1}{8 \times \text{number of threads per inch}}$$

N = Number of threads per inch

EXAMPLE: Find the pitch, depth, and minor diameter of a 1 in. − 8 N.C. thread.

$$\text{Major diameter} = 1.000 \text{ in.}$$

$$\text{Pitch} = \frac{1}{N}$$

$$= \frac{1}{8} \text{ in.}$$

$$\text{Depth of thread} = .61343 \times P$$

$$= .61343 \times \frac{1}{8}$$

$$= .077 \text{ in.}$$

$$\text{Minor diameter} = \text{Major diameter} - (D + D)$$

$$= 1.000 - (.077 + .077)$$

$$= .846 \text{ in.}$$

Calculations for ISO Metric Threads

P = Pitch of thread in millimeters
D = Depth of thread
$$= .54127 \times \text{pitch}$$

FC = Width of flat at the crest
$$= 0.125 \times \text{pitch}$$
FR = Width of flat at the root
$$= 0.25 \times \text{pitch}$$

EXAMPLE: What is the pitch, depth, minor diameter, width of crest, and width of root for a M 14 × 2 thread?

$$\text{Pitch} = 2 \text{ mm}$$

$$\text{Depth} = 0.54127 \times 2$$

$$= 1.082 \text{ mm}$$

$$\text{Minor diameter} = \text{Major diameter} - (D + D)$$

$$= 14 - (1.082 + 1.082)$$

$$= 11.84 \text{ mm}$$

$$\text{Width of crest} = 0.125 \times \text{pitch}$$

$$= 0.125 \times 2$$

$$= 0.25 \text{ mm}$$

$$\text{Width of root} = 0.25 \times \text{pitch}$$

$$= 0.25 \times 2$$

$$= 0.5 \text{ mm}$$

● THREADING EQUIPMENT AIDS

Knowing how to use the quick-change gearbox and the thread-chasing dial can help to produce work quickly and accurately.

To Set the Quick-Change Gearbox for Threading

The **quick-change gearbox**, Fig. 31-3, is designed to quickly set up the various gears to obtain the correct **pitch** for thread cutting. This unit has gear changes that transmit a direct motion or a ratio from the spindle to the lead screw.

1. Check the print for the pitch in the number of threads per inch (millimeters) required.

2. On the quick-change gearbox chart, find the *whole number* that represents the pitch in the number of threads per inch (millimeters).

3. Engage the tumbler lever in the hole at the bottom of the vertical column in which the number is located, Fig. 31-3.

4. Set the top lever in the proper position as indicated on the chart for the thread pitch required.

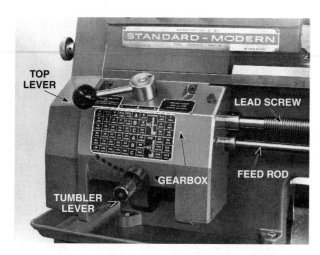

Fig. 31-3 The quick-change gearbox sets the proper ratio between the lathe spindle and the leadscrew to cut various thread pitches. (Courtesy of Standard-Modern Machine Co.)

5. Engage sliding gear in or out as required.

6. Turn the drive plate or chuck by hand and make sure that the lead screw revolves.

7. Recheck the complete setup before thread cutting.

Thread-Chasing Dial

The **thread-chasing dial** is an indicator with a revolving dial, that can be either fastened to the carriage or built into it, Fig. 31-4. The chasing dial shows the operator when to engage the split-nut lever in order to take successive cuts in the same groove or thread. It also indicates the relationship

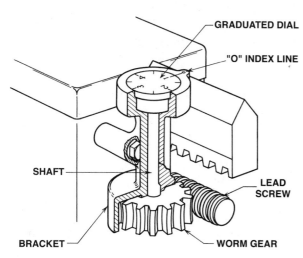

Fig. 31-4 The thread-chasing dial mechanism used for thread cutting. (Courtesy of Kelmar Associates)

between the ratio of the number of turns of the work and the lead screw with respect to the position of the cutting tool and the thread groove.

The thread-chasing dial is connected to a worm gear, which meshes with the threads of the lead screw. The dial is graduated into eight divisions, four numbered and four unnumbered, and it revolves as the lead screw turns. Fig. 31-9 on page 305 indicates when the split-nut lever should be engaged for cutting various numbers of threads per inch.

● THREAD CUTTING

Thread cutting on a lathe is a process of producing a helical ridge of uniform section by cutting a continuous groove around a cylinder. This is done by taking successive light cuts with a threading toolbit the same shape as the thread form. In order to produce an accurate thread, it is important that the lathe, the cutting tool, and the work be set up properly.

To Set up a Lathe for Threading 60° Threads

1. Set the lathe speed to about one-quarter of the speed used for turning.

2. Set the quick-change gearbox for the required pitch in the number of threads per inch or millimeters.

3. Engage the lead screw.

4. Secure a 60° threading toolbit, check the angle using a thread center gage, and mount it in a left-hand offset toolholder.

5. Set the compound rest at 29° to the right (to the left for a left-hand thread).

6. Mount the toolholder in the toolpost and set the point of the toolbit even with the dead center point.

7. Set the toolbit at right angles to the center line of the work, using a thread center gage, Fig. 31-5.

● **NOTE**

Never jam a toolbit into a thread center gage. This can be avoided by aligning only the cutting (leading) side of the toolbit with the gage. A piece of paper placed on the cross-slide under the gage and toolbit makes it easier to check the tool alignment.

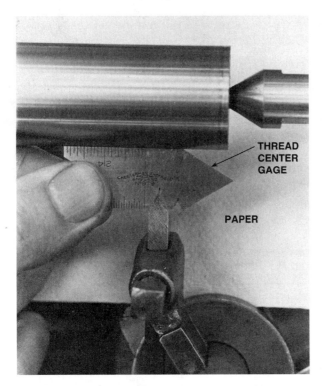

Fig. 31-5 Setting the threading tool square to the work with a thread center gage. (Courtesy of Kelmar Associates)

8. Set the apron feed lever in the neutral position and check the engagement of the split-nut lever.

To Cut a Thread

Thread cutting is a lathe operation that requires a great deal of attention and skill. It involves manipulation of the lathe parts, coordination of the hands, and strict attention to the operation being performed. Before proceeding to cut the thread, it is wise to take several trial passes without cutting (cutting air) in order to get the feel of the machine. To cut a thread, use the following procedure:

1. Mount the work in the lathe and check that the diameter to be threaded is .002 in. (0.05 mm) undersize.

2. With chalk, mark the drive plate slot that is driving the lathe dog, Fig. 31-6.

3. Mark the length to be threaded by cutting a light groove at this point with the threading tool while the lathe is revolving, Fig. 31-6.

4. Chamfer the end of the work with the side of the threading tool, Fig. 31-7.

Fig. 31-6 The drive-plate slot into which the lathe dog fits, should be marked with chalk for thread cutting. (Courtesy of Kelmar Associates)

Fig. 31-7 Chamfering the end of the work with the side of the threading tool. (Courtesy of Kelmar Associates)

The chamfer should be slightly deeper than the thread to be cut.

5. Move the carriage until the point of the threading tool is near the right-hand end of the work.

6. Turn the crossfeed handle until the threading tool is close to the diameter, but *stop* when the handle is at the 3 o'clock position, Fig. 31-8.

7. Hold the crossfeed handle in this position and set the graduated collar to zero.

8. Turn the compound rest handle until the threading tool *lightly marks the work*, and set the compound rest graduated collar to zero.

9. Move the carriage to the right until the tool-bit clears the end of the work.

Fig. 31-8 Thread cutting is easier when the crossfeed handle is set at the three o'clock position. (Courtesy of Kelmar Associates)

HANDLE SET AT 3 O'CLOCK POSITION

10. Feed the compound rest *clockwise* about .003 in. (0.07 mm).

11. Engage the split-nut lever on the correct line of the thread-chasing dial, Fig. 31-9, and take a trial cut along the length to be threaded.

12. At the end of the cut, turn the crossfeed handle *counterclockwise* to move the tool-bit away from the work and disengage the split-nut lever, Fig. 31-10.

13. Stop the lathe and check the number of threads per inch with a thread pitch gage, rule, or center gage, Fig. 31-11.

If the pitch in the number of threads per inch (millimeters) produced by the trial cut is not correct, recheck the quick-change gearbox setting.

THREADS PER INCH TO BE CUT	WHEN TO ENGAGE SPLIT NUT	READING ON DIAL
EVEN NUMBER OF THREADS	ENGAGE AT ANY GRADUATION ON THE DIAL 1 1½ 2 2½ 3 3½ 4 4½	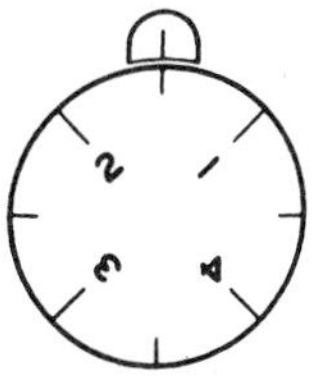
ODD NUMBER OF THREADS	ENGAGE AT ANY MAIN DIVISION 1 2 3 4	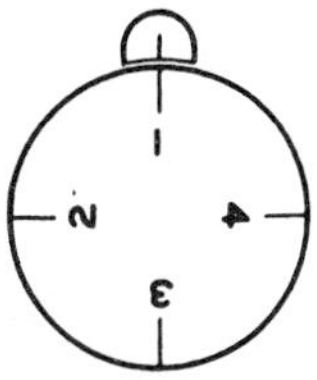
FRACTIONAL NUMBER OF THREADS	1/2 THREADS, E.G.,11 1/2 ENGAGE AT EVERY OTHER MAIN DIVISION 1 & 3, OR 2 & 4 OTHER FRACTIONAL THREADS ENGAGE AT SAME DIVISION EVERY TIME	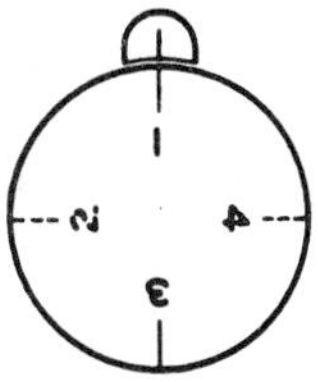
THREADS WHICH ARE A MULTIPLE OF THE NUMBER OF THREADS PER INCH IN THE LEAD SCREW	ENGAGE AT ANY TIME THAT SPLIT NUT MESHES	USE OF DIAL UNNECESSARY

Fig. 31-9 Split-nut engagement rules for cutting inch thread pitches. (Courtesy of Kelmar Associates)

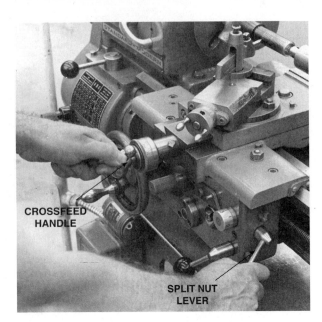

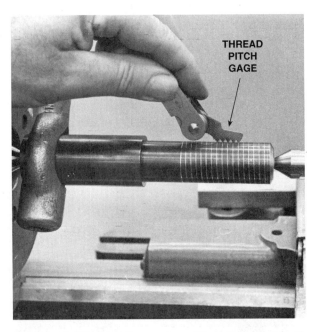

Fig. 31-10 At the end of the threaded section, <u>first</u> withdraw the toolbit, then disengage the split-nut lever. (Courtesy of Kelmar Associates)

Fig. 31-11 Checking the number of threads per inch with a thread pitch gage. (Courtesy of Kelmar Associates)

14. After each cut, turn the carriage handwheel to bring the toolbit to the start of the thread and return the crossfeed handle to zero.

15. *Set the depth of all threading cuts with the compound rest handle.* For National Form threads, use Table 31-2; for ISO metric threads, see Table 31-3.

Table 31-2 Depth settings when cutting 60° national form threads

	Compound Rest Setting		
tpi	0°	30°	29°
24	.027	.031	.031
20	.033	.038	.037
18	.036	.042	.041
16	.041	.047	.046
14	.047	.054	.053
13	.050	.058	.057
11	.059	.068	.067
10	.065	.075	.074
9	.072	.083	.082
8	.081	.094	.092
7	.093	.107	.106
6	.108	.125	.124
4	.163	.188	.186

When using this table for cutting National form threads, the correct width of flat (.125 P) must be ground on the toolbit point; otherwise the thread will not be the correct width.

Table 31-3 Depth settings when cutting 60° ISO metric threads

	Compound Rest Setting (mm)		
Pitch (mm)	0°	30°	29°
0.35	0.19	0.21	0.21
0.4	0.21	0.25	0.24
0.45	0.24	0.28	0.27
0.5	0.27	0.31	0.3
0.6	0.32	0.37	0.36
0.7	0.37	0.43	0.42
0.8	0.43	0.5	0.49
1	0.54	0.62	0.62
1.25	0.67	0.78	0.77
1.5	0.81	0.93	0.92
1.75	0.94	1.09	1.08
2	1.08	1.25	1.24
2.5	1.35	1.56	1.55
3	1.62	1.87	1.85
3.5	1.89	2.19	2.16
4	2.16	2.5	2.47
4.5	2.44	2.81	2.78
5	2.71	3.13	3.09
5.5	2.98	3.44	3.4
6	3.25	3.75	3.71

Fig. 31-12 Removing the burrs from the top of the thread with a fine file. (Courtesy of Kelmar Associates)

Fig. 31-13 Checking the accuracy of a thread with a master nut. (Courtesy of Kelmar Associates)

16. Apply cutting fluid and take successive cuts until the top (crest) and the bottom (root) of the thread are the same width.

17. Remove the burrs from the top of the thread with a file, Fig. 31-12.

18. Check the thread with a master nut and take further cuts, if necessary, until the nut fits the thread freely with no end play, Fig. 31-13.

To Convert an Inch-Designed Lathe to Metric Threading

Metric threads may be cut on a standard quick-change gear lathe by using a pair of change gears having 50 and 127 teeth respectively. Since the lead screw has inch dimensions and is designed to cut threads per inch, it is necessary to convert the pitch in millimeters into threads per inch. To do this, it is first necessary to understand the relationship of the inch and the metric systems of measurement, 1 in. = 2.54 cm.

Therefore the ratio of inches to centimeters is 1:2.54, or $1/2.54$

To cut a metric thread on a lathe, it is necessary to incorporate certain gears in the gear train which will produce a ratio of 1:2.54. These gears are:

$$\frac{1}{2.54} \times \frac{50}{50} = \frac{50}{127} \frac{\text{teeth}}{\text{teeth}}$$

In order to cut metric threads, two gears having 50 and 127 teeth must be placed in the gear train of the lathe. The 50-tooth gear is used as the spindle or drive gear, and the 127-tooth gear is placed on the lead screw.

To Cut a 2.5 mm Metric Thread on a Standard Quick-Change Gear Lathe

Use the following procedure and math conversion:

1. Mount the 127-tooth gear on the lead screw, Fig. 31-14.

2. Mount the 50-tooth gear on the spindle.

3. Convert the 2.5 mm pitch to threads per centimeter.

$$10 \text{ mm} = 1 \text{ cm}$$

$$\text{Pitch} = \frac{10}{2.5} = 4 \text{ threads/cm}$$

4. Set the quick-change gearbox to 4 threads /in.

By means of the 50- and 127-tooth gears, the lathe will now cut 4 threads/cm, or 2.5 mm pitch.

5. Set up the lathe for thread cutting. See the section entitled "To Set up a Lathe for Threading 60° Threads, page 303."

6. Take a light trial cut. At the end of the cut, back out the cutting tool and stop the machine.

Fig. 31-14 The gears that must be changed to convert an inch lathe to cut metric threads. (Courtesy of Kelmar Associates)

> Do not disengage the split nut.

7. Reverse the spindle rotation until the cutting tool has just cleared the start of the threaded section.

8. Check the thread with a metric screw pitch gage.

9. Cut the thread to the required depth.

> Never disengage the split nut until the thread has been cut to depth.

To Reset a Threading Tool

A threading tool must be reset whenever it is necessary to remove partly threaded work and finish it at a later time, to remove the threading tool for regrinding, or if the lathe dog slips on the work.

1. Mount the work and set up the lathe for thread cutting and for the proper number of threads to be cut.

2. With the threading toolbit clear of the work, start the lathe and engage the split-nut lever on the correct line for the pitch in the number of threads per inch (millimeters).

3. Allow the carriage to travel until the toolbit is opposite any portion of the unfinished thread.

4. Stop the lathe, but be sure to *leave the split-nut lever engaged.*

5. Feed the toolbit into the thread groove.

> Use ONLY the compound rest and crossfeed handles, until the right-hand side of the toolbit touches the right-hand side of the thread (left side for left-hand threads), Fig. 31-15.

6. Set the crossfeed graduated collar to zero *without moving the crossfeed handle,* and then set the compound rest feed collar to zero.

7. Back out the threading tool using the crossfeed handle, disengage the split-nut lever, and move the carriage until the toolbit clears the start of the thread.

8. Set the crossfeed handle back to zero and take a trial cut without setting the compound rest.

9. Set the depth of cut using the compound rest handle and take successive cuts to finish the thread.

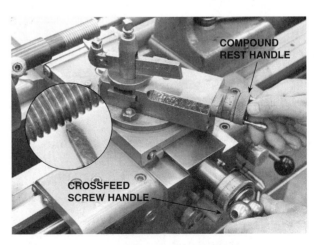

Fig. 31-15 The right side of the threading tool should be reset to the right side of a partially cut thread. (Courtesy of Kelmar Associates)

● THREAD MEASUREMENT

There are several methods of checking threads for depth, angle, and accuracy. Most commonly used are finished master hexagon nuts, thread micrometers, and thread gages.

A finished master hexagon nut can be used for checking all general-purpose threads. The thread should be cut deep enough to allow the nut to turn on freely with no end play.

Inch and metric thread micrometers can be used to check the pitch diameter of threads to an accuracy of .001 in. (0.02 mm). These micrometers are made to measure certain ranges of threads. Metric screw thread micrometers are available for checking thread diameters from 0 to 100 mm. Fig. 31-16 shows a thread micrometer used to check 14 to 20 threads/in. (60°).

A thread ring gage, Fig. 31-17, is often used in production work for testing threads. It is a hardened standard ring gage that can be adjusted to compensate for wear and tolerance.

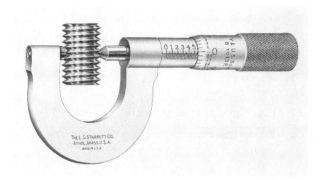

Fig. 31-16 A thread micrometer measures the pitch diameter of a thread. (Courtesy of The L.S. Starrett Co.)

Fig. 31-17 A thread-ring gage is used for checking the accuracy of threads on production work. (Courtesy of The Taft-Peirce Mfg. Co.)

SUMMARY

- Screw threads can be used for fastening, transmitting power, conveying material, and controlling movement.

- Threads are made in many forms; the ISO thread standard was created in order to standardize threads for international use.

- The quick-change gearbox and the thread-chasing dial are used to make the thread-cutting process efficient and accurate.

- Threading tools must be reset if partly completed work must be removed and completed later, if the tools need regrinding, or if the lathe dog slips on the work.

KNOWLEDGE REVIEW

Standard Thread Forms

1. State three purposes of threads.

2. Define a thread.

3. List four thread forms, give the angle of each thread, and state a use for each.

Thread Terms and Calculations

4. Define the following thread terms:
   a   major diameter
   b. lead
   c. pitch
   d. root

5. a. Calculate the depth, minor diameter, and width of flat for a 7/8 in.–9 N.C. thread.

   b. Calculate the pitch, depth, minor diameter, width of crest, and width of root for a M 16 × 3 thread.

To Set a Quick-Change Gearbox

6. What is the purpose of the quick-change gear box?

Thread-Chasing Dial

7. Explain the function of the thread-chasing dial.

8. What lines on the thread-chasing dial could be used for cutting 6, 11-1/2, and 16 threads/in.?

Thread Cutting

9. Explain how the threading toolbit is set up.

10. At what angle should the compound rest be set for cutting a 60° thread?

11. What should be the size of the diameter to be threaded?

12. At what position should the crossfeed handle be for threading?

13. What is the purpose of taking a light trial cut before threading the work?

14. Explain how the pitch in the number of threads per inch (millimeters) can be checked.

15. How is the depth of each threading cut set?

To Convert an Inch-Designed Lathe to Metric Threading

16. Describe how a standard quick-change gear lathe may be set up to cut a metric thread.

17. What precaution must be taken when cutting a metric thread on a standard quick-change gear lathe?

18. What is the pitch, depth, minor diameter, width of crest, and width of root for a M 20 × 2.5 thread?

To Reset a Threading Tool

18. State two reasons why it may be necessary to reset a threading tool.

20. Explain in point form how to reset a threading tool in a partially cut thread.

Thread Measurement

21. List three methods used to check a thread.

22. What instrument can be used to measure the pitch diameter of a thread?

Milling Machines

The milling machine is a machine tool used to produce accurately machined surfaces such as flats, angular surfaces, grooves, cams, contours, gear and sprocket teeth, helical grooves, and accurately sized holes. Milling operations are performed by feeding the workpiece into a revolving cutting tool called a milling cutter which could have one or more cutting edges. The shape of the milling cutter will determine the shape of the finished surface.

Figures courtesy of (clockwise from far left): Cincinnati Milacron Inc., Cincinnati Milacron Inc., Cincinnati Milacron Inc., Kelmar Associates, Bridgeport Machines Inc., Kelmar Associates.

UNIT
32

Horizontal Milling Machines and Accessories

The **horizontal milling machine** is a very versatile, highly efficient, and useful machine for machine shop work. Its spindle is generally in a horizontal plane, and the type of cut it makes is determined by the size and shape of the milling cutter. A vertical attachment is available for horizontal milling machines to make the machine more versatile for milling slots, pockets, contours, etc.

OBJECTIVES

After completing this unit, you should be able to:

- Describe the types and purposes of milling machines available.
- Identify and explain the purpose of the main operating parts.
- Recognize common cutters and explain their use.
- Observe the safety regulations when operating a milling machine.

KEY TERMS

climb milling   conventional milling   horizontal milling machine
milling cutter

TYPES OF MILLING MACHINES

The versatility of the milling machine makes it suitable for production, toolroom, job shops, and experimental and research work. The more commonly used milling machines are:

1. Plain knee and column
2. Universal knee and column, Fig. 32-1
3. Vertical knee and column
4. The manufacturing types
5. Automation type

The *universal knee* and column type is perhaps the most versatile milling machine and can be adapted to perform many jobs by the use of a variety of attachments. The *plain knee* and column milling machine is the same as the universal, except that it does not have the swivel table housing.

PARTS OF THE MILLING MACHINE

The following list describes the main parts of a milling machine:

The *base* gives support and rigidity to the machine and also acts as a reservoir for the cutting fluids.

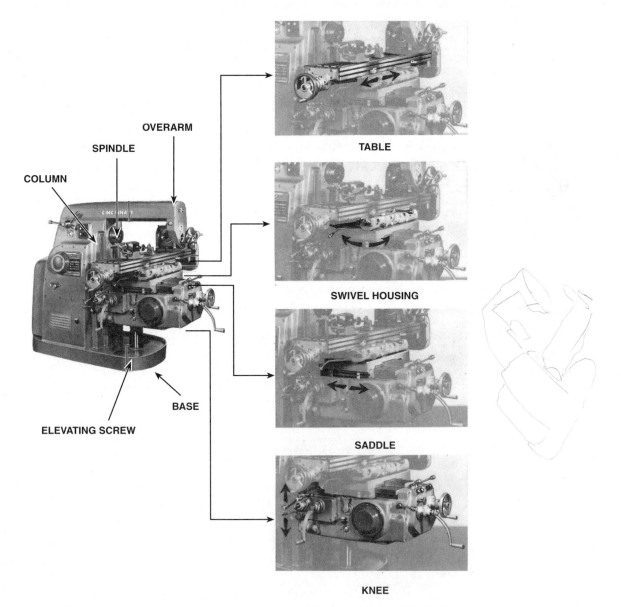

Fig. 32-1 The main parts of the universal knee and column milling machine. (Courtesy of Cincinnati Milacron Inc.)

The *column face* is a precision-machined and scraped section used to support and guide the knee when it is moved vertically.

The *knee* is attached to the column face and may be moved vertically on the column face either manually or automatically. It generally houses the feed mechanism.

The *saddle* is fitted on top of the knee and may be moved toward or away from the column manually by the crossfeed handwheel, or automatically by the crossfeed engaging lever.

The *swivel table housing*, fastened to the saddle on a universal milling machine, allows the table to be swiveled 45° to either side of the center line.

The *table* rests on guideways in the saddle and travels longitudinally in a horizontal plane. It supports work-holding devices and the work.

The *crossfeed handwheel* is used to move the table toward or away from the column.

The *table handwheel* is used to move the table horizontally left and right in front of the column.

The *feed dial* is used to set the rate of the table feeds.

The *spindle* provides the drive for arbors, cutters, and attachments used on a milling machine.

The *overarm* provides for correct alignment and support of the arbor and various attachments. It can be adjusted and locked in various positions, depending on the length of the arbor and the position of the cutter.

The *arbor support* is fitted to the overarm and can be clamped at any location on the overarm. Its purpose is to align and support various arbors and attachments.

The *elevating screw* is controlled by hand or an automatic feed. It gives an upward or downward movement to the knee and the table.

The *spindle-speed dial* is set by a crank that is turned to regulate the spindle speed. On some milling machines, the spindle-speed changes are made by means of two levers. When making speed changes, always check whether the change can be made when the machine is running, or if the machines must be stopped.

Most milling machines are equipped with two or more arbors on which differently-shaped cutters, Fig. 32-5, may be mounted. Most milling machines can be fitted with many other attachments, such as a dividing head, rotary table, vertical head, slotting attachment, rack-cutting attachment, and various special fixtures.

● MILLING CUTTERS

A **milling cutter** is a rotary cutting tool having equally spaced teeth around the periphery and sometimes on the end or sides. These teeth engage with the workpiece to remove the metal in the form of chips. Milling cutters are manufactured in a variety of shapes and sizes to produce many shapes or profiles. They are usually made of high-speed or special steel with tungsten-carbide teeth or inserts.

Coated Milling Cutters

Special coatings can be applied to the surface of milling cutters to prolong their tool life, increase productivity, and lower manufacturing costs. The most common coatings applied to cutting tools are titanium nitride and titanium carbonitride. These coatings are applied in a thin film, about .0001 in or 0.002 mm thick, with a hardness of over 80 Rc, Fig. 32-2. The ultra-thin, hard film protects the sharp edges of a cutting tool that are so important to their efficient cutting action and tool life. The titanium carbonitride coating, 30% harder than the titanium nitride coating, is applied to tools that are operated at higher speeds and feeds and used for machining abrasive or difficult-to-cut materials.

The prime function of a coating is to reduce abrasive wear and produce a fine finish on the work surface. *Abrasive wear* occurs as the teeth or cutting edges of a tool come into contact with the work material. With the higher hardness (over 80 Rc) that a coating provides, the cutting

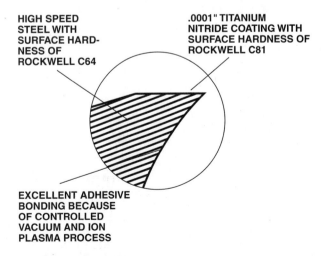

HIGH SPEED STEEL WITH SURFACE HARDNESS OF ROCKWELL C64

.0001" TITANIUM NITRIDE COATING WITH SURFACE HARDNESS OF ROCKWELL C81

EXCELLENT ADHESIVE BONDING BECAUSE OF CONTROLLED VACUUM AND ION PLASMA PROCESS

Fig. 32-2 Titanium nitride coating on a cutting tool reduces abrasive wear and prolongs the life of a tool. (Courtesy of Balzers Tool Coating Inc.)

edge is protected from abrasive wear and lasts longer. The added lubricity of a coating improves the chip formation and the material flow, to reduce chip welding and a built-up edge. The surface finishes of workpieces cut by an uncoated and coated tool are compared in Fig. 32-3.

In the uncoated tool, Fig. 32-3A, as the chip is separated from the metal and slides up the tool face, resistance to chip flow causes small particles of the metal to weld to the cutting-tool face. This *built-up edge* continually increases in size until it breaks off and takes some of the cutting tool material away with it. The result is a cratering action on the tool face which will eventually shorten the life of the cutting tool.

The lubricating properties of the coated tool, Fig. 32-3B, reduce the friction of the chip sliding up the tool face and change the shape of the chip. Because the chip slides up the tool face more quickly, there is less heat generated and less tendency for the chip to weld to the tool face. Since the coating keeps the cutting edges of the tool sharp longer and prevents built-up edges from forming, better surface finishes and more accurate work are produced.

Advantages

Coated cutting tools allow a manufacturer to increase the productivity of a machining operation without the expense of buying new machine tools. Some of the most common benefits of coated cutting tools are:

1. **Lower machining cost per part**. In most cases, the longer tool life, use of higher speeds and feeds, better surface finishes, and more accurate workpieces result in large cost savings per part. A 400% increase in tool life and higher material removal rates, Fig. 32-4, can result in up to a 50% increase in productivity.

2. **Higher tool life**. Titanium nitride-coated tools can run three to eight times longer than uncoated tools. As the number of parts produced by a tool increases, the cost per

HOW COATING IMPROVES SURFACE FINISH OF WORKPIECE

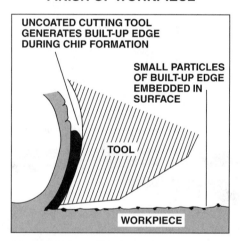

(A) UNCOATED TOOL

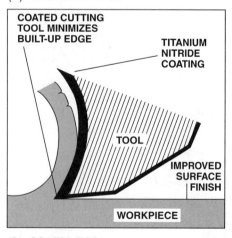

(B) COATED TOOL

Fig. 32-3 A comparison of the cutting action of an uncoated and coated cutting tool. (Courtesy of Balzers Tool Coating Inc.)

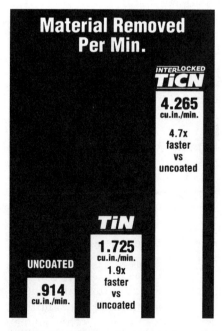

Fig. 32-4 A comparison of the material removal rates of uncoated, titanium nitride coated, and titanium carbonitride coated tools. (Courtesy of Niagara Cutter Inc.)

part can be reduced by 66% to 88%. This will depend on the type of material cut, the type and application of the coolant, and other factors.

3. **Improved surface finish**. The increased lubrication qualities of coated tools reduce built-up edges and provide a better chip flow along the tool face. Because chips flow faster over the tool face, they carry away heat faster, and the result is a better surface finish.

4. **Improved part quality**. Titanium-coated tools reduce the tendency for built-up edges to form by providing a smooth cutting action that results in improved surface finishes and greater work accuracy.

5. **Other improvements**. These include less downtime for tool changes and reconditioning, better ability to hold machining tolerances, and sometimes the elimination of two-step operations such as roughing and finishing cuts.

Types of Milling Cutters

Milling cutters are available in a wide range of shapes and sizes to suit the many types of machining operations performed on a milling machine. They can range from the plain milling cutter used to machine flat surfaces to the complex form cutters used to produce very intricate forms. Some of the most common cutters are as follows:

Plain milling cutters are the most commonly used types of milling cutters. They are usually wide cylindrical cutters with the teeth on the periphery. Plain milling cutters such as light-duty, Fig. 32-5A, light-duty helical, heavy-duty, Fig. 32-5B, and high-helix, are used for producing flat surfaces.

Side milling cutters are generally narrow cylindrical cutters with teeth on the sides and the periphery. The teeth may be straight, Fig. 32-5C, or staggered, where each tooth is alternately set to the right and to the left for better chip clearance. Side milling cutters are used for facing the edges of work and cutting slots. *Half-side milling cutters*, Fig. 32-5D, with teeth on the periphery and only one side, are used for straddle milling operations.

Angular cutters have teeth that are at an angle to the face and the axis of the cutter. Single angle cutters and double angle cutters are used for milling angular surfaces, grooves, and serrations, Fig. 32-5E.

Formed cutters are the exact shape of the part to be produced which permits duplication

of parts more economically than by most other means. Formed cutters may be of practically any shape such as concave, convex, and irregular. A convex cutter is shown in Fig. 32-5F, and a concave cutter is shown in Fig. 32-5G.

Gear cutters, Fig. 32-5H, are another type of formed cutter. They are manufactured in a wide range of sizes and contours, depending on the number of teeth and the pitch of the gear. They are generally used for special gear needs, since most gears are now mass produced by hobbing.

Shell end mills, Fig. 32-5I, have teeth on the periphery and on the end. They can be used for facing and peripheral cutting. They are held on a stub arbor which fits into and is driven by the milling machine spindle. The cutters are held on the arbor by a special type of cap screw and driven by a rear key.

● MILLING CUTTER USE

To get the best tool life and productivity out of a milling cutter, it is important that the cutter be used properly. Most common milling cutter problems are caused by excessive heat, abrasion, edges chipping, clogging, built-up edges, cratering, and work hardening of the workpiece. Before using any cutter, it is important to decide whether the operation will be performed by climb or conventional milling.

Climb milling is when the cutter rotation and the table (work) feed are in the same direction, Fig. 32-6A. When climb milling, it is important that the work be held securely, and the machine table must be equipped with a backlash eliminator.

Conventional milling is when the cutter rotation and the table (work) feed are going in opposite directions, Fig. 32-6B. The tooth enters the cut at zero chip thickness in an upward direction and progressively gets thicker. Conventional milling is used on workpieces where minimum shock is desirable when the cutter enters the work, or when the machine table is *not fitted with a backlash eliminator*. The forces in conventional milling try to lift the workpiece, therefore it is important that the work be held securely.

Basic Cutter Practice

The following are a few of the points that should be observed when using cutters:

1. Use a key when mounting cutters and saws. *Do not* depend on friction of the collar spacers to provide a good drive.

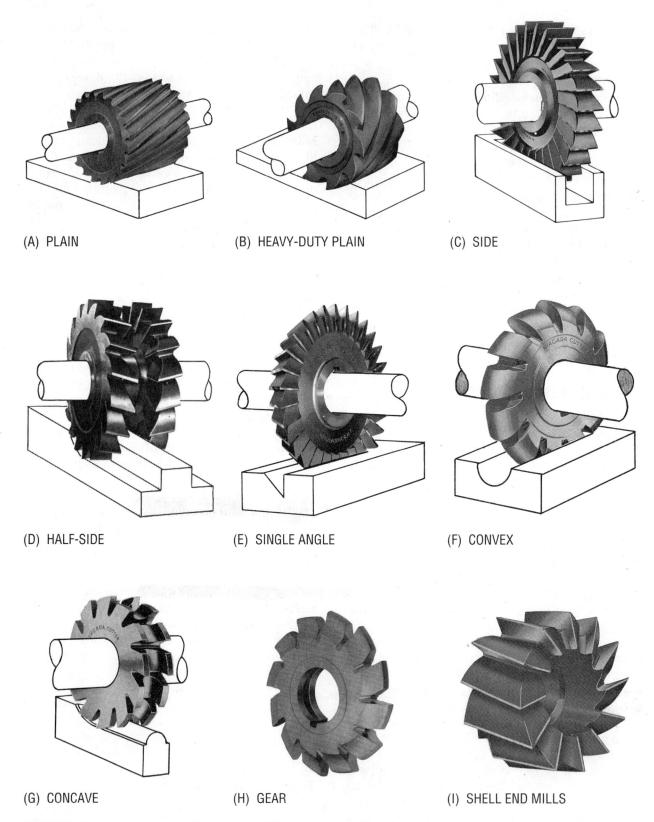

(A) PLAIN

(B) HEAVY-DUTY PLAIN

(C) SIDE

(D) HALF-SIDE

(E) SINGLE ANGLE

(F) CONVEX

(G) CONCAVE

(H) GEAR

(I) SHELL END MILLS

Fig. 32-5 Common cutters used for various milling operations. (Courtesy of Niagara Cutter Inc.)

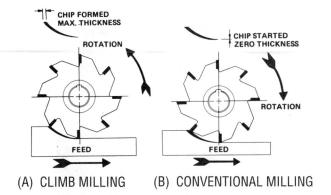

(A) CLIMB MILLING   (B) CONVENTIONAL MILLING

Fig. 32-6 Two common methods of milling are used. (Courtesy of Niagara Cutter, Inc.)

2. When mounting cutters, be sure that the arbor, collets, and cutters are clean and free of dirt and burrs that could cause misalignment and spring the arbor.

3. Check the arbor/cutter run-out and correct the problem. This could cause undue wear on a small section of the cutter circumference.

4. Remove as much backlash as possible from the table feedscrews to reduce play and dimensional errors while machining.

5. Use the low range of the recommended speeds and feeds on new applications. These can be increased gradually to increase productivity when suitable.

6. *Never cut with a dull cutter.* It is wise to determine just how many pieces a cutter might be expected to produce before it should be resharpened.

7. Use coolant wherever possible to reduce the machining friction and heat that will shorten the life of a cutter.

See Fig. 32-7 for a list of the common problems that may arise during milling and their suggested solutions.

CHECK LIST FOR CORRECTIVE ACTION →

PROBLEMS	INCREASE SPEED	DECREASE SPEED	INCREASE CHIP LOAD	DECREASE CHIP LOAD	CHECK CUTTER GEOMETRY, INCREASE CLEARANCE OR RAKE ANGLES	CHECK CUTTER GEOMETRY, DECREASE CLEARANCE OR RAKE ANGLES	CHECK COOLANT	CHECK LOOSE SPINDLE BEARINGS & WORN WAYS	CHECK OVERARM SUPPORT, USE ARM BRACES, USE 2 OVERARM SUPPORTS	CHECK OVERARM BEARING	CHECK BACKLASH ELIMINATOR, TIGHTEN GIBS	CHECK HORSEPOWER AT SPINDLE	MAKE WORK HOLDING DEVICE MORE RIGID	SUPPORT WORK IN DIRECTION OF CUTTING FORCES	CLIMB MILL	CONVENTIONAL MILL	DECREASE NO. TEETH TO MODIFY CUTTER FOR ADD. CHIP ROOM	INCREASE NO. TEETH	DECREASE DEPTH CUT	INCREASE DEPTH CUT (GET UNDER SURFACE SCALE)	USE LARGER ARBOR	CHECK ALIGNMENT, CUTTER RUN OUT, ARBOR COLLARS, ARBOR SHANK	CHECK SPEED & FEED RATES
ABRASION		●	●		●		●								●			●		●			
CRATERING		●		●	●		●									●							
CHIPPING	●		●		●	●	●				●		●					●	●		●	●	
BUILT-UP-EDGE	●		●		●		●																
EXCESSIVE CHATTER		●	●			●		●	●	●	●		●	●	●				●	●			
BREAKAGE			●	●			●	●	●	●	●	●		●			●		●			●	
CHIPS PACK IN GULLET			●	●			●										●		●				
CHIPS TORN OR CORRUGATED					●												●						●
SPINDLE STALLS IN CUT		●			●	●						●				●			●				
POOR FINISH	●				●	●		●			●		●	●				●	●		●		
PART SIZE VARIES OR TAPERS				●	●			●	●	●			●	●				●	●			●	

Recommended check list for reducing abrasion, cratering, chipping, built-up-edge, excessive chatter, breakage, chips packed in gullet, torn chips, spindle-stalls, finish, and part size variation problems.

Fig. 32-7 Milling cutter problems and suggested solutions. (Courtesy of Niagara Cutter Inc.)

● MILLING MACHINE SAFETY

The milling machine, like any other machine, demands the total attention of the operator and a thorough understanding of the hazards involved in its operation. The following rules should be observed when operating the milling machine:

1. Be sure that the work is mounted securely and that the cutter is revolving in the proper direction before taking a cut.

2. Always wear safety glasses and keep long hair under a cap or net, Fig. 32-8.

3. When mounting or removing milling cutters, always hold them with a cloth to avoid being cut, Fig. 32-9.

4. While setting up or measuring work, stop the machine spindle and move the table as far as possible from the cutter to avoid cutting your hands, Fig. 32-10.

5. *Never* attempt to mount, measure, or adjust work until the cutter is *completely stopped.*

6. Keep hands, brushes, and rags away from a revolving milling cutter *at all times.*

7. When using milling cutters, do not use an excessively heavy cut or feed.

Fig. 32-9 Use a cloth when handling milling cutters to prevent painful cuts. (Courtesy of Kelmar Associates)

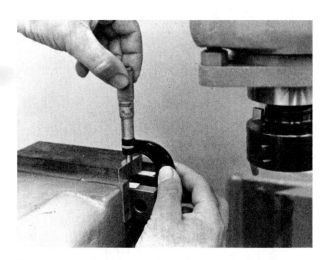

Fig. 32-10 Stop the machine spindle and move the work well clear of the cutter before taking measurements. (Courtesy of Kelmar Associates)

Fig. 32-8 Always wear safety glasses and keep long hair from getting caught in a machine. (Courtesy of Kelmar Associates)

 This can cause a cutter to break and the flying pieces could cause serious injury.

8. Always use a *brush*, not a rag, to remove the cuttings after the cutter has stopped revolving, Fig. 32-11.

9. Never reach over, or near a revolving cutter.

Fig. 32-11 Use a brush, not your hand, to remove sharp chips. (Courtesy of Kelmar Associates)

Fig. 32-12 Keep the floor around a machine clean and free of oil and hazards. (Courtesy of Kelmar Associates)

 Keep hands at least 12 in. (300 mm) from a revolving cutter.

10. Keep the floor around the machine free of chips, oil, and cutting fluid, Fig. 32-12.

SUMMARY

- Milling machines are used to produce a variety of machined surfaces including flats, angular surfaces, grooves, slots, gear and sprocket teeth, and holes.

- In milling, the workpiece is fed into a revolving milling cutter, that can have more than one cutting edge.

- Special coatings on the surfaces of milling cutters reduce abrasive wear, prolong tool life, and increase productivity. The most common coatings are titanium nitride and titanium carbonitride.

- In climb milling, the cutter rotation and the table (work) feed are in the same direction.

- In conventional milling, the cutter rotation and the table (work) feed move in opposite directions. This reduces the shock when the cutter enters the work.

KNOWLEDGE REVIEW

Milling Machines and Parts

1. List six types of work that can be produced on a vertical milling machine.
2. What is the difference between a plain horizontal and a universal milling machine?
3. What hand controls are used to move the table
   a. lengthwise?
   b. in or out?
   c. up or down?
4. From what type of steel are milling cutters made?
5. Name two types of coatings used on milling cutters.
6. State three ways that a coating reduces abrasive wear.
7. How is a built-up edge formed?
8. List five advantages of coated cutting tools.

Milling Cutters

9. For what purpose are the following cutters used?
   a. plain milling
   b. side milling
   c. formed
10. List six causes of milling cutter problems.

11. Explain the difference between climb and conventional milling.

12. What must a milling machine have in order to successfully climb mill?

13. What precaution should be taken when mounting cutters?

Milling Machine Safety

14. How should the machine be set when taking measurements?

15. What precaution should be taken when removing steel chips with a brush?

Milling Machine Setup

Due to the versatility of the milling machine and its attachments, it is possible to accurately machine parts of identical size and shape. For this to happen, the machine must be set up correctly and the workpiece must be aligned and held securely. Each job is different, therefore the correct cutter must be selected first and then mounted on the milling arbor.

OBJECTIVES

After completing this unit, you should be able to:

- Set up and operate the machine properly.
- Mount and remove milling arbors and cutters.
- Calculate and set the correct speeds and feeds for various milling operations and work material.

KEY TERMS

arbor   graduated collar   milling cutter

● MOUNTING AND REMOVING AN ARBOR

The milling **arbor**, Fig. 33-1, which is used to hold the cutter during the milling operation, is held in the machine spindle by the draw-in bar, Fig. 33-2. The cutter is driven by a key that fits into the keyways on the arbor and the cutter to prevent it from turning on the arbor. Spacer and bearing bushings hold the cutter in position on the arbor.

To Mount an Arbor

When mounting or removing an arbor, follow the proper procedure to preserve the accuracy of the machine:

1. Clean the tapered hole in the spindle and the taper on the arbor, using a clean cloth.

2. Check that there are no cuttings or burrs in the taper.

> Dirt, burrs, or metal chips will damage the machine and arbor tapers, and prevent the arbor from running true.

3. Mount the tapered end of the arbor into the spindle taper.

4. Place the right hand on the draw-in bar, Fig. 33-3, and turn the thread into the arbor approximately 1 in. (25 mm).

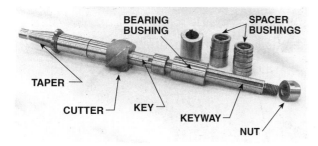

Fig. 33-1 The milling machine arbor is used to hold and drive the cutter. (Courtesy of Kelmar Associates)

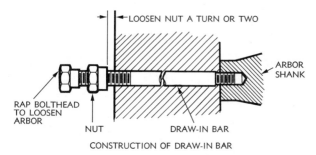

Fig. 33-2 The draw-in bar holds the arbor firmly in the machine spindle. (Courtesy of Kelmar Associates)

Fig. 33-3 Inserting an arbor into the milling machine spindle. (Courtesy of Kelmar Associates)

Fig. 33-4 Tightening the draw-in bar locknut holds the arbor securely in the spindle. (Courtesy of Kelmar Associates)

5. Tighten the draw-in bar locknut securely against the back of the spindle, Fig. 33-4.

To Remove an Arbor

An arbor must be removed occasionally from the machine spindle to accommodate larger-diameter arbors or special machine attachments. Care in this operation will preserve the life of the arbor. To remove an arbor, proceed as follows:

1. Remove the milling machine cutter.

2. Loosen the locknut on the draw-in bar approximately two turns, as shown in Fig. 33-2.

3. With a soft-faced hammer, hit the end of the draw-in bar until the arbor taper is free in the spindle, Fig. 33-5.

4. With one hand, hold the arbor, and use the other hand to unscrew the draw-in bar from the arbor, Fig. 33-6.

Fig. 33-5 Releasing the arbor taper by tapping the draw-in bar with a soft-faced hammer. (Courtesy of Kelmar Associates)

Fig. 33-6 Holding the arbor while unscrewing the draw-in bar. (Courtesy of Kelmar Associates)

5. Carefully remove the arbor from the tapered spindle to avoid damaging the spindle or arbor tapers.

6. Leave the draw-in bar in the spindle for further use.

● MOUNTING AND REMOVING A MILLING CUTTER

Milling cutters must be changed frequently, so it is important that the following sequence be followed in order not to damage the cutter, the machine, or the arbor.

To Mount a Milling Cutter

Use the following procedure to mount a **milling cutter:**

1. Remove the arbor nut and collars.

2. Clean all surfaces of cuttings and remove all burrs.

3. Check the direction of the arbor rotation.

4. Slide the collars on the arbor, positioning the bearing bushing as close to the cutter as the machining operation will allow, Fig. 33-7.

Whenever possible, locate the cutter close to the machine column or the arbor support to provide the best rigidity during the machining operation.

5. Hold the cutter with a cloth, having the teeth pointing in the direction of arbor rotation, and slide it on to the arbor fitted with a key.

6. Slide the arbor support in place and be sure that it is on the bearing bushing on the arbor, Fig. 33-8.

Fig. 33-7 Positioning the cutter on the arbor with the required number of arbor bushings. (Courtesy of Kelmar Associates)

Fig. 33-8 Sliding the arbor support into place over the bearing bushing. (Courtesy of Kelmar Associates)

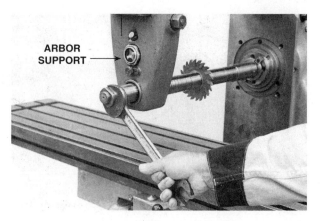

Fig. 33-10 Be sure the arbor support is locked in place over the bearing bushing before tightening the arbor nut with a wrench. (Courtesy of Kelmar Associates)

Fig. 33-9 Before taking a cut, be sure that the arbor support clears the work and vise. (Courtesy of Kelmar Associates)

7. Put on additional spacers, leaving room for the arbor nut.

TIGHTEN THE ARBOR NUT BY HAND.

8. Lock the overarm in position.
9. Tighten the arbor nut firmly with a wrench.

Do not tighten the arbor nut without the arbor support in place, it may bend the arbor.

10. Lubricate the bearing collar in the arbor support.
11. Make sure that the arbor and arbor support will clear the work, Fig. 33-9.

To Remove a Milling Cutter

Use the following procedure to remove a milling cutter:

1. Be sure that the arbor support is in place and supporting the arbor on a bearing bushing before using a wrench on the arbor nut, Fig. 33-10.

This will prevent bending the arbor.

2. Clean all cuttings from the arbor and cutter.

3. Use a wrench to loosen the arbor nut.
4. Loosen the arbor support and remove it from the overarm.
5. Remove the nut, arbor collars, and cutter. Place them on a wooden board, Fig. 33-11, not on the table surface.

This will preserve the accuracy of the cutter, arbor collars, and the machine table.

6. Clean the spacer and nut surfaces and replace them on the arbor.

Do not use a wrench to tighten the arbor nut.

7. Place the cutter in the proper storage.

Fig. 33-11 A piece of Masonite or wood can protect the machine table, cutter, and arbor collars from damage. (Courtesy of Kelmar Associates)

Fig. 33-12 Milling a workpiece being held in a fixture. (Courtesy of Cincinnati Milacron Inc.)

● WORK-HOLDING DEVICES

There are several devices used for holding work to be milled. The most commonly used are vises, V-blocks, strap clamps, angle plates, and fixtures.

The *vise* can be used for holding square, round, and rectangular pieces for the cutting of flat surfaces, angles, gear racks, keyways, grooves, and T-slots.

V-blocks usually have a 90° V-shaped groove and a tongue which fits into the table slot to allow proper alignment for milling special shapes, flat surfaces, or keyways in round work.

Angle plates are used for holding large work or special shapes when machining one surface square with another.

A *fixture*, Fig. 33-12, is a special holding device made to hold a particular workpiece for one or more milling operations on a production basis. It provides an easy setup method but is limited to the job for which it is made.

● CUTTING SPEEDS

The cutting speed for a milling cutter is the speed, in either feet per minute or meters per minute, that the periphery of the cutter should travel when machining a certain metal. The speeds used for milling-machine cutters are much the same as those used for any cutting tool. Several factors must be considered when setting the r/min to machine a surface. The most important are the following:

- material to be machined,
- cutter type and material,
- finish required,

Table 33-1   Milling machine cutting speeds

Material	High-Speed Steel Cutter		Carbide Cutter	
	Ft./Min	M/Min	Ft./Min	M/Min
Machine steel	70–100	21–30	150–250	45–75
Tool steel	60–70	18–20	125–200	40–60
Cast iron	50–80	15–25	125–200	40–60
Bronze	65–120	20–35	200–400	60–120
Aluminum	500–1000	150–300	1000–2000	150–300

- depth of cut,
- rigidity of the machine and the workpiece.

It is good practice to start from the calculated r/min using Table 33-1, and then progress until maximum tool life and productivity are reached.

Inch Cutting Speeds

The speed of a milling machine cutter is calculated by using the same formula as for a twist drill or a lathe workpiece.

$$r/min = \frac{CS \times 4}{D}$$

EXAMPLE: Find the r/min to mill a keyway 1 in. wide in a machine steel shaft, using a 4 in. diameter high-speed steel cutter (*CS* 80).

$$r/min = \frac{CS \times 4}{D}$$
$$= \frac{80 \times 4}{4}$$
$$= 80$$

EXAMPLE: Calculate the r/min required to end-mill a 3/4 in. wide groove in aluminum with a high-speed steel cutter (CS 600).

$$r/min = \frac{CS \times 4}{D}$$
$$= \frac{600 \times 4}{3/4}$$
$$= 3200$$

Metric Cutting Speeds

For metric cutters, the cutting speed is expressed in meters per minute. The formula used to determine the r/min is:

N = # of teeth on the cutter
CPt = chip load per tooth

Table 33-2 General recommendations for setting speeds

Use Higher Speeds for:	Use Lower Speeds for:
• Soft materials	• Hard, abrasive materials
• Better finishes	• Heavy cuts/rigid set-ups
• Light cuts	• High nickel or
• Excessive edge	manganese materials
chipping	• Sandy castings
• Small diameter	• Reducing excessive
tools/frail set-ups	tool wear

4" cutter 20 teeth 100cs
.005 chip load

$$r/min = \frac{CS \times 320}{D}$$

EXAMPLE: At what r/min should a 150 mm carbide-tipped milling cutter revolve to machine a piece of machine steel (CS 60)?

$$r/min = \frac{CS \times 320}{D}$$

$$= \frac{60 \times 320}{150}$$

$$= 128$$

Feed = $\frac{N \times CPt \times RPM}{10 \quad 20 \cdot .005 \quad 100}$

EXAMPLE: Calculate the r/min required to mill a groove 25 mm wide in a piece of aluminum using a high-speed end mill (CS 150).

$$r/min = \frac{CS \times 320}{D}$$

$$= \frac{150 \times 320}{25}$$

$$= 1920$$

See the general recommendations for setting speeds for milling in Table 33-2.

● MILLING FEEDS

Feed is the rate at which the work moves longitudinally (lengthwise) into the revolving cutter. It is measured either in inches per minute or millimeters per minute. *Chip per tooth* is the amount of material removed by each tooth of the cutter as it revolves and advances into the work. The *milling feed* is determined by multiplying the chip size (chip per tooth) desired, the number of teeth in the cutter, and the r/min of the cutter.

Inch Calculations

The formula used to find work feed in inches per minute is:

$$feed = N \times c.p.t. \times r/min$$

where:

N = number of teeth in the milling cutter

$c.p.t.$ = chip per tooth for a particular cutter and metal, as given in Table 33-3

r/min = number of revolutions per minute of the milling cutter

EXAMPLE: Find the feed in inches per minute using a 4 in. diameter, 12-tooth helical cutter to cut machine steel (CS 80). It would first be necessary to calculate the proper r/min for the cutter.

$$r/min = \frac{CS \times 4}{D}$$

$$= \frac{80 \times 4}{4}$$

$$= 80$$

$$Feed\ (in./min) = N \times c.p.t. \times r/min$$

$$= 12 \times .010 \times 80$$

$$= 9.6$$

$$= 10\ in./min$$

Metric Calculations

The formula used to find the work feed in millimeters per minute is the same as the formula used to find the feed in inches per minute, except that mm/min is substituted for in./min:

$$mm/min = N \times chip\ per\ tooth \times r/min$$

EXAMPLE: Calculate the feed in millimeters per minute for a 75 mm diameter, six-tooth helical milling cutter when machining a cast iron workpiece (CS 20).

First calculate the r/min of the cutter.

$$r/min = \frac{CS \times 320}{diameter\ of\ cutter}$$

$$= \frac{20 \times 320}{75}$$

$$= 85$$

$$Feed\ (mm/min) = N \times c.p.t. \times r/min$$

$$= 6 \times 0.18 \times 85$$

$$= 91.8$$

$$= 92\ mm/min$$

Table 33-3   Recommended feed per tooth (high-speed steel cutters)

Material	Side mills		End Mills		Plain Helical Mills		Saws	
	Inches	Millimeters	Inches	Millimeters	Inches	Millimeters	Inches	Millimeters
Machine steel	.007	0.18	.006	0.15	.010	0.25	.002	0.05
Tool steel	.005	0.13	.004	0.1	.007	0.18	.002	0.05
Cast iron	.007	0.18	.007	0.18	.010	0.18	.002	0.05
Bronze	.008	0.2	.009	0.23	.011	0.28	.003	0.08
Aluminum	.013	0.33	.011	0.28	.018	0.46	.005	0.13

The feed rates shown in this table are for industrial-size machines; use only half the recommended feed for training programs.

Table 33-4   General recommendations for setting feeds

Use More Feed for:	Use Less Feed for:
• Easy to machine materials • Controlling long continuous chips • Light depth of cut • Reducing chatter • Reducing excessive tool wear	• Deep slotting cuts • Better finishes • Reducing edge chipping • Milling thin walled parts

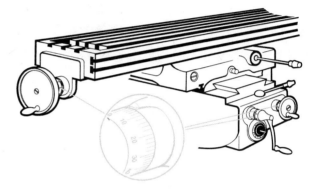

Fig. 33-13 The graduated collars permit accurate settings for horizontal, vertical, and longitudinal cuts. (Courtesy of Kelmar Associates)

See the general recommendations for setting feeds for milling in Table 33-4.

● GRADUATED COLLARS

Graduated collars are used on machine tools so that the workpiece or cutting tool can be accurately located in a horizontal or longitudinal position, and also set vertically for a depth of cut. Since work in a milling machine generally does not revolve, the amount of the collar movement is the amount that the machine table or cutting tool will move.

Fig. 33-13 shows the three important graduated collars required for most operations done on a milling machine. Each feed screw is fitted with a collar graduated in either thousandths of an inch or hundredths of a millimeter. This collar is free to revolve on a sleeve, but can be locked by a thumbscrew to a position aligned with the zero or index line as in Fig. 33-13. The graduated collars represent the amount of movement in either thousandths of an inch or hundredths of a mil-

limeter of the table, knee, or saddle. The number of graduations on the collar is directly related to the number of threads per inch or pitch in millimeters on the feed screw. For example, if the crossfeed screw on the saddle has five threads per inch, one revolution of the screw advances the saddle .200 in. A crossfeed screw with a pitch of 5 mm would advance the saddle 5 mm in a complete revolution.

SUMMARY

- A milling arbor is used to hold the cutter during the milling operation. When mounting an arbor, proper procedures must be followed to preserve the machine's accuracy.

- Arbors are removed from the machine spindle in order to accommodate larger-diameter arbors or other special machine attachments.

- Accessories used to hold work for milling include vises, V-blocks, strap clamps, angle plates, and fixtures.
- Factors to be considered when selecting the speed for a milling cutter include the material to be machined, the cutter type and material, required finish, depth of cut, and rigidity of the machine and workpiece.
- Milling feed rate is calculated from the number of teeth in the cutter, the desired chip per tooth, and the speed of the cutter.
- Graduated collars are used to accurately position workpieces and cutting tools.

KNOWLEDGE REVIEW

Mounting and Removing an Arbor

1. How is the cutter prevented from turning on the arbor?
2. Why is it necessary to clean the spindle and arbor tapers before mounting an arbor?
3. What is the purpose of the draw-in bar?
4. Explain in point form how to remove an arbor.

Mounting and Removing a Milling Cutter

5. In what position on an arbor should the cutter be placed?
6. What should be checked before starting to mill a surface?
7. Why is a wooden board used when changing a cutter?
8. Why is it important that the arbor support be in place before using a wrench on the arbor nut?

Work-Holding Devices

9. Name four commonly used holding devices.
10. What holding device can be used when many similar pieces are machined?

Cutting Speeds

11. Define cutting speed.
12. Name five important factors to consider when setting the r/min to machine a surface.
13. What r/min is required to mill a piece of cast iron (CS 60) using a 4 in. diameter high-speed cutter?
14. Calculate the r/min required to mill a piece of tool steel (CS 19 m/min) using a 75 mm carbide tipped cutter.

Milling Feeds

15. Define feed.
16. How is the milling feed determined?
17. What feed is recommended for milling machine steel (CS 100) using a 6 in. diameter, 24-tooth, side-milling cutter?
18. Calculate the feed for milling a piece of bronze (CS 30 m/min) using a 100 mm, 6-tooth, high-speed, plain helical milling cutter.

Graduated Collars

19. What feed screws are equipped with graduated collars?
20. What is the value of each division on the graduated collar?

Horizontal Milling Operations

The versatility of a **horizontal milling machine** and the number of attachments and cutting tools available make it possible to perform a wide variety of machining operations. These can range from milling a flat surface to producing intricate forms such as gears, cams, and helical forms. To prevent accidents and damage to the machine and workpiece, it is important that the work be accurately set up and held securely.

OBJECTIVES

After completing this unit, you should be able to:

- Set up and align vises and other work-holding devices.
- Machine flat and vertical surfaces to within .001 in. (0.02 mm) accuracy.
- Use the dividing head to machine geometric forms on a workpiece.

KEY TERMS

direct indexing   horizontal milling machine   index head
side milling   simple indexing

● ALIGNING A VISE

Whenever a workpiece is required to be machined to a layout or to have cuts square or parallel to an edge, the vise must first be aligned. The vise may be aligned by the following methods:

1. The simplest, but least accurate, method is to align the lines on the vise and the swivel base, Fig. 34-1.

2. The vise may be aligned at right angles to the table travel by placing the body of a steel square against the column face and the other edge against the solid jaw of the vise, Fig. 34-2.

When the full width of the vise jaw bears against the face of the square, the locknuts on the vise should be tightened.

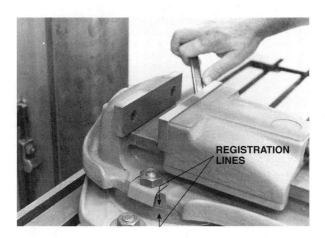

Fig. 34-1 Swivel the vise until the registration line on the vise aligns with the zero line on the base. (Courtesy of Kelmar Associates)

REGISTRATION LINES

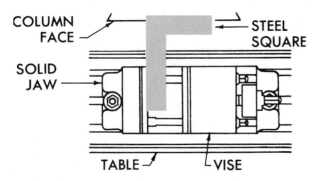

COLUMN FACE

STEEL SQUARE

SOLID JAW

TABLE

VISE

Fig. 34-2 Aligning the solid jaw of the vise with a square held against the column face. (Courtesy of Kelmar Associates)

Fig. 34-3 To accurately align a vise, mount an indicator on the milling machine arbor. (Courtesy of Kelmar Associates)

3. If greater accuracy is required, the vise should be aligned with a dial indicator mounted on the arbor. Use the following procedure:

   a. Mount the indicator on the arbor.
   b. Mount a parallel in the vise, Fig. 34-3.
   c. Raise the machine table until the indicator button is slightly below the top of the parallel.
   d. With the crossfeed handle, adjust the table until the indicator registers about .020 in. (0.5 mm) on the dial.
   e. Set the indicator bezel to zero, Fig. 34-4.
   f. Move the table longitudinally until the indicator is at the other end of the parallel, Fig. 34-5.

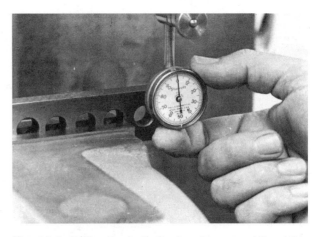

Fig. 34-4 Setting the indicator bezel to zero at the right-hand end of the parallel. (Courtesy of Kelmar Associates)

Fig. 34-5 The indicator reading at the left-hand end of the parallel shows how much the vise jaws are out of alignment. (Courtesy of Kelmar Associates)

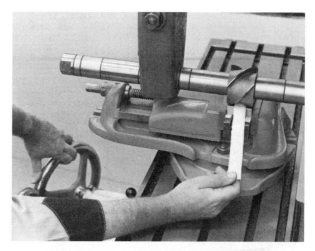

Fig. 34-6 Use a long strip of paper between the cutter and the work when setting a cutter to the work surface. (Courtesy of Kelmar Associates)

g. Loosen the vise locknuts and adjust the vise one-half the difference in the indicator readings.

h. Repeat steps (f) and (g) until the reading is the same at both ends of the parallel.

If the vise must be set at 90° to the table or the column, the same procedure may be followed and the parallel moved across the indicator with the crossfeed handle.

● SETTING THE CUTTER TO THE WORK SURFACE

Before setting a depth of cut, the operator should check that the work and the cutter are properly mounted and that the cutter is revolving in the proper direction.

To Set the Cutter to the Work Surface

1. Raise the work to within 1/4 in. (6.35 mm) of the cutter, and directly under it.

2. Hold a long piece of thin paper on the work surface, Fig. 34-6, or, a safer method is to soak a short piece of paper in coolant and paste it on the work.

When holding the paper, make sure it is long enough to prevent the fingers from coming close to the revolving cutter.

3. Start the cutter rotating and raise the table to bring the work up slowly until the cutter just touches the paper.

4. Stop the spindle and move the table until the work clears the cutter.

5. Move the knee up .002 in. (0.05 mm) for paper thickness, and set the graduated collar to zero.

6. Raise the table to the desired depth of cut.

If the knee is moved up beyond the desired amount, turn the handle backward one-half turn and come up to the required line. This will take up the backlash in the thread movement.

This method can also be used when setting the edge of a cutter to the side of a piece of work. In this case, the paper will be placed between the side of the cutter and the side of the workpiece.

Milling a Flat Surface

A milling machine vise can be used to hold square and rectangular work for machining. When milling a flat surface, use the following procedure:

1. Align the vise to the column face of the milling machine.

2. Remove all burrs from the workpiece with a file, Fig. 34-7.

3. Set the work in the vise, using parallels and paper feelers, to make sure that the work is seated properly on the parallels.

Fig. 34-7 Always remove any burrs from the work with a file to prevent inaccurate setups and painful cuts. (Courtesy of Kelmar Associates)

4. *Tighten the vise securely* on the workpiece.

5. Tap the work lightly at the four corners until the paper feelers are tight between the work and the parallels, Fig. 34-8.

6. Mount a plain helical cutter wider than the work to be machined. It should be large enough in diameter to allow clearance between the work and the arbor, Fig. 34-9.

Be sure that the teeth are pointing in the proper direction for the spindle rotation.

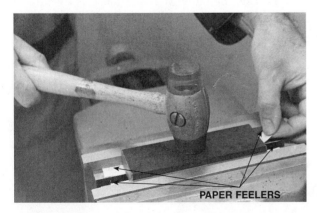

Fig. 34-8 With short paper feelers under each corner of the work, tap the work down until all feelers are tight. (Courtesy of Kelmar Associates)

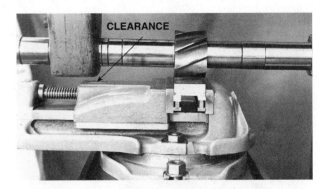

Fig. 34-9 Before any milling operation, check for clearance between the arbor support and the vise or work. (Courtesy of Kelmar Associates)

7. Set the proper spindle speed for the cutter diameter and the workpiece material, Table 33-1, page 327.

8. Set the feed to approximately .003 to .005 in. (0.07 to 0.12 mm) chip per tooth.

9. Start the cutter and raise the work, using a long paper feeler between the cutter and the work, as shown in Fig. 34-5.

10. Stop the spindle when the cutter just grips the paper.

11. Set the graduated collar on the elevating screw to zero, allowing for the paper thickness.

12. Move the work clear of the cutter and set the depth of cut using the graduated collar.

13. For roughing cuts, use a depth of not less than 1/8 in. (3 mm) and .010 to .025 in. (0.25 to 0.65 mm) for finish cuts.

14. Set the table dogs for the length of cut.

15. Engage the longitudinal feed and machine the surface.

16. Set up and cut the remaining sides as required.

See "Machining a Block Square and Parallel" in the Vertical Mill section, page 352, for the proper setup if the sides of the workpiece must be machined square and parallel.

● SIDE MILLING

Side milling is often used to machine a vertical surface on the sides or the ends of a workpiece, Fig. 34-10.

Fig. 34-10 Side milling a vertical surface. (Courtesy of Kelmar Associates)

Fig. 34-11 Mount the milling cutter as close as possible to the machine column for the best rigidity. (Courtesy of Kelmar Associates)

1. Remove all burrs from the workpiece with a file, as shown in Fig. 34-7.

2. Set up the work securely in a vise and on parallels.

> Be sure that the surface projects about 1/2 in. (12.7 mm) beyond the edge of the vise and the parallels to prevent the cutter, vise, or parallels from being damaged.

3. Mount a side milling cutter as close to the machine column as possible to provide maximum rigidity when milling, Fig. 34-11.

> Be sure that the cutter teeth are pointing in the proper direction for the cutter rotation.

4. Set the proper speed and feed for the diameter of the cutter being used.

5. Start the machine and move the table until the top corner of the work just touches the revolving cutter.

6. Set the crossfeed graduated collar to zero.

7. With the table handwheel, move the work clear of the cutter.

8. Set the required depth of cut with the crossfeed handle.

9. Take the cut across the surface, using the longitudinal automatic feed.

● CUTTING SLOTS AND KEYWAYS

Although keyways and slots are generally cut with an end mill, they may be cut on a horizontal mill using a side milling or slotting cutter.

Centering a Cutter to Mill a Slot

Use the following procedure to center the cutter.

1. Locate the cutter as close to the center of the work as possible.

2. Using a steel square and rule, adjust the work by using the crossfeed screw until the distance from the cutter to the blade of the square is exactly the same on both sides, Fig. 34-12.

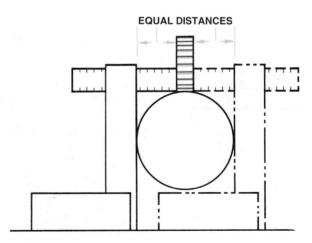

Fig. 34-12 Locating a cutter over the center of a round shaft. (Courtesy of Kelmar Associates)

3. Lock the saddle to prevent any movement during the cut.

4. Move the work clear of the cutter and set the depth of cut.

5. Cut the slot to length, using the same procedure as in milling a flat surface.

● THE INDEX OR DIVIDING HEAD

The **index head**, Fig. 34-13, is a device used to divide the circumference of a piece of work into any number of equal parts and to hold the work in the required position while the cuts are being made. The main parts of the index head are the worm and worm wheel, index crank, index plates, and sector arms, Fig. 34-14.

The index head consists of a 40-tooth worm wheel fastened to the index head spindle engaging with a single-start threaded worm attached to the index crank. Since there are 40 teeth in the worm wheel, one complete turn of the index crank will cause the spindle to rotate 1/40 of a turn. Therefore, 40 complete turns of the index crank revolve the spindle 1 complete turn, thus making a ratio of 40 to 1.

Work may be indexed by either **simple** or **direct indexing**. The formula for calculating the number of turns for simple indexing is:

$$N = \frac{40}{\text{(number of divisions to be cut)}}$$

= number of turns of the index crank

EXAMPLE: For 4 divisions,

$$\frac{40}{4} = 10 \text{ complete turns of the index crank.}$$

Fig. 34-13 An indexing or dividing head set. (Courtesy of Cincinnati Milacron Inc.)

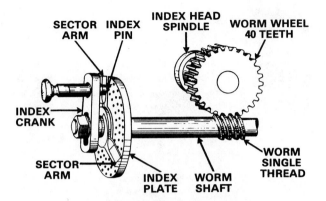

Fig. 34-14 The main parts and construction of the mechanism for simple indexing. (Courtesy of Kelmar Associates)

If it is required to cut a reamer with 8 equally spaced flutes, the indexing would be:

$$\frac{40}{8} = 5 \text{ complete turns.}$$

For six divisions it would be:

$$\frac{40}{6} = 6\frac{2}{3} \text{ complete turns.}$$

The six turns are easily made, but what of the 2/3 of a turn? This 2/3 of a turn involves the use of the index plate and sector arms.

The index plate is a circular plate having a series of hole circles. Each hole circle contains a different number of equally spaced holes into which the index crank pin can be engaged. The index plate and the crank are used along with the sector arms, which eliminates the need for counting a given number of holes each time the work is indexed or turned.

To get 2/3 of a turn, choose any circle with a number of holes that is evenly divisible by 3, such as 24, Table 34-1; then take 2/3 of 24 = 16 holes on a 24-hole circle. Thus, for the 6 divisions, we have 6 complete turns, plus 16 holes on a 24-hole circle.

To Set the Sector Arms

Use the following procedure to set the sector arm:

1. To carry out the example using the six turns and 16 holes on a 24-hole circle, place the beveled edge of a sector arm against the index crank pin, usually in the top hole, then count 16 holes on the 24-hole circle.

Table 34-1 Index plate hole circles

Brown & Sharpe	Cincinnati Standard Plate	
Plate 1 15-16-17-18-19-20	One Side	24-25-28-30-34-37-38-39-41-42-43
Plate 2 21-23-27-29-31-33		
Plate 3 37-39-41-43-47-49	Other Side	46-47-49-51-53-54-57-58-59-62-66

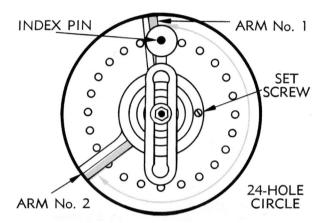

Fig. 34-15 The sector arms set for 16 holes on a 24-hole circle (2/3 turn). (Courtesy of Kelmar Associates)

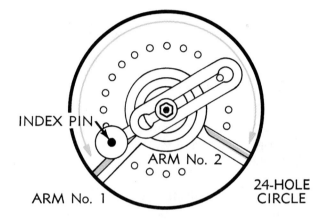

Fig. 34-16 The position of the sector arms after indexing 6⅔ turns. (Courtesy of Kelmar Associates)

> Do not include the hole in which the index crank pin is engaged, Fig. 34-15.

2. Move the other sector arm just beyond the sixteenth hole, and tighten the set screws.

3. After the first cut has been made and the cutter returned for the next cut, withdraw the index crank pin, make six complete turns *clockwise*, plus the 16 holes between the sector arms.

4. Stop between the last two holes, gently tap the index crank, and allow the pin to snap into place.

5. Move the sector arms around against the pin, ready for the next division, Fig. 34-16.

> If the pin is turned past the required hole, an error will appear in the spacing unless it is turned back at least one-half turn to eliminate backlash and then brought back to the proper hole.

To Mill a Hexagon

A hexagon, 1⅝ in. (40 mm) across flats is to be milled on a 2 in. (50 mm) diameter shaft, Fig. 34-17. To cut the hexagon, check the alignment of the index centers and then mount the work. The indexing for 6 divisions = 40/6 or 6⅔ turns.

The following procedure could be used for milling any hexagon—only the measurements would be changed:

1. Select a hole circle that the denominator 3 will divide into, such as 24.

2. Set the sector arms to 16 holes on 24-hole circle, as shown in Fig. 34-15

Do not count the hole the pin is in.

3. From the hole the pin is in, count 16 holes and adjust the other sector arm to just beyond the sixteenth hole, as shown in Fig. 34-15.

4. Start the machine and set the cutter to the top of the work by using a paper feeler, Fig. 34-6, page 331.

5. Set the graduated dial to zero.

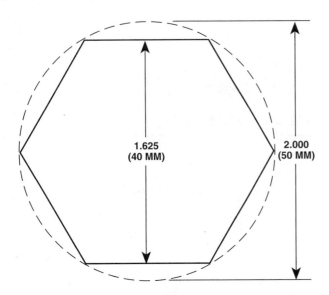

Fig. 34-17 A 1.625 in. (40 mm) hexagon to be cut on a 2.000 in. (50 mm) diameter shaft. (Courtesy of Kelmar Associates)

6. Calculate the depth of cut, Fig. 34-17, 2.000 in. − 1.625 in. = .375 in. (50 mm − 40 mm = 10 mm). As only one-half of the material is to be removed from each side, the depth of cut = .375/2 = .1875 in. (10/2 = 5 mm).

7. Move the work clear of cutter.

8. Raise the table .1875 in. (5 mm).

9. Lock the knee clamp.

10. Mill the first flat.

11. Index 20 turns ($3 \times 6\frac{2}{3}$) and mill the opposite side, Fig. 34-18.

12. With a micrometer, check the distance across the flats and adjust and recut, if necessary, Fig. 34-19.

13. Continue indexing $6\frac{2}{3}$ turns until all six sides are cut.

Fig. 34-18 Milling the second flat of the hexagon. (Courtesy of Kelmar Associates)

Fig. 34-19 Measuring the distance across the flats with a micrometer. (Courtesy of Kelmar Associates)

● PLAIN OR DIRECT INDEXING

Direct indexing is the simplest form of indexing, but it can only be used for milling divisions that are evenly divisible into 24, 30, or 36. The common divisions that can be obtained are listed in Table 34-2.

To Obtain Direct Indexing

Use the following procedure to obtain direct indexing:

1. Disengage the worm from the wheel by swinging the worm disengaging bracket upward, Fig. 34-20. This disengages the spindle from the index crank.

Fig. 34-20 The worm bracket must be moved to disengage the spindle from the index crank. (Courtesy of Kelmar Associates)

Table 34-2 Divisions for which plate can be used

Plate Hole Circles or Slots															
24	2	3	4	—	6	8	—	—	12	—	—	24	—	—
30	2	3	—	5	6	—	—	10	—	15	—	—	30	—
36	2	3	4	—	6	—	9	—	12	—	18	—	—	36

2. Mount the proper direct indexing plate with the holes or slots facing the plunger pin.

3. Mount the workpiece in the dividing head chuck or between centers.

4. Lock the spindle and cut the first surface or groove.

5. Disengage the plunger pin from the indexing plate and unlock the spindle, Fig. 34-21.

6. Turn the indexing plate the proper number of holes or slots.

7. Engage the plunger pin and lock the spindle.

8. Cut the remaining surfaces or grooves in the same manner.

EXAMPLE: What direct or plain indexing is required to mill four flats on a round shaft?

$$\text{Indexing} = \frac{24}{4} = 6 \text{ holes or slots in}$$
24-hole circle for each flat milled

OR

$$\frac{36}{4} = 9 \text{ holes or slots in}$$
36-hole circle for each flat milled.

SUMMARY

- A horizontal milling machine can mill flat surfaces as well as intricate forms.

- Side milling can be used to machine a vertical surface on the sides or ends of a workpiece.

- Keyways and slots can be cut on a horizontal mill using a side milling or slotting cutter.

- The index head is used to divide the circumference of a workpiece into a number of equal parts and to hold the work in the required position.

- Work can be indexed by either simple or direct indexing.

- Direct indexing is the simplest form of indexing, but it can only be used under certain conditions.

KNOWLEDGE REVIEW

Aligning a Vise

1. If the sides of the work do not have to be exactly square, how may the vise be aligned?

2. Name and briefly describe two other methods of aligning a vise on a milling machine.

Setting the Cutter to the Work Surface

3. Explain briefly how to set a cutter to the work surface before setting a depth of cut.

Milling a Flat Surface

4. State two methods by which a vise may be aligned.

Fig. 34-21 The direct indexing plate, attached to the dividing head spindle, is used for indexing a limited number of divisions. (Courtesy of Kelmar Associates)

PLUNGER PIN LEVER

PLUNGER PIN

5. What is the purpose of using paper feelers when setting up work on parallels?

6. What depth of cut is recommended for

   a. rough cuts?   b. finish cuts?

Side Milling

7. What is the purpose of side milling?

8. What precaution should be observed when setting up work for side milling?

9. Why should the side-milling cutter be mounted as close to the spindle as possible?

10. What graduated collar is used to set the depth of cut when side milling?

Centering a Cutter

11. List in point form how to center a 1/4 in. (6.35 mm) keyway cutter to a 1 in. (25.4 mm) shaft.

The Index or Dividing Head

12. What is the purpose of the dividing head?

13. What two types of indexing may be done on the dividing head?

14. What is the formula for calculating simple indexing?

15. Calculate the simple indexing for 24, 27, and 36 divisions using the Brown and Sharpe plate.

16. What procedure should be followed to set the sector arms for 15 holes in the 20-hole circle?

17. What will be the depth of cut required to mill a hexagon .750 in. across the flats from a 1.250 in. diameter piece of material?

18. After one side of a hexagon has been machined, what side should be milled next? Explain why.

19. What direct indexing is required to cut the following using a Brown and Sharpe index plate:

    a. a hexagon?   b. an octagon?

Vertical Milling Machines

A **vertical milling machine**, has basically the same parts as a horizontal plain milling machine. Instead of the cutter fitting into a horizontal spindle, it fits into a vertical spindle. On most machines, the head can be swiveled 90° to either side of the center line for the drilling of angular holes, and the milling of angular surfaces and slots. The vertical milling machine is especially useful for operations such as face and end milling, drilling, boring, jig boring and cutting keyways, dovetails, and T-slots.

OBJECTIVES

After completing this unit, you should be able to:

- Identify and state the purpose of the main vertical milling machine parts.
- Recognize and describe the purpose of common end mills.
- Select speeds and feeds for various types of end mills.

KEY TERMS

collets (solid, spring)   end mill (shell, solid)   flycutter

● VERTICAL MILL PARTS

The *base*, Fig. 35-1A, is made of ribbed cast iron. It may contain a coolant reservoir.

The *column* is cast in one piece with the base. The machined face of the column provides the ways for the vertical movement of the knee. The upper part of the column is machined to receive a turret on which the overarm is mounted.

The *overarm* may be round or of the more common dovetailed ram type. It may be adjusted toward or away from the column and swiveled to increase the capacity of the machine.

The *head* is attached to the end of the ram (or overarm), Fig. 35-1A. On universal type machines, the head may be swiveled in two planes, clockwise or counterclockwise, and toward or away from the column. The motor, which provides the drive to the spindle, is mounted on top of the head. Spindle speed changes may be made by means of gears and V-belts or variable speed pulleys on some models. The spindle mounted

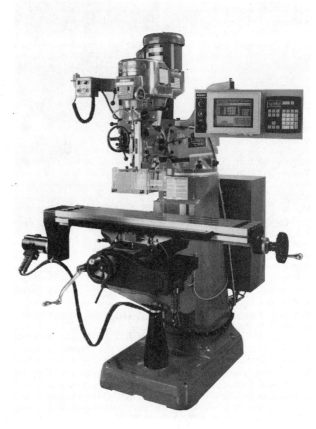

Fig. 35-1B The EZ-TRAK SX 2-axis control vertical milling machine can store thousands of operations in its memory. (Courtesy of Bridgeport Machines, Inc.)

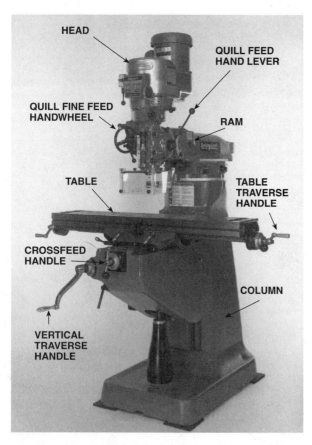

Fig. 35-1A A vertical milling machine can be used for face and end milling operations. (Courtesy of Bridgeport Machines Inc.)

in the *quill* may be fed by means of the quill feed hand lever, the quill fine feed handwheel, or by automatic power feed.

The *knee* can be moved up and down the face of the column and supports the saddle and table. On most machines, all table movements are controlled manually. Automatic feed control units for any table movements are usually added as accessories.

The *2-axis control vertical mill*, Fig 35-1B, is a step between the standard milling machine and the CNC machining center. It contains a digital readout (DRO) which can store thousands of individual operations in its memory even though no computer is involved and no CNC programming is necessary. The machine can be taught to produce any part simply by manually moving the table to a position and pushing the *ENTER* button at the end of each move to store the machine position in memory. This 2-axis control machine is capable of machining angles and radii without a rotary table, routine milling of arc, slots, pockets, and drilling bolt-hole circles.

● MILLING CUTTERS AND COLLETS

Most machining on the vertical mill is done with either an end mill, a shell end mill, or a fly-cutter.

End mills have cutting teeth on the end as well as on the periphery, and are fitted to the machine spindle by a suitable collet or adapter. They may be of two types, the solid end mill, Fig. 35-2, or the shell end mill, which is fitted to a separate arbor.

Solid end mills may have two or more flutes.

- The *two-flute end mills*, Fig. 35-2A, are used for general-purpose milling because they provide good chip clearance and remove metal quickly. They are end-cutting and can be used for milling slots, keyways, plunge cutting, and drilling shallow holes.

- The *three-flute end mills*, Fig. 35-2B, provide good chip clearance, while at the same time minimize chatter. They are used for the same purpose as two-flute end mills.

- The *four-flute end mills*, Fig. 35-2C, are generally used for finish cutting and are available as center-cutting and non center-cutting types. The non center-cutting type requires a starting hole before milling a slot in the center of a workpiece. When plunging with a center-cutting, four-flute end mill, downfeed it slowly to prevent the end flutes from loading or clogging.

- *Roughing end mills*, Fig. 35-2D, are used for general-purpose milling to remove metal quickly. Their design breaks chips into small pieces, thereby reducing heat, friction, and the amount of horsepower required to remove metal.

The most common form end mills are the ball and the corner-rounding type.

- The *ball end mill*, Fig. 35-2E, is used for milling a concave radius on the bottom of slots, fillets, etc. The teeth on these end mills are cut to the center, allowing them to drill into the material at the start of a cut.

- The *corner-rounding end mill*, Fig. 35-2F, is used to produce a convex form on the corner of a workpiece. Standard corner-rounding end mills are available with a radius of 1/32 in. to 1 in. (1 mm to 25 mm).

(A) TWO-FLUTE

(B) THREE-FLUTE

(C) FOUR-FLUTE

(D) ROUGHING

(E) BALL END

(F) CORNER ROUNDING

(G) T-SLOT

(H) DOVETAIL

Fig. 35-2 Common types of end mills. (Courtesy of Niagara Cutter, Inc.)

- The *T-slot cutter*, Fig. 35-2G, is used to mill T-slots in the tables and beds of machine tools. A vertical slot must first be cut in the workpiece to allow the neck and shank of the cutter to enter.

- The *dovetail cutter*, Fig. 35-2H, is used for milling dovetail slides into a workpiece. They are available as standard cutters with 45° and 60° included angles.

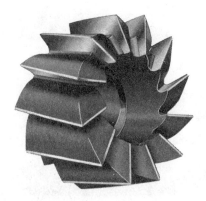

(A) SHELL END MILL

(B) FLYCUTTER

Fig. 35-3 Common cutters used for the removal of larger work surfaces. (A) (Courtesy of Niagara Cutter, Inc.), (B) (Courtesy of Kelmar Associates)

(A) UNCOATED

(B) TITANIUM NITRATE COATED

Fig. 35-4 Coated cutting tools improve chip flow and reduce the possibility of a built-up edge. (Courtesy of Balzers Tool Coating Inc.)

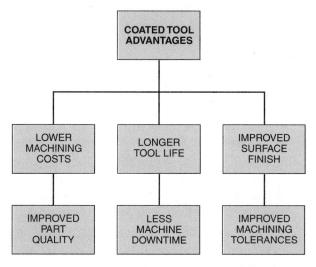

Fig. 35-5 The main advantages of coated cutting tools. (Courtesy of Kelmar Associates)

The most common methods of machining large surfaces on a vertical milling machine are with a shell end mill or a flycutter. *Shell end mills*, Fig.35-3A, are multi-tooth cutters with teeth on the face and periphery. They are usually held on a stub arbor and can be changed quickly when they are worn or must be replaced. The **flycutter**, Fig. 35-3B, can hold two or more single-pointed cutting tools. They provide an economical way of machining large work surfaces.

Coated End Mills

Special coatings can be applied to the surface of end mills to prolong their life, increase productivity, and lower manufacturing costs. The most common coatings are titanium nitride and titanium carbonitride. These ultra-thin coatings improve the lubricity of the end mill's cutting edge and face to allow the chip to slide freely and resist a buildup of work material, Fig. 35-4. The coated end mill lasts longer because

less heat is created at the cutting edge which improves surface finish and work accuracy, even at higher speeds and feeds. For a more detailed explanation of coated tools, see Unit 32.

The main advantages of coated cutting tools are shown in Fig. 35-5.

Types of Collets

There are two main types of **collets**—the spring type and the solid type. Both are driven by a key in the spindle bore and a keyway on the outside of the collet.

The *spring collet*, Fig. 35-6A, holds and drives the cutter by means of friction between the collet and cutter. On heavy cuts, the cutter may move up in the collet if it is not tightened securely.

The *solid collet,* Fig. 35-6B, is more rigid and holds the cutter securely. Solid collets may be driven by a key in the spindle and a keyway in the collet, or by two drive keys on the spindle.

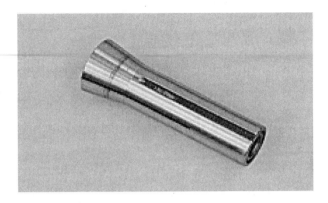

(A) SPRING

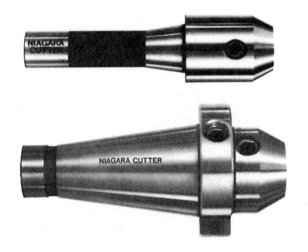

(B) SOLID

Fig. 35-6 Types of vertical milling machine collets. (A) Spring (Courtesy of Kelmar Associates), (B) Solid (Courtesy of Niagara Cutter, Inc.)

The cutter is driven and prevented from turning by one or two setscrews in the collet, which bear against flats on the cutter shank.

● MILLING GUIDELINES

Several factors must be considered when setting up a milling job. These include the type of milling operation, speeds, feeds, depth of cut, and safety.

Conventional or Climb Milling

Before using any end mill, it is important to decide whether the machining operation will be performed by conventional or climb milling. Since most vertical milling machines do not have a backlash eliminator, climb milling, where the work and cutter are traveling in the same di-

rection, can be performed by taking lighter cuts and snugging up the machine table slides. See Unit 32 (page 318) for a more detailed explanation of conventional and climb milling.

Speeds, Feeds, Depth of Cut

The speeds and feeds used for vertical milling operations are basically the same as those listed for milling cutters in Unit 33. Since end mills are generally smaller and not quite as sturdy as milling cutters, it is wise to use a little slower feedrate with end mills. A good rule of thumb is that the maximum depth of cut should not be greater than the diameter of the end mill.

Safety Precautions

Modern metal-cutting operations generally use high spindle speeds, high temperatures, and cutting forces. It is very important to protect yourself from flying chips, especially when using flycutters. All guards for the machine must be in place and safety glasses worn at all times. A safe machining operation must also consider the rigidity of the setup, cutter spindle speed, the size, shape and overhang of the end mill, and the condition of the machine. See Unit 32 for a more detailed section on Milling Machine Safety, page 320.

● END MILL PERFORMANCE

An end mill will give the best machining results when the teeth are sharp and the end mill is used properly. Too often, end mills fail prematurely because the operator does not recognize the factors that affect tool life. Some of the main factors are heat, abrasion, chipping of cutting edges, clogging, built-up edges, work hardening, and cratering.

Heat

Excessive heat, Fig. 35-7, created during machining is one of the main causes of cutting-edge failure and shortened tool life. The heat is caused by the cutting edges rubbing on the work material and the chip sliding along the tooth face. When the heat becomes too high, it affects the hardness of the cutting edges and they become dull. Keep the heat generated to a minimum by using sharp end mills and applying a good supply of cutting fluid.

Abrasion

Abrasion, Fig. 35-8, is a wearing-away action caused by the microstructure of the material

Fig. 35-7 Excessive heat is one of the main causes of tool failure and shortened life. (Courtesy of the Weldon Tool Co.)

Fig. 35-9 Too heavy a load on the cutting edge will cause it to chip. (Courtesy of the Weldon Tool Co.)

Fig. 35-8 Abrasion will dull the cutting edges and create wear lands. (Courtesy of the Weldon Tool Co.)

Fig. 35-10 Clogging reduces the chip space and can cause a cutter to break. (Courtesy of the Weldon Tool Co.)

being cut. It wears away the sharp cutting edges and creates wear lands which get bigger as the wear continues. This will produce more heat during machining and dull the cutting edges quickly.

Chipping of Cutting Edges

Small fractures or chipping occur on cutting edges when the cutting forces used in machining are greater than the cutting edges can stand, Fig. 35-9. The main causes of cutting-edge chipping

are feedrate too high, chattering, built-up edge breakaway, brittleness of end mill, and running cutters backward.

Clogging

Clogging occurs when material with a "gummy" composition clogs or jams in the flutes of the end mill, Fig. 35-10. To reduce clogging, take lighter cuts, use finer feeds, use end mills with fewer teeth to provide more chip space, and apply coolant under pressure to flush out the chips.

Fig. 35-11 Built-up edges will cause a poor cutting action and produce rough surface finishes. (Courtesy of the Weldon Tool Co.)

Fig. 35-12 Work hardening of the material surface will increase cutter wear and cause cutter failure. (Courtesy of the Weldon Tool Co.)

Fig. 35-13 A good supply of cutting fluid can help to reduce cratering. (Courtesy of the Weldon Tool Co.)

Built-Up Edges

Built-up edges, Fig. 35-11, occur when particles of the work material cold-weld themselves to the faces of the teeth. This buildup affects the cutting action, requiring more power for cutting and producing a poor surface finish on the work. When the built-up edge gets too large, it breaks away taking with it a portion of the cutting edge. Reduce the built-up edge by reducing the feed, depth of cut, and apply cutting fluid.

Work Hardening

Work hardening of some material, Fig. 35-12, is caused by the action of the cutting edges deforming or compressing the work surface which changes the microstructure and increases the surface hardness. This condition can occur when machining high-temperature and high-strength superalloys, austenitic steels, and many of the high-alloyed carbon tool steels.

To prevent or reduce work hardening, keep a constant feed on the cutter, use proper cutting speeds, climb mill wherever possible, and use a good supply of cutting fluid.

Never allow a cutter to rub on the work to prevent work hardening. KEEP IT CUTTING AT ALL TIMES.

Cratering

Cratering, Fig. 35-13, is caused by the high heat and abrasion of the chips sliding on the tooth face next to the cutting edge. The sliding and curling of the chips create a narrow hollow or groove in the tooth face which keeps getting larger, eventually resulting in tool failure. Cratering can be reduced by applying a good supply of cutting fluid and using coated end mills.

SUMMARY

- Vertical milling machines use a variety of cutters such as solid end mills, shell end mills, and flycutters to perform many machining operations.

- Special coatings applied to end mills can prolong tool life, increase productivity and reduce costs.

- Spring collets and solid collets are used to hold and drive milling cutters.

- Proper use of an end mill requires consideration of the type of milling (conventional or climb), speed, feed, depth of cut, and safety issue.

- The main factors affecting tool life include heat, abrasion, chipping of cutting edges, clogging, built-up edges, work hardening, and cratering.

KNOWLEDGE REVIEW

Vertical Milling Machine

1. How does the vertical mill differ from the horizontal mill?

2. Name five operations that can be performed on a vertical mill.

3. State the purpose of the following vertical mill parts:
   a. column
   b. overarm
   c. knee

Milling Cutters and Collets

4. Name four types of end mills used on a vertical mill.

5. What type of end mill should be used for:
   a. plunge cutting?
   b. finishing?
   c. roughing?
   d. concave radius?
   e. convex radius?

6. For what purpose are shell end mills and flycutters used?

7. Name two types of collets used in a vertical mill.

8. How can climb milling be performed if the machine does not have a backlash eliminator?

8. What is the maximum depth of cut recommended for an end mill?

10. List four safety factors that should be considered for any machining operation.

End Mill Performance

11. List six main factors that can affect tool life.

12. How can the heat generated during a milling operation be kept to a minimum?

13. Name five causes of cutting edge chipping.

14. How does a built-up edge affect the cutting action?

15. Explain how work-hardening occurs.

16. How can work-hardening be prevented or reduced?

UNIT
36

Vertical Milling Machine Operations

The **vertical milling machine** is one of the most versatile and useful machines in a school or manufacturing shop. The many types of cutters and attachments that are available for this machine allow machining operations such as end and surface milling, radius and cam milling, drilling, reaming, boring, cutting slots and keyways, etc. to be performed.

OBJECTIVES

After completing this unit, you should be able to:

• Mount and remove end mills and cutters.
• Machine the four sides of a workpiece square and parallel.
• Drill and ream holes in a vertical mill.
• Mill slots and keyways.

KEY TERMS

dial indicator   vertical head   vertical milling machine

● MOUNTING AND REMOVING CUTTERS

The vertical milling machine permits the use of a wide variety of cutting tools. These tools may be held in the spindle by a spring collet or an adapter which is held in the spindle by means of a draw-in bar, Fig. 36-1.

Mounting a Cutter in a Spring Collet

Use the following procedure to mount a cutter in a spring collet:

1. Shut off the electric power to the machine.

2. Place the proper cutter, collet, and wrench on a piece of masonite on the machine table.

3. Clean the taper in the spindle.

4. Place the draw-in bar into the hole in the top of the spindle.

5. Clean the taper and keyway on the collet.

6. Insert the collet into the bottom of the spindle, press up, and turn it until the keyway aligns with the key in the spindle.

7. Hold the collet up with one hand and with the other, thread the draw-in bar into the collet for about four turns.

8. Hold the cutting tool with a rag and insert it into the collet for the full length of the shank.

9. Tighten the draw-in bar into the collet (clockwise) by hand.

10. Hold the spindle-brake lever and tighten the draw-in bar securely with a wrench, Fig. 36-2.

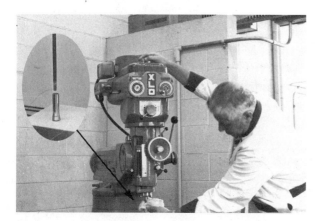

Fig. 36-1 Mounting a spring collet in a vertical mill spindle. (Courtesy of Kelmar Associates)

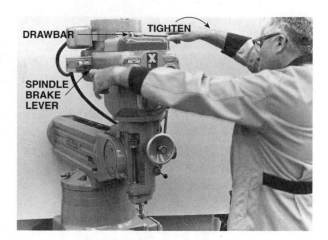

Fig. 36-2 Tightening the draw-in bar holds the collet and cutter in the spindle. (Courtesy of Kelmar Associates)

Removing a Cutter from a Spring Collet

Use the following procedure to remove a cutter from a spring collect:

1. Shut off the electric power to the machine.

2. Place a piece of masonite on the machine table to hold the necessary tools.

3. Pull on the spindle brake lever and loosen the draw-in bar with a wrench (counter-clockwise).

4. Loosen the draw-in bar by hand, only about three turns.

Do not unscrew the draw-in bar from the collet, otherwise the thread on the draw-in bar and collet could be damaged.

5. With a soft-faced hammer, strike down sharply on the head of the draw-in bar to break the taper contact between the collet and spindle.

6. With a cloth, remove the cutter from the collet.

7. Clean the cutter and replace it in its proper storage place.

Mounting a Cutter in a Solid Collet

Use the following procedure to mount a cutter in a solid collet:

1. Shut off the electric power to the machine.

2. Place the cutter, collet, and necessary tools on a wooden board on the machine table.

3. Slide the draw-in bar through the top hole in the spindle.

4. Clean the spindle taper and the taper on the collet.

5. Align the keyway or slots of the collet with the keyway or drive keys in the spindle, and insert the collet into the spindle.

6. Hold the collet up in the spindle and thread the draw-in bar clockwise with the other hand.

7. Pull on the brake lever and tighten the draw-in bar securely with a wrench.

8. Insert the end mill into the collet until the flat(s) align with the setscrew(s) of the collet and tighten the screws.

9. To *remove* an end mill from a solid collet, loosen the screws, hold the end mill with a cloth, and pull down.

● ALIGNING THE VERTICAL HEAD

Alignment of the head of a vertical milling machine is very important when performing most operations on this machine. If the head is not set 90° to the table, any surfaces machined or holes drilled will not be square with the work surface. When face milling, the machined surface will be stepped if the head is not square with the table.

Three methods may be used to align the vertical head (spindle):

- Aligning the graduations on the head.

- Setting a square on the machine table and checking two sides of the spindle sleeve at 90° to each other.

- The dial indicator method, which is the most accurate.

Procedure for the Dial-Indicator Method

Use the following procedure to align the vertical head using the dial-indicator method:

1. Fasten an indicator clamp unit onto a suitable rod held in the spindle, Fig. 36-3A.

2. Position the indicator so that when it is revolved it will be close to the front and back edges of the table.

3. Mount a large convex-contact point on the end of the indicator plunger.

Fig. 36-3A The indicator dial being set to zero (0) at the right-hand end of the table. (Courtesy of Kelmar Associates)

4. Lower the spindle until the indicator contact point touches the table and the dial indicator registers about one-quarter of a revolution, then set the bezel to zero (0), Fig. 36-3A.

5. Lock the spindle in this location.

6. Slowly rotate the vertical mill spindle 180° by hand until the contact point is at the opposite end of the table, Fig. 36-3B. Compare the two indicator readings.

7. If there is any difference in the readings, loosen the swivel-mount locknuts and adjust the head until the indicator registers one-half the difference between the high and low reading. Tighten the locking nuts.

8. Recheck the accuracy of the head and adjust if necessary.

Fig. 36-3B A double-exposure view showing the indicator being used to check both ends of the machine table. (Courtesy of Kelmar Associates)

Fig. 36-3C A double-exposure view showing the spindle rotated 90° to check the two edges of the table. (Courtesy of Kelmar Associates)

9. Turn the vertical mill spindle 90° and set the dial indicator as in step 4.

10. Rotate the machine spindle 180° and check the reading at the other side of the table, Fig. 36-3C.

11. If the two readings are not the same, repeat step 7 until both readings are the same.

12. Tighten the swivel-mount locknuts.

13. Recheck the readings and adjust if necessary.

To prevent the indicator from being damaged in the T-slots of the table, it is advisable to work from the high reading first and then rotate to the low reading. For best accuracy when aligning the head, always use as long a clamp rod as possible so that a large table area can be covered.

● TO MACHINE A FLAT SURFACE

Use the following procedure to machine a flat surface:

1. Check that the vertical head is at right angles to the table in both directions, so that a flat surface is produced.

2. Mount a suitable flycutter in the machine spindle, Fig. 36-4A.

3. Set the machine spindle speed for the size of the flycutter and material being milled.

4. Remove all burrs from the workpiece.

5. Clean the work and the vise.

6. Mount the work in the vise, on suitable parallels and with paper feelers under each corner.

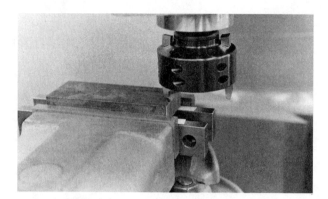

Fig. 36-4A Machining a flat surface with a flycutter. (Courtesy of Kelmar Associates)

7. Start the machine and raise the table until the cutter just touches the surface of the work near the right-hand end.

8. Move the table so that the cutter clears the end of the work.

9. Set the vertical feed graduated collar to zero.

10. Raise the table .010 in. (0.25 mm) and take a trial cut approximately 1/4 in. (6 mm) long.

11. Stop the spindle and measure the work thickness.

12. Raise the table the desired amount and mill the surface to size.

● MACHINING A BLOCK SQUARE AND PARALLEL

In order to machine the four sides of a piece of work so that its sides are square and parallel, it is important that each side be machined in a definite order, Fig. 36-4B. It is very important that dirt and burrs be removed from the work, vise, and parallels every time the work is reset, because dirt and burrs can cause inaccurate work.

Machining Side #1:

1. Check that the vertical head is at right angles to the table in both directions.

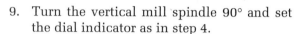

NOTE

A square on the machine table with its blade against the side of the spindle sleeve is a fairly accurate method of checking for squareness.

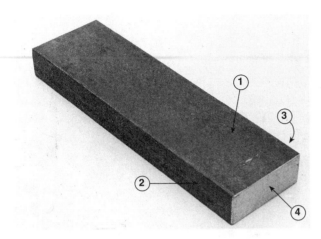

Fig. 36-4B The sequence for machining the sides of a rectangular workpiece square and parallel. (Courtesy of Kelmar Associates)

Fig. 36-6 Raise the table until the revolving cutter just scratches the right-hand end of the work. (Courtesy of Kelmar Associates)

2. Remove all burrs from the workpiece.

3. Clean the work and the vise.

4. Mount the work in the center of the vise on parallels, with its largest side (#1) up and use paper feelers under each corner, Fig. 36-5.

5. Mount a flycutter in the milling machine spindle.

6. Set the machine to the proper speed for the size of the cutter and the material to be machined. (See Table 33-1 on page 327.)

7. Start the machine, and raise the table until the cutter just touches near the right-hand end of the side #1, Fig. 36-6.

8. Move the work clear of the cutter.

9. Raise the table about .060 in. (1.5 mm) and machine side #1 using a steady feed rate, Fig. 36-7.

10. Take the work out of the vise and remove all burrs from the edges with a file.

Machining Side #2:

11. Clean the vise, work, and parallels thoroughly.

12. Place the work on parallels, if necessary, with side #1 against the solid jaw and side #2 up, Fig. 36-8.

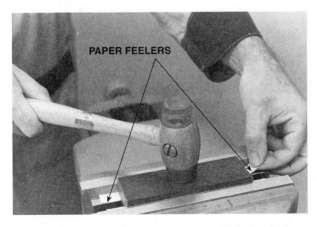

Fig. 36-5 Side #1 (the largest surface side) should be facing up. (Courtesy of Kelmar Associates)

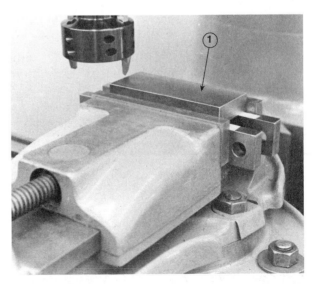

Fig. 36-7 Only a clean-up cut is required on side #1. (Courtesy of Kelmar Associates)

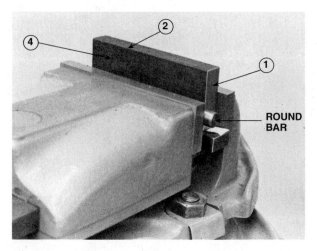

Fig. 36-8 The finished side (#1) is placed against the solid jaw of the vise when setting up to machine side #2. (Courtesy of Kelmar Associates)

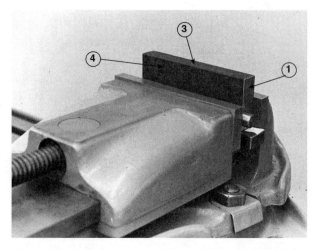

Fig. 36-9 The setup required to machine side #3. The width of the part should be to size after this operation. (Courtesy of Kelmar Associates)

13. Place a round bar between side #4 and the movable jaw.

> The round bar must be in the center of the amount of work being held inside the vise jaws to hold the work squarely against the solid jaw.

14. Tighten the vise securely and tap the work down until the papers feelers are tight.

15. Follow steps 7 to 10 and machine side #2.

Machining Side #3:

16. Clean the vise, work, and parallels thoroughly.

17. Place side #1 against the solid vise jaw with side #2 resting on parallels, if necessary, Fig. 36-9.

> Move the parallel to the left so that about 1/4 in. (6 mm) of the work extends beyond the end of the parallel. This permits the work to be measured while in the vise.

18. Place a round bar between side #4 and the movable jaw.

> The round bar must be in the center of the amount of work being held inside the vise jaws.

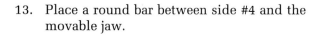

19. Tighten the vise securely and tap the work down until the paper feelers are tight.

20. Start the machine and raise the table until the cutter just touches near the right-hand end of side #3.

21. Move the work clear of the cutter and raise the table about .010 in. (0.25 mm).

22. Take a trial cut about 1/4 in. (6 mm) long, stop the machine, move the work clear of the cutter, and measure the width of the work, Fig. 36-10.

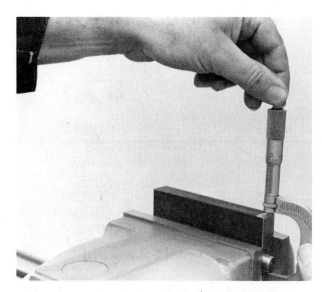

Fig. 36-10 Measuring the size of the work after the trial cut. (Courtesy of Kelmar Associates)

Fig. 36-11 The setup for machining side #4. (Courtesy of Kelmar Associates)

23. Raise the table the required amount and mill side #3 to the correct width.

24. Remove the work and file off all burrs.

Machining Side #4:

25. Clean the vise, work, and parallels thoroughly.

26. Place side #1 down on the parallels with side #4 up and tighten the vise securely, Fig. 36-11.

With three finished surfaces, the round bar is not required when milling side #4.

27. Tap the work down until the paper feelers are tight.

28. Follow steps 20 to 23 and machine side #4 to the correct thickness.

● MACHINING THE ENDS OF A BLOCK OR WORKPIECE

The ends of the workpiece may be machined by two methods, depending on the length and shape of the workpiece. Short pieces (no more than about 3½ in. or 90 mm long) may be held upright in the center of the vise and machined with a flycutter or shell end mill. Long workpieces must be gripped in the vise, which is set parallel to the table travel and machined with an end mill.

Flycutter or Shell End Mill Method

Use the following procedure to machine short pieces:

1. Remove the burrs from the workpiece.

2. Clean the vise and the work.

3. Set the work in the center of the jaws with one end on the base of the vise and tighten the vise lightly.

4. Tap the work down on the base and square it with the solid vise jaw as shown in Fig. 36-12.

5. Tighten the vise securely.

6. Using the crossfeed handle, center the workpiece with the cutter and take a cleanup cut.

7. Remove the work from the vise, remove the burrs, and check for squareness.

8. Clean the vise and the work thoroughly, and place the machined end on paper feelers on the base of the vise.

9. Tighten the vise securely and tap the work down onto the feelers so that the work is seated properly.

10. Take a cleanup cut off the end.

11. Remove the burrs and measure the height of the work, Fig. 36-13.

12. Raise the table the required amount and machine the work to length.

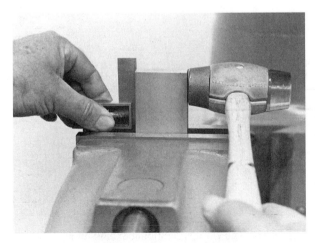

Fig. 36-12 Squaring a short workpiece before machining the end. (Courtesy of Kelmar Associates)

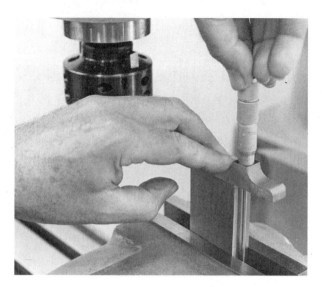

Fig. 36-13 Measuring the length of the workpiece with a micrometer after a clean-up cut. (Courtesy of Kelmar Associates)

End Mill Method

Use the following procedure to machine long pieces:

1. Set the toolhead square (90°) in both directions.
2. Align the vise parallel to the table travel.
3. Remove the burrs from the workpiece.
4. Clean the workpiece and the vise thoroughly.
5. Mount the work in the vise on suitable parallels and on paper feelers, Fig. 36-14.

Fig. 36-14 The workpiece should extend about 1/4 in. (6 mm) past the vise jaws and parallels for machining the ends. (Courtesy of Kelmar Associates)

Fig. 36-15 Mount the end mill and check the direction of spindle rotation. (Courtesy of Kelmar Associates)

> The end of the work should extend past the vise jaws by at least 1/4 in. (6.35 mm).

6. Tighten the vise securely; then tap the work onto the parallels until all the paper feelers are tight.
7. Mount a suitable end mill in the spindle, Fig. 36-15.
8. Set the proper spindle speed (r/min) and check that the cutter is revolving in a clockwise direction.
9. Center the cutter with the work and move the table until the *revolving* end mill just cuts along paper feeler between the work and the cutter, Fig. 36-16.

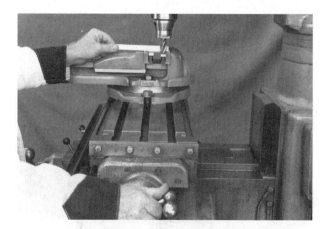

Fig. 36-16 A long piece of paper should be used when touching up a cutter to the work. (Courtesy of Kelmar Associates)

Fig. 36-17 The depth of cut when squaring an end is set with the longitudinal feed handle. (Courtesy of Kelmar Associates)

10. Move the work clear of the cutter, using the crossfeed handle.

11. Set a .030 to .060 in. (0.8 to 1.5 mm) depth of cut with the longitudinal table feed handle, Fig. 36-17, and tighten the table traverse lock.

12. Take the cut across the end of the work, using the crossfeed handle.

13. Remove the work, remove the burrs, and check for squareness.

14. Repeat the procedure on the other end and machine the work to length.

● PRODUCING AND FINISHING HOLES

Drilling, tapping, reaming, counterboring, countersinking, and boring operations can be done accurately on a vertical mill. The spindle is adaptable for the use of the tools for these operations, and the feed screw graduated collars provide accurate locating of the holes to be machined.

The vertical head must be at right angles (90°) to the table before doing any of these operations.

To Drill on a Vertical Mill

Use the following procedure to drill on a vertical mill:

1. Mount the work in a vise or clamp it to the table.

Work must be supported by parallels which are positioned so that they will not interfere with the drill.

2. Mount a drill chuck in the spindle.

3. Mount a center finder in the drill chuck.

4. Adjust the table until the center-punch mark of the hole to be drilled is in line with the tip of the rotating center finder.

5. Tighten the table and saddle clamps.

6. Stop the machine and remove the center finder.

7. Place a center drill in the chuck and drill the center hole until about 1/4 of the countersink section enters the work.

8. Mount the proper size drill in the drill chuck.

9. Set the machine to the proper speed for the size of the drill being used.

10. Use the quill hand feed lever and drill the hole, using cutting fluid.

To Ream on a Vertical Mill

After the hole has been drilled, the table should *not* be moved, so that the reamer and the drilled hole stay aligned.

1. Mount the reamer in the spindle or drill chuck.

2. Set the speed and feed for reaming (approximately 1/4 of the drilling speed).

3. Apply cutting fluid as the reamer is fed into the hole with the quill hand feed lever.

4. Stop the machine spindle.

5. Remove the reamer from the hole. (Do *not* turn the reamer backwards.)

To Mill Slots and Keyways

Keyways, keyseats, and slots can be easily and quickly cut in a vertical milling machine by using an end milling cutter.

1. Lay out the keyseat and the end of the shaft as in Fig. 36-18.

2. Place the work in the vise, or in V-blocks for longer work.

3. Using a square, align the end layout line, which will set the keyseat to proper position on the top of the shaft, Fig. 36-19.

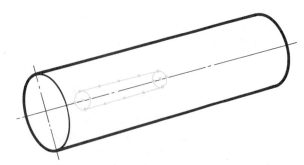

Fig. 36-18 The layout required for a keyseat on a shaft. (Courtesy of Kelmar Associates)

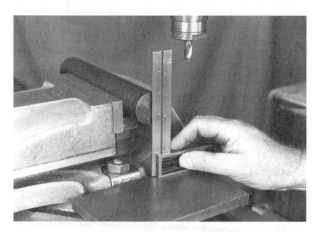

Fig. 36-19 Setting up the work to position the keyseat for milling. (Courtesy of Kelmar Associates)

Fig. 36-20 Using a long piece of paper to set the cutter to the side of the work. (Courtesy of Kelmar Associates)

Fig. 36-21 The keyseat machined to the proper length. (Courtesy of Kelmar Associates)

4. Fasten the vise securely.

5. Mount a two- or three-fluted end mill. The diameter of the end mill must be the width of the keyseat.

6. Center the workpiece by touching the revolving end mill to a piece of paper held against the shaft, Fig. 36-20.

7. Set the crossfeed graduated collar to zero.

8. Lower the table until the revolving end mill clears the work.

9. Move the table over a distance of half the diameter of the work plus half the diameter of the cutter and lock the table.

10. Tighten the saddle lock to stop the table from moving during the machining operation.

11. Adjust the table until the end mill is in line with one end of the keyseat.

12. Feed the table up until the end mill cuts to its full diameter on the shaft.

13. Set the graduated collar on the vertical traverse screw shaft to zero.

14. Raise the knee until the depth of cut equals one-half the thickness of the key or the diameter of the cutter.

15. Lock the knee clamp and machine the keyseat to the proper length, Fig. 36-21.

SUMMARY

• The vertical head must be aligned to produce surfaces and holes that are square with the work surface. The dial-indicator method of alignment is the most accurate.

- When machining a piece of work so that its four sides are square and parallel, each side must be machined in a definite sequence.

- A vertical mill can be used for drilling, boring, tapping, and reaming operations, the vertical head must be at right angles to the table.

KNOWLEDGE REVIEW

Mounting and Removing Cutters

1. Why is it important that the draw-in bar be loosened only three turns?

Machining a Flat Surface

2. When machining a flat surface, why must the vertical head be at right angles (90°) to the table?

Milling a Block Square and Parallel

3. What effect would burrs and dirt on the work, parallels, or vise have?

4. How is the work set up to mill side #1?

5. How is the work set up to mill side #2?

6. Where should the round bar be placed when milling sides #2 and #3?

7. How is the work set up for milling side #4?

Machining the Ends of a Block

8. Why it is important to remove the burrs from a workpiece before mounting it in a vise?

9. What is the purpose of the paper feelers?

10. Briefly describe how the end of a short block may be squared.

11. How is longer work held for squaring the end?

Producing and Finishing Holes

12. Explain briefly how the following operations can be performed:

    a. drilling   b. reaming

To Mill Slots and Keyways

13. Explain how to set up a shaft in a vise to mill a keyseat.

14. Briefly describe how to center the cutter after the shaft has been set up.

Grinders

Grinding is a metal-removal process that uses an abrasive cutting tool to produce a high surface finish and bring the workpiece to an accurate size and shape. In the grinding process, a revolving grinding wheel or an abrasive belt is brought into contact with the surface of a workpiece. Each abrasive grain on the periphery of the grinding wheel or on the surface of the abrasive belt, is a cutting tool, and as it contacts the work surface it removes a minute chip of metal. The modern grinding machine is capable of finishing soft or hardened workpieces to tolerances of .0002 in. (0.005 mm) or less on high-production runs, while at the same time producing a very high surface finish.

Figures courtesy of (clockwise from far left): Carborundum Abrasives of North America, Cincinnati Milacron Inc., GE Superabrasives, AVCO Bay State Abrasives, Do ALL Co., Do ALL Co.

Bench and Abrasive Belt Grinders

There are various types of grinding machines used in the machine tool trade to suit the sizes and shapes of a wide variety of workpieces. *Bench and pedestal* grinders are used for the sharpening of cutting tools and the rough grinding of metals. *Abrasive belt grinders* are used for finishing flat and contour work. There are many other special types of grinders used for specific purposes in the machine tool trade.

OBJECTIVES

After completing this unit, you should be able to:

- Rough grind metals safely.
- Sharpen a lathe toolbit and a standard twist drill.
- Use a belt grinder to finish flat and contour-shaped workpieces.

KEY TERMS

belt grinder	bench grinder	dressing
offhand grinding	pedestal grinder	truing

BENCH AND PEDESTAL GRINDER

Bench and pedestal grinders are used for the sharpening of cutting tools and the rough grinding of metal. Because the work is usually held in the hand, this type of grinding is commonly called **offhand grinding**.

The **bench grinder** is mounted on a bench while the **pedestal grinder**, Fig. 37-1, being a larger machine, is fastened to the floor. Both types consist of an electric motor with a grinding wheel mounted on each end of the spindle. One wheel is usually a coarse-grained wheel for the fast removal of metal, while the other is a fine-grained wheel for finish grinding. The U-shaped work rests provide a rest for either the work or the hands while grinding.

Always keep the work rests adjusted within 1/16 in. (1.5 mm) of the wheel to prevent work being jammed between the rest and the wheel.

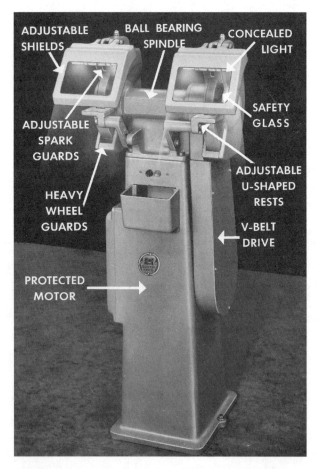

Fig. 37-1 The main parts of a pedestal grinder. (Courtesy of South Bend Lathe Corp.)

 The safety glass eyeshields provide eye protection for the operator, but it is good safety practice to wear safety glasses while grinding, even though the machine is equipped with eyeshields.

GRINDING WHEELS

Aluminum oxide and *silicon carbide* are the two types of grinding wheels generally used on bench and pedestal grinders. These manufactured abrasives are superior to natural abrasives such as emery, sandstone, corundum, and quartz, because they contain no impurities and are of a more uniform grain structure.

Aluminum oxide, Fig. 37-2, is made in an arc-type electric furnace by charging bauxite, ground coke, and iron borings. Grinding wheels made from aluminum oxide are used to grind high-tensile strength materials such as hard or soft carbon and alloy steels, tough bronze, etc.

Silicon carbide, Fig. 37-3, is manufactured in a resistance-type electric furnace by charging silica sand and coke with small amounts of sawdust and salt. Grinding wheels made of silicon carbide are used to grind low-tensile strength materials such as aluminum, copper, ceramics, cast iron, etc.

Grinder Safety

Because grinding wheels operate at very high speeds and the grinding particles are very

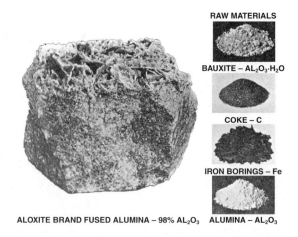

RAW MATERIALS

BAUXITE – $AL_2O_3 \cdot H_2O$

COKE – C

IRON BORINGS – Fe

ALOXITE BRAND FUSED ALUMINA – 98% AL_2O_3 ALUMINA – AL_2O_3

Fig. 37-2 Aluminum oxide is manufactured from bauxite ore, ground coke, and iron borings. (Courtesy of Carborundum Abrasives of North America)

RAW MATERIALS

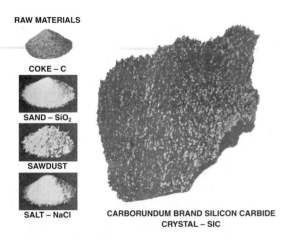

COKE – C

SAND – SiO₂

SAWDUST

SALT – NaCl

CARBORUNDUM BRAND SILICON CARBIDE
CRYSTAL – SIC

Fig. 37-3 Silicon carbide is manufactured from silica sand, coke, sawdust, and salt. (Courtesy of Carborundum Abrasives of North America)

fine, it is important to observe the following safety precautions.

1. Always wear approved safety glasses when operating a grinder.

2. Use a screwdriver handle or wooden mallet to ring test every wheel before mounting. (See Fig. 38-9.)

3. Always stand to one side of the wheel when starting a grinder.

 Never stand in line with a grinding wheel, in case the wheel breaks during the start-up, Fig. 37-4

4. Allow a new wheel to run for about one minute before using.

Fig. 37-4 Always stand to one side of a wheel when starting a grinder. (Courtesy of Kelmar Associates)

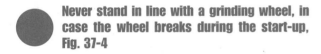 **If a wheel is going to break, it will break within the first minute.**

5. Always use a wheel guard that covers at least one-half of the grinding wheel.

6. Never run a grinding wheel faster than the speed recommended on its blotter.

7. Do not grind on the side of a wheel unless it is designed for this purpose.

8. Never force a grinding wheel by jamming work into it.

9. Be sure that the grinder workrest is within 1/16 in. (1.5 mm) of the grinding wheel face to prevent work from jamming between the wheel and the rest.

10. Always remove burrs produced by grinding with a file, or use an abrasive stone on hardened material, Fig. 37-5.

Dressing and Truing a Wheel

When a grinding wheel is used, several things happen to it:

1. Small metal particles embed themselves in the wheel, causing it to become loaded or clogged, Fig. 37-6A.

2. The abrasive grains become worn smooth and the wheel loses its sharp cutting edges.

3. Grooves become worn in the face of the wheel.

Any one of these conditions requires the immediate *dressing* or *truing* of a wheel. If a

Fig. 37-5 Remove the burrs and sharp edges produced by grinding as soon as possible. (Courtesy of Kelmar Associates)

Fig. 37-6A The face of a grinding wheel that has been severely loaded. (Courtesy of Kelmar Associates)

loaded wheel is not dressed, it will not cut properly, and it will start to heat, burn, and distort the workpiece. **Dressing** is the process of reconditioning the wheel to make it cut better, Fig. 37-6B. **Truing** refers to shaping a wheel to a desired shape and to make its grinding surface run true with its axis. Both truing and dressing may be done at the same time using a mechanical wheel or an abrasive stick dresser.

Fig. 37-6B The face of a grinding wheel that has been properly dressed. (Courtesy of Kelmar Associates)

(A) MECHANICAL

(B) ABRASIVE STICK

Fig. 37-7 Types of grinding wheel dressers. (Courtesy of Norton Co.)

A mechanical wheel dresser, Fig. 37-7A, commonly called a star dresser, is usually employed to dress an offhand grinding wheel. It consists of a number of hardened, pointed disks mounted loosely in a handle. The abrasive dressing stick, Fig. 37-7B, is used occasionally to give clogged grinding wheels a light dressing to keep them cutting freely.

To Dress and True a Wheel

To dress and true a wheel, use the following procedures:

1. Adjust the grinder work rest so that when the lugs of the mechanical dresser are against its edge, the dresser rolls just touch the face of the wheel, Fig. 37-8.

 Never make any adjustments on a grinder unless the wheel is stopped.

2. Wear an approved pair of safety glasses.
3. Stand to one side of the wheel and then start the grinder.
4. Hold the dresser down firmly with its lugs against the edge of the work rest.

Fig. 37-8 Dressing a grinding wheel using a mechanical dresser. (Courtesy of Norton Co.)

 DO NOT have the dresser rollers contact the revolving wheel at this time.

5. Move the dresser across the face of the wheel in a steady motion for a trial pass.

6. After each pass, tilt the holder up slightly to advance the dresser disks into the wheel.

7. When the wheel is dressed, stop the grinder and adjust the work rest back to within 1/16 in. (1.5 mm) of the wheel face.

● GRINDING A LATHE TOOLBIT

All lathe toolbits, regardless of shape, must have relief angles and side rake in order for them to cut properly. Since it is impossible to cover the grinding of all types of lathe toolbits, only the general-purpose toolbit will be explained in detail.

A toolbit grinding gage, Fig. 37-9, should be used to check all angles and clearances.

Fig. 37-9 Checking the end-relief angle with a toolbit grinding gage. (Courtesy of Kelmar Associates)

To Grind a Lathe Toolbit

Use the following procedure to grind a lathe toolbit:

1. Hold the toolbit firmly while supporting the hands on the grinder work rest.

2. Tilt the bottom of the toolbit in toward the wheel and grind the 10° side-relief angle and form required on the left side of the toolbit, Fig. 37-10A.

3. Grind until the side cutting edge is about 1/2 in. (12.7 mm) long and the point is over about 1/4 of the width of the toolbit, Fig. 37-10B.

> While grinding, move the toolbit back and forth across the face of the wheel. This helps to grind faster and prevents grooving the wheel.

4. High-speed steel toolbits must be cooled frequently.

(A)

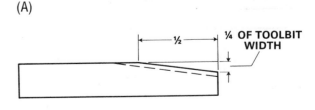

(B)

Fig. 37-10 Grinding the side-cutting edge and the side-relief angle on a general-purpose toolbit. (Courtesy of Kelmar Associates)

Never overheat a toolbit.

5. Hold the back end of the toolbit lower than the point and grind the 15° end-relief angle on the right side, Fig. 37-11A. At the same time, the end cutting edge should form an angle of 70 to 80° with the side cutting edge, Fig. 37-11B.

6. Hold the toolbit about 45° to the axis of the wheel, Fig. 37-12A, tilt the bottom of the toolbit in, and grind the 14° side rake on the top of the toolbit, Fig. 37-12B.

7. Grind the side rake for the entire length of the side cutting edge, but do not grind the top of the cutting edge below the top of the toolbit.

8. Grind a slight radius on the point, being sure to keep the same end and side relief angles.

9. Use an oilstone to hone the point and cutting edge of the toolbit to remove sharp edges and improve its cutting action.

(A)

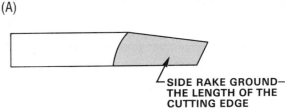

(B)

Fig. 37-12 Grinding the side rake on top of the toolbit. (Courtesy of Kelmar Associates)

To Sharpen a General-Purpose Toolbit

A general-purpose lathe toolbit, ground to the shape and dimensions shown in Fig. 37-13, can be quickly sharpened by grinding only the end cutting edge. It is important to maintain the same shape and end relief angle when sharpening a toolbit.

After the worn portion is removed, grind a slight radius on the point and hone the cutting edge. When the side-cutting edge becomes too short, after repeated sharpening, regrind the whole toolbit to the original shape and dimensions.

(A)

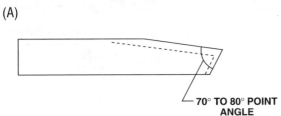

(B)

Fig. 37-11 Shaping the point and grinding the end-relief angle. (Courtesy of Kelmar Associates)

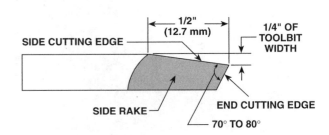

Fig. 37-13 The top view of a general-purpose toolbit showing its shape and dimensions. (Courtesy of Kelmar Associates)

● GRINDING A DRILL

Before using any drill, it is always wise to examine its condition. To cut properly and efficiently, a drill should have the following characteristics:

- The cutting edges should be free from wear or nicks.
- Both cutting edges should be the same angle and the same length.
- The margin should be free of wear.
- There should be a proper amount of lip clearance.

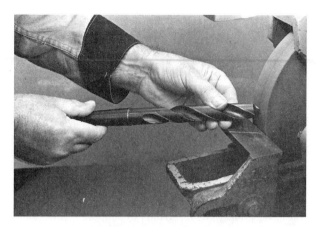

Fig. 37-15 The shank of the drill should be held a little lower than the point when sharpening. (Courtesy of Kelmar Associates)

To Sharpen a Drill

A general-purpose drill has an included point angle of 118° and lip clearance from 8 to 12°, Fig. 37-14A&B. Use the following procedure for sharpening:

1. Hold the drill near the point with one hand; with the other hand, hold the shank of the drill slightly lower than the point, Fig. 37-15.

2. Move the drill so that it is 59° to the face of the grinding wheel, Fig. 37-16.

A line scribed at 59° on the grinder work rest will help to keep the drill at the correct angle.

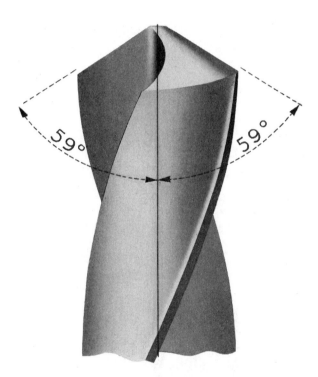

(A) POINT ANGLE

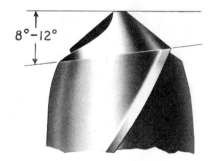

(B) LIP CLEARANCE

Fig. 37-14 The correct angle and lip clearance of a general- purpose drill. (Courtesy of Kelmar Associates)

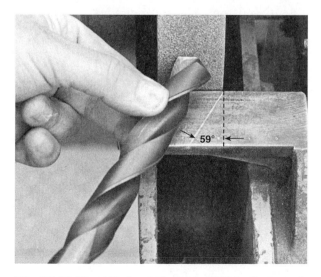

Fig. 37-16 The drill should be held at 59° to the face of the grinding wheel. (Courtesy of Kelmar Associates)

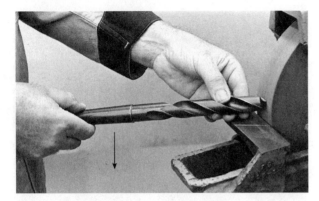

Fig. 37-17 Keep the lip against the grinding wheel and lower the drill shank. (Courtesy of Kelmar Associates)

3. Bring the drill close to, but not touching the grinding wheel.

4. Rotate the drill in the fingers until the lip or cutting edge of the drill is parallel to the grinder work rest.

5. Bring the lip of the drill against the grinding wheel and slowly lower the drill shank, Fig. 37-17.

● **DO NOT TWIST THE DRILL.**

6. Back the drill out, so that it just clears the grinding wheel.

7. Without moving the position of the body or hands, rotate the drill one-half turn, until the other lip or cutting edge is parallel to the work rest.

8. Grind the other cutting edge in the same manner as the first.

9. Check the angle of the drill point with a drill point gage, Fig. 37-18.

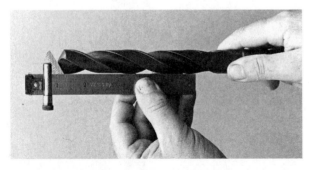

Fig. 37-18 Checking the point angle and the lip length with a drill point gage. (Courtesy of Kelmar Associates)

10. Repeat operations 5 to 8 until the cutting edges are sharp and the lands are *free from wear.*

● ABRASIVE BELT GRINDERS

Abrasive belt grinders have provided industry with a fast, easy, and economical method of finishing flat or contour work. In most cases, this machine has replaced the old hand method of using a file and abrasive cloth. Work that would take hours to finish by hand can be finished in a few minutes on an abrasive belt grinder.

The abrasive **belt grinder**, whether horizontal or vertical, consists of a motor, a contact wheel, an idler wheel, and an endless abrasive belt, Fig. 37-19.

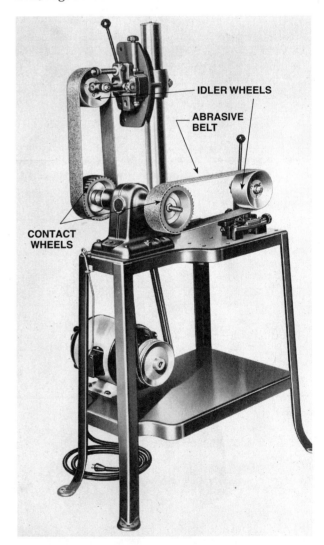

Fig. 37-19 The main parts of an abrasive belt grinder. (Courtesy of Delta International Machinery Corp.)

● ABRASIVE BELTS

Aluminum oxide and *silicon carbide* abrasive belts are used on belt grinders. Aluminum oxide abrasive belts should be used for grinding high-tensile strength materials (all steels and tough bronze). Silicon carbide abrasive belts should be used for grinding materials with low tensile strength (cast iron, aluminum, brass, copper, glass, plastic). A 60- to 80- grit belt may be used for general-purpose work. For fine finishes, a 120- to 220- grit belt is recommended.

Safety Precautions

1. Always wear eye protectors when grinding.
2. Run the abrasive belts in the direction indicated by the arrows stamped on their backs.
3. Never grind on the up side of an abrasive belt (i.e., with the belt rotating toward rather than away from you).
4. If much grinding or polishing is required on a workpiece, cool it frequently in a suitable medium.
5. Sharp corners or edges should be brought in contact with the belt *lightly*, otherwise these rough edges will tear the belt.

● FINISHING FLAT SURFACES

Whenever *flat* surfaces are to be finished, a hard, flat platen, Fig. 37-20, should be mounted on the underside of the belt. This platen will pre-

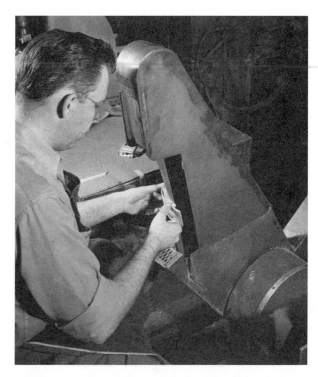

Fig. 37-21 A soft or formed contact wheel should be used for finishing contour surfaces. (Courtesy of Grinding Wheel Institute)

vent the belt from giving, thereby ensuring flat work.

● FINISHING CONTOUR SURFACES

Concave and *convex* surfaces are easily finished by mounting a soft or formed contact wheel on the machine. The work is then held against the contact wheel which conforms to the shape of the part, Fig. 37-21. This method makes it possible to finish intricate forms or sharp radii.

Fig. 37-20 A flat platen under an abrasive belt allows flat surfaces to be finished on a belt grinder. (Courtesy of Norton Co.)

SUMMARY

- Bench and pedestal grinders are used to sharpen cutting tools and rough grind metals.
- The grinding-wheel materials, aluminum oxide and silicon carbide, are manufactured abrasives that are superior to natural abrasives.
- Abrasive belt grinders use aluminum oxide and silicon carbide belts to finish flat and contour work.

- Dressing is the process of reconditioning a grinding wheel to make it cut better; truing is the process of making a grinding wheel run true with its axis.

KNOWLEDGE REVIEW

Bench and Pedestal Grinders

1. Define "offhand grinding."

2. Why is a coarse and a fine-grained wheel usually found on bench or pedestal grinders?

3. How close should the work rests be set to the wheel? Explain why.

Grinding Wheels

4. Name the two types of grinding wheels used on bench or pedestal grinders.

5. Name four natural abrasives and explain why they are rarely used in wheel manufacture.

6. What types of materials are ground with

   a. aluminum oxide wheels?

   b. silicon carbide wheels?

Safety Precautions

7. What precaution should be observed when starting a grinder?

8. Why should a new wheel be left running for about one minute before using?

9. Why should burrs produced by grinding be removed?

Dressing and Truing a Wheel

10. Name three reasons for dressing and truing a wheel.

11. Define "dressing" and "truing."

12. Name two types of wheel dressers.

13. Explain how to dress and true a grinding wheel.

Grinding a Lathe Toolbit

14. Why are the end- and side-relief angles required on a general-purpose lathe toolbit?

15. What instrument can be used to check the toolbit angles and clearances?

16. Why is it recommended to move the toolbit back and forth over the face of the grinding wheel?

17. Why should the cutting edge of a toolbit be honed after grinding?

18. How can a general-purpose toolbit be resharpened quickly and easily?

Grinding a Drill

19. List the characteristics of a correctly ground drill.

20. What is the lip clearance and point angle of a general-purpose drill?

21. At what angle is the drill held to the face of the wheel?

22. How should the cutting edge be held in relation to the grinder work rest?

Abrasive Belt Grinders

23. Why are abrasive belt grinders used in industry?

Abrasive Belts

24. Name two types of abrasive grinding belts.

25. For what purpose is each type used?

Safety Precautions

26. In what direction should abrasive belts be run?

27. How should sharp edges of work be brought in contact with the belt? Explain.

Finishing Surfaces

28. Why is a platen used when finishing flat surfaces?

29. How can contour surfaces be finished?

UNIT 38

Surface Grinder Wheels and Operations

The most common surface grinder is the horizontal spindle grinder with a reciprocating table, Fig. 38-1, which consists of a revolving abrasive wheel mounted on a horizontal spindle and a rectangular table that moves back and forth under the wheel. A magnetic chuck, mounted on the table, provides a fast and easy method of holding the work while grinding.

OBJECTIVES

After completing this unit, you should be able to:

- Select the proper grinding wheel to suit the work material and the grinding operation.
- Mount and prepare a wheel for grinding.
- Grind a flat surface within .001 in. (0.02 mm) accuracy.

KEY TERMS

abrasive grain   grade   surface grinder

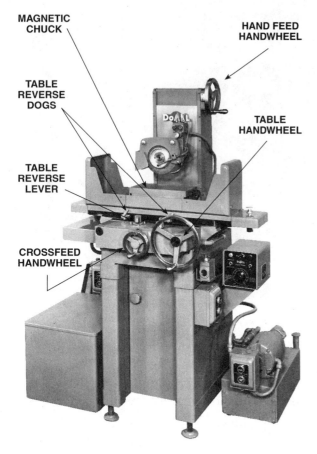

Fig. 38-1 The main operative parts of a horizontal spindle, reciprocating surface grinder. (Courtesy of DoALL Co.)

● SURFACE GRINDER PARTS

The following list describes the main parts of a surface grinder.

The *wheel feed handwheel* moves the grinding wheel up or down to set the depth of cut.

The *table traverse handwheel* moves the table back and forth (longitudinally) under the grinding wheel. The table can be operated by hand or automatically.

The *crossfeed handwheel* is used to move the table in or out (transversely). Either a hand or automatic crossfeed can be used.

The *table reverse dogs* are used to set the length of table travel.

The *table traverse reverse lever* is used to reverse the direction of the table travel .

Safety Precautions

1. Never run a grinding wheel faster than the maximum speed recommended on the wheel blotter.

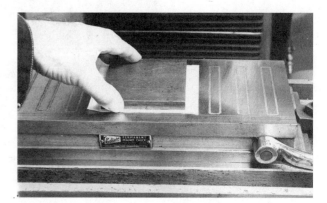

Fig. 38-2 Before starting a grinder, try to remove the workpiece to make sure the magnetic chuck is holding. (Courtesy of Kelmar Associates)

● **A wheel that is run faster than recommended can break and damage the machine or cause an accident.**

2. Always have the wheel guard covering at least one-half of the grinding wheel.

3. Before starting a grinder, *always* make sure that the magnetic chuck has been turned on by trying to remove the work from the chuck, Fig. 38-2.

4. See that the wheel clears the work before starting the grinder.

5. Stand to one side of the wheel before starting the machine.

6. Never attempt to clean the magnetic chuck, or mount and remove work, until the wheel has completely stopped.

7. *ALWAYS* wear safety goggles while grinding, Fig. 38-3.

Fig. 38-3 Always wear safety glasses while in a machine shop, especially when grinding. (Courtesy of Kelmar Associates)

● GRINDING WHEELS

Grinding wheels are made up of a large number of sharp-edged, abrasive crystals that are bonded (cemented) together in the form of a grinding wheel. Each abrasive crystal is a cutting tool that removes minute (very small) chips, similar to a chip produced on a lathe or milling machine. The most common grinding wheels are made of aluminum oxide or silicon carbide.

With the development of superabrasives in the mid 1950's, diamond and cubic boron nitride (CBN) grinding wheels are finding wide acceptance in metalworking operations because of their super-hard, super wear-resistant qualities.

The following list describes the major grinding-wheel materials and their applications:

- *Aluminum oxide* crystals, found in approximately 75% of the grinding wheels manufactured, are used to grind high-tensile strength materials and most ferrous metals.

- *Silicon carbide* abrasives are used for grinding low-tensile strength materials such as cast iron, aluminum, brass, bronze, and cemented carbides.

- *Diamond abrasives*, four times harder and three and one-half times more wear-resistant than aluminum oxide, are used for grinding ultra-hard materials such as tungsten carbide, ceramics, and space-age alloys. *They should never be used on ferrous metals.*

- *Cubic boron nitride (CBN)* abrasives have two- and one-half times the hardness and wear-resistance of aluminum oxide. They have found wide use in grinding hardened cutting tools and very hard ferrous metals and superalloys.

Wheel Composition

The materials that go into the manufacture of grinding wheels are the abrasive grains and the bond. Variations of these two materials can produce wheel characteristics such as grade and structure, which are important factors in the selection and use of grinding wheels.

Abrasive Grain

An important factor in the selection of a grinding wheel is the **abrasive grain** size, Fig. 38-4A,B,C. The size of the abrasive grain will affect the following:

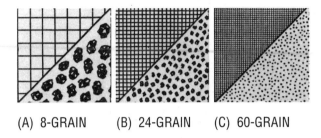

(A) 8-GRAIN   (B) 24-GRAIN   (C) 60-GRAIN

Fig. 38-4 Relative abrasive grain sizes. (Courtesy of Carborundum Abrasives of North America)

- *Type of surface finish produced.* Use coarse grains for rough grinding, fine grains for finish grinding.

- *Material being ground.* Use coarse grains on soft materials, fine grains on hard materials.

- *Amount of material removed.* Coarse grains remove metal quickly, fine grains remove metal more slowly.

- *Contact between wheel and work.* Use a coarse-grain wheel where the contact area is large, a fine grain-wheel where the contact area is small.

Grade

The **grade** of a grinding wheel generally refers to the strength with which the bond material holds the abrasive grains together in a wheel. Weak bond posts, Fig. 38-5A, release abrasive grains rapidly, therefore the wheel is classified as soft grade. Strong bond posts, Fig. 38-5C, are classified as a hard grade because they hold the abrasive grains more tightly in the wheel.

The grades of grinding wheels range from A (softest) to Z (hardest). The grade of grinding wheel selected depends on the hardness of the

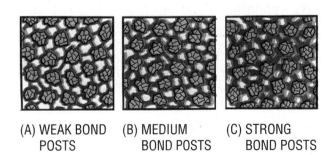

(A) WEAK BOND POSTS   (B) MEDIUM BOND POSTS   (C) STRONG BOND POSTS

Fig. 38-5 Grinding wheel grades. (Courtesy of Carborundum Abrasives of North America)

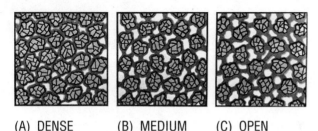

(A) DENSE   (B) MEDIUM   (C) OPEN

Fig. 38-6 Grinding wheel structure. (Courtesy of Carbarundum Abrasives of North America)

material, area of contact, machine condition, wheel speed, and the feed rate.

Structure

The structure of a grinding wheel depends on the space taken up by the grain and the bond compared to the voids (spaces) between them. If the spacing of the grains is close, the structure is dense, Fig. 38-6A; the structure is open when the grain spacing is fairly wide, Fig. 38-6C. Wheels with open structure provide better chip clearance and remove metal faster than those with a dense structure.

The structure of grinding wheels is indicated by numbers ranging from 1 (dense) to 15 (open). When selecting the structure of a grinding wheel, the following factors must be considered: work material, contact area, surface finish, and coolant application.

Wheel Identification and Selection

The most common wheel used for surface grinding operations is the Type 1 aluminum oxide straight wheel, Fig. 38-7. The standard marking system, Fig. 38-8, is used by manufacturers to identify all the specifications of an aluminum-oxide and silicon-carbide grinding wheel. These specifications include abrasive type, grain size, grade, structure, and bond. This information, printed on the blotters of most grinding wheels, allows the user to select the proper wheel to suit the workpiece material and the type of grinding operation.

When selecting a wheel for a surface grinding operation, consider the following factors:

- *Abrasive*—Whenever steel must be ground, use an aluminum oxide wheel.
- *Grain Size*—A medium grain, 46 to 60 grit, can be used for most surface grinding operations.

Fig. 38-7 A Type 1 straight grinding wheel. (Courtesy of Norton Company)

- *Grade*—Use a medium-grade (J) wheel because it will break down and present new sharp cutting edges when grains become dull.
- *Structure*—For most steels of medium hardness, use a medium-hard (#7) wheel.
- *Bond*—For most steel grinding, use a vitrified-bond wheel.

Grinding Wheel Care

The proper care of grinding wheels is important to the operator's safety and the life of the wheel. A wheel carelessly handled can break and damage the machine and possibly cause an accident. Three ways to ensure safe grinding wheels are by inspection, handling, and storage.

Inspection

After each wheel is received, it should be visually checked for cracks and chips, and then ring-tested, Fig. 38-9. A good vitrified wheel, tapped with a wooden mallet or screwdriver handle, should give a good, clear, metallic tone. If the wheel is cracked or damaged, it will produce a dull sound. Should there be any doubt about the wheel's condition, *do not use the wheel;* return it to the manufacturer for speed testing.

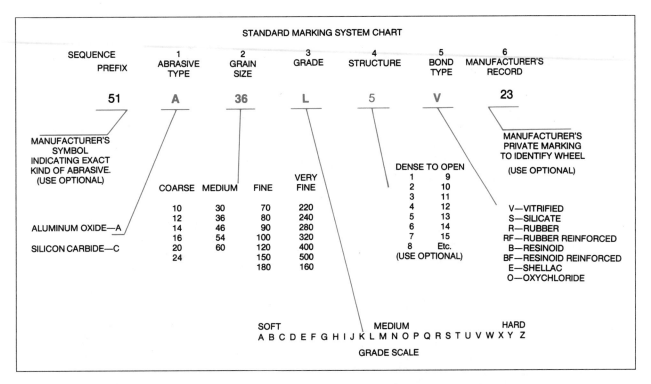

STANDARD MARKING SYSTEM CHART

SEQUENCE	1	2	3	4	5	6
PREFIX	ABRASIVE TYPE	GRAIN SIZE	GRADE	STRUCTURE	BOND TYPE	MANUFACTURER'S RECORD
51	A	36	L	5	V	23

MANUFACTURER'S SYMBOL INDICATING EXACT KIND OF ABRASIVE. (USE OPTIONAL)

MANUFACTURER'S PRIVATE MARKING TO IDENTIFY WHEEL (USE OPTIONAL)

ALUMINUM OXIDE—A

SILICON CARBIDE—C

COARSE	MEDIUM	FINE	VERY FINE
10	30	70	220
12	36	80	240
14	46	90	280
16	54	100	320
20	60	120	400
24		150	500
		180	160

DENSE TO OPEN
1	9
2	10
3	11
4	12
5	13
6	14
7	15
8	Etc.

(USE OPTIONAL)

V—VITRIFIED
S—SILICATE
R—RUBBER
RF—RUBBER REINFORCED
B—RESINOID
BF—RESINOID REINFORCED
E—SHELLAC
O—OXYCHLORIDE

SOFT   MEDIUM   HARD
A B C D E F G H I J K L M N O P Q R S T U V W X Y Z
GRADE SCALE

Fig. 38-8 Standard grinding wheel marking system. (Courtesy of Grinding Wheel Institute)

Fig. 38-9 Supporting the wheel on one finger for ring testing. (Courtesy of Carborundum Abrasives of North America)

Handling

Careful handling of grinding wheels is important to the life of the grinding wheel and its safe performance. The following suggestions regarding grinding wheels are offered by the American National Standards Institute and wheel manufacturers:

- Do not drop or bump grinding wheels; they can be damaged easily.
- Do not roll any grinding wheel on its edge.
- Stack wheels carefully to prevent damage when moving wheels.
- Do not lay tools or materials on the top of a grinding wheel.

Storage

Store grinding wheels in a dry area and in a proper rack for safekeeping. A typical storage rack, Fig. 38-10, shows how various types of wheels should be properly stored. Thin wheels should be stored on a flat horizontal surface to prevent warping, while small-cut and mounted wheels can be stored in boxes. It is recommended that special wheels and high-speed grinding wheels be stored away from regular wheels to avoid interchanging them.

● MOUNTING A GRINDING WHEEL

A type No. 1 aluminum oxide straight wheel, Fig. 38-11, is generally used for most sur-

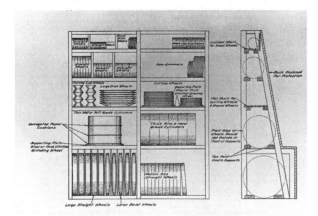

Fig. 38-10 A typical rack for storing various types of grinding wheels. (Courtesy of Grinding Wheel Institute)

Fig. 38-11 A Type 1 straight grinding wheel. (Courtesy of Norton Co.)

face grinding operations. Before a wheel is mounted on a machine, it should be checked to make sure it is not defective. Suspend the wheel by slipping one finger through the hole and gently tap the side with the handle of a hammer or screwdriver, as was shown in Fig. 38-9. A good wheel will give a sharp, clear ring.

Care should be used when handling or mounting a grinding wheel to prevent it from being damaged. A wheel that has been misused or damaged may shatter, cause damage to the grinder, or possibly cause a serious accident.

To Mount a Wheel on a Straight Spindle

Use the following procedure to mount the grinding wheel:

1. Check that the wheel is not cracked, by tapping it at four points about 90° apart with a plastic or wooden handle screwdriver.

> A good wheel will give a sharp, clear ring.

2. Clean the inner flange on the machine, the spindle, and the hole in the grinding wheel, Fig. 38-12.

3. Check the wheel to make sure there is a wheel blotter on each side.

> If not, place two blotters the same size as the flanges and place one on each side of the wheel.

4. Slide the wheel on the grinder spindle. The wheels should go on freely without binding.

> Never force a wheel onto the spindle.

5. Clean and place the outer flange against the wheel.

6. Hold the wheel with a rag and tighten the spindle nut against the flange enough to hold the wheel firmly, Fig. 38-13.

> DO NOT exert excessive pressure while tightening, or strains may be set up in the wheel that may cause it to break.

7. Replace the grinding wheel guard on the machine.

Fig. 38-12 Clean the bore of the wheel, the machine spindle and flange face before mounting a wheel. (Courtesy of Kelmar Associates)

Fig. 38-13 Tightening the spindle nut to hold the grinding wheel firmly on the spindle. (Courtesy of Kelmar Associates)

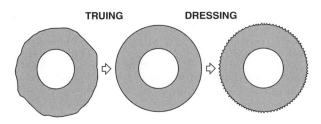

TRUING   DRESSING

Fig. 38-14 Truing makes the wheel surface run true to its axis; dressing sharpens a wheel. (Courtesy of Kelmar Associates)

● TRUING AND DRESSING A WHEEL

Truing is the operation of making a wheel run true, or altering the face to give it a desired shape, Fig. 38-14.

Dressing is the operation of removing dull grains or metal particles to make the wheel cut better. A dull or loaded wheel should be dressed for several reasons:

- To keep down the heat generated between the wheel and the work.
- To reduce the strain on the grinding wheel and machine.
- To improve the surface finish of the work.

On a surface grinder, an industrial diamond mounted in a holder is used to dress and true the wheel.

To Dress and True a Surface Grinder Wheel

Use the following procedure to dress and true the wheel:

1. Check the diamond for wear, and when necessary, turn it in the holder to expose a new point.

2. The diamond is canted in the holder at a 10° angle.

> This helps to prevent chattering and the tendency to dig in during the dressing operation.

3. Clean the magnetic chuck thoroughly with a cloth, and then wipe over it with the palm of the hand.

4. Place the diamond holder on the last two magnetic poles on the left end of the magnetic chuck to stop abrasive particles from pitting the chuck surface during truing.

> Paper should be placed between the diamond and the chuck to prevent scratching or marring the chuck surface when removing the diamond holder, Fig. 38-15.

5. The point of the diamond should be offset about 1/2 in. (12.7 mm) to the left of the grinding wheel centerline, Fig. 38-16.

6. Energize the magnetic chuck by turning the lever to the *ON* position, Fig. 38-15.

7. Make sure the diamond clears the wheel; then start the grinder.

8. Lower the wheel until it touches the diamond.

9. Move the diamond *steadily* across the face of the wheel. A fast feed produces a rough

Fig. 38-15 Setting the diamond dresser on the left-hand end of the magnetic chuck. (Courtesy of Kelmar Associates)

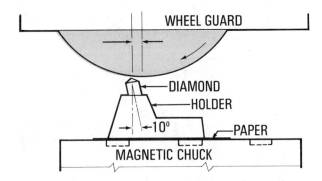

Fig. 38-16 The diamond should be offset about 1/2 in. (12.7 mm) to the left of the wheel centerline for truing and dressing. (Courtesy of Kelmar Associates)

Fig. 38-17 Use a fine honing stone to remove any nicks or burrs from the magnetic chuck face. (Courtesy of Kelmar Associates)

wheel surface that removes metal quickly. A slow feed produces smooth surfaces for high surface finishes.

10. Take light cuts .001 in. (0.02 mm) in a smooth, steady motion until the wheel is clean, sharp, and running true.

11. Take a finish pass of .0005 in. (0.01 mm) across the face of the grinding wheel.

● GRINDING A FLAT SURFACE

The most common operation performed on a surface grinder is grinding a flat or horizontal surface. To obtain the best results, the correct type of wheel, properly dressed, should be used.

To Grind a Flat Surface

Use the following procedure to grind a flat surface:

1. Thoroughly clean the magnetic chuck with a cloth and then wipe it with the palm of the hand.

2. Use a hone to remove any burrs from the face of the magnetic chuck, Fig. 38-17.

3. File off any burrs on the surface of the work that is to be placed on the magnetic chuck.

4. Place a piece of paper *slightly* larger than the workpiece in the center of the chuck.

5. Place the work on the paper and turn on the magnetic chuck.

Try to remove the work to make sure it is held securely, Fig. 38-18.

6. Set the table reverse dogs so that the center of the grinding wheel clears each end of the work by approximately 1 in. (25 mm), Fig. 38-19.

7. Set the crossfeed to advance approximately .030 to .040 in. (0.8 mm to 1 mm) at every table reversal.

8. Turn the crossfeed handwheel until the edge of the work overlaps the edge of the grinding wheel by about 1/8 in. (3 mm), Fig. 38-20.

9. Turn the wheel feed handwheel until the grinding wheel is about 1/32 in. (0.8 mm) above the work surface.

Fig. 38-18 Energize the magnetic chuck to hold the workpiece securely. (Courtesy of Kelmar Associates)

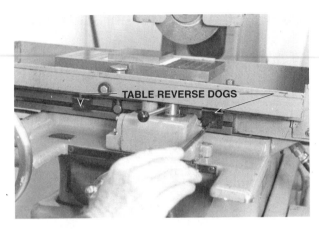

Fig. 38-19 Set the table reverse dogs so that the grinding wheel center clears each end of the work by 1 in. (25.4 mm). (Courtesy of Kelmar Associates)

Fig. 38-21 A properly trued and dressed wheel should produce a good surface finish. (Courtesy of Kelmar Associates)

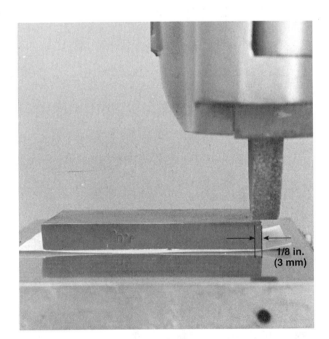

1/8 in. (3 mm)

Fig. 38-20 Setting the grinding wheel to the surface of the work near the edge. (Courtesy of Kelmar Associates)

10. Start the grinder and lower the wheelhead until the wheel just sparks the work.

11. Raise the wheel about .005 in. (0.12 mm).

> The wheel may have been set on a low spot of the work.

12. Start the table traveling automatically and feed the entire width of the work under the wheel to check for high spots.

13. Lower the wheel .002 to .003 in. (0.05 to 0.07 mm) for every cut until the surface is completed.

> Cutting fluid should be used whenever possible to aid the grinding action and keep the work cool.

14. For the final pass, take a .0005 in. (0.01 mm) depth of cut.

> A clean-cutting wheel should produce a good surface finish on the work, Fig. 38-21.

SUMMARY

- Grinding wheels are made up of aluminum oxide and silicon carbide crystals that are bonded together. Superabrasives are used to grind ultra-hard materials.

- The grade of a grinding wheel refers to the strength that the abrasive grains are bonded together.

- The structure of a grinding wheel refers to the spacing of the abrasive grains. A more open structure provides more chip clearance and removes metal faster.

- Proper inspection, handling, and storage prolong grinding wheel life and enhance safety.

- When grinding a flat surface, use cutting fluid whenever possible to assist the cutting action and reduce heat buildup.

KNOWLEDGE REVIEW

Surface Grinder

1. Why is the surface grinder considered an important machine tool?

2. Name the most common surface grinder.

3. What device is generally used to hold work while grinding?

Surface Grinder Parts

4. Describe the purpose of:
   a. the crossfeed handwheel
   b. the table traverse handwheel
   c. the wheel feed handwheel

Safety Precautions

5. Before starting the grinder, how can the magnetic chuck be tested for holding power?

6. List any three important safety precautions for a surface grinder.

Grinding Wheels

7. Name and state the purpose of the two common wheels used in machine shop work.

8. Name and state the purpose of the two superabrasive materials.

9. What materials go into the composition of a grinding wheel?

10. Define:
    a. wheel grade
    b. wheel structure

Wheel Identification and Selection

11. What wheel specifications are identified by a standard marking system?

12. Name the five factors which must be considered when selecting a wheel.

Mounting a Grinding Wheel

13. Explain the procedure for testing a wheel to make sure it is not defective.

14. List the steps to mount a grinding wheel.

15. Why should excessive pressure not be used when tightening the spindle nut?

Truing and Dressing a Wheel

16. Define: truing, dressing.

17. Why should a dull or loaded wheel be dressed?

18. Where should the diamond holder be placed on the magnetic chuck?

19. Why is paper used between the chuck and the diamond holder?

20. How should the diamond be located in relation to the wheel?

Grinding a Flat Surface

21. Explain the procedure for mounting work on a magnetic chuck.

22. How long should the table travel be in relation to the work length?

23. Explain the procedure for setting the wheel to the work surface.

24. Why should cutting fluid be used whenever possible?

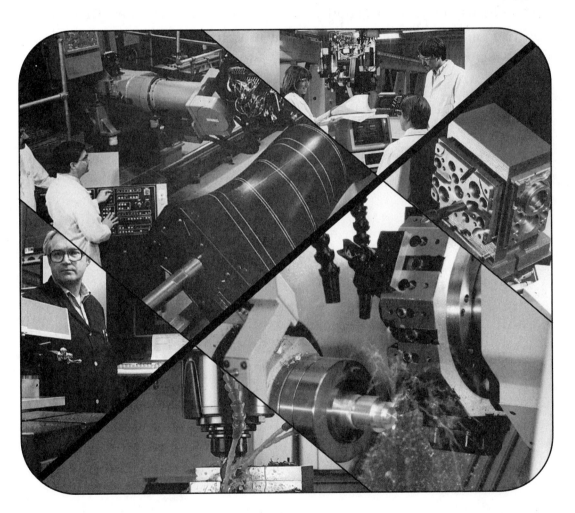

Computer Age Machining

The development of the computer has made changes in our everyday life from retail sales, banking, and medicine to communications, transportation, science, and manufacturing. No other invention in history has had such an impact on humanity, in such a short period of time, as the computer. The computer has made possible the exploration of space, world-wide television, improved health care, quality-controlled manufacturing, robots, flexible manufacturing systems, and many other benefits to humanity. Since our present-day computer is considered to be in its infancy, it is hard to imagine what effect the development of newer and more powerful computers will have. One thing is certain—this revolutionary invention will drastically change our lifestyles and the world in general.

Figures courtesy of (clockwise from far left): AMT (The Association for Manufacturing Technology): AMT, AMT, Cincinnati Milacron Inc., Hardinge Brothers, Inc., Fadal Engineering Co., Inc.

U N I T
39

Computer Numerical Control

Numerical Control (NC) is a big change from the conventional machine tool operation where the operator guides a cutting tool from information on the part print. The accuracy of the part produced depends upon the skills of the operator or machinist. With NC, information on a print is converted into a part program which guides the machine to produce the part.

The code numbers and symbols of a NC part program must represent the information on the part print and contain every movement, path, or action a machine may require to accurately produce the part. The cutting action of NC machine tools is similar to conventional machining, however the machine movements now depend upon the accuracy of the programmer.

OBJECTIVES

After completing this unit, you should be able to:

- Understand the role of the modern-day computer in manufacturing.
- Compare conventional and CNC machining procedures.
- Recognize the common turning and machining center axes.
- State six main advantages of CNC.

KEY TERMS

computer numerical control (CNC)

machine control unit (MCU)

mainframe computer

numerical control (NC)

coordinate measuring machine (CMM)

machining center

microcomputer

minicomputer

● BACKGROUND

Ever since the beginning of mankind, some type of device or system has been used to count and perform calculations. Primitive people used their fingers, toes, and stones to count, Fig. 39-1. The abacus, Fig. 39-2, developed in the Orient around 4000 B.C., was really the first primitive computer and can still be found in use today in some Oriental businesses.

The first mechanical calculator, that could only add and subtract, was developed in 1642. This was followed by improvements in calculators that could add, subtract, multiply, and divide. The punched-card system, the first method of data processing, was introduced in 1804, and the first simple computer was developed in Germany during the 1930s. The world's first electronic digital computer was introduced in 1946 and in 1971—the microprocessor, which contained the entire central processing unit in a single one-quarter inch square chip, was developed, Fig. 39-3. In 1993, the Pentium chip, designed for 32-bit computers, was introduced. It contains almost 3.1 million transistors and is capable of processing 166 million computer instructions per second. The P6 chip, released in the fall of 1995, contains about 5.5 million transistors and is capable of processing 250 to 300 million com-

Fig. 39-3 Many thousands of transistors can be contained on a tiny 1/4 in. square silicon chip. (Courtesy of Delta International Machinery Corp.)

puter instructions per second. In 1997, the P7 will be introduced; it will contain 10 million or more transistors and will be much more powerful than the P6 chip. Continual developments will make computers smaller, more powerful, and user-friendly.

● WHAT IS NUMERICAL CONTROL?

Numerical control (NC) is the operation of machine tools by a series of coded instructions consisting of numbers, letters of the alphabet, and symbols, Fig. 39-4. This means that numerical codes, converted to a form that can be understood and used by machine tools, direct some or all of the fundamental operations of a given machine tool. Numerical commands range all the way from controlling the position of the spindle in relation to the workpiece (the most important function) to the speed, feed, and even the flow of the coolant. The numbers are coded instructions which refer to specific distance, position, motion, or function that will be followed by the machine tool as it produces a workpiece.

Advantage of NC

Numerical control (NC) or machine control units with hard-wired controls were replaced in

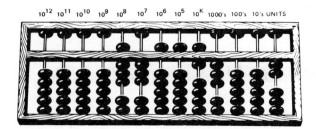

Fig. 39-1 The methods primitive people used to count. (Courtesy of Kelmar Associates)

TOES AND FINGERS   STONES   BEADS

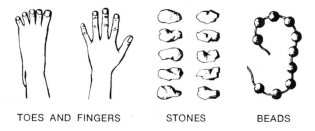

10^{12} 10^{11} 10^{10} 10^{9} 10^{8} 10^{7} 10^{6} 10^{5} 10^{K} 1000's 100's 10's UNITS

Fig. 39-2 The abacus, the first computer, was developed around 4000 B.C. and is still used in some Far East countries. (Courtesy of Kelmar Associates)

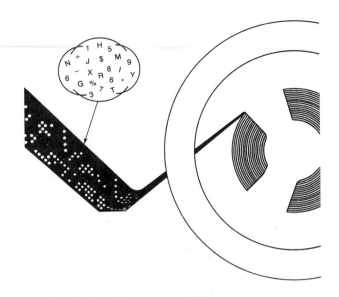

Fig. 39-4 Coded instructions of letters, numbers, and symbols are used to operate CNC machine tools. (Courtesy of Kelmar Associates)

the mid-1970s by **computer numerical control (CNC)** because of the many advantages its software-logic, microprocessor-based controls offered. The following list describes several of these advantages:

1. Programs can be stored on floppy disks, tapes, or cassettes which can be loaded into the **machine control unit (MCU)** memory. This same program can be recalled as many times as required from the memory.

2. The MCU has random-access memory (RAM) which allows programs to be edited, and also allows for on-board programming (manual data insert—MDI).

3. An external device (computer) can be used to make the program, which is then uploaded to a computer disk or directly to the MCU.

4. CNC controls can compensate for size and tool wear, inspect parts, and communicate with other computers and robots. The tool-path can be displayed on the screen (CRT) and errors in the program can be corrected before machining begins.

CNC versus Conventional Machining

Computer numerical control is a complete departure from the conventional machine tool operation where an operator or machinist studies an engineering drawing or print and guides a cutting spindle based on the print information. The successful operation of a conventional machine tool depends in large part on the skill of the operator or machinist developed over many years of practice. With CNC, the drawings are studied and the information about the workpiece or part is converted to a series of appropriate code numbers by the part programmer, Fig. 39-5. The code numbers represent every movement, path, or action the machine tool must take to properly machine the workpiece described in the engineering drawing. The complete series of codes necessary to produce a single workpiece on a machine tool is known as a *part program.*

The movements and functions of the machine tool have now become the *programmer's* responsibility, while the role of the operator is drastically changed. Thus the basic functioning of the machine does not rest with the person who is standing by monitoring the operation, but with the one who developed the part program in a coded form which guides and directs the work-

From the drawing to the workpiece

Fig. 39-5 The steps that are taken in CNC from the engineering drawing to the finished part. (Courtesy of Deckel Maho, Inc.)

piece machining. Machine tools cannot read numbers, however, the coded instructions are read by a program reader which is part of the overall Machine Control Unit often called the MCU. The function of the MCU is to act upon the coded instructions to convert them into movement and operation of the machine tool. Normally the control unit output signals are converted by electronic, hydraulic or mechanical servomechanisms.

It is important to know that CNC is not a mechanical brain and cannot think, evaluate, judge, discern, or show any real adaptability. Sometimes a choice of alternatives or limits can be set up in the program so that incorrect or faulty instructions will not be followed by overriding the programmed instructions; however, all of these conditions with their possible choice of alternatives have to be included in the program.

CNC Programming and Operation

CNC programming and operation can be learned in a few days by anyone who has a knowledge of basic machine tool operations and machining procedures. The same basic machining procedures are used whether producing a part with conventional machine tools or with CNC machine tools. The basic difference between the two is that, with conventional machine tools, the operator uses machine handwheels to manually control the table or spindle movements to produce the part, Fig. 39-6A. With CNC machine tools, the programmer or machine operator programs the machine control unit (MCU) through the use of symbols, letters, and numbers (coded instructions) which automatically control the machine tool movements to produce the desired part, Fig. 39-6B.

A CNC machine tool can be thought of as a servant, a very obedient servant, but one that cannot make decisions alone. It will do exactly what it has been told to do and will follow instructions with speed, accuracy, and reliability. The factor that limits its operation is a combination of the capacity of the machine tool and the machine control unit. For a CNC machine to function properly, it must have instructions in the form it can understand and accept. Incorrect instructions will result in either a rejection of the instructions or a faulty machining operation. Provided with good instructions within the machine capacity, CNC offers the programmer complete confidence that the program will be carried out

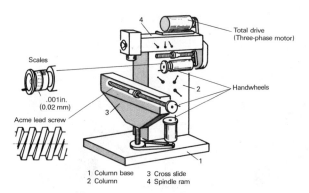

(A) CONVENTIONAL MILLING MACHINE

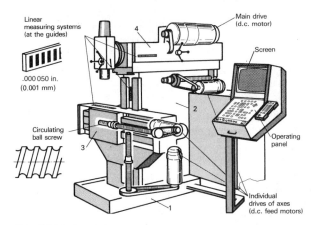

(B) CNC MILLING MACHINE

Fig. 39-6 Comparison of a conventional and a CNC milling machine. (Courtesy of Deckel Maho, Inc.)

exactly, either once or one thousand times. The program will be carried out at the best speeds, with the accuracy built into the machine and the machine control unit, and in a predictable amount of time.

● COMPUTERS IN MANUFACTURING

The first computer, developed in the early 1950s, was large and had many breakdowns because of the heat created by the many vacuum tubes it contained. As the electronics developed from vacuum tubes to transistors, solid-state components, integrated circuits (IC), and then microprocessors, computers became smaller, more reliable, and less expensive. Today a computer no larger than a typewriter is common in industry and in many homes.

The computer, Fig. 39-7, is a tool which can perform many tasks with amazing speed, accuracy, and reliability. It does not have a brain and

Fig. 39-7 The computer is a valuable tool for CNC machining because it can perform many tasks with speed, accuracy, and reliability. (Courtesy of The Association for Manufacturing Technology)

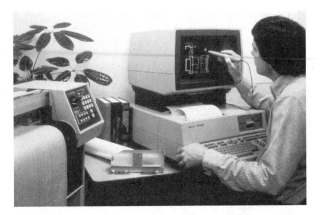

Fig. 39-9 The mainframe computer is a large-scale data processing system which is generally a company's host or central computer. (Courtesy of Hewlett-Packard Co.)

therefore cannot think for itself. The computer is only an extension of a person's brain and must be told exactly, in language that it understands, what it should do.

The computer is the building block for the entire concept of computer assisted manufacturing (CAM). A computer is a machine that consists of an arithmetic logic unit (ALU), a memory, a control unit, and input/output (I/O) devices that, through the use of stored programs, can process data, Fig. 39-8. In manufacturing, the computer is used to collect, store, process, and transmit data, and the computer is chosen for a particular purpose depending upon the data-handling job involved.

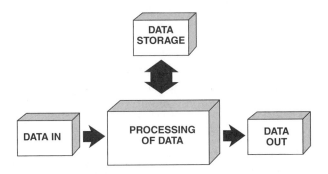

Fig. 39-8 The functions of a computer are to collect, store, process, and transmit data. (Courtesy of *Modern Machine Shop Magazine*, copyright 1994, Gardner Publications Inc.)

Types of Computers

There are a number of types of computers used in industry, many of which are designed for specific applications.

- The **mainframe computer**, Fig. 39-9, is a large-scale data processing system made up of one or more separate devices to perform each of the basic functions (i.e., central processing unit or CPU, main memory, I/O devices such as disk and tape drives and printers, etc.). Such large systems can usually store millions of bytes (characters) of information in its main memory and can process millions of instructions per second. It is often a company's main computer and the one that performs general-purpose data processing, such as CNC part programming, payroll, cost accounting, inventory, and many other applications. The mainframe computer generally has a number of individual keyboard terminals connected to it, and all of these can feed information to the mainframe computer at the same time.

- The **minicomputer**, Fig. 39-10, is smaller in size and capacity than the mainframe computer. It is generally the dedicated type, which means it will only perform a specific task such as:

1. Producing CNC part programs.
2. Sending program data to various CNC machines.
3. Controlling the movements of a single CNC machine.

Fig. 39-10 The minicomputer is a dedicated computer designed to perform specific tasks. (Courtesy of Ingersoll Cutting Tool Co.)

4. Managing inventory, scheduling, and recordkeeping.

The minicomputer's CPU is generally designed to accept a maximum word length of 32 bytes; however, some of the more recent models can accept 64 bytes.

- The **microcomputer**, Fig. 39-11, generally contains one microprocessor or chip that contains at least the arithmetic-logic and the control-logic functions of the CPU. The microprocessor is generally designed for simple applications and must be accompanied by other electronic devices (usually on a printed circuit board) for more complex ap-

Fig. 39-11 The microcomputer, used for simple applications, contains only one chip. (Courtesy of Hewlett-Packard Co.)

plications. The CPU is usually designed to accept 8, 16, and 32 bytes; however, there are also 64-byte microprocessors available.

Memory and Storage Devices

A computer system usually has both internal and external storage (memory). The *internal memory* is used to temporarily store data that is being processed. There are several types of internal memory systems, such as the main memory, control memory (for program control), and local memory (in high-speed registers or buffers). The *external memory* is used for storing data that is about to be processed or for recording data for future use.

Four types of computer memories can be used for either internal or external storage:

1. *Random-Access Memory (RAM).* This is an internal memory where data can be stored and recalled from specific addressable locations. RAM is high-speed memory, usually consisting of high-density semiconductor integrated circuit (IC) devices, which can have access times from tens to hundreds of nanoseconds (billionths).

2. *Read-Only Memory (ROM).* This is an internal memory where the data is stored permanently and can only be read out. ROM is used primarily as internal memory to store small programs that can be recalled frequently at high speeds. They are usually made of semiconductors similar to those in RAMs, but the data is entered during the manufacturing process.

3. *Direct-Access Storage Device (DASD).* This is an external memory where large amounts of data can be stored and recalled randomly, but a combination of direct access and sequential searching is required to reach the specific storage location. DASD memories are usually stored on hard or soft magnetic disks. The data-transfer rates of such systems often range in the hundreds of thousands of characters per second. Large, high-performance DASD systems store billions of bytes (gigabytes) of data.

4. *Sequential-Access Storage (SAS).* This is an external memory where large amounts of data are stored and recalled in sequence, not randomly. This memory is used to store large amounts of data as a permanent record or backup that does not have to be accessed

directly by the CPU. Magnetic tape devices are the main memories of this type and are like audio tapes, available in reels, cartridges, or cassettes.

Functions of a Computer in CNC

Computers in CNC are used in three main areas:

1. Almost all machine tools built today include a computer in their control unit or use a computer in their operation.

2. Some of the part programming for CNC machine tools is done by off-line computers.

3. Many machine tools are controlled by computers in a control room, or even in another plant. This is commonly known as Direct Numerical Control (DNC).

The computer has provided many benefits to the manufacturing process by increasing productivity, producing high-quality parts, and reducing manufacturing costs. It is used for such things as part design (CAD), testing, inspection, quality control, planning, inventory control, gathering of data, work scheduling, warehousing, and many other functions in manufacturing. Computers will have a strong influence on the manufacturing of the future.

● CNC MACHINE TOOLS

Machine tools were originally designed to be single-purpose units. There were machines for milling, drilling, boring, tapping, grinding, turning, sawing, etc. When numerical control was first developed, it was applied to the single-purpose machine concept. The first major breakthrough came in the late 1950s with the development of the machining center that combined rotating spindle operations including milling, drilling, tapping, boring, counterboring and spot facing. This was possible because of the automatic tool changer which supplied various tools to the spindle as they were required.

In the 1990s, both rotating-spindle and turning operations have been incorporated into many newer machine models; some machines also include grinding capabilities. When all these features are combined with faster speeds and the newer cutting-tool materials, the multi-purpose CNC units combine the best of all machining operations in a single setup. If drilling, boring, and milling operations can be done on one machine in one setup with one operator instead of making three separate setups on three machines with three operators, productivity increases and costs are decreased.

CNC technology is being used on all types of machine tools, from the simplest to the most complex. Some of the more common machine tools are the turning and chucking centers, horizontal and vertical machining centers, coordinate measuring machines, and electro-discharge machines.

Turning and Chucking Centers

Turning and chucking centers, Fig. 39-12A, were developed in the mid-1960s because about 40% of all metal cutting operations were performed on lathes. These computer numerically controlled machines are capable of greater accuracy and higher production rates than are possible on the conventional engine lathe. The basic turning or chucking center operates on only the X and Z axes, Fig. 39-12B.

- The X axis controls the cross motion of the turret head (toward or away from the workpiece).

- The Z axis controls the lengthwise travel of the turret head (toward or away from the headstock).

Machining Centers

Horizontal and vertical **machining centers**, Fig. 39-13A, were developed in the 1960s so that parts did not have to be moved from machine to

Fig. 39-12A A chucking center increases productivity and reduces manufacturing costs. (Courtesy of Emco Maier Corp.)

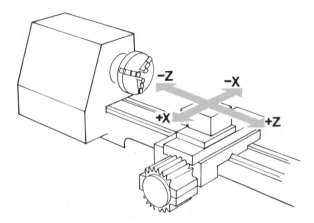

Fig. 39-12B The main axes of a lathe or turning center. (Courtesy of Emco Maier Corp.)

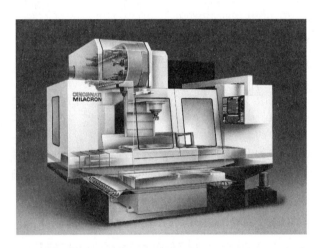

Fig. 39-13A Vertical machining centers can perform many different operations on a part in one setup. (Courtesy of Cincinnati Milacron Inc.)

machine in order to perform various operations. This increased productivity because more operations could be performed on a workpiece in one setup. There are two main types of machining centers, the vertical and the horizontal spindle types.

a. The vertical machining center, Fig. 39-13B, operates on three axes:

- The X axis controls the table movement (left or right).

- The Y axis controls the table movement (toward or away from the column).

- The Z axis controls the spindle movement (up or down).

Axes and directions of axes when working with vertical spindle

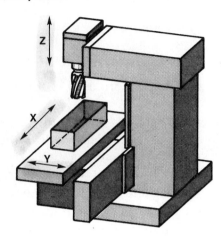

Fig. 39-13B The main axes of a vertical machining center. (Courtesy of Deckel Maho, Inc.)

b. The horizontal machining center, Fig. 39-13C, operates on three axes:

- The X axis controls the table movement (left or right).

- The Y axis controls the vertical movement of the spindle (up or down).

- The Z axis controls the horizontal movement of the spindle (in or out).

Directions of axes when working with horizontal spindle

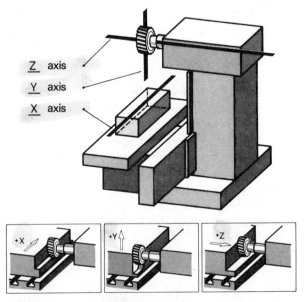

Fig. 39-13C The main axes of a horizontal machining center. (Courtesy of Deckel Maho, Inc.)

Fig. 39-14 Coordinate measuring machines are used to inspect parts produced on other CNC machines. (Courtesy of Giddings & Lewis—Sheffield Measurement)

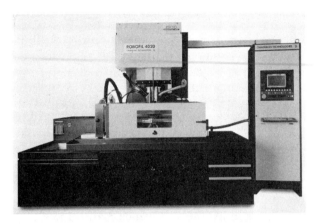

Fig. 39-15 The shape of the part produced on a wire-cut EDM machine is controlled by the CNC program in the machine control unit. (Courtesy of Charmilles Technologies Co.)

Coordinate Measuring Machines (CMM)

Coordinate measuring machines, Fig. 39-14, were developed because of the need for higher quality and increased productivity. The inspection techniques that catch defects only *after* parts have been made are no longer acceptable. This shift from detection to prevention is taking inspection to the manufacturing floor to provide timely feedback for process control and correction. CMMs are designed to check the accuracy of the part as it is being manufactured, based on the information about the part programmed into its control unit.

Wire-Cut Electro-Discharge Machines (EDM)

The wire-cut EDM machine, Fig. 39-15, is an electro-discharge machine that uses CNC-controlled movement to produce the desired contour or shape on a part. It uses a continuous traveling wire under tension as the electrode to follow a programmed path along the X and Y axes to produce the contour required.

● CNC PERFORMANCE

CNC has made great progress since NC was first introduced in the mid-1950s as a means of guiding machine tools through various motions automatically, without human assistance. These advances came as the result of the development of transistors, solid-state circuitry, integrated cir-

cuits (ICs), and the computer chip, which made it possible to program machines to produce complex forms thought to be impossible to machine just a few years ago. Not only have the machine tools and machine control units been greatly improved, but their cost has been continually dropping. CNC machines can now be easily purchased by small manufacturing shops and educational institutions. Their wide acceptance throughout the world has been a result of their accuracy, reliability, repeatability, and productivity, Fig. 39-16.

Accuracy

CNC machine tools were accepted by industry because they were capable of machining to very close tolerances and increasing manufacturing productivity. Modern CNC machines are capable of producing parts which are accurate to within a tolerance of .0001 to .0002 in. (0.0025 to 0.005 mm). The machine tools have been built better, and the machine control units ensure that parts within the tolerance allowed by the engi-

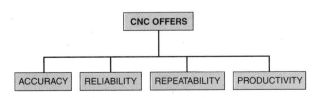

Fig. 39-16 CNC is widely used by industry because of its effect on product quality and productivity. (Courtesy of Kelmar Associates)

neering drawing will be maintained. The accuracy, which formerly depended on the machinist's many years of skill, is now being exceeded by reliable CNC control systems and better machine tool construction.

Reliability

Consumers throughout the world are demanding better and more reliable products; therefore there is a need for equipment that can produce high-quality goods time and time again. Machine tools have been greatly improved with better machine slides, bearings, ball screws, and machine tables to make sturdier and more reliable machines. New machine control systems which can compensate for cutting tool wear ensure that accurate parts will be produced every time.

Repeatability

The repeatability of a machine tool means checking parts produced on that machine to see how they compare to others for size and accuracy. The repeatability of a CNC machine should be at least one-half the smallest tolerance allowed for the part. Machine tools capable of greater accuracy and repeatability will naturally cost more, but this increase in cost will be quickly offset by reduced scrap and increased productivity.

Productivity

Industry is always striving to produce better products at competitive or lower prices to gain a bigger share of the market. To meet competition throughout the world, manufacturers must continue to lower manufacturing costs and produce better-quality products. They must get greater output per worker, greater output per machine, and greater output for each dollar of capital investment. These factors provide strong justification for using CNC to remain competitive.

● ADVANTAGES OF CNC

CNC has grown at an ever-increasing rate, and its use will continue to grow because of the many advantages it has to offer industry. Some of the most important advantages of CNC are illustrated in Fig. 39-17.

1. *Less Scrap*—Because of the accuracy of CNC systems and the elimination of most human errors, scrap has been greatly reduced or eliminated.

2. *Reduced Production Lead Time*—The program preparation and setup for CNC machines is usually short. Many jigs and fixtures formerly required are not necessary. The program can be stored in a small space and used again as required.

3. *Less Human Error*—The CNC eliminates the need for an operator to take trial program cuts, make trial measurements, or make table positioning movements. The operator does not have to change cutting tools

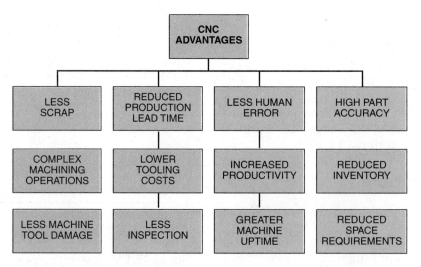

Fig. 39-17 Some of the major advantages CNC offers manufacturing. (Courtesy of Kelmar Associates)

except during setup and when tools become dull.

4. *High Part Accuracy*—CNC ensures that all parts produced will be accurate and uniform in quality. The improved accuracy of the parts produced by CNC assures the interchangeability of parts.

5. *Complex Machining Operations*—Complex operations can be done quickly, easily, and accurately with CNC and electronic measuring equipment.

6. *Lower Tooling Costs*—CNC generally does not require complex holding fixtures, therefore the cost of fixtures may be reduced by as much as 70%.

7. *Increased Productivity*—Because CNC controls all the machine functions, parts are produced faster, with less setup and lead time.

8. *Reduced Inventory*—An inventory of spare parts is no longer necessary, since duplicate parts can be made to the same accuracy when the same program is used.

9. *Less Machine Tool Damage*—The damage to machine tools as a result of operator error or carelessness is almost eliminated because there is less need for operator intervention.

10. *Less Inspection*—Because CNC produces parts of uniform quality, less inspection time is required. Many operations or parts may be inspected while they are being machined.

11. *Greater Machine Uptime*—Because there is less time required for setup and operator adjustments, production rates could increase as much as 80 percent.

12. *Reduced Space Requirements*—CNC requires less floor space for machines because of their high productivity, and less storage space for fewer jigs and fixtures.

SUMMARY

- Numerical Control (NC) is the control of machine tool movements through coded instructions consisting of letters, numbers, and symbols.

- Computer Numerical Control (CNC) replaced NC in the mid-1970s because the onboard computer stores information, has random access memory, can compensate for size and tool wear, and inspect parts.

- In conventional machining, the operator manually moves the machine slides to produce a part, while in CNC machining, the part program automatically moves the slides to produce a part.

- Computers are the building blocks to Computer Assisted Manufacturing (CAM) because they can perform many tasks quickly, accurately, and reliably.

- The most common CNC machines are the turning center and the machining center. These machines increase productivity and reduce manufacturing costs.

KNOWLEDGE REVIEW

Computer Numerical Control

1. Define numerical control.
2. Name three ways primitive people used to count.
3. Trace the development of the computer, starting with the first mechanical calculator.

What Is Numerical Control?

4. State four advantages of CNC over NC.
5. What is the function of the machine control unit.
6. Compare conventional machining with CNC machining.

Computers in Manufacturing

7. Name and state the purpose of three computers used in manufacturing.
8. State the purpose of the following:
   - internal memory
   - external memory
   - RAM
   - ROM

CNC Machine Tools

9. State the purpose of each axis on:
   a. turning and chucking centers
   b. vertical machining centers

CNC Performance

10. Name four qualities that have been responsible for the wide acceptance of CNC throughout the world.

Advantages of CNC

11. List six of the most important advantages of CNC.

UNIT 40

How CNC Controls Machines

A French mathematician and philosopher, M. Rene Descartes, is credited for establishing the Cartesian or rectangular coordinate system. The location data (information) for any absolute point in space may be described very accurately by using rectangular coordinates. CNC systems use coordinates when programming the distance and direction the machine tool table must move to produce the part described on the print.

A standard Cartesian or rectangular coordinate system for CNC purposes may be likened to a grid arrangement of several square blocks of a community that may be used as reference points for giving directions. Each block functions as a unit of measurement in describing the location of one point in the community in relation to another point. Assume that someone needs instructions to get from the present location to the their destination. By using the grid in Fig. 40-1 as a small sectional map showing the person's present location, exact directions can be given from the present point to the destination point as listed to the right of the map. Rectangular coordinates are often used in everyday living by expressing directions in two-dimensional coordinates.

OBJECTIVES

After completing this unit, you should be able to:

- Describe the Cartesian coordinate system and how it is used to locate various part locations.
- Use the absolute and incremental systems and the various codes used for CNC programming.
- Explain how the Machine Control Unit (MCU) uses various types of programmed information to produce a part.

KEY TERMS

absolute system
continuous-path positioning
Cartesian coordinate system
incremental system
modal code
nonmodal code
point-to-point positioning

● CARTESIAN COORDINATE SYSTEM

The **Cartesian coordinate system** works on a grid system, similar to graph paper, where reference lines run at 90° to each other. If the center of the graph is considered to be the point where X and Y equal zero, then all lines run parallel to either the X (horizontal) axis or Y (vertical) axis. If a point is plotted to the right of the X zero, the dimension would be a plus move or X+; to the left, it would be a minus move or X−. If a point is plotted above the Y zero, the dimension would be a plus move or Y+; below, it would be a minus move or Y−.

In Fig. 40-2, the graph paper has been divided into four quadrants, with the X and Y axes in the center. The location of the three points, A, B, and C, is as follows:

- Point A would be X2 and Y2.

> Plus dimensions do not have to be indicated with a plus (+) sign, they are assumed.

- Point B would be X1 and Y−2.

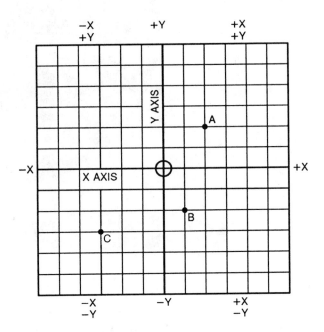

Fig. 40-2 The four quadrants formed when the X and Y axes cross allow accurate locations of points from the XY zero, or origin point. (Courtesy of Allen-Bradley Co.)

> The minus (−) sign must be included, otherwise the CNC control would assume that it is a plus, move and locate the point above the X axis.

- Point C would be X−3 and Y−3.

Machine tools fit the coordinate system very well because they generally have three primary axes of motion, the X (longitudinal), Y (cross), and Z (vertical), Fig. 40-3. The X motion

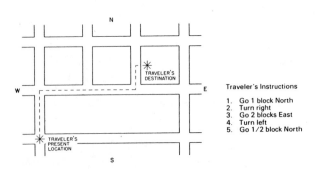

Fig. 40-1 A city map containing a number of square blocks is a form of the coordinate system. (Courtesy of The Superior Electric Co.)

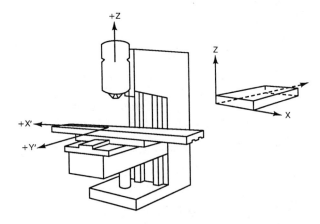

Fig. 40-3 The X, Y, and Z axes on a vertical milling machine are the same as those on a vertical machining center. (Courtesy of Electronic Industries Association)

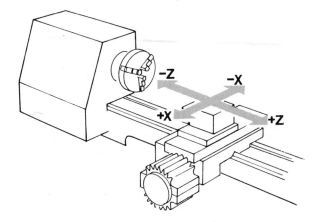

Fig. 40-4 The primary axes on CNC turning centers are the X and Z axes. (Courtesy of Emco Maier Corp.)

on a CNC vertical machining center is the longitudinal movement of the table right or left. The Y motion is for table cross-movement, toward or away from the column, and the Z motion is for the vertical movement (up or down) of the spindle or knee. There are only two primary axes on a turning center, the X and Z. The X axis, which controls the cross-slide, moves the cutting tool toward or away from the workpiece. The Z axis, which controls the carriage, moves the cutting tool lengthwise, toward or away from the workpiece, Fig. 40-4.

● BINARY NUMBERS

Electronic computers and electronic machine control units do have one thing in common: they use binary notation for numerical values. Knowledge of binary is not essential since manual input to both computers and MCUs is done with standard Arabic symbols, but a better appreciation of computer and control-unit function is possible with an understanding of binary.

The familiar decimal system of numbering is based on the power of ten. Early humans had ten fingers and ten toes, and to them ten may have seemed a law of nature. "Base ten" is what we have used all our lives. Binary uses only two digits, a zero and a one. On a punched tape, a hole represents one, no hole represents zero. Binary notation is used in numerical (digital) control because electrical circuits and devices are stable in the following conditions: charged or discharged, on or off, conductor or nonconductor, positive or negative, and so on.

Within a computer logic and processing system, a binary zero (0) might be a positive charge and a binary one (1) might be a negative charge. Numerical values can be added, subtracted, multiplied and divided, the same as in our familiar decimal (Arabic) system, as shown in Fig. 40-5.

Binary was the key that unlocked the development of the electronic computer. Small electrical cores or films that make up the memory elements of a computer can be changed from one electrical condition to another without moving or changing the physical position. Current electronic memory elements can switch from a charge to a no-charge condition and back over five million times per second.

The computer programmer or CNC part programmer does not make the conversion from the Arabic to binary. This is done by the computer or tape-punch unit itself after the data has been entered. A conversion is made back automatically to the Arabic system for the final data output. It is important to understand that accurate data must be placed in the computer, as well as accurate instructions as to how the data shall be used, in order to obtain the desired output.

● PROGRAMMING MODES

There are two types of programming modes used in CNC work, the absolute and incremental systems. There are advantages to both systems, and CNC controls are capable of handling both at any time, providing the correct code has been programmed.

The **absolute system** gives the dimensions and locations of any spot on a job or machine from a single zero or reference point. For example, using the absolute system, a person driving a car from New York to Chicago and making a number of stops along the way might calculate the mileage as follows:

New York to Albany	—155 miles
New York to Buffalo	—453 miles
New York to Detroit	—647 miles
New York to Chicago	—816 miles

All mileage is calculated from the starting point.

The **incremental system** gives the dimensions and locations of any spot on a job or machine from the *previous* reference point. As the

Arabic	Binary	Powers of 2
0 0		
1 1	. . . .	2^0
2 10	. . . .	2^1
3 11		
4 100	. . . .	2^2
5 101		
6 110		
7 111		
8 1000	. . . .	2^3
9 1001		
10 1010		
11 1011		
12 1100		
13 1101		
14 1110		
15 1111		
16 10000	. . . .	2^4
17 10001		
18 10010		
19 10011		
20 10100		
21 10101		
22 10110		
23 10111		
24 11000		
25 11001		
26 11010		
27 11011		
28 11100		
29 11101		
30 11110		
31 11111		
32 100000	. . . .	2^5
64 1000000	. . . .	2^6
128 . . . 10000000	. . . .	2^7

Value of 182 expressed in	1×10^2 =	100
Arabic Numbers	8×10^1 =	80
	2×10^0 =	2
1 8 2		182

Value of 182 expressed in	1×2^7 =	128
Binary Numbers	0×2^6 =	0
	1×2^5 =	32
	1×2^4 =	16
Remember:	0×2^3 =	0
Any number raised to	1×2^2 =	4
the zero power equals 1.	1×2^1 =	2
	0×2^0 =	0
1 0 1 1 0 1 1 0		182

Fig. 40-5 Binary numbers are important to the rapid and accurate operation of computers and control systems. (Courtesy of *Modern Machine Shop Magazine*, copyright 1994, Gardner Publications Inc.)

machine tool moves from one point to another, it remembers and uses each previous point as a reference.

In the *incremental system* using the same example, the mileage would be as follows:

New York to Albany	—155 miles
Albany to Buffalo	—296 miles
Buffalo to Detroit	—269 miles
Detroit to Chicago	—271 miles

All mileage is calculated from the previous point.

Modern controls are capable of handling mixed data (absolute and incremental) in a given block of data. The workpiece shown in Fig. 40-6 has been dimensioned in the absolute mode, with all locations shown from the same zero or reference point. In Fig. 40-7, all dimensions are

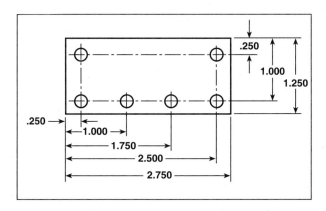

Fig. 40-6 A workpiece dimensioned in the absolute mode with all dimensions taken from the same reference point. (Courtesy of Kelmar Associates)

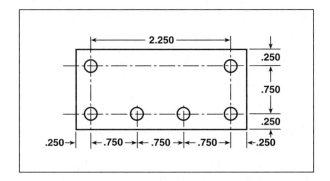

Fig. 40-7 A workpiece dimensioned in the incremental system where all sizes are given from the previous point. (Courtesy of Kelmar Associates)

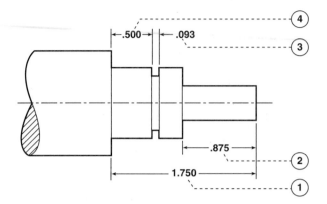

Fig. 40-8 Using a combination of absolute (#1 and #2) and incremental (#3 and 4) dimensions for programming a part. (Courtesy of Emco Maier Corp.)

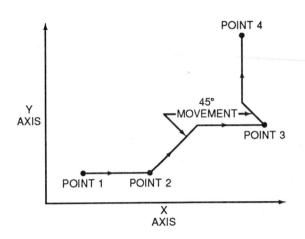

Fig. 40-10 The path which point-to-point positioning uses to reach straight-line location points. (Courtesy of Kelmar Associates)

shown in the incremental mode, with each location given from the previous location.

No system is either right or wrong all the time, and it may be advantageous to use a combination of both during the programming of a job, Fig. 40-8.

● CNC POSITIONING SYSTEMS

The CNC positioning systems most commonly used are *point-to-point* and *continuous path*, Fig. 40-9. Most control units can handle both of them, and it is important that the applications for each are understood.

Point-to-Point Positioning

Point-to-point positioning or rapid traverse is used to locate the spindle, or the workpiece fastened on the machine table, at a specific location to perform operations such as drilling, reaming, boring, and tapping, Fig. 40-9. **Point-to-point positioning** (G00) consists of positioning from

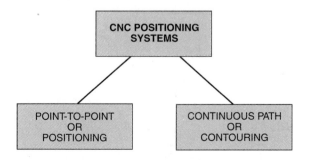

Fig. 40-9 The two types of CNC positioning systems. (Courtesy of Kelmar Associates)

one coordinate (XY) position or location to another, performing the machining operation, clearing the tool from the work, and moving to the next location, until all the operations have been completed at the programmed locations. Drilling machines, or point-to-point machines, are ideally suited for positioning the machine tool to an exact location or point.

Point-to-point machining moves the spindle or workpiece from one point to another as fast as possible (rapids) while the *cutting tool is above the work surface*. The rate of rapid travel usually ranges between 200 and 800 in./min (5 and 20 m/min). Both X and Y axes move simultaneously and at the same rate along a 45° angle line until one axis is reached, and then there is a straight line movement to the other axis, Fig. 40-10.

Continuous-Path Positioning (Contouring)

Continuous path machining, or contouring, is where the cutting tool is always in contact with the workpiece as it travels from one programmed point to the next. **Continuous-path positioning** is the ability to control motions on two or more machine axes simultaneously to keep a constant cutter-workpiece relationship. The CNC program information must accurately position the cutting tool from one point to the next and follow a predefined accurate path at a programmed feed rate in order to produce the form or contour required, Fig. 40-11.

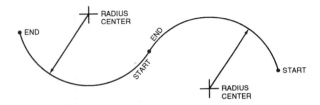

Fig. 40-11 Complex forms on two axes can be generated by circular interpolation. (Courtesy of Kelmar Associates)

● PROGRAMMING FORMAT

Word address is the most common type of programming format used for CNC programming systems. This format contains a large number of different codes (preparatory and miscellaneous) to transfer program information from the part print to machine servos, relays, micro-switches, etc., to manufacture a part. These codes, which conform to EIA (Electronic Industries Association) standards, are put together in a logical sequence called a *block of information*. Each block should contain only enough information to perform one step of a machining operation.

Word Address Format

Every program for any part to be machined, must be put in a format that the machine control unit can understand. The format used on any CNC machine is built in by the machine tool builder and is based on the type of control unit on the machine. A *variable-block format* which uses words (letters) is most commonly used. Each instruction word consists of an address character, such as X, Y, Z, G, M, or S. Numerical

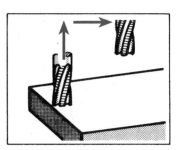

**G00
RAPID TRAVERSE**

**G01
LINEAR INTERPOLATION
(STRAIGHT LINE MOVEMENT)**

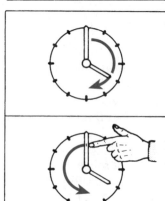

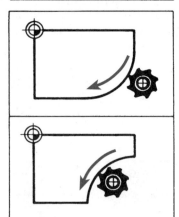

**G02
CIRCULAR INTERPOLATION
(CLOCKWISE)**

**G03
CIRCULAR INTERPOLATION
(COUNTERCLOCKWISE)**

Fig. 40-12 The functions of a few common G-codes. (Courtesy of Deckel Maho, Inc.)

data follows this address character to identify a specific function such as the distance, feed rate, or speed value.

The address code G90 in a program tells the control that all measurements are in the absolute mode. The code G91, tells the control that measurements are in the incremental mode.

Codes

The most common codes used when programming CNC machine tools are *G*-codes (preparatory functions), and *M*-codes (miscellaneous functions). Other codes, such as *F, S, D,* and *T*, are used for machine functions such as feed, speed, cutter diameter offset, tool number, etc.

G-codes, are sometimes called *cycle codes*, because they refer to some action occurring on the X, Y, and/or Z axis of a machine tool, Fig. 40-12. The G-codes are grouped into categories such as group 01, containing codes G00, G01, G02, and G03, which cause some movement of the machine table or the head. Group 03 includes either absolute or incremental programming, while group 09 deals with canned cycles.

- A G00 code rapidly positions the cutting tool or workpiece from one point to another point on a job. During the rapid traverse movement, either the X or Y axis can be moved, or both axes can be moved at the same time. Although the rate of rapid travel varies from machine to machine, it ranges between 200 and 800 in./min (5 and 20 m/min).

- The G01, G02, and G03 codes move the axes at a controlled feedrate.
  - G01 is used for straight-line movement (linear interpolation).
  - G02 (clockwise) and G03 (counterclockwise) are used for arcs and circles (circular interpolation).

Some G-codes are classified as either modal or nonmodal. **Modal codes** stay in effect in the program until they are changed by another code from the same group number. **Nonmodal codes** are in effect for only one operation, and must be reprogrammed whenever required. Using group 01 as an example, only one of the four codes in this group can be used at any one time. For example, if a program begins with a G01 and a G00 is entered next, the G01 is canceled from the program until it is entered again. Entering a G02 or G03 code into the program, will cancel the G01

code. Figure 40-13A shows many of the common G-codes that conform to the EIA Standards.

M or miscellaneous codes are used to either turn ON or OFF different functions which control certain machine tool operations, Fig. 40-13B. M-codes are not grouped by categories, although several codes may control the same types of operations such as M03, M04 and M05 which control the machine tool spindle.

- M03 makes the spindle turn clockwise.

- M04 makes the spindle turn counterclockwise.

- M05 turns the spindle off.

All three of these codes are considered modal, because they stay operational until another code is entered in their place. Figure 40-13B shows the most common M-codes used on CNC machine tools.

On a turning center, the function of some G-codes and M-codes may differ from the those on a

Group	Code	Function
01	G00	Rapid positioning
01	G01	Linear interpolation
01	G02	Circular interpolation clockwise (CW)
01	G03	Circular interpolation counterclockwise (CCW)
06	G20	Inch input (in.)
06	G21	Metric input (mm)
	G24	Radius programming (**)
00	G28	Return to reference point
00	G29	Return from reference point
	G32	Thread cutting (**)
07	G40	Cutter compensation cancel
07	G41	Cutter compensation left
07	G42	Cutter compensation right
08	G43	Tool length compensation positive (+) direction
08	G44	Tool length compensation minus (-) direction
08	G49	Tool length compensation cancel
	G84	Canned turning cycle (**)
03	G90	Absolute programming
03	G91	Incremental programming

(**)—Refers only to CNC lathes and turning centers

Fig. 40-13A Some of the most common G-codes (preparatory) used in CNC programming. (Courtesy of Electronic Industries Association)

Code	Function
M00	Program stop
M02	End of program
M03	Spindle start (forward CW)
M04	Spindle start (reverse CCW)
M05	Spindle stop
M06	Tool change
M08	Coolant on
M09	Coolant off
M10	Chuck - clamping (**)
M11	Chuck - unclamping (**)
M12	Tailstock spindle out (**)
M13	Tailstock spindle in (**)
M17	Toolpost rotation normal (**)
M18	Toolpost rotation reverse (**)
M30	End of tape and rewind
M98	Transfer to subprogram
M99	End of subprogram

(**)—Refers only to CNC lathes and turning centers.

Fig. 40-13B Some of the most common M-codes (miscellaneous) used in CNC programming. (Courtesy of Electronic Industries Association)

machining center and these are marked with (**) in Fig. 40-13 A and B. Other codes not listed for machining centers are used exclusively for turning centers.

Fig. 40-14 shows the functions of some common M-codes.

Block of Information

CNC is generally programmed with only the information needed to complete the desired operation. Each word conforms to the EIA standard and they are written on a horizontal line. If multiple codes or same group codes are mixed, the machine control unit (MCU) will not recognize what to perform.

Using the example shown in Fig. 40-15, the words are as follows:

N 001	represents the sequence number of the operation.
G01	represents linear interpolation.
X1.2345	will move the table 1.2345 in. in a positive direction along the X axis.
Y.6789	will move the table .6789 in. along the Y axis.
M03	Spindle on CW.

INPUT AND OUTPUT MEDIA

There are different methods that can be used for putting information into a computer and receiving the processed output. Punched tape, magnetic tape, floppy disk, and hard disk are all methods used to transfer print information to a CNC machine, Fig. 40-16. Computer systems have a keyboard, similar to a typewriter keyboard, that is used to program information into the computer. Magnetic tape, also used as input, usually comes from some other computer or a conversion of some other input media. Input can come from numerous specially designed sensors that record data and convert it to binary signals that are compatible with the computer, Fig. 40-16.

The floppy disk or diskette is a small, flexible disk similar to a phonograph record. It has tracks or channels that will store magnetic bits of information. The bits can be sensed by a reading head and function similarly to magnetic tape, except that it is much faster to access data from disk. One disk, depending on its density, can store the equivalent of 2000 to 8000 feet or more of punched tape.

Output may be signals which are sent to units that control all types of CNC equipment. For example, computer data may run the servomechanisms of the machine tool table and spindle drives. Output can also be in the form of punched tape, print-out sheets, magnetic tape, or even data transferred to direct storage and used later for further processing.

With the introduction of computers, many of the problems associated with these methods were eliminated because the programming data was sent directly to the CNC machine tool's central processing unit (CPU). Today the computer is the main method of transferring data from the job print to the CNC machine.

The program input medium used depends on what each MCU was designed to read. Except for control units designed for manual data input (MDI) of the part program at the shop-floor level, almost all MCU's have some type of media input method which can use magnetic tape cassettes or floppy disks.

Tape Format

For many years, the 1 in. wide, 8-track punched tape was the industry standard for transferring data from the programmer to the NC machine tool. Even with the advent of computers and CNC soft-wired control units, perforated or

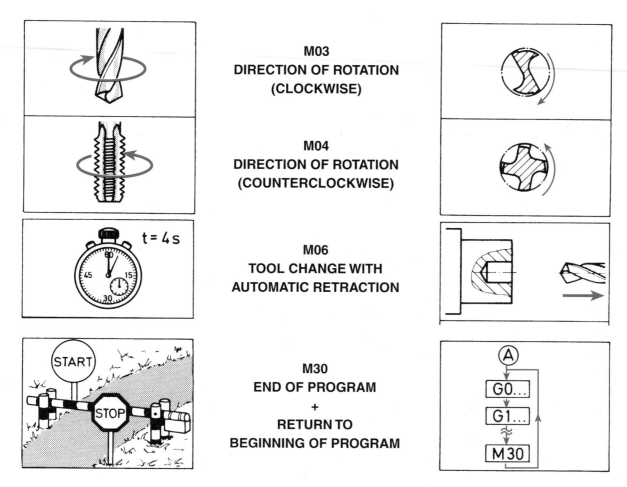

Fig. 40-14 The functions of a few common M-codes. (Courtesy of Deckel Maho, Inc.)

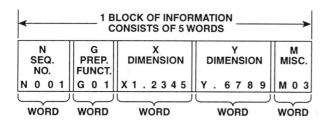

Fig. 40-15 A block contains all the information required to complete an operation. (Courtesy of Kelmar Associates)

punched tape is still in use, but rapidly being replaced by the computer. Several tape formats have been in use since the introduction of NC, but the trend has been toward control units that are manufactured to the Electronics Industries Association standards RS-244 and RS-358 which define variable-block tape format for positioning and straight-cut controls, and for contouring and positioning controls.

EIA RS-244 and RS-358

The Electronics Industries Association RS-244 standard binary coded decimal was becoming the method of coding individual digits or symbols when the ASCII (American Standard Code for Information Interchange) was introduced. Considerable progress has been made in resolving the differences between the old EIA code and the newer ASCII code. The EIA now has a new standard RS-358, which is a subset of ASCII character codes used in numerical control work. The two EIA character coding systems are used interchangeably for almost all CNC-related activities involving the perforated tape. Notice the similarity of the coded tape formats for EIA RS-244 and RS-358 shown in Fig. 40-17.

Most modern control systems can recognize and accept tape in either EIA standard format. Most of the newer CNC units can identify whether tape code RS-244 or RS-358 is being used through the parity check. Even though

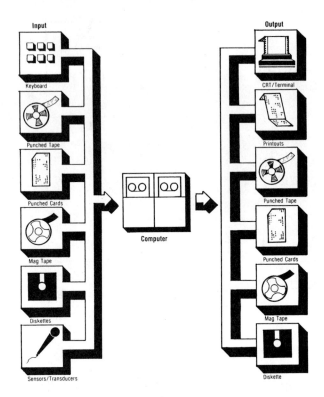

Fig. 40-16 Various methods of putting information into a computer (input) and receiving information (output). (Courtesy of *Modern Machine Shop Magazine*, copyright 1994, Gardner Publications Inc.)

RS-358 is intended to replace RS-244, many companies are still using the older tape code because of existing tape preparation equipment which cannot be economically altered. Once a company invests in equipment to suit a certain standard, it is often reluctant to change because of the cost involved.

● INPUTTING DATA AT THE MACHINE

Every machine control unit (MCU) has a mode selection switch or control which allows the operator to put in or change data at the machine, Fig. 40-18. This switch is the heart of the CNC machine and the operator must first set the switch for the proper mode before any function can be performed. The switch can usually be set for various modes such as manual, manual data input, edit, program operation, and tape.

Manual Mode

In the manual mode, a CNC machining center acts like a manual milling machine; a CNC

turning center acts like an engine lathe. In the manual mode, the CNC machine operator can press jog buttons or use the pulse generator handwheel to move a selected axis, and turn on switches to perform various machine functions.

Manual Data Input Mode

The mode switch has two positions, the edit, and the manual data input (MDI) position. Both mode switch positions require entering data through the keyboard. In the edit position of the mode switch, a program can be entered or changed and in the MDI position, CNC commands can be entered and executed.

In edit mode, an operator can enter CNC programs into the control memory. This can also be done by loading the program from some outside device such as a computer or tape reader. In the MDI mode, CNC commands are entered through the keyboard and display screen manually and can be run once or stored. If the same command is required a second time, the operator must enter the MDI command again.

Edit Mode

In the edit mode, the operator can do two basic things, enter CNC programs into the control memory and change current programs. Any good CNC control allows the operator to insert new information into the program, change current information in the program, and delete information from the program.

Program Operation Mode

The operation mode allows the operator to actually run the programs. The two positions of the mode switch are memory (or auto) and tape. The memory (or auto) mode is used to run programs from the control's memory. The operator usually has the choice of running the full program in the auto mode, or one line at a time in the single step mode for each time the cycle start button is depressed. While the program is being run, the operator can see one full page of the program on the control's display screen.

Tape Mode

With the ever-increasing use of the computer, the need for the tape mode switch has decreased. Many of the newer CNC controls no longer have the ability to run programs from the

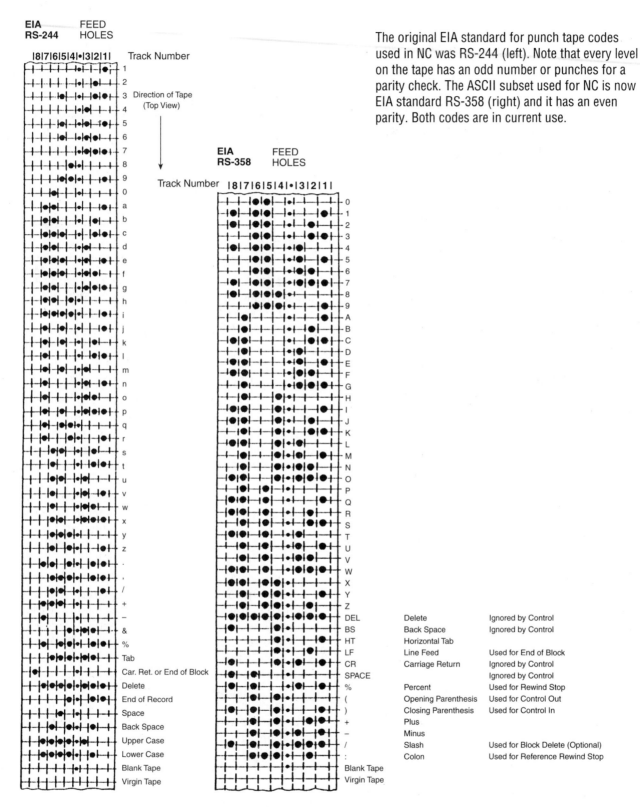

The original EIA standard for punch tape codes used in NC was RS-244 (left). Note that every level on the tape has an odd number or punches for a parity check. The ASCII subset used for NC is now EIA standard RS-358 (right) and it has an even parity. Both codes are in current use.

Delete	DEL	Ignored by Control
Back Space	BS	Ignored by Control
Horizontal Tab	HT	
Line Feed	LF	Used for End of Block
Carriage Return	CR	Ignored by Control
SPACE		Ignored by Control
Percent	%	Used for Rewind Stop
Opening Parenthesis	(	Used for Control Out
Closing Parenthesis	)	Used for Control In
Plus	+	
Minus	–	
Slash	/	Used for Block Delete (Optional)
Colon	:	Used for Reference Rewind Stop

Fig. 40-17 A comparison of the differences in the EIA standard RS-244 and RS-358 codes for punched tape. (Courtesy of Modern *Machine Shop Magazine*, copyright 1994, Gardner Publications Inc.)

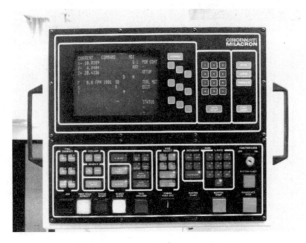

Fig. 40-18 The CNC machine control unit allows the operator to input new programs or change existing programs. (Courtesy of Cincinnati Milacron Inc.)

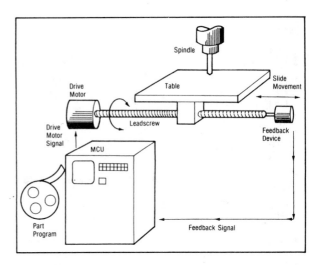

Fig. 40-20 A CNC machine's slides are positioned by the information on the part program in the MCU. (Courtesy of *Modern Machine Shop Magazine*, copyright 1994, Gardner Publications Inc.)

tape. The tape position of the mode switch allows the operator to see only one or two commands of the program at a time on the display screen. In memory mode, the operator can monitor a whole page of the program.

● HOW THE MCU CONTROLS THE MACHINE

The MCU (**machine control unit**) controls the machine tool movements and positions a machine slide to the desired location. The conventional manual machine has a handwheel connected to a leadscrew, Fig. 40-19. Each revolution of the screw moves the table slide by an amount equal to the lead of the table screw. For example, if the screw has 10 threads per inch, the table movement would be .100 inch. If the handwheel dial is graduated into 100 equal divisions, an op-

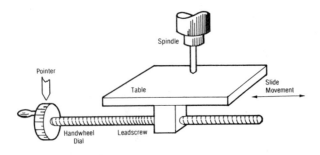

Fig. 40-19 A conventional machine tool's slides are moved by the operator turning the handwheels to position the workpiece. (Courtesy of *Modern Machine Shop Magazine*, copyright 1994, Gardner Publications Inc.)

erator can position the table to an accuracy of .001 inch by counting the number of turns of the handwheel and the number of graduations on the dial. To perform the same function without the operator, the electronics in the MCU are required, and a drive motor and a feedback device must be added to the machine, Fig. 40-20.

The instructions can be entered into the MCU by a computer, a program tape, or by manual data input. The MCU takes the input dimensions and compares them to the output of the feedback device that gives the current slide position. If there is any difference between the input signals and the table position, a signal is sent to the drive motor to rotate it in the proper direction to cancel the difference. When the two signals are the same or match, all slide motion is stopped. If the table slide is forced out of position by the cutting forces, the feedback device sends a signal to the MCU which recognizes this condition and the MCU sends a signal to the drive motor to move the slide back into position. What an operator did visually, the MCU does electronically and to a much higher degree of accuracy and speed.

Feedback Devices

In a closed-loop system, various types of feedback devices for table slide movements are used—rotary resolvers, synchros, linear scales, optical encoders, precision potentiometers, and lasers. Rotary resolvers or linear encoders are most commonly used for slide position feedback, and optical encoders are used to a lesser extent.

Rotary resolvers are usually mounted directly to the drive motor or to the end of the leadscrew, and can make very accurate position measurements. In most cases, the resolver's analog output signal is fed back to the MCU and processed by a resolver-to-digital converter that counts the pulses generated. These pulses, representing the angular displacement of the rotor and therefore the leadscrew, can be used to measure the leadscrew revolutions, and as a result, the position of the machine slide.

A linear-scale feedback device consists of two magnetically-coupled parts, a scale and a slider, and operates like a multiple resolver. The scale is usually mounted to a fixed-position machine member such as the machine column, and the slider is mounted to the related sliding part. As the carrier travels along the column, the slider moves along the scale. The slider printed-circuit tracks produce a wave-form pattern similar to that of a track on the scale, but with one track shifted one-quarter of a cyclic pitch. When the voltage energizes the scale, it induces an output voltage in proportion to the slider's linear displacement. This output signal can be used to produce results similar to the rotary resolver.

MCU Features

There are many types of MCUs available, each with specific features for a CNC machine which provide flexibility and advantages. A manufacturer of controls generally designs an MCU for a specific application or machine and these must be built into the MCU. On a hard-wired control, the options are usually built into the hardware by the manufacturer. On a CNC unit, many of the features are software and must be included in the executive program by the control builder in consultation with the machine builder. These features are permanently installed in the MCU's memory.

Some of the more common and more important features include:

- *Part Program Storage:* An advantage of newer CNC units is the large-capacity computational and data storage combination. This may hold a number of complete part programs which can be stored in the computer memory or on tape and then read into the CNC unit. Each use of the program is

from the stored memory rather than reading and rereading the tape.

- *Manual Data Input:* Most CNC units have some degree of MDI which enables special or revised data, or even a complete part program, to be entered right at the machine tool. Therefore, if the original program has errors or must be revised in any way, it is possible to make revisions right at the machine. Today, MDI is usually made by keyboard input, which may identify the part program to be called from memory. It can also be used to communicate with a central computer if the program data is stored away from the machine tool.

- *Program Edit with Program Copy:* This feature allows part programs to be corrected or modified at the CNC. Program copy permits the stored part program and the edit data to be copied as a corrected program into another storage area of memory. The data is then cleared from edit memory, freeing it for further corrections.

- *Multi-Program Storage:* This allows more than one part program to be stored as long as the capacity of the MCU is not exceeded. This can be done with either greater capacity in the standard part program memory, or the use of diskettes.

SUMMARY

- Cartesian coordinates are an accurate method of locating any point on a job. The primary axes on a vertical machining center are X, Y, and Z; on a turning center, X and Z.

- In the absolute programming mode, all dimensions and locations are given from a fixed zero or origin point. In the incremental programming mode, all dimensions are given from the previous location.

- Point-to-point positioning is used to rapidly locate the cutting tool while it is above the work surface. Continuous path is to position the tool from one point to another while it is in contact with the work.

- G-codes (preparatory) cause some action on the axes of a machine, while M-codes (miscellaneous) control functions of a machine. Some codes are modal and stay in effect until replaced by another code of the same group, while others are nonmodal and stay in effect for only one operation.

- Input media to a computer can consist of punched tape, magnetic tape, floppy disk, or hard disk.

- The EIA standard codes for punched tapes are RS-244 and RS-358, which most control units can recognize and use.

- The MCU controls tool movements and slide positions, while feedback devices compare and correct the information sent from the MCU with the table position.

KNOWLEDGE REVIEW

Cartesian Coordinate System

1. Explain how the Cartesian coordinate system and its four quadrants are used for dimensional locations.

2. Name the three primary axes of motion of a vertical machining center.

3. Name the two primary axes of motion on a turning center and state the motion of each.

Binary Numbers

4. Compare decimal numbers with binary numbers.

Programming Modes

5. Name and describe the two types of programming modes.

Positioning Systems

6. Name the two positioning systems and describe the use for each.

Programming Format

7. Explain the word-address programming format.

8. What do the following codes represent? G, G00, G01, G02, G03, F, S, D, T.

9. Define modal and nonmodal codes.

10. What do the following codes represent? M, M03, M04, M05.

Input and Output Media

11. List four types of media that computers use to transfer data to a CNC machine.

12. Name four types of output data.

13. What punched tape was previously used as an industry standard?

14. Name the two EIA standards for variable-block tape format.

Inputting Data at the Machine

15. What input modes may be used to input data at the machine?

16. What input mode can only be used to change existing programs?

How the MCU Controls the Machine

17. Explain how the machine table is moved to various positions by conventional and CNC machining.

18. What is the purpose of the feedback device in closed-loop systems?

19. Name the two most commonly used devices for slide position feedback.

MCU Features

20. List four important MCU features.

Preparing for Programming

Accurate programming of the part is very important to the accuracy of the part being produced. Part programming is becoming easier because of advanced computer technology and better CNC software. Computer assistance in programming is widely used in industry; however, parts are still being programmed manually in many plants where machines have not been replaced. For the less-complex part geometries, especially where two-axes positioning machines are applicable, manual programming is quite common.

Regardless of whether programming is done with a computer or manually, a knowledge of the basics is still very important. The programmer must decide how to process the part, what operations in what sequence are required, and what tools to use. The programmer must be able to effectively communicate with the machine control unit.

OBJECTIVES

After completing this unit, you should be able to:

- Program a part using linear and circular interpolation.
- Set and shift workpiece and program zero to suit the part being programmed.
- Compensate for variations in tool length, cutter diameter, or nose radius in order to produce accurate parts.
- Program parts or operations using the X and Y coordinates.

KEY TERMS

absolute positioning	circular interpolation	compensation
contouring	helical interpolation	incremental positioning
linear interpolation	zero-reference point	

● PROGRAMMING FOR POSITIONING

Before a person starts to program, it is important to become familiar with the part to be produced. Using the engineering drawings and specifications, the programmer should be able to plan the machining sequences required to produce the part. Visual concepts must be put into a written manuscript as the first step in developing a part program, Fig. 41-1. It is the part program that will be supplied to the machine control unit by the computer, tape, diskette, or some other medium.

The programmer must first establish a reference point for aligning the workpiece and the machine tool for programming purposes. The manuscript must include this along with the types of cutting tools and work-holding devices required, and where they are to be located.

Dimensioning Guidelines

The system of rectangular coordinates is very important to the successful operation of CNC machines. Certain guidelines should be observed when dimensioning parts for CNC machining. The following seven guidelines will help to insure that the dimensioning language means exactly the same thing to the design engineer, the technician, the programmer, and the machine operator.

1. Define part surfaces from three perpendicular reference planes.

2. Establish reference planes along part surfaces that are parallel to the machine axes.

3. Dimension from a specific point on the part surface.

4. Specify the tolerances allowed for each part dimension.

5. Allow for surface irregularities on contours.

6. Dimension the part clearly so that its shape can be understood without making mathematical calculations or guesses.

7. Define the part so that a computer numerical control cutter path can be easily programmed.

From the drawing to the workpiece

1. Reading drawing

2. Programming

3. Inputting program

4. Manufacturing

Fig. 41-1 The first step in producing a CNC program is to take the information from the print and produce a program manuscript. (Courtesy of Deckel Maho, Inc.)

● BLOCK FORMATS

The structure of CNC programs consists of a number of well-ordered sequences of commands, Fig. 41-2. These program commands can consist of:

- Program-technical commands such as program number and block numbers N1, N2, etc. Blocks are labeled N.

- Geometrical commands, such as G01, for distances in the X, Y, and Z axes.

- Technological commands or miscellaneous functions such as speed, feed, spindle ON or OFF, etc.

The following is a brief description of some of the more common program commands used in CNC work:

1. *Block number*—consists of the letter followed by up to 5 digits; N − − − − −.

2. *G-address* (preparatory functions)—consists of the letter followed by up to 2 digits, G − −.

3. *X, Y, Z addresses* (dimensioning codes)—have the letter, plus or minus (±) signs, and 4 or 5 digits.

EXTRACT FROM PROGRAM:

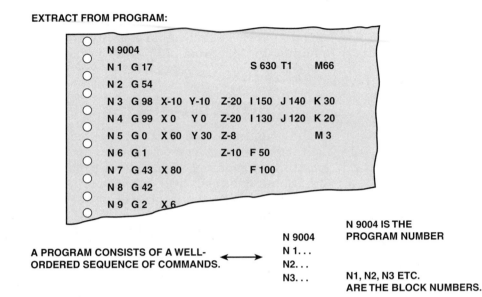

N 9004

N 1   G 17   S 630 T1   M66

N 2   G 54

N 3   G 98   X-10 Y-10   Z-20   I 150   J 140   K 30

N 4   G 99   X 0   Y 0   Z-20   I 130   J 120   K 20

N 5   G 0   X 60 Y 30   Z-8   M 3

N 6   G 1   Z-10   F 50

N 7   G 43   X 80   F 100

N 8   G 42

N 9   G 2   X 6

A PROGRAM CONSISTS OF A WELL- ⟷   N 9004   N 9004 IS THE
ORDERED SEQUENCE OF COMMANDS.   N 1...   PROGRAM NUMBER
                                     N2...
                                     N3...   N1, N2, N3 ETC.
                                               ARE THE BLOCK NUMBERS.

STRUCTURE OF A PROGRAM:

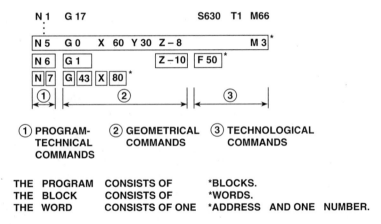

① PROGRAM-   ② GEOMETRICAL   ③ TECHNOLOGICAL
  TECHNICAL   COMMANDS   COMMANDS
  COMMANDS

THE PROGRAM CONSISTS OF   *BLOCKS.
THE BLOCK   CONSISTS OF   *WORDS.
THE WORD   CONSISTS OF ONE *ADDRESS AND ONE NUMBER.

Fig. 41-2 A summary of program structure and block size. (Courtesy of Deckel Maho, Inc.)

Vertical machine: X ± 5 digits, Y ± 4
                       digits, Z ± 5 digits

Horizontal machine: X ± 5 digits, Y ± 5
                       digits, Z ± 4 digits.

Turning machine: X ± 5 digits, Z ± 5
                       digits

4. *F (feed) address*—The feed rate of the cutting tool, stated as a letter and up to three
   digits; F – – –.

5. *S (speed) address*—generally refers to spindle speed, consists of a letter and up to four
   digits, S – – – –.

6. *M-address* (miscellaneous functions)—consists of a letter followed by up to 2 digits; M
   – –.

7. *T-address* (tool number)—consists of a letter followed by up to three digits; T – – –.

8. *J,K—addresses* (circle parameters)—consist
   of a letter followed by two digits, J – –, K – –.

Most CNC manufacturers generally follow the word-addresses outlined in the ANSI-EIA standard RS-274-D for positioning and contouring machines.

A – Rotation around X-axis

B – Rotation around Y-axis

C – Rotation around Z-axis

F – Feed rate

G – Preparatory functions

I – Circular interpolation X-axis arc center

J – Circular interpolation Y-axis arc center

K – Circular interpolation Z-axis arc center

M – Miscellaneous functions

N – Sequence or block number

R – Arc radius

S – Spindle speed

T – Tool number

X – X-axis data

Y – Y-axis data

Z – Z-axis data

% – Program start (rewind)

See the example of the block format and size shown in blocks N3 and N5 of Fig. 41-2 on how various addresses are used in a CNC program. There are differences in programming CNC turning centers and CNC machining centers, and it is important to know the programming required for each machine.

MEASUREMENT SYSTEMS

Most CNC control units and some hardwired units can provide both inch and metric data. Either can be used by setting a switch or by entering a specific code (G20 for inch and G21 for metric) in the part program. A control is basically inch or metric, depending primarily on the feedback system and how slide drives are geared. When it is switched from one to the other, all software and internal processing of data must also switch to agree with the input. The display monitor will show the data in inches or millimeters depending on the information supplied to the control. Some control units may only be able to convert to metric from inch and vice versa.

When programming, all dimensions, tools, etc., must be in either inch or metric sizes and should not be mixed to prevent errors. To change an inch dimension into millimeters, for example, multiply by 25.4; from millimeters to inch, divide by 25.4.

POSITIONING OR CONTOURING

Computer numerical control programming falls into two main types: positioning or point-to-point, and contouring or continuous-path. Modern control units can easily handle a mixture of positioning and contouring, however it is wise to have a good understanding of the two operations and their programming capabilities.

Positioning

Positioning or point-to-point programming is used to direct a tool that is not touching the workpiece to a specific location on a workpiece where some operation will be performed. It can be used in operations such as drilling, where the drill moves over and across the workpiece to a specific location and performs the programmed operation there, Fig. 41-3A. It may be locating the tool for operations such as drilling, reaming, tapping, and boring. The process is repeated until all machining is completed at the programmed locations on the workpiece.

While moving from one coordinate position to the next position, the cutting tool is *not* in contact with the workpiece. The spindle may move above the workpiece or the table may move under the spindle. It is only after the spindle and workpiece reach the programmed location that the spindle advances the tool into the workpiece.

Linear Interpolation

Linear interpolation, programmed points connected by straight lines, is used for straight-line moves, whether the length between the two points is short or long. In the early days of NC, it was used to produce curves that had to be broken into a very large number of short, straight-line programmed points to produce the required shape. Most modern CNC controls have circular interpolation that can produce contour and curved shapes with very few programmed steps.

Most CNC machine tools can also mill straight lines along either the X or Y axis, Fig. 41-3B, or angular lines in two axes with no change in the depth of tool along the line of travel, Fig. 41-3C. The coordinate positions of the beginning

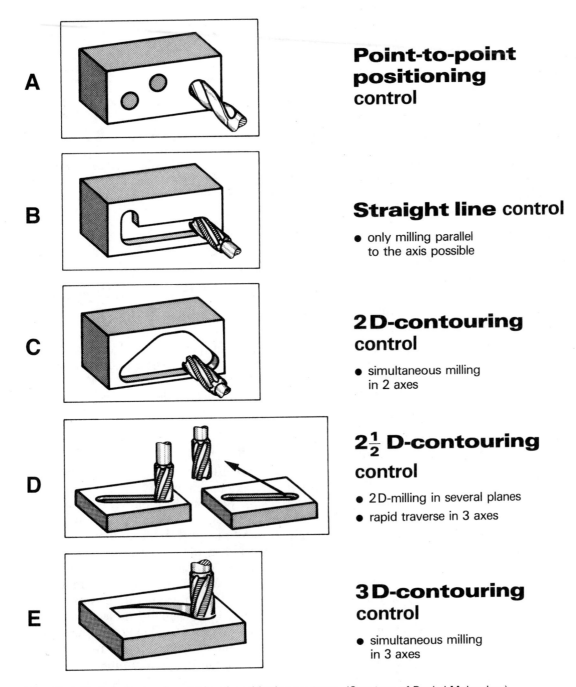

Point-to-point positioning control

Straight line control
- only milling parallel to the axis possible

2 D-contouring control
- simultaneous milling in 2 axes

2½ D-contouring control
- 2 D-milling in several planes
- rapid traverse in 3 axes

3 D-contouring control
- simultaneous milling in 3 axes

Fig. 41-3 Various types of control and positioning systems. (Courtesy of Deckel Maho, Inc.)

and end points of the line are programmed along with a feed rate, Fig. 41-3D (left view). The feed rate is the key since a pure positioning program involves absolutely no contact between tool and workpiece. In positioning, the tool is moving from one position to the next at a rapid traverse rate, which could be hundreds of inches per minute. A milling cut must be programmed at a feed rate to suit the material being cut and the type and diameter of the cutter.

Contouring

Contouring or continuous-path operation is where the cutting tool is always in contact with the workpiece as the coordinate movements are being made. Therefore, a contour path is being formed as the programmed coordinate movements are being made, Fig. 41-3E. The most common contouring operations use milling cutters and lathe tools to machine and shape the workpiece.

The path that the cutting tool follows in relation to the workpiece is very important and must be accurately programmed. The main difference between a positioning and a contouring control is the interpolation features, that coordinate the axes movements and allow the programmer to program a specific feed rate.

Programming for Contouring

There are similarities and differences in programming for contouring (or continuous path) and programming for positioning. The similarity is that dimensioning for both uses rectangular coordinates. In all CNC work, there must be reference points or planes on the workpiece that relate to machine axes.

There is one very important difference—interpolation. The function of interpolation is to store the programmed information and continually monitor and direct the progress of the machine axis movements to maintain a straight-line motion between the coordinate points at a defined vector feed rate. Interpolation is the main feature that simplifies contour programming.

The interpolator in the control calculates the movements required to make the tool follow the programmed path at the programmed vector feed rate. The function of the interpolator is to calculate or generate the many coordinate positions required to produce the desired shape or contour. The contouring principle can be illustrated with a simple circle, defined as many very small straight lines joined end-to-end to form a closed plane, with each line being the same distance from a center point. It takes many thousands of very short straight lines within the working framework of rectangular coordinates to describe a circle as shown in Figure 41-4. Circles, arcs, curves or shapes can be machined by connecting the points with a series of tool movements. It is possible to program the most complex form if enough coordinate points can be defined that accurately describe the surface, Fig. 41-5.

The five interpolation methods used to define coordinate points to generate curved shapes are: linear, circular, helical, parabolic, and cubic. Most modern contouring controls have linear and circular interpolation, and usually helical.

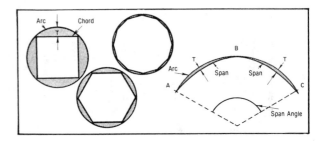

Fig. 41-4 The types of forms which can be obtained through continuous path machining. (Courtesy of *Modern Machine Shop Magazine*, copyright 1994, Gardner Publications Inc.)

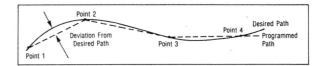

Fig. 41-5 Linear interpolation is used to program a curve which cannot readily be described mathematically. (Courtesy of *Modern Machine Shop Magazine*, copyright 1994, Gardner Publications Inc.)

Circular Interpolation

Circular interpolation is very valuable for programming complete circles, approximations of circles, or portions of a circle. Normally, all that must be programmed for an arc are the coordinate locations of the start and end points of the arc, the radius of the circle or arc, the coordinate locations of the center of the circle, and the direction in which the cutter is to travel, Fig. 41-6. The circular interpolator in most MCUs breaks up the circular span into a series of the smallest increments of movement caused by a single output pulse, usually .0001 inch. The interpolator automatically computes enough of these pulses to describe the circular cut and then generates the signals that move the cutting tool to produce the cut.

The benefit of circular interpolation can be quickly seen when comparing the linear and circular programming required to produce circular cuts. Thousands of data blocks, each defining a short span, may be necessary with linear interpolation to produce the same circle as compared to a few blocks of programmed data with circular interpolation. This MCU feature automatically calculates what had to be described previously in the smallest detail.

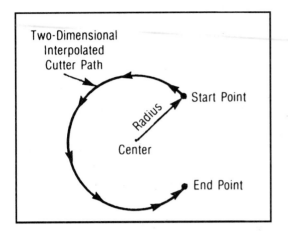

Fig. 41-6 To program any portion of a circle requires the start and end point of the arc, the center location, the radius, and the direction of cut. (Courtesy of *Modern Machine Shop Magazine*, copyright 1994, Gardner Publications Inc.)

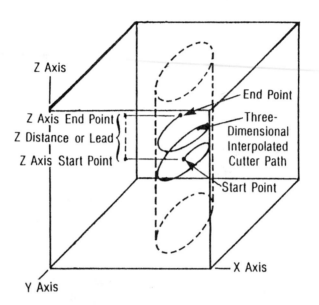

Fig. 41-8 Helical interpolation uses circular interpolation for two axes along with a linear movement in the third axis. (Courtesy of *Modern Machine Shop Magazine*, copyright 1994, Gardner Publications Inc.)

Many other curves and free-form shapes can be closely approximated with a series of arcs, using circular interpolation, Fig. 41-7. In many cases a true circle (or series of circles) can be fitted to the desired contour, using less data and providing a smoother surface than with linear interpolation.

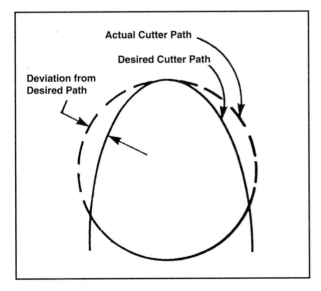

Fig. 41-7 Circular interpolation can be used to come close to producing a curve on surfaces that are not true arcs. (Courtesy of *Modern Machine Shop Magazine*, copyright 1994, Gardner Publications Inc.)

Helical Interpolation

The third most common form of interpolation is helical. As shown in Fig. 41-8, **helical interpolation** combines the use of two-axes circular interpolation with a linear movement of the third axis. All three axes move simultaneously to produce a helical spiral cutter path. This method is most commonly used in milling large internal-diameter threads, and requires minimal input data.

● MACHINE AND WORK COORDINATES

Most CNC machine tools have a default coordinate system (home or zero position). This is generally a fixed position, usually located at the tool change position, which is built into the machine by the manufacturer. It is used as a reference point when locating work-holding devices and workpieces. Most MCUs have two position registers, the absolute position and the command position, Fig. 41-9. The absolute position register shows the position of the machine table from the absolute zero or home position at all times. The command accumulator register shows the table axis position in relation to the programmer's or operator's XY zero or reference point.

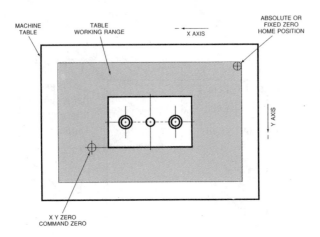

Fig. 41-9 The command (part) zero location can be moved from the fixed or home position to any suitable point in the table working range. (Courtesy of Kelmar Associates)

When a part is being programmed, the machine's absolute or zero position and the part zero are rarely the same. The programmer can choose any location on the part, work-holding device, or machine table as the program zero.

Machine Zero Point

For convenience in programming, the machine zero point can be set by three methods—by the operator, manually by a programmed absolute zero shift, or by work coordinates—to suit the holding fixture or the part to be machined.

- *MANUAL SETTING*—The operator can use the MCU controls to locate the spindle over the desired part zero and then set the X and Y coordinate registers on the console to zero.

- *ABSOLUTE ZERO SHIFT*—The absolute zero shift is a method of changing the coordinate system by a command in the CNC program. The programmer first sends the machine spindle to home zero position by a G28 command in the program. Then another command (G92 for absolute zero shift) tells the MCU how far from the home zero location the coordinate system origin is to be positioned, Fig. 41-10.

The sample commands may be as follows:

N1 G28 X0 Y0 Z0 (sends spindle to home zero position)

N2 G92 X4.000 Y5.000 Z6.000 (the position the machine will reference as part zero)

Stored zero shifts (G54 ... G59)
Programmed zero shift (G92)

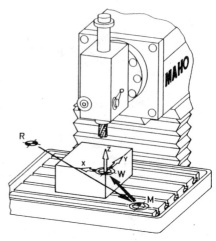

⊕ **R = Reference point** (maximum travel of machine)

⊕ **M = Machine zero point** (X0, Y0, Z0) of machine coordinate system

⊕ **W = Part zero point workpiece** coordinate system

Under G54 ... G59 the actual machine coordinates of part zero are stored in the stored zero offsets memory, and activated in the part program.

Under G92 the actual machine coordinates are inserted and used on the G92 line of the part program.

Fig. 41-10 The relationship between the part zero and the machine system of coordinates. (Courtesy of Deckel Maho, Inc.)

If more than one fixture or part needs to be located on a machine table, the programmer can use a different zero location for each part or fixture. This can be done by inserting a G28 X0 Y0 Z0 command to return to home zero, followed by a G92 command with new coordinates for each part zero location required.

Full-Floating Part Zero

Full-floating zero machines have no fixed reference or zero point, therefore a new zero can be used for each setup. When using controls having full-floating capabilities, the part may be located at any convenient place on the machine table. Once the workpiece is set up, the operator must check the program manuscript for the proper alignment positions and these must be entered into the control. When the cycle button is pressed, the machine will rapid to the previous work setup; then the operator manually moves the machine table zero location to match the new setup's zero location. Once the new workpiece location is set, the machine can be locked and matched with the control unit.

Full-floating zero can increase machine up-time by reducing the amount of time spent on setting up the workpiece.

● TOOL SETTINGS AND OFFSETS

All types of CNC machine tools require some form of tool setting and offsets (**compensation**) to allow the programmer or operator to meet unexpected problems relating to tooling. In many cases it will be impossible to predict the exact tooling problems that may come up on a job. Tool compensation enables changes to be made to the program to overcome the tooling problems that may arise.

● SETTING ZERO POINT AND TOOL AXIS

Before any part is machined, it is very important to set the cutting tool to the zero locations of the X, Y, and Z axes. Each machine tool has a home or **zero reference point** that is built into it by the manufacturer and cannot be changed. In Fig. 41-9, previously shown on page 418, the XY zero (home) position of the machine is in the top right-hand corner of the working range of the table. However, the programmer can set any reference point which best suits the part, operation, or tool change position. In most cases, this reference point will *not* be the same as the machine reference point.

Setting X and Y Axes to Zero

After the part to be machined has been fastened to the table and aligned, the center of the machine spindle should be aligned with the X and Y edges. This can be done by:

* Setting an edge finder to left edge of the part, Fig. 41-11.
* Raising the spindle or lowering the table.
* Moving the table one-half the diameter of the edge finder along the X axis.

If the diameter of the contact point on the edge finder is .200 in., a movement of .100 in. in the direction of the arrow will align the spindle center with the edge of the workpiece.

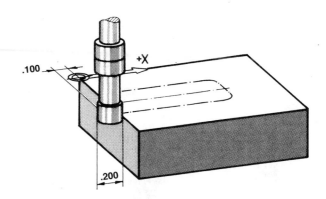

Fig. 41-11 Setting the spindle center to zero on the X axis with an edge finder. (Courtesy of Deckel Maho, Inc.)

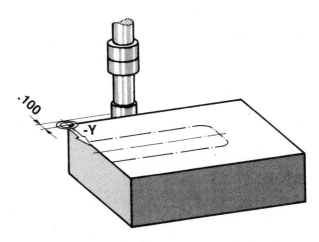

Fig. 41-12 Setting the spindle center to zero on the Y axis with an edge finder. (Courtesy of Deckel Maho, Inc.)

* Setting the X axis register to zero.

The same procedure should be used to set the Y axis, Fig. 41-12, and then setting the Y axis register to zero. The *zero shift* on the machine control unit (MCU) allows the XY zero location to be easily shifted to the one chosen by the programmer.

Setting the Z Axis Register to the Tool Length Offset

To set the end of the milling cutter to the surface of the workpiece in the Z axis, Fig. 41-13:

* Bring the cutter over the surface of the part.
* Place a thin piece of paper between the surface of the work and the end of the cutter.
* Raise the table or lower the spindle until a slight drag is felt on the paper.

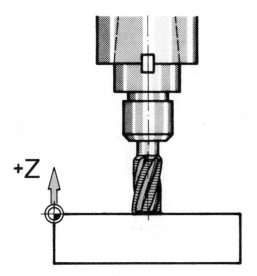

Fig. 41-13 Setting the bottom of the cutter to the work surface in the Z axis. This method can only be used when machining with one tool. (Courtesy of Deckel Maho, Inc.)

- After clearing the workpiece, raise the table .002 in. to allow for paper thickness.
- Set the Z axis register to zero.

● WORK SETTINGS AND OFFSETS

All CNC machine tools require some form of tool setting, work setting, and offsets (compensation). Compensation allows the operator or programmer to make adjustments for unpredictable tooling and setup conditions. In some cases, it is almost impossible to predict certain conditions and compensation is especially designed to handle such problems.

Work Coordinates

In **absolute positioning**, work coordinates are generally set on one edge or corner of a part and all programming is taken from this work coordinate. In Fig. 41-14, the part zero is used for all positioning for hole locations #1, 2, and 3.

In **incremental positioning**, the work coordinates change because each location is the zero point for the move to the next location, Fig. 41-15.

On some types of parts, it may be desirable to change from absolute to incremental, or vice versa, at a certain point in the job. This can easily be done by inserting the G90 (absolute) or the G91 (incremental) command into the program at the point where the change is to be made.

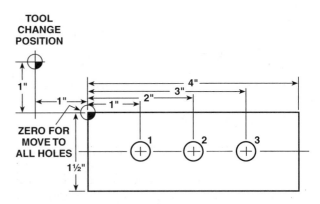

Fig. 41-14 In absolute programming, all dimensions must be taken from the XY zero at the top left-hand corner of the part. (Courtesy of Kelmar Associates)

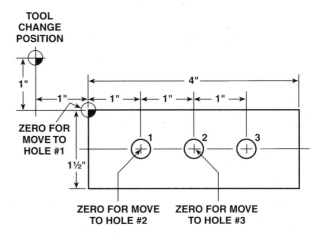

Fig. 41-15 In incremental programming, all dimensions are taken from the previous point. (Courtesy of Kelmar Associates)

Work or R Plane

The word-address letter R refers to either the work surface or the rapid-traverse distance (often called the work plane) programmed. The R work surface is set a specific height or distance above the work surface and it is commonly used with canned or fixed cycles. This setting is referred to as R0, or the reference dimension, and all programmed depths for cutting tools and surfaces to be machined are taken from the R0 surface.

The R0 work plane is generally set at .100 in. above the highest surface of the workpiece, Fig. 41-16, which is also known as gage height. Some manufacturers build a gage height distance of .100 in. into the MCU, and whenever the feed motion in the Z axis is called for, .100 in. will automatically be added to the depth programmed.

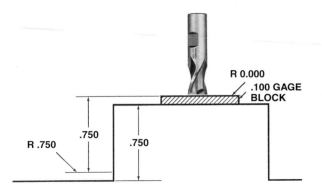

Fig. 41-16 Using a .100 in. gage block to set the gage height or R0 on the work surface. (Courtesy of Kelmar Associates)

When setting up cutting tools, the operator generally places a .100 in. thick gage on top of the highest surface of the workpiece. Each tool is lowered until it just touches the gage surface and then its length is recorded on the tool list. Once the gage height has been set, it is not generally necessary to add the .100 in. to all future depth dimensions since most MCUs will do this automatically.

● TOOL SETTINGS AND OFFSETS

For every part, the programmer must consider the step-by-step operations required to machine the part. At the same time, a list of cutting tools used for each operation must be included. This *tool list* must include the type of each cutting tool, its diameter, and length. Since a wide variety of cutting tools of various lengths and diameters are generally used for machining a part, it is important to be able to compensate for the differences in their diameters and lengths to ensure that the part will be machined accurately. This compensation involves working with offsets so that the machine control unit knows exactly how to adjust for differences in tool diameters and lengths.

All forms of compensation work with offsets. CNC offsets can be thought of as the memory on an electronic calculator. If the calculator has a memory, a constant value can be stored into each memory and used whenever required for a calculation. This keeps from having to enter the same number over and over again when it is required. Like the file locations in the memory of an electronic calculator, offsets in the CNC control are storage locations into which numerical values for each tool can be placed.

Tool Lengths

The programmer must specify a minimum length, based on the maximum depth to which each cutting tool must travel. Each tool is then programmed as though it had zero length, and the spindle/tool gage line and the tool point were the same. The amount of Z-axis travel is programmed from the R plane (.100 above the work surface) at a specified feed rate, Fig. 41-17. The actual length of each tool is entered into the MCU either by NC tape, computer, or manually by the machine operator.

Some MCUs are equipped with semiautomatic tool compensation to make allowances for differences in cutting tool lengths. The tool-length compensation feature provides for as many as sixty-four or more tool length offsets. The programmer uses the same basic tool length for all tools. When the machine is being set up, the operator mounts each tool, starting with the longest, jogs it against a fixed reference stop, and presses a tool offset button on the MCU. This automatically records the tool's setting into memory. All cutting tools for a job are set against the

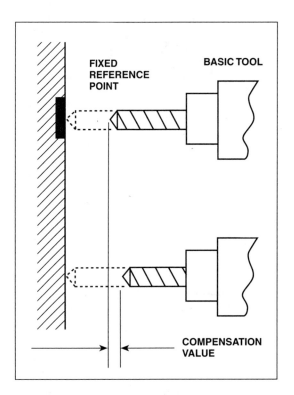

Fig. 41-17 Semiautomatic tool compensation is used to quickly figure tool lengths for all tools at setup time. (Courtesy of *Modern Machine Shop Magazine*, copyright 1994, Gardner Publications Inc.)

SETTING ZERO POINT IN THE TOOL AXIS

The length of the various tools to be used for each operation must be recorded and entered into the tool memory.

In actual practice **two modes of operation** are used.

1. Machine with **one** tool

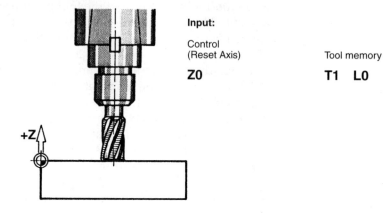

Input:

Control
(Reset Axis)

Z0

Tool memory

T1   L0

2. Matching with **several** tools

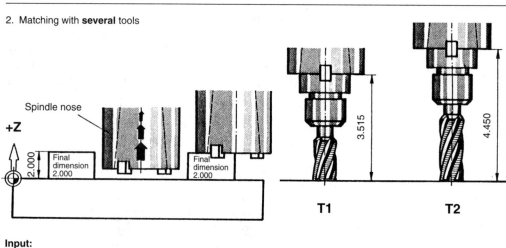

Input:

Control
(Reset Axis) **Z2.000**   Tool memory   **T1   L3.515**
                                          T2   L4.450

Fig. 41-18 Two methods of setting and recording tool lengths in the MCU memory. (Courtesy of Deckel Maho, Inc.)

same reference stop, and the difference between each tool's actual length and the basic tool length is entered into memory as a compensation value, Fig. 41-18. With this information, the control's computer changes all program steps to take into account the length and diameter of each tool.

Most modern CNC machines use preset tools which have been prepared in the toolroom. Their type, diameter, and length have all been recorded on the tool assembly drawing and used when the program is being prepared. Each tool is assigned a specific number which is stored in the

MCU, and can be recalled by a command in the CNC program any time that specific tool is required. For example, a T1 M06 code will place tool #1 into the spindle on a machine equipped with an automatic tool changer (ATC).

Cutter-Diameter Compensation

Cutter-diameter compensation (CC) changes a milling cutter's programmed centerline path to compensate for a small difference in cutter diameter. On most modern MCUs, it is effective for most cuts made using either linear or circular interpolation in the X-Y axis, but does not affect the programmed Z-axis moves. Compensation is usually in increments of .0001 in. up to +1.0000 in., and usually most modern controls have as many CDCs available as there are tool pockets in the tool storage matrix.

The advantage of the CC feature is that it:

1. Allows the use of cutters that have been sharpened to a smaller diameter.

2. Permits the use of a larger or smaller tool already in the machine's storage matrix.

3. Allows backing the tool away when roughing cuts are required due to excessive material present.

4. Permits compensation for unexpected tool or part deflection, if the deflection is constant throughout the programmed path.

The basic reference point of the machine tool is never at the cutting edge of a milling cutter, but at some point on its periphery. If a 1 in. end mill is used to machine the edges of a workpiece, the programmer would have to keep a 1/2 in. offset from the work surface in order to cut the edges accurately, Fig. 41-19. The 1/2 in. offset represents the distance from the centerline of the cutter or machine spindle to the edge of the part. Whenever a part is being machined, the programmer must calculate an offset path, which is usually half the cutter diameter.

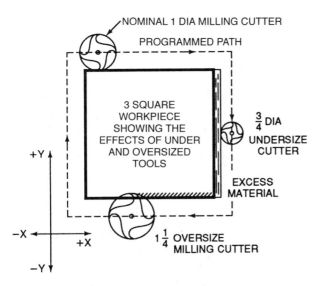

Fig. 41-19 Cutter-diameter compensation must be used when machining with various size cutters. (Courtesy of Kelmar Associates)

Modern MCUs, which have part surface programming, automatically calculate centerline offsets once the diameter of the cutter for each operation is programmed. Many MCUs have operator-entry capabilities which can compensate for differences in cutter diameters; therefore an oversize cutter, or one that has been sharpened, can be used as long as the compensation value for oversize or undersize cutters is entered.

Programming Cutter-Radius Compensation

The use of cutter-radius compensation may vary with different controls. Each control generally has a set of rules that specify how cutter radius compensation is used and canceled on that control. The basics of how a control is programmed are shown in the following example. Be sure to refer to the CNC control manufacturer's manual for the machine control unit being used.

Most controls use three G-codes with cutter radius compensation. G41 is used to activate a cutter-left condition (climb milling with a right hand cutter). G42 is used to activate a cutter-right condition (conventional milling). G40 is used to cancel cutter-radius compensation. The principle of cutter-radius offset is illustrated in Fig. 41-20.

To easily determine whether to use G41 or G42, look in the direction the cutter is moving during machining.

- If the cutter is on the left, use G41.

- If the cutter is on the right, use G42.

Figure 41-21 shows some examples that should help in deciding when to use the G41 or G42 code. Once cutter-radius compensation is properly activated, the cutter will be kept on the same side of all surfaces until the G40 command is used to cancel the compensation.

CNC Lathe Tooling

The parts produced on a lathe or turning center require only a limited number of cutting tools. Single-point lathe tools are used for some ID (internal diameter) operations and almost all OD (outside diameter) operations such as turning, facing, grooving, etc. Form tools are not commonly used because of the contouring capability of modern MCUs.

The workpieces, when machined with commercial-grade tools, tend to vary in size from tool to tool depending on the class of insert. Gener-

THE PRINCIPLE OF THE TOOL RADIUS OFFSET

The cutter radius offset makes it possible to program the **workpiece dimensions.**

PRINCIPLE:

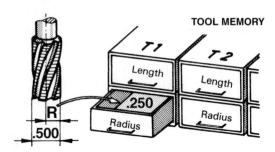

The workpiece dimensions are programmed, the control calculates the tool center path by means of the respective radius value R.

Input into the tool memory:

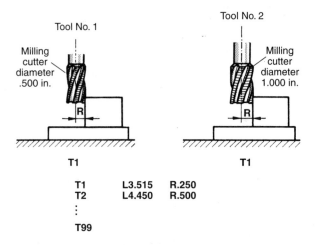

T1	L3.515	R.250
T2	L4.450	R.500
⋮		
T99		

Fig. 41-20 The principle of cutter radius offset. (Courtesy of Deckel Maho, Inc.)

ally, a class which provides ± .005 on the tool tip would be used for rough cuts, while a class ± .0002 would be used for precise finish cuts. After a looser tolerance insert has been indexed, more time has to be to be spent in compensating for these variations if they are used for finish cuts.

With preset tooling, a variation of insert-type tooling, the toolholder generally has some adjustment for length or other characteristic. Tool adjustments to the exact dimensions shown in the part program are made in a special measuring machine or presetting device. With presetting, carbide or ceramic inserts can be set to very close tolerances.

G41 / G42,G40

In order to enable the control to calculate from the program data and from the tool memory data the center path of the milling cutter, it is necessary to communicate to the control **where** the tool has to mill.

For this there are 3 G-functions:

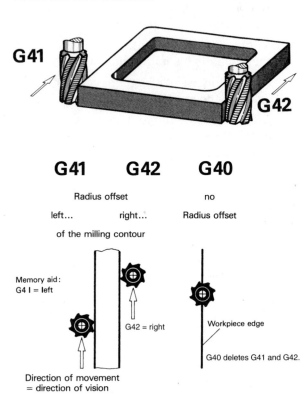

Fig. 41-21 Methods of programming cutters for radius compensation. (Courtesy of Deckel Maho, Inc.)

Qualified tooling has become almost universally accepted by most CNC operators and programmers. Machine builders are designing for such tooling, and a specific machine may have slots provided in the turret face to accommodate certain size tools. Two tools of the same designation are going to cut identically, and a tool will cut the same after each index of the insert.

Tool Nose-Radius Compensation

Just as cutter-radius compensation allows the programmer to program milling work surface coordinates, so does tool nose-radius compensation in lathe-type work. Figure 41-22 shows why tool nose-radius compensation is used to prevent deviations from the work surface on circular and angular cuts.

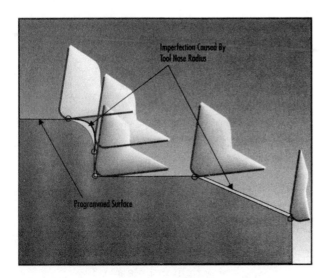

Fig. 41-22 The tool nose radius will cause variations in the work surface when cutting tapers and contours. (Courtesy of *Modern Machine Shop Magazine*, copyright 1994, Gardner Publications Inc.)

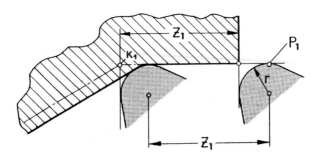

Fig. 41-23A The final form is produced by P_1 of the tool nose radius, therefore the full length programmed will not be cut. (Courtesy of Emco Maier Corp.)

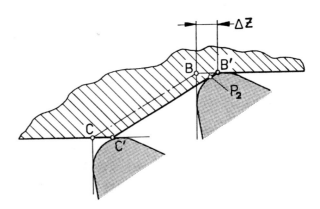

Fig. 41-23B When taper turning, P_2 of the tool nose radius produces the final taper outline. (Courtesy of Emco Maier Corp.)

Usually the nose radius on a lathe tool is small (1/64, 1/32, 3/64, or 1/16 in.), meaning that the variations on the work surface caused by the tool radius will also be small. For the chamfering of corners to break sharp edges, tool nose-radius compensation may not be required. However, if the surfaces being machined are very important (contours and radii), compensation must be provided for the radius of the tool. *Tool nose-radius compensation should only be required on finishing cuts.*

The tool nose radius does not affect the workpiece outline when cutting in the XZ direction, Fig 41-23A. When cutting in the Z direction, machining occurs at the P_1 of the cutting tool; however, when the tool starts to cut the taper, Fig 41-23B, the cutting changes to P_2 of the tool nose. Even though the Z1 dimension was programmed, the taper will not be cut at the correct position because the cutting occurs at different positions on the tool nose radius. The small difference can be calculated mathematically and the program adjusted accordingly. Modern MCUs do this automatically when the tool information is entered in the program and called into use with the correct code.

To decide whether to use the G41 or G42 command, look in the direction the tool is moving during the cut and note which side of the workpiece the tool is on. For a turning center with tools on the back side of the centerline, the

tool is on the left, use G41 (e.g. boring toward the chuck); if the tool is on the right, use G42 (turning toward the chuck). Once this is determined, include the proper G code in the tool's first approach to the workpiece. Once tool nose-radius compensation is entered, it remains in effect until canceled by the G40 code.

Programming Coordinate Dimensions

To provide the basics of the programming required to locate edges or hole locations on actual workpieces, a few examples follow. At this point it is helpful to review the two methods of positioning, absolute and incremental. **Absolute positioning** is where all dimensions are taken from one fixed point or location. **Incremental positioning** is where the dimension for the next location is given from the previous point.

When locating points on a workpiece in the coordinate system, there are two lines at right angles to each other, one vertical and one horizon-

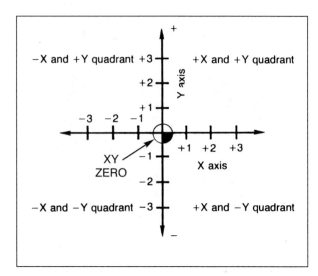

Fig. 41-24 The intersecting lines, at right angles to each other, form the four quadrants of the coordinate system. (Courtesy of Allen-Bradley)

tal, Fig. 41-24. These are called axes, and where they intersect is called the origin or zero point. The horizontal line is called the X axis; the vertical line is called the Y axis.

- Any point to the *right of the Y axis* would be an *X+* (plus) or positive move.

- Any point to the *left of the Y axis* would be an *X−* (minus) or negative move.

- Any point *above the X axis* would be a *Y+* move.

- Any point *below the X axis* would be a *Y−* move.

Drilling Operations

The hole locations on the *absolute* example shown in Fig. 41-25 will be programmed, assuming that the center of the tool or machine spindle is located over the XY zero at the top left-hand corner of the part. After locating at hole #3, return the tool to XY zero.

Hole #1 = X 1.000 (a positive move to the right of XY zero)

= Y−.750 (a negative move below the XY zero)

Hole #2 = X 2.000 (2.000 to the right of XY zero)

= Y−.750 (generally does not have to be programmed since the Y location is the same as for hole #1)

Hole #3 = X 3.000 (3.000 to the right of XY zero)

XY Zero = X0 Y0 (tool returns to start position)

In the *incremental* example, Fig. 41-26, the XY zero is at the bottom left-hand corner of the part. Program the center location of all holes in numerical sequence, and after hole #3, return the tool to the XY zero position.

Hole #1 = X 1.000 (a positive move to the right of XY zero)

= Y .750 (a positive move above XY zero)

Hole #2 = X 1.000 (a positive move to the right of hole #1)

= Y0 (this indicates no change in the Y location)

Hole #3 = X 1.000 (a positive move to the right of hole #2)

= Y0 (no change)

XY Zero = X−3.000 (a negative move from hole #3)

= Y−.750 (a negative move from hole #3)

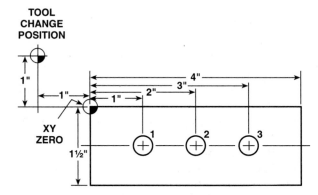

Fig. 41-25 A workpiece using absolute (datum) dimensioning. (Courtesy of Kelmar Associates)

Milling Operations

In milling operations, it is not the center of the cutter that does the machining, but generally the periphery. The cutter path would not be the contour or outer surface, but a distance equal to the radius of the cutter away from the part. For example, if a .750 in. diameter end mill is being used to machine the edges of the parts shown in Figs. 41-25 and 41-26, the center of the machine spindle (or cutter) must be programmed (offset) .375 in. away from all edges.

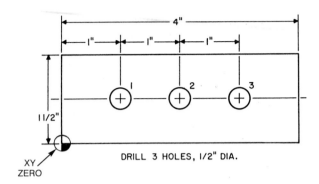

Fig. 41-26 A workpiece using incremental (delta) dimensioning. (Courtesy of Kelmar Associates)

To program for machining the edges of the absolute part shown in Fig. 41-25, the start point will be the XY zero at the top left-hand corner of the part. The machining will be in a clockwise (climb milling) direction around the edges of the part.

1. The .750 in. diameter cutter must be positioned .375 in. to the left of XY zero, or in a minus direction, and .375 in. above the XY zero in a plus direction.

    Start position = X−.375, Y .375

2. Milling the top edge

    X 4.375   (the −.375 from the left of the part to .375 past the right-hand edge)

    Y .375   (remains the same, and on most controls, does not have to be programmed again)

3. Milling the right-hand edge

    X 4.375   (does not have to be programmed because there is no move in the X direction)

    Y−1.875   (the cutter center moves .375 in. below the bottom edge)

4. Milling the bottom edge

    X−.375   (the cutter center moves .375 in. past the left-hand edge of the part)

    Y−1.875   (this position will stay at −1.875 in.)

5. Milling the left-hand edge

    X−.375   (the cutter will remain at −.375 in. or the start point)

    Y .375   (the cutter returns to the start point)

To program for machining the edges of the incremental part shown in Fig. 41-26, the start point will be the XY zero at the bottom left-hand corner of the part. The machining, with a .750 in. diameter cutter, will be in a clockwise (climb milling) direction around the edges of the part.

1. The .750 in. diameter cutter must be positioned .375 in. to the left of the XY zero, or in a minus direction, and .375 in. below the bottom edge, or a minus direction.

    Start position = X−.375, Y−.375

2. Milling the left-hand edge

    X 0   (the cutter will remain at −.375 in., or the start point)

    Y 2.250   (the cutter will move to .375 in. above the top edge)

3. Milling the top edge

    X 4.750   (the cutter will move to .375 in. past the right-hand edge)

    Y 0   (the cutter will remain at .375 in. above the part)

4. Milling the right-hand edge

    X 0   (the cutter will remain at .375 in. past the right-hand edge)

    Y−2.250   (the cutter will move to .375 in. below the bottom edge)

5. Milling the bottom edge

    X−4.750   (the cutter will return to the start point)

    Y 0   (the cutter will remain at .375 in. below the bottom edge)

SUMMARY

-   The manuscript for a part program must contain specifications about the part, sequence of operations, types of cutting tools, and work-holding methods.

-   A program block can consist of technical, geometrical, and technological commands.

-   Positioning, or point-to-point positioning, is used to quickly locate a tool at a specific point while the tool is above the work surface.

-   Linear interpolation is used to make straight-line cuts between two points at a programmed feed rate.

- Circular interpolation is used to machine circles, arcs, or contours, at a programmed feed rate.

- Machine and work coordinates are used as reference points when locating workpieces, work-holding devices, and for tool-change positions.

- Offsets or compensation are used to make adjustments in the program for tool length, diameter, nose radius, resharpening, and deflection.

- All absolute coordinate dimensions are taken from one fixed point.

- Each incremental coordinate dimension is taken from the previous point.

KNOWLEDGE REVIEW

Programming for Positioning

1. What information should a program manuscript contain?

Block Formats

2. Name three types of program commands.

3. Define the following word-addresses: G, I, M, N, R, T.

Positioning or Contouring

4. What is the purpose of
   a. positioning?   c. contouring?
   b. linear interpolation?   d. interpolator?

5. What four pieces of information are required to program an arc?

Machine and Work Coordinates

6. Define the purpose of
   a. an absolute position register
   b. a command accumulator register

7. What is the purpose of
   a. absolute zero shift?
   b. full-floating zero?

Tool Settings and Offsets

8. In point form, explain how to set the X and Y axes to zero with an edge finder.

9. Briefly explain how to set the Z axis to zero.

Work Settings and Offsets

10. Where is the R plane generally set in relation to the work surface?

Tool Settings and Offsets

11. Explain how semi-automatic tool compensation works.

12. State three advantages of cutter diameter compensation.

13. State the purpose of the following cutter radius compensation codes: G41, G42, G40.

14. What effect does tool nose radius have on angular and contour surfaces?

UNIT 42

Linear Programming

Linear interpolation involves straight-line movement of a machine tool slide between any two programmed points, whether they are close together or far apart. There are two primary codes, G00 and G01, used for straight-line moves between two points. The G00 command, a rapid-traverse mode, is used for the rapid point-to-point positioning of a cutting tool between two points. The G01 command, with a programmed feed rate, is used when machining straight-line cuts between two points parallel to the X and Y axes, Fig. 42-1.

Straight-line angular cuts can be made with the G01 command when the X and Y coordinates at the beginning and end of the line are given along with a programmed feed rate, Fig. 42-2.

OBJECTIVES

After completing this unit, you should be able to:

- Prepare two-axes programs to position a cutting tool for drilling or tapping operations.
- Write programs that combine positioning and machining operations.
- Prepare programs using the X, Y, and Z axes.
- Write a program to produce angular cuts and surfaces.

KEY TERMS

canned cycle   continuous-path control   gage height
two-axes programming

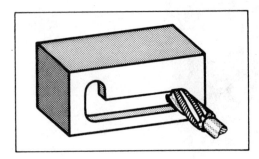

Fig. 42-1 Straight-line control is used when machining cuts parallel to the X and Y axes. (Courtesy of Deckel Maho, Inc.)

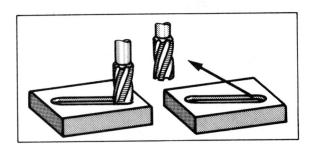

Fig. 42-2 Straight-line angular cuts can be made with linear interpolation. (Courtesy of Deckel Maho, Inc.)

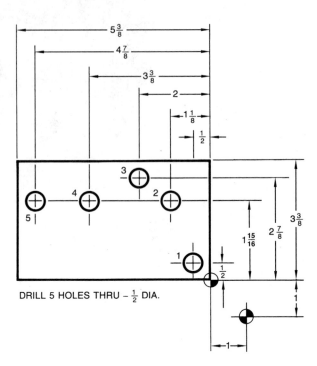

DRILL 5 HOLES THRU – $\frac{1}{2}$ DIA.

Fig. 42-3 Drilling example #1 requires the positioning for hole locations in the absolute mode. (Courtesy of The Superior Electric Co.)

● TWO-AXES PROGRAMMING

In the following drilling and milling examples, only cutting tool positioning in the X and Y axes will be used for programming. This **two-axes programming** is to keep the programs easy to understand while some G and M codes, sequences, speeds, tool number, etc., are gradually introduced.

● TWO-AXES DRILLING

Positioning to only the X and Y axis location for drilling is a very practical application for a two-axes control machine. Once at the coordinate location, the spindle is manually controlled by the operator to drill a hole.

Drilling Example #1 — Absolute (Inch)

Fig. 42-3 illustrates an absolute, inch-dimensioned part that will be used for programming hole locations only. In practice, the programming for the hole locations and the drilling operations would be combined in the same program. A G-code readable controller, such as a Fanuc or Fanuc

compatible, will be used for the programming examples.

The codes introduced in this example are:

G00 – rapid traverse mode

G20 – inch program mode

G90 – absolute positioning

G92 – absolute program command to reset axis position register coordinates

M00 – program stop; operator must restart cycle

M06 – tool change command

M30 – end of data; resets control and/or machine

S – spindle speed

T – tool number

% – program start code (beginning of program); rewind stop code (end of program)

Program Notes

1. The machine spindle must be first located at the tool-change (start) point.

2. All programming will start at the XY part zero at the lower right corner of the part.

3. Use absolute positioning for all hole locations.

4. Program all moves in rapid traverse (G00).

5. Program holes in the numerical sequence shown on the print.

6. After hole #5, return to the start point.

7. Use 800 r/min. speed for the 1/2 in. diameter stub drill.

The Program

```
%
   – program start code
O421
   – program number; most controllers use
     the letter O as the first digit
N05   G20   G92   X1.000   Y–1.000
   G20   – inch program mode
   G92   – position register reset
         – this positions the tool at the
           tool-change position 1.000 to the
           right and 1.000 below the part
           zero
N10   G90   T01   M06
   G90   – absolute positioning mode
   T01   – tool #1, 1/2 in. dia. stub drill
   M06   – tool change command
N15   G00   X–.500   Y.500   S800   M03
   G00   – tool rapids to Hole #1 position
   X–.500 – tool moves .500 in. to left of part
           zero
   Y.500  – tool moves .500 above part zero
   S800   – spindle speed 800 r/min.
   M03   – spindle ON clockwise
N20   M00
   – program stop command for drilling hole
N25   X–1.125   Y1.9375
         – tool rapids to Hole #2 position
   X–1.125 – tool moves 1.125 to left of part
             zero
   Y1.9375 – tool moves 1.9375 above part
             zero
N30 M00
   – program stop at Hole #2
N35   X–2.000   Y2.875
   – tool rapids to Hole #3
   X–2.000 – tool moves 2.000 to left of part
             zero
   Y2.875  – tool moves 2.875 above part
             zero
```

```
N40   M00
   – program stop at Hole #3
N45   X–3.375   Y1.9375
   – tool rapids to Hole #4
   X–3.375 – tool moves 3.375 to left of part
             zero
   Y1.9375 – tool moves 1.9375 above part
             zero
N50 M00
   – program stop at Hole #4
N55   X–4.875   Y1.9375
   – tool rapids to Hole #5
   X–4.875 – tool moves 4.875 to left of part
             zero
   Y1.9375 – tool stays on same line as Hole
             #4
N60 M00
   – program stop at Hole #5
N65   X1.000   Y–1.000
   – tool returns to tool-change or start
   position
N70   M30
   – end-of-program command; resets
     program to start (rewind)
%
   – end-of-program symbol
```

Drilling Example 2 — Absolute (Metric)

In Fig. 42-4, the same basic inch part has been dimensioned in metric (inch to millimeters, multiply by 25.4; millimeters to inch, divide by 25.4). It will be used to program hole locations only and pause at each location as it would if the holes were to be drilled. The G92 command has been omitted from this example because all programming begins at the XY zero location on the part. It is being replaced by the G00 command, along with the proper XY axes locations, to take the spindle back to the tool-change position. Since this example is only used for positioning and no drilling will occur, the M00 commands have been omitted from this and following examples.

The code introduced in this example is:

G21 – metric program mode

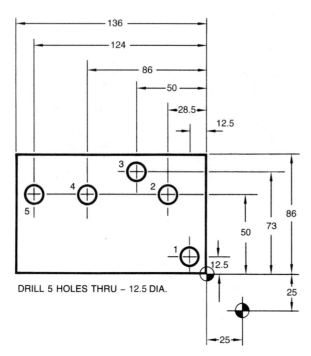

DRILL 5 HOLES THRU – 12.5 DIA.

Fig. 42-4 Drilling example #2 with metric dimensions requires hole positioning in the absolute mode. (Courtesy of The Superior Electric Co.)

Program Notes

1. Locate the machine spindle at the tool-change (start) point.

2. Start all programming at the XY part zero (lower right corner of part).

3. Use absolute programming for all locations.

4. Program holes in numerical sequence.

5. Program all moves in rapid traverse (G00).

6. After Hole #5, return to start point.

7. Use 800 r/min for the 12.5 mm diameter stub.

The Program

%
 – program start code

O422
 – program number

N05 G21 G00 X25 Y–25
 G21 – metric program code
 G00 – rapid traverse move to tool-change position, 25 mm to the right and below part zero

N10 G90 T01 M06
 G90 – absolute positioning
 T01 – 12.5 mm diameter stub drill
 M06 – tool change command

N15 G00 X–12.5 Y12.5 S800 M03
 – tool rapids to Hole #1 position
 X–12.5 – tool moves 12.5 mm to left of part zero
 Y12.5 – tool moves 12.5 mm above part zero
 S800 – 800 r/min. spindle speed
 M03 – spindle ON clockwise

N20 X–28.5 Y50
 – tool rapids to Hole #2
 X–28.5 – tool moves 28.5 mm to left of part zero
 Y50 – tool moves 50 mm above part zero

N25 X–50 Y73
 – tool rapids to Hole #3
 X–50 – tool moves 50 mm to left of part zero
 Y73 – tool moves to 73 mm above part zero

N30 X–86 Y50
 – tool rapids to Hole #4
 X–86 – tool moves 86 mm to left of part zero
 Y50 – tool moves to 50 mm above part zero

N35 X–124 Y50
 – tool rapids to Hole #5
 X–124 – tool moves 124 mm to left of part zero
 Y50 – tool stays at 50 mm above part zero

N40 X25 Y–25
 – tool rapids to tool-change or start position
 X25 – tool moves 25 mm to right of part zero
 Y–25 – tool moves 25 mm below part zero

N45 M30
 – end of program; rewind code

%
 – end-of-program symbol

Drilling Example #3 — Incremental (Inch)

Fig. 42-5 shows a part dimensioned incrementally that will be used for programming hole locations only. In incremental programming, the last location is the XY zero for the next location. In practice, a cutting tool is moved to the desired location and the machining operation is performed before moving to the next location.

The code introduced in this example is:

G91 — incremental positioning

Program Notes

1. Locate the machine spindle at the tool-change (start) point.

2. Start programming at the XY zero at the bottom left corner of the part.

3. Use incremental programming for all hole locations.

4. Program all moves in the rapid mode (G00).

5. The tool path should follow the holes in numerical sequence.

6. After Hole #5, return to the start point.

7. Use 800 r/min. for the 1/2 in. diameter stub drill.

The Program

```
%
   – program start code
O423
   – program number
N05  G20  G90  G00  X–1.000  Y–1.000
   G20  – inch programming
   G00  – positions the spindle at the tool-
          change (start) point
N10  G91  T01  M06
   G91  – incremental programming
   T01  – 1/2 in. diameter stub drill
   M06  – tool change point
N15  X2.000  Y2.000  S800  M03
          – tool moves to Hole #1 location
   S800 – spindle speed 800 r/min.
N20  X.250  Y1.250
   – tool moves to Hole #2 location
N25  X0  Y.750
   – tool moves to Hole #3 location
N30  X1.750  Y0
   – tool moves to Hole #4 location
N35  X0  Y–2.000
   – tool moves to Hole #5 location
N40  X–4.000  Y–2.000
   – tool moves back to tool-change position
N45  M30
   – end of program; rewind symbol
%
   – end-of-program symbol
```

Drilling Example #4 — Incremental (Metric)

The same part used in Fig. 41-5 has been dimensioned in metric sizes for Fig. 42-6. (inch to millimeters, multiply by 25.4; millimeters to inch, divide by 25.4). It will be used to program hole locations and pause at each location where a hole should be drilled.

Program Notes

1. Locate the spindle at the tool-change (start) point.

2. Start programming at the XY zero at the bottom left corner of the part.

3. Use incremental programming for all locations.

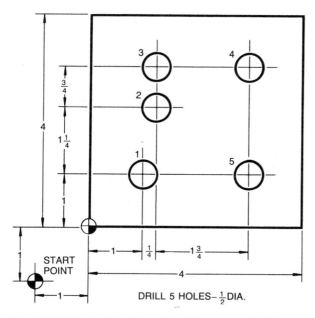

Fig. 42-5 The hole locations in drilling example #3 are to be programmed in the incremental mode. (Courtesy of The Superior Electric Co.)

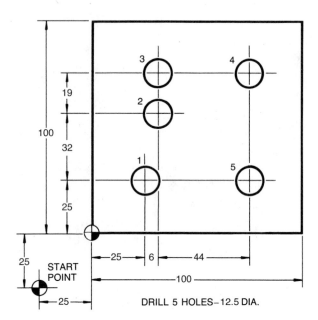

Fig. 42-6 The hole locations in the metric drilling example #4 are to be programmed in the incremental mode. (Courtesy of The Superior Electric Co.)

4. Move to all locations in the rapid traverse mode (G00).

5. The tool path should follow the holes in numerical sequence.

6. After Hole #5, return to start point.

7. Use 800 r/min speed for the 12.5 mm diameter stub drill.

The Program

%
   — program start code
O424
   — program number
N05 G21 G90 G00 X–25 Y–25
   G21 — metric programming
   G00 — tool rapids to tool-change (start) point
N10 G91 T01 M06
   G91 — incremental positioning
   T01 — 12.5 mm diameter stub drill
   M06 — tool change command
N15 G00 X50 Y50 S800 M03
   — tool rapids to Hole #1 location
   S800 — spindle speed 800 r/min.
N20 X6 Y32
   — tool moves to hole #2 location

N25 X0 Y19
   — tool moves up to Hole #3 location
N30 X44 Y0
   — tool moves across to Hole #4 location
N35 X0 Y–51
   — tool moves down to Hole #5 location
N40 X–100 Y–50
   — tool returns to tool-change position
N45 M30
   — end of program; rewind code
%
   — end-of-program symbol

● TWO-AXES MILLING

In drilling, the center of the cutting tool (drill) was positioned in the center of the hole location. In most milling operations, it is not the center of the tool that does the cutting, but some point on its circumference. Let us assume that a 1.000 in. diameter end mill will be used to machine the edges of a part, then the tool path must be the center of the end mill. In other words, the cutter must be programmed .500 in. or its radius away from all edges to be cut. This is referred to as cutter-diameter compensation, and for more information on this see Unit 41.

● CLIMB AND CONVENTIONAL MILLING

To get the best tool life and productivity out of a milling cutter, it is important that the cutter be used properly. Most common milling cutter problems are caused by excessive heat, abrasion, edges chipping, clogging, built-up edges, cratering, and work hardening of the workpiece. Before using any cutter, it is important to decide whether the operation will be performed by climb or conventional milling.

Climb milling is when the cutter rotation and the table (work) feed are in the same direction, (see Fig. 32-6A). When climb milling, it is important that the work be held securely, and the machine table must be equipped with a backlash eliminator. Climb milling is not a problem on a CNC machine tool because they are generally equipped with a recirculating ball screw. The advantages of climb milling are: increased tool life, improved surface finishes, easier chip removal, less edge breakout, and lower energy requirements. Climb milling is not recommended for machining parts that have a hard or abrasive surface.

Conventional milling is when the cutter rotation and the table (work) feed are going in opposite directions, (see Fig. 32-6B). The tooth enters the cut at zero chip thickness in an upward direction and progressively gets thicker. Conventional milling is used on workpieces where minimum shock is desirable when the cutter enters the work, or when the machine table is *not fitted with a backlash eliminator.* The forces in conventional milling try to lift the workpiece, therefore it is important that the work be held securely.

● CUTTING SPEEDS

The cutting speed for a milling cutter is the speed, in either feet per minute or meters per minute, that the periphery of the cutter should travel when machining a certain metal. The speeds used for milling-machine cutters are much the same as those used for any cutting tool. Several factors must be considered when setting the r/min to machine a surface. The most important are the following:

- material to be machined,
- cutter type and material,
- finish required,
- depth of cut,
- rigidity of the machine and the workpiece.

It is good practice to start from the calculated r/min using Table 33-1, and then progress until maximum tool life and productivity are reached.

Inch Cutting Speeds

The speed of a milling machine cutter is calculated by using the same formula as for a twist drill or a lathe workpiece.

$$r/min = \frac{CS \times 4}{D}$$

EXAMPLE: Find the r/min to mill a keyway 1 in. wide in a machine steel shaft, using a 4 in. diameter high-speed steel cutter (*CS* 80).

$$r/min = \frac{CS \times 4}{D}$$
$$= \frac{80 \times 4}{4}$$
$$= 80$$

EXAMPLE: Calculate the r/min required to end-mill a 3/4 in. wide groove in aluminum with a high-speed steel cutter (CS 600).

$$r/min = \frac{CS \times 4}{D}$$
$$= \frac{600 \times 4}{3/4}$$
$$= 3200$$

Metric Cutting Speeds

For metric cutters, the cutting speed is expressed in meters per minute. The formula used to determine the r/min is

$$r/min = \frac{CS \times 320}{D}$$

EXAMPLE: At what r/min should a 150 mm carbide-tipped milling cutter revolve to machine a piece of machine steel (CS 60)?

$$r/min = \frac{CS \times 320}{D}$$
$$= \frac{60 \times 320}{150}$$
$$= 128$$

EXAMPLE: Calculate the r/min required to mill a groove 25 mm wide in a piece of aluminum using a high-speed end mill (CS 150).

$$r/min = \frac{CS \times 320}{D}$$
$$= \frac{150 \times 320}{25}$$
$$= 1920$$

See the general recommendations for setting speeds for milling in Table 33-2.

● MILLING FEEDS

Feed is the rate at which the work moves longitudinally (lengthwise) into the revolving cutter. It is measured either in inches per minute or millimeters per minute. *Chip per tooth* is the amount of material removed by each tooth of the cutter as it revolves and advances into the work. The *milling feed* is determined by multiplying the chip size (chip per tooth) desired, the number of teeth in the cutter, and the r/min of the cutter.

Inch Calculations

The formula used to find work feed in inches per minute is:

$$feed = N \times c.p.t. \times r/min$$

where:

N = number of teeth in the milling cutter

$c.p.t.$ = chip per tooth for a particular cutter and metal, as given in Table 33-3

r/min = number of revolutions per minute of the milling cutter

EXAMPLE: Find the feed in inches per minute using a 4 in. diameter, 12-tooth helical cutter to cut machine steel (CS 80). It would first be necessary to calculate the proper r/min for the cutter.

$$r/min = \frac{CS \times 4}{D}$$

$$= \frac{80 \times 4}{4}$$

$$= 80$$

$$Feed\ (in./min) = N \times c.p.t. \times r/min$$

$$= 12 \times .010 \times 80$$

$$= 9.6$$

$$= 10\ in./min$$

Metric Calculations

The formula used to find the work feed in millimeters per minute is the same as the formula used to find the feed in inches per minute, except that mm/min is substituted for in./min:

$$mm/min = N \times chip\ per\ tooth \times r/min$$

EXAMPLE: Calculate the feed in millimeters per minute for a 75 mm diameter, six-tooth helical milling cutter when machining a cast iron workpiece (CS 20).

First calculate the r/min of the cutter.

$$r/min = \frac{CS \times 320}{diameter\ of\ cutter}$$

$$= \frac{20 \times 320}{75}$$

$$= 85$$

$$Feed\ (mm/min) = N \times c.p.t. \times r/min$$

$$= 6 \times 0.18 \times 85$$

$$= 91.8$$

$$= 92\ mm/min$$

See the general recommendations for setting feeds for milling in Table 33-4, (Unit 33).

Since the calculated feed rates are generally for industrial-size machines, on lighter machines or in training programs, it is wise to start with half the calculated feed rate and gradually increase it to suit the machine and type of operation.

Milling Example #1 - Absolute (Inch)

The four edges on the part shown in Fig. 42-7 are to be machined with a 1.000 in. diameter four-flute end mill. On this part, the machining will occur on the X or Y axis at any one time. Therefore if the machine is already positioned on an axis from the previous cut, it is not necessary to program that coordinate axis again.

Program Notes

1. Locate the machine spindle at the tool-change (start) position.
2. Start the programming at the XY zero at the bottom left corner of the part.
3. Program in the absolute mode.
4. Program in a clockwise direction (climb milling).
5. Use a 1.000 in. diameter, four-flute end mill and program a roughing cut (allow .020 in.) on all four edges.
6. Program a finishing cut around the part.
7. Return to the tool-change (start) position. The codes introduced in this example are:

F – feed rate

G01 – linear interpolation

M03 – spindle *ON* clockwise rotation

M05 – turns spindle *OFF*

M06 – tool change command

The Program

```
%
   – program start code
O426
   – program number
N05  G21  G90  G00  X–1.000  Y–1.000
   G00  – sets end mill to tool-change
          (start) position.
```

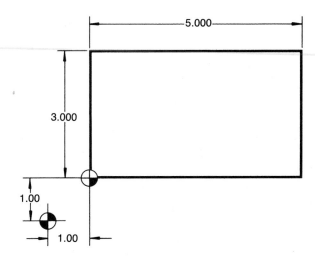

Fig. 42-7 The edges of milling example #1 are to be programmed in the absolute mode. (Courtesy of Kelmar Associates)

G21   – inch programming
G90   – absolute positioning

N10 T01 M06
T01   – 1.000 dia. end mill
M06   – tool change command

N15 S400 F5.0 M03
S400 – 400 r/min.
F5.0 – 5 in./min. feed rate
M03   – spindle on clockwise

Roughing Cut

N20 G01 X–.520 Y0
– positions end mill .020 in. away from left edge (X axis) and on the bottom edge (Y axis)

N25 Y3.520
G01   – linear interpolation at 5 in. feedrate
– cuts left edge leaving .020 in. for finish cut
Y3.520 – feeds in Y axis to .020 in. above top edge

N30 X5.520
– cuts top edge leaving .020 in. for finishing
– tool moves .020 in. past the right edge

N35 Y–.520
– cuts left edge leaving .020 in. for finishing
– tool moves .020 in. below bottom edge

N40 X–.500
– cuts bottom edge leaving .020 in. for finishing
– tool moves to finish size at the left edge

Finish Cut

N45 Y3.500
– finish cut on left edge; stops at finish size on top

N50 X5.500
– finish cut on top edge; stops at finish size on right edge

N55 Y–.500
– finish cut on right edge; stops at finish size at the bottom

N60 X–.500
– finish cut on bottom edge

N65 G00 X–1.000 Y–1.000
– tool rapids to tool-change position

N70 M05
– turns spindle OFF

N75 M30
– end of program; rewind code

%
– end-of-program symbol

Milling Example #2 — Incremental

Fig. 42-8 shows the same part that was used in example #1, but which will now be programmed in the incremental mode. Although the part is the same, the tool-change (start) position and the XY part zero have been changed. In practice, the tool-change position and the XY part zero are generally on the bottom left-hand side. However, since these two positions can be changed at any time to suit the machine, workpiece, fixture, or operations, it is wise to learn to program from a variety of locations.

Program Notes

1. Locate the machine spindle at the tool-change (start) position.

2. Start programming at the XY zero at the top left corner of the part.

3. Program in the incremental mode.

4. Program clockwise (CW).

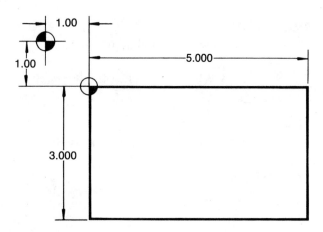

Fig. 42-8 The edges of milling example #2 are to be programmed in the incremental mode. (Courtesy of Kelmar Associates)

5. Use a 1.000 in. diameter, four-flute end mill, and program around all edges.

6. Return to the tool-change (start) position.

The Program

%
— program start code

O427
— program number

N05 G20 G90 G00 X–1.000 Y1.000
G20   — inch programming
       — sets end mill to tool-change (start) position

N10 G91
G91   — incremental positioning

N15 T01 M06
T01   — 1.000 dia. end mill
M06   — tool change code

N20 S400 F5.0 M03
S400   — 400 r/min.
F5.0   — 5 in./min. feed rate
M03   — spindle on clockwise

N25 G00 X.470 Y–.470
— rapids end mill close to top edge and left side

N30 G01 X.030 Y–.030
— feeds end mill to top edge and left side

N35 X6.000
— machines top edge and positions the edge of the end mill on the right edge

N40 Y–4.000
— machines right edge and positions the edge of the end mill on bottom edge

N45 X–6.000
— machines bottom edge and positions the edge of the end mill on left edge

N50 Y4.000
— machines left edge to top of part

N55 G00 X–.500 Y.500
— rapids back to tool-change (start) position

N60 M05
— turns spindle off

N65 M30
— end of program command; rewind code

%
— end-of-program symbol

● TWO-AND-A-HALF AXES PROGRAMMING

Two-and-a-half axes programming is where there is motion in the X, Y, and Z axes, but not all at the same time. This type of programming uses the X and Y axes to position the tool, and the Z axis to perform some type of operation such as drilling, tapping, etc. It is the most common form of CNC programming and is suitable for contouring operations in only two axes simultaneously. Three-axes programming involves motion on all three axes at the same time.

Milling and Drilling Example — Absolute (Inch)

In the example shown in Fig. 42-9, the programming for milling the edges will be combined with drilling the holes. This will involve the use of some new codes that should be understood:

G40 — cancels cutter compensation/offset

G43 — tool length compensation

G49 — tool length compensation cancel

G54 — reference point offset (selects work coordinate system #1)

G80 — cancels any canned cycle

G81 — starts canned drilling cycle

G99 — returns tool to R0 (gage height) after hole has been drilled

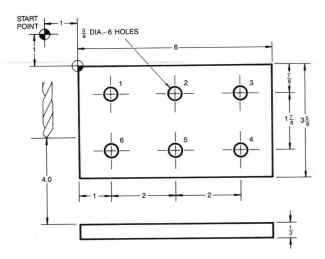

Fig. 42-9 The edges of this part are to be milled to size and the holes drilled in the absolute mode. (Courtesy of The Superior Electric Co.)

H – tool length compensation code

R0 – reference or gage height used for setting tool lengths. (usually .100 in. above work surface)

Program Notes

1. Locate the machine spindle at the tool-change (start) point.

2. Start programming at the XY zero at the top left corner of the part.

3. Use absolute programming.

4. Use a .500 in. diameter, two-flute end mill to machine the four edges clockwise.

5. Return to start position to change the tool.

6. Drill the six 3/8 in. diameter holes in numerical sequence and return to the start point.

The Program

%
– program start code

O4210
– program number

N05   G54   G20   G90   G40
G54 – calls the distance from the machine zero to part zero, and is stored in the offset page.

G20 – inch

G90 – absolute
G40 – cancels cutter compensation

N10   T01   M06
T01 – .500 in. diameter two-flute end mill
M06 – tool change code

N15   G43   H01   X–1.0   Y1.0   Z.1
G43 – tool length offset
H01 – offset value

N20   S800   M03
S800 – speed 800 r/min.
M03 – spindle on clockwise

N25   Z–.750
Z–.750 – tool point set to .750 below work surface

N30   G00   X–.250   Y.250
– tool feeds close to edge of part in X axis and on the top edge in Y axis

Milling the Edges

N35   G01   X6.250   F3.0
– top edge of work machined and tool stops at the finish size at the right edge
F3.0 – 3 in./min. feed rate

N40   Y–3.875
– right edge is cut and tool stops at finish size at bottom edge

N45   X–.250
– bottom edge is machined and the tool stops at the finish size at the left edge

N50   Y.250
– left edge is machined and the tool clears the top edge

N55   G00   X–1.0   Y1.0   Z1.0   G49   M05
– tool rapids to tool-change position and 1.0 above work surface
G49 – cancels G43 code (length compensation)
M05 – spindle turns OFF

N60   T02   M06
T02 – .375 stub drill
M06 – tool change code

Drilling Holes

N65   S1066   M03
S1066 – 1066 r/min.
M03 – spindle ON clockwise

N70 G43 H02
G43 — tool length compensation
H2 — compensation value

N75 G00 X1.0 Y-.875 Z.1
— tool rapids to Hole #1 position .100 in. above work surface.

N80 G81 G99 Z-.750 R.1 F3.0
G81 — canned cycle; drill Hole #1. Fig. 42-10
G99 — tool rapids out of hole to R (gage) height
Z-.750 — tool to clear bottom of work by 1/4 in.
R.1 — gage height .100 above work surface
F3.0 — 3 in. feed per minute

N85 X3.0
— Hole #2 drilled

N90 X5.0
— Hole #3 drilled

N95 Y-2.750
— Hole #4 drilled

N100 X3.0
— Hole #5 drilled

N105 X1.0
— Hole #6 drilled

N110 G80
— cancels drill cycle

N115 G00 X-1.0 Y1.0 Z1.0 M05
— tool rapids back to tool-change position 1.0 above work surface.
— spindle turns OFF

N120 M30
— end of data

%
— end-of-program symbol

● ANGULAR PROGRAMMING

Linear interpolation involves moving the cutting tool from one position to another in a straight line. With this type of programming, any straight-line section can be machined, including all tapers or angular surfaces. When linear moves are programmed, the coordinates (XY axes) for the beginning and end of each line must be given. Most straight-line moves are parallel to the X and Y axes; however, there may be a need for a straight-line angular move.

Almost all modern CNC controls have a **continuous-path control** system. This means that the drive motors of the X and Y axes can operate

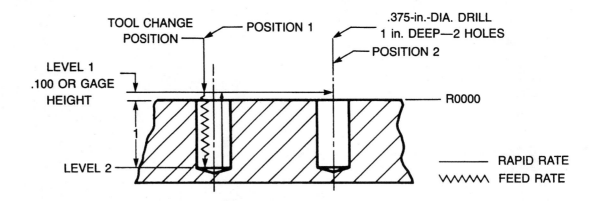

```
N80   G81   X6.0   Y4.0   Z-.75   R.1   F3.0
N85
```

Fig. 42-10 A G81 drill cycle is used to drill the two .375 in. diameter holes to 1 in. depth. (Courtesy of Cincinnati Milacron Inc.)

at different rates of speed. When milling angles, the MCU knows the start coordinate position and can calculate the difference between it and the end coordinate position. It quickly calculates the difference between the X and Y coordinates, and automatically produces the cutter centerline or vector path, Fig. 42-11.

When movement is required along two axes (X and Y), the axes move simultaneously along a vector path. The rate of travel along the vector path is set automatically by the MCU so that it is equal to the programmed feed rate. Since an angle is a straight line connecting a start point and end point, linear interpolation can be used to cut angular and taper surfaces of any length. After the two points have been programmed, along with a vector feed rate, the MCU stores this information in its memory until the end point of the line is reached. Therefore the function of interpolation is to store information and constantly compare and correct the machine axes movement to keep a straight-line movement between the start point and end point coordinates at a specified vector feed rate.

The workpiece shown in Fig. 42-12 requires a 30° angular slot .250 in. wide by .125 in. deep to be cut. Before this can be programmed, it is necessary to calculate the coordinate positions (XY) of the start point and the end point of the angular slot. If this will be programmed in the absolute positioning mode, the calculations are as follows:

Start point	X1.500 Y2.000
End point	X3.548 Y.566
Feed rate	5 in./min

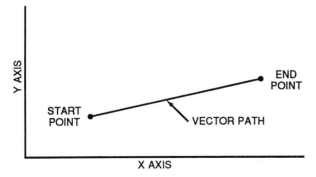

Fig. 42-11 Angular surfaces can be programmed by linear interpolation when a start point, end point, and feed rate are specified. (Courtesy of Kelmar Associates)

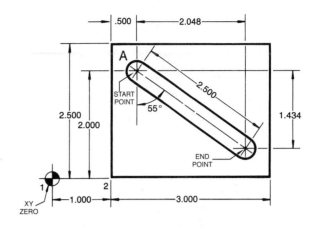

Fig. 42-12 A workpiece that requires an angular slot to be programmed. (Courtesy of The Superior Electric Co.)

The program to machine the workpiece would be as follows:

The Program

%
  – program start code
O4213
  – program number
N05   G20   G90   G00   X–1.000   Y0   T1   M06
  – inch/absolute
  – spindle rapids to tool-change position
  – .250 in. 2-flute end mill to be placed in spindle
N10   G40   S1000   M03
        – cancel cutter compensation
  M03   – spindle ON clockwise
N15   G00   X1.500   Y2.000   Z0.100
  X--Y-- – positions XY coordinates at point A
  Z.100 – tool rapids to .100 in. above work surface
N20   G01   Z–.125   F5
  Z–.125 – tool feeds to .125 in. below work surface at 5 in. feed rate
N25   X3.548   Y.566
  – groove cut from start to end point
N30   G00   Z.100   M05
  – tool rapids out of hole to .100 in. above the work surface
  – spindle is turned off

N35   X0   Y0
— tool rapids to tool-change position

N40   M30
— end of data

%
— end-of-program symbol

In Fig. 42-12, the coordinate dimensions of the start point and end point of the angular surface were shown on the part print. In Fig. 42-13, a five-sided figure (pentagon) is to be machined, and the print does not define the coordinate locations of the start and end point of each straight line. The XY coordinate locations for the start point and end point of each angular line can be calculated by trigonometry, Fig. 42-14.

The coordinate locations of the various points of the pentagon are as follows:

- Points 1 and 6 = X0 Y−1.000

The XY locations for points 2, 3, 4, and 5 can be calculated mathematically.

- Point 2 = X−1.902   Y−2.382
- Point 3 = X−1.1755 Y−4.618
- Point 4 = X1.1755   Y−4.618
- Point 5 = X1.902   Y−2.382
- Point 6 = X0   Y−1.000

Programming the Pentagon

A five-sided (pentagon) groove 1/2 in. wide and .250 in. deep must be machined on the part shown in Fig. 42-13. A two-flute end mill should be used because it is center-cutting and can be used to plunge-cut to depth. The following absolute program is required to machine the part.

The Program

%
— program start code

O4214
— program number

N05   G20   G90   G00   X0   Y0   T1   M06
— inch/absolute
— spindle rapids to XY0
— .250 in. 2-fluted end mill to be placed in the machine spindle

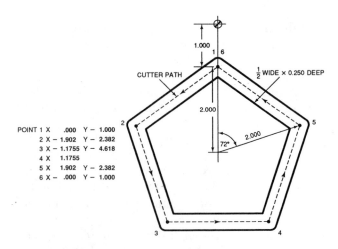

Fig. 42-13 A part that requires a five-sided (pentagon) groove to be programmed. (Courtesy of The Superior Electric Co.)

POINT 1 X   .000   Y − 1.000
  2 X − 1.902   Y − 2.382
  3 X − 1.1755 Y − 4.618
  4 X   1.1755
  5 X   1.902   Y − 2.382
  6 X − .000   Y − 1.000

N10   G40   S1000   M03
— cancel cutter compensation
M03   — spindle ON clockwise

N15   G00   X0   Y−1.000   Z.100
X−Y−   — positions spindle at Point #1
Z.100   — tool rapids to .100 in. above work surface

N20   G01   Z−.250   F3
Z−.250 — tool feeds .250 in. below work surface at 3 in. feed rate

N25   X−1.902   Y−2.382
— groove is cut from Point 1 to 2

N30   X−1.1755   Y−4.618
— groove is cut from Point 2 to 3

N35   X1.1755
— groove is cut from Point 3 to 4

N40   X1.902   Y−2.382
— groove is cut from Point 4 to 5

N45   X0   Y−1.000
— groove is cut from Point 5 to 6

N50   G00   Z2.000   M05
— tool rapids to 2.000 in. above work surface
— spindle is turned OFF

N55   X0   Y0
— tool rapids back to start point

N60   M30
— end of data

%
— end-of-program symbol; rewind

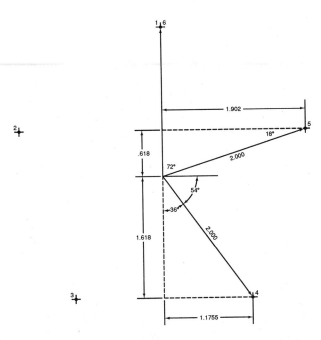

Fig. 42-14 The mathematical measurements required to locate the coordinates on the five-sided figure. (Courtesy of Kelmar Associates)

- The machine spindle can be located at the tool-change position by a G92 or G00 code.

- Absolute programming is where all coordinate locations are taken from one fixed XY zero point.

- Incremental programming is where the last coordinate position becomes the XY zero for the next location.

- To convert from inch to millimeters dimensions, multiply by 25.4; from millimeters to inch, divide by 25.4.

- Cutter-diameter compensation makes it possible to change a programmed cutter-tool path to allow for differences between the actual and the programmed cutter diameter.

- A **canned cycle** is a preset sequence of events which is started by a single G code. For example, the G81 drilling code locates the position of the hole, rapids to **gage height**, drills the hole to depth, and rapids back to gage height.

- Angular or taper (straight-line) surfaces can be programmed by linear interpolation if

the coordinates for the start point, the end point, and a feed rate are given.

1. Define linear interpolation

Two-Axes Programming

2. Use absolute positioning to program the three hole locations and return to start point.

%

O0002

N05_____

N10_____

N15_____

N20_____

N25_____

N30_____

N35_____

%

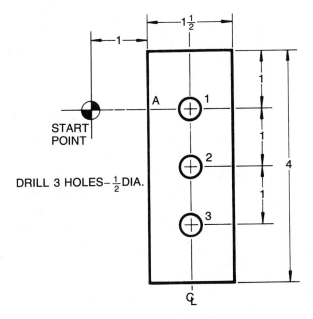

DRILL 3 HOLES—$\frac{1}{2}$ DIA.

3. Use incremental positioning to program the three hole locations and return to start point.

%

O0003

_____

_____

_____

_____

_____

_____

_____

%

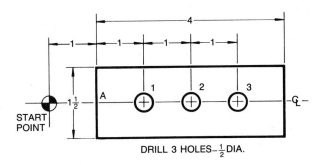

DRILL 3 HOLES—$\frac{1}{2}$ DIA.

Climb and Conventional Milling

4. Explain the difference between climb and conventional milling, and the relationship between cutter rotation and table travel.

Cutting Speeds

5. List four important factors that must be considered when setting the r/min for a machining operation.

6. Calculate the r/min required for a 3/4 in. diameter end mill to cut tool steel (70 CS).

7. Calculate the r/min required for a 100 mm diameter milling cutter to cut aluminum (200 CS).

Milling Programs

8. Use the part shown in question #2 to program the part path in the absolute mode; the first program point should be at the top left corner. Ignore cutter size and tool compensation.

9. Use the part shown in question #3 to program the edges in the incremental mode; the first program point should be the top left corner.

Milling/Drilling Program

10. Program the mill/drill exercise in the absolute mode using a .500 diameter end mill for machining the edges and a .375 diameter drill with G81 codes for the holes. Start all programming at the XY part zero (top left corner). The machine steel part (CS 100) is .500 thick.

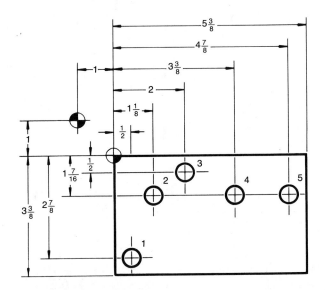

Angular Programming

11. What three pieces of information are necessary to program an angular cut?

12. Calculate the absolute coordinate locations for the four holes from the start point, drill the holes, and return to the start point.

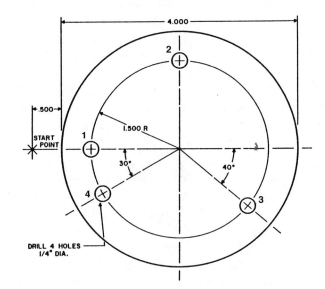

Circular Interpolation

Contouring, sometimes called continuous-path machining, refers to where the tool is always in contact with the workpiece as it travels from one programmed location to the next. It is not only important for the machine to accurately position the tool from one location to the next, but also to follow a predefined, accurate, programmed path. The accuracy of this path is as important as positioning to the correct programmed location. Because the tool is always in contact with the part from one location to the next, it is actually contouring or shaping the part throughout the machining cycle.

OBJECTIVES

After completing this unit, you should be able to:

- State the purpose of circular interpolation and give the basic programming requirements.
- Write programs to produce arcs by various methods.
- Produce simple programs to machine full circles.
- Use polar coordinates to locate rotary or angular positions.

KEY TERMS

center-point programming
parabolic interpolation
polar reference line
radius programming
vector radius

circular interpolation
polar coordinates
quadrant
vector angle
X axis

● INTERPOLATION

The method used to move contouring machines from one point to the next is called interpolation. This feature combines the individual axis commands into a predefined tool path. There are five types of interpolation, the most common of these being linear, circular, and helical. All contouring controls have linear interpolation, and most of the modern controls have both linear and circular interpolation. Only the most complex controls use cubic or parabolic interpolation.

Circular Interpolation

Circular interpolation was developed to eliminate the many calculations required with early controls and to simplify the programming of arcs and circles. It allows a programmer to make the cutting tool follow any circular path ranging from a small arc segment to a full 360° circle. In order to program an arc or circle on some MCUs, the tool must first cut a path to the starting location of the arc or circle. This information is followed by a program line containing the direction of cut, end point coordinates of the arc or circle, and generally the radius of the circle or distance from the arc start point to the arc center point, Fig. 43-1. The circular interpolation of the MCU automatically breaks up the arc into very small (minute) linear moves, generally .0001 or .0002 in. (0.0025 or 0.005 mm) each, depending upon the resolution of the servo motors, to describe the circular path. The MCU then generates the controlling signals to move the cutting tool to produce the desired arc or circle. If the same arc or circle were to be programmed in linear interpolation, hundreds or even thousands of coordinates,

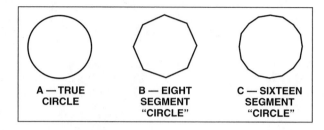

Fig. 43-2 The MCU breaks up a circle into many thousands of very small (.0002 in.) straight-line segments. (Courtesy or Allen-Bradley Co.)

each defining a very small span, would have to be calculated and programmed, Fig. 43-2.

There are a variety of machine control units available; some generate only one quadrant (90°) at a time, while others can generate a full circle (360°). Some models of MCUs can only do circular interpolation in a two-axis plane at a time, such as XY, XZ, or YZ axes, while others can interpolate circular movements for three axes at the same time. Circular interpolation can also be used to generate second- and third-degree curves and free-form shapes which can be closely described with a series of arcs or circles.

When circular interpolation is programmed, four pieces of information are necessary:

1. The start point of the arc (XY coordinates).

2. The direction of the cutter travel (preparatory function).

3. The end point of the arc (XY coordinates).

4. The center point of the arc (I J coordinates) or the arc radius (R).

The standards for *direction of cutter travel* are defined by the EIA preparatory function codes for circular interpolation. G02 is the code for circular movement in a clockwise direction (CW). G03 represents circular movement in a counterclockwise direction (CCW). These codes must be programmed in the block of information where circular interpolation starts, and they are modal (remain in effect) until a new preparatory (G) code is programmed, namely G00 or G01.

The *start point of the arc* is usually the end point of a line or the end point of a previous arc. The start point locates the cutting tool to the start of the machining position of the arc and is generally given as XY and/or Z coordinate dimensions.

The *end point of the arc* (XY and/or Z coordinates) is the last point where the cutter path

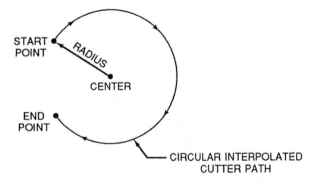

Fig. 43-1 To machine a full circle on some controllers, the start and end point of each quadrant (90°) must be programmed. (Courtesy of Kelmar Associates)

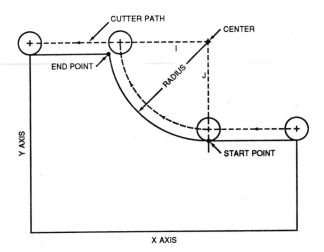

Fig. 43-3 Two-dimensional circular interpolation requires a start point, end point, the coordinate location of the arc center, and the direction of travel. (Courtesy of Kelmar Associates)

center line completes the circular path. Whenever an arc uses more than one 90° quadrant, the point where it crosses into the next quadrant must be programmed as the end point, when using single quadrant programming. The MCU assumes that this is also the start point for the next quadrant; therefore it is only necessary to program the end point for the next quadrant.

The *center point of the arc* (XY and/or Z coordinates) is the center of the circle or arc and is described by I (X coordinate value), J (Y coordinate value), and K (Z coordinate value), Fig. 43-3. Generally the I, J, and K words are signed, incremental values regardless of whether they have been programmed in the absolute or incremental mode. The I and J coordinate values are always taken *from the start point of each arc or 90° segment to the center point of the radius.*

Parabolic Interpolation

Another method of programming complex contour forms is **parabolic interpolation,** which is especially suited to free-form designs such as mold work and automotive die sculpturing. Parabolic interpolation positions the machine between three non-straight-line positions in a movement that is either a complete parabola or a portion of one. Its advantage lies in the ability to closely approximate curved sections with as much as 50:1 fewer points than with linear interpolation. It very easily adapts to free-form shapes

that have eye appeal rather than a mathematical description.

● CONTOUR PROGRAMMING

The programmer must follow the same steps and use the same principles as in point-to-point positioning. The part drawing must be studied, the manuscript written, and the program produced. The contouring program requires more detail than the point-to-point program, where positioning takes place at maximum speeds. When in the interpolation mode, the milling feed rate into the workpiece must be provided.

Machine feed rates and tool geometry are two requirements that the programmer must consider for contouring. Not only must the cutting path be plotted, but an appropriate feed rate must be given. Factors such as type of cutting tool, material of the workpiece, and depth of cut all must be considered. For example, a machine may only be able to negotiate a corner at slow speed while fairly high speeds can be used on a straight-line cut. At the point of acceleration or deceleration, a tool mark may result if the change is too sharp.

One major point that the programmer must consider is allowing for the *cutting tool offset.* Every CNC machine tool has a basic machine reference point which is usually the centerline of the spindle and the end of the spindle nose. Naturally, the cutting tool may extend a few inches beyond the basic reference point, but it is not difficult for the programmer to compensate for tool length.

The CNC machine that uses a milling cutter for sculpturing or contouring presents other factors that must be considered. It is not the center of the cutter, but normally a point on its periphery, that actually does the machining. If a ball-nose cutter is used, different positions of the cutter will be machining, depending on where the cutter contacts the workpiece surface. If a cutter is one inch in diameter and if it is being used for profiling, the programmer must use a one-half inch offset because that represents the distance between the spindle centerline and the edge of the cutter.

The contouring example shown in Fig. 43-4 is fairly simple because all machining is done parallel to the X or Y axes. However, the part shown in Fig. 43-5 is more difficult to program. If the same example were contoured in all three axes, the programming can be quite complex. It

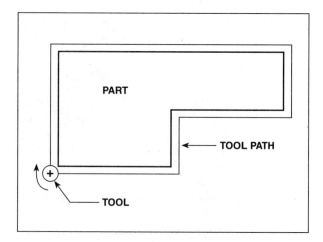

Fig. 43-4 In simple contouring, the machining occurs parallel to either the X or Y axis of the workpiece. (Courtesy of Allen-Bradley Co.)

is important to remember that the offsets for cutting tool geometry must always be made perpendicular to the surface being machined, regardless of the direction that the surface faces.

Contour programming can involve parts ranging from the very simple to the very complex. A CNC turning center that has linear and circular interpolation may require only about fifty input steps to turn several diameters and radii. The same is true for CNC machining centers where the machining is done parallel to the X or Y axes. On the other hand, a complex aircraft part may require several hundred thousand programming coordinates. For example, the simple part in Fig.

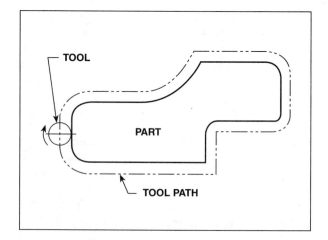

Fig. 43-5 In three-axes contouring, it is possible to do straight-line and curved shapes. (Courtesy of Allen-Bradley Co.)

43-4 only requires six cutting moves. With the addition of arcs in Fig. 43-5, the same basic-shaped part requires fourteen cutting moves.

● PROGRAMMING AN ARC

The two most common methods of programming an arc are by **center-point programming** and by **radius programming.** Some MCUs will generate an arc if it is defined by the end point of the arc and the arc center point (I & J coordinates). Other MCUs will generate an arc if it is defined by the end point of the arc and the arc radius. Before an arc or circle is programmed, it is necessary to determine what information is needed for the particular MCU being used.

There are three questions which must be answered before an arc is programmed. Place a pencil point at the start point of the arc and answer the following questions before the pencil point is moved.

1. Which way?—Clockwise (G02) or counterclockwise (G03) direction from the start point of the arc.

2. Where to?—The X and Y coordinates of the end point of the arc.

3. How far?—The I and J values from the start point of the arc to the center of the circle.

The information required for center-point programming is shown in Fig. 43-6:

1. G-code—G02 for circular interpolation clockwise, G03 for circular interpolation counterclockwise.

2. End point—The X and Y coordinates of the end point of the arc, or signed incremental distances.

3. Center point—The signed coordinates of the center point of the arc. The letters I (X axis) and J (Y axis) are used to define the point.

The information required for radius programming is as follows:

1. G-code—G02 for circular interpolation clockwise, G03 for circular interpolation counterclockwise .

2. End point—The X and Y coordinates of the end point of the arc or signed incremental distances.

3. Radius—The radius of the arc preceded by the letter address R.

THE CIRCLE CENTER COORDINATES I AND J

When the programming of circles and arcs with the address R is not possible, the circle center has to be programmed. This is done with the addresses I and J:

I is a coordinate parallel to the   X axis
J is a coordinate parallel to the   Y axis.

G 17

Vertical Machining Center

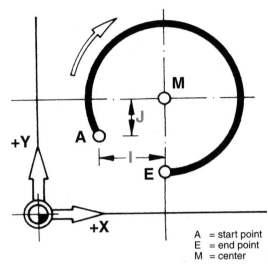

A = start point
E = end point
M = center

are therefore used as circle center coordinates

In the plane

G 17
I and J

Fig. 43-6 The information necessary to program arcs and circles. (Courtesy of Deckel Maho, Inc.)

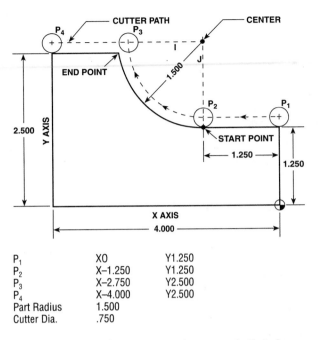

P₁	X0	Y1.250
P₂	X–1.250	Y1.250
P₃	X–2.750	Y2.500
P₄	X–4.000	Y2.500
Part Radius	1.500	
Cutter Dia.	.750	

Fig. 43-7 An absolute programming example that shows the relationship of the I and J (center-point) addresses to the X and Y axes. (Courtesy of Kelmar Associates)

Center-Point (IJ) Method

The I (X distance) and J (Y distance) are always taken incrementally from the start point of each arc or segment to the center point of the radius. This method is very common on many controllers.

Figure 43-7 shows absolute programming examples for center-point (IJ) and radius (R) programming of a partial arc. These coordinates take into account that the centerline of the tool is programmed. Depending on the location of the tool, the .375 in. radius of the tool has either been added to, or subtracted from the absolute dimensions.

The codes introduced in this example are:

I – X axis center point

J – Y axis center point

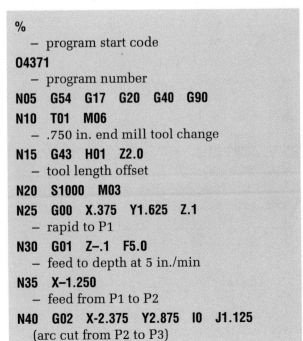

```
%
– program start code
04371
– program number
N05   G54   G17   G20   G40   G90
N10   T01   M06
– .750 in. end mill tool change
N15   G43   H01   Z2.0
– tool length offset
N20   S1000   M03
N25   G00   X.375   Y1.625   Z.1
– rapid to P1
N30   G01   Z–.1   F5.0
– feed to depth at 5 in./min
N35   X–1.250
– feed from P1 to P2
N40   G02   X-2.375   Y2.875   I0   J1.125
(arc cut from P2 to P3)
```

N45 G01   X–4.0
 – feed from P3 to P4

N50   G00   Z1.0   G49   M05

N55   X1.0   Y0
 – tool change position

H60   M30
%
 – end-of-program symbol

Radius Method

Most modern controllers do not require the I and J word addresses but use the R address code for the arc radius along with the XY coordinates dimensions.

%
 – program start code
O4372
 – program number
N05   G54   G17   G20   G40   G90

N10   T01   M06
 – .750 end mill tool change

N15   G43   H01   Z2.0
 – tool length offset

N20   S1000   M03

N25   G00   X.375   Y1.625   Z.1
 – rapid to P1

N30   G01   Z–.1   F5.0
 – feed to depth at 5 in./min

N35   X–1.250
 – feed from P1 to P2

N40   G02   X–2.375   Y2.875   R1.125
 – arc cut from P2 to P3

N45   G01   X–4.0
 – feed from P3 to P4

N50   G00   Z1.0   G49   M05

N55   X1.0   Y0
 – tool change position

N60   M30
%
 – end-of-program symbol

The example in Fig. 43-8 shows the milling of arcs up to 180° in both a clockwise (G02) and counterclockwise (G03) direction. Note the difference in the programming; the G42 code in the programming example on the bottom left means

cutter diameter compensation on the right side of the surface being cut. In the example on the bottom right, the G41 code means cutter diameter compensation on the left side of the work surface being cut.

Figure 43-9 on page 452 shows the programming required to mill a part containing an entry slot and a part circle using the center-point (IJ) method. A summary of the various methods that can be used to program arc and circles on vertical machining centers is shown in Fig. 43-10 on page 453.

● TO PROGRAM AN ARC

Circular interpolation makes it possible to program an arc or a full circle depending on the job requirements. As stated previously, arcs and circles can be programmed by two methods, the center-point method and the radius method. Since many controllers do not have the capabilities of the radius method, the example with the two 90° arcs shown in Fig. 43-11 on page 454 will be programmed using the center-point method.

Program Notes

1. All programming begins at the XY zero at the bottom left of the part.

2. Use absolute programming.

3. Use a .375 in., two-flute end mill to machine the circular grooves clockwise .200 in. deep.

4. The material is aluminum (CS 500).

5. Start machining at Point 1 and when slot two has been cut return to XY zero.

The Program

%
 – program start code
O4311
 – program number
N05   G54   G17   G20   G90   G40
 – spindle offsets to XY zero
 – inch/absolute/cancel cutter
  compensation

N10   T01   M06
 – .375 in. end mill/tool change code

N15   G43   H01   Z 1.0
 – tool length offset/offset information/tool
  1 in. above work surface

(cont'd p. 454)

MILLING ARCS OF CIRCLES UP TO 180°

Arcs of circles/circles are programmed via the functions G2 or G3.

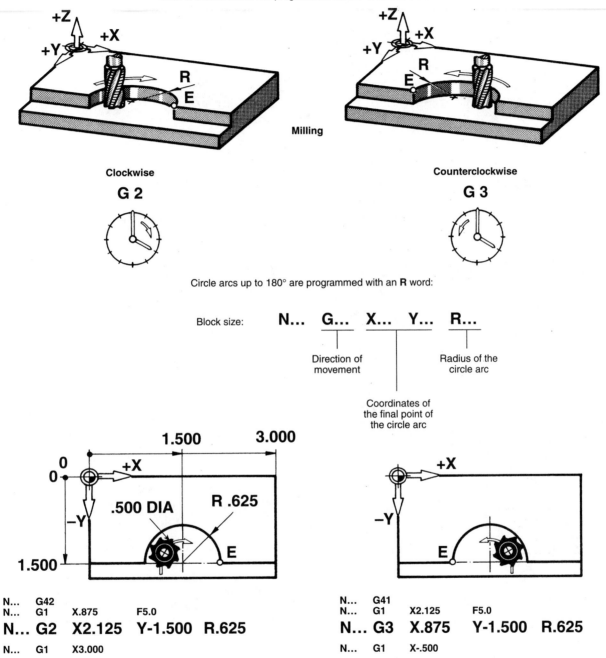

Milling

Clockwise

G 2

Counterclockwise

G 3

Circle arcs up to 180° are programmed with an **R** word:

Block size: **N...   G...   X...   Y...   R...**

Direction of movement

Coordinates of the final point of the circle arc

Radius of the circle arc

N...	G42		
N...	G1	X.875	F5.0
N...	**G2**	**X2.125**	**Y-1.500 R.625**
N...	G1	X3.000	

N...	G41		
N...	G1	X2.125	F5.0
N...	**G3**	**X.875**	**Y-1.500 R.625**
N...	G1	X-.500	

Fig. 43-8 Using G02 and G03 codes to climb and conventional mill two arcs. (Courtesy of Deckel Maho, Inc.)

MILLING ARCS OF CIRCLES WITH ABSOLUTE DIMENSIONS

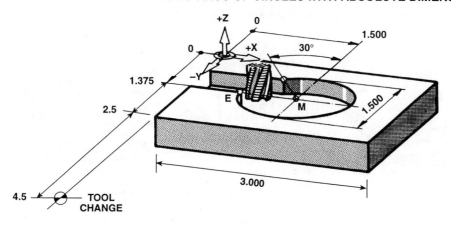

Circle arcs ≦ 180° **can**, circle arcs > 180° **must** be programmed with the center coordinates I and (J)

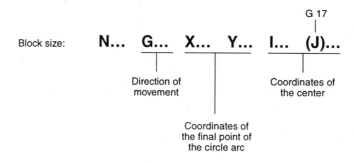

The control calculates the radius value from the coordinates of A and M.

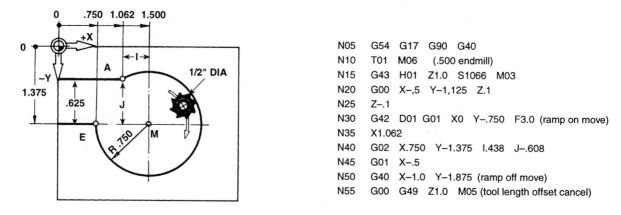

N05	G54	G17	G90	G40
N10	T01	M06	(.500 endmill)
N15	G43	H01	Z1.0	S1066	M03
N20	G00	X−,5	Y−1,125	Z.1
N25	Z−.1
N30	G42	D01	G01	X0	Y−.750	F3.0	(ramp on move)
N35	X1.062
N40	G02	X.750	Y−1.375	I.438	J−.608
N45	G01	X−.5
N50	G40	X−1.0	Y−1.875	(ramp off move)
N55	G00	G49	Z1.0	M05	(tool length offset cancel)

Fig. 43-9 A program for milling arcs and circles in the absolute mode. (Courtesy of Deckel Maho, Inc.)

MILLING ARCS OF CIRCLES IN THE PLANE G17 (SUMMARY)

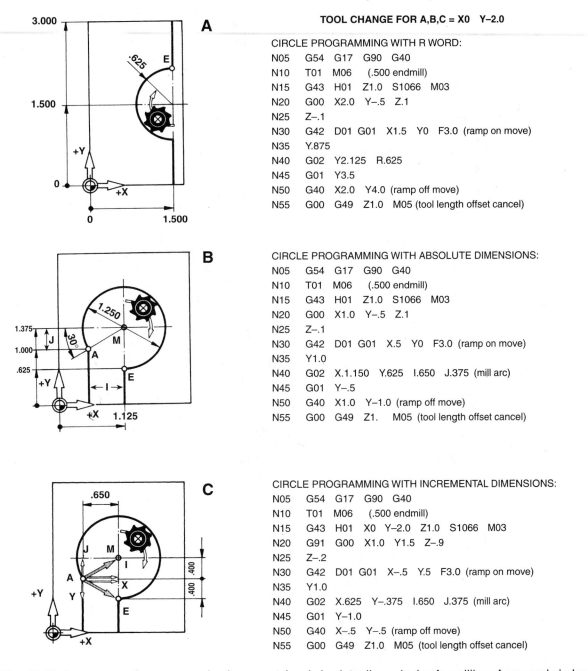

TOOL CHANGE FOR A,B,C = X0   Y–2.0

CIRCLE PROGRAMMING WITH R WORD:

N05	G54	G17	G90	G40
N10	T01	M06	(.500 endmill)	
N15	G43	H01	Z1.0	S1066 M03
N20	G00	X2.0	Y–.5	Z.1
N25	Z–.1			
N30	G42	D01	G01 X1.5 Y0 F3.0 (ramp on move)	
N35	Y.875			
N40	G02	Y2.125	R.625	
N45	G01	Y3.5		
N50	G40	X2.0	Y4.0 (ramp off move)	
N55	G00	G49	Z1.0 M05 (tool length offset cancel)	

CIRCLE PROGRAMMING WITH ABSOLUTE DIMENSIONS:

N05	G54	G17	G90	G40
N10	T01	M06	(.500 endmill)	
N15	G43	H01	Z1.0	S1066 M03
N20	G00	X1.0	Y–.5	Z.1
N25	Z–.1			
N30	G42	D01	G01 X.5 Y0 F3.0 (ramp on move)	
N35	Y1.0			
N40	G02	X.1.150	Y.625 I.650 J.375 (mill arc)	
N45	G01	Y–.5		
N50	G40	X1.0	Y–1.0 (ramp off move)	
N55	G00	G49	Z1. M05 (tool length offset cancel)	

CIRCLE PROGRAMMING WITH INCREMENTAL DIMENSIONS:

N05	G54	G17	G90	G40
N10	T01	M06	(.500 endmill)	
N15	G43	H01	X0 Y–2.0 Z1.0 S1066 M03	
N20	G91	G00	X1.0 Y1.5 Z–.9	
N25	Z–.2			
N30	G42	D01	G01 X–.5 Y.5 F3.0 (ramp on move)	
N35	Y1.0			
N40	G02	X.625	Y–.375 I.650 J.375 (mill arc)	
N45	G01	Y–1.0		
N50	G40	X–.5	Y–.5 (ramp off move)	
N55	G00	G49	Z1.0 M05 (tool length offset cancel)	

Fig. 43-10 A summary of programs using incremental and absolute dimensioning for milling of arcs and circles. (Courtesy of Deckel Maho. Inc.)

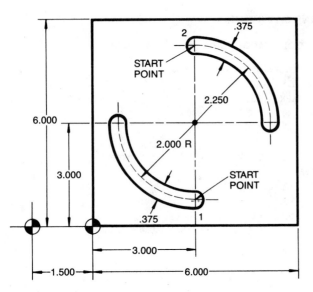

Fig. 43-11 The I and J (center point) locations are taken from the arc start point to the center point of rotation. (Courtesy of The Superior Electric Co.)

N20   S5333   M03
   – speed 5333 r/min./spindle ON clockwise

N25   G00   X3.0   Y1.0   Z.1
   – spindle rapids to Point 1 on the first arc and .100 in. above work surface

Milling the Arcs

N30   G01   Z–.2   F3.0
   – spindle feeds to .200 in. below work surface

N35   G02   X1.0   Y3.0   I0   J2.0
   – circular interpolation clockwise/first arc cut

N40   G00   Z.1
   – cutter rapids to .100 above the work surface

N45   X3.0   Y5.250
   – rapids to start of the second arc, position #2

N50   G01   Z–.2
   – the cutter feeds to depth

N55   G02   X5.250   Y3.0   I0   J–2.250
   – same as sequence N35

N60   G00   Z.1
   – same as sequence N40

N65   Z1.0   G49   M05
   – tool rapids back to XY zero/spindle shuts OFF and raises to the full retract position

N70   X–1.5   Y0
   – rapid to tool change position

N75   M30
   – program end and rewind

%
   – end-of-program. symbol

To Program a Circle

A 360° full circle will be programmed using two different methods. In the center-point method, the full circle will be treated as four 90° arcs, and since many MCUs generate only one quadrant at a time, each arc must be programmed. Since the end point of one arc automatically becomes the start point of the next arc, a full circle requires that the *first start point* and *four end points* be programmed.

When using the full-circle programming method, either the I-word or J-word can command full 360° circular movement. The selection of the I- or J-word is determined by the starting location of the circle. If starting at 12 o'clock or 6 o'clock position on the circle the J-word is used. If starting at the 3 o'clock or 9 o'clock position the I-word is used. These methods should suit most controllers.

The center-point program for the circular groove shown in Fig. 43-12 is as follows:

Program Notes

1. All programming begins at XY zero above the top center of the part.

2. Use absolute programming.

3. Use a .375 in., two-flute end mill to machine the circular groove clockwise .200 in. deep.

4. The material is machine steel (CS 100).

5. Start machining at Point 1 and when Point 5 is reached, return to XY zero.

The Program

%
   – program start code

04312
   – program number

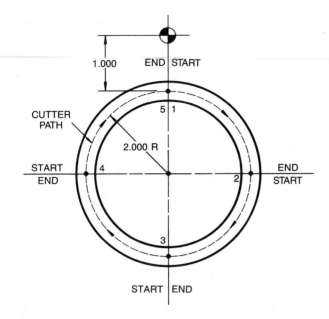

Fig. 43-12 A full circle to be machined, each quadrant at a time, requires the start point and end point of each quadrant. (Courtesy of The Superior Electric Co.)

N05   G54   G17   G20   G90   G40
— spindle located at work coordinate position XY zero/inch/absolute/cancels cutter compensation

N10   T01   M06
— .375 in. end mill/tool change code

N15   G43   H01   Z1.0
— tool length offset/offset information/tool 1 in. above work surface

N20   S1066   M03
— speed 1066/spindle ON clockwise

N25   G00   X0   Y–1.0   Z.1
— rapids to position #1, 1 in. above work surface

N30   G01   Z–.2   F3.0
— tool feeds into work .200 in. at 3 in./min.

N35   G02   X2.0   Y–3.0   I0   J–2.0
— circular groove cut from Point 1 to 2

N40   X0   Y–5.0   I–2.0   J0
— circular groove cut from Point 2 to 3

N45   X–2.0   Y–3.0   I0   J2.0
— circular groove cut from Point 3 to 4

N50   X0   Y–1.0   I2.0   J0
— circular groove cut from Point 4 to 5

N55   G00   Z1.0   G49   M05
— tool rapids to 1 in. above work surface/spindle shuts OFF

N60   X0   Y0
— tool rapids to XY zero

N65   M30
— program end and rewind

%
— end-of-program symbol

Full-circle programming written as a J command for the same part shown in Fig. 43-12 is as follows:

The Program

%
— program start code

O4312
— program start number

N05   G54   G17   G20   G90   G40
— spindle offsets to XY zero/inch/absolute/cancels cutter compensation

N10   T01   M06
— .375 in. end mill/tool change code

N15   G43   H01   Z1.0
— tool length offset/offset information/tool 1 in. above work surface

N20   S1066   M03
— speed 1066/spindle ON clockwise

N25   G00   X0   Y–1.0   Z.1
— rapids to position #1, .1 in. above work surface

N30   G01   Z–.2   F3.0
— tool feeds into work .200 in. at 3 in./min

N35   G02   J–2.0

N40   G00   Z1.0   G49   M05
— tool rapids to 1 in. above work surface

N45   X0   Y0

N65   M30
— program end and rewind

%
— end-of-program symbol

● POLAR COORDINATES

The polar-coordinate concept makes it possible to quickly calculate and define rotary and angular movements. If the radius and angle of movement is supplied, most modern controllers can calculate the points necessary to accurately position the cutting tool. This eliminates much of the trigonometry calculations that were required in the past to change polar coordinates into rectangular coordinates. With rectangular coordinates, a point can be located by its distance from the two axis lines (X and Y) that intersect at the XY zero, Fig. 43-13A. With **polar coordinates,** the position of a point is located by its radial distance and direction from a fixed reference point or origin, (XY zero).

Polar coordinates basically follow the same principles as rectangular coordinates. A polar reference grid is a series of invisible circles around a reference point, the circle center, Fig 43-13B. A point can be located by the radius distance from the center of the grid and the angle from a base line extending from the center. This base line, commonly called the **polar reference line** (PRL), is usually the **X plus axis,** which divides the top and bottom, as well as the left and right portions of the cartesian coordinate system. On some controls the PRL can be shifted to suit the programmers preference. A line drawn from the origin to any point in the circle is the vector; the length of the line is the **vector radius,** and the angle formed by the vector and the polar reference line is the polar or **vector angle.** Any point in a plane may be located when given the radius and the vector angle. Both positive and negative angles are used for polar coordinates. Any angle starting at the polar reference line and going counterclockwise is a positive circle; a negative circle goes clockwise from the polar reference line, Fig. 43-14.

Just as plane selection is required when using circular interpolation, the same is required when using polar coordinates. An additional code, G16, is required to turn the polar coordinate system ON. When returning back to the rectangular system, a G15 code is used to turn the polar coordinate system OFF. Polar coordinates

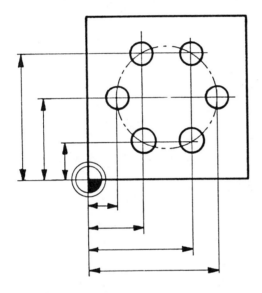

(A) RECTANGULAR COORDINATES

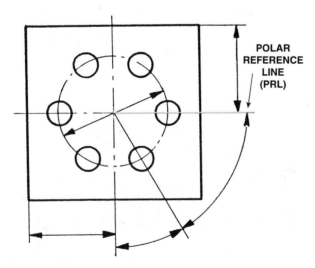

(B) POLAR COORDINATES

Fig. 43-13 A bolt circle dimensioned using two methods. (Courtesy of Emco Maier Corp.)

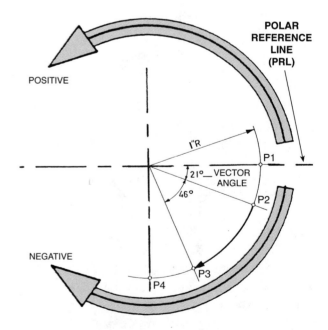

Fig. 43-14 Point locations are either positive or negative in relation to the polar reference line. (Courtesy of Kelmar Associates)

can be programmed in either absolute or incremental positioning.

A designer may find it convenient to use polar coordinates to show angular relationships of two or more holes. For example, the center of a bolt circle should be located by X and Y coordinate dimensions from the workpiece zero point, and the polar axis of the circle should be parallel with the X axis of the workpiece. It is not necessary to have a full circle of holes; the angular relationship of just two holes establishes part of a circle.

The polar principle is used in CNC programming for calculations required for contouring, especially when it is necessary to program a complex offset cutter-centerline path. The Woodworth circular tables in the Appendix of Tables at the end of this book can simplify the conversion of polar coordinates into rectangular coordinates for many types of work.

Using Polar Coordinates

Polar coordinates, often called polar rotation, can be thought of as rotating the rectangular coordinate system. They are formed by constructing a line whose slope is not the same as the X or Y axis, see P2 and P3 in Fig. 43-14. It is important for the programmer to check the programming manual for the controller being used, because the methods and codes may vary from manufacturer to manufacturer. Even the EIA has not designated codes for polar rotation, and the codes used in the following examples are used by several manufacturers, but not all.

Regardless of the differences in controllers, the MCU needs the following information in order to properly locate polar coordinate locations:

1. The code to turn ON the polar rotation system.

2. The *radius,* with the I and J, to establish the polar center of rotation parallel to the XY coordinates.

3. The radial distance from the center of rotation to the location required, given as an X value.

4. The *vector angle* measured from the polar reference line; positive rotation is counterclockwise, negative is clockwise. In Fig. 43-14, the angle from P1 to P2 is negative 21°.

5. Any additional G codes required to perform particular operations, such as G81 to drill hole locations or G02 to cut an arc.

6. The *number of rotations* required. Assume six equally spaced holes were required on a bolt circle; each hole would be 60° apart, therefore the number of rotations, not counting the first hole would be five.

7. The code to turn OFF the polar coordinate system.

There are three common ways that polar coordinates are used for programming full or partial arcs.

1. A complete circle (360°) can be programmed in one block of information, Fig. 43-15A.

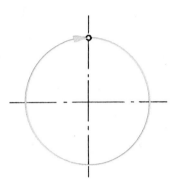

(A)

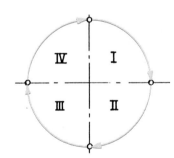

(B)

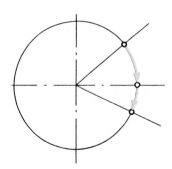

(C)

Fig. 43-15 Three common ways that polar coordinates can be used to program arcs and full circles. (Courtesy of Emco Maier Corp.)

2. A circle is divided into four 90° **quadrants** and a separate block of information is required for each quadrant, Fig. 43-15B. An arc cannot extend beyond the quadrant.

3. Arcs which extend into two quadrants, Fig. 43-15C, require a separate block of information for each quadrant.

The polar coordinate word address format used for Fig. 43-16 is only an example and this format could vary with each controller.

G16   X1.000 Y1.000 R.75 A60 D60 L6

G16   − the code to initiate polar rotation

X1.0 − the X axis coordinate of the center of rotation

Y1.0 − the Y axis coordinate of the center of rotation

R.75 − the radius from the center of rotation to the bolt circle

A60   − the starting index angle measured from the polar reference line (+X axis)

D60   − other rotations measured in degrees

L5   − the number of rotations to be performed

G15   − the code to cancel polar rotation

Partial Quadrant Arcs

Partial arcs within a quadrant can be programmed providing that the controller has this capability. In Fig. 43-17, the location of Point 1

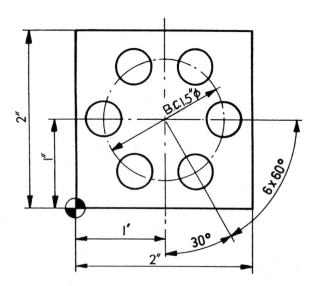

Fig. 43-16 Six holes on a bolt circle can easily be programmed using polar coordinates. (Courtesy of Emco Maier Corp.)

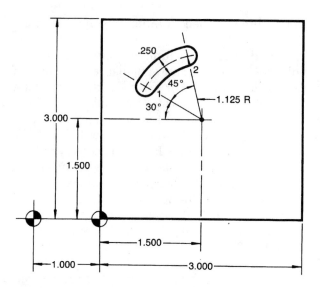

Fig. 43-17 Partial arcs can be programmed by the start-point coordinates, the radius, and the degree of rotation. (Courtesy of The Superior Electric Co.)

and Point 2 have to be defined giving the radius first and then the angle.

- Point 1 radius is 1.125 and the angle is 30° from the X axis or −210° from the polar reference line.

- After locating to Point 1, it requires a 45° polar rotation or −255° from the polar reference line to reach Point 2.

Any arc that extends into another quadrant must be treated as two separate arcs and programmed in two separate blocks.

SUMMARY

- Interpolation is the method used to move a cutting tool from one point to another along a predefined path.

- For circular interpolation, the MCU requires the arc start- and end-point coordinates, the circle radius, the location of the arc or circle center, and the direction in which the cutter is to travel.

- Some MCUs use the center-point method and can only generate one quadrant (90°) at a time, and require I and J coordinates. Modern MCUs can generate a full circle (360°) and require the coordinates for the arc/circle center and its radius.

- Polar coordinates are used on modern MCUs to quickly calculate and define rotary and angular movements if the arc radius and the angle of movement are provided.

- Polar locations are located from the polar reference line (X+ axis) with the movements in a counterclockwise direction being positive, and those clockwise being negative.

KNOWLEDGE REVIEW

Interpolation

1. Name and state the purpose of three types of interpolation.

Contour Programming

2. List five factors that a programmer must consider regarding the cutting tool and machine.

3. For what purpose is simple contouring used?

Programming an Arc

4. Name two methods used to program arcs and circles.

5. What three questions must be answered before an arc is programmed?

6. List the information required for center-point programming.

7. What three pieces of information are required for radius programming?

Center-Point Method

8. What is the purpose of the word addresses I and J?

Radius Method

9. What address code is used in the radius method?

10. Name the two pieces of information that controllers require for machining arcs and circles by the radius method.

Programming Arcs/Circles

11. Write an absolute program to cut the two arcs, 1/8 in. deep in machine steel (CS 100).

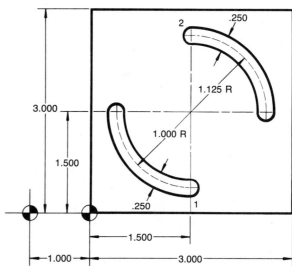

12. Write an absolute program to cut the 3/8 in. wide circular groove, .150 in. deep in aluminum (CS 500).

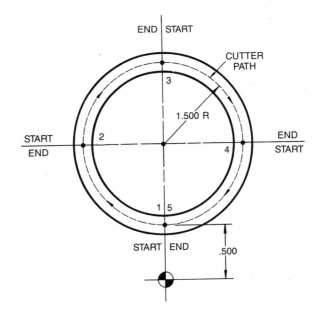

Polar Coordinates

13. What information is required to locate any point by polar coordinates?

14. Why is the polar reference line important?

15. What is a vector angle?

16. Define positive circle, negative circle.

17. List the six pieces of information the MCU requires to locate multiple polar coordinate positions.

UNIT 44

Subroutines or Macros

Program flow is the logic path that the control follows as it moves through a program. **Linear logic** is when a program goes from the beginning command and follows each program step in sequence directly to the end. Within a CNC program, it is possible to use branching logic and logic statements that shorten and simplify the program.

Using logic is sometimes the only way a geometry can be generated. Once this possibility is understood, there are many different ways to change a program to produce the same form on the machine. There are different kinds of program flow statements; loops and subroutines, or short programs within the main program that can also contain other logic statements called *nesting* logic. Figure 44-1 shows how a main program can have subprograms nested within it, and these subprograms could in turn, still have others, and so on.

Also, there are short programs called fixed cycles, or canned cycles, which are built into the control to simplify groups of program commands that are frequently used. These may be repetitive operations such as drilling, boring, tapping, etc. Fixed cycles can be incorporated into loops and subroutines.

OBJECTIVES

After completing this unit, you should be able to:

- Describe the purpose of subroutines or macros and know how they are used in programming.
- Describe loops and state the difference between smart and dumb loops.
- Recognize and use canned or fixed cycles in programming where they are applicable.

KEY TERMS

canned cycle　　　　GOTO command　　　　linear logic
loop　　　　　　　　macro

Basics of using macros (sub programs)

Macros are separate programs that are stored in a separate memory.

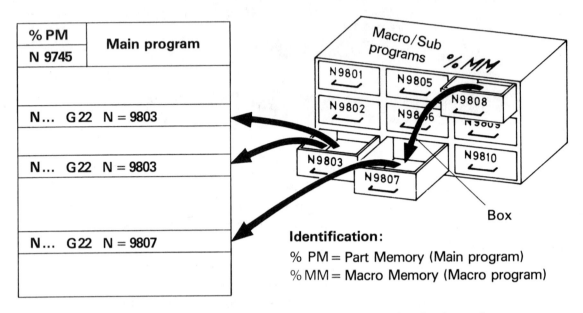

Identification:
% PM = Part Memory (Main program)
% MM = Macro Memory (Macro program)

Subroutines (macros) are used to simplify program with returning factions of program.

Run-off program

Can be nested 8 times.

Fig. 44-1 Macros are separate programs that are stored in a separate memory and can be called into the main program at any time. (Courtesy of Deckel Maho, Inc.)

● SUBROUTINES OR MACROS

The easiest logic statement to understand is the subroutine, a miniprogram sometimes called a subprogram or **macro.** Subroutines are called up within the body of the main program, but are actually a separate group of commands that are usually outside the main program. A subroutine, sometimes called a "program within a program," is used to store frequently used data sequences (one block or a number of blocks of information), that can be recalled from memory as often as required by a call statement in the main program, Fig. 44-1. Subroutines are usually stored as separate programs, but in the same general memory area as a main part program. An example of a subroutine could be a drilling cycle where a series of 3/8 in. (9.5 mm) diameter holes 1 in. (25.4 mm) deep must be drilled in a number of locations on a workpiece.

A subroutine is a miniprogram used within a main program, which would have to be repeated a number of times on a workpiece. It is a group of instructions or data that is permanently stored in memory, and can be recalled as a group to *solve recurring problems* such as bolt-circle locations, drilling and tapping cycles, and other frequently used routines. The subroutine must be written as a separate program and can be identified and stored with an O block and four digits, the same as a main program. Within the main program, the subroutine can be activated by using a M98 followed by the letter P and the same four digits that were used to identify the subroutine. An example of a subroutine would be the XY locations of the three slots in the workpiece illustrated in Fig. 44-2.

Programming a Subroutine

When the subroutine (three slots) in Fig. 44-2 is programmed, the following points should be kept in mind:

1. The subroutine program is usually stored under a program number, such as O4421, and recalled into the main program by a call statement, such as M98 P4422.

2. The sequence numbers of the subroutine should be large enough so that they do not conflict with the sequence numbers of the main program.

3. If the absolute programming mode (G90) is used in the main program, the G90 code must be used at the end of the subroutine in order to return to the programming mode of the main program, if the subroutine was using the incremental programming mode (G91).

4. The M99 code should be used at the end of the subroutine to return back to the main program at the line directly after the call statement.

The Subroutine for the Three Slots (Stored as O4422)

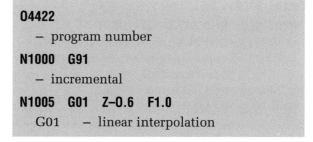

O4422
— program number

N1000   G91
— incremental

N1005   G01   Z–0.6   F1.0
G01   — linear interpolation

(cont'd)

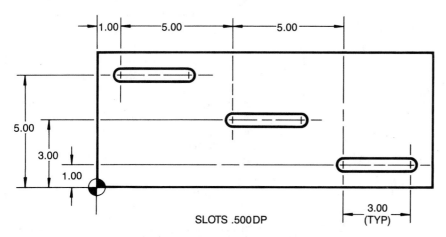

Fig. 44-2 A macro can be used to machine the three slots; it would simplify programming and save computer memory. (Courtesy of Kelmar Associates)

Z-.600 — the cutter is fed down into the work .500 + .100 gage height.

F1.0 — the feed rate is 1 in./min.

N1010 X3.0 F3.0

X3.0 — a slot is cut 3.000 in. along the X axis

F3.0 — the feed rate for milling the slot is 3 in./min.

N1015 G00 Z.6

the cutter rapids out of the workpiece .600 in. or back to gage height (.100 above the work surface)

N1020 G90

— absolute to return to main program

N1025 M99

— end of subroutine/return to main program

The Main Program for the Three Slots

O4421

N05 G90 G00 G17 G20 G54

— absolute

N10 T01 M06 G43 H01

T01 — number and type of cutting tool

H01 — tool length offset

N15 S1066 M03

N20 G00 X1.0 Y5.0 Z.1 M08

G00 — table rapids to position #1

Z0.1 — spindle rapids to within .100 of work surface

M08 — coolant is turned on

N25 M98 P4422

— subroutine recalled to mill slot #1

N30 X6.0 Y3.0

— table rapids to position #2

N35 M98 P4422

— subroutine mills slot #2

N40 X11.0 Y1.0

— table rapids to position #3

N45 M98 P4422

— subroutine mills slot #3.

N50 X0 Y0 Z2.0 G49 M05

— table rapids back to program XY zero

Z2.0 — spindle raises 2.0 in. above work surface

M05 — stops spindle

N55 M30

— program rewinds to start for next part

● LOOPING LOGIC

A **loop** in a program is designed to repeat a group of commands as often as they are required in a program. It may be used for locating purposes, such as moving from one hole location to another if all the moves are the same distance from each other. Loops can save valuable programming time and steps, since they require only one statement in the program.

Loops, which use incremental values, can be in the main program or in a subroutine, or both. The commands inside a loop are no different than regular program commands. The two categories of loops used are dumb loops that never stop, or smart loops that can repeat themselves a specific number of times.

Dumb Loops

Dumb loops are rarely used in CNC programming because once they are started, they never stop. They are often used to demonstrate the full range of motions of a machine tool at a machine tool show. Dumb loops cause the machine to repeat the same motions in the program until someone stops the machine manually.

Figure 44-3 shows a dumb loop which is started at program step N35 and sends the program back to step N05 to repeat the main program steps. This programming technique is called a jump, a skip, or a GOTO command. The **GOTO command** causes the program to return to a certain line in the program to repeat the same commands. It causes the machine to run through the same X, Y, and Z motions of steps N05 to N35.

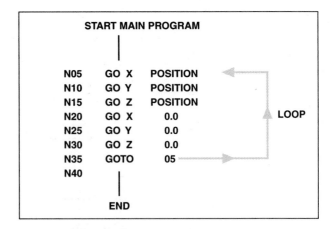

Fig. 44-3 Dumb loops return to a specific line in the main program and continue to loop until they are stopped manually. (Courtesy of Kelmar Associates)

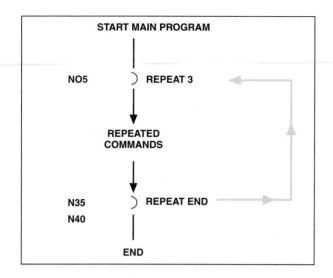

```
           START MAIN PROGRAM

                    |
   NO5        )  REPEAT 3

                    ↓

              REPEATED
              COMMANDS

                    ↓

   N35        )  REPEAT END
   N40             |

                  END
```

Fig. 44-4 A simple smart loop repeats three times between lines N05 to N35 and then continues on to the main program. (Courtesy of Kelmar Associates)

Smart Loops

A smart loop is used to repeat a series of commands a certain number of times. Each time the loop repeats itself, it is recorded by the MCU, and when the specific numbers of loops in the command are reached, the control moves on to the block of information outside the end of the loop. In Fig. 44-4, the loop would repeat program steps N05 to N35 three times and then move on to program step N40.

● CANNED OR FIXED CYCLES

A **canned cycle,** sometimes called a fixed cycle, is a preset combination of programmed commands that causes the machine axis movement and/or the machine spindle to perform operations such as drilling, boring, tapping, turning a diameter, milling a pocket, etc. All modern CNC controls contain a variety of canned cycles which are hard wired into the control and cannot be erased. A canned cycle shortens and simplifies the programming required for repetitive operations because the programmer only has to enter the cycle code to program that operation. Each operation may involve as many as seven machine movements, and a G81 to G89 code in the program will make the machine perform a particular operation. Control units with this feature can save up to 50% in programming time, one-third the data-processing time, and reduce the length of program required.

Modern CNC turning centers (lathes) that have turning cycles, will program a series of moves such as turning a diameter to size by simply entering the specifications of the part to be machined. Using a straight turning of a diameter to size on a lathe as an example of a turning cycle shown in Fig. 44-5, the program would have to:

1. Rough cut to a certain diameter (X move) and to a certain length (Z move).

2. Rapid back to the start position for a second or semi-finish cut, if necessary.

3. Feed in and turn to the specified diameter and length.

4. Rapid back to the start position for the finish cut.

5. Feed in and finish turn to diameter and length.

On most CNC turning centers (lathes), four movements are possible with the G84 turning cycle code; two for outside diameter turning—A towards the headstock and B towards the tailstock, and two for internal boring—C towards the headstock and D towards the tailstock, Fig. 44-5A. The information required for programming a turning cycle would be the: block number, G84 code, X value (depth of cut), Z value (length of cut), and the feed rate.

In Fig. 44-5A the first and fourth movements are in rapid traverse, while the second and third movements are at the programmed feed rate. The part required is shown in Fig. 44-5B and the tool setup is shown in Fig. 44-5C. The program for cutting the two diameters is shown in Fig. 44-5D and the tool paths are shown in Fig. 44-5E.

Some of the factors that would change from part to part with a turning cycle would be the:

- size of the rough stock,
- depth of cut taken for each pass,
- distance to the end of the cut (Z and X),
- start point of the cuts (Z and X),
- amount of material left for the finish cut.

Once this information is entered, a canned cycle will complete a particular operation, such as rough, semi-finish, and finish turn a shaft.

Most manufacturers of CNC machines and control systems use the standard EIA numbering system for canned or fixed cycles. A CNC drilling machine will be used to illustrate and describe some of the more common canned cycles.

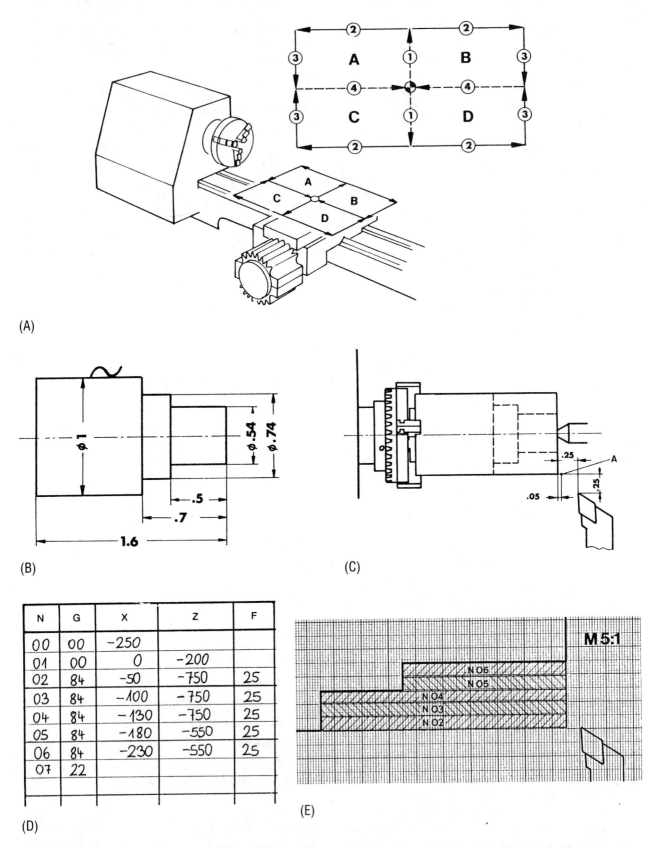

(A)

(B)

(C)

N	G	X	Z	F
00	00	−250		
01	00	0	−200	
02	84	−50	−750	25
03	84	−100	−750	25
04	84	−130	−750	25
05	84	−180	−550	25
06	84	−230	−550	25
07	22			

(D)

(E)

Fig. 44-5 A diameter turning cycle (G84) can be used to take a series of cuts to rough and finish turn a diameter. (Courtesy of Emco Maier Corp.)

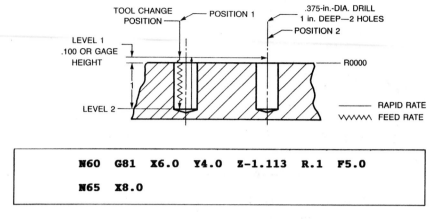

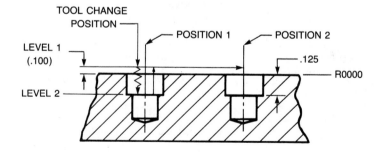

Fig. 44-6 A drill cycle (G81) will rapid to hole location, drill the hole, and rapid to the next location and repeat the drill cycle. (Courtesy of Cincinnati Milacron Inc.)

Drill Cycle (G81)

In Fig. 44-6, sequence number (N60), a drill cycle (G81) is used to drill a 3/8 in. diameter, 1 in. deep hole (Z1.113) at 5 in./min (F5.0) at position #1.

1. The spindle rapids to gage height (level 1); then it feeds to Z1.113 depth (level 2).

2. The spindle rapids back up to gage height (level 1).

3. In sequence N65, the table moves (X8.0) along the X axis to position #2 and repeats the drilling cycle.

To calculate the programmed Z depth (level #2) for a 118° included angle drill point, use the following formula:

$$Z = \text{Full body depth} + \text{drill point length}$$
$$= 1.000 + (.300 \times \text{drill diameter})$$
$$= 1.000 + (.300 \times .375)$$
$$= 1.000 + .113$$
$$= 1.113$$

Drill/Dwell Cycle (G82)

In Fig. 44-7, sequence N70, code G82 calls for a drill/dwell cycle. This cycle is the same as

Fig. 44-7 The drill/dwell cycle (G82) goes through the same sequence as the G81 drill cycle, but pauses for a few seconds at the bottom of the hole to produce a good finish. (Courtesy of Cincinnati Milacron Inc.)

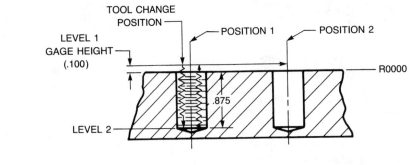

Fig. 44-8 The tap cycle (G84) is used to thread a series of holes. (Courtesy of Cincinnati Milacron Inc.)

a drill cycle with a dwell time added when the drill reaches the programmed depth.

1. The spindle feeds from the gage height (level 1) to the depth at level 2.

2. It stops at the depth for a period of time which is preset by a P word of 0500, which is equal to one-half second and then it completes the cycle.

3. In sequence N75, the table moves along the X axis (X8.0) to position #2.

4. The drill/dwell cycle is repeated at position #2.

NOTE

The dwell cycle is often used for counterboring and spot facing operations where a smooth surface is desirable.

Tap Cycle—Right Hand (G84)

In Fig. 44-8, sequence N80, code G84 calls for a right-hand tap cycle to thread a hole .875 in. deep (Z1.000) at 31.25 in./min (16 pitch thread at 500 r/min.).

1. The spindle feeds from gage height at level #1 to the depth at level #2.

2. At .875 in. depth, the spindle reverses and feeds back up to gage height (level #1).

3. The spindle reverses direction again.

4. In sequence N85, the table moves (Y9.0) along the Y axis to position #2, and the next hole is tapped.

Bore Cycle (G85)

In Fig. 44-9, sequence N90, code G85 calls for a bore cycle to bore a .750 in. hole at 2 in./min.

1. The table moves to position in the X and Y axes, and the spindle rapids to level #1.

2. The boring tool is fed to depth (1.250) at level #2.

3. The feed direction reverses, and the boring tool returns to level #2.

4. In sequence N95, the table moves along the X and Y axes to position #2, and the bore cycle is repeated.

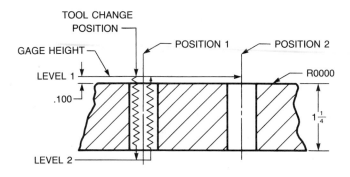

```
N90  G85  X28.0  Y18.0  Z-.125  R.1  F5.0
N95  X24.0  Y14.0
```

Fig. 44-9 The bore cycle (G85) is used to bring a hole to accurate size and location. (Courtesy of Cincinnati Milacron Inc.

SUMMARY

- A subroutine or macro, sometimes called a program within a program, is used to store frequently used data that can be recalled into the main program by a call statement.

- Subroutines are programmed separately and are identified and stored the same as a main program. They can be recalled into the main program by a M98 P0000 call statement.

- Loops, which save programming time and steps, repeat a group of commands as often as they are required in the main program. Dumb loops repeat a group of commands continuously until stopped manually. Smart loops repeat a group of commands a specific number of times, and then continue with the next block of information in the main program.

- Canned cycles, sometimes called fixed cycles, are preset combinations of programmed commands that are initiated by a single command to cause some machine axis movement or to perform specific machining operations.

- Canned cycles are available on modern machining centers and turning controls for repetitive machining operations such as milling a pocket, turning a diameter, thread cutting, etc.

KNOWLEDGE REVIEW

1. Define linear logic.
2. Name three different types of programs or logic flow statements.

Subroutines or Macros

3. What are macros or subroutines, and for what purpose are they used?
4. How should macros be programmed?
5. Under what code should a macro be stored, and what call statement is used to call it into the main program?
6. What must be used at the end of a macro program to return it to the programming mode of the main program?

Looping Logic

7. Why are loops valuable in programming?
8. Name two types of loops and state the purpose of each.

Canned or Fixed Cycles

9. State the purpose of canned or fixed cycles.

10. Describe the actions that occur in the four steps of the A-movement of a turning cycle shown in Fig. 44-5A.

11. List the five factors that could change from part to part in a turning cycle.

12. Define the following cycle codes: G81, G82, G84, G84 (lathe), G85.

UNIT 45

CNC Machining Centers

The first NC machines, introduced in the early 1960s, were drilling machines capable of point-to-point positioning (straight-line motions). Numerical Control (NC) quickly developed into Computer Numerical Control (CNC), and CNC machine tools, because of their many advantages, became widely accepted in the world so that by the mid 1990s, 90 percent of the machine tools manufactured were CNC controlled and only 10 percent were manual machines.

The most common machine tools used by industry are the machining center, turning center, electro-discharge (EDM) machines, and **coordinate measuring machines** (CMM). Most of these machine tools can be combined into a **flexible manufacturing system** (FMS) or flexible manufacturing cell (FMC) for the automated manufacturing of parts. Only the machining center will be covered in this unit, as other topics such as EDM, CMM, FMS, and FMC are covered in the Technology section of this book.

OBJECTIVES

After completing this unit, you should be able to:

- Describe the uses of the main types of CNC machining centers used in industry.
- State eight main advantages of machining centers.
- Program basic operations on two types of machining centers.

KEY TERMS

coordinate measuring machines
modular tooling
VARIAX

flexible manufacturing cells
tombstone fixture

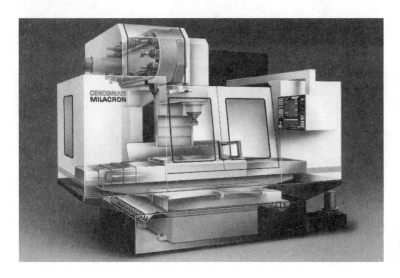

Fig. 45-2A A vertical machining center with an automatic tool changer. (Courtesy of Cincinnati Milacron Inc.)

Fig. 45-2B A horizontal machining center performing operations on the sides of a component. (Courtesy of Giddings & Lewis, Inc.)

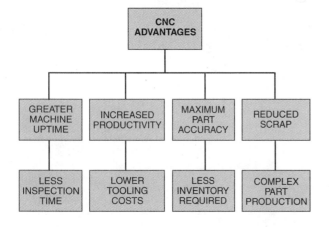

Fig. 45-1 The advantages of CNC are available to anyone associated with the metalworking industry. (Courtesy of Kelmar Associates)

● CNC ADVANTAGES

CNC machines do not require as much operator intervention, and once the machine has been set up, it will machine without stopping until the end of the program is reached. Some of the main advantages that CNC gives a manufacturing shop are greater machine uptime, increased productivity, maximum part accuracy, reduced scrap, less inspection time, lower tooling costs, less inventory, and complex part production, Fig. 45-1.

● MACHINING CENTERS

There are four types of machining centers used in the machine tool industry. They are the

Fig. 45-2C A universal machining center whose spindle can be programmed to a horizontal or vertical position. (Courtesy of Deckel Maho, Inc.)

vertical machining center, Fig. 45-2A, horizontal machining center, Fig. 45-2B, universal machining center, Fig. 45-2C, and the VARIAX, Fig. 45-3.

- *Vertical machining centers* have the cutting tool held in a vertical position. They are generally used to perform operations on flat parts that require motion in the X, Y, and Z axes.

- *Horizontal machining centers* have the cutting tool held in a horizontal position. They can machine parts on more than one side in one clamping, and find wide use in flexible manufacturing systems.

- *Universal machining centers* allow the spindle to be programmed for either vertical or horizontal machining anytime in the program. This combines the features of both the vertical and horizontal machining centers, allowing the machining of all sides of a part in one setup.

- *VARIAX™—The Machine Tool of the Future,* developed by the Giddings & Lewis Inc., is a radical departure from the conventional design of machining centers, Fig. 45-3A. The advanced technology design of the machine uses some of the most basic physical laws of nature. The triangles formed by its six legs connect the upper and lower platforms and contribute to the machine's impressive rigidity. The spindle, virtually floating in space, can move at very high speeds. The **VARIAX** has the potential to change and revolutionize CNC machine tools as we know them today.

    Even though many of the components consist of standard field-proven parts such as ball screws, servomotors, and spindles, the way they are designed and assembled mechanically has developed the VARIAX into an unparalleled and unique machine tool.

- The VARIAX, Fig. 45-3B, consists of a lower platform housing a pallet to hold the part for machining, and an upper platform containing what appears to be a free-floating spindle to drive the cutting tool.

Fig. 45-3A The VARIAX, the machine tool of the future, seems to have a spindle floating in space. (Courtesy of Giddings & Lewis, Inc.)

Fig. 45-3B The VARIAX has a lower platform for holding the part and an upper platform containing the spindle. (Courtesy of Giddings & Lewis, Inc.)

Fig. 45-3C The two platforms are held together by self-aligning bearings and legs that contain ball screws. (Courtesy of Giddings & Lewis, Inc.)

- It has six legs that connect the bed to the head; these legs can be extended or retracted to change the position of the spindle to allow for six-axis machining.

- The platforms are held together by gimbals (self-aligning bearings), and the legs that contain ballscrews are driven by servo motors to position the upper platform.

The main advantages of the VARIAX are:

1. The VARIAX can move and position in almost any direction giving it 6-axis contouring capabilities.

2. The rigidity of the triangulated crossed-leg structure is five times greater than that of a traditional machining center.

3. It can contour five to ten times faster than a conventional machining center.

4. The VARIAX is two to ten times more accurate than a traditional machine tool.

5. The load and movement are shared with all the actuating components. This reduces the stress and wear in any one axis and provides *five times* the acceleration rate; for example, contouring as shown in Fig. 45-3C, can be performed at 2600 in/min.

This unique mechanical design could not function without the VARIAX's computer and software capabilities, which permit CMM (coordinate measurement machine) accuracy. The VARIAX can accept CNC programs written in standard program format and make the transition into any shop quite easy.

● VERTICAL MACHINING CENTERS

Vertical machining centers are the most common machines used in industry; therefore, this section will only deal with these machines. For teaching purposes, two types of CNC Bench-Top machines will be used because they use the same basic programming features as industrial machines, and can perform the same operations using lighter cuts. Students feel at ease on a smaller machine and they are relatively inexpensive. References will be made to industrial machines throughout this unit because bench-top machines are so similar, especially those with Fanuc compatible controls. Since programming codes do vary slightly with manufacturers, it is always wise to consult the programming manual for each specific machine to avoid crashes or scrap work.

The 3-axes bench-top CNC milling machine with the Fanuc-compatible controller, Fig. 45-4, is ideal for teaching the basics of CNC programming. It can be set for vertical and horizontal machining, and includes all important G and M codes, milling cycles, subroutines, etc. It can be programmed in inch or metric dimensions in both incremental and absolute programming. Some models are equipped with a graphics display which allows the operator to test-run the program on the screen without cutting a part. This is a safe way to check the accuracy of a program to prevent crashes and scrap work without actually running the machine.

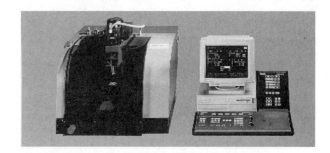

Fig. 45-4 The CNC Bench-Top teaching milling machine is ideal for basic CNC programming and operations. (Courtesy of Emco Maier Corp.)

Fig. 45-5A The main parts of a CNC machining center. (Courtesy of Cincinnati Milacron Inc.)

Fig. 45-5B A CNC machining center with a tilting contouring spindle provides the machine with an A axis. (Courtesy of Cincinnati Milacron Inc.)

Machining Center Parts

The main operating parts of CNC machining centers are the bed, column, saddle, table, servomotors, ball screws, spindle, tool changer, and the machine control unit (MCU), Fig. 45-5A.

- BED—The bed, usually made of high-quality cast iron, provides a rigid machine capable of performing heavy-duty machining and maintaining high precision. Hardened and ground ways on top of the bed provide rigid support for all linear axes movements.

- COLUMN—The column, mounted to the saddle, is designed to prevent distortion and deflection during machining. The column provides the machining center with the Z axis vertical linear movement.

- SADDLE—The saddle, mounted on the hardened and ground bedways, provides the machining center with the X axis longitudinal linear movement.

- TABLE—The table, mounted on the bed, provides the machining center with the Y axis cross linear movement.

- SERVO SYSTEM—The servo system consists of servo drive motors, ball screws, and position feedback encoders to provide fast, accurate movement and positioning of the XYZ axes slides.

- SPINDLE—The spindle, programmable in one-r/min increments, can have speed ranges from 20 to 6000 r/min. or higher.

The spindle can be a fixed position (horizontal) type, or a tilting/contouring spindle, Fig 45-5B, to provide the Z axis vertical movement.

- TOOL CHANGERS—There are two types of tool changers, the vertical tool changer, Fig. 45-6, and the horizontal tool changer, Fig. 45-7. The tool changer can store a number of preset tools that can be called for use by a command in the part program. Tool changers are usually bi-directional and take the shortest travel distance to randomly access a tool. Modern tool change time is usu-

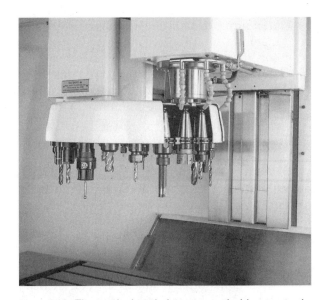

Fig. 45-6 The vertical tool changer can hold many tools that can be recalled as required by a call statement in the program. (Courtesy of Fadal Engineering Co., Inc.)

Fig. 45-7 The tools in a horizontal tool changer can be recalled by a call statement in the program. (Courtesy of Cincinnati Milacron Inc.)

ally only 3 to 5 s, which improves machine uptime.

- MCU—The machine control unit allows the operator to perform operations such as programming, machining, diagnostics, tool and machine monitoring, etc. MCUs vary according to manufacturers' specifications and modern MCUs are making machine tools more reliable and the machining operations less dependent on human skills.

● WORKHOLDING DEVICES

When a workpiece is set up, it must be securely fastened, and the setup must be rigid enough to withstand the forces created during the machining operation. If the workpiece or the holding device becomes loose during machining, damage can result to the tooling, the machine, and the workpiece. Many workholding devices used with conventional machines can also be used on CNC machines. However, in order to increase productivity, many quick-change devices, which may be air- or hydraulic-powered are now available to hold parts accurately and securely.

The machine operator should be sure that all workholding devices are free from chips and burrs before using. The workholding devices, generally specified by the programmer, should be located in the proper position on the machine table. Failure to follow these instructions may result in operator injury, damage to the machine, or scrapped workpieces.

Types of Workholding Devices

- PRECISION VISES—Fig. 45-8, which may be air- or hydraulic-powered, are keyed directly to the table slots and make positioning and clamping of parts fast and accurate. When multiple, identical parts must be machined and each held in a separate vise, a matched set of vises can be used. These vises can also be used to support a long part on both ends to maintain parallelism.

- VISE JAW SYSTEMS can add versatility to, and increase the flexibility of a precision vise for holding workpieces in single station and double-station vises. A set of master jaws is placed in the vise and parallels, modular workstops, angle plates, V-jaws, and machinable soft jaws can be snapped into position as required, Fig. 45-9.

- ANGLE PLATES—Fig. 45-10, are L-shaped pieces of cast iron or hardened steel finish machined to a 90° angle. They are made in a variety of sizes and have holes or slots that provide a means for fastening the workpiece.

- V-BLOCKS—Fig. 45-11, are generally used in pairs to support round work. A U-shaped

Fig. 45-8 A precision vise provides for quick and accurate changeover of parts. (Courtesy of Kurt Manufacturing)

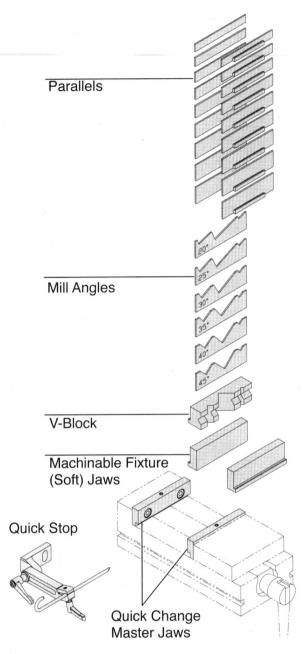

Parallels

Mill Angles

20°
25°
30°
35°
40°
45°

V-Block

Machinable Fixture
(Soft) Jaws

Quick Stop

Quick Change
Master Jaws

Fig. 45-9 Interchangeable vise jaws can include parallels, modular workstops, angle plates, V-jaws, and soft jaws. (Courtesy of Toolex Systems, Inc.)

Fig. 45-10 Angle plates, machined to an accurate 90° angle, allow workpieces to be clamped or bolted to it for machining. (Courtesy of Kelmar Associates)

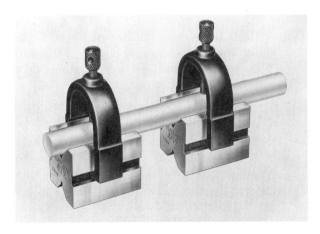

Fig. 45-11 V-blocks are generally used to hold round parts for machining. (Courtesy of The L.S. Starrett Co.)

clamp may be used to fasten the work in a V-block.

- STEP BLOCKS—Fig. 45-12, are used to provide support for strap clamps when work is being fastened to the table or workholding device.
- CLAMPS OR STRAPS—Fig. 45-13, are used to fasten work to the table, angle plate, or

Fig. 45-12 Step blocks provide support for strap clamps when work is fastened to the machine table. (Courtesy of Northwestern Tools, Inc.)

Fig. 45-13 Clamps and straps are used to fasten parts to the machine table, angle plate, or fixture. (Courtesy of J. H. Williams & Co.)

fixture. They are available in a variety of sizes and are usually supported at the end by a step block and bolted to the table by a T-bolt.

- SUPPORT JACKS, Fig. 45-14, are used to support the workpiece to prevent distortion of the workpiece during clamping.
- PARALLELS—Fig. 45-15, are flat, square, or rectangular pieces of hardened metal bars used to support the workpiece for setup.

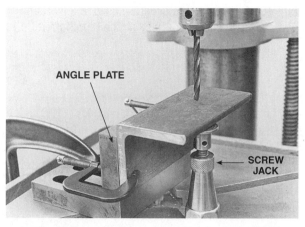

Fig. 45-14 Support jacks are used to prevent thin sections from distorting when they are clamped and machined. (Courtesy of Kelmar Associates)

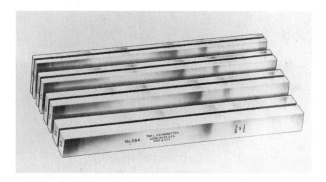

Fig. 45-15 Parallels are used to raise and support a part when held in a vise or clamped to the machine table. (Courtesy of L. S. Starrett Co.)

- SUBPLATES—are generally flat plates that may be fitted to the machine table to provide quick and accurate location of workpieces, workholding devices, or fixtures. The fixturing holes in these subplates are accurately located and, when set up on the machine table in relation to the machine datum, provide the programmer with known locating positions.

● FIXTURES

CNC eliminates many of the expensive jigs and fixtures that were necessary to hold and locate a workpiece on conventional machine tools. CNC fixtures, Fig. 45-16, are used to accurately locate a part and hold it securely for machining operations. Fixtures should be kept as simple as possible to keep the time required to load and unload a part as short as possible. Since part setup is nonproductive time, the savings from the use of proper fixturing will result in correspond-

Fig. 45-16 Various styles of tombstone fixtures that are used on CNC machining centers for holding parts. (Courtesy of Mid-State Machine Products, Inc.)

ing savings in the cost of producing a part. When designing a fixture to hold a part, the following points must be considered:

1. *Positive location:* The fixture must hold a workpiece securely enough to stop the workpiece from moving in the X, Y, and Z axes.

2. *Repeatability:* Identical parts should always be located in exactly the same location for every part change.

3. *Rigidity:* The workpiece must be held securely enough to prevent any movement due to the forces created by the machining operation.

4. *Design:* Modular fixtures using standard components are less costly than custom fixtures and can be quickly modified to accommodate different parts.

5. *Low profile:* Parts of the fixture and the necessary clamping devices should be designed so that there is free movement for the cutting tool at any point in the machining cycle.

6. *Part loading/unloading:* The fixture and its clamping devices should not interfere with the rapid loading or unloading of a part.

7. *Part distortion:* The fixture design should prevent stress from being applied to the part by the clamping forces, otherwise the machined part will distort when the clamping forces are removed.

Tombstone Fixtures

Tombstone fixtures, which may be square, hexagonal, or octagonal, are available in vertical or horizontal models for use on machining centers to hold parts for machining, Fig. 45-16. Parts can be loaded on one side, or any of the available sides of the fixture for machining parts in one setup, if possible. They are especially useful in palletized machining where work is loaded off the machine and the pallet readied for machining operations. On machines that have indexing capabilities, being able to machine parts fastened to more than one side of a tombstone fixture reduces setup time and increases machine uptime.

There are many variations of tombstone fixtures, some standard and some specifically designed to suit a specific part. Some tombstone fixtures have locating features that provide accurate alignment at specific positions on the ma-

Fig. 45-17 Various parts clamped to a tombstone fixture ready for the carousel of the machining center. (Courtesy of Prohold Workholding Inc.)

chine. Figure 45-16 on the previous page, shows the plain square, square tapped, and the hexagonal tapped tombstones on which parts can be fastened. Figure 45-17 shows various types of parts fastened on the sides of a tombstone ready for the pallet to deliver it to the machining center. Figure 45-18 shows a modular workholding station that consists of a precision tombstone having ways into which various vises and workholding fixtures can be quickly and accurately attached. This type of system is very flexible because the vise jaws, fixture, and accessories can be attached or changed to suit different shapes or sizes of workpieces. A SnapLock system allows jaws, fixtures, and accessories to be attached, removed, or indexed in the matter of seconds to a high degree of accuracy.

Clamping Hints

1. Always place the bolt as close to the work as possible.

2. Place a piece of soft metal ("packing") be-

Fig. 45-18 A precision modular workholding station consisting of a tombstone and interchangeable workholding fixtures. (Courtesy of Toolex Systems, Inc.)

Fig. 45-19 Tooling inserts are available in many types, sizes, and grades. (Courtesy of Kaiser Precision Tooling, Inc.)

tween the clamp and the workpiece to spread the clamping force over a wider area and prevent damage to the workpiece.

3. Make sure the packing does not extend into the machining path of the cutting tool.

4. Use two clamps whenever possible to hold a part.

5. Parts that are not flat should be shimmed to prevent distortion when the work is clamped.

6. Apply equal pressure on clamping bolts to prevent workpiece distortion.

● CNC TOOLING

The proper selection of cutting tools for each machining operation is essential to producing an accurate part. Generally, not enough thought and planning go into the selection of cutting tools for each job. The CNC programmer must have a thorough knowledge of various tooling available in order to program a part for the maximum accuracy and productivity.

Machining centers use a variety of cutting tools to perform various machining operations. These tools may be conventional high-speed steel, cemented-carbide inserts, CBN (cubic boron nitride) inserts, or polycrystalline dia-

mond insert tools, Fig. 45-19. Some of the more common tools used are end mills, drills, taps, reamers, boring tools, flycutters, form tools, etc.

Studies show that a machining center uses about twenty percent of the time milling, ten percent boring, and seventy percent hole-making in an average machine cycle. On conventional milling machines, the cutting tool is removing metal about twenty percent of the time, while on machining centers it can be as high as seventy-five percent.

Modular Tooling

Modular tooling is a complete tooling system providing the flexibility and versatility to build a series of tools necessary to produce a part. The modular tooling system combines rigidity, accuracy, and quick-change capabilities to increase the productivity of CNC machine tools. A **modular tooling** system has a basic clamping unit that fits into a spindle or turret, and holds a variety of cutting tool units or cutting tool carriers, Fig. 45-20A. The modular tooling system's building blocks, Fig. 45-20B, consist of almost any combination of accessories, adapters, extensions, inserts, reducers, shanks, and toolholders that can be interchanged for a wide range of machining operations.

The main advantages of modular tooling are:

1. FLEXIBILITY—Shops can create their own tools and reduce the time and cost of obtaining them from suppliers.

CLAMP SCREW

POCKETS FOR THE FLOATING CROSS-BOLT

LARGE CLAMPING SURFACE

SELF-CENTERING GROOVE

AIR ESCAPE

FLOATING DRIVE PIN

Fig. 45-20A Precise and simple assembly of components is the main advantage of modular tools. (Courtesy of Kaiser Precision Tooling, Inc.)

2. PRESETTING—Allows for the accurate off-line setup of tools in order to produce quality parts.

3. TOOL SUPPLY—In a short period of time, a shop will accumulate a supply of reusable tools that have the flexibility to be designed for most machining jobs.

4. PRODUCTIVITY INCREASE—Operators can quickly change tooling from one operation to the next in a matter of seconds, thereby increasing machine uptime. This is especially important for rapid adaptability often caused by engineering changes.

5. REDUCED COSTS—The investment in modular tooling results in large savings in tool-change time, tool-setup time, time required for trial cuts, and reduced scrap.

6. RIGIDITY AND REPEATABILITY—Most of the modular tooling is as rigid as solid tooling, and it provides positive location for length and diameter to within .00008 in.

● STANDARD TOOLING

Most of the tooling used on conventional milling machines can also be used on CNC machining centers. The most common are:

Center Drills—Center drills, Fig. 45-21, are used to accurately spot hole locations for a drill that is to follow. The disadvantage of using center drills is that the small pilot drill can break easily unless care is used. An alternative to the center drill is the spotting tool that has a 90° included angle and is widely used for spotting hole locations.

Drills—Conventional as well as special drills are used to produce holes, Fig. 45-22. Always choose the shortest drill that can produce a hole of the required depth. As drill diameter and length increase, so does the error in hole size and location. Stub drills are recommended for drilling on machining centers, wherever possible.

Indexable Insert Drills—Spade drills, which are available in a wide range of diameters and lengths, are generally used when drilling large holes in tough materials, Fig. 45-23. TiN (titanium nitride) coated HSS and carbide grade inserts can be used in spade drills. Insert drills with helical flutes use a length to diameter ratio to provide good rigidity.

End Mills—End mills and shell end mills, Fig. 45-24, are widely used on machining centers for operations such as face, pocket, and contour milling, spotfacing, counterboring, and roughing and finishing of holes using circular interpolation. Short stub end mills are stronger and can remove metal faster.

Taps—Machine taps, Fig. 45-25, are designed to withstand the torque required to thread a hole and clear the chips out of the hole. Tapping is one of the most difficult machining operations to perform because of the following factors:

- clearing chips from the hole,
- getting a good supply of cutting fluid into the hole,
- speed and feed of threading operations being governed by the lead of the thread,
- depth of thread required.

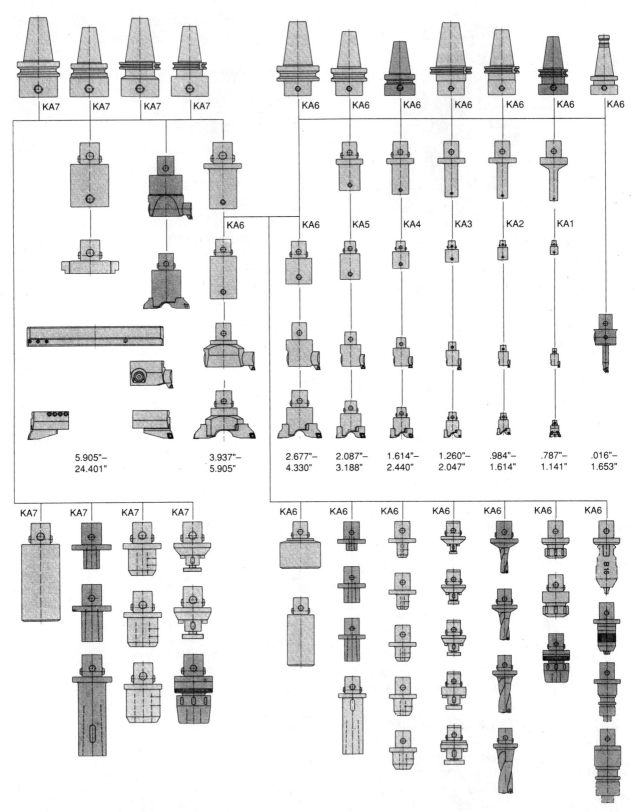

Fig. 45-20B The large variety of components in a modular tooling system makes it possible to suit any type of machine spindle. (Courtesy of Kaiser Precision Tooling, Inc.)

Fig. 45-21 Center drills can be used to accurately spot a hole location before a larger drill is used. (Courtesy of Cleveland Twist Drill Co.)

Fig. 45-23 Replaceable indexable insert drills are rigid, accurate, and can be removed very quickly. (Courtesy of Kaiser Precision Tooling, Inc.)

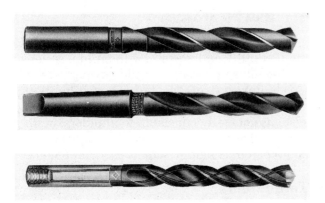

Fig. 45-22 Stub drills are generally used on machining centers because they are rigid and in most cases more accurate than longer drills. (Courtesy of Cleveland Twist Drill Co.)

Fig. 45-24 Short stub end mills are used whenever possible on CNC machines because they are stronger and remove metal faster. (Courtesy of Niagara Cutter, Inc.)

Reamers—Reamers are available in a variety of designs and sizes, Fig. 45-26. A reamer is a rotary end cutting tool used to accurately size and produce a good surface finish in a hole previously drilled or bored.

Boring Tools—Boring is the operation of enlarging a previously drilled, bored, or cored hole to an accurate size and location with a desired surface finish. For smaller holes, a single-point boring tool is generally used, Fig. 45-27A. When producing larger, extremely accurate holes, a multiple boring operation may be required. Twin cutter boring heads, Fig. 45-27B, which balance the cutting pressures to produce the best roundness and accuracy are used for a rough bore pass, which helps to eliminate semifinishing boring in

many applications. This is generally followed by a single-point boring tool for the final pass. When selecting a boring bar, use the largest diameter and the shortest length possible to provide the best rigidity and produce the best accuracy.

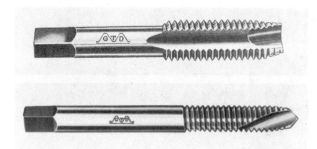

Fig. 45-25 Two- and three-flute taps are generally used on machine tools to thread a hole under power. (Courtesy of Greenfield Industries, Inc.)

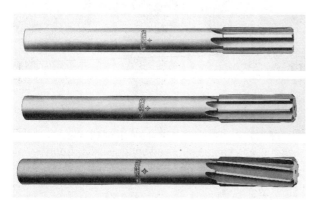

Fig. 45-26 Reamers are used to bring a hole to an accurate size and produce a smooth finish. (Courtesy of Cleveland Twist Drill Co.)

Fig. 45-27A Precision single-point boring tools are available in many different sizes. (Courtesy of Kaiser Precision Tooling, Inc.)

Fig. 45-27B Many precision twin-cutter boring tools are available for various machining operations. (Courtesy of Kaiser Precision Tooling, Inc.)

● TOOLING SYSTEMS

A complete tooling system must provide the flexibility and versatility to build a series of cutting tools necessary to build any part. A modular-tooling system must incorporate rigidity, accuracy, and quick-change capabilities. Tooling systems generally consist of toolholders, cutting tools, and adapters that allow a variety of tools to be prepared using the same basic components.

Toolholders

The machining center uses a wide variety of tools such as drills, taps, reamers, end mills, face mills, boring tools, etc., to perform various machining operations on a workpiece. For these cutting tools to be inserted into the machine spindle quickly and accurately, all must have the same taper shank toolholders to suit the machine spindle. The most common taper used in CNC machining center spindles is the No. 50 taper, which is a self-releasing taper. The toolholder must also have a flange or collar for the tool-change arm to grab, and a stud, tapped hole, or some other device for holding the tool securely in the spindle by a power drawbar or other holding mechanism, Fig. 45-28A and B.

When preparing for a machining sequence, examine the tool assembly drawing and select all the cutting tools required to machine the part. Each cutting tool should be assembled off-line in a suitable toolholder and preset/premeasured to the correct length, diameter, and adjusted for tool runout with a precision tool presetter, Fig. 45-28C. The presetter can be used for fast resetting of a tool during production or for preparing a tool

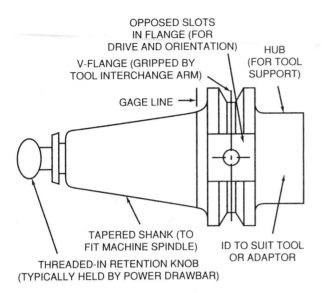

Fig. 45-28A The construction of a toolholder designed to be used in a No. 50 taper CNC machining center spindle. (Courtesy of Hertel Carbide Ltd.)

Fig. 45-28B A variety of cutting tools, each in their own toolholder, can be preset for use in machining centers. (Courtesy of Hertel Carbide Ltd.)

Fig. 45-28C A precision tool presetter is used to set tool lengths, diameters, and to adjust tool runout. (Courtesy of Kaiser Precision Tooling, Inc.)

for the next machining operation. Once all the cutting tools are assembled and preset, they are loaded into specific pocket locations in the machine's tool-storage magazine where they can be selected as required by the part program.

Tool Identification

CNC machine tools use a variety of methods to identify the various cutting tools that are used for machining operations. The most common methods of identifying tools are:

1. TOOL POCKET LOCATIONS—Tools for early machining centers are assigned a specific pocket location in the tool-storage magazine, and each tool is called up for use by a command in the part program.

2. CODED RINGS ON TOOLHOLDERS—A special interchange device reader is used to identify tools by special coded rings on the toolholder.

3. TOOL ASSEMBLY NUMBER—Most modern MCUs have a tool identification feature allowing the part program to recall a tool from the tool-storage magazine pocket by using a five- to eight-digit tool assembly number.

   • Each tool assembly number may be assigned a specific pocket in the tool-storage magazine in the computer program by the operator using the MCU or by a remote tool management console.

Cutting Tool Management

In order to get the best productivity from any machine tool, it is important to have a sound program covering all of the cutting tools and tooling. The best CNC machine tool can only come close to its productivity potential if the best tools for each operation are available for use when they are required. The tool-management program must include tool design, standard coding system, purchasing, good tooling practices, part programming that is cost effective, and the best use of cutting tools on the machine.

A good cutting-tool policy must include the following:

1. STANDARD POLICY

    -   A standard policy regarding cutting tools must be established.

    -   Everyone must clearly understand the policy: the tool engineer, part programmer, setup person, and machine tool operator.

    -   The role each person has in selecting the proper tools must be clearly defined.

2. TOOL DIMENSIONAL STANDARDS

    -   All cutting tools must conform to established cutting tool dimensional standards.

    -   When it is necessary to recondition cutting tools, they should be ground to the next CNC standard.

    -   The part programmer must use the tool standards when programming.

3. RIGID CUTTING TOOLS

    -   Select the shortest cutting tool possible for each job to ensure locational accuracy and rigidity.

    -   Cutting toolholders should be of one-piece construction to provide rigidity.

4. TOOL PREPARATION

    -   There must be a rigid policy on tool setting, compensation, and regrinding that everyone understands.

    -   Clearly define who has the responsibility for each function so that there is no misunderstanding.

5. INDEXABLE INSERT TOOLS

    -   Use cemented carbide insert-type tooling wherever possible because of their wear resistance, productivity, and dimensional accuracy.

    -   Borazon (CBN) inserts should be used on hard ferrous metals where cemented carbides are not satisfactory.

    -   Polycrystalline diamond (PCD) inserts should be used for machining nonferrous materials.

The success or failure of a tool-management program depends on the knowledge of the programmer. To be most effective, the programmer must have a good knowledge of machining practices and procedures, and the type of cutting tool required for each operation. Most modern CNC MCUs have features or software programs available to make any tool-management policy more effective.

● PROGRAMMING PROCEDURE

Most modern bench-top CNC teaching mills (machining centers) come equipped with Fanuc or Fanuc-compatible machine control units (MCUs). Many older CNC bench-top teaching mills were equipped with the EIA RS 274D standard programming controls. Since both types can be used for teaching purposes, both programming examples will be covered for each part to be programmed. Although there are some differences in programming, they will both produce the same part, however the Fanuc-compatible control is widely used in industry.

The part shown in Fig. 45-29 will be used to cover the programming procedures with each code and step explained in detail for easy understanding. The following program notes for the EIA RS 274D standard control will be helpful in the programming and machining procedures.

1. Program in the absolute mode.

2. Begin programming at *XY zero reference point* (lower left of part); set X and Y axes registers at zero at this point.

3. Use a two-flute, 1/4 in. diameter end mill for all operations, including the holes.

4. Set the end mill to the top of the work surface and set the Z axis register at zero.

5. Program operations in numerical sequence, #1 to 13.

6. Return to XY zero when job is complete.

7. Material is .500 × 3.000 × 3.000 in. aluminum (CS 500 sf/min.)

8. Machine—Bench-top CNC mill.

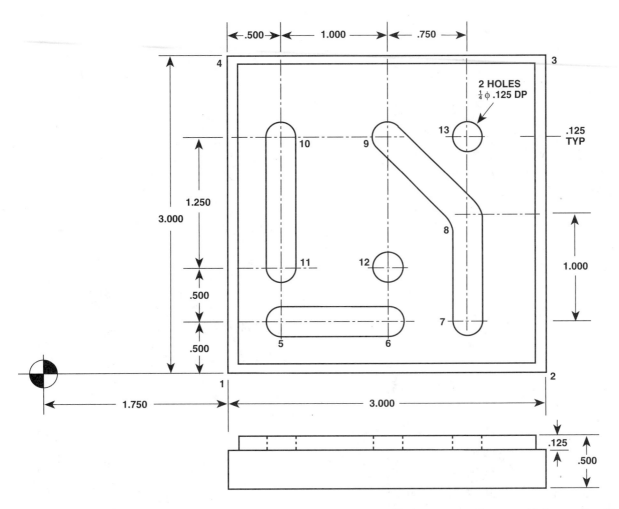

Fig. 45-29 A programming exercise for machining the edges, slots, and holes on a part for a machining center. (Courtesy of The Superior Electric Co.

Programming Sequence (EIA RS 274D Control)

000	**G92**	**X–1750 Y00 Z2000**
	G92	– programmed offset of reference point
	X–1750	– tool set 1.750 in. to left of Point 1
	Y00	– tool center on bottom edge of part
	Z2000	– end of cutter is 2.000 in. above part
N05	**G20**	**G90**
	G20	– inch data input
	G90	– absolute programming mode
N10	**M06**	**T01**
	M06	– tool change command

	T01	– tool No. 1 (.250 in. dia., 2-flute end mill)
N15	**S2000**	**M03**
	S2000	– spindle speed set at 2000 r/min
		– spindle ON clockwise
N20	**G00**	**X–125 Y00 Z2000**
	G00	– rapid traverse rate
	X–125	– tool located .125 in. to left of Point 1
	Y00	– center of tool on lower left edge of part
	Z2000	– end of cutter 2.000 in. above work surface
N25	**G43**	**X00 Y00 H1 Z100**
	G43	– tool No. 1 length offset
	Z100	– tool rapids down to within .100 in. above work surface

N30　G01　X00　Y00　Z–125　F40
　　G01　　　– linear interpolation
　　Z–125　– tool feeds to .125 in. below part
　　　　　　　surface
　　F40　　　– feed rate set for 4 in./min.

N35　G01　X3000　Y00　Z–125　F40
　　X3000　– tool feeds right 3.000 in. along
　　　　　　　X axis from Point 1 to 2
　　Y00　　　– Y axis does not move

N40　G01　X3000　Y3000　Z–125　F40
　　X3000　– X axis does not move
　　Y3000　– tool feeds up 3.000 in. along Y
　　　　　　　axis from Point 2 to 3

N45　G01　X00　Y3000　Z–125　F40
　　　　　　　– tool feeds 3.000 in. along X
　　　　　　　axis from Point 3 to 4

N50　G01　X00　Y00　Z–125
　　　　　　　– tool feeds 3.000 in. along Y
　　　　　　　axis from Point 4 to 1

N55　G00　X00　Y00　Z100
　　　　　　　– tool rapids .100 in. above work
　　　　　　　surface

N60　G00　X500　Y500　Z100
　　　　　　　– tool rapids to Point 5

N65　G01　X500　Y500　Z–125　F40
　　　　　　　– tool feeds .125 in. into the part

N70　G01　X1500　Y500　Z–125　F40
　　X1500　– tool feeds 1.000 in. from Point
　　　　　　　5 to 6

N75　G00　X1500　Y500　Z100
　　　　　　　– tool rapids to .100 in. above
　　　　　　　part surface

N80　G00　X2250　Y500　Z100
　　　　　　　– tool rapids to Point 7

N85　G01　X2250　Y500　Z–125　F40
　　　　　　　– tool feeds .125 in. into part

N90　G01　X2250　Y1500　Z–125　F40
　　Y1500　– tool feeds 1.000 in. along Y
　　　　　　　axis from Point 7 to 8

N95　G01　X1500　Y2250　Z–125　F40
　　　　　　　– tool feeds along angular line
　　　　　　　from Point 8 to 9

N100　G00　X1500　Y2250　Z100
　　　　　　　– tool rapids to .100 in. above
　　　　　　　part surface

N105　G00　X500　Y2250　Z100
　　　　　　　– tool rapids to left 1.000 in. to
　　　　　　　Point 10

N110　G01　X500　Y2250　Z–125　F40
　　　　　　　– tool feeds .125 in. into part

N115　G01　X500　Y1000　Z–125　F40
　　　　　　　– tool feeds 1.250 in. along Y
　　　　　　　axis from Point 10 to 11

N120　G00　X500　Y1000　Z100
　　　　　　　– tool rapids to .100 in. above
　　　　　　　part surface

N125　G00　X1500　Y1000
　　　　　　　– tool rapids 1.000 in. to hole
　　　　　　　position at Point 12

N130　G82　Z–125　F30　P1.0　R100
　　G82　　　– fixed boring cycle
　　Z–125　– tool feeds .125 in. into part
　　P1.0　　– tool pauses at bottom of hole
　　R100　　– tool retracts to .100 in. above
　　　　　　　part surface

N140　G00　X2250　Y2250　Z100
　　　　　　　– tool rapids to hole Position 13

N145　G82　X2250　Y2250　Z–125
　　　　　　　– repeats G82 cycle

N150　G80　G00　X2250　Y2250　Z2000
　　G80　　　– cancels boring cycle
　　Z2000　– tool rapids to 2000 above part
　　　　　　　surface

N155　G00　X–1750　Y00　Z2000
　　　　　　　– tool rapids back to XY zero
　　　　　　　(start position)

N160　M05
　　　　　　　– turns spindle OFF

N165　M30
　　　　　　　– end-of-program code

The programming for the Fanuc-compatible control is basically the same as for the RS 274D control, however there are some differences. A few of the main differences are:

1. The G54 code is used to set the programmed offset of the reference point.

2. Codes are modal and do not have to be repeated in every sequence line.

3. All dimensions are entered as decimals.

Using the same job as in Fig. 45-27, the programming for a Fanuc-compatible control would be as follows:

Programming Sequence
(Fanuc-compatible control)

```
%
O4528
N5   G54   T1   M06
     – work coordinate system #1
N10   G43   H01
N15   M03   S2000
N20   G00   X–.5   Y0   Z.2
N25   G01   Z–.125   F5.0
N30   X3.0
N35   Y3.0
N40   X0
N45   Y0
N50   Z.1
N55   G00   X.5   Y.5
N60   G01   Z–.125
N65   X1.5
N70   Z.1
N75   G00   X2.25
N80   G01   Z–.125
N85   Y1.5
N90   X1.5   Y2.25
N95   Z.1
N100   G00   X.5
N105   G01   Z–.125
N110   Y1.0
N115   Z.1
N120   G00   X1.5
N125   G81   Z–.125   R.1
N130   X2.25   Y2.25
N135   G00   X–1.75   Y0 Z.2
N140   M30
%
```

● **MACHINE AND WORK COORDINATES**

Before running any program, refer to Unit 41 for:

1. Setting the zero point and tool axis.

2. Tool settings and offsets.

3. Work settings and offsets.

4. Cutter-diameter compensation.

SUMMARY

- Over 90 percent of the machine tools manufactured in the world are CNC controlled, while only 10 percent are conventional, manually operated machines.

- CNC machine tools increase productivity, improve part accuracy, reduce scrap, and lower manufacturing costs.

- There are vertical, horizontal, and universal CNC machining centers, and modern machines with graphic displays to allow a part to be run on the screen before it is machined.

- CNC fixtures, such as tombstone, are used to accurately locate and hold parts for machining. They reduce the amount of time taken to load and unload parts, and as a result, increase productivity.

- Modular tooling is a complete tooling system providing flexibility, rigidity, repeatability, and presetting. It can result in savings in tool-change time, setup time, trial cuts, and scrap.

- CNC cutting tools can be identified by tool-pocket locations, coded rings on toolholders, and the tool assembly number.

- Most of the programming for CNC machine tools is in the absolute programming mode.

KNOWLEDGE REVIEW

1. Name six types of machines or systems that use computer numerical control.

CNC Advantages

2. List six of the most important advantages of CNC.

Machining Centers

3. Describe and state the purpose of the following:

a. vertical machining center

b. horizontal machining center

c. universal machining center

Vertical Machining Centers

4. List three reasons why Bench-Top machines are ideal for teaching basic CNC programming.

5. Why is it always wise to consult the manual for each different machine when programming?

6. Name three reasons why it is important to test-run a program without actually cutting the part.

7. What is the purpose of the servo system?

8. Name two types of tool changers and state their purposes.

Workholding Devices

9. Describe and state the purpose of vise jaw systems.

Fixtures

10. Name six important features that a CNC fixture should have.

11. Name three types of tombstone fixtures and explain why they are important on CNC machines with indexing capabilities.

CNC Tooling

12. What is modular tooling and what is its purpose?

13. List six advantages of modular tooling.

Tooling Systems

14. Describe a toolholder to suit a CNC machining center spindle.

15. Name three methods used to identify cutting tools used for machining operations.

16. List the five characteristics a good cutting management policy must include.

Programming Procedures

17. What two types of controls can be found on CNC bench-top teaching machines?

18. Prepare a program to cut the 1/4 in. wide rectangular groove in a piece of aluminum.

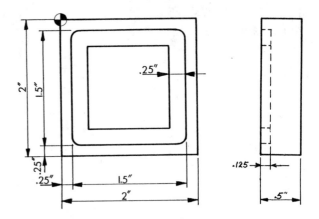

PROGRAM NOTES

- Program in the absolute mode for Fanuc-compatible control.
- The tool change position is X–2.0 Y0.
- Start at the top left corner of the groove and machine clockwise.
- When the groove is complete, return to the tool change position.

UNIT
46

Turning Centers

NC turning centers were developed during the 1960s to improve the productivity and accuracy of the lathes because about forty percent of all metal-cutting operations were performed on these machines. The early NC turning centers, that were standard lathes retrofitted with controls and measuring systems, were a great improvement over the conventional and tracer lathes in use at that time. Through continual improvement, a variety of CNC turning machines were developed to meet the needs of industry and the competition from many other manufacturers.

OBJECTIVES

After completing this unit, you should be able to:

- Describe a chucking and turning center and state the purpose for which each is used.
- Recognize and use the proper programming codes for chucking and turning centers.
- Write a simple program for a bench-top machine and a Fanuc-compatible machine.

KEY TERMS

centrifugal force   chucking center   dynamic gripping force
in-process gaging   material-handling system   static gripping force
tool-monitoring system   turning center

Fig. 46-1 Turning centers are used mainly for machining shaft-type workpieces. (Courtesy of Cincinnati Milacron Inc.)

● TYPES OF CNC TURNING MACHINES

The main types of CNC turning machines used in industry are the turning center, the chucking center, and the combination turning/milling center.

The **turning center,** Fig. 46-1, is mainly designed to handle shaft-type workpieces which are usually supported by a chuck at one end and a heavy-duty tailstock center at the other end. Some turning centers are equipped with two programmable indexing turrets mounted on separate cross-slides, one above and one below the centerline of the work. Each turret can be fitted with eight different types of tools, and can cut on both sides of the work at the same time. Dual turrets are usually used for operations such as:

* rough and finish cuts in one pass,
* different diameters machined at the same time,
* turning and threading operations simultaneously.

The **chucking center,** Fig. 46-2, is designed mainly for machining round work held in a chuck. These machines can also be equipped with dual turrets, one for external operations and the other for internal machining operations at the same time. However, if the workpiece allows, both turrets can be used for external operations at the same time, or both for internal operations.

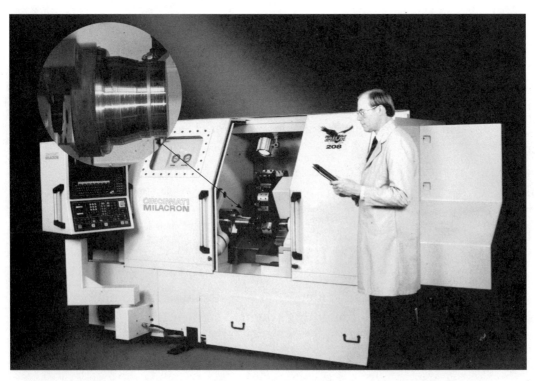

Fig. 46-2 Chucking centers are used for machining work held in a chuck or holding fixture. (Courtesy of Cincinnati Milacron Inc.)

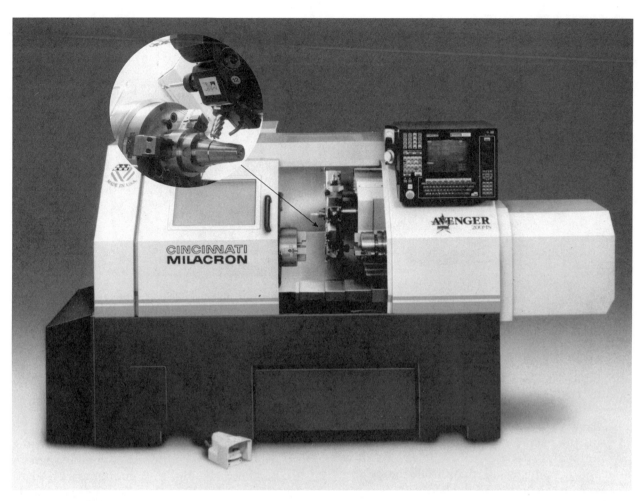

Fig. 46-3 Turning and milling centers, equipped with a motorized (live) tooling turret, can perform turning and milling operations on a part. (Courtesy of Cincinnati Milacron Inc.)

The **combination turning/milling center**, Fig. 46-3, was developed to perform additional operations such as milling, drilling, tapping, etc., on machined round parts. This is possible because this type of machine has a special tool turret containing pockets with their own drive for the cutting tools required. Previous to this development, the round part was machined in a turning or chucking center and then moved to another machine for the additional operations.

In order to perform operations such as milling slots, flats, drilling and tapping holes, etc., the machine must have a contouring spindle that can be indexed and locked at exact locations around the work circumference. For example, if four holes were required at 90° intervals around the part circumference, the programmed procedure would be as follows:

- The spindle would be located at a 90° position.
- The revolving drill would be moved to the correct length along the Z axis.
- The hole would be drilled to depth along the X axis and then retracted from the hole.
- The spindle would index 90° and the cycle would be repeated until all four holes were drilled.

● MAIN PARTS OF CHUCKING AND TURNING CENTERS

The main parts of CNC turning and chucking centers are the bed, headstock, cross-slide, carriage, turret, tailstock, servomotors, ball screws, hydraulic and lubrication systems, and the machine control unit (MCU).

Fig. 46-4 The slant-bed design provides good rigidity and clears the chips from the machining area. (Courtesy of Cincinnati Milacron Inc.)

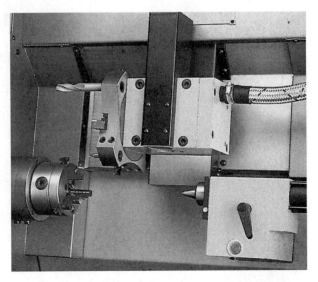

Fig. 46-6 The manual tailstock is standard equipment for turning centers. (Courtesy of Emco Maier Corp.)

- BED—The bed, usually made of high-quality cast iron, can absorb the shock of heavy cuts. Most turning and chucking centers have a slant-bed design, Fig. 46-4, providing the operator easy access for loading and unloading parts. It also allows the chips and coolant to fall away from the cutting area to the bottom of the bed.

- HEADSTOCK—The headstock of the CNC chucking and turning centers, Fig. 46-5, contains the driving mechanism for the machine spindle. These machines are equipped with a variety of motor sizes ranging from 5 to 75 horsepower, and spindle speeds from 32 to 5500 revolutions per minute (r/min).

- TAILSTOCK—CNC chucking and turning centers can be equipped with either a manual tailstock, Fig. 46-6, similar to a standard engine lathe, or an automatic programmable tailstock, Fig. 46-7. The tailstock travels on its own hardened and ground bearing ways.

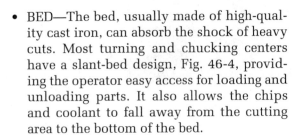

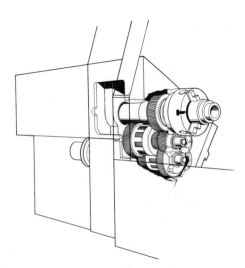

Fig. 46-5 The driving mechanism for the machine spindle is contained in the headstock. (Courtesy of Cincinnati Milacron Inc.)

Fig. 46-7 The movements of a programmable tailstock are controlled by a call statement in the program. (Courtesy of Cincinnati Milacron Inc.)

This allows the carriage to move past the tailstock when a short shaft is being held.

The *automatic programmable tailstock* can be moved by a program call statement, or manually by the operator using the switches on the MCU. Positioning, clamping, and unclamping of the tailstock are usually done by hydraulic pressure. The tailstock is protected against collision with the indexing tools by a contact sensor that immediately stops the indexing motion on contact.

The *swing-up tailstock* can provide flexibility and versatility to the chucking or the turning center. It can swing up to support workpieces for external machining, and then swing away to allow the machine to perform internal work such as drilling and boring.

- TURRETS—The type and number of turrets on a CNC chucking or turning center can vary with the size of the machine and the manufacturer. The more common types are the drum turret, disk turret, and the square multi-tool turret. The turret can hold eight or more tools for external and internal machining operations, Fig. 46-8. Most turrets are capable of bidirectional indexing, and the slides on which the turrets are positioned can travel at a rapid traverse rate of approximately 400 in./min (100 m/min), which reduces noncutting time.

Fig. 46-8 Some drum-type turrets can contain live and stationary tooling. (Courtesy of Cincinnati Milacron Inc.)

Fig. 46-9 The MCU is used to enter, edit, display, and store CNC programs. (Courtesy of Cincinnati Milacron Inc.)

- SERVO SYSTEM—The servo system, consisting of servo drive motors, ball screws, and rotary resolvers, provides fast, accurate positioning of the X and Z axes slides. The rotary resolvers provide the system with unidirectional slide-positioning repeatability of +.0002 in. (0.005 mm). The rapid movement of the slides is approximately 200 to 400 in./min (5000 to 10000 mm/min), and the feed rates are programmable.

- MCU—The MCU (machine control unit), Fig. 46-9, allows the operator to enter a program, edit a program, graphically display programs, store programs into memory, perform diagnostics on the machine, run a program manually or automatically, and perform many more functions and operations. These controls are designed and manufactured using the latest "state-of-the-art" technologies and features which may vary with manufacturers.

● WORKHOLDING DEVICES

The need to increase productivity to meet the demands of industry has resulted in higher metal-removal rates and higher spindle speeds. This has created a need for high-performance chucks with quick-change jaws that can grip the workpiece quickly, accurately, and more securely.

The most common workholding device on turning and chucking centers is some type of chuck. There is a wide variety of chucks, such as self-centering, counter-centrifugal, and collet, to suit various workpieces and machining conditions.

Self-Centering Chuck

Self-centering chucks are designed to move all jaws simultaneously to center the part in the chuck. Self-centering chucks generally have higher gripping forces and are more accurate than other chuck types. These chucks are recommended for bar stock or turned parts that are located from the gripping diameter. The self-centering chuck can hold collet pads in the master jaws for bar stock operations.

Counter-Centrifugal Chuck

The counter-centrifugal chuck provides good gripping pressure even at high speeds when centrifugal force tries to move the jaws outward. The counter-centrifugal chuck reduces the centrifugal force developed by the high r/min; counterweights pivot so that the centrifugal force tends to increase the gripping pressure, thus offsetting the outward forces developed by centrifugal force of the chuck jaws.

Collet Chuck

The collet chuck can accurately hold square, hexagonal, and round bar stock. The collet assembly consists of a drawtube, a hollow cylinder with master collets, and collet pads. Master collets are available with three or four gripping fingers and are often referred to as either three-split or four-split design. The four-split design has better gripping power, but is not quite as accurate. Most collet chucks are front-actuated, and the collet pads are sized for the diameter they are to hold.

Chuck Clamping Force

Two devices can be used to measure the clamping force being put on the workpiece. One measures the **static gripping force,** Fig. 46-10, which is the force per jaw exerted by the chuck on the workpiece when the spindle is stopped. The **dynamic gripping force** is the force per jaw exerted by the chuck on the workpiece when the spindle is running.

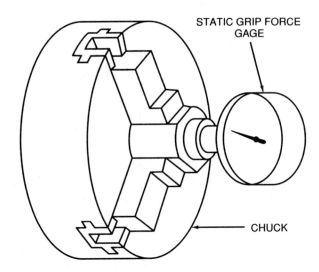

Fig. 46-10 The amount of force exerted by the jaw on a stationary workpiece is called static gripping force. (Courtesy of Cincinnati Milacron Inc.)

Chuck Jaw Pressures

When setting the chuck jaw clamping pressure, make sure it does not exceed the maximum pressure stamped on the warning plate. If greater pressure is used, high stress forces can be created within the chuck, resulting in possible damage to the chuck or to the machine, which may result in personal injury. Front-actuated chucks are usually operated between 200 and 500 psi. Operating the chuck below 200 psi will not provide enough clamping force on the part to hold it securely. Always use the maximum chuck clamping pressure unless the amount of pressure can damage the part.

Centrifugal Force

The amount of centrifugal force that can be tolerated imposes speed limits upon all types of chucks. **Centrifugal force,** which increases as the speed of rotation increases, tries to move the chuck jaws outward. This tends to reduce the amount of the clamping force on the part. Every chuck is affected by the high internal stresses caused by centrifugal force; therefore, all chucks have a maximum rotation speed, which should never be exceeded.

Operating chucks at spindle speeds that are higher than their rated maximum speed results in higher internal stresses and reduces the clamping force on the part. Centrifugal force also increases as the jaws are moved outward from

WORKPIECE-HANDLING SYSTEM

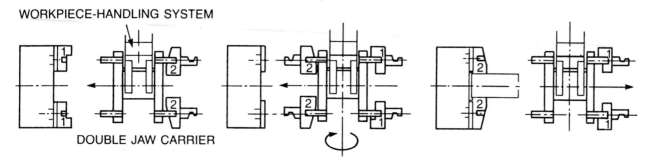

DOUBLE JAW CARRIER

Fig. 46-11A A quick jaw-changing system can replace chuck jaws in one minute or less. (Courtesy of Rohm Products of America)

the center line of rotation and as the jaws are made heavier. It is wise to reduce the spindle speed when using special jaw tooling shapes.

Chuck Jaw Mounting

When it is necessary to change the chuck jaws frequently for different-size workpieces, or when machining operations require a different method of holding the workpiece, a quick jaw-changing system can be used. A quick jaw-changing system, Fig. 46-11A, can reduce the changing time from the usual 30 minutes to about 1 minute or less, reducing downtime and increasing productivity.

Turning centers can also be equipped with an automatic jaw- or chuck-changing system, Fig. 46-11B. When the automated jaw-changing system is called up by the CNC program, it moves into position in front of the chuck, jaws are re-

moved and returned to the magazine, and the next set of jaws is mounted into the chuck.

Some automated systems change one jaw at a time, while others can change all three jaws at once. To manually change chuck jaws could take an operator 20 to 30 minutes, while the quick-change designs change them in 1 minute or less.

Steadyrest/Follower Rest

When long, thin workpieces or shafts are machined, chatter usually occurs. The chatter or vibration can be reduced or eliminated by the use of a steadyrest or a follower rest, Fig. 46-12.

The steadyrest is mounted to the bed of the turning center and can be programmed to open or close automatically, providing *feed-through* capabilities for the workpiece. The steadyrest uses constant hydraulic pressure applied to the support rollers, allowing it to adjust to the work-

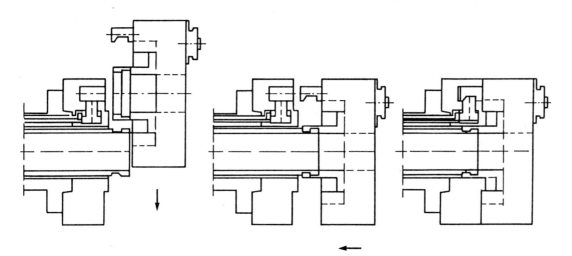

Fig. 46-11B An automatic chuck-changing system can reduce machine downtime and increase productivity. (Courtesy of Forkardt, Inc.)

Fig. 46-12 Follower rests are used to support long shafts as close as possible to the cutting tool. (Courtesy of Cincinnati Milacron Inc.)

piece diameter. The support provided by the steadyrest reduces chatter in the workpiece and allows the machine to operate at higher and more efficient speeds and feeds.

When a four-axis chucking or turning center is used, a follower rest can be mounted in the lower turret. The turret can be programmed to move at the desired rate of travel, providing constant support to the workpiece, while the cutting tool in the upper turret does the machining.

● TOOLING SYSTEMS AND CUTTING TOOLS

A wide variety of tooling systems are available for turning and chucking centers. It is important to remember that the success of any machining operation will depend on the accuracy of the tooling system and the cutting tools being used. A typical tooling system, Fig. 46-13, can consist of toolholders, boring bar holders, facing and turning holders, and drill sockets.

Cutting Tools

There is a variety of types and styles of indexable insert tooling available for chucking and turning centers, Fig. 46-14. This tooling allows the operator to change the indexed inserts at the machine tool instead of removing the cutting tool for resharpening and replacing it with another tool.

Carbide Inserts

Because of the wide variety of carbide inserts available, Fig. 46-15, it is important to select the correct grade and shape for the type of material and machining application. Two main factors should be kept in mind when selecting inserts:

- Is the insert capable of cutting the required contours?
- Does it have enough strength to complete the operation?

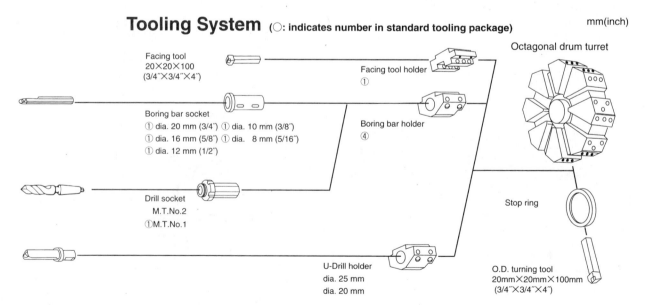

Fig. 46-13 A typical tooling system for turning and chucking centers. (Courtesy of Mazak Quick Turn 10 Tooling System)

Fig. 46-14 A variety of cutting tools and toolholders used on turning and chucking centers. (Courtesy of Kennametal, Inc.)

There is a basic relationship between the cutting condition, speed and feed, and the tool life. Some of the factors that affect tool life are:

- workpiece material,
- insert nose radius,
- insert grade,
- depth of cut,
- insert geometry,
- coolant.

Fig. 46-15 Carbide-insert cutting tools help to reduce machining time and increase productivity. (Courtesy of Carboloy, Inc.)

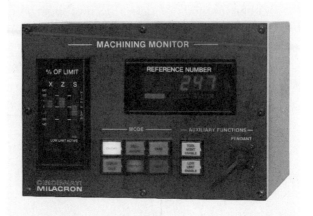

Fig. 46-16 A tool-monitoring system alerts the operator when a tool becomes dull and needs replacing. (Courtesy of Cincinnati Milacron Inc.)

Tool Monitoring

When machining on a turning center, the operator must pay continuous attention to tool wear and breakage. A **tool-monitoring system,** Fig. 46-16, can signal to the operator that tools should be replaced because they are worn and broken.

There are many types of tool-monitoring systems and the way each detects tool wear varies with the manufacturer. A worn or dull tool requires more power to machine a workpiece than a sharp tool; therefore, the most common method of determining wear measures the power or force it takes to drive the cutting tool. The tool-monitoring system measures the load on the main spindle drive motor in two stages:

1. When the machine is set up and ready to produce a workpiece, the normal machining cycle is run with new tools and all speeds and feeds at 100 percent of programmed rate.

2. When the cycle is completed, a second cycle is made without contacting the workpiece.

Using these two cycles, the system can calculate the net machining forces and torque for every part of the program where monitoring may be desirable. Once the limits have been set, the monitoring system signals the operator when machining forces and torque exceed acceptable limits and, in some cases, automatically reduces the speed or feed to compensate for the dullness of the cutting tool.

Other methods of detecting tool wear and breakage involve measurements of the following:

- sound,
- electrical resistance,
- vibration,
- radioactivity,
- measuring heat,
- optical magnification.

The tool-wear-monitoring system provides such benefits as:

- broken-tool detection,
- reduced operator attention,
- machine protection,
- worn tool detection,
- improved productivity.

In-Process Gaging

An **in-process gaging** system, Fig. 46-17, is a way of checking what is happening to the workpiece and tools during machining. It can compensate for tool wear and thermal growth, determine tool offsets, locate workpieces, and be used for inspection purposes.

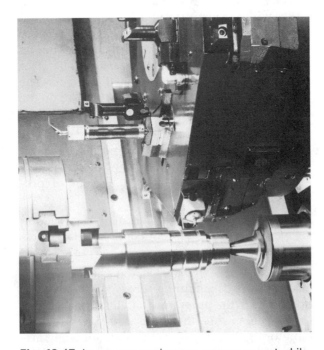

Fig. 46-17 In-process gaging can measure a part while it is being machined and automatically adjust the machine to compensate for tool wear. (Courtesy of Cincinnati Milacron Inc.)

The probe sends a signal to the machine control when it is deflected in any direction once contact is made with the tool or workpiece. In-cycle gages are omnidirectional (all directions), which means that the sensing probes will sense any +X, +Y, or +Z direction. Once the probe stylus has been deflected, a signal is automatically sent to the control where the data can be acted on.

On turning and chucking centers, the probes can be set either in a toolholder or a turret. The probes can be selected the same way a cutting tool is selected by using a call statement in the program. The use of in-process gaging helps to reduce operator errors in setup and allows for inspection of parts while they are being machined.

● SPEEDS AND FEEDS

There are many speed and feed tables available that can be used as guidelines for setting surface speed, feed, and depth of cut, Fig. 46-18. When choosing a surface speed, look at the conditions that relate to the workpiece being machined. Interrupted cuts, deep cuts, long continuous cuts, surface scale, high feed rates, no coolant, and rigidity of setup or workpiece all indicate that there should be a reduction in the recommended surface speed. Uninterrupted cuts, light feed and depth of cuts, short length of cuts, smooth prefinished materials, flood coolant, and rigid setup permit using the recommended surface speeds and still maintain good tool life.

The correct feed should be used for all tools because incorrect feed rates can produce problems. Too slow a feed rate can cause problems with chip control and reduce cutter life. Too fast a feed rate can cause insert chipping and breakage, and can reduce cutter life.

● MATERIAL-HANDLING SYSTEMS

There are a number of options or accessories that can be added to the CNC chucking and turning centers to increase performance and productivity. Collectively, they make up a **material-handling system.** Some of the common accessories are the bar feeder, parts catcher, parts loader/unloader, chip conveyor, and robot loader.

Bar Feeder

The bar feeder, Fig. 46-19, is capable of handling 6 ft (2 m) and 12 ft (4 m) bar lengths to

		Turning								
		High-speed steel tool					Carbide tool			
		Depth of cut		Speed	Feed		Speed sfpm (CS)		Feed	
Material	Finish*	in.	mm	sfpm (CS)	ipr	mm	Brazed	Throw-away	ipr	mm
Aluminum alloys	R	.150	3.81	600	.015	0.38	1100	1500	.020	0.5
Wrought	F	.025	0.63	800	.007	0.17	1400	1800	.010	0.25
Brass	R	.150	3.81	400	.015	0.38	800	925	.020	0.5
330-340-353	F	.025	0.63	480	.007	0.17	960	1100	.007	0.17
Cast iron	R	.150	3.81	145	.015	0.38	500	550	.020	0.5
Soft	F	.025	0.63	185	.007	0.17	650	725	.010	0.25
Cast iron	R	.150	3.81	80	.015	0.38	300	340	.015	0.38
Hard	F	.025	0.63	120	.007	0.17	360	410	.007	0.17
Carbon steel (Lo)	R	.150	3.81	120	.015	0.38	400	485	.020	0.5
1010–1020	F	.025	0.63	160	.007	0.17	475	625	.007	0.17
Carbon steel (Med)	R	.150	3.81	75	.015	0.38	300	375	.020	0.5
1030–1055	F	.025	0.63	105	.007	0.17	385	475	.007	0.17
Carbon steel (Hi)	R	.150	3.81	65	.015	0.38	275	345	.015	0.38
1060–1095	F	.025	0.63	85	.007	0.17	360	440	.007	0.17
Alloy steel (Med C)	R	.150	3.81	90	.015	0.38	300	400	.020	0.5
4130–4140	F	.025	0.63	120	.007	0.17	400	500	.007	0.17
Tool steel (HS)	R	.150	3.81	60	.015	0.38	250	290	.015	0.38
M-3, M-4, M-7	F	.025	0.63	65	.007	0.17	275	320	.007	0.17
Stainless steel	R	.150	3.81	105	.015	0.38	425	475	.015	0.38
300 Series	F	.025	0.63	125	.007	0.17	475	520	.007	0.17
Stainless steel	R	.150	3.81	150	.015	0.38	475	525	.015	0.38
400 Series	F	.025	0.63	170	.007	0.17	525	590	.007	0.17

*R designates rough cut depths; F designates finish cut depths.

Note: These cutting speeds can be used to calculate spindle speeds; however, such speeds are approximate and should be adjusted for type and condition of machine, exact dimensions of operation, and type of material being machined.

Fig. 46-18 Suggested cutting speeds and feeds for high-speed steel and carbide cutting tools. (Courtesy of Kelmar Associates)

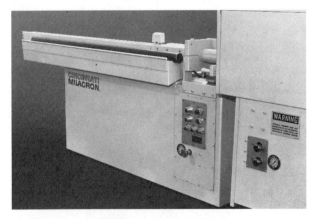

Fig. 46-19 The bar feeder handles long bars to eliminate the loading of individual part blanks. (Courtesy of Cincinnati Milacron Inc.)

eliminate the loading of individual part blanks. This reduces the amount of time spent on loading parts, and as a result, increases productivity.

Parts Catcher

The parts catcher, Fig. 46-20, takes the machined parts and deposits them outside of the machine. When the bar feeder and parts catcher are used, the loading and unloading time is reduced.

Parts Loader/Unloader

The parts loader/unloader, Fig. 46-21, allows individual part blanks of .75 to 2 in. (20 to 50 mm) in diameter and 1 to 2 in. (25 to 50 mm) in length, to be loaded and unloaded in approximately 6 s.

Fig. 46-20 The parts catcher takes finished parts and stores them outside the machine. (Courtesy of Cincinnati Milacron Inc.)

Fig. 46-21 The parts loader/unloader places a blank into the machine and removes the finished part. (Courtesy of Cincinnati Milacron Inc.)

Fig. 46-22 The chip conveyor system collects the chips from machining and moves them to the storage area. (Courtesy of Cincinnati Milacron Inc.)

Chip Conveyor

The chips produced during machining by the cutting cycle fall freely onto the chip conveyor track because of the slant bed design of the machine. The chip conveyor, Fig. 46-22, takes all the chips and transports them by the conveyor system out of the bottom of the machine into containers for storage and recycling.

Robot Loaders

Robot loaders, Fig. 46-23, are controlled by the MCU and can perform many different operations, such as loading and unloading parts, stor-

Fig. 46-23 Programmable robots can be used to load and unload turning and chucking centers. (Courtesy of Cincinnati Milacron Inc.)

ing and retrieving parts from pallets, transporting parts to inspection stations, and changing chuck jaws. Dedicated robot loaders represent the major trend in automated workhandling for turning centers.

● FUNCTION CODES

Many preparatory and miscellaneous function codes are the same for turning centers, machining centers, and wire-cut electrical discharge machines (EDMs). However, because different types of work are produced on each machine, there are certain differences in the coding systems to suit each machine. In order to properly program each machine, it is very important for the programmer to recognize that differences do exist. The use of an incorrect code might result in scrapped work, damage to the machine or cutting tool, or no response from the MCU. The common codes used on turning centers are listed as follows:

Preparatory Codes for Turning Centers

G00	Rapid positioning
G01	Linear interpolation
G02	Circular interpolation clockwise (CW)
G03	Circular interpolation counterclockwise (CCW)
G04	Dwell
G20*	Inch data input
G21*	Metric data input
G27	Zero return check
G28	Zero return
G29	Return from zero
G32	Thread cutting
G40	Tool tip radius compensation cancel
G41	Tool tip radius compensation left
G42	Tool tip radius compensation right
G50	Absolute coordinate preset
G52	Local coordinates preset
G54–G59	Work coordinates system selection
G70	Finish cycle
G71	OD, ID rough cutting cycle
G72	End surface rough cutting cycle
G73	Rough-turn cutting cycle
G74	End surface cutting off cycle
G75	OD, ID cutting off cycle
G78	Thread-cutting cycle

G90	Absolute positioning
G91	Incremental positioning
G92	Maximum spindle speed
G94	Feed in in./min
G95	Feed in r/min
G96	Constant surface speed
G97	Constant surface speed cancel

*On some machines and controls, these may be G70 (inch) and G71 (metric)

Miscellaneous Functions for Turning Centers

M00	Program stop
M01	Optional stop
M02	End of program
M03	Spindle rotation — forward
M04	Spindle rotation—reverse
M05	Spindle stop
M08	Coolant on
M09	Coolant off
M10	Chuck — clamping
M11	Chuck — unclamping
M12	Tailstock spindle out
M13	Tailstock spindle in
M17	Toolpost rotation normal
M18	Toolpost rotation reverse
M21	Tailstock forward
M22	Tailstock backward
M23	Chamfering on
M24	Chamfering off
M30	Reset and rewind
M31	Chuck bypass on
M32	Chuck bypass off
M41	Spindle speed — low range
M42	Spindle speed — high range
M73	Parts catcher out
M74	Parts catcher in
M98	Call subprogram
M99	End subprogram

● CNC BENCH-TOP TEACHING MACHINES

For teaching and programming purposes, a CNC bench-top teaching lathe will be used because it has the same standard programming features found on industrial-size machines. It can perform the same machining operations with lighter cuts, it is relatively inexpensive, and stu-

Fig. 46-24 The CNC bench-top lathe can be used for the same basic, but lighter, operations as a standard turning center. (Courtesy of Emco Maier Corp.)

dents seem more at ease on this smaller machine. Because some programming codes may vary with manufacturers, it is important to refer to the programming manual for each machine to avoid crashes and scrap work.

The CNC bench-top lathe, Fig. 46-24, is ideal for teaching the basics of CNC lathe programming. It uses the EIA RS274D standard control and includes all the standard G- and M-codes that can be programmed in inch or metric dimensions in both absolute and incremental programming. This machine also features canned cycle processing and canned cycle thread cutting. Some models are equipped with a graphics display which can test-run a program on the CRT screen without the machine actually cutting the part. This allows the operator to test-run the program, checking it for accuracy, preventing crashes, and avoiding scrap parts.

The basic difference between CNC bench-top lathes and conventional CNC turning centers is the position of the cutting tools. On CNC turning centers, the cutting tool is at the back of the workpiece, while on bench-top machines, the cutting tool is at the front. Since G-code standards are based on industrial-type machines, to cut a radius or contour counterclockwise, a G03 command is required, Fig. 46-25. On bench-top machines with the cutting tool in front, a G02 or

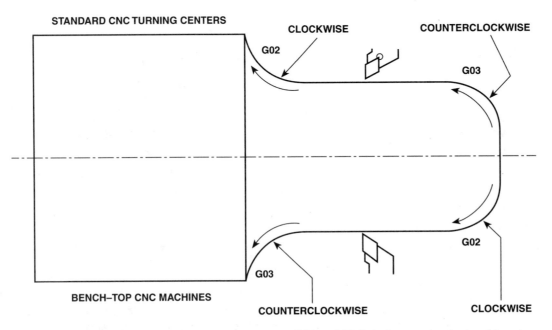

Fig. 46-25 A comparison of the circular interpolation (G02 and G03) codes on standard and bench-top turning centers. (Courtesy of Kelmar Associates)

clockwise command is required to cut the same radius as in Fig. 46-25.

Programming Procedure

The part shown in Fig. 46-26 will be used to cover the programming procedures, with each code and step explained in detail for easy understanding. The following program notes will be helpful in the programming and machining procedures.

1. Program in the absolute mode.

2. All programming begins at the XZ zero reference point at the right edge of the part.

> Touch tool point to diameter and set X register to zero; touch tool point to right end of part and set Z register to size.

3. Use diameter programming code; this default code is automatic on startup.

4. A cemented-carbide insert tool will be used for machining.

5. Set the tool reference point .200 in. from work diameter and .100 in. to the right of the right end of work.

6. Rough cut all diameters only, to .020 in. oversize.

7. Finish cut diameters, radius, chamfer, and taper.

8. Material—1.000 in. diameter aluminum, 3.00 in. long (CS 500 sf/min).

9. Machine—bench-top CNC lathe.

Program Sequence (EIA RS274D Control)— ROUGH CUT

%	
rewind stop code/parity check	
N05 G90	
	– absolute programming mode
N10 G92 X1200 Z100	
G92	– reference point offset
X1200	– tool point located .100 in. off the outside diameter (diameter 1.000 plus .100 on both sides of work)
Z100	– tool point located .100 in. to right of part face (Z0)
N15 S2000 M03	
S2000	– spindle speed set at 2000 r/min.
M03	– spindle ON clockwise (speed of 2000 r/min may have to be set manually)
N20 G00 X945 Z050	
G00	– rapid traverse rate
X945	– tool point located for first cut (Point 1)
Z050	– tool point located .050 in. from part face
N25 G01 X945 Z–2050 F40	
G01	– linear interpolation (straight-line movement)
X945	– .925 diameter will be cut to .945

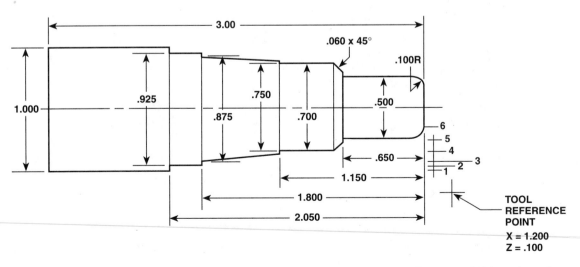

Fig. 46-26 A typical round part used for CNC programming and machining. (Courtesy of Kelmar Associates)

Z-2050 – the diameter will be cut 2.050 in. from end of part.

F40 – feed rate is 4 in./min.

N30 G00 X945 Z050
– tool rapids to Point 2

N35 G00 X895 Z050
– tool rapids to Point 3

N40 G01 X895 Z-1800 F40
– the .875 in. diameter will be cut to .895

N45 G00 X895 Z050
– tool rapids to Point 3

N50 G84 X720 Z-1150 F40 H050
G84 – fixed turning cycle
X720 – diameter turned to .720 in. diameter in incremental cuts of .050 in. (H.050) each
Z-1150 – the diameter turned 1.150 in. from end of part

N55 G00 X720 Z050
– tool rapids to Point 4

N60 G84 X520 Z650 F40 H050
G84 – fixed turning cycle with .500 in. diameter turned to .520 for .650 in. length

N65 G00 X520 Z050
– tool rapids to Point 5

Finish Turning

N70 G00 X300 Z0
– tool rapids to Point 6, the start of the .100 radius and on the edge of the work

N75 G03 X500 Z-100 F40
– tool moves clockwise (CW) to the end of the radius, .100 in. from part face

N80 G01 X500 Z-650 F40
– the .500 in diameter is finish turned for .650 in. length

N85 G01 X580 Z-650 F40
– shoulder is faced to start of the 45° chamfer

N90 G01 X700 Z-710 F40
– the .060 × 45° chamfer is cut

N95 G01 X700 Z-1150 F40
– the .700 diameter is turned to 1.150 length

N100 G01 X750 Z-1150 F40
– shoulder is faced to small diameter of the taper

N105 G01 X875 Z-1800 F40
– the taper section is cut

N110 G01 X925 Z-1800 F40
– shoulder is faced to the start of the .925 diameter

N115 G01 X925 Z-2050 F40
– the .925 diameter is cut to 2.050 length

N120 G01 X1200 Z-2050 F40
– shoulder is faced so tool clears the OD of the part

N125 G00 X1200 Z100
– tool rapids back to start position Point 1

N130 M30
– end-of-program code

%
rewind/stop code

● FANUC COMPATIBLE PROGRAMMING

The programming for the Fanuc-compatible control is basically the same as for the EIA RS274D control, however, there are some differences. A few of the main differences are:

1. The G28 code is used to set the programmed offset of the reference point.

2. Codes are modal and do not have to be repeated in every sequence line.

3. All dimensions are entered as decimals.

Using the same job as in Fig. 46-26, the programming for a Fanuc-compatible control would be as follows:

Programming Sequence

%

N05 G20 G90 G40
G40 – cancels tool radius compensation

N10 G95 G96 S2000 M03
G95 – feed rate per revolution
G96 – constant surface speed

N15 T0202
– tool number and offsets

N20 G00 X1.2 Z.1

N25 G73 U.05 R.05
G73 – rough turning cycle
U.05 – allowance on diameter for depth of cut

```
    R.05    – pull-off distance
N30 G73 P35  Q95  U.025  W.005  F.008
    P35     – start block of rough contour
              cycle
    Q95     – end block of rough contour
              cycle
    U.025   – allowance for finish cut
    W.005   – shoulder allowance for finish
              cut
N35  G00  X.3
N40  Z.05
N45  G01  Z0
N50  G03  X.5  Z–.1  R.1
N55  G01  Z–.65
N60  X.58
N65  X.7  Z–.71
N70  Z–1.15
N75  X.75
N80  X.875  Z–1.8
N85  X.925
N90  Z–2.05
N95  X1.05
N100 G00  X1.2  Z.1
N105 G72  P35  Q95  F.005
    G72     – finish turn cycle
N110 G00  X2.0  Z.5
N115 M30
%
```

SUMMARY

- Turning centers are used to machine shaft-type workpieces; chucking centers machine work held in a chuck, while turning/milling centers can perform turning and milling operations.

- The most common turrets that can hold a number of external and internal machining tools can be of the drum, disk, or square multi-tool type.

- Modern workholding devices are designed to grip the workpiece quickly, accurately, and securely to reduce setup time and increase productivity.

- Most modern workholding devices can be controlled to operate by a code in the part program.

- Tool monitoring systems are used to detect tool wear or breakage by measuring the load on the main spindle drive motor.

- Material-handling systems are used to load and unload parts automatically to reduce machine downtime and increase productivity.

- Bench-top CNC teaching machines are excellent for teaching purposes because they use standard programming features to teach basic fundamentals, and they are more affordable by educational institutions.

KNOWLEDGE REVIEW

1. Why were NC turning centers developed?

Types of CNC Turning Machines

2. Name and state the purposes of three types of CNC turning machines.

Main Parts

3. List three types of tailstocks that can be used on chucking and turning centers.

4. Name three types of turrets and state the purposes they serve.

5. What is the purpose of the servo system?

Workholding Devices

6. State three reasons why centrifugal chucks are used on chucking and turning centers.

7. Define static and dynamic gripping force.

8. Name two types of chuck jaw-mounting systems.

9. What is the purpose of a steady or follower rest?

Tooling Systems and Cutting Tools

10. List six factors that can affect the life of a cutting tool.

11. What five benefits can a tool-wear-monitoring system provide?

12. For what five purposes is in-process gaging used?

Material Handling Systems

13. State the purpose of the following:

    a. bar feeder

    b. parts catcher

    c. chip conveyor

    d. robot loaders

Function Codes

14. Define the following preparatory codes: G01, G20, G40, G73, G78, G94.

15. Define the following miscellaneous codes: M03, M05, M23, M30, M98.

CNC Teaching Machines

16. Compare the position of the cutting tool on CNC bench-top lathes and conventional CNC turning centers.

17. What codes would be used to cut a .250 in. radius on the right-hand end of a part?

    a. on a bench-top CNC lathe

    b. on a conventional CNC turning center

18. Prepare a program to turn the following part using a Fanuc compatible control.

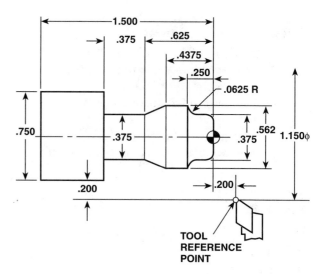

PROGRAM NOTES

- Program in the absolute mode
- The material is aluminum, 1.150 diameter
- Use diameter programming code
- Start at the tool reference point
- Machine—Bench-top CNC lathe.

SECTION

16

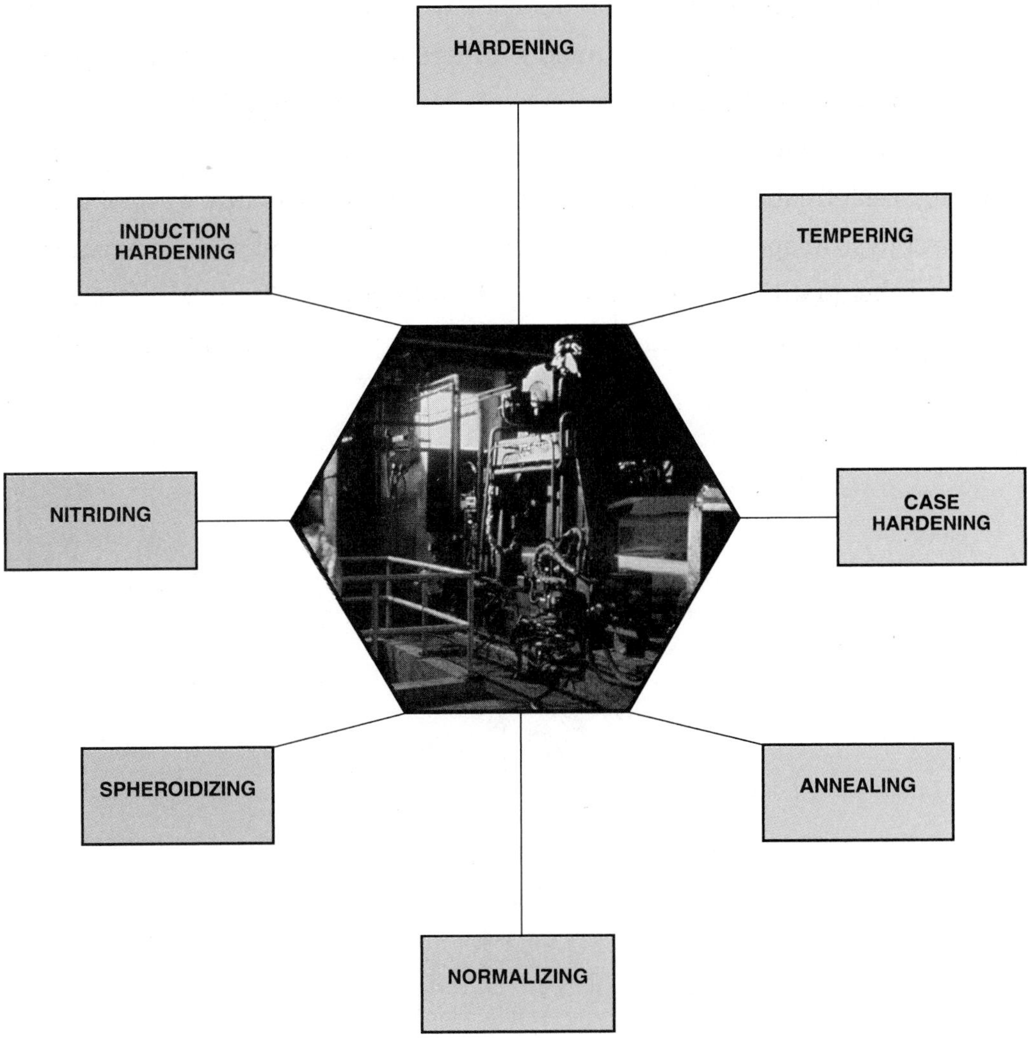
HARDENING
TEMPERING
CASE HARDENING
ANNEALING
NORMALIZING
SPHEROIDIZING
NITRIDING
INDUCTION HARDENING

Heat Treating

Many centuries ago, it was known that heating and cooling iron and steel greatly changed their properties. The village blacksmith, who hammered the chisel to shape, knew that metal got harder when it was cooled quickly. The development of the microscope enabled workers to observe that the inner structure of the metal had assumed a different grain structure. This process was eventually called heat treatment, which is a term applied to a variety of procedures used to change the physical characteristics of metal by heating and cooling. Heat treatment is used to improve the microstructure of steel to meet certain physical specifications. Toughness, hardness, and wear resistance are a few of the qualities obtained through heat treatment. To obtain these characteristics, operations such as hardening, tempering, annealing, and case-hardening are necessary.

Figure courtesy of Kelmar Associates; inset courtesy of Nucor Corp.

Heat Treatment of Steel

Understanding the properties and heat treatment methods for various steel has become very important to the proper functioning of a part. New alloys have resulted in an increase in the strength and wear-resistance of metals. The selection of the proper steel and its subsequent heat treating operations will ensure that the manufactured part will perform properly in use.

OBJECTIVES

After completing this unit, you should be able to:

- Explain the changes that occur in the microstructure of steel when it is heat treated.
- Harden, temper, and anneal a piece of high-carbon tool steel.
- Caseharden a piece of low-carbon steel.

KEY TERMS

annealing
hardening
normalizing

casehardening
heat treatment
tempering

flame hardening
induction hardening

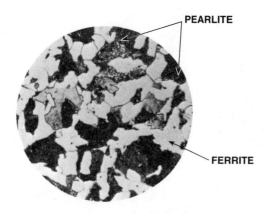

Fig. 47-1 Steel is made up of ferrite (iron) and pearlite (carbon). (Courtesy of Kelmar Associates)

● TYPES OF STEELS

There are many different types of steels available for wide-ranging uses such as paper clips, bridges, springs for automobiles, and cutting tools. All of these steels have one thing in common: they all contain iron and carbon, the main elements in steel, Fig. 47-1. Carbon is the element which may have the greatest effect on the steel's properties because it is the hardening agent. As the percentage of carbon in steel increases over 0.83%, the hardness, hardenability, tensile strength, and the wear-resistance of steel are increased.

There are four distinct categories of steels, low-carbon, medium-carbon, high-carbon, and alloy steels, Fig. 47-2.

- *Low-carbon steel* contains from 0.02 to 0.30% carbon, and because of the low carbon content, cannot be hardened. The surface of this steel can be casehardened by

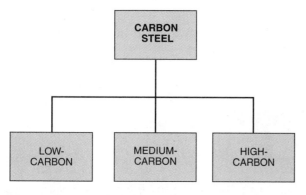

Fig. 47-2 Types of carbon steels. (Courtesy of Kelmar Associates)

heating and increasing its carbon content. Low-carbon steel is used for shafts, nuts, washers, sheet metal, etc.

- *Medium-carbon steel* contains from 0.30 to 0.60% carbon, can be hardened (toughened) which increases its tensile strength. It is widely used for forgings, wrenches, hammers, screwdrivers, etc.

- *High-carbon steel,* also called tool steel, contains over 0.60% and as high as 1.7% carbon. It can be fully hardened and is used for cutting tools such as drills, taps, reamers, milling cutters, lathe toolbits, etc.

- *Alloy steels* are generally some form of high-carbon steel to which certain alloying elements have been added to provide it with qualities such as increased hardness, toughness, tensile strength, wear and corrosion resistance.

Tool Steels

There are many types of tool steels produced by manufacturers under their own trade names. These steels are generally classified as water-hardening, oil-hardening, air-hardening, or high-speed steels to suit a variety of uses. To avoid confusion with trade names, it is wise to order steel by its AISI (American Iron and Steel Institute) or SAE (Society of Automotive Engineers) number. Be sure to consult the steel manufacturer's handbook for the selection and heat treatment of each type of steel. The most common tool steels used in industry are shown in Table 47-1.

- *High-speed steels,* may have a tungsten or molybdenum base, and are used primarily for cutting tools such as drills, reamers, milling cutters, and lathe tools. They provide good wear-resistance and red hardness.

- *Hot-work steels,* may have a chromium, tungsten, or molybdenum base, and provide good wear-resistance, toughness, and excellent red hardness. These steels resist heat checking and cracking during heat treatment, and give good life in forging tools, piercing dies, and die-casting dies.

- *Cold-work steels,* generally a high-carbon, high-chromium base, provide deep hardenability, size stability, and outstanding wear-resistant parts, tool and die applications, and cutting tools.

Table 47-1　Common types of tool steels (Courtesy of Kelmar Associates)

Group	Code	Hardening Depth	Toughness	Wear Resistance
1　High speed				
– molybdenum	M	deep	low	very high
– tungsten	T	deep	low	very high
2　Hot work				
– chromium base	H	deep	good	fair
– tungsten base	H	deep	good	fair–good
– molybdenum base	H	deep	medium	high
3　Cold work				
– high carbon–high chrome	D	deep	poor	best
– air hardening	A	deep	fair	good
– oil hardening	O	medium	fair	good
4　Shock resisting	S	medium	best	fair
5　Mold Steels	P	shallow	high	low–high
6　Special Purpose				
– low alloy	L	medium	high	medium
– carbon tungsten	F	shallow	low–high	low–high
7　Water hardening	W	shallow	good	fair–good

- *Shock-resistant steels,* provide good toughness and hardness for chisels, pneumatic tools, heavy-duty dies and punches, and general purpose applications.

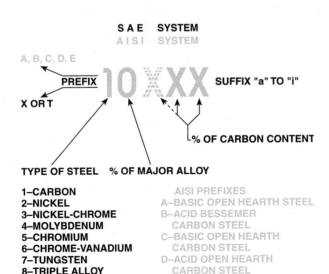

TYPE OF STEEL　% OF MAJOR ALLOY

1–CARBON
2–NICKEL
3–NICKEL-CHROME
4–MOLYBDENUM
5–CHROMIUM
6–CHROME-VANADIUM
7–TUNGSTEN
8–TRIPLE ALLOY
9–SILICON-MANGANESE

AISI PREFIXES
A–BASIC OPEN HEARTH STEEL
B–ACID BESSEMER
CARBON STEEL
C–BASIC OPEN HEARTH
CARBON STEEL
D–ACID OPEN HEARTH
CARBON STEEL
E–ELECTRIC FURNACE STEEL

Fig. 47-3　AISI and SAE steel numbering system. (Courtesy of Kelmar Associates)

HEAT TREATMENT SPECIFICATIONS FOR STEELS

There are two standard steel identifications: SAE (Society of Automotive Engineers) and AISI (American Iron and Steel Institute), Fig. 47-3. The coding system for various types of steels is shown in Table 47-2. To obtain the heat-treatment specifications for various steels, use a standard reference book and look at the AISI or SAE tables of steels, or refer to the steel manufacturer's specifications.

HEAT-TREATMENT TERMS

To understand heat treating better, it is useful to know that steel is a combination of many elements, mainly iron (ferrite) and iron carbide (pearlite). These elements, along with heat treating methods, can be varied to produce steel with different characteristics and properties. **Heat treatment** falls into two main categories—one where there is no change in the composition of the steel, the other involves a change in the steel's surface through the absorption of carbon, nitrogen, or a combination of both.

Table 47-2 Steel coding system (Courtesy of Kelmar Assocaites)

Types of steel	Coding
Carbon	1XXX
– plain	10XX
– free cutting	11XX
– high manganese	13XX
Nickel	2XXX
– 3.50% nickel	23XX
– 5.00% nickel	25XX
Nickel chromium	3XXX
– 1.25% ni., .60% cr	31XX
– 3.50% ni., 1.50% cr	33XX
Molybdenum	4XXX
– 1.0% chrome	41XX
– 5% cr, 1.8% ni	43XX
– 3.5% nickel	48XX
Chromium	5XXX
– low chrome	51XX
– medium chrome	52XX
Chromium-vanadium	6XXX
Triple Alloy	8XXX
Silicon-Manganese	92XX

To provide a better understanding of the heat treating process, it is important to be familiar with some common metallurgy and heat-treating terms.

Metallury

AUSTENITE—a solid solution of carbon in iron existing between the lower and upper critical temperatures.

CEMENTITE—a carbide of iron that is the hardening agent in steel.

FERRITE—iron in its pure form.

MARTENSITE—the structure of fully hardened steel obtained when austenite is quenched. Martensite is characterized by its needle-like pattern, Fig. 47-4B.

TEMPERED MARTENSITE—the structure obtained after martensite has been tempered, Fig, 47-4C.

PEARLITE—a saturated mixture of ferrite (iron) and cementite (iron carbide); usually the condition of steel before heat treatment, Fig. 47-4A.

(A) PEARLITE: STEEL BEFORE HEAT TREATING

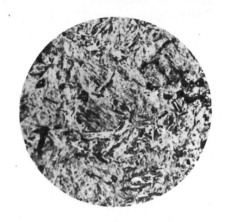

(B) MARTENSITE: STEEL AFTER HARDENING

(C) TEMPERED MARTENSITE: STEEL AFTER TEMPERING

Fig. 47-4 The effects of heating and cooling on the grain structure of 0.83% carbon steel. (Courtesy of Kelmar Associates)

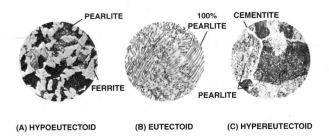

(A) HYPOEUTECTOID   (B) EUTECTOID   (C) HYPEREUTECTOID

Fig. 47-5 Types of eutectoid steels. (Courtesy of Kelmar Associates)

EUTECTOID STEEL—steel containing just enough carbon to dissolve completely in the iron when the steel is heated to its critical range. Eutectoid steel contains from 0.80% to 0.85% carbon and forms austenite at the lower critical temperature, Fig. 47-5B.

HYPEREUTECTOID STEEL—steel containing more carbon than will completely dissolve in the iron when the steel is heated to the critical range. This steel has over 0.90% carbon, Fig. 47-5C.

HYPOEUTECTOID STEEL—steel containing less carbon than can be dissolved by the iron when the steel is heated to the critical range. This steel has less than 0.90% carbon, Fig. 47-5A.

Heat Treating

The most common heat treating terms are illustrated in Fig. 47-6.

HEAT TREATMENT—the heating and cooling of metals to produce certain mechanical properties.

DECALESCENCE POINT—the temperature at which carbon steel, when heated, transforms

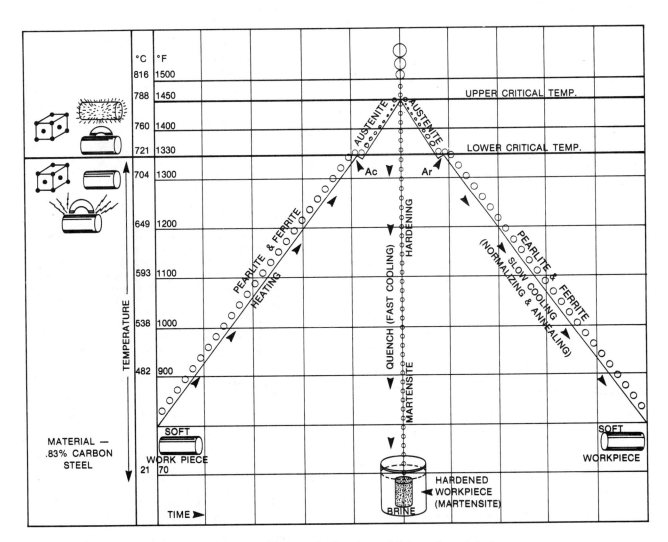

Fig. 47-6 Some common heat treating terms illustrated. (Courtesy of Kelmar Associates)

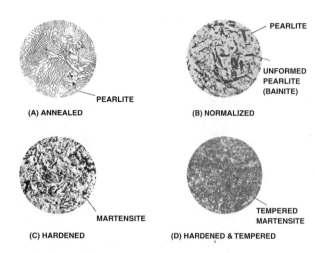

(A) ANNEALED — PEARLITE

(B) NORMALIZED — PEARLITE, UNFORMED PEARLITE (BAINITE)

(C) HARDENED — MARTENSITE

(D) HARDENED & TEMPERED — TEMPERED MARTENSITE

Fig. 47-7 The effects of various heat-treating operations on the grain structure of 0.83% carbon steel. (Courtesy of Kelmar Associates)

from pearlite to austenite, point *Ac* (generally about 1330° or 721° for 0.83% carbon steel).

RECALESCENCE POINT—the temperature at which carbon steel transforms from austenite to pearlite, point *Ar*.

LOWER CRITICAL POINT—the lowest temperature at which steel may be quenched in order to harden it.

UPPER CRITICAL POINT—the highest temperature at which steel may be quenched in order to attain maximum hardness and the finest grain structure.

CRITICAL RANGE—the temperature range between the upper and lower critical temperatures.

HARDENING—the heating of steel above its lower critical temperature, followed by quenching in water, oil, or air to produce a martensite structure, Fig. 47-7C.

TEMPERING (Drawing)—reheating hardened steel to a point below its lower critical temperature, followed by a rate of cooling, Fig. 47-7D. Tempering removes the brittleness and toughness in the steel.

ANNEALING—the process of relieving internal stresses and strains and softening steel by heating it above its critical temperature and allowing it to cool slowly in a closed furnace, ashes, lime, or asbestos, Fig. 47-7A.

NORMALIZING—heating steel to just above its upper critical temperature and cooling it in still air. Normalizing improves the grain structure and removes stresses and strains, Fig. 47-7B.

SPHEROIDIZING—the heating of steel to just below the lower critical temperature for a certain period of time followed by cooling in still air. This process, that improves the machinability of the metal, works best on high-carbon steel.

● HEAT-TREATING FURNACES

There are three types of furnaces used in various heat treating operations. These are the low-temperature or drawing furnace, Fig. 47-8A, the high-temperature furnace, Fig. 47-8B, and the pot-type furnace, Fig. 47-8C. The type of heat treating operation will determine the type of furnace used. For hardening carbon and alloy steels, and for pack carburizing machine steel, the high-temperature furnace is used. For preheating and tempering operations, the low-temperature furnace is used. Casehardening or liquid carburizing is carried out in the pot furnace. These furnaces may be heated by gas, oil, or electricity. The furnace temperature in all these furnaces is indicated and accurately controlled by a thermocouple and pyrometer.

The thermocouple, which is made of two dissimilar metals, is mounted inside the furnace and is connected by leads to the pyrometer, generally mounted on a wall near the furnace, Fig. 47-9. Any changes in the furnace temperature show on the pyrometer, which may be set to shut

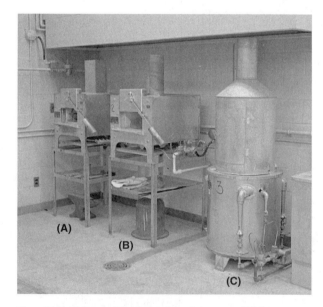

Fig. 47-8 Types of heat treating furnaces. (A) Low-temperature (B) High-temperature (C) Pot-type. (Courtesy of Kelmar Associates)

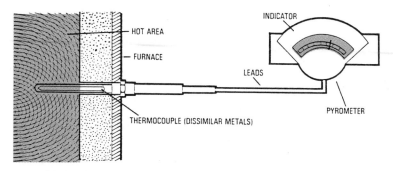

Fig. 47-9 The pyrometer and thermocouple control indicate the furnace temperature. (Courtesy of Kelmar Associates)

off the heat to the furnace when a preset temperature is reached.

Another method used to determine the temperature of steel is to note its color. This method is not as accurate as the pyrometer, but it can be used when heat treating small parts and tools. In Table 47-3, various heat colors and their approximate temperatures are given.

Table 47-3 Chart giving heat colors and their approximate temperatures (Courtesy of Kelmar Associates)

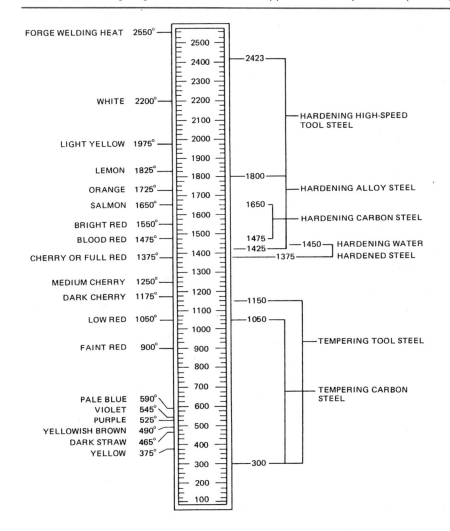

● SAFETY PRECAUTIONS

During any heat treating operation, the following safety precautions should be observed:

1. Know and understand the operation of the heat treating system.

2. Check that all the safety devices, such as automatic shut-off valves, air switches, exhaust fan, etc., are working properly before starting the furnace.

3. Follow the manufacturer's instructions on lighting the furnace.

4. Stand to one side when lighting a gas- or oil-fired furnace, in case of a blow back.

5. Always wear a face shield, gloves, and approved protective clothing when working with hot metal, Fig. 47-10.

6. Use the proper tongs for the job, and be sure the tongs are dry (heated above the dew point) before removing any work from a liquid carburizing pot.

 Any moisture (water or oil) coming in contact with this liquid may cause an explosion and seriously burn the operator.

7. Never inhale the fumes from a liquid carburizing solution.

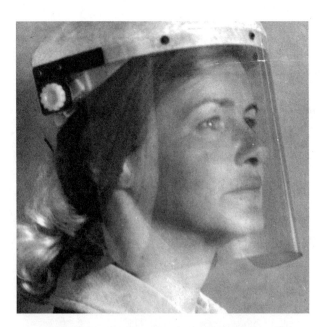

Fig. 47-10 Face shields offer the best protection when performing heat treating operations. (Courtesy of Kelmar Associates)

 These solutions may contain CYANIDE OF POTASSIUM, A DEADLY POISON.

● METCALF'S EXPERIMENT

This experiment shows the effect of temperature on the grain structure, hardness, and strength of tool steel, Fig. 47-11.

1. Select a piece of SAE 1090 (tool steel) about 1/2 in. (13 mm) diameter and about 4 in. (100 mm) long.

2. With a sharp pointed tool, cut shallow grooves approximately 1/2 in. (13 mm) apart.

3. Number each section as in Figure 47-11.

4. Heat the bar with an oxyacetylene torch, bringing number 1 section to a white heat.

5. Keep number 1 section at white heat, and heat sections 4 and 5 to a cherry red color.

DO NOT apply heat to sections 6 to 8.

6. Quench in cold water or brine.

7. Test each section with the edge of a file to compare it for hardness.

8. Break off the sections and examine the grain structure under a microscope. The results will indicate that:

    *   Sections 1 and 2 have been overheated. They break easily and the grain structure is very coarse.

Fig. 47-11 Metcalf's experiment shows the effects of over or underheating on the grain structure of the steel. (Courtesy of Kelmar Associates)

- Section 3 requires more force to break and the grain structure is somewhat finer.

- Sections 4 and 5 have the greatest strength and resistance to shock. These sections have the finest grain structure.

- Sections 6 to 8, where the metal was underheated, require the greatest force to break, and bending occurs. The grain structure becomes coarser toward section 8. This section is the original structure of unheated steel (pearlite).

HARDENING CARBON STEEL

Hardening is the process of heating metal uniformly to its proper temperature and then quenching or cooling it in water, oil, air, or in a refrigerated area.

Carbon tool steels may be hardened by heating them to a bright cherry red color, approximately 1450 to 1500° (790 to 830°C), and quenching in water or oil. When the steel is heated to this temperature, a chemical and physical change takes place, with the carbon combining with the iron to form a new structure called *austenite*. When the steel is quenched, the austenite changes to a hard, brittle, fine-grained structure called martensite, Fig. 47-12.

To Harden Carbon Tool Steel

1. Start the furnace and preheat the steel slowly.

2. Heat the steel to the temperature recommended by the manufacturer.

3. Quench in water, brine, or oil, depending on the type of steel or manufacturer's recommendation.

Long, slender pieces should be quenched vertically to avoid warping.

4. Move the work about in the quenching medium in a "figure 8" motion, to allow the steel to cool quickly and evenly.

5. Test for hardness with a hardness tester or the edge of a file.

TEMPERING

After hardening, the steel is brittle and may break with the slightest tap, due to the stresses

(A) BEFORE HARDENING

(B) AFTER HARDENING

Fig. 47-12 The grain structure of carbon tool steel. (Courtesy of Kelmar Associates)
(A) Austenite   (B) Martenite

caused by quenching. To overcome this brittleness, the steel is tempered; that is, it is reheated until it is brought to the desired temperature or color, and then quenched. **Tempering** toughens the steel and makes it less brittle, although a little of the hardness is lost. As steel is heated it changes in color, and these colors indicate various tempering temperatures, Table 47-4.

To Temper Carbon Tool Steel

1. Clean all the surface scale from the work with abrasive cloth.

2. Select the temperature or color desired, Table 47-4.

3. Heat the steel slowly and evenly.

4. When the steel reaches the correct temperature or color, quench it quickly in the same cooling medium used for hardening.

Table 47-4 Tempering colors and approximate temperatures for carbon steel

Color	Temperature		Use
	°F	°C	
Pale yellow	430	220	Lathe tools, planer tools
Light straw	445	230	Milling cutters, drills, reamers
Dark straw	475	245	Taps and dies
Brown	490	255	Scissors, shear blades
Brownish-purple	510	265	Axes and wood chisels
Purple	525	275	Cold chisels, center punches
Bright blue	565	295	Screwdrivers, wrenches
Dark blue	600	315	Woodsaws

(A) BEFORE CASEHARDENING

● **ANNEALING**

Annealing is the process of relieving internal strains and softening steel by heating it above its critical temperature (see Fig. 47-6), and allowing it to cool slowly in a closed furnace, or in ashes, lime, or asbestos.

● **CASEHARDENING**

Casehardening is a method used to harden the outer surface of low-carbon steel while leaving the center or core soft and ductile, Fig. 47-13. Because carbon is the hardening agent, some method must be used to increase the carbon content of low-carbon steel before it can be hardened. Casehardening involves heating the metal to its critical temperature while it is in contact with some carbonaceous material. Three common methods used for casehardening are the pack method, the liquid bath method, and the gas method.

The *pack method,* also called *carburizing,* consists of packing the steel in a closed box with a carbonaceous material and heating it to a temperature of 1650 to 1700°F (900 to 927°C) for a period of 4 to 6 hours. The steel may then be removed from the box and quickly quenched in water or brine. To avoid excessive warping, it is

SOFT CORE

HARD CORE

(B) AFTER CASEHARDENING

Fig. 47-13 The grain structure of low-carbon steel. (Courtesy of Kelmar Associates)

sometimes better to allow the box to cool, remove the steel, reheat to 1400 to 1500°F (760 to 815°C), and then quench.

In school shops, *Kasenit,* a non-poisonous coke compound, is often used for casehardening (pack method).

1. Heat the piece of steel to about 1450 to 1500°F (790 to 815°C), which will produce a bright cherry red color.

2. Remove from furnace, and cover the work with powdered Kasenit.

3. Replace metal in furnace and leave it there until the Kasenit appears to boil; then quench it in cold water. This will only give from .010 to .015 in. (0.25 to 0.4 mm) penetration.

In the *liquid bath method,* the steel to be casehardened is immersed in a liquid cyanide bath containing up to 25% sodium cyanide. Potassium cyanide may also be used, but its fumes are dangerous. The temperature is held at 1550°F (845°C) for 15 minutes to 1 hour, depending on the depth of case required. At this temperature the steel will absorb both carbon and nitrogen from the cyanide. The steel is then quickly quenched in water or brine.

 Never inhale the poisonous cyanide fumes or allow any water to come in contact with the cyanide; explosive spattering may occur.

With the *gas method,* carburizing gases are used to caseharden low-carbon steel. The steel is placed in a furnace and sealed. Carburizing gas is introduced and the furnace is held between 1650 to 1700°F (900 and 927°C). After a predetermined time, the carburizing gas is shut off and the steel allowed to cool. The steel is then reheated to between 1400 to 1500°F (760 to 815°C) and quickly quenched in water or brine.

● NORMALIZING

Normalizing is performed on metal to remove internal stresses and strains, and to improve machinability.

Procedure

1. Set the pyrometer approximately 30°F (16.6°C) above the upper critical temperature of the metal (as shown in Fig. 47-6 on page 515) and start the furnace.

2. Place the part in the furnace and after the required temperature has been reached, allow the part to remain at that temperature one hour for each inch (25.4 mm) of thickness.

3. Remove the part from the furnace and allow it to cool slowly in still air.

Thin workpieces may cool too rapidly and may harden if normalized in air. It may be necessary to pack them in lime to slow the cooling rate.

● SPHEROIDIZING

Spheroidizing is a process of heating metal for an extended period to just below the lower critical temperature. This process produces a special grain structure where the cementite particles become spherical in shape. Spheroidizing is generally done on high-carbon steel to improve machinability.

Procedure

1. Set the pyrometer approximately 30°F (16.6°C) below the lower critical temperature of the metal (as shown in Fig. 47-9 on page 517) and start the furnace.

2. Place the part in the furnace and allow it to soak for several hours at this temperature.

3. Shut down the furnace and let the part cool slowly to about 1000°F (537.7°C).

4. Remove the part form the furnace and cool it in still air.

● NITRIDING

Nitriding is used on certain alloy steels to provide maximum hardness. Most carbon alloy steels can be hardened to only about 62 Rockwell C by conventional means; however, readings of 70 Rockwell C may be obtained on some vanadium and chromium alloy steels using a nitriding process.

Gas Nitriding

The parts to be nitrided are placed in an airtight drum, which is heated to a temperature of 900 to 1150°F (482.2 to 621.1°C). Ammonia gas is circulated through the chamber and decomposes into nitrogen and hydrogen. The nitrogen penetrates the outer surface of the workpiece and combines with the alloying elements to form hard nitrides. Gas nitriding is a slow process requiring approximately 48 hours to obtain a case depth of .020 in. (0.5 mm). Because of the low operating temperatures used in this process, and since no quenching of the part is required, there is little or no distortion.

Salt Bath Nitriding

Nitriding may also be carried out in a salt bath containing nitriding salts. The hardened part is suspended in the molten nitriding salt, which is held at a temperature from 900 to 1100°F (482.2 to 593.3°C) depending on the application. Parts such as high-speed taps, drills, and reamers are nitrided to increase surface hardness, improving wear resistance.

● INDUCTION HARDENING

In **induction hardening,** the part is surrounded by a coil through which a high-frequency electric current is passed. The current heats the surface of the steel to above the critical temperature in a few seconds. An automatic spray of water, oil, or compressed air is used to quench and harden the part, which is held in the same position as for heating. Since only the surface of the metal is heated, the hardness is localized at the surface. The depth of hardness is determined by the frequency of the electrical current and the length of the heating cycle.

Induction hardening may be used for the selective hardening of gear teeth, splines, crank shafts, camshafts, and connecting rods.

● FLAME HARDENING

Flame hardening is used extensively to harden ways on lathes and other machine tools, as well as gear teeth, splines, crankshafts, etc.

The surface of the metal is heated very rapidly to above the critical temperature by an oxy-acetylene torch and is hardened quickly by a quenching spray. Large surfaces, such as lathe ways, are heated by a special torch that is moved automatically along the surface, followed by a quenching spray. Smaller parts are placed under the flame, and spray quenched automatically.

Flame-hardened parts are generally tempered by a special low-temperature torch which follows the quenching nozzle as it moves along the work. The depth of flame hardening varies from 1/16 to 1/4 in. (1.58 to 6.35 mm), depending on the speed at which the surface is brought up to the critical temperature.

● TESTING STEEL FOR HARDNESS

Hardness in steel may be defined as the property it has to resist penetration and deforma-

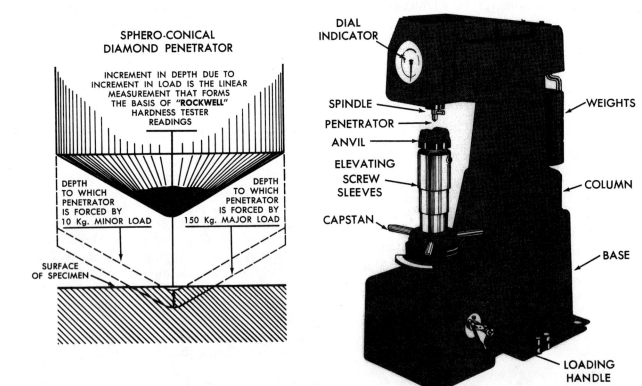

Fig. 47-14 The hardness tester and the principle on which it operates. (Courtesy of Delta International Machinery Corp.)

tion. The harder the steel, the greater its resistance to penetration and deformation. A common method of testing the hardness of steel is by trial and error, using a file, often called the File Test. Modern science has introduced many up-to-date methods that will test the hardness of metals very accurately. Some of these hardness testers are:

ROCKWELL—The Rockwell hardness tester measures the amount of penetration caused by a diamond point being forced into the metal, Fig. 47-14. The greater the penetration, the softer the metal. The penetration is recorded on a visible dial, and the resultant figure is called the Rockwell hardness number.

BRINELL—This tester uses a 10 mm hardened steel ball that is forced into the work surface. The diameter of the impression is measured by a special microscope. The higher the Brinell number, the harder the metal.

SHORE SCLEROSCOPE—The test is done with a diamond-tipped hammer that is dropped on the metal from a given height. The amount of rebound indicates the degree of hardness.

SUMMARY

- Heat treatment consists of heating and cooling steels according to specified procedures to improve their physical properties.
- Heat treatment changes the microstructure of steel to produce desired properties.
- Heat treatment operations include annealing, hardening, normalizing, spheroidizing, and tempering.
- Heat treatment is used to develop hardness, softness, toughness, machinability, and wear resistance.
- Metal hardness can be measured using a Rockwell tester, Brinell tester, or Shore scleroscope.

KNOWLEDGE REVIEW

1. Define the term heat treating.
2. Name three qualities obtained through heat treating.
3. List three heat treating operations.

Types of Steels

4. What are the common elements in all steels?
5. State the carbon content for:
   a. low-carbon steel
   b. high-carbon steel
6. Name five qualities that alloying elements provide in steels.
7. State the purpose of:
   a. high-speed steels
   b. shock-resistant steels

Heat Treatment Terms

8. Define the following metallurgy terms.
   a. austenite
   b. cementite
   c. martensite
9. Define the following heat-treating terms.
   a. decalescence point
   b critical temperature
   c. normalizing
   d. eutectoid steel

Types of Furnaces

10. Name the three common types of heat treating furnaces, and state where each is used.
11. Describe a thermocouple and explain its use.
12. If a pyrometer is not available, what other method can be used to check the temperature of the workpiece?

Safety Precautions

13. State five precautions to observe when heat treating steel.

Metcalf's Experiment

14. What knowledge can be obtained about heat treating temperatures from Metcalf's experiment?

Hardening and Tempering Carbon Steel

15. Define hardening.
16. What change occurs to steel when it is hardened?

17. Define martensite.

18. How should long slender pieces be quenched? Explain why.

19. Why is it necessary to temper steel after hardening?

Annealing

20. Define annealing.

Casehardening

21. What type of steel must be casehardened? Explain why.

22. Describe a piece of casehardened steel.

23. Name three methods of casehardening.

24. What is the purpose of the following heat-treating processes?

    a. normalizing

    b. nitriding

    c. induction hardening

Testing Steel for Hardness

25. Name three methods of testing steel for hardness.

26. How does a Rockwell tester operate?

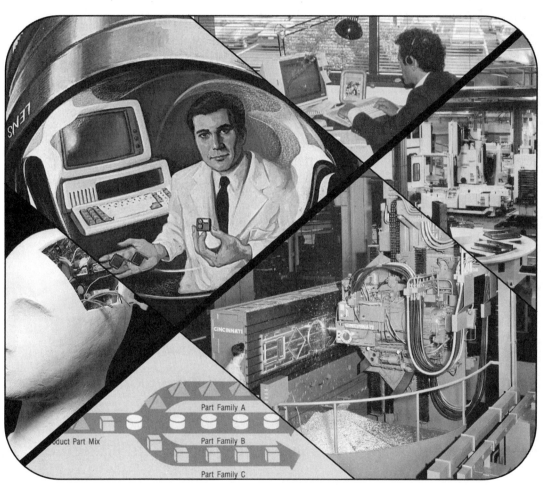

Part Family A

Product Part Mix

Part Family B

Part Family C

CINCINNATI

LENS

Manufacturing Technologies

The development of new technologies, mainly due to the increasing use of computers in the machine tool trade, has created a wide variety of activities for those who wish to explore the exciting and rewarding field of manufacturing. The rapidly changing technology in manufacturing requires that a student or tradesperson be versatile enough to learn about new technology as it is introduced, to be able to gain the advantages each offers. The importance of keeping up with technological developments was a lesson we painfully learned over the past thirty years with the loss of world markets and a reduction of the workforce. Education and training in manufacturing related courses are the key to the survival of any manufacturing nation.

Figures courtesy of (clockwise from far left): Manufacturing Engineering Magazine, Manufacturing Engineering Magazine, *Autodesk, Inc., Cincinnati Milacron Inc., Cincinnati Milacron Inc.*, Modern Machine Shop Magazine.

UNIT
48

Artificial Intelligence

There has always been a great challenge to make machines intelligent so that they could think, reason, and react as humans, Fig. 48-1. Over the years, this has evolved into the field of a computer science called **Artificial Intelligence,** which is a term applied to machine functions that normally require some human understanding. Artificial Intelligence, primarily concerned with symbolic reasoning and problem solving, is a new tool for manufacturing which is having a great effect in the areas of artificial vision, expert systems, robotics, natural language understanding, and voice recognition. It is also finding applications in product design, diagnostics, inspection, planning, and scheduling.

OBJECTIVES

After completing this unit, you should be able to:
- List the basic elements of an artificial-intelligence system.
- Describe the primary industrial applications of artificial intelligence.
- Explain the uses of expert systems in manufacturing.
- Describe how vision systems function and their uses in manufacturing.

KEY TERMS

artificial intelligence   expert system   knowledge-base building blocks
natural language   robotics   vision system

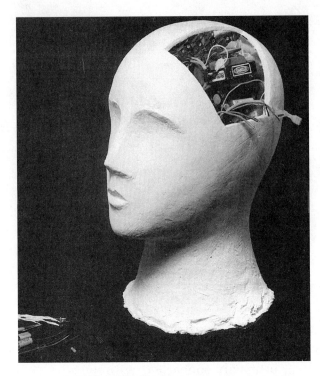

Fig. 48-1 Artificial Intelligence systems are trying to come as close as possible to the capabilities of the human brain. (Courtesy of *Manufacturing Engineering Magazine*)

LEADERSHIP AND MOTIVATION

Education and training of the workforce and those entering the workforce in the new manufacturing technologies are very important to a strong manufacturing nation and the country's standard of living. People in a leadership position, such as a teacher or supervisor in industry, will find it necessary to be a force, stimulus, and motivating influence to challenge students to the exciting field of manufacturing. In order to create an interest in manufacturing careers, instructors must interest, excite, inspire, and challenge students to expand and stretch their minds. This may involve some strain and some pain to make gains, however, students must be aware that only through life-long learning will it be possible to keep up with technological advances. The key to survival of students or workers in industry will be their versatility and willingness to adapt to changes in job descriptions, the skills and knowledge required, and the ability to cope with changes in general.

Some of the most progressive manufacturing countries in the world recognize the link between education and industry, as well as the effects on the country's economy and its standard of living. In these countries, it is not unusual for the brightest students to choose a career in manufacturing. Unfortunately, in the Western World, the importance of a strong link between the school and industry is not very well understood.

BASIC ELEMENTS

Artificial Intelligence (AI) requires:

- a large database on a subject, gathered by taking the knowledge of a number of experts in the field,
- new approaches to computer hardware and software,
- special programming languages to handle symbols rather than numbers,
- data and parallel processing capabilities.

The knowledge base of an expert system consists of a large body of information that has been assembled, widely shared, and accepted by experts in the field, and a diagnostic computer software package that the operator uses when seeking an answer to a specific problem. For example, assume that a manufacturer machining titanium discovers tiny stress cracks, which if ignored, can lead to the failure of the product. Is this a new problem that has arisen when machining this type of titanium, or have others had the same experience before? Computer software uses stored knowledge from leading titanium engineers and metallurgists to guide the manufacturer through a series of questions about stress appearances and location. Questions, responses, and results from others machining titanium have also been added to give the database a wider, more practical approach.

The software program examines a number of conditions that range from very small stress marks to tears in the metal. It can provide a number of digitally scanned metallurgical photographs of unusual stress cracks, as well as display photographs and schematics of the microstructure. From the answers, the knowledge-based system diagnoses the probable cause. Each problem is stored in a library of case studies to allow others access to answers and information when faced with a similar problem in the future.

While artificial intelligence is composed of knowledge-based fixed rules, it also deals with continuously changing production problems by using fuzzy logic. This is a method by which the fixed rules of artificial intelligence may be ignored in part, while giving preference to other rules, utilizing fuzzy sets to deal with the uncertainty.

AI systems do not operate as conventional computer systems, but use association, reasoning, and decision-making processes similar to how the brain would solve the problem. Because this system has the ability to handle data with some level of meaning and understanding, it can be used as an intelligent problem solver.

● AI USES

AI can perform many human tasks, but *as yet cannot create.* It is being used in the four primary industrial applications shown in Fig. 48-2.

1. EXPERT SYSTEMS—These provide decision-making capabilities for specific problems that require expert knowledge.

2. NATURAL LANGUAGE—This generally refers to whether a machine can understand and interpret a **natural language.**

3. ARTIFICIAL SENSES—These are special systems which simulate one or more of the human senses such as sight, hearing, touch, and smell.

4. ROBOTICS—The term **robotics** relates to a robot that uses adaptive control, sensing, and learning capabilities to perform physical tasks.

● EXPERT SYSTEMS

An **expert system** is a form of AI that uses a knowledge base to solve problems formerly requiring some form of human expertise. This system must obtain the knowledge from experts and store it in the form of rules under knowledge

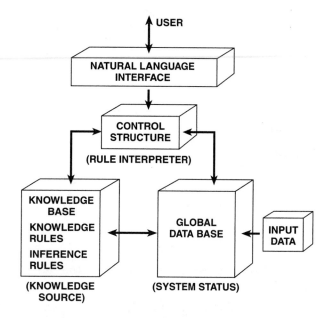

Fig. 48-3 The basic elements of an expert system. (Courtesy of *Manufacturing Engineering Magazine*)

base, Fig. 48-3. Expert systems use **knowledge-base building blocks** containing a number of advanced techniques, Fig. 48-4.

1. LOGICAL PROCESS—This consists of a number of logical (thinking) tools which perform problem-solving tasks such as

   • symbolic programming: uses symbols which represent objects and relationships.

   • proportional or predicate (base) calculus: a form of logic used to arrive at conclusions.

   • search methods: seeks solution from a number of alternate sources or approaches.

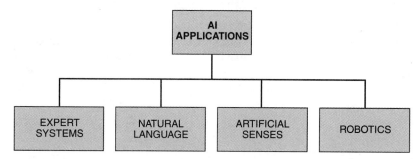

Fig. 48-2 The primary application of AI in manufacturing. (Courtesy of Kelmar Associates)

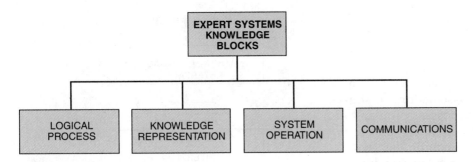

Fig. 48-4 The knowledge blocks contained in an expert system. (Courtesy of Kelmar Associates)

- heuristics (discovery): uses rules coming from past experiences to guide the thinking process.

2. KNOWLEDGE REPRESENTATION—Knowledge is stored in forms that include constraints, assertions, rules, and certainty factors. This information is used in the problem-solving process by eliminating alternatives and associating facts.

3. SYSTEM OPERATION—Expert systems organize and control their problem-solving activities. They will offer intermediate results and explanations of their conclusions so that the user can interact and change the reasoning path, if necessary.

4. COMMUNICATIONS—Expert systems must have the hardware and software to be able to communicate efficiently with the user, experts, knowledge engineers, databases, and other computers.

● AI MANUFACTURING APPLICATIONS

AI knowledge-based expert systems were initially used in the fields of medicine and science, but are now finding wide use in manufacturing. There are generally two types of systems used, one to solve complex problems usually performed by humans and the other to perform simpler tasks. Expert systems have been successfully used in the following manufacturing applications:

1. EQUIPMENT DIAGNOSING AND MAINTENANCE—Expert systems can simplify the job of troubleshooting complex equipment by referring to the knowledge of equipment designers and the experience of expert maintenance personnel.

2. ORDERING AND SCHEDULING—Expert systems use the knowledge of bills of materials and ordering rules to accurately schedule the production of product lines. They can also be used to track the product through the production line.

3. MODELING AND SIMULATION—Potential problems in layout and scheduling of production lines can be avoided by modeling the line and simulating the flow of product through the manufacturing process. It will show potential problem areas and allow changes in design without actually running the production lines.

4. SOFTWARE DEVELOPMENT—Developing software for manufacturing applications is very time consuming and expensive. Expert systems use automated programming techniques to reduce and simplify this type of programming, resulting in greater productivity.

5. COMPUTER-AIDED DESIGN (CAD)—In order to minimize the time and cost and to avoid errors in the design of complex products, expert systems have proven to be very valuable. They allow operators to interact with the design process, make design changes, and try out alternatives, Fig. 48-5.

6. TEST—As the complexity of a product increases, so does the testing operation to ensure that the product will perform satisfactorily. Expert systems use a knowledge-base of design and test information to reduce the time and effort required to test a product or system.

7. COMPUTER-AIDED INSTRUCTION (CAI)—Expert systems can provide a large base of education programs because they have a

Fig. 48-5 Complex parts can be quickly designed, tested, and altered on a CAD system. (Courtesy of Reprinted with the permission from and under the copyright of © 1995 Autodesk, Inc.)

broader knowledge base and have the ability to simulate and practice learning situations.

8. PROCESS CONTROL—Expert systems can assist skilled operators to monitor manufacturing operations and equipment to quickly find changes in the process or potential problems. The intelligent control system can store thousands of previous process measurements and use its knowledge base of rules and its expert operator experience to advise human operators.

● VISION SYSTEMS

AI technology combined with computers, software, television cameras, and other optical sensors, allows machines to perform jobs usually done by humans, Fig. 48-6. **Vision systems** are able to analyze and interpret 2D (two-dimensional) camera images into 3D (three-dimensional) objects by:

1. IMAGE FORMATION—The camera image is broken down electronically into very small picture elements or pixels. Each pixel is graded and coded by its light intensity to form an electronic image in the computer memory.

2. IMAGE ANALYSIS—The stored image is then described by the computer in terms that can be related and compared to known objects. One method is by *edge detection,* which divides the image into regions and uses lines to define the outline of an object.

Fig. 48-6 Vision systems combine computers, software, and television cameras to perform jobs usually done by humans. (Courtesy of *Manufacturing Engineering Magazine*)

3. IMAGE INTERPRETATION—The object, described in outline form in the computer's memory, is compared to other models in the knowledge base. The object can be recognized by its form in comparison to a template image, or by its dimensions such as area, length, etc.

Types of Systems

The use of AI technology has provided industry with a valuable manufacturing tool. Combined with AI computers, software, television cameras, and other optical sensors, it allows machines to perform many tasks normally done by humans. One of these tasks is that of gaging (measurement) and inspection of manufactured parts. Dimensional gaging systems generally fall into four categories.

1. TRIANGULATION—Used extensively on some types of coordinate measuring machines, triangulation-type optical sensors, Fig. 48-7, are used to accurately measure the distance between the probe and the surface of the part. With the part location and probe location known, a signal from the optical sensor completes the three-point dimensioning. A spot on the surface to be measured is illuminated by a controlled light

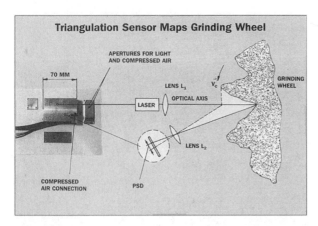

Fig. 48-7 A non-contact optical sensor uses triangulation to measure the grinding wheel's cutting life. (Courtesy of *Manufacturing Engineering Magazine*)

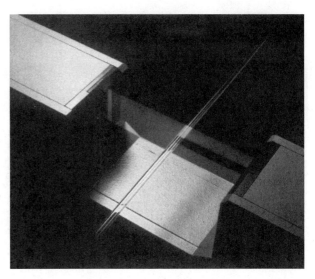

Fig. 48-8 A scanning laser beam system being used to measure the diameter of a glass rod. (Courtesy of *Manufacturing Engineering Magazine*)

source, and some of the diffused light is picked up by a detector. Any change in the location of the reflected light is quickly noted by the sensors to identify the new position of the surface.

2. SCANNING LASERS—The laser (light amplification by stimulated emission of radiation) is another technology that is used in production optical gaging. A low-powered laser is formed into a thin beam of light that is sent through a rotating mirror to scan a linear scale. The linear scale, called a holographic scale, uses a laser generated grating rather than a conventional ruled glass scale, and scans this area at a constant speed. As the part being measured interrupts this beam of light, the ultra fine graduations are picked up by a laser reader. This system can produce results with a resolution of 1/2 millionth of an inch (0.01 micron) throughout its measurement range. Current applications include non-impact printing, laser radar, electronic mail, barcode reading, optical character recognition, robotic vision, and quality control.

3. TV CAMERAS—A TV camera is one of the most sensitive visual systems ever invented. Not only can a high resolution black-and-white camera intensify any close-up, but it also digitizes the object at the same time. This non-contact vision system allows the digitized video signal to be processed by a computer to electronically produce measurement. Enormous amounts

of information are rapidly produced in its field of view, and this process is fast, accurate, and very reliable.

4. LINEAR ARRAY SYSTEMS—A linear array image sensor is used for non-contact measurement and inspection. The sensor system consists of a light source, that sends out a parallel light beam from one side of the object to be measured, and a sensing head located on the opposite side, Fig. 48-8. The system has no moving parts but contains a compact controller that can be programmed for counting, positioning, gaging, sorting, orienting, and configuring.

As an example, the diameter of an object is measured by the controller counting the number of elements of the array that are blocked off. Large diameters can be measured by using the lights of two linear arrays, one placed on each side of the part. The speed of the system is limited only by the rate of reading as the part is electronically scanned and processed by a minicomputer.

● ARTIFICIAL INTELLIGENCE IN THE FUTURE

The applications of AI in industry are only in the beginning stages. AI research will continue toward making computer-based systems more powerful, easier to use, and less expensive. Future AI software and systems will produce machines that can adapt to a changing environment and learn from their experiences. The future areas of growth will be in deep-knowledge systems that will be able to reason, learning subsystems that will learn from their own experiences, conceptual networks that can make decisions on the meaning of words, and voice recognition.

SUMMARY

- The basic requirements of an AI system are a large data base, computer hardware and software, special programming language, and data and parallel processing capabilities.

- Artificial Intelligence is used for four primary industrial applications: expert systems, natural language, artificial senses, and robotics.

- Expert systems use a knowledge base of building blocks to solve problems formerly done by humans.

- The knowledge-base building blocks contain advanced techniques such as logic process, knowledge representation, system operation, and communications.

- Vision systems, using television cameras, computers, software, and other optical senses, are able to analyze and interpret 2D camera images into 3D objects.

KNOWLEDGE REVIEW

1. Define the term Artificial Intelligence.

AI Uses

2. Name the four primary industrial applications of AI.

3. List six manufacturing applications of expert systems.

AI Manufacturing Applications

4. What are some benefits of using an expert system for modeling and simulation?

5. Why is CAD valuable?

Vision Systems

6. What type of system can interpret 2-D camera images into 3-D? Why is this important?

Types of Systems

7. List the four categories of dimensional gaging.

8. For what purpose is a linear array system used?

UNIT 49

Computer Manufacturing Technologies

The introduction of computers into manufacturing has created a revolution in manufacturing technology and processes. As more powerful computers became available, they found their way into numerous manufacturing applications such as part design and analysis, production scheduling and control, inventory and cost accounting, process control, optimization, and machinability data analysis. Computer networks have brought the computer to design engineers, process planners, production control personnel, and into the manufacturing shop to machine operators. Every phase of manufacturing has been affected, and it seems as though more and more manufacturing processes will be assisted by the computer in the future, Fig. 49-1.

The earliest uses of computers in manufacturing were in the general accounting area (job control, inventory control, shipping/receiving, payroll, etc.). Engineering applications came later in the general areas of:

- design and drafting (computer-aided design or CAD),
- manufacturing planning and control (computer-assisted manufacturing or CAM) and
- computer-integrated manufacturing (CIM), a high-technology approach to more efficient manufacturing.

Each area will be covered to provide basic information about how each fits into the manufacturing process.

OBJECTIVES

After completing this unit, you should be able to:

- Describe the components and uses of computer-aided design (CAD).
- List the advantages of computer-assisted manufacturing (CAM).
- Explain how computer-integrated manufacturing (CIM) improves machine-shop productivity.

KEY TERMS

computer-aided design
computer-aided manufacturing
computer-integrated manufacturing
digitizing

● COMPUTER-AIDED DESIGN (CAD)

Computer-aided design (CAD) systems are the computerized version of conventional design tools such as pencils, erasers, compasses, drafting machines, French curves, and even calculators. CAD allows the operator to produce and modify drawings with a light pen or digitizer faster and more accurately than by traditional drafting. The process allows complex designs to be produced five to twenty times faster than was formerly done by hand, Fig. 49-2. A typical system consists of a computer or central processing unit (CPU), memory for design-data storage, a hard-copy unit for checking revised drawings, a plotter for producing drawings, and an interactive terminal, Fig. 49-3. The principal function of a CAD system is to produce engineering drawings; however, the system is also able to assist in all phases of the production-design cycle. CAD systems are used in the fields of mechanics, architecture, medicine, and engineering. Civil engineers use the CAD system to develop roads, bridges, and highways. Architects are using CAD systems to design homes, buildings, and sports arenas. Medical doctors use computer art as aids in facial reconstruction and developing body joints for bone implants. Designers of mechanical structures and components use the CAD system

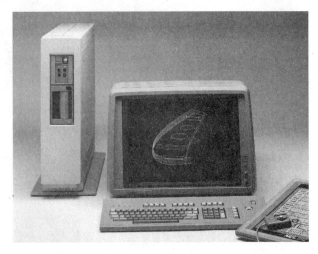

Figure 49-2 CAD being used to design a complex part. (Courtesy of Numerical Control Computer Sciences)

Figure 49-3 A complete CAD system can be used to design complex parts and the printer produces the engineering drawings. (Courtesy of Ingersoll Cutting Tool Co.)

extensively to create a wide variety of products. There are many CAD software programs available to suit specific engineering and manufacturing applications. It is important that the program loaded into the computer suits the purpose for which it will be used.

CAD Operating Components

The main operating components of a CAD system are the interactive graphics terminal, digitizer, and computer.

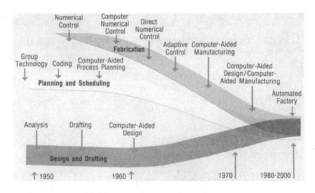

Fig. 49-1 The effect computers have on manufacturing technology. (Courtesy of *Modern Machine Shop Magazine*, copyright 1994, Gardner Publications Inc.)

- INTERACTIVE GRAPHICS TERMINAL—The operator supplies information about the design of a product to the interactive graphics terminal by means of a cursor on the CRT (cathode-ray tube) screen with a light pen, joy stick, or cursor control arm, and a keyboard for entering text and information. Some terminals are equipped with a digitizer tablet to convert line drawings and other graphics into digital data for analysis, storage, or computation.

  The operator usually describes or modifies the design of a product on the CRT screen with a keyboard and cursor. With light-pen systems, many drafting symbols, functions, lines, arcs, text, cross-hatching, and dimensions are displayed on the *menu,* Fig. 49-4. When a symbol or command is touched with the light pen or mouse, this activates the corresponding information in the computer memory and displays it on the CRT. Any design change made on one view is automatically added to other views. Images on the CRT screen can be rotated, inverted, enlarged, examined in close-ups by zooming in on a specific area, or displayed from different views. Most systems today can display three-dimensional objects in motion to show the relationships between moving parts.

- DIGITIZER—Terminals equipped for **digitizing** usually have a sensitized table for the drawing to be digitized. Tracing the drawing with a cursor or light pen displays the drawing on the CRT screen, and also enters the digitized drawing into the computer memory.

- COMPUTER—The computer, while controlling various system components, performs many valuable functions in a CAD system.

  - It provides a database from which information can be recalled and used for various needs.

  - It can calculate volume, mass, center of gravity, and moment of inertia in relation to the principal axes.

  - It can perform stress analysis and generate bills of material for the component being designed.

  - It can provide English/metric conversions.

  - It can produce a CNC program for machining the part.

  - It can communicate, over telephone lines, with other systems that may wish to access the same information.

● COMPUTER-ASSISTED MANUFACTURING (CAM)

The uses of computers in production are in manufacturing planning (the process) and manufacturing control (production) areas. CAM is a natural followup to CAD where the product has been designed. Since this is a logical progression, CAD/CAM is often thought of as one continuous process, that of designing and producing a part with the help of computers. However, CAD covers all the functions involved in getting ready to produce a product, while CAM is applied to all the functions actually involved in producing the finished product.

There are three steps to producing a product,

- first, coming up with an idea (conceptualization),

- second, designing the product (CAD),

- third, producing the product (CAM), Fig. 49-5.

Computer-assisted manufacturing (CAM) generates computer data for the machining/manufacturing processes from the CAD program. The availability of low-cost desk top CAM systems makes it possible for small shops to develop a new product quickly, reducing the time and cost to bring the product to the market. Where companies

Figure 49-4 The icons on the menu tablet can be selected and transferred to the computer drawing simply by pressing the button on the cursor. (Courtesy of Reprinted with the permission from and under the copyright of © 1995 Autodesk, Inc.)

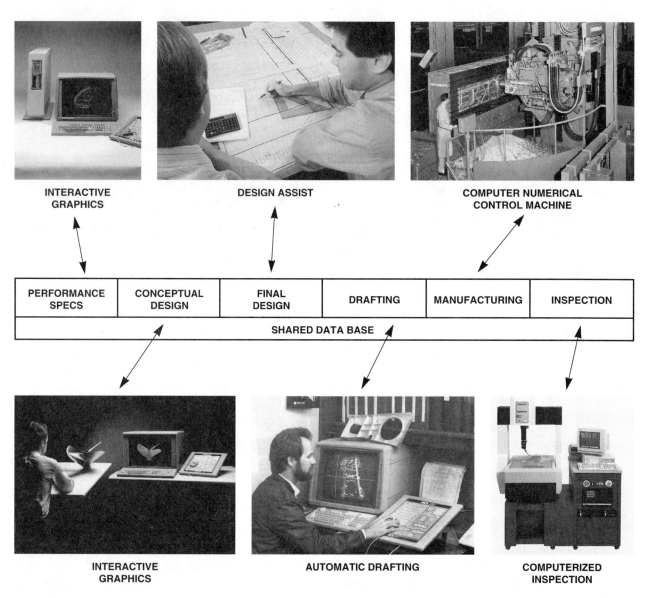

Figure 49-5 A fully integrated CAD/CAM system. (Courtesy of Cincinnati Milacron Inc., Giddings & Lewis, Inc., Kelmar Associates, Manufacturing Engineering, Numerical Control Computer Sciences)

once had to invest large sums of money to develop new products physically, CAM systems now quickly convert CAD drawings into finished parts.

Normally, it required a team of tool engineers, design engineers, machine shop modelmakers, and diemakers to produce a product. Much time and effort was spent to debug and refine the product to produce an efficient manufacturing operation. CAM systems allow engineers to recognize and remedy problems early in the manufacturing process. By collecting, analyzing, transmitting and storing large amounts of data, CAM systems bring great efficiency and effectiveness to manufactur-

ing. They organize and direct manufacturing engineers and production workers by automating processing plans and scheduling production resources. The CAM system can also generate a CNC machine tool program and the cutting tool path from the CAD data.

CAD/CAM Components

The most important component of a CAD/CAM system is its database. The database includes the drawings and alpha-numeric information such as a bill of materials, results of analysis routines, and text placed on drawings. All designs begin with a concept (idea) of a three-dimensional part that is to be produced. Usually, engineering drawings (2-D representations) had to be re-interpreted and produced as a three-dimensional part. A CAD/CAM system can use the three-dimensional part description created during the design and drafting processes, Fig. 49-6. Therefore, all design and manufacturing analysis can be done using the stored information, saving valuable time and reducing the chances for errors.

At one time, NC programmers had to be able to use trigonometry and mathematical calculations to develop the cutting tool path (the steps a cutting tool takes to perform an operation, or cut a part to size). CAM software can use the information from the CAD design to outline the part geometry and produce a CNC cutting tool program for the machine to follow, Fig. 49-7. This eliminates much of the time spent with paper drawings and time-consuming mathematical calculations.

CAM systems offer other advantages such as machining routines like hole drilling and pocket machining. For example, if it is required to machine a pocket, rather than defining every move, all that has to be done is to define the size of the pocket and the tool that will do the machining. Every move required to machine the pocket is created automatically by the CAM system. All types of canned routines or macros are available that make it easy to perform common machining operations quickly and accurately.

Generally, a CAM setup is located in one area of a manufacturing shop and consists of a workstation or a PC, a printer or plotter, as well

Figure 49-7 A complex part being machined from information supplied by the CAD system. (Courtesy of Cincinnati Milacron Inc.)

as CAD and CAM software, or both. With the manufacturing engineer specifying which cutting tool to use, the CAM software is used to define the cutting tool's path by generating a machine-code file that will drive the machine tool. It is important that any CAM package (software) used is compatible with the CNC machine tools it will be used on. The CAM-generated machine codes are carried to the machine tool by means of a floppy disk, or directly from the computer. CAM can also interface with a Coordinate Measuring Machine (CMM) to allow sample parts to be digitized into a CAM system.

● CAM ADVANTAGES

The main reason for implementing computer-assisted manufacturing may be the need to be competitive with other industries. It seems that the saying today is "automate, emigrate, or evaporate." The industries who have taken advantage of the benefits of the new technologies have been able to improve their productivity and remain competitive, while those resisting change have gone out of business.

Some of the main benefits of implementing CAM are:

- lower manufacturing costs,
- increased productivity,

Figure 49-6 The design and specifications of a part are included in the CAD design. (Courtesy of Hewlett-Packard Co.)

- improved product quality,
- lower inventory levels,
- shorter manufacturing cycle times,
- reduced space requirements,
- shared use of design and manufacturing data,
- faster introduction of new products,
- consistency and improvement in data collection,
- flexibility in design and manufacturing operations.

● COMPUTER INTEGRATED MANUFACTURING (CIM)

Technology is the most important factor that produces quality products and increases productivity. Productivity may be defined as the ratio of labor, capital, raw material, energy, and time (input) to usable product (output). What numerical control (NC) did to increase the productivity of the individual machine tool, **computer-integrated manufacturing** (CIM) can do for the entire shop. The purpose of CIM is to link automated manufacturing functions such as computer numerical control (CNC), process planning, material planning, and CAD/CAM, and integrate these with data processing functions such as accounting, inventory control, and payroll. CIM must be thought of as an information system that controls the flow of data throughout the engineering and manufacturing stages, and also involves the areas of marketing and finance, Fig. 49-8. CIM promises

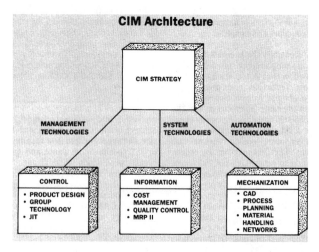

Figure 49-8 The CIM architecture must include management, system, and automation technologies. (Courtesy of *Manufacturing Engineering Magazine*)

to be one of the most advanced, far-reaching, and surest method of improving plant and corporate productivity.

The System

As more and more sectors of a manufacturing plan become integrated, CIM continues to offer more benefits at each step of the operation. The key to all manufacturing operations is the creation, storage, analysis transmission, modification, and the processing of data. CIM should be thought of as an information concept where data can be used throughout a company so that all aspects of engineering and manufacturing can use timely information to help them do a better job. Some of the major benefits of using the same database are:

- INFORMATION USAGE—The centralized CIM database allows product information to be reused, rather than creating it again for another machine tool or process. This lowers manufacturing costs for machine tool programming, tool and fixture engineering, material handling, and quality control.

- ELIMINATION OF ERRORS—Because the same data can be used for each separate manufacturing function, formatting errors are eliminated. This means that product information is more accurate, resulting in better control of product quality and reliability.

- IMPROVED COMMUNICATIONS—The barriers to effective communication between design manufacturing and other parts of the company are eliminated, and communication is greatly improved.

● INFORMATION EXCHANGE

A hierarchy (systematic grouping) of computers at all levels of the manufacturing operation is necessary to successfully implement CIM. Three major tiers (levels) of computers should be in this grouping.

1. MAINFRAME—The first of the group must be the large, mainframe-based data management system, which provides the on-line storage capabilities to store and use the large amounts of information required by a CIM system, Fig. 49-9. This top tier provides the capabilities to:

   - Store large amounts of manufacturing data such as drawings, three-dimensional

Figure 49-9 The mainframe computer is capable of storing large amounts of information required by a CIM system. (Courtesy of Hewlett-Packard Co.)

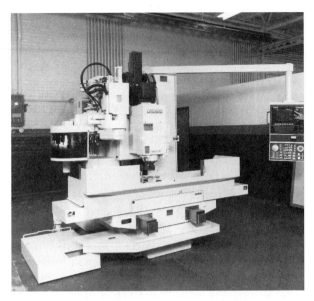

Figure 49-10 Computers on machine tools can access the central database stored in the mainframe computer. (Courtesy of Cincinnati Milacron Inc.)

part models, robot and CNC machine tool programs, process plans, and tooling and fixturing information,

- Perform tracking functions when data is released from engineering to manufacturing,
- Provide data transmission to output devices such as plotters and printers, and parts classification and process planning for group technology procedures,
- Combine alphanumeric MIS management information system databases with graphic CAD/CAM databases.

2. CAD/CAM—The middle tier is composed of the automated engineering and design technologies such as part design, analysis, and graphic programming for computer numerical control and robotic programs. Whenever there is a need to change the design of any component, the computer searches out the entire file this component may affect and makes the necessary changes in the complete file. These changes are immediately passed on to the central database, and the revised information is available to everyone immediately.

3. FLOOR LEVEL—The desktop engineering workstations provide access to the central database. This allows engineers, designers, quality control personnel, and machine operators to use and expand the central database at the same time, Fig. 49-10.

CIM ADVANTAGES

The use of common computer data produces higher quality products, shorter cycle times between design and production, efficient production of small lot numbers, and faster design changes, which in turn respond to market demands for quality, flexibility, and delivery. The following benefits have been reported by industries using CIM:

- 15 to 30% reduction in engineering design costs,
- 30 to 60% reduction in product lead time,
- 30 to 60% reduction in the work in process,
- 5 to 20% reduction in labor costs,
- 40 to 70% increase in final assembly productivity,
- 200 to 300% increase in equipment productivity.

Computers make it possible to bring all functions of manufacturing into a single, smooth operating manufacturing system. CIM occurs in manufacturing when:

1. All the processing and related management functions are expressed in the form of data.

2. This data is in a form that can be generated, transformed, used, moved, and stored.

3. The data can move freely between all parts of a system, for the life of the product, to make the entire company operate at maximum efficiency.

SUMMARY

Computer-integrated manufacturing is changing the way products are designed, manufactured, and serviced. The following are reasons why manufacturers must consider implementing CIM technology into their manufacturing operations:

- The system provides great flexibility in the range of products a company can produce. It allows faster and more customized designs to be made because of the computer analysis.

- There can be faster response time to market needs and demands, because of the ability to change the product mix, production, and the use of manufacturing equipment.

- There is better control of all phases of the manufacturing process which results in higher quality products and more efficient use of the machine tools.

- There is less scrap and waste because of better use of materials, equipment, and personnel.

- Every phase of manufacturing can be closely tracked, and information on every operation can be stored in the central database for everyone's use.

- Faster product turnaround is possible because of the best use of manufacturing equipment, using the same database, reducing inventories, and avoiding missing parts or material that cause work stoppages.

KNOWLEDGE REVIEW

CAD Operating Components

1. List the components of a computer-aided design (CAD) system.

Computer-Assisted Manufacturing

2. For what does the acronym CAM stand?

3. CAD/CAM is thought of as one continuous process; explain how they differ.

CAM Advantages

4. What are the benefits of using a CAM system to develop new products?

5. Define the meaning of cutting tool path.

Computer-Integrated Manufacturing

6. What information system controls the flow of data throughout the engineering and manufacturing stages and integrates them with the areas of marketing and finance?

7. Name three major benefits of using the same database.

Information Exchange

8. List the three major levels of computers required to successfully implement CIM.

U N I T
50

Coordinate Measuring Systems

The productivity and speed of CNC machine tools brought about a need for inspection systems that could keep up with the CNC machine tools. Accurate measurement of locations and size dimensions was usually dependent upon the skills of the inspector, using conventional measuring tools. These could not keep up with the productivity of CNC machines and were time-consuming, costly, and subject to human error.

The first coordinate measuring system, a two-axes machine that could locate accurately to .001 in. (0.025 mm), was introduced in 1959 at the International Tool Show in Paris. It was the computer that allowed the coordinate measuring machines (CMM) to be fully automated, greatly reducing errors made by human operators.

OBJECTIVES

After completing this unit, you should be able to:

- Explain the operating principles of a coordinate measuring machine (CMM).
- Recognize the two main types of coordinate measuring machines.
- List the benefits of using coordinate measuring machines in inspection and quality-control operations.

KEY TERMS

coordinate measuring machine
index grating

direct computer control
reference scale

● COORDINATE MEASURING MACHINES

A **coordinate measuring machine** (CMM) is an advanced, multi-purpose quality control system used to quickly inspect components and keep pace with the productivity of CNC machine tools, Fig. 50-1. It replaces long, complex, and inefficient conventional inspection methods with simple procedures that are up to twenty times faster, and much more accurate. A CMM can reduce or eliminate CNC machine downtime, reduce scrap or rework, and is easy to operate.

Purpose

A CMM is a versatile machine that can measure almost any dimension on a part, at any stage of production, regardless of its shape. It can check the dimensional and geometric accuracy of parts ranging from engine blocks and circuit boards, to sheet metal parts and picture tubes. Some of the typical uses are:

- inspection of the first part produced by a CNC machine,
- random sampling of parts during a production run,
- preventive maintenance checks of dies, jigs, and fixtures,
- qualifying or receiving inspection of machined parts,
- measuring parts at any stage of production, quickly and accurately,
- installation as a station on CNC machines to allow inspection of critical parts.

Operating Principles

The CMM consists of an indicating probe supported on three perpendicular (X, Y, & Z) axes. The measuring system on each axis may vary from the stainless steel reference scale (.0001 in. or 0.0025 mm accuracy) to the fiber-optic laser measuring system (.000,080 in. or 2 micrometers) for ultra-accuracy CMMs. The most common measuring system used on CMMs is the reference scale.

- Each axis has a built-in stainless steel reference scale or spar which is etched with 41-2/3 lines per millimeter, Fig. 50-2A.

- An **index grating** with the same line structure is placed at a slight angle on top of the **reference scale** so that as it moves an integrated interference or moire fringe pattern is produced, Fig. 50-2B.

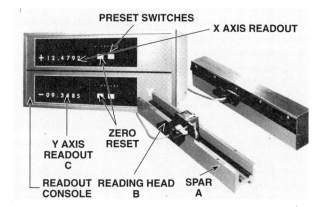

Fig. 50-2A The components of a coordinate measuring unit. (Courtesy of Giddings & Lewis—Sheffield Measurement)

Fig. 50-1 A coordinate measuring machine is used for inspection and quality control purposes. (Courtesy of Giddings & Lewis—Sheffield Measurement)

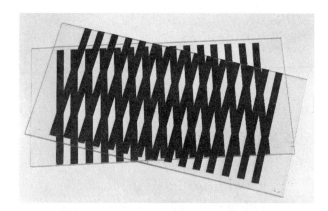

Fig. 50-2B The principle of the moire fringe pattern which produces accurate measurements. (Courtesy of Giddings & Lewis—Sheffield Measurement)

- Ultra-accuracy CMMs usually have a fiber-optic laser measuring system for greater accuracy.

- Photocells in a non-contact reading head convert this pattern into electrical signals, that are converted by a processor into a digital reading on the display.

- The digital readout instantly shows any changes in direction, providing accurate information on the probe's movement and position.

- Any point on the job can be located in relation to another point in all three axes in one check.

● TYPES OF CMMs

CMMs combine the conventional inspection methods of the surface plate, micrometer, and vernier instruments into one machine capable of high accuracy and great speed. They usually have a granite surface table, operating on air bearings, and a bridge-like structure that carries the ram holding the indicating probe. This probe can be moved vertically (up and down) in the Z axis for taking measurements. They are available in vertical and horizontal CMM systems to suit many different sizes and shapes of workpieces.

Vertical CMMs

Vertical CMMs, Fig. 50-3, are designed with a wrap-around ring bridge that surrounds the measuring area with a very rigid frame. This ensures that the machine will be stable, accurate, and able to precisely repeat measurements. The accuracy of the CMM will vary from .000,080 in. (2 micrometers) for models with a fiber-optic laser measuring system to .0001 in. (0.0025 mm) for models with reference scales. The bridge-type construction allows access from five sides to position various-sized workpieces.

These machines can be operated manually or can use standard CMM software programs for automatic operation. The main features of CMM systems are the superior accuracy, repeatability, and inspection speed.

Horizontal CMMs

Horizontal CMMs are designed for large, bulky workpieces that cannot normally be handled on vertical arm machines, Fig. 50-4. They

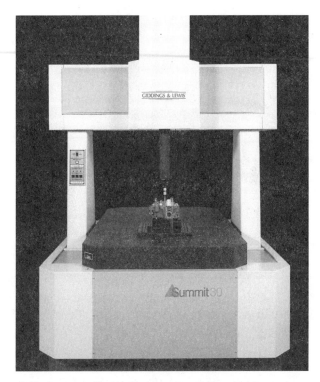

Fig. 50-3 The main features of a vertical coordinate measuring machine are superior accuracy, repeatability, and speed. (Courtesy of Giddings & Lewis, Inc.)

are especially well-suited to measure large gear cases and engine-block type parts, where high precision alignment and geometry measurements are required. A rotary table is usually supplied with horizontal machines so that five sides of a part can be inspected in one setup. This reduces the time usually spent on part handling, and greatly speeds up the inspection process. The accuracy and repeatability of horizontal CMMs is usually in the .000080 in. (2 micrometers) range for all four axes.

● CMM ADVANTAGES

Ever since CMMs were introduced, they have been able to greatly reduce inspection costs, improve inspection processes, reduce scrap and rework, and increase productivity. A CMM can help an industry control quality during a manufacturing process, not after, as is generally the case, by keeping checks on the manufacturing system. It can also supply data on factors such as tool wear, part accuracy, rework values, and manufacturing processes for each machine. The

Fig. 50-4 The horizontal coordinate measuring machine is used to inspect large, bulky workpieces. (Courtesy of Giddings & Lewis, Inc.)

Fig. 50-5 The large digital display helps to prevent measurement errors. (Courtesy of Giddings & Lewis—Sheffield Measurement)

following are some of the benefits CMMs can offer industry:

1. One-time setup—Unlike surface plate techniques where an inspector must make multiple setups to measure all axes of a part, the CMM is set up only once. This eliminates the cost of making multiple setups.

2. Reduction of errors—Even the very best inspectors are bound to make some mistakes, and it is very difficult to exactly repeat measurements every time. Once the CMM is set up or programmed, it is almost impossible to make errors.

3. Flexibility—A CMM can be used for almost any part that fits within its measuring range. It eliminates the time and expense required to build specific gages and fixtures for special parts. Because of its flexibility, it can be used in a production line which may change product frequently.

4. Digital readouts—CMM measurement readings are displayed brightly and clearly in easy-to-understand digits, Fig. 50-5. In most cases, a hard copy of these readings can be produced on a printer for permanent records, or for quality control purposes.

5. Geometric calculations—A measurement processor-equipped CMM can perform the most complex calculation. For example, by touching three points on a circle, it will immediately give you its diameter.

6. Statistical calculations—Random or systematic sampling of product helps to reduce inspection time, and also identify potential problems before they result in scrap work.

7. **Direct computer control** (DCC) capability—The complete inspection process can be automated. The computer will not only make all the necessary mathematical calculations, but also move the probe to various locations according to the pre-programmed instructions.

See Table 50-1 for a summary of the advantages of manual, measurement-processor equipped and direct computer controlled CMMs.

Table 50-1   The comparative advantages of measurement-equipped and direct computer controlled CMMs (Courtesy of Giddings & Lewiss, Inc.)

Profit and Payback Elements	CMM with Measurement Processor	CMM with Direct Computer Control
Inspection cost savings	✔	✔
Reduced manufacturing downtime and scrap	✔	✔
Increased inspection and manufacturing capacity	✔	✔
Improved accuracy	✔	✔
Minimization of operator error	✔	✔
Reduced operator skill requirements	✔	✔
Reduced inspection fixturing and maintenance costs	✔	✔
Increased versatility through reduced dedication to specific tasks (unlike go/no go gages)	✔	✔
Uniform inspection quality	✔	✔
Reduction in calculating and recording time and errors	✔	✔
Reduction in set-up time and fixturing costs through automatic compensation for misalignment	✔	✔
Provision of a permanent record for process control and traceability of compliance to specifications	✔	✔
Reduction in off-line analysis time which can often be longer than required inspection time	✔	✔
Simplified inspection procedures, especially when geometric dimensioning and tolerancing are involved	✔	✔
Possibility of reduction of total inspection time through use of statistical data analysis techniques (random or systematic sampling)	✔	✔
Even greater productivity — up to 20 *times* faster than open set-up methods		✔
Improved accuracy and reproducibility — operator influence is eliminated and low gage force improves results		✔
Operable by less skilled personnel		✔
Elimination of repeated inspections		✔

SUMMARY

- A coordinate measuring machine is an advanced multi-purpose quality control system that can be programmed to automatically inspect machined parts and keep pace with the productivity of CNC machine tools.

- The CMM measuring system consisting of stainless-steel reference scales, and a reading head that provides readings to an accuracy of .0001 in. (0.0025 mm). Fiber-optic laser measuring systems measure to within an accuracy of .000 080 in. (2 micrometers).

- CMMs reduce inspection costs, improve the inspection process, reduce scrap and rework, increase productivity, and provide industry with quality control during a manufacturing process.

KNOWLEDGE REVIEW

Coordinate Measuring Machines

1. What is a coordinate measuring machine (CMM)?

2. List four main uses of a CMM.

3. What is the most common measuring system used on CMMs?

4. What measuring system is used on ultra-accuracy CMMs?

Types of CMMs

5. Name two types of CMMs.

CMM Advantages

6. List four important advantages of a CMM.

UNIT 51

Electrical Discharge Machining (EDM)

Electrical discharge machining, commonly known as EDM, is a process that is used to remove metal by means of electric spark erosion. Machining is performed by the action of an electrical discharge of short duration and high-current density between the electrode (tool) and the workpiece, Fig. 51-1. EDM was initially used for removing broken taps, but has now become a powerful tool for producing intricate dies and molds in hardened material. There are two types of EDM machines, the ram or die- sinking EDM, and the wire-cut EDM.

OBJECTIVES

After completing this unit, you should be able to:
- Describe the operation of electrical discharge machines.
- Explain the purposes and functions of dielectric fluids.
- Recognize the differences between the ram and wire-cut types of electrical discharge machines.
- List the advantages of electrical discharge machining.

KEY TERMS

electrical discharge machining   dielectric fluid

servomechanism   spark erosion

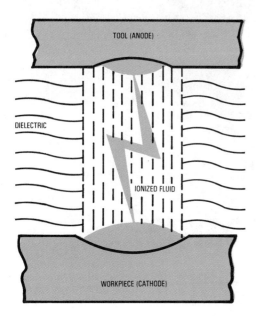

Fig. 51-1 A controlled spark discharge removes a very small particle of metal during each electrical discharge. (Courtesy of Cincinnati Milacron Inc.)

RAM OR DIE-SINKING EDM

The ram or die-sinking EDM, Fig. 51-2, has a cutting tool (electrode) shaped to the form of the cavity, mounted in the ram of the machine. The electrically conductive workpiece is fastened to the machine table below the electrode. The DC power supply produces a series of short, high-frequency electrical arc discharges between the electrode and the workpiece. This action removes (erodes) tiny particles of metal from the workpiece and as the process continues, the electrode reproduces its form in the workpiece. Fig. 51-3 illustrates a number of products that have been produced by molds or dies manufactured on EDM ram-type machines.

PRINCIPLE OF RAM OR DIE-SINKING EDM

The principle of **spark erosion** is very simple. The workpiece and the tool (electrode) are placed in a working position in such a way that

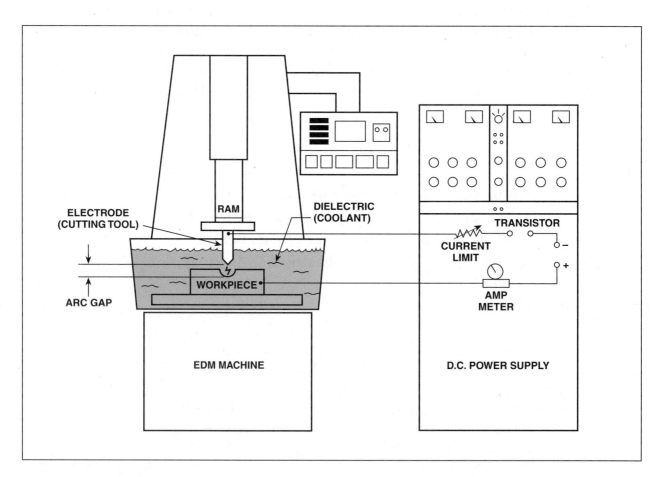

Fig. 51-2 Ram-type EDMs plunge a tool, shaped to the form of the cavity required, into a workpiece. (Courtesy of Kelmar Associates)

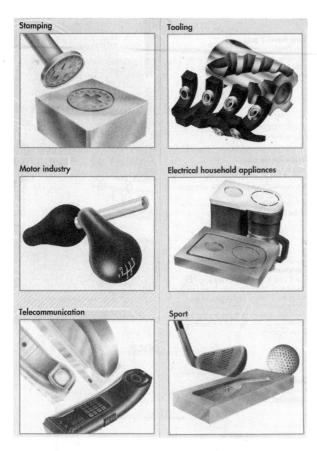

Fig. 51-3 Types of forms and cavities produced by ram EDM machines. (Courtesy of Charmilles Technologies Co.)

they do not touch each other. The cutting process takes place in a tank where the tool and work are separated by a gap that is filled with an insulating fluid (dielectric), Fig. 51-2. The workpiece and tool are both connected to a DC power supply with a cable. When the switch on one cable lead is closed, an electrical potential is applied between the tool and work. Initially, no current flows because the dielectric fluid between the tool and work is an insulator. As soon as the gap between them is reduced, a spark jumps across the gap and highly heats a very small area of the work material. The current is then automatically shut off and the discharge channel collapses very quickly. The molten metal on the surface of the material evaporates explosively, forming a small crater. The metal particles are washed away by the dielectric fluid and the process is repeated many times per second. As one discharge follows another, the crater continues to get bigger, taking the shape of the electrode.

● EDM COMPONENTS

Because EDM removes metal electrically, the hardness of the material does not determine whether or not it can be machined. One of the major benefits of the EDM process is that a part can be cut to the proper form in its hardened state. In this way, the damage or distortion that could occur during the heat treating operation is eliminated. The main components of an EDM machine are the electrode, dielectric fluid, servomechanism, and power supply.

Electrode

In EDM, a relatively soft graphite or metallic electrode can be used to machine hardened steel, or even carbide. The electrode can be of a special EDM graphite, copper, brass, copper tungsten, steel, or other electrically conductive material. The EDM process always produces a cavity slightly larger than the electrode, and because of the *overcut,* the electrode must be made a little smaller than the actual size required. Once the amount of overcut is determined, it will be repeated accurately, unless there is a change in the DC current supply.

Dielectric Fluid

The workpiece and the electrode are submerged in dielectric oil, an electrical insulator that helps to control the arc discharge. The oil also acts as a coolant, and is pumped through the arc gap to wash away the chips (swarf). A **dielectric fluid** performs three important functions:

1. It insulates the electrode/workpiece gap and prevents a spark from forming until the gap and voltage are correct. When this happens, the oil ionizes and allows the discharge to occur.

2. It must cool the work, the electrode, and the molten metal particles. Without coolant, the electrode and the workpiece would become dangerously hot.

3. It must flush the metal particles out of the gap. Poor flushing will cause erratic metal removal, poor machining conditions, and increase machining time and costs.

In order to be effective, the dielectric fluid must be circulated under constant pressure. There are a number of flushing methods used to carry away the metal particles efficiently while assisting in the machining process. Too much fluid pressure will remove the chips before they can assist in the cutting action, resulting in slower metal removal. Too little pressure will not remove the chips quickly enough and may result in short-circuiting the erosion process.

There are four methods generally used to circulate the dielectric fluid:

- DOWN THROUGH THE ELECTRODE, Fig. 51-4A. A hole (or holes) is/are drilled through

the electrode and the dielectric fluid is forced through the electrode to the spark erosion zone. This quickly washes away swarf or metal particles from the cutting area. A small standing slug or core is generally left in the cavity and must be removed after the machining operation.

- UP THROUGH THE WORKPIECE, Fig. 51-4B. Dielectric fluid can also be circulated up through holes in the workpiece. Die

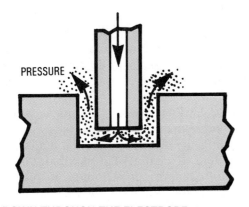

(A) DOWN THROUGH THE ELECTRODE

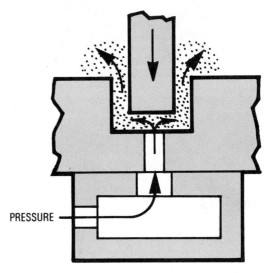

(B) UP THROUGH THE WORKPIECE

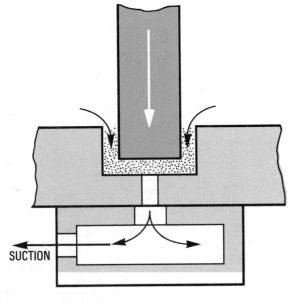

(C) BY VACUUM FLOW

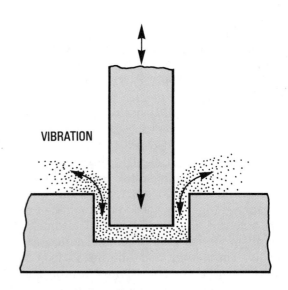

(D) BY VIBRATION

Fig. 51-4 Methods of circulating dielectric fluid. (Courtesy of Cincinnati Milacron Inc.)

blocks and larger cavities are flushed this way to remove the metal particles.

- VACUUM FLOW, Fig. 51-4C. A vacuum (negative pressure) is produced in the gap causing the dielectric to flow through the normal .001 in. (0.02 mm) gap to remove metal particles from the cutting zone. The flow can be either down through a hole in the workpiece, or through a hole in the electrode. The advantages of vacuum flow over other methods are improved machining efficiency, reduced smoke and fumes, and sometimes reduced or eliminated taper in the workpiece.

- VIBRATION, Fig. 51-4D. Vibration or jet flushing is used when the coolant cannot be run through either the work or the electrode. A pumping and sucking action is used to cause the dielectric to carry away the chips from the spark gap. The vibration method is especially valuable for very small holes, deep holes, or blind cavities where it would be impractical to use other methods. Jumping the electrode (rapid up-and-down movements) is used with this method of flushing.

● THE SERVO MECHANISM

The **servomechanism** works with the power supply unit to control the incoming voltage level and the arc gap distance between the electrode and the workpiece. This is very important because it helps to break down the insulating quality of the dielectric. As a spark is generated, it has to jump the arc gap and remove material. If a spark does not happen, the servo feeds the electrode closer to the workpiece, maintaining a gap of about .0005 to .002 in. (0.0l to 0.04 mm). If this gap gets any smaller than .0005 in. (0.01 mm), it will short-circuit the system. During the machining process, the servo senses and corrects any shorted conditions by rapidly retracting and returning the tool. If the gap is too large, ionization of the dielectric fluid does not occur and machining cannot take place. Precise gap control is made possible by the use of *fuzzy logic,* a process where the computer/control can evaluate and adjust all conditions (amperage, on/off time, voltage, etc.) to resume or maintain stable EDM.

When chips in the spark gap reduce the voltage below a critical level, the servomechanism causes the tool to withdraw until the chips are flushed away by the dielectric fluid. In addi-

tion to the vertical movement of the electrode, the servo control mechanism can be used on the table motion for work requiring horizontal movement of the electrode.

● MATERIAL REMOVAL RATES

Metal removal rates for EDM are slower than for conventional machining methods. The amount of metal removed with each pulse is directly in relation to the energy of the pulse. This pulse energy is determined by the on-time and peak current values set on the power supply. The rate at which metal is removed depends on the amount of energy in each spark and the number of sparks produced per second. This energy (electrical current) is measured in amperes (current). In Fig. 51-5A, the spark contains only a certain amount of energy which removes only a certain amount of material. In Fig. 51-5B, the current is doubled, resulting in double the material removed. The energy in Fig. 51-5C is again doubled, and compared to A, four times the amount of material is removed. The rate of metal removal is dependent upon the following factors:

1. Electrode material; graphite electrodes produce higher material removal rates than other electrodes.

2. Workpiece material; the harder the material, the slower the metal removal rate.

3. Power supply used; the higher the pulse energy, the higher the material removal rate.

4. Amount of current in each discharge; higher current removes material faster.

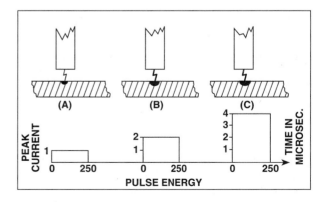

Fig. 51-5 Metal-removal rates increase with the amount of energy per spark. (Courtesy of Kelmar Associates)

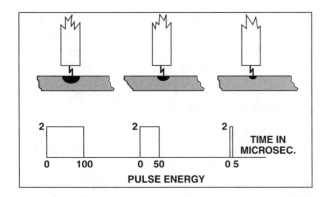

Fig. 51-6 The surface finish improves as the amount of spark energy decreases. (Courtesy of Kelmar Associates)

5. Frequency of the discharge; the more sparks produced in one second, the more material is removed.

6. Dielectric flushing conditions; good flushing conditions increase the metal removal rate.

The normal metal removal rate is approximately 1 in.^{3} (16.38 cm^3) of material per hour for every 20 amps. of machining current. Removal rates in tool steel with a graphite electrode are approximately .0008 in.^{3} (0.01308 cm^3) per minute per amp. As a general rule, the faster the metal is removed using higher current, the rougher the surface finish that will be produced, Fig. 51-6.

● **ADVANTAGES OF EDM**

Some of the newer electrical discharge machines can be programmed for vertical, directional, orbital, vectorial, cylindrical, conical, and helical machining cycles. Electrical discharge machines have many advantages over conventional machine tools:

1. Any material that is electrically conductive, especially cemented carbides and the new supertough space-age alloys, can be cut, regardless of their hardness. These are extremely difficult to cut by conventional machine tools.

2. Work can be machined in a hardened state which eliminates the deformation caused by hardening.

3. Intricate shapes can be produced because of vertical, rotary, and orbital motions, Fig. 51-7A & B.

Fig. 51-7A Types of forms produced by ram EDM. (Courtesy of Cincinnati Milacron Inc.)

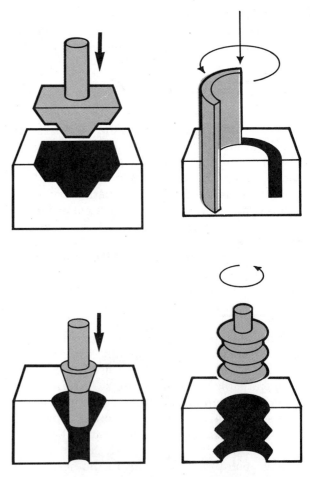

Fig. 51-7B Tool movements can produce other types of shapes. (Courtesy of Cincinnati Milacron Inc.)

4. The process is automatic because the servo mechanism advances the electrode into the work as the metal is removed.

5. Once the program is prepared and the work set up, no operator attention is required.

6. Dies and molds can be produced more accurately and at lower costs.

7. A die punch can be used as the electrode to reproduce its shape in the matching die plate, and the necessary clearance is produced by the overcut.

8. Stresses are not created in the work material since the tool (electrode) never comes in contact with the work.

9. The process is burr-free, eliminating any deburring operations.

10. Thin, fragile sections can be machined without deforming.

● WIRE-CUT EDM

Wire-cut EDMs are designed for different workpiece forms than the ram-type machine. It uses a vertically travelling wire that follows a horizontal path through the workpiece in the same manner as a bandsaw. Movement of the wire is controlled continuously and simultaneously on two to five axes, in minimum increments of .00001 in. Any contour within the maximum worktable movements can be cut with great accuracy, including tapers up to 20°.

Wire-cut EDMs, Fig. 51-8, use CNC programmed movements to produce almost any part of a desired size or shape within the capacity of the machine. It uses a thin, vertically-traveling wire that cuts a narrow groove to any programmed shape such as tapers, ellipses, parabolas, involutes, etc., through electrically-conductive material, Fig. 51-9. Wire-cut EDMs can machine the hardest of all metals such as tungsten carbide, space-age alloys, difficult-to-machine material, pure molybdenum, polycrystalline diamond, and polycrystalline cubic boron nitride.

Wire EDM Process

The wire-cut EDMs use CNC programmed instructions to move the workpiece in the X and Y axes, in a horizontal plane, to cut a variety of shapes, Fig. 51-10. The vertically moving wire never contacts the workpiece, but operates in dielectric fluid (usually deionized water) that is in

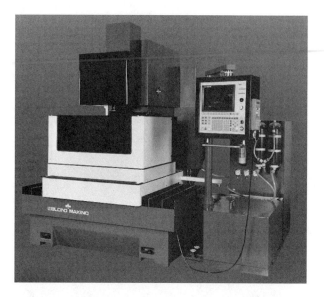

Fig. 51-8 Wire-cut EDMs use a wire as the electrode and can cut almost any shape or form. (Courtesy of Makino, Inc.)

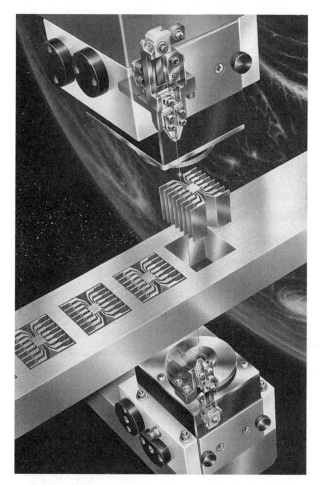

Fig. 51-9 An intricate form being produced by wire-cut EDM. (Courtesy of Makino, Inc.)

Fig. 51-10 Examples of work that can be produced by wire-cut EDM. (Courtesy of Makino, Inc.)

the spark area between the work and the wire electrode. In operation, the dielectric fluid breaks down and forms a gas that allows the spark to jump the gap between the work and the wire. The small metal particles created by the spark are then washed away by the dielectric fluid.

Movement of the wire is continuously controlled by the servomechanism in increments of .00001 in. (0.0002 mm) in all axes. The head can be tilted up to 30° along the XY axis for cutting tapered sections, and can be raised or lowered in the Z axis to suit the thickness of the workpiece. Any programmed contour or shape, within the capacity of the machine, can be cut very accurately, Fig. 51-10.

● OPERATING SYSTEM

The main parts of the wire-cut EDMs are the electrode, dielectric fluid, servomechanism, and the machine control unit.

Electrode

The electrode (wire) used for EDM systems may be made of brass, copper, zinc, molybdenum, steel, manganese, or tungsten. Its thickness generally ranges from .002 to .012 in. (0.05 to

0.30 mm) in diameter, however, thickness as small as .0005 and .001 in. (0.01 and 0.02 mm) is available for special jobs. The wire electrode travels in a continuous loop from the supply spool to the takeup spool, always keeping new wire in the spark area, Fig. 51-11. This ensures the accuracy of the cut because the wire passing through the spark area is always the same size and not worn.

Good wire electrodes must have the following characteristics:

1. Be good electrical and thermal conductors.
2. Provide efficient metal removal rates.
3. Possess high tensile strength and a high melting point.

Dielectric Fluid

The dielectric fluid plays an important role in the EDM process because it creates the spark gap and removes metal particles from the spark erosion area. Flushing the metal particles out quickly creates good cutting conditions, while poor flushing will cause erratic cutting and poor metal-removal rates. The recommended flushing is a high-pressure jet stream directed along the wire, flushing both the top and bottom of the work.

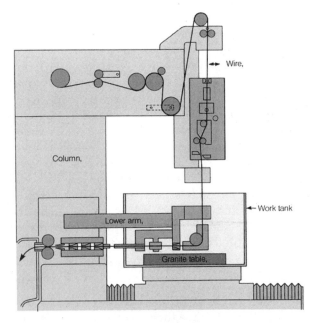

Fig. 51-11 The path of the wire from the upper supply spool, through the workpiece, and onto the lower takeup spool. (Courtesy of Makino, Inc.)

Deionized water is generally the dielectric used in the EDM wire-cut process. This is generally distilled water that has been passed through an ion-exchange resin to make it a good insulator. Untreated tap water is not suitable for EDM wire-cut operations because it is an electrical conductor.

To be effective, a dielectric fluid must serve the following functions:

1. Provide a path for the spark between the wire (electrode) and the work material.

2. Act as an insulator between the wire and work to prevent shorting.

3. Flush away the metal particles from the wire and the work material.

4. Cool both the wire and the workpiece.

Servomechanism

The EDM servomechanism controls the discharge current, the feed rate of the table motors, the speed of the wire, and the distance in the spark gap. It automatically keeps a constant gap of about .001 to .002 in. (0.02 to 0.05 mm) between the wire and the work to prevent any contact that could cause a short and break the wire. As the servomechanism feeds the work into the wire, it senses the work/wire spacing, and slows or speeds up the drive motors to maintain the proper arc gap. Accurate control of the gap width is very important to the EDM process. Too large a gap will not allow the spark to be conducted between the wire and work, while too small a gap will cause the wire to short and break.

Machine Control Unit

The machine control units of modern EDMs are big improvements over those on former machines. They are more user-friendly and contain many programs and cycles that make it easier to operate the EDM machines, Fig. 51-12. A few of the more common features of EDM controls are:

• ADAPTIVE OR "FUZZY LOGIC" CONTROLS—Automatically adjust all machining conditions (ON/OFF, servo-voltage, amperage, gap-width, etc.) to reduce wire breakage.

• AUTOMATIC JUMP CONTROL—A program to rapidly retract and return the electrode to the cutting position. This allows the dielectric fluid to wash away the chips from the spark erosion area.

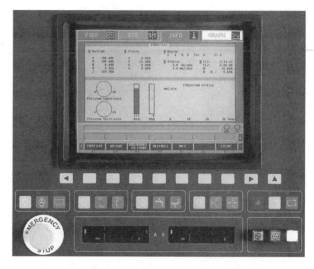

Fig. 51-12 The machine control unit contains many features to assist the operator in programming and machining. (Courtesy of Charmilles Technologies Co.)

• MIRROR IMAGING—An exact copy of a program can be made in any three quadrants to reproduce left- and right-hand parts, Fig. 51-13.

• PATTERN ROTATION—Sometimes called a rotation copy, is used for cutting repetitive forms. Assume a gear is to be cut, the program for one tooth is created and the first tooth is cut. The work is then indexed for the next tooth and uses the same program until all the teeth have been cut.

• MIRROR FINISHING—Made possible by advanced circuitry, can eliminate most secondary finishing operations.

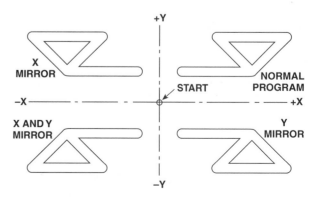

Fig. 51-13 Mirror imaging allows left- and right-hand parts to be reproduced accurately in any quadrant from one program. (Courtesy of The Superior Electric Co.)

Some features of new control units are:

1. Automatic wire threading and rethreading.

2. The control unit can be linked directly to a computer.

3. A new program can be prepared while another part is being machined.

4. Canned cycles are available for a number of standard operations.

5. Programming and operation sequence assistance.

   - Once the operator provides the following information about a job:
     - the thickness and type of work material,
     - the accuracy and surface finish required,
     - the maximum taper angle,
     - the smallest radius to be cut,
     - the type and quality of flushing,
   - The control system will select:
     - the type and diameter of the wire,
     - generator settings for the roughing and finishing passes,
     - calculate the offsets required, and
     - generate a complete operational command file.

SUMMARY

- Electrical discharge machining is a process that erodes metal by an electric spark of short duration and high-current density between the electrode and the work material.

- Two common types of EDM machines are used in industry; the ram or die-sinking machine has an electrode the shape of the cavity required, and the wire-cut EDM that uses a thin, vertically-moving wire to cut almost any shape in the metal.

- The servomechanism maintains a gap distance between the electrode and the material being cut to prevent a short-circuit and to provide the best material-removal rates.

- The dielectric fluid acts as an electrical insulator to control the electrical discharge, as a coolant, and as a means of washing away the chips (swarf).

KNOWLEDGE REVIEW

1. Briefly describe the process of electrical discharge machining.

Ram or Die-Sinking EDM

2. Name the two types of EDM machines.

EDM Components

3. List the main components of an EDM machine.

4. What are three important functions of a dielectric fluid?

5. What methods can be used to circulate dielectric fluid?

The Servomechanism

6. What is the purpose of the servomechanism?

7. What are the characteristics of a good wire electrode?

8. Why is tap water not a suitable dielectric fluid for EDM wire operation?

UNIT
52

Flexible Manufacturing Systems

There is a world-wide demand for a good selection of high-quality products that are reasonably priced and available whenever the customer wants them. To meet these requirements, flexible manufacturing systems (FMS) were developed to enable manufacturers to quickly change their manufacturing operation to produce any product, at any time, and still maintain an economical operation. Flexible manufacturing systems are one of the most efficient ways of reducing or eliminating manufacturing problems. FMS is a business-driven solution that leads to improved profitability due to reduced lead times and inventory levels, rapid response to market changes, lower labor costs, and improved manufacturing productivity as a result of operational flexibility, predictability, and control.

OBJECTIVES

After completing this unit, you should be able to:
- Describe the factors that influence the design of flexible manufacturing systems.
- Explain how the concept of modularity is applied to flexible manufacturing systems.
- List the characteristics of an effective flexible manufacturing system.

KEY TERMS

automatic identification system
flexible manufacturing system

automatic tool changing
manufacturing cell

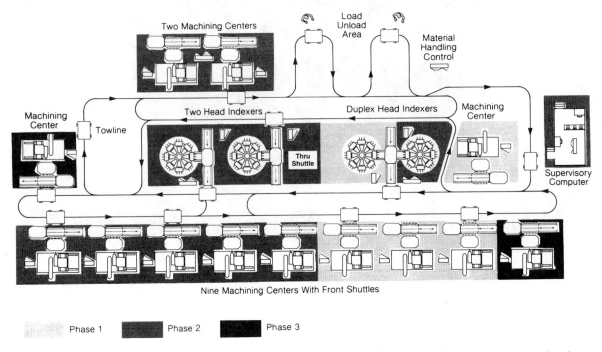

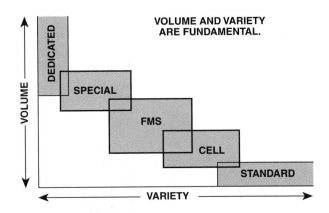

Fig. 52-1 FMS consists of a number of CNC machine tools serviced by a material-handling system under the control of computers. (Courtesy of Giddings & Lewis, Inc.)

● ADVANTAGES

A **flexible manufacturing system** generally consists of a number of CNC machine tools serviced by a material-handling system under the control (supervision) of one or more dedicated supervisory (executive) computers, Fig. 52-1. A typical flexible manufacturing system can automatically and completely process the members of one or more part families on a continuing basis without human intervention. FMS brings flexibility and responsiveness to manufacturing to produce a part when it is required by the market and not when it is most suitable for production. It provides additional flexibility to suit changing market conditions and product type without expenditures for other equipment. FMS uses equipment in off-hours, such as second and third shifts when it is not normally used. This is a major stepping stone to unmanned manufacturing, the automated factory of the future.

● MANUFACTURING APPLICATIONS

Today's ever-changing manufacturing requirements and the need for flexibility make part variety and volume factors in selecting the basic design of the manufacturing system required, Fig. 52-2. The greatest need for flexible manufac-

Fig. 52-2 The manufacturing method selected depends on the variety of the product mix and the number of parts required. (Courtesy of Giddings & Lewis, Inc.)

turing systems seems to be in the areas of mid-volume and mid-variety part production. New technology including machine design, material handling methods, and control techniques make it possible to meet the ever-changing requirements of part volume and variety. System modules and phasing techniques are the keys that provide the on-going adaptability to meet changes in product mix and production volumes.

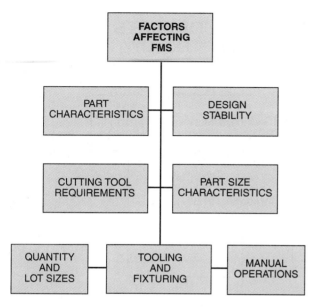

Fig. 52-3 The main factors that affect the design of a flexible manufacturing system. (Courtesy of Kelmar Associates)

System Design

Although production volume per part number and production variety greatly define the areas of application, there are other factors that will affect and modify system design.

Visualize a series of transparencies. The lifting of the first transparency (that illustrates volume and variety) uncovers more underlying factors (for example, part size, assembly in a common unit, workpiece accuracy and shape). Each layer is more definitive, but not necessarily more important.

Further system design is determined by the general characteristics of the part. The size, shape, common features, and life cycle will determine how the part is processed, what tooling and fixtures may be needed, and the type of machinery required in the system, Fig. 52-3.

Part Size

Large parts require more automated methods for systems work, while smaller parts are more "forgiving" and may not require as high a degree of automation.

Workpiece Accuracy and Shape

Because of the accuracy required, a very critical series of tolerances often determines the manufacturing method and the design of the machine. Part shape is closely associated with accuracy; very complex parts may require special machine design and system layout.

Assembly

A series of parts is often required to be produced in a one-to-another relationship because they are assembled as a common unit. This "manufacturing-to-assembly" often becomes a critical variable in system design.

Product Life Cycle

The longer the product design life cycle, the more desirable it is to invest in dedicated or less flexible approaches to manufacturing. On the other hand, the shorter the life cycle of the product design, the more advantageous it is to have a flexible system that allows rapid changes.

Planning for the Future

Planning is the key factor in meeting present and future goals. A prime characteristic of a system is a phased implementation of equipment—machinery, material handling, and controls. The time span for phasing can be months or even years.

During this time, a system can evolve from a single manufacturing cell into an FMS as production increases. Further requirements may lead to process specialization of a special system. This forward planning allows complete control over manufacturing needs, both through expansion and contraction. Even part changes or totally different parts can be adapted; production adaptability to changing requirements is a key advantage. The system must eliminate the need for a total and immediate investment, based on a market forecast projection of 10 years or more, where there is little or no room for adaptability to market changes once the commitment is made.

Productivity versus Flexibility

The conflict between productivity and flexibility can be resolved by the application of a mid-volume, mid-variety manufacturing systems concept, Fig. 52-4. The productivity versus flexibility arrows show the different paths of each objective. The system allows the optimum level of each objective so that they are in harmony, thereby enhancing each other's qualities. Flexibility has three levels—complete flexibility, process specialization, and machine specialization. Each defines the degree of commitment to a process. Complete flexibility relies on universal

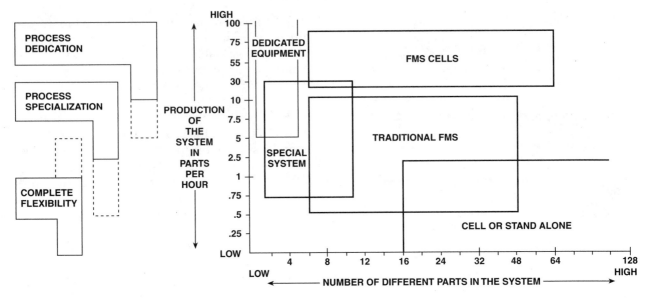

Fig. 52-4 A mid-volume, mid-variety manufacturing systems concept suits both productivity and flexibility requirements. (Courtesy of Giddings & Lewis, Inc.)

machinery, tooling, fixtures, and processes. In the next level, process specialization shifts the emphasis to tooling, fixturing, and processing to increase productivity. The machine specialization level is not as flexible because machinery is generally designed for its application like the tooling and fixturing of the previous level.

Productive capacity starts at stand-alone machinery and ends with dedicated equipment. It should be noted that high-volume production does not necessarily result in high efficiency. High machine efficiency is generally the most efficient use of machine tools, material handling, and control systems technologies.

● TYPES OF SUPPORT SYSTEMS

In order to complete short runs of products, flexible manufacturing must combine several automation systems. Previously, traditional flexible manufacturing involved long setup times and frequent changeovers. On the other hand, traditional automation involved the use of dedicated or single-purpose machine tools set in line to complete or produce a specific component or product in large volume at high speed.

Flexible manufacturing systems combine the best of both features. While in-line conveyors maintain the speed of fixed automation, flexible

machining or assembly is introduced with the use of the following:

- **AUTOMATIC IDENTIFICATION SYSTEM—**Data is automatically collected and produced by means of advanced technology such as laser-read bar coding, machine vision, and radio frequency identification.

- COMPUTER WORK DIRECTORS—In response to component or product design changes, computing systems intercept and direct parts/assemblies to work centers.

- MULTI-FUNCTIONAL TOOLS—Multi-purpose machine tools are required that include quick-change tooling, fixturing, as well as assembly machines all controlled by computers.

- AUTOMATED MATERIAL HANDLING—A system of storage and retrieval of materials as work is transported between work cells by automatic guided vehicles, Fig. 52-5.

- ROBOTICS—A work system to execute computer commands and parts distribution for different operations, Fig. 52-6.

The availability of better graphics and computer software for product design, lower costs of intelligent machines and computers, better product description (Group Technology), and im-

Fig. 52-5 CNC machines are supplied with tools, fixtures, and materials by automated guided vehicles. (Courtesy of Giddings & Lewis, Inc.)

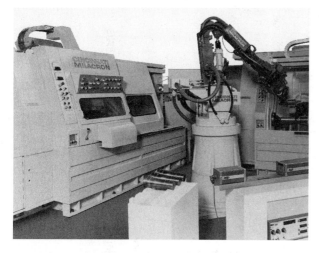

Fig. 52-6 A robot being used to load and unload parts on a CNC chucking and turning center. (Courtesy of Cincinnati Milacron Inc.)

proved communications and networking between systems are making it easier and more desirable for companies to invest in flexible automation. Along with the implementing of manufacturing technologies, the continuous retraining and cross-training of employees will make flexible manufacturing systems work, and make it possible for companies to meet the changes in market trends.

● **THE FUNDAMENTAL BUILDING BLOCKS**

The modular adaptability of the technological building blocks establishes a good foundation for a manufacturing system that is flexible to meet changing production needs. Machine mod-

ules, material handling modules, and control modules can be individually selected to suit present needs, yet have the ability to interrelate with each other for future needs, Fig. 52-7.

The machining modules are designed for stand-alone manufacturing and future system integration requirements. The material-handling modules tie the machine modules together to work with inspection, wash modules, and load/unload areas. The system control modules are the management tools required to monitor and control all the modules in a system. The type and number of each of these modules, as well as the shape and physical design of the system, must meet the manufacturer's objectives, method of production, size of parts, etc.

There are three main manufacturing concepts used to provide productivity and flexibility to suit product volume and variety. These are the stand-alone manufacturing center, **manufacturing cell**, and the flexible manufacturing system:

1. The *stand-alone manufacturing center* consists of a multi-pallet system working with a machining center, Fig. 52-8A. It can work on several types of parts at the same time, selecting them at random from a pre-loaded queue (line) of parts on pallets. This type of system has found wide use because of the multi-pallet design and the minimum attention it requires during production. When production requirements increase beyond the capability of the stand-alone manufacturing center, it can easily be fitted (integrated) into a flexible manufacturing system module, Fig 52-8B.

2. The *manufacturing cell* contains a group of processing modules combined to handle a family of parts and complete all the manufacturing operations before the part leaves the cell. There are three types of manufacturing cells generally used to suit various production purposes—the palletized cell, the robot or automated cell, and the FMS cell.

   • THE PALLETIZED CELL, Fig. 52-9—is generally used for high-variety, low-volume production problems. Material handling joins a group of flexible general purpose machine tools using a common pallet design with pre-fixtured parts on pallets.

   Palletized cells are different from the FMS concept because part flow usually limits them to two or three stations.

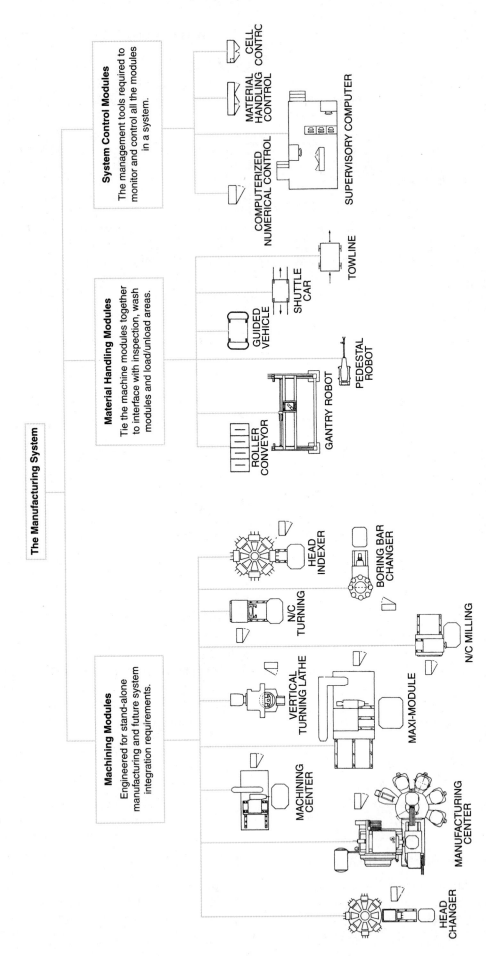

The Manufacturing System

Machining Modules
Engineered for stand-alone manufacturing and future system integration requirements.

Material Handling Modules
Tie the machine modules together to interface with inspection, wash modules and load/unload areas.

System Control Modules
The management tools required to monitor and control all the modules in a system.

HEAD CHANGER

MANUFACTURING CENTER

MACHINING CENTER

MAXI-MODULE

VERTICAL TURNING LATHE

N/C MILLING

N/C TURNING

BORING BAR CHANGER

HEAD INDEXER

ROLLER CONVEYOR

GANTRY ROBOT

PEDESTAL ROBOT

GUIDED VEHICLE

SHUTTLE CAR

TOWLINE

COMPUTERIZED NUMERICAL CONTROL

SUPERVISORY COMPUTER

MATERIAL HANDLING CONTROL

CELL CONTROL

Fig. 52-7   The fundamental building blocks of a manufacturing system. (Courtesy of Giddings & Lewis, Inc.)

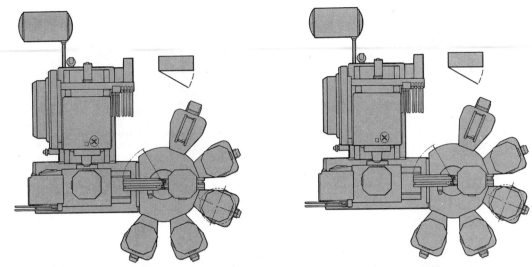

Two Manufacturing Centers

Fig. 52-8A The stand-alone manufacturing centers find more use due to multiple pallets and minimum operating attention. (Courtesy of Giddings & Lewis, Inc.)

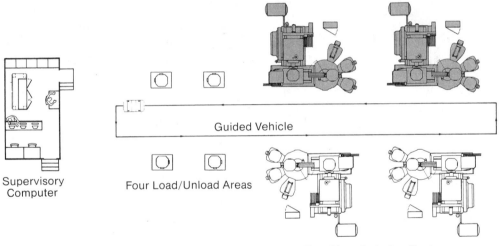

Supervisory Computer

Four Load/Unload Areas

Guided Vehicle

Four Manufacturing Centers

Fig. 52-8B When higher production is required, stand-alone manufacturing cells can be added to an FMS. (Courtesy of Giddings & Lewis, Inc.)

These cells do not have central computer control with real-time routing, load balancing, and production scheduling logic.

- THE ROBOT OR AUTOMATED CELL, Fig. 52-10—consists of a group of flexible machines, linked with robotics or specialized material handling. Its typical application is to high-volume production of a small, well-defined, homogeneous family of parts.

Cells generally have a fixed process and an orderly flow of parts between operations. The cell becomes a fully-automatic process through the application of robotics, power clamping of parts, special tools, and other forms of automation.

- THE FMS CELL, Fig. 52-11—is a distinct machine group within an FMS. The distinguishing characteristics of this cell are the automated flow of raw material to the

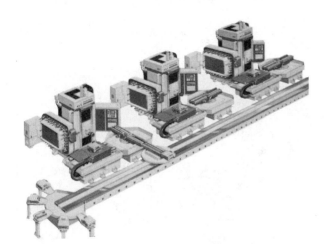

Fig. 52-9 The palletized cell is suitable for the production of high-variety, low-volume parts. (Courtesy of Giddings & Lewis, Inc.)

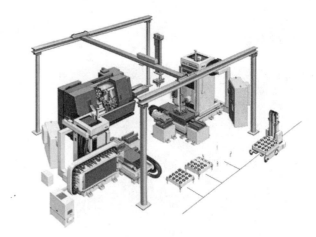

Fig. 52-11 The FMS cell supplies raw material to the cell, machines the part completely within the cell, and then removes the finished part. (Courtesy of Giddings & Lewis, Inc.)

Fig. 52-10 The robot or automated cell contains a group of flexible machines linked with robotics and a material handling system. (Courtesy of Giddings & Lewis, Inc.)

cell, total machining of the part across the machines within the cell, and then the removal of the finished part.

The FMS cell is a station in a larger automated processing network, where different types of cells can be linked by material handling devices. These ideas approach the concept of a computer-controlled factory.

3. The *FMS module*, Fig. 52-12, combines the features beyond the standard design to assure that it can be integrated into a full FMS. The features of these modules are increased control capabilities, material handling capacity, tooling and processing facilities, axis travel, thrust requirements, and spindle designs.

● FMS SYSTEM PLANNING

In order to achieve optimum efficiency in manufacturing, various components must be combined to make an effective flexible manufacturing system, Fig. 52-13. If possible, to justify the large capital investment, the system should operate 24 hours every day. Factories of the future will operate 8760 hours per year, produce a wide variety of products, and have very little human intervention.

Machines will be grouped together in small clusters or cells with automatic handling of all materials in, and between different cells. All machines will be computer controlled, which will greatly improve the accuracy and consistency of the production system. **Automatic tool changing** will be common on all machines to speed the machining process and reduce machine downtime. Tool management will ensure that the right tool arrives at the right place, at the right time.

In order to satisfy production needs, several cells may be set up, each connected with the others. These in turn will be connected with areas of the plant such as material storage, transport, and data communications, all under the control of the executive and mainline computers.

The inspection system of an FMS must be accurate and reliable enough to detect variations in dimensional accuracy and initiate the system to make changes to correct the error by an adjustment, or by replacing the cutting tool. All types of material-handling systems such as ro-

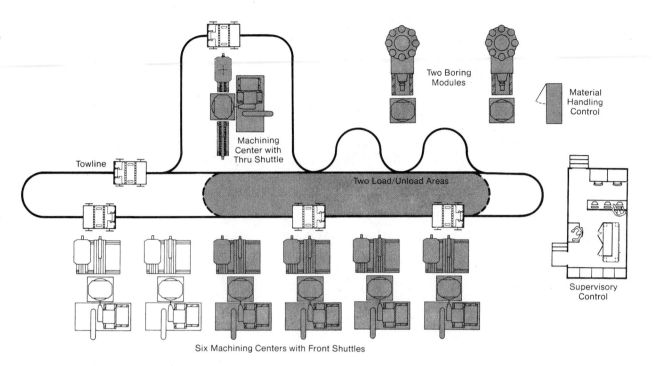

Fig. 52-12 FMS modules have increased control capabilities, material handling, and tooling and processing facilities. (Courtesy of Giddings & Lewis, Inc.)

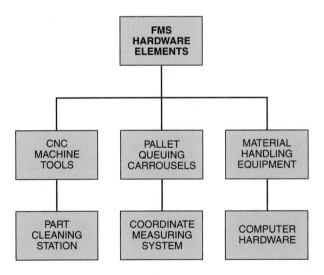

Fig. 52-13 The main elements of a flexible manufacturing system. (Courtesy of Kelmar Associates)

bots, automated guided vehicles, etc., will ensure that the right material, fixture, tooling, and whatever is required for production, is available when it is needed.

To make the system complete, all functions in the manufacturing process, that is, design, planning, inventory control, scheduling, and shop floor control, must be able to communicate with each other automatically. By networking, they will share the common data base and all other current information. Each of the cell systems can take any independent action required, based on this shared information. The mainframe or control computer will report to the staff and call attention to problems that may require human attention.

SUMMARY

- Flexible manufacturing systems should be considered whenever productivity and flexibility are essential for a manufacturing system.

- FMS provides manufacturers with a way to reduce many hidden manufacturing costs such as scrap, in-process inventory, labor costs, material and overhead costs, and to improve machine up-time.

- The direct reduction in costs can vary from a low of 30% in labor to a high of 90% in machine up-time. This allows manufacturers to be competitive on the world market while still being profitable.

KNOWLEDGE REVIEW

1. Why were Flexible Manufacturing Systems (FMS) developed?

Manufacturing Applications

2. List the three main manufacturing concepts used to provide productivity and flexibility to suit product volume and variety.

The Fundamental Building Blocks

3. List three types of manufacturing cells.

FMS Planning

4. Why is automatic-tool changing advantageous?

5. What is the purpose of tool management?

6. What are some factors that determine system design?

UNIT
53

Group Technology

Group technology is a system where a large number of parts are classified into various part families on the basis of similarities in their shape, configuration, holes, threads, and machining operations, to improve manufacturing productivity, Fig. 53-1. This also involves the clustering of machines into cells where automated work-handling devices provide an efficient flow and processing of parts through the machines.

OBJECTIVES

After completing this unit, you should be able to:
- Define group technology.
- Explain the importance of a well-designed parts classification and coding system.
- Discuss how load balancing relates to group technology.
- Describe the computer-aided process planning procedure.

KEY TERMS

cellular manufacturing
group technology
part families

computer-aided process planning
load balancing
work-in-process

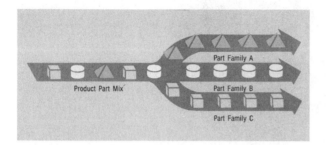

Fig. 53-1 Group technology classifies parts into part families on the basis of their size, shape, and machining operations. (Courtesy of *Modern Machine Shop Magazine*, copyright 1994, Gardner Publications Inc.)

● PURPOSE AND ADVANTAGES

Group technology (GT) is a system where the features of many parts are studied and examined through the use of interactive computers, to identify and take advantage of the similarities that exist in parts and their manufacturing processes. While the shape of parts may vary in size, they may contain the same features as smaller parts, Fig. 53-2. Parts such as nuts, washers, gears, and similar components, are easily grouped into classes of similar parts because they are relatively small and have the same features. By grouping parts together into families, productivity can be increased because all parts require similar operations or manufacturing steps, Fig. 53-3. This improves process planning and the scheduling of machines and equipment.

Part-coding and standardization of the manufacturing process can improve shop floor management and result in more efficient use of machine tools and related equipment. This is ac-

Fig. 53-2 A variety of parts that may be found in many manufacturing plants. (Courtesy of Computervision Corp.)

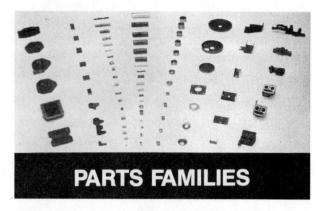

Fig. 53-3 The same parts as in Fig. 53-2 which have been grouped into part families on the basis of similarities. (Courtesy of Computervision Corp.)

complished through the use of interactive computer programs that classify and assign codes to the parts based on the design features and manufacturing operations.

GT can be very valuable for product design because it can greatly reduce lead times associated with engineering, production planning, and fabrication creating cost savings ranging from 10% or higher. Where it formerly took an engineer many hours to look over a few thousand prints searching for a part that would lend itself to a particular operation, using GT this job can be completed in a few minutes by studying the reports of a computerized group technology system. As the process is fully developed, other departments such as purchasing, testing, and quality control can begin using the same information contained in the classification and coding system to reduce their costs.

Benefits

Group technology can provide advantages in the standardization of components, reliability of estimates, effective machine operation, productivity, costing accuracy, customer service, and order potential. The parts within a family can be analyzed to find common elements and identify non-standard design features. In Fig. 53-4, the darker bars show the non-standard variations in some of the parts. Since there are very few parts involved, it may be possible to change these dimensions to standard dimensions which would reduce manufacturing costs and increase productivity. At the same time, it can reduce planning effort, paperwork, setup time, machine downtime, **work-in-process,** work movement,

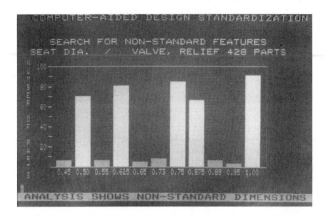

Fig. 53-4 Group technology can show variations in part design that might be changed to reduce manufacturing costs and increase productivity. (Courtesy of Computervision Corp.)

overall production times, finished parts stock, and overall costs. Some of the benefits are as follows:

- 52% in new part design,
- 10% in the number of drawings through standardization,
- 30% in new shop drawings,
- 60% in industrial engineering time,
- 20% in production floor space required,
- 42% in raw material stock,
- 69% in setup time,
- 70% in throughput time,
- 62% in work-in-process inventory,
- 82% in overdue orders.

In addition to these advantages, group technology provides better working conditions that improve the quality of life in a manufacturing environment. Because machines are grouped to accommodate part families, the workers are able to see their contribution to the final product, which results in increased worker satisfaction.

● **PART CLASSIFICATION**

The classification and coding method is the key to the application of group technology principles. Although many systems have been developed, there does not seem to be any system that has been universally accepted. Most use a system where parts are classified, coded, and assigned symbols that represent important information

about the part. Interactive computers take this information and assign each part a digit or letter, and a digit code to describe its characteristics.

The GT coding system has been simplified to include the similarities in parts according to key manufacturing features. One system uses a 30-digit geometric code that allows hundreds of thousands of individual items to be identified and grouped together. If the first digit of the code is #1, it identifies the major classification and digits 2 to 30 describe the details of the part. If the first digit is #2, the part is in a different family and digits 2 to 30 have another set of meanings.

Advantages

A well-designed classification and coding system has many benefits and results in group technology applications in many areas of a manufacturing operation. The major benefits of a well-designed classification and coding system are as follows:

- assists in the formation of part families and machine groups,
- allows quick retrieval of design, drawings, and production information,
- reduces design duplication and costs,
- provides valuable workpiece statistics,
- allows accurate estimation of machine tool requirements, logical machine scheduling, and makes the best use of capital expenditures,
- provides information for tooling setups, reductions in setup time and overall production time,
- results in improved tool design,
- improves production planning and scheduling procedures,
- improves cost accounting and estimation procedures,
- makes better use of machine tools, workholding devices, and labor,
- improves the effective use of CNC machine tools.

All departments may use a classification and coding system in one way or another, however the two major users are in the design/engineering and manufacturing areas. Therefore, a system should meet all the requirements and provide all the information needed by these two

Table 53-1 Basic information of classification and coding systems

Design/Engineering Area	Manufacturing Area
Main shape	Major operation
	Minor operation
Shape element	Major dimension and size ratio
Material	Rough shape and size
	Machine tool
Rough shape and size	Workholding devices
Major dimensions	Cutting tools
	Lot size
Minor dimensions	Setup time
Tolerances	Production time
	Operation sequence
Functions	Surface roughness
Assembly	Special treatment
Drawing size, etc.	Inspection
	Assembly, etc.

areas. Table 53-1 shows some of the information and data that should be included in a good classification and coding system.

● STANDARDIZATION PROCESS

Once the coding numbers have been established, the data should be analyzed so that the family of parts can be better defined. By examining the part characteristics and how often they occur, it is possible to identify the major **part families.** For example, families may be defined as flat or cylindrical parts that fall within certain shapes, diameter ranges, and dimensional tolerances, Fig. 53-5.

After the major family has been defined, a further breakdown is needed to break the family into smaller and smaller groups. Each breakdown further defines the part, and code numbers describe the part in greater detail, as well as describing its material, shape, dimensions, and accuracy. By analyzing the product, the difference between parts can be identified and separated into families. How successfully group technology will work depends on the combined cooperation of engi-

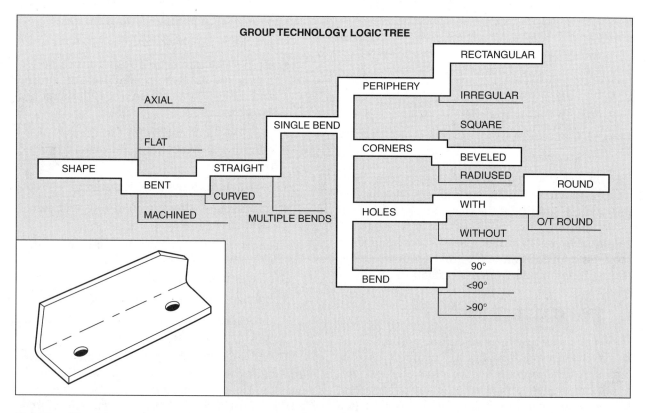

GROUP TECHNOLOGY LOGIC TREE

Fig. 53-5 The characteristics of this metal part are clearly defined by the various stages of a good coding system. (Courtesy of *Manufacturing Engineering Magazine*)

neers, designers, supervisors, and shop personnel. Their main purpose must be to eliminate waste and duplication, and find better ways to manufacture that will include the best use of personnel, time, materials, and equipment.

● LOAD BALANCING

In order to distribute the workload so that manufacturing takes place efficiently on available machine tools, it is important that coded information be as clear and simple as possible. The information should include the number of parts to be produced and the number of parts manufactured yearly. A comparison can then be made to see how each family, and each part within a family, relates to production (load) demand over a period of time. When all the information about parts and available machine tools is known, they can be grouped and positioned on a machine tool capability worksheet. The efficient distribution of this workload is the purpose of **load balancing.**

Since the code number listed on the worksheet identifies the features of a part, a user of the system can also use the same code number to identify the characteristics of the part (sheet metal, turned part, etc.). The rest of the coded numbers would have to be examined to determine exactly what machine tools and equipment are needed to machine the number of parts required. By using the group technology database, the computer can analyze the information and show the exact relationship between the parts required and the machine tools available, Fig. 53-6.

By analyzing this information, it will show when to purchase new equipment, and when it may be possible to retire some existing machines.

● DESIGN AND PART RETRIEVAL

Preparing for the manufacture of a part can be costly because of the many factors that must be considered such as tool design, process planning, and CNC program. It is important to avoid designing a new part if a part is already available in a part family. A part should only be considered new if there has been a design change that does not fit the family of parts described in the GT database.

Standardization and part design retrieval are two very important functions in group technology classification and coding. A design retrieval system eliminates the duplication of previous designs stored in the database, and also reminds engineers of company standards, Fig. 53-7. It has been estimated that as much as 15% savings in design costs can be had by using an efficient design retrieval system. Even greater savings can be made in the manufacturing area by grouping shop orders that will reduce setup

Fig. 53-7 The interactive computer analyzes information about a part to identify common elements and find inconsistencies. (Courtesy of Deere & Company Mfg.)

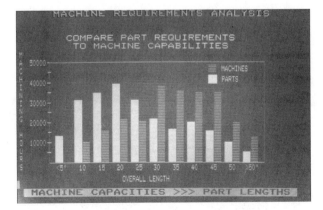

Fig. 53-6 Group technology can compare part manufacturing requirements and determine when additional machine tools are required. (Courtesy of Computervision Corp.)

times with part family tooling. In order to achieve the greatest savings, it is important that design and manufacturing engineers completely understand each other's needs.

● GROUP TECHNOLOGY AND CNC RELATIONSHIP

Group technology allows CNC programs to be prepared quickly and more accurately than by previous methods. The code number for each family of parts appears on the worksheet of machine tool capability listing. All the design and manufacturing information of the part is contained within the grouping of code numbers. Each member of a family falls within a certain design category, and by using this information, a CNC program can be written to produce the parts within the family. Only a small amount of information may have to be changed, such as a bore diameter, etc., to make these programs suit various pieces within the same family.

● GROUP TECHNOLOGY AND CELLULAR MANUFACTURING

Computer-aided process planning (CAPP) uses much of the same information contained in the GT file to convert a part design into a manufactured product. The main steps in process planning include:

- selection of raw material,
- sequence of machining operations,
- selection of workholding devices,
- tools and tooling required,
- CNC program and machinery,
- machining requirements—speeds, feeds, tool-change, etc.

There are two types of CAPP systems, the variant system and the generative system.

In the *variant system,* a process is recalled by part number, code number, or some other key, as long as the plan fits the same part being manufactured. If the part differs in any way, the recalled plan can be varied to meet the different requirements.

In the *generative system,* that relies heavily on artificial intelligence, a part is described to a computer, which then produces a plan to manufacture the part based on stored manufacturing logic. The computer looks at the part to be man-

ufactured, and using decision rules programmed into the system, proceeds to produce a manufacturing plan.

To improve manufacturing productivity, industry has turned to **cellular manufacturing** where machining processes, machinery, and people are grouped together to manufacture a specific family of parts. These cells are designed to handle all the operations of a family of parts, ranging from receiving the raw material to producing the finished product within the cell.

The group technology database can be used to analyze the manufacturing requirements and show when cellular manufacturing cells would be justified, Fig. 53-8. The analysis could also show that by proper use and placement of the machines in the cell, productivity could be increased, while at the same time reducing the floor space required, Fig. 53-9.

In order for cell manufacturing to work most efficiently, several factors have to be considered in the planning or manufacture of any component.

1. The physical characteristics of a part will affect the efficiency of the cell, or the machines within that cell.

2. At least one or more cells should be capable of performing all the operations required to produce the part.

3. Some machines within a cell should be flexible enough to perform multiple operations.

4. Wherever possible, any job assigned to a cell should be completed within that cell.

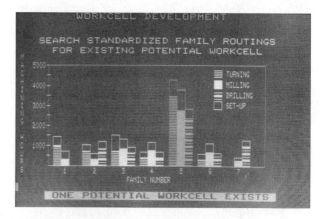

Fig. 53-8 Group technology can search existing part-family routines for potential work cells. (Courtesy of Computervision Corp.)

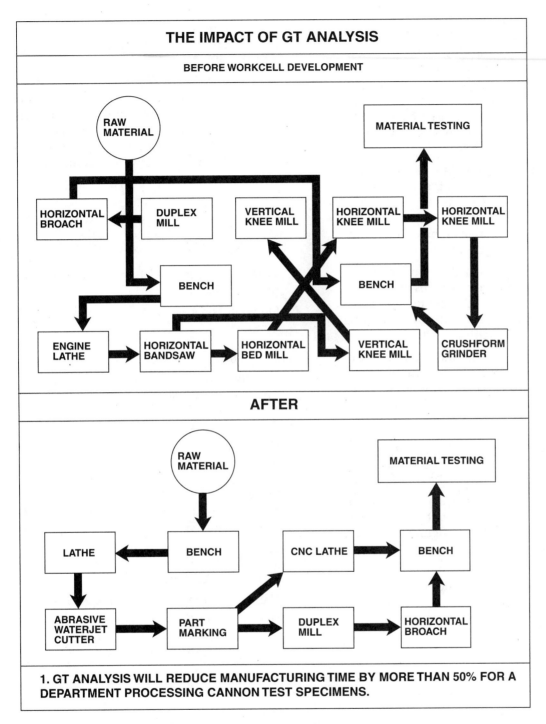

Fig. 53-9 Group technology analysis can reduce manufacturing time by more than 50% with the proper cell organization. (Courtesy of *Manufacturing Engineering Magazine*)

● DATA HANDLING SYSTEMS

The entire manufacturing should be controlled by its data processing operation which includes creating, transmitting, storing, analyzing, and modifying data. In order for data to be an effective manufacturing tool, it must be accurate, organized, and easily accessible to anyone involved in manufacturing.

- Product data describes the part to be designed and manufactured, that includes information about its size, shape, etc. The data also contains printouts describing materials, finishes, tolerances, heat treating instructions, and part numbers.

- Process data includes the manufacturing operations and procedures necessary to complete the fabrication. Also included is information concerning the cutting or forming processes, their machines, or tools.

- Rules data controls the design and manufacture of the product by describing in detail the procedures, specifications, analysis, and production standards.

SUMMARY

- Group Technology is a system that classifies large numbers of different parts by their characteristics (shape, configuration, holes, threads, size, etc.) before creating a family of parts.

- GT also involves the clustering of machines into cells for the most efficient flow of parts between machines and operations.

- The grouping of parts into families increases machining productivity, process planning, and the efficient planning and scheduling of machines and tools.

- Parts are classified, coded, and assigned symbols that provide important information about the part. These codes can be read by sensors located in the manufacturing process to avoid errors by machining the wrong part.

KNOWLEDGE REVIEW

1. What system is used to classify parts into part families?

2. On what basis are parts classified into part families?

Purpose and Advantages

3. List three advantages of classifying parts into parts families.

Part Classification and Coding

4. Describe the GT coding system for a 30-digit geometric code.

Design and Part Retrieval

5. Name three advantages of an efficient design-retrieval system?

GT and Cellular Manufacturing

6. How does CAPP operate?

7. Name the two types of CAPP systems.

8. Describe cellular manufacturing.

UNIT 54

Just-In-Time Manufacturing

Just-In-Time (JIT) manufacturing is a concept that was developed because manufacturers were looking for ways to improve productivity, reduce costs, reduce scrap and rework, overcome the shortage of machines, reduce inventory and work-in process (WIP), and utilize manufacturing space efficiently, Fig. 54-1. It originated with the Toyota automobile company, and its initial concept was to supply or produce parts only when they were required for the manufacturing or assembly process. Through careful planning, it was found that waste could be eliminated, plant space was better utilized, and overall productivity improved. Its success led to a revolution in production planning and inventory control.

OBJECTIVES

After completing this unit, you should be able to:

- Define Just-In-Time (JIT) manufacturing and its impact on waste in manufacturing operations.
- Identify the elements of a successful JIT purchasing plan.
- Describe the issues and obstacles involved in setting up a JIT manufacturing program.

KEY TERMS

Just-In-Time   zero-defect manufacturing

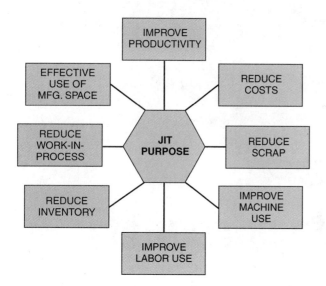

Fig. 54-1 JIT strategy is to eliminate waste through inventory reduction, shorter lead times, product and process design, education, and other related programs. (Courtesy of Kelmar Associates)

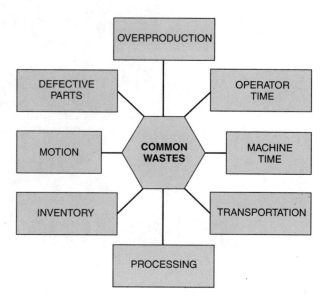

Fig. 54-2 Common wastes that must be eliminated if a JIT program is to be successful. (Courtesy of Kelmar Associates)

JIT PHILOSOPHY

The basic philosophy of JIT manufacturing is the elimination of all waste in materials, machines, labor, time, manufacturing space, etc. to improve manufacturing efficiencies. This is an approach to business management and manufacturing with the prime purpose of serving the customer better. To achieve these goals, it is important to work on eliminating the following common wastes, Fig 54-2:

1. OVERPRODUCTION—It is important to produce only what is required, when it is required; making more than required is a waste of time, money, and material.

2. TIME—Some of the common things that do not make effective use of time are operators waiting due to long setup times, slow material movement, shortage of parts and tools, slow processes and procedures.

3. TRANSPORTATION—Moving of material from one place to another is a costly waste of time. The manufacturing process should be designed to eliminate or greatly reduce the handling and movement of material.

4. PROCESSING—One of the objectives of JIT is to eliminate products and processes through redesigning the product or process. It is often found that through redesigning,

some products are not needed and should not be made.

5. STOCK—Keeping inventory at any stage, from raw materials to work-in process (WIP) is costly, wasteful, and the sign of an ineffective and inefficient manufacturing operation.

6. MOTION—Any motion that does not add value to the product or service is wasteful and should be eliminated.

7. DEFECTIVE PRODUCTS—Parts or products that must be scrapped or reworked, are a costly waste that must be eliminated by changing the manufacturing process for zero defects.

JIT ADVANTAGES

When JIT techniques are properly applied, there is a great reduction in all waste resulting in decreased manufacturing costs and improved productivity. The savings are evident in most areas involved in the manufacturing operation, as can be seen by the following:

• 20% reduction in direct labor,

• 45% reduction in indirect labor,

• 40% reduction in manufacturing space,

• 50% increase in productivity,

Fig. 54-3 To succeed, any JIT implementation must have the full support of top management. (Courtesy of The Association for Manufacturing Technology)

- 60% reduction in quality defects,
- 90% reduction in production time,
- 38% reduction in capital expenditures,
- 58% reduction in inventory.

Most companies that implemented JIT, found that the primary benefit at the start of the program was the reduction of inventory. The inventory tied up a great amount of money that could have been used to update manufacturing processes, buy new equipment, or reduce the company debt.

Even though there are many benefits of JIT and its philosophy is fairly simple, implementing it is sometimes difficult and time-consuming. The implementation depends on the product demand, flexibility of the supplier (subcontractor), quality of the parts and material, and the total involvement and teamwork of management and employees, Fig. 54-3. The difficulties in implementing JIT are generally due to changes in supplier relationships, purchasing policies, total quality management, and management philosophy.

● JIT PURCHASING

Any material that is purchased from suppliers (sub-contractors) must meet quality standards and be delivered to the production line as it is required for use. Material that arrives too early takes up valuable floor space, while material that is late in arriving holds up production. JIT purchasing departments are concerned whether suppliers can meet the following requirements:

- reduced order quantities,
- frequent and reliable delivery schedules,
- reduced and predictable lead times,
- consistent high quality for all materials.

Features

The key to a successful JIT purchasing plan is the continuous long-term relationship with suppliers who are committed to, and can adhere to all JIT principles. This is one of the key changes to former purchasing policies, because the number of suppliers is limited, and each will be evaluated on their ability to consistently supply quality products.

1. LIMITING SUPPLIERS—The number of suppliers for any particular product in JIT is reduced to a very few. This allows companies to improve their relationship with the suppliers, help them improve their part quality, and reduce the problems of working with many suppliers.

2. LONG TERM RELATIONSHIPS—Companies are working toward establishing better bonds with suppliers and treating them as partners, and not competitors. They expect suppliers to initiate a certification program where they take total responsibility for the quality of their products.

3. COOPERATIVE PRODUCT DESIGN—Many companies are looking to use the supplier's expertise in the design of product components. Since the supplier knows best its own equipment and processes, and will be re-

Fig. 54-4 Input from suppliers, who must make the product, is important when designing a product. (Courtesy of The Association for Manufacturing Technology)

sponsible for producing the part, its input is very valuable in product design, Fig. 54-4.

4. JIT DELIVERIES—The characteristics of JIT deliveries are small orders, consistent order quantities, and firm order schedules. It is quite possible for a supplier to make several deliveries in a day to feed the production requirements. Deliveries in some companies range anywhere from three to twelve times per month.

5. GEOGRAPHIC LOCATION—Part of JIT is to try to use suppliers who are close to the company, or encourage them to set up an operation nearby. Many suppliers have established service centers near their customers to better support the JIT requirements.

6. SUPERIOR QUALITY PRODUCTS—A key issue in the JIT program is the continual and never-ending improvement of product quality. The goal of the JIT purchasing program is to eliminate the inspection of incoming goods entirely because the supplier's production processes are so reliable. Almost all world-class manufacturers work with suppliers by offering training and assistance to improve the supplier's production processes and the quality of their product. The eventual goal is to continually improve quality until it achieves the goal of **zero-defect manufacturing.**

7. SUPPLIER FEEDBACK—It is important for the company to regularly provide suppliers with information on how their product compares with that of their competitors in quality, technology, responsiveness, dependability, and cost. It is the responsibility of JIT purchasing to see that any errors in procedures or product quality are corrected by the supplier before further purchases are made.

The purchasing process of companies working with JIT has undergone many changes in order to meet competition. Suppliers are now looked at as being just a part of the company's production process, and the need to cooperate with each other is essential for the success of both.

● **JIT IMPLEMENTATION**

The implementation of a full JIT program takes a long time, and there are many obstacles that must be faced. Unless management is fully

committed to the program, it will be very difficult, if not impossible, to implement a JIT program because it takes a long time for results to be noticeable.

Time Required to Implement and Show Results

In many cases, top management underestimates the time it takes to make improvements that prove financially that the system is working. It was easier to implement JIT gradually into various company areas and correct problems as they arose. The full implementation of a JIT manufacturing in a company can take three years or longer, with visible benefits within the first year.

Changes Are Required

JIT will affect all levels of an organization from management to the worker, and also include the relationship with the suppliers. The support of workers is essential to the successful implementation of a JIT system. The following are some of the points that management must consider in gaining worker support:

1. Management and workers must share the same goal, a strong desire to see the company succeed.

2. There must be a mutual trust that the implementation of JIT will not lead to job losses.

3. Workers must understand that their intelligence and experience in solving problems and suggesting changes are welcomed and appreciated.

4. Workers must receive training to make them flexible enough to handle many jobs, Fig. 54-5.

5. Management must always keep in mind the human dignity of the workers.

JIT must be applied from the ground up, and major changes must be made in both the purchasing and manufacturing processes. It is important to everyone for changes to be made in order to reap the benefits JIT will surely bring. Some of the more common changes are:

- Fundamental philosophies and business ground rules used for decades will be challenged.

- A high level of cooperation and interdependence is required between all departments of a company.

Fig. 54-5 Employee retraining and involvement is essential for a successful JIT program. (Courtesy of Fadal Engineering Co., Inc.)

- Systems information, such as inventory management, order entry, purchasing, and order tracking, must be obtained.
- JIT manufacturers and suppliers must adopt and adhere to strict statistical quality control standards.
- Manufacturers may have difficulty with manufacturing capabilities and production flow because JIT achieves superior quality, volume, and product flexibility, with a small work-in-process inventory, Fig. 54-6.
- The quality of purchased goods must be of the highest quality, otherwise production stops. This means that suppliers will re-

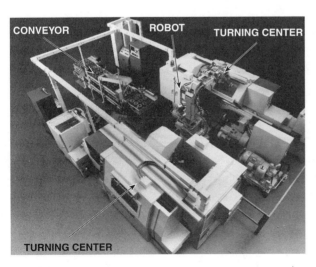

Fig. 54-6 The proper design of manufacturing cells can improve product quality and increase productivity. (Courtesy of Cincinnati Milacron Inc.)

quire specific quality control training programs that often take time and slow JIT implementation.

- Quality must be engineered into the production process, which can take time to train workers to identify defects and change processes to correct them.
- The setup time and cost of running small batches require engineering and retraining time to avoid production delays.

The number of obstacles for implementing JIT can be overcome if there is support from top management for the long period of time it takes for the change to JIT. Once the obstacles of support and time can be overcome, JIT is very achievable and will increase productivity, provide high-quality parts, and reduce manufacturing costs.

SUMMARY

JIT is a philosophy of getting the right material to the right place at the right time. Its main purpose is to reduce waste at all levels and improve manufacturing efficiency. The following points must be considered when implementing JIT.

- Use group technology effectively.
- Make operations simple, yet flexible.
- Eliminate or greatly reduce setup time.
- Strive for zero inventory in all operations.
- Use machines and equipment effectively.
- Control quality at the source (suppliers).
- Emphasize product performance.
- Monitor and control processes and the results.
- Establish a total preventive maintenance program.
- Stress never-ending improvement in processes and the product.
- Encourage management to be actively involved.
- Encourage and allow worker involvement.
- Assure employment security for workers.
- Establish long-term relationships with suppliers.
- Assist suppliers in producing and supplying quality products.

KNOWLEDGE REVIEW

1. What is the basic philosophy or goal of JIT?

JIT Philosophy

2. List seven areas where it is important to eliminate waste.

JIT Advantages

3. What have companies found to be the primary benefit of implementing a JIT program?

JIT Purchasing

4. List four requirements that JIT suppliers must meet.

5. Why is a supplier's input valuable in the design of a product?

UNIT 55

Lasers

In industry today, the need for flexibility and soft-tooling (non-contact) are vital to producing products for this dynamic and ever-changing world. Laser systems have been found to be very useful in highly competitive industries such as automotive, aerospace, and advanced manufacturing because they can solve part manufacturing and identification problems. Laser systems are generally used for material processing (cutting, welding, drilling, and mold work) and product identification (marking), Fig. 55-1.

The **laser,** the acronym of **L**ight **A**mplification by **S**timulated **E**mission of **R**adiation, is a unique tool that can cut, weld, drill, produce cavities, heat treat, and mark a wide variety of materials. Lasers have found many applications in material processing, surveying, identification and tracking, medicine, spectroscopy, communications, and precision measurement. It has often been referred to as a tool that is flexible, can be set up quickly, and needs no hard tooling or dies. It responds quickly to programmed commands, does not wear, reduces or eliminates finishing operations, and does not create excessive noise, dust, fumes, or chips.

OBJECTIVES

After completing this unit, you should be able to:

- Explain the operating principles of a laser.
- Describe the differences between solid, gas, and liquid lasers.
- Identify areas of laser application in manufacturing processes.

KEY TERMS

bar coding

laser (excimer)

laser

lasercaving

Fig. 55-1 Lasers are used to quickly and accurately make complex cuts in sheet metal parts. (Courtesy of *Manufacturing Engineering Magazine*)

LASER PRINCIPLE

The laser beam is a very narrow, intense beam of coherent (united) light that can be controlled over a wide range of temperatures at the point of focus, from slightly warm to several times hotter than the surface of the sun. There are three main types of lasers commonly used for material processing purposes; they are the CO_2 (carbon dioxide), the Nd:YAG (neodymium-doped yttrium aluminum garnet) and the excimer lasers.

The general principle that most lasers operate on is as follows:

- Electrons or molecules in a lasing medium, a gas or solid core, are excited to higher energy levels by photons from another energy source.

- As they return to their original unexcited state, they release a burst of energy in the form of a photon.

- The energy source for solid-state lasers is a high-intensity light, while for gas lasers, it is an electrical discharge.

- The photons are bounced between two high-polished mirrors within the laser unit.

- When a photon passes near another excited particle in the same wavelength, that particle is stimulated to emit a photon.

- Each new photon is capable of initiating the release of other photons, thereby forming a chain reaction.

- One of the mirrors in the laser unit is partially transparent, and when enough photons are created, an intense column of coherent light, the laser beam, comes through the partially transparent mirror.

TYPES OF LASERS

There are several types of lasers used for manufacturing purposes such as solid, gas, liquid, and semiconductor. Since the solid, gas, and liquid lasers are the most common and basically use the same principle, only these will be covered. The main parts of any laser, Fig. 55-2, includes a power supply, a lasing medium (suitably contained), and a pair of precisely aligned mirrors. One of the mirrors has a totally reflective surface, while the other is only partially reflective (about 96%).

SOLID LASER

The power source charge fires the large flash tube similar to a photographic electronic flash. The flash tube produces a burst of light sending or "pumping" energy into the lasing medium (ruby rod), Fig. 55-2. This excites the chromium atoms in the ruby rod to a high energy level. As these atoms return to their initial state, they produce heat and bundles of light energy called photons. The photons produced then strike other atoms causing more photons of identical wavelengths to be produced, Fig. 55-3. These photons are reflected back and forth (oscillate) along the rod by the mirrors on each end

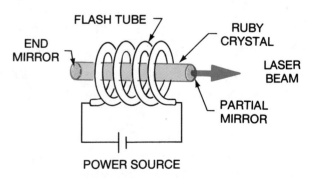

Fig. 55-2 The solid ruby laser has found uses in surgery, measurement, welding, and drilling. (Courtesy of Kelmar Associates)

WAVE GROWTH BY SIMULATED EMISSION

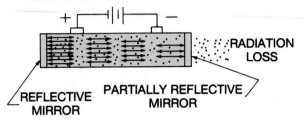

Fig. 55-3 Light waves traveling along the ruby rod grow as excited photons are stimulated. (Courtesy of Kelmar Associates)

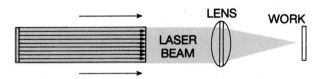

Fig. 55-4 Parallel light waves pass through the partially reflective mirror and are focused by the lens to a contact point on the work surface. (Courtesy of Kelmar Associates)

of the rod. As a result, the number of photons continues to increase, producing more energy until some of them pass through the partially reflective mirror producing a laser beam. This laser beam then passes through a lens onto the point of focus on the work, Fig. 55-4.

Solid lasers have found wide applications in the fields of surgery, atomic fusion, drilling diamond dies, measurement, and spot-welding.

● GAS LASERS

Almost 60% of the lasers used in industry are CO_2 (carbon dioxide) systems because they provide more power than YAG systems. Though the lasing medium is often a mixture of gases, CO_2 or helium-neon provides the molecular action required for photon generation. The lasing medium is contained in a resonator cavity (glass tube) through which gas flows to replenish the molecules used during the electrical discharge which stimulates the lasing medium, Fig. 55-5. The length of the resonator determines the amount of effective laser power, therefore pre-

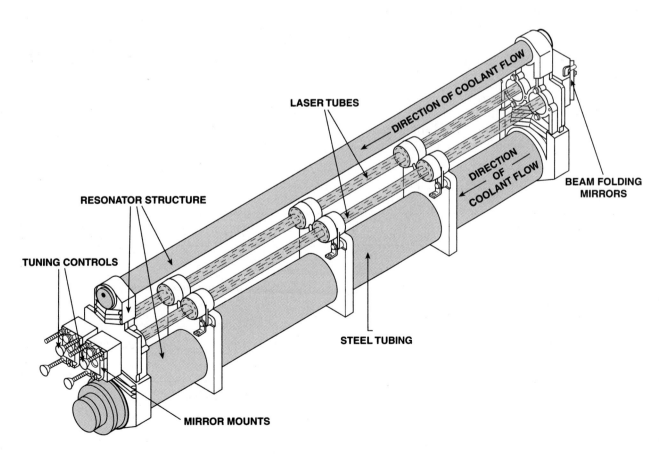

Fig. 55-5 The construction of a CO_2 gas laser resonator. (Courtesy of *Manufacturing Engineering Magazine*)

cisely aligned mirrors are used to bend the beam through more sections of the resonator. In this way, the length of the optical path can be increased without increasing the overall length of the laser.

CO_2 lasers are used for sheet metal cutting operations because they are faster or better than conventional machining. They can cut sheet metal faster, more accurately, with less distortion, thereby producing a better quality product. They are especially useful for prototype parts and where small quantities are required. CO_2 lasers also find use in cutting organic materials such as leather and rubber, for welding, drilling, perforating, etc.

● YAG LASERS

YAG solid-state lasers, although not as powerful as CO_2 lasers, are being used in manufacturing for cutting, welding, drilling, and heat-treating operations. Their power ranges from a few milliwatts to more than 400 watts, with the most common being in the 100 watt range. The power range is necessary to cut various thicknesses of material. Low-power levels are used in the manufacture of integrated circuit-mounted resistors.

High-power YAG lasers are being used in cutting, welding, and drilling operations:

1. CUTTING—Sections of stainless steel up to 1/2 in. (12.7 mm) thick have been cut with less than 1° taper per side, minimal thermal damage, and good edge quality. These lasers are especially suited to cutting materials with high thermal conductivity and reflectivity such as aluminum because of their high-energy, high-repetition rate, and short pulse widths.

2. WELDING—YAG lasers are being used to weld ferrous and nonferrous metals due to their longer pulse widths and pulse-shaping capabilities. It might eventually replace electron-beam, resistance, and gas tungsten arc welding.

3. DRILLING—YAG lasers produce high-quality holes in turbine blades for jet engines. They are also used for drilling cooling holes in automotive components, ceramics, and other exotic materials.

See Table 55-1 for advantages and disadvantages of CO_2 and YAG Lasers for various applications.

● EXCIMER LASERS

Excimers (pulsed-gas lasers), the fastest growing class of lasers, emit light in the ultraviolet (UV) region of the spectrum. The lasing gases are mixtures of a noble gas (those that have no tendency to combine with other elements and are found in the atmosphere), such as argon, krypton, or xenon, and a halogen gas.

Under normal conditions, the atoms of these gases do not react chemically, however, when they are exposed to the excited environment of an electrical discharge, a noble gas atom and a

Table 55-1 Advantages and disadvantages of CO_2 and YAG lasers for various applications

Advantages	Disadvantages
Drilling	
• Can drill high-alloy steels • Can drill angled and difficult-to reach holes • No tool cost or wear • Fast setup times	• Deep hole quality not good • Cannot drill blind holes to accurate depth • Material must be removed from exit holes • Trepanning slows producing large holes
Cutting	
• Able to cut high-alloy metals • Can process small lot sizes where die costs would be too high • Able to cut complex shapes • Edge quality is very good	• Uneconomical for high volumes • Thickness is limited due to taper of cut

halogen gas atom combine to form a dimer (double the molecular formula) such as xenon chloride or argon fluoride. The word excimer comes from **EXCI**ted di**MER** which is the excited combination of a noble and halogen gas. As the electrical discharge decays, the dimers split apart and emit UV light.

Excimer lasers interact with materials via ablation (erosion or wearing away), which happens when the photon (radiant) energy of the light absorbed by the material is great enough to break chemical bonds. The fragments from the material expand explosively away from the machined area. The entire process takes only a few hundred nanoseconds (one-billionth) and most of the laser energy and the heat created are carried away by the ejected fragments.

Applications

The excimer's biggest advantage is being able to produce high-quality edges on parts with almost no microcracking or thermal damage. Ablation (erosion) removes a very small slice of material with each laser pulse for accurate depth control within fractions of a micron (one thousandth of a millimeter).

The Excimer Process

A polymide layer mask, of the form required, is deposited on the surface of the material. The laser beam then scans over the part and machines the sample through the open areas of the mask, Fig 55-6. Sometimes the mask is put in front of the lens and the mask pattern is imaged onto the material. The high resolution of the imaging process, combined with the erosion process, virtually eliminates any heat-affected zone and allows machining fine features in polymer films, ceramics, and semiconductors.

Excimers are excellent for:

- removing small amounts of material precisely,
- high-speed marking of parts,
- photochemical applications that use high-efficiency, high-power UV laser light,
- hardening and annealing thin surfaces,
- micromachining of materials,
- semiconductor processing in the submicron range.

LIQUID LASERS

Liquid lasers use an organic dye in a solvent as the lasing medium, Fig. 55-7. The liquid is circulated by the pump and the flash tube excites the dye molecules to high energy levels to produce photons. When a photon passes another excited particle, that particle is stimulated and produces another photon, creating a chain reaction. The process continues as the adjustable external mirrors create a feedback, and when enough photons are created in the tube, an intense light passes through the partial mirror as a laser beam. Liquid lasers are best suited to chemical analysis because of the tuning prism which can be rotated, and in turn produce a choice of several different colors and corresponding wavelengths.

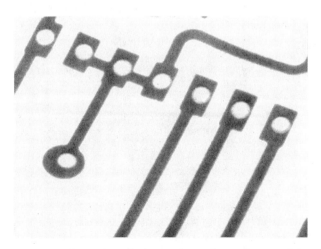

Fig. 55-6 An excimer laser produces forms in material through a polymide layer mask of the desired shape. (Courtesy of *Manufacturing Engineering Magazine*)

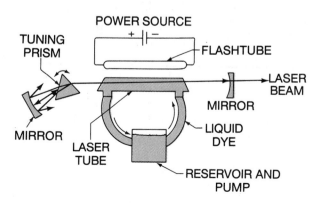

Fig. 55-7 Liquid lasers are well suited for chemical analysis. (Courtesy of Kelmar Associates)

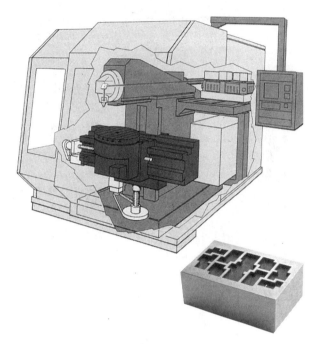

Fig. 55-8 Lasercaving makes it possible to produce intricate forms in the hardest materials with no machining forces or tool wear. (Courtesy of *Modern Machine Shop Magazine*, copyright 1994, Gardner Publications Inc.)

● LASERCAVING

Lasercaving, a relatively new application for lasers, makes it possible to machine cavities in work which is too hard to mill, or cannot be EDM'd because the material is nonconductive. Lasercaving is especially effective in materials such as ceramics, composites, carbides, quartz glass, and titanium-based alloys.

The control of where the laser beam cuts is similar to the milling process. Pockets are created by running the cutting tool (laser) back and forth through the area to be machined. Because the laser cutter is very small, the material-removal rates are relatively slow. Although the tool is small, it is possible to produce very accurate forms and sharper corners, Fig. 55-8. The computer part programs used by lasercaving are similar to those used for stereolithography, but in reverse. The geometric computer model of the part is sliced up into cross-sections, the thickness of which corresponds to the laser's depth of cut. This technology could be very beneficial to die and mold makers who are constantly faced with producing and finishing intricate forms and shapes.

● INDUSTRIAL APPLICATIONS

Measurement and Inspection

Gas discharge lasers, having mirrors with high reflectivity, are widely used in metrology applications such as laser micrometers, Fig. 55-9, and image recognition systems. The laser beam is directed from a fixed mirror toward a rotating mirror that is mounted on the end of a motor shaft. The rotating mirror flashes the beam quickly across the transmitter lens which causes the beam to scan through the work area in a linear or parallel fashion. This makes it possible to take accurate measurements of stationary parts, or parts moving at high speed. When used with a microprocessor-based controller, laser-based measurements can be made in a fully automatic manufacturing system.

Laser Welding

In laser welding, the weld area can be located away from the laser source. By using laser optics, the depth and width of bead can be controlled. Any material that can be welded using conventional methods, can generally be welded faster and better with a laser. Because of the rapid heating, heat transfer to other parts of the workpiece is kept to a minimum, thereby reducing distortion and keeping clean-up operations to a minimum, Fig. 55-10.

Some of the most common advantages of laser welding are:

1. The ability to weld dissimilar metals at consistent quality.

2. Thin material can be welded at a high speed.

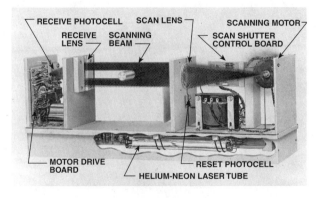

Fig. 55-9 The LaserMike can make accurate, non-contact measurements of parts while they are in production. (Courtesy of LaserMike Div. Techmet Co.)

Fig. 55-10 On some lasing machines, the laser beam can be orbited to produce a wider weld bead. (Courtesy of *Manufacturing Engineering Magazine*)

3. Thin and thick materials can be easily welded.

4. It can be readily automated.

5. Can be used for glass, quartz, and plastic applications.

● LASER HARDENING

The high-speed transformation hardening of carbon steels, over 0.2% carbon, is a major application of lasers. The advantages laser heat treating has over conventional methods are:

- It reduces or eliminates workpiece distortion.

- Induces residual compressive stresses into the part.

- Selected areas can be hardened such as ring gears, helical gears, gear racks, internal bores, etc.

- High speed hardening of various part thicknesses.

- The process can be automated in almost any environment.

Before steel shafts are laser hardened, they must be stripped of any rust-protective coatings. Then a coating that improves the absorption of the laser radiation into the material is applied only to the areas requiring hardening. The heating process is immediately followed by a water and/or oil quench. The heating changes the carbon steel into austenite, and the quenching transforms it to martensite.

● LASER GAGING

Lasers are playing an ever-increasing, important role in gaging and inspection so necessary for statistical process control. Lasers are able to do what quality control standards require, that is, to perform 100% inspection with 100% accuracy at full production speeds. Some of the more common purposes of lasers in gaging/inspection are:

1. DIMENSIONAL MEASUREMENT—Measuring a part as it is being made, to see if it is right, and adjusting the process on the spot, if necessary.

2. LASER PROBES—Coordinate-measuring machines, using laser scanning probes with triangulation to improve the machine's accuracy, can scan contoured surfaces 50 times faster than noncontact probes.

3. NONDESTRUCTIVE TESTING—The use of lasers has created a new field of nondestructive testing known as holographic interferometry. This is the process which has been used to inspect machine tools, tires, computer components, automobiles, and aircraft parts to find defect strains, unbonded areas, and cracks that are invisible to the eye. The use of hologram (interference patterns created with the aid of a beam of laser light from which three-dimensional images can be reconstructed) allows precise measurements to be made in a manufacturing environment.

● LASER MARKING/BAR CODING

The initial use of laser-based systems for on-line product marketing was in food packaging, where it was used to monitor shelf life and trace products in case of recall. It quickly found applications in electronic industries for the noncontact coding of small fragile electronic components because it was two to twenty times faster than other methods. The laser marks do not have any effect on the product being marked, and the marks do not deteriorate.

Laser marking is being used in many industries because of the drive for zero defects, inventory control, and the need to identify defects in the manufacturing process, Fig. 55-11. Laser machining is able to mark identification codes on parts, particularly those of nontraditional materials or complex shapes.

Fig. 55-11 A YAG laser system was used to identify this piston for processing and inventory purposes. (Courtesy of *Manufacturing Engineering Magazine*)

Bar Coding

Bar coding, one of the best data collection systems, can be used on almost anything that can be counted (taking inventory), packaged (supermarket labeling), and tracked (letters and packages in courier services). It can also be used for monitoring quality control on assembly lines, and collecting information through multiple work stations.

The most common application of bar coding is in the retail stores, where printed labels consisting of a series of dark lines against a white background are used to code certain information about the product. The code could be used for pricing, tracking, marketing, and inventory purposes. There are as many as 16 digits of numerical information in a bar code, and by varying the width of the lines and spaces, the information can be altered for each product, Fig. 55-12.

Industry is using bar coding for many of the same purposes as in retail stores. Components

Table 55-2 Selected laser applications

Type of Process	Typically Used Lasers
Metal cutting	Pulsed and continuous-wave CO_2, ruby, Nd:YAG
Metal welding	Pulsed and continuous-wave CO_2, Nd:YAG, Nd:glass, ruby
Metal drilling	Pulsed CO_2, Nd:YAG, Nd:glass, ruby
Metal annealing/ heat treating	Continuous-wave CO_2
Plastic cutting	Continuous-wave CO_2
Ceramic cutting	Pulsed CO_2
Soldering	Pulsed Nd:YAG
Ceramic scribing	Pulsed CO_2
Metal marking	Pulsed CO_2, Nd:YAG
Plastic sealing	Continuous-wave CO_2
Photoresist removal	Excimer
Metal surface alloying	Continuous-wave CO_2, Nd:YAG
Transformation hardening	Continous-wave CO_2
Metal surface texturizing	Pulsed CO_2
Resistor trimming	Nd:YAG, CO_2
Semiconductor fabrication	Excimer, Nd:YAG
Plastic drilling	Excimer
Plastic/ceramic marking	Excimer

are identified with bar codes so that when they are brought to a machine for processing, the laser reader can quickly identify the part to be sure only correct components are machined. See Table 55.2.

SUMMARY

- A laser beam is a very narrow, intense beam of coherent (united) light which can be controlled at the point of focus over a wide range of temperatures. Lasers are generally used for material processing and product identification applications.

- Lasers can cut, drill, weld, and harden a wide range of materials that would be difficult to process by other means.

- Lasercaving is used to machine cavities in materials such as ceramics, composites, carbides, quartz, and titanium-based alloys

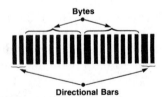

1 1 0 1 0 1 0 0 0 0 0 0 1 0 0 1

Fig. 55-12 A basic bar code format used for identification. (Courtesy of Hewlett-Packard Co.)

that may be too hard or impractical for a milling cutter. This process is useful for die and mold applications.

- Lasers are especially useful for measurement applications and image-recognition systems.

KNOWLEDGE REVIEW

1. Name two manufacturing areas where lasers are generally used.

Laser Principle

2. Describe a laser beam.

Types of Lasers

3. List the main parts of a laser.

4. List the three types of lasers and state where each is used.

Excimer Lasers

5. Name four important uses of excimer lasers.

Lasercarving

6. For what materials is lasercarving effective?

Industrial Applications

7. Name three advantages of laser welding.

Laser Hardening

8. List five advantages of laser heat treating.

Laser Marking/Bar Coding

9. What are three applications of bar coding data-collection systems?

UNIT
56

Robotics

The first American industrial **robot** was installed in the early 1960s in an automotive die casting department. Thousands of other robots quickly followed and they worked many hours in dangerous or boring environments that were not-suitable for humans, without tiring, complaining, or breaking down. Robots are excellent welders and painters, good pickers and placers, however, they can perform many more tasks.

As robot design improved, many of the problems associated with early robots, such as misalignment of parts and the inability to lift heavy parts, were eliminated. Human-like senses of vision, touch, and hearing, were added that enabled robots to perform more complicated and sensitive tasks. Early robots were taught by leading them by hand through the operation to be performed and saving the program on magnetic wire memories or cassette tape. Current robot programs can be created in an off-line computer using 3-D simulation software, Fig. 56-1.

OBJECTIVES

After completing this unit, you should be able to:
- Describe the components that make up an industrial robot.
- Identify many manufacturing applications where robots are used successfully.
- List the advantages of using industrial robots.

KEY TERMS

automated-guided vehicle   gripper (end effector)
robot   sensor

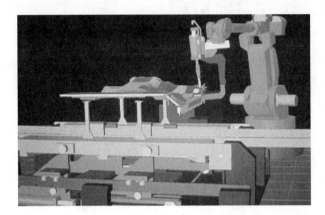

Fig. 56-1 As much as 70% programming time may be saved using an off-line programming system. (Courtesy of Tecnomatix Technologies, Inc.)

ROBOT REQUIREMENTS

In order for an industrial robot to be most effective, it must meet the following criteria: it must be adaptable for many applications, reliable, easy to teach, safe to operate, and be capable of working in hazardous environments.

APPLICATION FLEXIBILITY

The ideal industrial robot should be capable of performing many different tasks and operations. There are two main manufacturing areas where robots are widely used:

1. HANDLING APPLICATIONS—Robots used in handling materials must be equipped with some type of device, such as a pair of gripping fingers or vacuum cups, to pick up or move the material. Robots can be used to load or unload machine tools, forging presses, injection-molding machines, and even use inspection or gaging equipment. They are also used to handle material (moving it from one station to another), retrieving parts from storage areas or conveyor systems, packaging, and palletizing (placing parts on a pallet for transporting to other stations).

2. PROCESSING APPLICATIONS—Robots in this category carry out operations such as spot welding, seam welding, spray painting, metallizing, cleaning, applying sealants, or any other operations where the robot can operate a tool to carry out a manufacturing process.

ROBOT DESIGN FEATURES

An industrial robot is basically a single-arm device with a pair of **grippers (end-effectors)** which are used to move something or perform some task. It must be able to reach all locations in the work area with speed and fluid motion. The robot system must also have high accuracy during its path and positioning, and must be able to direct the robot through a series of directions in sequence. Versatility, high repeatability, ease of programming, compactness, and quick change-over capabilities are key requirements for robots.

Industrial robots are equipped with a manipulator arm that may be able to move in six or more axes, including rotary motion to imitate the human arm, Fig. 56-2. Some spray painting robots have as many as twelve-axes movements to enable the arm to be positioned to suit the part surface being painted. Most of the robots used in industry today are powered electrically because they are fast, accurate, and easy to maintain. Electric robots are widely used because of their low power consumption, smooth motion, and reliability. Special electric robots can be found operating in hazardous environments that would be dangerous for humans.

TYPES OF ROBOTS

Robots are being used for many applications in industry, science, research, engineering, and medicine. The floors of many factories and warehouses are embedded with steel tracks to guide the AGV, **automated-guided vehicle,** Fig. 56-3, used for moving tools, materials, and parts. Astronauts use robots in space exploration for storing cargo and retrieving objects in space. In the private sector, personal robots have been developed with sensory perceptions and artificial intelligence to work with people in service functions. One type of mobile robot is used in hospitals to transport material automatically to various hospital wards. These robots can be programmed to aid nurses by carrying light meal trays, glass supplies, lab specimens, mail or pharmaceuticals to wherever they may be required. Equipped with ultrasonic sensors and microprocessors, these robots are capable of going around poles or abutments that may be in their path.

END-OF-ARM TOOLING

Generally, industrial robots are designed to work with grippers (end-effectors), specially de-

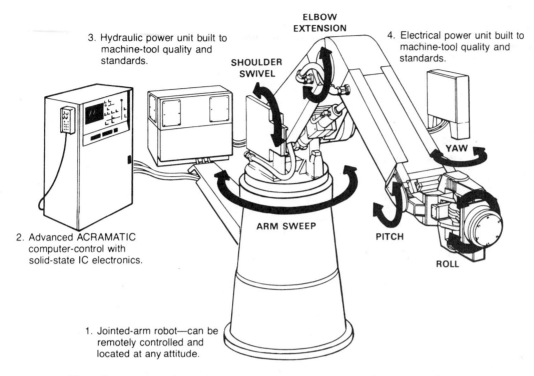

3. Hydraulic power unit built to machine-tool quality and standards.

ELBOW EXTENSION

SHOULDER SWIVEL

4. Electrical power unit built to machine-tool quality and standards.

YAW

ARM SWEEP

PITCH

ROLL

2. Advanced ACRAMATIC computer-control with solid-state IC electronics.

1. Jointed-arm robot—can be remotely controlled and located at any attitude.

Fig. 56-2 The robot arm may have six or more axes of movement. (Courtesy of Cincinnati Milacron, Inc.)

Fig 56-3 Automated guided vehicles (AGV) are used in flexible manufacturing systems to move tools, equipment, and materials. (Courtesy of Giddings & Lewis, Inc.)

Fig. 56-4 A gripper fitted with a pneumatic toggle mechanism to hold radiator cores for brazing. (Courtesy of *Manufacturing Engineering Magazine*)

signed fixtures, or suction cups. Grippers used for handling workpieces are one form of end-effectors, Fig, 56-4. Grippers are usually designed for holding a paint gun, a spot welder, or for a specific purpose. They may also be equipped with gaging, optical sensors, or probes for measuring, inspection, or location purposes. Some grippers collect data that is relayed back to a cen-

tral data processing point that enables the robot to make adjustments for missing or improperly positioned parts. While grippers have been developed that resemble human hands, much work remains to be done to get these mechanical hands to move and act like human hands.

TYPES OF SENSORS

Many robots use some form of sensory system to collect data, to enable the control system to predict or make decisions, to locate and position components, and to regulate the operation of the robot.

In order for a robot to feel or fit parts properly, tactile (touch) **sensors** are located in the gripper, and/or in combination with other sensors, in the wrist, Fig. 56-5. These are used to control the degree and direction of squeezing pressure. Sensory systems enable the robot to identify the type and rotational position of the object. Force or torque sensors may also be located in the gripper or the robot's wrist to indicate the resistance force (weight) of the object being held, and adjust gripping pressure to avoid crushing the object.

ROBOT VISION

Photoelectric diodes play an important role in the robot's arm contact sensing. These diodes trigger on/off signals such as gripper closed or part jammed. Solid state, closed circuit, black-and-white video cameras are used in robot vi-

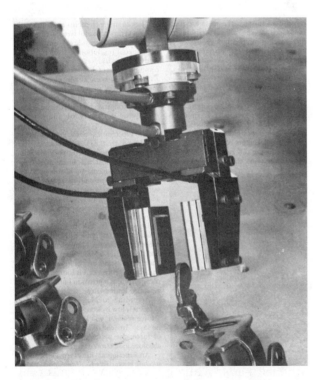

Fig. 56-5 Tactile sensors mounted on a robot gripper, are used during the assembly of carburetors. (Courtesy of *Manufacturing Engineering Magazine*)

sion, and by a system of template matching, the images are processed by computer analysis. Robot vision can also evaluate grasp orientation and automatically find deviations in the part by comparing them with stored images in the computer memory.

The video system relies on one of two light sources, ambient or structural. Ambient light depends on surrounding natural light; the structural light source is much faster and provides special templates of illumination that are stored for comparison in the computer's memory. Ambient light compares corresponding points on two images by triangulation, taking measurements from different viewing locations. The robot's movements are controlled by the speed with which its visual information is processed.

ARTIFICIAL INTELLIGENCE

How fast industrial robots can move depends upon the command and control languages and software that works with the robot's sensory systems. All this activity is stored in the computer program for the robot.

There are three ways of programming robots—manual, teach, and automatic.

1. MANUAL MODE—In the manual mode, a robot can be taught by leading the arm through the necessary movements by an operator. The robot learns from this experience and then forms its own pattern of movements based on its stored knowledge. In this way, the robot can perform the task and also evaluate itself. Manual-mode programming is suited for performing operations that are difficult to do, or difficult to explain.

2. TEACH MODE—In the teach mode, the axes of the robot arm can be moved in a coordinated manner by the use of the teach pendant, Fig. 56-6. Movement within the coordinate system is controlled by the positional pushbuttons on the pendant. The pushbuttons on the teach pendant make it very simple for the human operator to teach the robot to perform specific tasks using any coordinate system.

3. AUTOMATIC MODE—In the automatic mode, the robot movement is controlled by the computer program. There are various types of control languages used for programming robots, the most common being the VAL and KAREL formats. The VAL

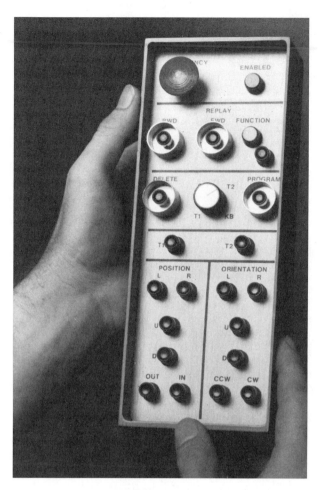

Fig. 56-6 The teach pendant makes it easy to teach a robot various tasks. (Courtesy of Cincinnati Milacron Inc.)

(versatile assembly language) and KAREL (named after the person who introduced the term robot), generally cover the range of motion, grippers, input, output, etc.

While teaching by example is still widely used, it seems that in the near future, it will be replaced by artificial intelligence. With the tremendous technological changes taking place, it will be difficult to know exactly to what extent robots will be used in the future.

● SAFETY PRECAUTIONS

Robots, like other pieces of machinery, must be handled carefully in order to prevent accidents. Always observe the same basic safety rules as with other machines, and also the following that apply specifically to working with robots:

1. The robot working area should be enclosed by some form of barrier to prevent people from entering while the robot is working. Entrance gates to the enclosure should have controls that automatically stop the robot if anyone enters accidentally.

2. Emergency stop buttons, that stop the robot and cut off the power, should be available outside the working range of the robot.

3. Robots should be programmed, serviced, and operated only by fully trained workers who understand the robot's action.

4. Extreme care should be used during the programming cycle, especially if it is necessary to have humans within the robot's working area.

5. The work done by a robot should not be a health or safety hazard to human workers in nearby areas.

6. Hydraulic and electrical cables should be placed so that they cannot be damaged by the operation of the robot or any related equipment.

● INDUSTRIAL APPLICATIONS

The industrial robot is finding many applications where jobs are monotonous, physically difficult, or unhealthy for human workers. Because of the flexibility of the robot arm and the ability to teach it new tasks, the robot is continually finding new industrial applications. The most common applications are:

1. LOADING AND UNLOADING MACHINE TOOLS—Robots are widely used for loading and unloading machine tools, forging presses, and injection-molding machines, Fig. 56-7. It is quite common to have one robot loading and unloading two or more machines in one work cycle.

2. WELDING—Computer-controlled industrial robots can spot-weld car bodies as they are moving on a conveyor. Unique tracking capabilities enable the robot to place the welds accurately on the car body without stopping the line.

Seam welding is another application where robots are finding wide use, Fig. 56-8. The Cincinnati Milacron Miltrac seam-tracking device inspects the weld path *through the arc,* and adjusts the robot's po-

Fig. 56-7 A robot can load or unload two machine tools at the same time. (Courtesy of Cincinnati Milacron Inc.)

Fig. 56-8 Robots are very well suited for welding and spot-welding operations. (Courtesy of The Association for Manufacturing Productivity)

sition to compensate for variations in the start point or seam path in order to consistently produce accurate weld weaves.

3. MOVING HEAVY PARTS—Robots are being used to move heavy parts from an input station, through various machining and gaging stations, to an exit conveyor. They can also bring materials from storage areas, package and palletize, inspect, sort, and assemble.

4. SPRAY PAINTING—The painting of automobile bodies is an example of a job that would be unhealthy for human workers. The robot can be taught (programmed) to follow the contour of an automobile body or part and apply a consistently even coat of paint to the surface, saving as much as 20% of the materials that humans use.

Fig. 56-9 A robotic workcell being used for the assembly of wire harnesses. (Courtesy of *Manufacturing Engineering Magazine*)

5. ASSEMBLY—An assembly job such as stringing a wire harness around a series of pegs is an example of a job that is very monotonous for the human worker, Fig. 56-9. The robot arm is fitted with a spool of wire and a feeding mechanism, and then taught to proceed around pegs and through closely spaced slots to produce a harness.

6. MACHINING OPERATIONS—Some repetitive machining operations such as drilling, are ideally suited for the computer-controlled industrial robot.

● ADVANTAGES

The computer-controlled industrial robot has many advantages that make it a very important industrial manufacturing tool:

1. The computer controls the position, orientation, velocity, and acceleration of the path of the tool center point (TCP) at all times.

2. The robot can easily be taught because it can be moved to different coordinate locations with the teach pendant. The console display provides communication between the teacher and the control.

3. The computer can store system software and taught data in its memory. This allows editing of the taught program at any time, especially when changes are necessary in the program to suit a different application.

4. The program in the computer's memory can be easily expanded to add more taught points or extra functions.

5. The robot can be interfaced with other equipment or machine tools, and can receive and interpret signals from various sensor devices. It can also provide valuable data to a supervisory or executive computer on manufacturing conditions, quantities, and problems.

6. The system can be easily modified to suit production changes. The robot computer can be linked to a supervisory computer that can direct the robot to convert to other programmed operations as required.

7. The computer is able to monitor the robot and the equipment in service, and displays diagnostic messages on the console screen to indicate various error conditions. If necessary, the computer may direct the robot to stop an operation.

SUMMARY

- A robot, a computer-controlled device, consists of three basic components: the machinery or mechanical parts, the controller/computer system, and the software program.

- The industrial robot can perform various combinations of programmed movements, manipulations, and actions involving the movement, placement, or assembly of parts.

- Sensors in the robot wrist enable it to identify the type and position of an object, feel or fit parts properly, and adjust the gripping pressure to suit the part being handled.

- Robots are widely used in industrial applications where jobs are monotonous, physically difficult, or for operations that are unhealthy for humans such as spray painting and welding.

KNOWLEDGE REVIEW

1. What qualities must an industrial robot have in order to be effective?

Application Flexibility

2. List two main manufacturing areas where robots are used and give one application for each.

3. Name four key requirements for an industrial robot.

Robot Design Features

4. What type of sensors are located in the gripper or robot's wrist?

Artificial Intelligence

5. List the three methods of programming a robot.

6. In what mode is a robot's movement controlled by a computer program?

7. Why is the manual mode important?

Safety Precautions

8. Why should a robot's working areas be enclosed by a barrier?

UNIT 57

Statistical Process Control

The need to remain competitive in international commerce forces all nations to control manufacturing processes so that only the highest quality goods are produced. This involves the improvement of the entire production system. The customer must be the most important part of the production line, and quality should be aimed at meeting the needs of the consumer.

The key to quality improvement rests with management. Management must drive the quality system and match its behavior to the objectives of the organization. Some companies think that quality can be improved simply by sending people to the right quality training seminars—but that alone does not work. Quality must spread throughout everything a company does, at all levels of the organization. You cannot buy a solution to all quality problems, nor is there a single, right approach to excellence.

The two key people largely responsible for the move toward quality manufacturing were Dr. W. Edwards Deming and Dr. Joseph Juran. Their work in Japan in the 1950s had a great influence on Japan becoming a leading manufacturing nation.

Dr. W. Edwards Deming stresses simple, straightforward, quantitative methods that prevent quality slippage rather than after-the-fact repair. Although simple in concept, his process control statistics have the potential to revolutionize the management of any process, not just manufacturing and assembly operations. Central to Deming's message is that competitive quality cannot be obtained with traditional inspection methods. He regards statistics as abstracted models of how all systems should function and feels that this theoretical base is essential for quality improvement.

Deming also believes strongly that "quality on the shop floor can be no better than the intent of management which is made in the board room," and maintains that until executives and managers actively adopt quality methods into their own decision making, in effect model the correct attitudes and actions, a quality effort will not work.

Dr. Joseph Juran, whose concepts are very similar to Deming's, uses technical approaches instead of statistics. He focuses on product designs that are both high in quality and able to be manufactured to consistently high quality standards. Juran feels it is critical to have participation not only in engineering and manufacturing, but in all parts of the organization. Managerial review systems set up to measure performance against quality goals, and on-site top management involvement in implementing a quality program are crucial.

Juran insists that quality goals be specific. The statement, "Quality is priority number one," is unacceptable. Instead, a properly specified goal would be "to raise quality levels to at least the levels of competitors" by a certain date, or to "cut costs of poor quality by 50 percent in five years."

OBJECTIVES

After completing this unit, you should be able to:

- List the factors that determine the quality of a manufactured product.
- Compare quality control in traditional and progressive manufacturing processes.
- Explain the principles of statistical process control.
- Understand the use of X-bar charts, R-charts, and histograms used in quality control.

KEY TERMS

histogram
quality control
statistical control chart
statistical process control

● QUALITY CONTROL

Statistical process control can be used to evaluate the stability and predictability of a process, and then use this information as a means to improve the process. The three major steps in a sound **quality control** process involve:

1. The specification of what is wanted.
2. The production of things to meet the specifications.
3. The inspection of things produced to see that they have met the specifications.

The purpose of a quality-control program is to examine manufactured products with the following objectives:

1. The level of quality must meet customer satisfaction (design performance).
2. The part must be within the boundaries of design limitation (production quality).
3. The cost of the manufactured part must be competitive on the world marketplace (cost reduction).

A good quality-control program must be set on a foundation of building blocks that would include planning, sampling, inspection, and correction, Fig. 57-1. All the steps are necessary, and since the quality-control program requires all of them working together, no one part can be eliminated. Since a company is judged by its performance in relation to quality, volume output, cost and delivery, all the steps must be repeated until acceptable products are made. As quality improves, a chain reaction occurs resulting in the benefits shown in Fig. 57-2.

Poor-quality goods result in short-term and long-term costs such as warranty repairs, legal liabilities, product recalls, lost sales, and even bankruptcy, Fig. 57-3. A quality-control program does not cost money, it lowers manufacturing costs since for every dollar spent on eliminating

Fig. 57-1 The building blocks of a good quality control program. (Courtesy of Kelmar Associates)

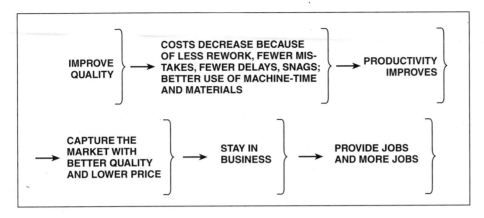

Fig. 57-2 The chain reaction that occurs as quality improves. (Courtesy of Kelmar Associates)

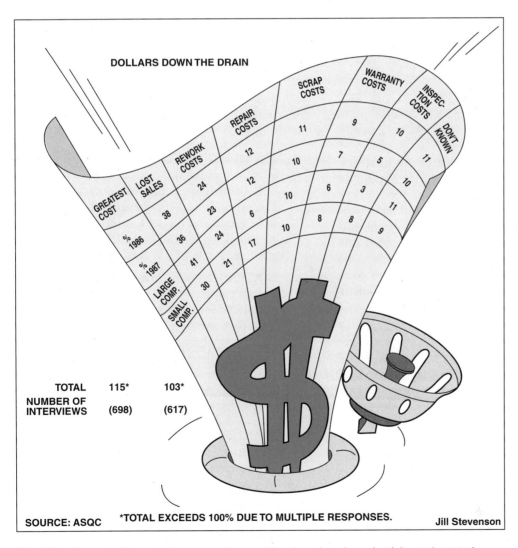

Fig. 57-3 Poor-quality work is very costly, resulting in reduced productivity and wasted money. (Courtesy of *Manufacturing Engineering Magazine*)

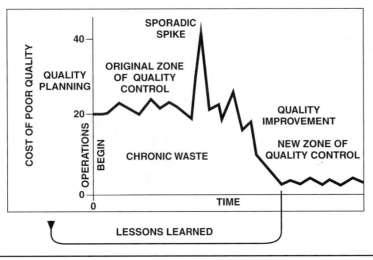

THE JURAN TRILOGY

Fig. 57-4 The Juran Trilogy shows that quality improvement saves manufacturing costs and increases productivity. (Courtesy of Juran Institute)

defects, a saving of five to ten dollars results. Fig. 57-4 shows a comparison of the cost of poor quality in relation to the reduction of waste as a result of quality improvement.

Industry soon realized that there must not be a separation between manufacturing and quality control. It was felt that by the time production got to quality, especially in the final inspection stage, it was too late. It was necessary to catch the defects during the manufacturing cycle and not in quality control, since quality, or the lack of it, occurs in the manufacturing cycle. Quality control could not make the part good or bad; all it could do was to accept or reject the part.

● QUALITY ASSURANCE

Manufacturing control requires quality assurance to make sure the decisions manufacturing control makes are correct. They must determine whether the equipment is capable of producing the required accuracy, if it is being used properly, and whether the correct statistical tool is being used.

● TOTAL QUALITY CONTROL (TQC)

The purpose of total quality control is to produce high-quality products, at low cost, so they may reach the marketplace at the right time and meet customer satisfaction. A number of factors determine the quality of the product manufactured, such as:

1. PRODUCT DESIGN—The design process should involve the customer who should be able to provide valuable information about the product and its eventual use. This will assure that the final product will satisfy the needs of the customer and at the same time make best use of the company's facilities.

2. MATERIAL—The raw material must be of suitable quality to produce the quality of product required. Using poor raw material usually results in a poor product.

3. PROCESS DESIGN—This step in the planning sequence is to develop the best process, system, procedures, etc., for producing the product to specifications and within cost restraints.

4. MANUFACTURING METHOD—The manufacturing process must be error-free, otherwise product error will result.

5. MANUFACTURING EQUIPMENT—The equipment must be dependable and produce the quality product required. Old, worn machines breaking down constantly cannot be expected to produce top quality products.

6. WORKFORCE SKILL—The workforce must be properly trained to handle new machine tools and manufacturing processes.

The quality of a manufactured product depends on all factors working together through a well-planned, coordinated effort by management and everyone in the company. All departments must work toward a common goal for total quality control to be effective.

- Workers should be quality-oriented and understand how they can affect customer satisfaction.

- Every department should understand the role it plays in producing a quality part, and how to work effectively to achieve customer satisfaction.

- Departments should work together in a cooperative spirit and assist any department that is having difficulty.

Quality must begin with management and must be translated by engineers and others into plans, specifications, tests, and production. They are responsible for the material, components, methods of processing, and the method of inspection.

In the traditional manufacturing process, Fig. 57-5A, suppliers brought raw materials, components, and sub-assemblies to a receiving

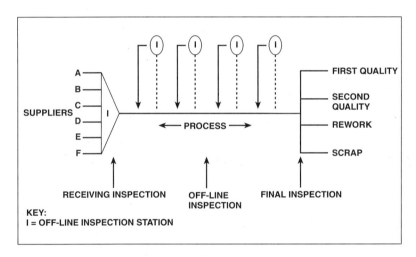

(A) TRADITIONAL MANUFACTURING INSPECTION PROCESS

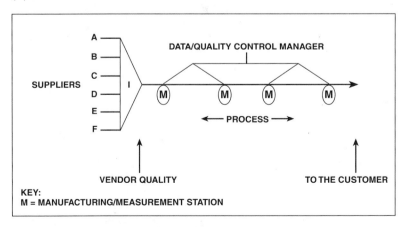

(B) PROGRESSIVE MANUFACTURING INSPECTION PROCESS

Fig. 57-5 A comparison between the traditional and progressive manufacturing inspection processes. (Courtesy of *Manufacturing Engineering Magazine*)

inspection area where they had to be checked for quality. This took up someone's time sorting the good from the bad, and also tied up inventory. The quality control function, primarily an off-line inspection, inspected a few parts after they were machined, while the manufacturing process continued. By the time the results of the inspection were known, it was quite possible that other bad parts were made and mixed with good parts. This meant that the final inspection had to separate the good from those that could be salvaged by rework, and the rest were scrapped. This was very costly from the standpoint of time, material, and rework and resulted in poor quality products.

● NOTE

The real goal must be to manufacture quality into each product. For this to happen, inspection must be an integral part of the manufacturing process, and each person must be responsible for the quality of the part he or she produces.

The progressive manufacturing process, Fig. 57-5B, requires suppliers to use statistical methods to ensure that top quality raw materials, sub-assemblies, and parts are being supplied. In the manufacturing process, inspection information is collected in statistical form by those actually operating the equipment. Should there be a variation in the statistics, it is important to identify and correct all special causes before production resumes. The operator must be aware of changes in part quality and when tools should be changed or adjusted. As can be seen by the process in Fig. 57-5B, the chances of a bad or defective product reaching a customer are greatly reduced.

To assure that company standards are met and maintained, regular meetings must be scheduled by an appointed committee to oversee the obligations of the different departments and personnel. The committee must be assigned the responsibility to recommend, modify, or change procedures that are not effective.

● STATISTICAL METHODS

Line graphs, control diagrams, and checklists are used in industry to check the quality of the parts being produced, as well as those al-

ready produced. The data collected should be in statistical form so everyone can analyze it. Dr. Juran coined the phrase, the *Pareto Principle*, named after an economist, Vilfredo Pareto, who studied the normal distribution of wealth in Italy. What Pareto discovered many years ago was that a relatively small number of the population accounted for a majority of the wealth. That wealth was concentrated in its distribution indicates that it is not normally distributed. Juran concluded that the same principle of common distribution applied to many things.

Under the Pareto analysis, a workpiece, cell, or operation, can be divided into many small classes and examined to see what effect each has on part quality and cost. In Fig. 57-6, the assembly being manufactured showed five problem classes which kept occurring and affected the operation of the engine. Out of the 59 defects found, 47 were in categories 1 and 2; therefore, to improve product quality and reduce costs, improvement efforts should be directed to these areas first.

For many years, Dr. W. Edwards Deming, a quality expert, claimed that only 15% of the problems a company may have are directly related to the workers on the shop floor. He also stated that most of these could be handled by the worker, if he or she were given the proper tools, such as training, equipment, and improved processes.

The remaining 85% of the problems are related to management, because these can only be solved by actions they can take. This may involve changes in scheduling and processes, more and better training programs, and the purchase of new tools and equipment. These decisions can only be made by a well-informed management that is fully aware of problems through the use of statistical process control data analysis.

● STATISTICAL PROCESS CONTROL (SPC)

To understand how SPC works it may be wise to examine how production lots were examined before the use of SPC. For example, if a company had to produce one hundred pieces, a sampling of 13 might have been inspected. If a defective part was found in the sampling and it was established to accept on zero, or reject on a 1 criteria for 13 samples, that was a 1 point AQL (acceptable quality level). A part not meeting specifications was disposed of in a number of ways: scrap, sort, rework, return to supplier, or sell at a discount.

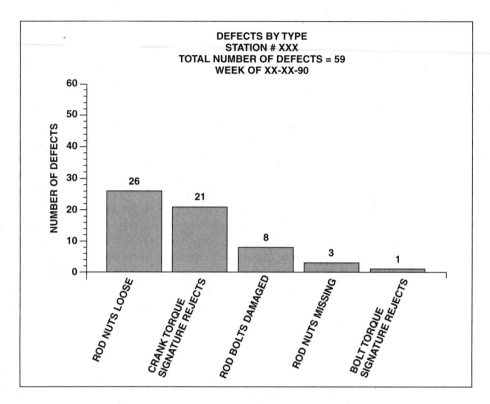

Fig. 57-6 The Pareto analysis can be used to find the most commonly occurring or most expensive problem. (Courtesy of GM Powertrain Group)

SPC uses data from measuring instruments recorded on a graph or chart. To be of value in analyzing the data, it is important that it be accurate and as current as possible. When arriving defects are in statistical control, the only choices are 0% or 100% acceptance inspection, Fig. 57-7. If measurements start running near the high or low quality limit, the process must be changed

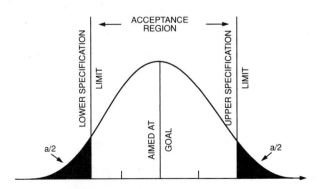

Fig. 57-7 The Bell Curve or normal distribution shows that out of 100 parts inspected, approximately 68% will fall within the acceptance region, and a total of 32% will fall outside. (Courtesy of Juran Institute)

quickly before the product quality is unacceptable. SPC looks at the statistical differences of the part or process and does not accept or reject a product. Analysis alone cannot improve a process. It is when SPC is used actively to build process knowledge, with integrated action based on that knowledge, that process improvement results.

SPC takes data supplied by measuring instruments and records it on control charts, that help to identify problem areas. SPC can provide data that can result in corrective action to bring the problem into statistical control.

● **THE STATISTICAL CONTROL CHART**

The **statistical control chart** provides an operational definition of the state of statistical control of a process. The chart contains points plotted in time sequence; the points are either individual measurements or summary values obtained from process data. A center line and control limits *derived from process data* are drawn on the chart in order to detect signals of the existence of special causes from the pattern of plotted points. The control chart provides a means:

- to study and assess variation,
- to judge its state of statistical control,
- to maintain a continual record of the process so that the results of changes to the process or its inputs can be seen,
- to receive signals of change from the process, and
- to reveal through patterns of variation, coupled to various data collection and subgrouping strategies, information about causes of variation.

● X-BAR AND R-CHARTS

The kind of control chart to be used in a given situation depends upon the kind of data to be studied. X-bar and R-charts are used to analyze variable data organized into subgroups of two or more measurements. There should be some rational basis for organizing the data into subgroups. The X-bar chart must be considered together with the R-chart because transitional control limits for the X-bar chart cannot be compared unless the R-chart is in statistical control. Use of a control chart does not depend on a process being in statistical control, it is the method by which it may be achieved, and by which one may see whether a process is in statistical control.

Using the data in the X-bar control chart in Fig. 57-8, let us examine the included data:

1. There were 25 subgroups (measurement) over the time period covered by the data.

2. The center line (X) on the X-bar chart is the average reported value of the master using this measurement.

3. The average reported value is not equal to the accepted value assigned by a standard method.

4. Since this method of measurement results in a different average, the measurement process is said to be biased.

5. The amount of bias, or systematic error, would be $\overline{\overline{X}}$ (double bar) – accepted value.

When studying a process and learning about the variations of that process over a period of time, it is important to understand the reason for the variations. A point outside the control limits is a signal of *special cause.* There are supplemental rules such as eight (8) points above or below the center line may indicate other special causes. In any case, action on the process needs sound diagnosis of the cause.

● HISTOGRAM

The **histogram** is a quality descriptive control tool, normally used as a final inspection for lot-size parts with the purpose of seeing what the lot looks like. Histograms show the shape and spread created when a number of measurements are stacked on a measurement scale. Histograms indicate when a process is holding tolerance, and where it lies in relation to the minimum and maximum limits, Fig. 57-9A. They can also show the variation of product beyond the minimum and maximum limits, Fig. 57-9B.

Histograms generally have a three-part focus:

1. The *centering* of the histogram defines the aim of the process.

2. The *width* of the histogram defines the variability about the aim.

3. The *shape* of the histogram shows the deviations from the normal pattern.

Histograms can indicate several reasons why some processes are not capable of holding tolerances:

1. The variability of the process is too large.

2. The process is misdirected and adjustments should be made to the process.

3. The instrumentation is inadequate.

4. There is a process drift that may require a resetting of the process.

5. There are cyclical changes in the process that must be identified and corrected.

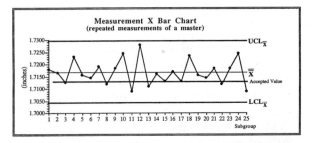

Fig. 57-8 Control charts show stability (statistical control) of a process over a period of time covered by the data. (Courtesy of GM Powertrain Group)

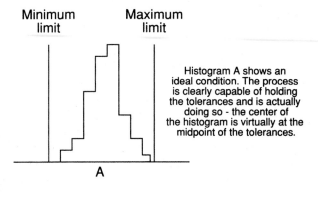

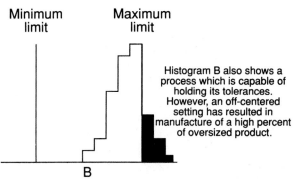

Fig. 57-9 Histograms show both the shape and spread of a collection of data for a manufacturing process. (Courtesy of Juran Institute)

6. The process is erratic and appropriate action must be taken.

● ADVANTAGES

The advantages of a good SPC program are:

1. The process has an identity, its performance is predictable, and it has measurable qualities.

2. Costs are predictable.

3. Production output is regular.

4. Productivity is at a maximum and costs are at a minimum.

5. Vendor relationships are in statistical control and costs are at a minimum.

6. The effects of changes in the system can be measured quickly and accurately.

SUMMARY

● Every company must set an acceptable level for its product or service. This quality level must agree with what the customer defines as an acceptable level of quality.

● Once the quality level has been set, it must become a company-wide policy. At this time, the cost of quality is set, and the only choice management must make is how and when it is to be incurred.

● Supervisors must decide if they want to incur preventative costs to assure quality now, or pay the price later for costs to repair defects and satisfy customers.

KNOWLEDGE REVIEW

1. Name two people whose work had a great influence on Japanese manufacturing in the 1950s.

Quality Control

2. List the three major steps in a quality control process.

3. What are three objectives of a quality control program?

4. Name five short-term and long-term costs of producing poor-quality goods.

Total Quality Control

5. What is the purpose of total quality control?

6. Why must the supply of raw materials be of good quality?

7. Who collects the statistical inspection information in the progressive manufacturing process?

8. What is used to provide an operational definition of the state of statistical control of a process?

Histogram

9. For what purpose is a histogram used?

UNIT
58

Stereolithography

In the highly competitive world of today, rapid time-to-market for new products is critical for manufacturers to be successful. Three-dimensional prototype models are a key and essential step before the manufacture of most industrial and consumer products and parts. Therefore, if a company can reduce the product developmental time by making fast, accurate prototypes, masters, and patterns, it would have an edge on the competition. **Stereolithography** (SL), one of the most widely used **rapid prototyping and manufacturing** technologies (RP&M), can literally go from art to part. It can quickly and accurately take a CAD or CAM design or a computer-scanned image, and produce solid three-dimensional objects in a matter of hours, Fig. 58-1. This technology allows manufacturers to reduce time-to-market by as much as 80%, reduce product development costs, and respond quickly to market changes while improving product quality.

OBJECTIVES

After completing this unit, you should be able to:

- Explain the benefits of stereolithography and rapid prototyping in manufacturing.
- Describe how stereolithography is used to produce a part.
- Discuss the factors to be considered in evaluating the potential of stereolithography and rapid prototyping in a manufacturing shop.

KEY TERMS

rapid prototyping and manufacturing

stereolithography

611

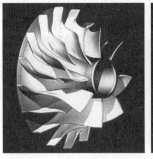

Fig. 58-1 A part, initially designed in wire frame, was easily converted to a solid model on an SLA system. (Courtesy of *Manufacturing Engineering Magazine*)

● PREPARING DESIGNS AND MODELS

Stereolithography Apparatus (SLA) uses an ultraviolet laser beam to scan a single cross-section of a CAD design into a vat of liquid photopolymer resin. The laser changes the resin from liquid to solid in thin, very accurate layers using the CAD design information. This computer-controlled process is repeated until the three-dimensional solid object is complete.

SL allows designers to touch, hold, and evaluate a physical model within hours of design. It allows them to test their designs to see if they would function properly in use. To correct any design error or make changes, the design can be quickly changed on the CAD drawing and a new model made within a few hours. The SL process can be very valuable to many phases of manufacturing, Fig. 58-2. Designers have time to think, create, and improve their designs. Engineers can test designs and locate any error early in the manufacturing cycle. Tooling and manufacturing departments can reduce their costs by purchasing

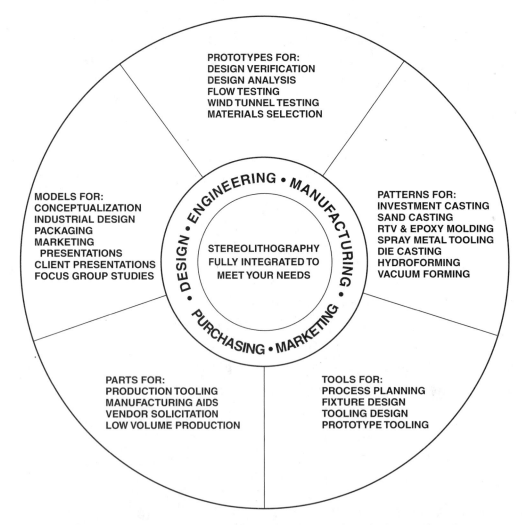

Fig. 58-2 Stereolithography can be useful to departments such as design, engineering, purchasing, and marketing. (Courtesy of 3D Systems)

the best tooling and developing the best process to produce the part. Sales and marketing people can do a better job because they can show potential customers the exact model of the product.

● BENEFITS OF SL AND RP&M

Although this technology is fairly new, being first introduced commercially in 1988, it has found applications in design engineering, product development, building prototypes, mold making, and pattern making to list just a few. The advantages that SL and RP&M offer industry are as follows:

1. VISUALIZATION—It transforms a complex CAD design into a three-dimensional part so that a designer can see exactly how the part looks. It has often been said that "A picture is worth a thousand words," which those using SL have updated by saying "A real prototype is worth a thousand pictures." Being able to see a real prototype with complex interior sections and curved surfaces can eliminate many errors later in manufacturing because all features of the part can be checked for accuracy.

2. VERIFICATION—An RP&M prototype can be quickly produced and its design verified that it contains all the features required of the final product. This system allows design changes to be made, or to correct design errors quickly, preventing costly mistakes on the final product.

3. REDESIGNING—Before RP&M, developing a prototype was a very costly and time-consuming operation. In many cases, small design errors were allowed to go through to the final product because of the time and money involved in making changes. RP&M technology produces a much better quality product because design changes are not so costly and a new prototype can be made and tested within a few hours, Fig. 58-3.

4. OPTIMIZATION—RP&M technology allows a design engineer to try new designs in order to make a better product without spending a lot of time or money. It gives the designer the potential of making a better product by optimizing the design.

5. FABRICATION—SL has expanded the role of rapid prototyping to where the manufacture of real parts is becoming more com-

Fig. 58-3 A prototype of an intake manifold being tested on a running engine to evaluate its performance. (Courtesy of 3D Systems)

mon. The type of material used should have the qualities required for the product, or the prototype can be used as a mold or pattern to produce the part.

● THE STEREOLITHOGRAPHY PROCESS

SL translates a CAD design into a solid object through a combination of laser, photochemistry, optical scanning, and computer software technology. The RP&M process begins with a three-dimensional CAD design of the part or component required, (see Fig. 58-1 on page 612).

- CAD DESIGN
  - Stereolithography begins with data from a CAD system.
  - A good CAD designer or engineer is very important to successful prototyping.
  - A good CAD system is a must to ensure success.
- TRANSLATOR
  - The CAD file must go through a CAD to an RP&M translator so that the data can be handled by the SL apparatus.
  - The SL file format appears to be the standard use for RP&M.
- PROCESS
  - The SLA builds physical models one layer at a time, Fig. 58-4.

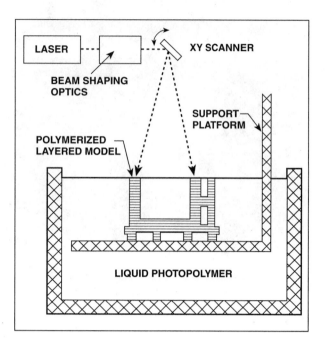

Fig. 58-4 An ultraviolet laser builds a part layer by layer, drawing each layer onto a vat of photopolymer which hardens when exposed to radiation. (Courtesy of 3D Systems)

- The CAD data of the product design is first sliced into very thin cross-sections.

- The slice unit is the value that the software program uses in place of CAD units, and can range in thickness from .003 in. (0.07 mm) to .015 in. (0.38 mm).

- A laser beam of ultraviolet light is then focused onto the surface of a vat of liquid photopolymer.

- The laser beam traces the thin cross-section of the part, turning a thin layer of liquid plastic to solid.

- The layer thicknesses (slice units) can range from .003 in. (0.07 mm) to .015 in. (0.38 mm), however, if a very accurate form is required, layers of .005 in. (0.13 mm) or less are recommended.

- The model is then lowered into the vat and recoated with liquid photopolymer and the laser traces the next slice on top of the previous slice.

- The process continues layer by layer until the part is complete.

- Unattended the SLA works around the clock, without tooling, machining, or cutting, and a 3-dimensional part is built within a few hours.

- The final part is removed from the vat and begins a curing process.

- After curing, the part may be sanded, plated, and painted.

- The SL prototype can be used in secondary manufacturing processes to produce the final product.

● PREPARING FOR SL

Although RP&M is an important technology that may improve product quality and get it to the market quicker, it may not be suitable for every industry. To evaluate its potential for a manufacturer, it is important to consider the following variables:

1. FINISHED PART SIZE—SL systems can only handle parts within a specific size, however, it is possible to make parts in sections and bond them together.

2. VOLUME—RP&M systems must be used for at least 200 prototypes (models) per year, or run half of each workday to justify the expense. If several million parts were required, one prototype could justify the system.

3. SPEED—Most systems take about a minute to produce each layer, therefore making a prototype is not very rapid. It could take as much as 90 hours to produce a full-size V-6 cylinder block.

4. PART FINISH—Slice thickness used to make the model will determine the surface finish. If a fine finish is required, the slice thickness should be small, and will take longer to produce. If surface finish is not too important, larger slice thicknesses can be used.

5. MATERIALS—If a model must only last for a few hours to "show and tell," it can be made from relatively weak materials. If it must undergo rigorous testing, it requires stronger material, especially where temperature is a prime consideration. The best RP&M model materials can withstand a constant temperature of 150°F (65.5°C).

6. ACCURACY—The accuracy of the model depends on the end use of the product. If high-volume plastic parts are to be produced, a fraction of a cent saving on each

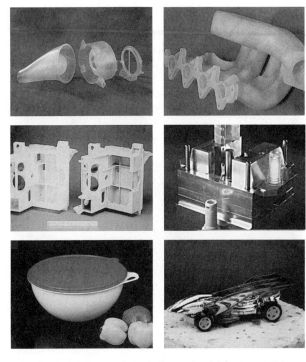

Fig. 58-5 A variety of models produced by stereolithography to improve quality and reduce time-to-market. (Courtesy of 3D Systems)

part could be critical, therefore it is important to know exactly how the RP material will perform.

7. COST—The cost of the SL machine is only one factor; the software required to run the system may be more expensive than the machine. Therefore, when considering this type of equipment, consider how the system will improve your operation and product quality. Despite the high up-front cost, the long term savings from design and production improvements can be as high as 30%.

There is a wide variety of part models that can be produced by SL; a few samples are shown in Fig. 58-5.

● SL ADVANTAGES

Stereolithography, a relatively new technology, allows manufacturers to quickly develop a new product and reduce the time it takes to bring it to market. Some of the major advantages of this technology are:

- Produces good quality prototypes quickly.
- Permits quality design changes to be made easily.

- Allows the model of the actual part to be seen and handled.
- Provides a reliable method of determining if designed parts will fit and function properly.
- Lowers developmental costs and improves part quality.
- Reduces time-to-market dramatically.
- Improves overall productivity and competitiveness.
- Produces physical models for market research.

SUMMARY

- Stereolithography uses a combination of lasers and photosensitive liquid plastic to create a three-dimensional model from a CAD/CAM design or computer-scanned image.
- This process uses an ultraviolet laser beam to scan a single cross-section of a CAD design into a vat of liquid photopolymer resin and change it from liquid to solid in very thin, accurate layers.
- Stereolithography reduces the product development time by making fast, accurate prototypes of a product and, if necessary, by making changes to the design before starting to manufacture.

KNOWLEDGE REVIEW

1. Why is it advantageous for a company to make fast, accurate prototypes, masters, and patterns?
2. Name one of the most widely used rapid prototyping and manufacturing technologies and describe what it does.
3. How can this process benefit the sales and marketing staff? Engineers?

Benefits of SL and RP&M

4. List five advantages that SL and RP&M offer industry.

The Stereolithography Process

5. How does SL translate a CAD design into a solid object?

6. Describe the role of laser beams in SL.

Preparing for SL

7. List the variables which should be considered before implementing a RP&M system?

SL Advantages

8. List three of the most important advantages of SL.

UNIT
59

Superabrasive Technology

The superabrasives, diamond and cubic boron nitride, possess properties unmatched by conventional abrasives such as aluminum oxide and silicon carbide. The hardness, abrasion resistance, compressive strength, and thermal conductivity of superabrasives make them logical choices for many difficult grinding, sawing, lapping, machining, drilling, wheel dressing, and wire drawing applications. **Superabrasives** can cut and grind the hardest materials known, making difficult material-removal applications routine operations. Superabrasive cutting tools are designed to increase productivity, produce better quality products, and reduce manufacturing costs.

OBJECTIVES

After completing this unit, you should be able to:

- State the specific physical properties that make superabrasives useful in manufacturing.
- Describe the manufacture and application of manufactured diamond and cubic boron nitride in grinding operations.
- List the advantages of using superabrasive tools.
- State the uses of polycrystalline cutting tools and the factors to consider in optimizing their performance.

KEY TERMS

abrasion resistance	cubic boron nitride	dressing
microcrystalline	monocrystalline	superabrasives
thermal conductivity	truing	

BACKGROUND

In 1954, after many years of research, GE produced manufactured diamond in the laboratory. Carbon and a catalyst, such as iron, chromium, cobalt, nickel, etc., were subjected to tremendous heat and pressure to form diamond crystals, Fig. 59-1. In 1957, after more research and testing, GE began the commercial production of these diamond abrasives. Because the temperature, pressure, and catalyst-solvent can be varied, it is possible to produce diamond abrasives of various sizes, shapes, and crystal structure, to suit a range of grinding applications on nonferrous and nonmetallic materials.

In 1969, GE introduced an entirely new material, cubic boron nitride (CBN). **Cubic boron nitride** is made from hexagonal boron nitride and a catalyst using the same high-pressure, high-temperature technology perfected in the manufacture of diamond. CBN, next to diamond in hardness, is used for the grinding of hard alloy steels and other difficult-to-grind ferrous metals.

PROPERTIES OF SUPERABRASIVES

Superabrasives can penetrate the hardest industrial and structural materials and make difficult material-removal operations into routine grinding operations. They are used to grind very hard, abrasive, and difficult-to-grind materials at higher rates, and are capable of improving grinding efficiency, reducing scrap, and increasing product quality.

The main physical properties of superabrasives, which make them superior to conventional abrasives, are shown in Fig. 59-2.

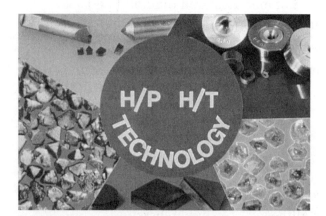

Fig. 59-1 A combination of high pressure and high temperature are necessary for diamond growth. (Courtesy of GE Superabrasives)

Hardness

An abrasive must be harder than the workpiece in order for it to cut; the harder the abrasive, the more easily it will cut. Fig. 59-2A compares the hardness of manufactured diamond and CBN with the conventional abrasives, aluminum oxide and silicon carbide. Diamond is the hardest substance known and has a hardness range between 7,000 and 10,000 on the Knoop hardness scale. CBN, with a Knoop hardness of 4700, is second only to diamond in hardness.

Abrasion Resistance

Superabrasives maintain their keen cutting edges much longer than conventional abrasives because of their superior **abrasion resistance.** Fig. 59-2B shows that diamond has about three times the abrasion resistance of silicon carbide, while CBN has about four times that of aluminum oxide. This high abrasion resistance makes them exceptional tools for grinding hard, tough materials at high material-removal rates, and for increasing productivity.

Compressive Strength

The high compressive strength of superabrasives gives them excellent qualities to withstand the forces created during high material-removal rates and the shock of severe interrupted cuts. Fig. 59-2C shows diamond with about 19 times the compressive strength of silicon carbide, while CBN has two and one-half times the compressive strength of aluminum oxide.

Thermal Conductivity

Thermal conductivity in cutting tools is defined as the ability to dissipate (transfer) the heat created at the chip-tool interface, to prevent thermal damage to the workpiece. Fig. 59-2D shows that both diamond and CBN have excellent thermal conductivity qualities for transferring the heat generated when grinding hard, abrasive, or tough materials, at high material-removal rates.

MANUFACTURED DIAMOND

Diamond is used for truing and dressing grinding wheels and for the manufacture of diamond wheels. The need for a reliable source of diamond during World War II was realized when natural diamond was not readily available.

(A) HARDNESS COMPARISON

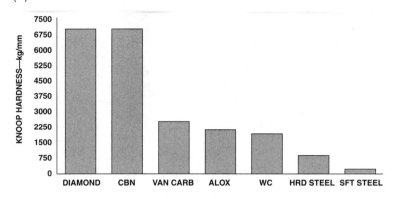

(B) ABRASION RESISTANCE COMPARISON

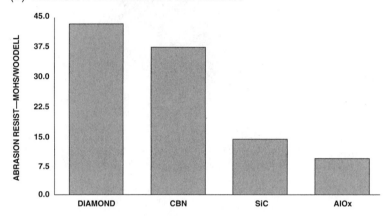

(C) COMPRESSIVE STRENGTH

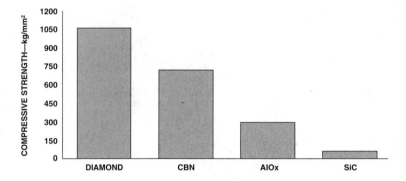

(D) THERMAL CONDUCTIVITY

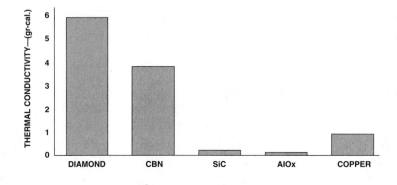

Fig. 59-2 The properties that make superabrasives, super-hard, super wear-resistant cutting tools. (Courtesy of GE Superabrasives)

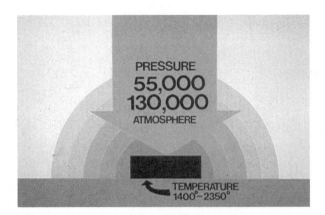

Fig. 59-3 The high-pressure, high-temperature belt apparatus used for manufacturing diamond. (Courtesy of GE Superabrasives)

To produce diamond by a manufacturing process, the conditions of pressure and temperature found far below the earth's surface had to be duplicated. This required a high-pressure, high-temperature belt apparatus capable of reproducing the conditions necessary for diamond growth. Graphite (a form of carbon) and a catalyst (such as iron, chromium, cobalt, nickel, etc.) were subjected to high temperatures (2550° to 4260°F, or 1400° to 2350°C) and high pressures (800,000 to 1,900,000 lbs./sq. in. of 55,000 to 130,000 atmospheres) to form diamond crystals, Fig. 59-3. Under these conditions, the graphite is transformed into diamond and remains that way when it is cooled and the pressure is removed.

Types of Manufactured Diamond

There are many different types of manufactured diamond to suit various grinding applications. Manufactured diamond is available for grinding cemented carbides, carbide/steel combinations, nonferrous and nonmetallic materials, and many products such as natural stone, concrete, and masonry. The three main manufactured diamond are:

- TYPE RVG DIAMOND—Is an elongated, friable crystal with rough edges, Fig. 59-4A, that consists of thousands of tightly bonded small crystals that make up each abrasive grain. Type RVG (resin and vitrified) wheels are used to grind ultra-hard materials, such as tungsten carbide, and tough, abrasive nonmetallic and nonferrous materials.

- TYPE CSG-11 DIAMOND, Fig. 59-4B, is designed to grind cemented-carbide brazed

Fig. 59-4A RVG diamond is used to grind hard, abrasive materials. (Courtesy of GE Superabrasives)

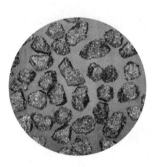

Fig. 59-4B CSG-11 diamond is used to grind carbide and steel combinations. (Courtesy of GE Superabrasives)

Fig. 59-4C MBG-11 diamond performs well on cemented carbides, glass, and ceramics. (Courtesy of GE Superabrasives)

tools where it may be necessary to grind both the carbide and some of the steel shank supporting the carbide insert.

- TYPE MBG-II DIAMOND—The wheels with MBG (metal-bond grinding) abrasive have a metal bond to hold the tough dia-

Fig. 59-4D MBS diamond is used primarily for cutting stone, marble, concrete, and masonry products. (Courtesy of GE Superabrasives)

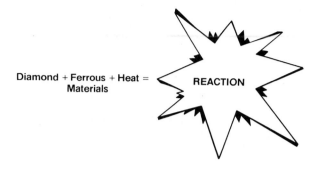

Diamond + Ferrous + Heat = Materials   REACTION

Fig. 59-5 Diamond tools react chemically, under the proper temperature and pressure conditions, when cutting ferrous materials. (Courtesy of GE Superabrasives)

mond crystals in the wheel. Type MBG-II diamond, Fig. 59-4C, is used for grinding glass, ceramics, and carbides.

- TYPE MBS DIAMOND—The Type MBS (metal-bond sawing) diamond, Fig. 59-4D, is used in metal-bond saws to cut granite, concrete, marble, and a variety of masonry and refractory materials.

Metal Coatings

The RVG diamond abrasive can be coated to prevent the diamond crystals from being pulled out of the resin bond. Coatings, such as nickel and copper provide better retention (holding power) for the RVG crystal in the wheel bond.

- TYPE RVG-W—(Resin, Vitrified, Grinding—Wet) is an RVG diamond with a special nickel coating, that sticks to all surfaces of the crystal, provides a better holding or bonding surface for the resin bond, and results in much longer grinding wheel life.

- TYPE RVG-D—(Resin, Vitrified, Grinding—Dry) is an RVG diamond with a special copper coating that improves the bonding strength of the diamond in the wheel and controls its fracturing under the stresses of grinding.

Under the pressure and temperatures created when grinding ferrous metals, diamond will react chemically and result in excessive diamond wear. Diamond is not recommended for grinding ferrous metals, Fig. 59-5.

CUBIC BORON NITRIDE

A major breakthrough in the precision high-production grinding of hard, difficult-to-grind ferrous metals was the discovery and manufacture of cubic boron nitride. CBN is twice as hard as aluminum oxide and its performance on hardened steels is far superior. CBN is cool cutting, chemically resistant to inorganic salts and organic compounds, and can withstand grinding temperatures up to 1832°F (1000°C) before breaking down. Because of the cool-cutting action of CBN wheels, there is little or no thermal (heat) damage to the surface of the part being ground. The main benefits of grinding wheels made of CBN abrasive are shown in Fig. 59-6.

Manufacture

CBN is synthesized in crystal form from hexagonal boron nitride, sometimes referred to as white graphite. Hexagonal boron nitride, composed of boron and nitrogen atoms along with a solvent catalyst, is converted into cubic boron nitride through the application of heat (3000°F or 1650°C) and pressure (up to 1,000,000 lbs./sq. in., or 68,500 atmospheres). The combination of high temperature and high pressure causes each nitrogen atom to donate an electron to a boron atom, which uses it to form another chemical bond to the nitrogen atom. This produces a strong, hard, blocky, crystalline structure similar to diamond.

CBN Types

There are various types of CBN available to suit a variety of steel grinding applications; they do not perform well on nonferrous or nonmetal-

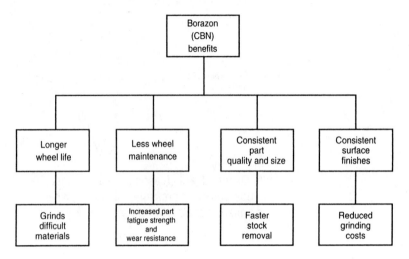

Fig. 59-6 The main benefits of CBN grinding wheels. (Courtesy of Kelmar Associates)

lic materials. Two main classes of CBN abrasive are monocrystalline and microcrystalline:

- MONOCRYSTALLINE CBN—**Monocrystalline** CBN abrasive contains a large number of cleavage (break) planes along which a fracture can occur. This macrofracture (large break) is necessary for the abrasive grains to resharpen themselves when they become dull, Fig. 59-7.

- MICROCRYSTALLINE CBN—**Microcrystalline** CBN abrasive consists of thousands of micron-size crystalline regions tightly bonded to each other to form a 100% dense particle. When the grains dull and the grinding pressure increases, they resharpen themselves by microfracturing (creating very small breaks), Fig. 59-8.

Table 59-1 lists the various types of abrasives and the workpiece materials for which each is best suited.

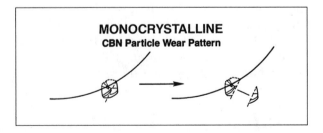

Fig. 59-7 Monocrystalline CBN crystals macrofracture (large break) under high grinding forces and expose sharp cutting edges. (Courtesy of GE Superabrasives)

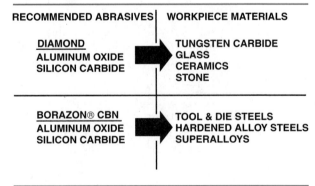

Fig. 59-8 Microcrystalline CBN crystals microfracture (very small break) to resharpen the wheel and promote long wheel life. (Courtesy of GE Superabrasives)

Table 59-1 Abrasive-workpiece profile (Courtesy of GE Superabrasives)

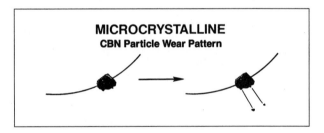

RECOMMENDED ABRASIVES	WORKPIECE MATERIALS
DIAMOND **ALUMINUM OXIDE** **SILICON CARBIDE**	**TUNGSTEN CARBIDE** **GLASS** **CERAMICS** **STONE**
BORAZON® CBN **ALUMINUM OXIDE** **SILICON CARBIDE**	**TOOL & DIE STEELS** **HARDENED ALLOY STEELS** **SUPERALLOYS**

● GRINDING WITH SUPERABRASIVES

The advantages that superabrasive tools offer the metalworking industry more than offset their high initial costs. Primarily designed to grind or machine very hard and difficult-to-cut metals, superabrasive tools are capable of greatly improving efficiency, reducing scrap parts, and increasing the quality of the product. Due to the tough, hard microstructure of these tools, their cutting edges last longer and efficiently remove material at higher rates that would cause conventional tools to break down rapidly.

The advantages of superabrasive tools that can be used for grinding, turning, and milling operations are:

- LONG TOOL LIFE—Superabrasive tools resist chipping and cracking, provide uniform hardness and abrasion resistance, and consistently outperform other tools by 10 to 700 percent. Reduced tool wear results in closer tolerances on workpieces, and fewer tool adjustments keep machine downtime to a minimum.

- HIGH MATERIAL-REMOVAL RATES—Because these tools are so hard and resist abrasion so well, cutting speeds in the range of 250 to 900 feet per minute (ft./min.) or 274 m/min. are possible. This results in higher material-removal rates (as much as three times that of carbide tools) with less tool wear reducing the total machining cost per piece .

- CUTS HARD, TOUGH MATERIALS—These cutting tools are so hard that they are capable of efficiently machining all ferrous materials with a Rockwell C hardness of 45 and above, and for machining cobalt-base and nickel-base high temperature alloys with a Rockwell C hardness of 35 and above. Many of these materials were so hard and abrasive that grinding was previously the only practical way to machine them.

- HIGH QUALITY PRODUCTS—Because the cutting edges of superabrasive tools wear very slowly, they produce high quality parts faster, and at a lower cost per piece than conventional cutting tools. The need for the inspection of parts is greatly reduced, as is the adjustment of the machine tool to compensate for cutting tool wear or maintenance.

- UNIFORM SURFACE FINISH—Surface finishes in the range of 20 to 30 microinches during rough cuts, and single-digit numbers on finish cuts are possible with superabrasive tools.

- LOWER COST PER PIECE—Superabrasive tools stay sharp and cut efficiently through long production runs. This results in consistently smoother surface finishes, better control over workpiece shape and size, fewer tool changes, less inspection, and increased machine uptime.

- REDUCED MACHINE DOWNTIME—Since these tools stay sharp much longer than conventional cutting tools, there is less time required to index, change, or recondition the cutting tool.

- INCREASED PRODUCTIVITY—A combination of all the advantages that superabrasive tools offer, such as increased speeds and feeds, long tool life, longer production runs, consistent part quality, and savings in labor costs, all have an effect on the overall production rates and the manufacturing cost per piece. A company's reputation can be improved by producing quality products at competitive prices, while poor-quality products hurt a company's reputation.

All these advantages come together to produce high-quality parts faster and at a lower cost per piece. When machining time on many operations can be reduced by 50%, and the need to replace worn cutting edges reduced from three times a day to once a month, superabrasive tools are very cost effective and can easily be justified.

● GRINDING MACHINES

The performance of a superabrasive grinding wheel depends on the capabilities of the machine. Trying to use superabrasive grinding wheels to make up for poor machine conditions will be doomed to failure right from the beginning. In order for superabrasive wheels to work effectively, they should be used on grinders having the following characteristics:

- Tight spindle bearings and snug machine slides to prevent vibration and chatter that could shorten the wheel life, produce a poor surface finish, and inaccurate work, Fig. 59-9.

- Consistent spindle speeds to handle the torque required for high metal-removal rates are necessary to keep superabrasive wheels operating at best efficiency. Loss of spindle speed reduces the efficiency of the cutting action and shortens the life of the wheel.

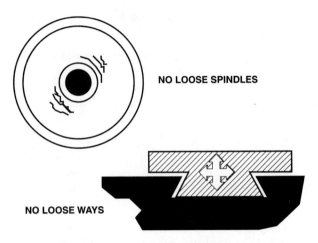

NO LOOSE SPINDLES

NO LOOSE WAYS

Fig. 59-9 Tight spindle bearings and snug machine slides are important when using superabrasive tools. (Courtesy of GE Superabrasives)

GRINDING WHEELS

The first application of the superabrasives, diamond and cubic boron nitride, was in grinding wheels. Although the applications for diamond and cubic boron nitride grinding wheels are very different, these two superabrasives contain four main properties cutting tools must have to cut extremely hard or abrasive materials at high metal removal.

Diamond grinding wheels are used to grind a variety of nonferrous and nonmetallic materials. In the metalworking industries, diamond wheels are widely used to manufacture and regrind tungsten carbide tools.

Cubic boron nitride (CBN) grinding wheels are recognized worldwide as superior cutting tools for grinding difficult-to-machine ferrous-based metals. From their initial use in toolroom and cutter grinding applications, CBN wheels have made their presence felt in production grinding operations worldwide, where high-technology CNC machines are revolutionizing the metalworking industry.

- Grinding wheels containing CBN abrasives last longer, provide more accurate parts, and require little or no conditioning after initial truing and dressing.

- These wheels are more productive when grinding hardened steels, tool steels, and superalloys, because of the properties of the CBN crystals.

- Not only is the CBN crystal an extremely hard abrasive, it is able to withstand the machining pressures and the heat of production grinding better than other abrasives.

- The CBN abrasive crystal has the toughness to match its hardness, so that its cutting edges stay sharp longer, and the crystal re-sharpens itself to stay free cutting.

WHEEL SELECTION

The type of wheel selected, and how it is used, will affect the metal-removal rate (MRR) and the life of a grinding wheel. The selection of a CBN grinding wheel can be a complex task, and it is always wise to follow the manufacturer's suggestions for each type of wheel. They have the experience in designing and applying wheels for specific jobs, therefore, their suggestions usually result in selecting the best wheel for each job.

Superabrasive wheel selection is affected by the following factors:

- type of grinding operation,
- grinding conditions,
- surface finish requirements,
- shape and size of the workpiece,
- type of workpiece material.

All four types of CBN wheels (resin, vitrified, metal, and electroplated) are highly effective, however, they are designed for specific applications and must be selected accordingly. There is no one type of CBN wheel that is suitable for all grinding operations, therefore, for the best grinding performance, the characteristics of the wheel abrasive must be matched to the requirements of the specific grinding job. Abrasive characteristics such as concentration, size, and toughness must be considered when selecting a wheel because they affect metal-removal rates, wheel life, and surface finish of any grinding operation.

TRUING AND DRESSING CBN WHEELS

It is not unusual for superabrasive wheels to outperform conventional grinding wheels by a margin of 250:1. Simple, but essential, truing, dressing, and coolant-application methods will ensure that superabrasive wheels live up to their reputation and justify their investment. Proper truing and dressing of the grinding wheel is the *most important factor* in making a CBN wheel

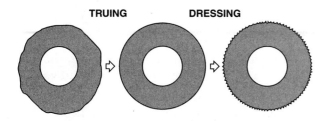

Fig. 59-10 Truing makes a wheel run true, and dressing sharpens the wheel. (Courtesy of Kelmar Associates)

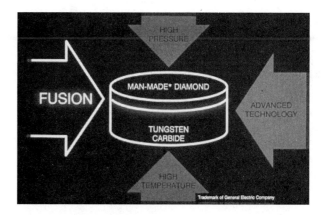

Fig. 59-11 Polycrystalline tools consist of a thin layer of diamond or CBN bonded to a cemented tungsten carbide substrate. (Courtesy of GE Superabrasives)

work, Fig. 59-10. An out-of-round wheel will pound the workpiece, reduce wheel life, and produce poor surface finishes. When properly conditioned, these wheels have the capability of increasing productivity, reducing overall grinding costs, consistently producing parts to close tolerances, and leaving the workpiece surface with no metallurgical damage.

● TRUING

Truing is the process of making a grinding wheel round and concentric with its spindle axis, to produce the required form or shape on the wheel, Fig. 59-10. This procedure involves grinding or wearing away a portion of the abrasive section of a grinding wheel in order to produce the desired form or shape. Superabrasive wheels should always first be trued with a dial indicator, to within .001 in.(0.02 mm) accuracy before using any truing devices.

A properly trued wheel will grind with minimum or no chatter, produce a straight cylinder, flat surface, or accurate profile on the workpiece, provided that it is also dressed properly.

● DRESSING

The truing process usually leaves the working surface of the wheel smooth, with little or no abrasive crystal protrusion or clearance for chip generation and removal. A wheel in this condition would burn the workpiece and remove little or no work material. **Dressing** is the process of removing some of the bond material from the surface of a trued wheel to expose the abrasive crystals and make the wheel grind efficiently.

● MACHINING WITH POLYCRYSTALLINE TOOLS

With the development of polycrystalline diamond (PCD) and polycrystalline cubic boron nitride (PCBN) blanks and inserts, a new generation of superabrasive cutting tools became available. These tool blanks and inserts consist of a .025 in. (0.63 mm) layer of diamond or cubic boron nitride bonded to a cemented carbide base, Fig. 59-11.

- Polycrystalline diamond (PCD) tools are used to cut abrasive nonferrous and nonmetallic materials.
- Polycrystalline cubic boron nitride (PCBN) tools are used to cut hard, ferrous metals, superalloys, and certain cast irons.

The cutting edges of polycrystalline cutting tool blanks and inserts are more wear-resistant than the cutting edges of conventional tools such as cemented carbides, even when removing material at much higher rates. Therefore inserts need to be indexed or changed far less often than carbide and ceramic tools, which reduces the amount of machine downtime and increases productivity.

● MACHINING

Polycrystalline cutting tool blanks and inserts are being used worldwide because they are able to machine tough and abrasive materials at higher rates, with longer tool life, than carbides and ceramics. Turning and milling operations on hard-to-machine metals lend themselves to the use of these cutting tools, and in many cases, eliminate the finishing grinding operation. Polycrystalline tool blanks are available in a wide variety of shapes and sizes to suit many machining applications.

● TYPES OF MATERIALS CUT

Polycrystalline diamond (PCD) tools are used for turning and milling nonferrous or nonmetallic materials, especially where the workpiece is hard and abrasive. The largest group of nonferrous metals is generally soft, but can have hard particles in them, such as silicon suspended in aluminum or glass fibers in plastic. These hard abrasive particles destroy the cutting edge of conventional tools. PCD tools often have a wear life of 100 times more than cemented carbide tools in such an abrasive machining application.

Polycrystalline cubic boron nitride (PCBN) tools are used for turning and milling operations on abrasive and difficult-to-cut (DTC) ferrous materials materials. PCBN tools can remove material at much higher rates than conventional cutting tools, with far longer tool life. Wherever PCBN cutting tools were used to replace a grinding operation, machining time was greatly reduced because of the higher metal-removal rate.

The best applications for PCBN cutting tools are on materials where conventional cutting-tool edges of cemented carbides and ceramics are breaking down too quickly. Their long-lasting cutting edges are capable of transferring the accuracy of computer controlled machine tools and flexible manufacturing systems, thereby producing accurate parts, increasing productivity, and reducing expensive machine downtime.

● MACHINING EXAMPLES

Superabrasive tools are widely used in automotive, aerospace, and other manufacturing industries for turning and milling operations of hard, abrasive, and difficult-to-cut materials. Industry has found that these superabrasive cutting tools are among the most effective tools for production cost reduction and product improvement. In terms of the number of pieces per cutting edge, downtime and overall productivity, superabrasive tools have proven to be the most cost efficient tools available today.

General Applications Guidelines

To obtain the best tool performance and the most number of parts per cutting edge, the following guidelines should be closely followed:

- Use PCD cutting tools to machine only nonferrous and nonmetallic materials.
- Select a rigid machine with enough horsepower to maintain the cutting speed where PCD tools perform best.
- Use a speed three times faster than for a cemented tungsten carbide tool.
- Set speed and feed rates which give a good balance between productivity and long tool life.
- Use rigid toolholders and keep the tool overhang as short as possible to avoid deflection, chatter, and vibration.
- Use positive rake angles and the largest nose radius possible for better surface finishes and to spread the cutting force over a wider area.
- Establish the life of each cutting edge or tool (usually after a certain number of pieces are cut) and change tools at the first sign of dullness.
- Use coolant wherever possible to reduce heat, promote free cutting, and flush away the abrasive chips from the finished work surface.

SUMMARY

- The superabrasives, manufactured diamond and cubic boron nitride (CBN), possess exceptional qualities of hardness, abrasion resistance, compressive strength, and thermal conductivity that are unmatched by conventional abrasives.
- Manufactured diamond is used to machine and grind hard, abrasive nonferrous and nonmetallic materials and is rarely used on ferrous materials. CBN is used to machine and grind hard, abrasive ferrous metals 35C Rockwell or higher.
- Diamond and CBN grinding wheels must be properly trued and dressed to avoid wheel damage and to grind effectively.

- Diamond and CBN cutting tools must have the proper rakes and angles to suit the work material and the machining operation.
- For the best cutting tool performance, machines should have tight spindle bearings, snug slides, and consistent spindle speeds.

KNOWLEDGE REVIEW

1. Name the two superabrasives.

Background

2. Briefly describe how manufactured diamond is produced.

Properties of Superabrasives

3. Name the four main properties of super-abrasives.

Manufactured Diamond

4. Name the manufactured diamonds that can grind carbide and steel.

5. What is the purpose of the coatings applied to diamond crystals?

Cubic Boron Nitride

6. Name the two main classes of CBN abrasive crystals.

Grinding with Superabrasives

7. List four main advantages of superabrasive tools.

Grinding Machines

8. Why are tight spindle bearings and snug machine slides necessary when using superabrasive tools?

Wheel Selection

9. List five factors that should be considered when selecting a superabrasive wheel.

Truing and Dressing CBN Wheels

10. Define truing and dressing, and state the purpose of each.

UNIT 60

The World Of Manufacturing

Scientific management of production began around the turn of the twentieth-century. It was the brain child of an engineer, Frederick Winslow Taylor, and was based on the specialization of people and machinery. New products and tools developed by research laboratories resulted in new methods and machines that became the foundation of mass production. In less than fifteen years, American industry produced 76 percent more goods than it had at the start of the century. While there were many more machine tools in factories devoted to mass production, the processes were not flexible and the workers had limited skills.

OBJECTIVES

After completing this unit, you should be able to:

- Trace the evolution of manufacturing during the twentieth century.
- Describe the aims and objectives of agile manufacturing.
- Explain the importance of the computer in producing quality products at reduced manufacturing costs.

KEY TERMS

agile manufacturing
customized manufacturing
flexible manufacturing
rapid response

automation
continuous corporate renewal
mass production
strategic planning

FACTORY FOCUS

Since the early 1900s, the focus of factories has been under continual change as a result of the development of more accurate machine tools, computer controls, and manufacturing processes. All these factors helped industry to increase productivity, improve product quality, and reduce manufacturing costs. The changes projected for the future will continue to amaze everyone associated with manufacturing.

Mass Production

Specialized and single-purpose machines were developed between 1900 to 1960 for the **mass production** of identical parts. Many different transfer-type machines, each designed to produce a specific product or perform a specific machining operation, were used in manufacturing. This process was not very flexible and as many as 150 different machines were required to produce a limited number of finished products in the early 1900s, Fig. 60-1. Many of these prod-

ucts required rework because the accuracy of the machine tool was not too reliable and because of human error. With the gradual improvement in machine tools, the number of machines required by 1960 dropped to about sixty, the number of products increased and their quality improved.

Flexible Manufacturing

The development of **flexible manufacturing** processes has enabled manufacturers to operate more efficiently and to be more responsive to customer needs. The introduction of numerical control (NC) in the early 1960s, that quickly evolved into computer numerical control (CNC), produced a major change in manufacturing methods. CNC machine tools were very accurate and also flexible enough to produce many different products. Therefore, as better and more versatile machine tools became available, the number of machines required dropped from 50 to 30, while the number of products increased from 100 to 1000, Fig. 60-1. The rework required was reduced due to the accuracy of the CNC ma-

The Evolution of Manufacturing			
Industry will soon be able to customize products for consumers **Factory Focus**	**Mass Production**	**Flexible Manufacturing**	**Customized Manufacturing**
	1900–1960	**1961–2000**	**2001–2020**
Number of machine tools	150–60	50–30	25–20
Products manufactured	10–15	100–1000	Unlimited
Products reworked	25% or more	0.02% or less	Less than 0.0005%

Fig. 60-1 In the 20th century, manufacturing has evolved from mass production, where many identical goods were made, to customized manufacturing where products are made to customer specifications. (Courtesy of Giddings & Lewis, Inc.)

chine tools, the manufacturing systems, and the reduction of human error.

Customized Manufacturing

The newest form of manufacturing, **customized manufacturing**, combines state-of-the-art manufacturing and product-delivery technologies with the craftsman's aim to please. In the 1990s, the introduction of smart machine tools, improved software, intelligence systems, expert systems, communication networks, and intelligent manufacturing systems makes it possible to build products to customer's specifications as fast and as cheaply as mass-produced products. The number of machine tools required will drop to between 25 to 20, the number of products manufactured will be unlimited, and the rework on parts will be almost eliminated, Fig. 60-1.

● AGILE MANUFACTURING

Manufacturing is quickly entering a new age of industrial excellence and mass customization. The newest form of manufacturing is **agile manufacturing**, combining state- of-the-art fabrication and product delivery technologies to custom-make products suiting each customer's taste, specifications, and budget. It will soon be possible to produce a wide variety of goods from

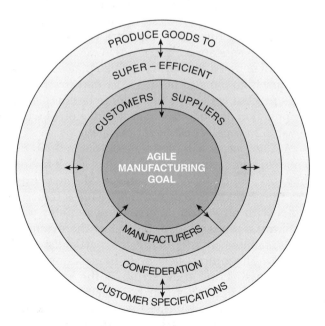

Fig. 60-2 The goal of agile manufacturing is to involve customers, suppliers, and manufacturers to produce custom-made parts. (Courtesy of Kelmar Associates)

clothing to computers, to automobiles, and many other goods as fast, and as cheaply as mass-produced products.

This is not just fantasy; more than 200 companies are already part of the Agile Manufacturing Enterprise Forum created by Lehigh University's Iaccoca Institute in 1991. This was in response to a call for action from the American Congress and the Department of Defense who wanted a competitive manufacturing enterprise for the 21st century. It includes companies such as Chrysler, AMF, Texas Instruments, Westinghouse, Honeywell, Milliken, and many others. To show how committed the United States is to agile manufacturing, the National Science Foundation, along with the Pentagon's Advanced Research Projects Agency, helped to set up three Agile Manufacturing Research Institutes—the University of Illinois, University of Texas, and Rensselaer Polytechnic Institute. Agile manufacturing has already brought back to the United States work that was formerly made in Korea and Mexico. Companies found that the time it took to move products through the international pipeline cost more than the labor savings were worth. The goal of agile manufacturing is to link customers, suppliers, and manufacturers into a kind of super-efficient confederation, Fig. 60-2. What the factory produces one day will be driven by yesterday's retail sales, or an order received moments ago from an on-line partner. Agile manufacturing is coming faster than anyone expected, and by the year 2000, it will be the norm.

If the philosophy of mass production is based on treating everyone the same, agile manufacturing is based on the philosophy of making things in quantity according to what each customer wants. This concept can be applied to almost any production system such as clothing, publication, service, manufacturing, government, educational, etc. Manufacturing on demand, whether it be for a single piece or a thousand pieces, will be part of a customized agile manufacturing program.

● CONTINUOUS CORPORATE RENEWAL

Successful manufacturing in the 21st century will include continuous changes, rapid response, and evolving quality standards. It is predicted that the products made on any given day will be costly and obsolete in fabrication when compared to products built two weeks later. This will happen due to technologies that were not in existence when the products were first produced. Such will be the extraordinary daily changes that will transform manufacturing. Newer and better methods will be consistently coming; a fixed-plan process will not be able to exist. Management of change and vision will be the way factories will run, with survival depending upon meeting unexpected customer changes quickly and economically.

The management of continuous improvements, while at the same time developing the next-generation technology that's designed to leapfrog past the competition using current systems, is referred to as **continuous corporate renewal.** Continuous renewal will include five key points, Fig. 60-3.

1. Future Vision—An important consideration is direction. Where is the company headed and where must it be in five years in order to survive?

2. Options—An alternate plan for setting the company in the right direction must be drawn up.

3. **Strategic Planning**—A listing of existing skills, resources, and capabilities should be made so that they can be compared against those that are required.

4. Partnerships—A business plan listing the resources required and a technology cooperative to fill the voids that exist should be developed.

5. Workforce Skills—The constant retraining and upgrading of workforce skills will create a life-long learning environment.

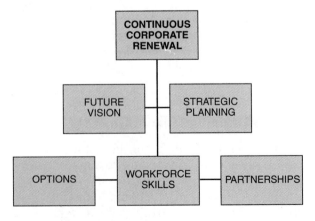

Fig. 60-3 The five key points of continuous corporate renewal. (Courtesy of Kelmar Associates)

● RAPID RESPONSE

In order for any healthy company to respond rapidly to a manufacturing challenge, it must be adaptable and know its strengths. It must be able to get involved in the particular activity at hand, have control of the technology, and be able to provide the right production facilities. Whether the company is involved in a solo effort or working with others to form a "virtual" enterprise, they must know when to dissolve the operation, freeing the participants to go into other projects. Ideally, each company or organization would have to contribute what it does best. One company might be good at designing the product, another at developing the prototype, and still another company would be good at marketing.

Rapid response involves working and cooperating with a number of companies where each company gives or takes something unique from the other. Sharing, teaming, and cooperation can only take place when it is based on trust. The sharing of fundamental knowledge, especially proprietary information among alliances, will become a major ethical challenge.

Rapid response can only be achieved through a network of information technologies and the knowledge and services that allow a company to satisfy the customer. In some cases, communication will be extended to individual machine tools on the shop floor. With a phone call, customers can discuss their particular needs with company engineers. In turn, the engineers enter the customer's specifications into a CAD/CAM system to design a one-of-a-kind article, which then activates the machine tools to produce the part overnight. This process will allow the order to be completed in a fraction of the time and cost a fraction of what it would normally cost by former methods.

Computers will play a major role in rapid response manufacturing. Through a nationwide network, computers will be able to "discuss" production schedules and routine business transactions with other computers. The computer will determine the workload and select the machine tools capable of producing the part most efficiently. The rapid collection of information, both in processing and transferring, is the key to the agility concept. This can be illustrated by the laser scanners used at the supermarket checkout. With one sweep of the scanner, the supermarket manager can monitor sales, shelf life, and trace products in case of a recall. Similar technology is used in industry where laser scanners are used in metrology applications. Accurate measurements can be made in a fully automatic manufacturing system, whether the parts are stationary or moving through production at a very high rate of speed.

● EVOLVING QUALITY

In the future, quality will be primarily based on satisfying the customer. While the current standard of quality is said to be almost defect-free, this standard will just be considered entry-level quality in the future. The prime consideration will be on the appreciation of value, the customer's insight and feeling of how good the product is for the price paid. Quality must be designed into the product and the manufacturing process, not built into it later with inspection, rework, or retrofit in the final assembly. Quality is free; what costs money are the poor-quality things that result from not doing the job right the first time. Quality products will be expected to be made the first time without inspection, scrap, or rework time. Defective work is very expensive due to the added costs that are involved, Fig. 60-4.

Competition in the mass individualized world will make it necessary for companies to communicate with their operators and floor people in a way that allows the workers to understand what they are doing. They must be part of a team that produces something, as opposed to just following instructions. Workers will be re-sponsible for quality fabrications. They will be required to understand what is happening in production or assembly and have the power to stop the operation if they feel something is wrong. A strong emphasis will be placed on teamwork, and it is quite likely that a team that works together one day may be mixed the next day. This will provide each member of the team with experience on all phases of manufacturing, and also adjust them to working with many different people. If a team of workers does not coordinate, they are going to build a terrible product. **Automation** will still be used but the assembly team will go with the product. Automation will still be delivering the right equipment and tools to the right station. It will be the team's responsibility to assemble a quality product. It is not their function to become experts in inserting components, such as a bumper or a windshield on a car, for example. The team that comprises the assembly group will have to work well and think for themselves, or the whole operation will suffer.

Self-directed and empowered work teams will make the master scheduler a thing of the past.

SUMMARY

- The focus of the factory has changed from single-purpose machine tools of the early 1900s to smart flexible machine tools of the late 1900s.

- Computer-controlled machine tools have made it possible to build products to customer specifications as fast and cheaply as mass-produced goods.

- Agile manufacturing combines state-of-the-art fabrication and product technology to produce products to suit each individual customer.

- Continuous corporate renewal, which includes continuous change, rapid response, and evolving quality standards, will be necessary for industries to survive in the future.

- Companies must respond rapidly to manufacturing challenges by sharing, teaming, and cooperating with other companies.

- Each worker must be a team player, responsible for producing defect-free products to satisfy customer needs.

Fig. 60-4 The manufacture of defective parts results in a poorer-quality product and a higher cost to the consumer. (Courtesy of Kelmar Associates)

KNOWLEDGE REVIEW

1. What became the foundation for mass production?

Factory Focus

2. Describe mass production from the standpoint of:
   a. number of machine tools
   b. number of products produced
   c. amount of rework

3. Describe flexible manufacturing from the standpoint of:
   a. number of machine tools
   b. number of products produced
   c. amount of rework

4. Describe customized manufacturing from the standpoint of:
   a. number of machine tools
   b. number of products produced
   c. amount of rework

Agile Manufacturing

5. What is the purpose of agile manufacturing?

6. When and by what institute was the Agile Manufacturing Enterprise Forum created?

7. What is the goal of agile manufacturing?

Continuous Corporate Renewal

8. Name three factors that successful manufacturing will include in the 21st century.

9. Name the five key points of continuous corporate renewal.

Rapid Response

10. Name three important factors necessary for rapid response when working with other companies.

11. List three important benefits of rapid response.

Evolving Quality

12. On what will the future of product quality depend?

13. Why is teamwork important to the quality of a product?

Appendix of Tables

Table 1   Decimal inch, fractional inch, and millimeter equivalents

Decimal inch	Fractional inch		Millimeter	Decimal inch	Fractional inch		Millimeter
.015625		1/64	0.397	.515625		33/64	13.097
.03125	1/32		0.794	.53125	17/32		13.494
.046875		3/64	1.191	.546875		35/64	13.891
.0625	1/16		1.588	.5625	9/16		14.288
.078125		5/64	1.984	.578125		37/64	14.684
.09375	3/32		2.381	.59375	19/32		15.081
.109375		7/64	2.778	.609375		39/64	15.478
.125	1/8		3.175	.625	5/8		15.875
.140625		9/64	3.572	.640625		41/64	16.272
.15625	5/32		3.969	.65625	21/32		16.669
.171875		11/64	4.366	.671875		43/64	17.066
.1875	3/16		4.762	.6875	11/16		17.462
.203125		13/64	5.159	.703125		45/64	17.859
.21875	7/32		5.556	.71875	23/32		18.256
.234375		15/64	5.953	.734375		47/64	18.653
.25	1/4		6.35	.75	3/4		19.05
.265625		17/64	6.747	.765625		49/64	19.447
.28125	9/32		7.144	.78125	25/32		19.844
.296875		19/64	7.541	.796875		51/64	20.241
.3125	5/16		7.938	.8125	13/16		20.638
.328125		21/64	8.334	.828125		53/64	21.034
.34375	11/32		8.731	.84375	27/32		21.431
.359375		23/64	9.128	.859375		55/64	21.828
.375	3/8		9.525	.875	7/8		22.225
.390625		25/64	9.922	.890625		57/64	22.622
.40625	13/32		10.319	.90625	29/32		23.019
.421875		27/64	10.716	.921875		59/64	23.416
.4375	7/16		11.112	.9375	15/16		23.812
.453125		29/64	11.509	.953125		61/64	24.209
.46875	15/32		11.906	.96875	31/32		24.606
.484375		31/64	12.303	.984375		63/64	25.003
.5	1/2		12.7	1.	1		25

Table 2 Inch/Millimeter Conversion

Inches to Millimeters						Millimeters to Inches					
Inches	Milli-meters	Inches	Milli-meters	Inches	Milli-meters	Milli-meters	Inches	Milli-meters	Inches	Milli meters	Inches
.001	0.025	.290	7.37	.660	16.76	0.01	.0004	0.35	.0138	0.68	.0268
.002	0.051	.300	7.62	.670	17.02	0.02	.0008	0.36	.0142	0.69	.0272
.003	0.076	.310	7.87	.680	17.27	0.03	.0012	0.37	.0146	0.7	.0276
.004	0.102	.320	8.13	.690	17.53	0.04	.0016	0.38	.0150	0.71	.0280
.005	0.127	.330	8.38	.700	17.78	0.05	.0020	0.39	.0154	0.72	.0283
.006	0.152	.340	8.64	.710	18.03	0.06	.0024	0.4	.0157	0.73	.0287
.007	0.178	.350	8.89	.720	18.29	0.07	.0028	0.41	.0161	0.74	.0291
.008	0.203	.360	9.14	.730	18.54	0.08	.0031	0.42	.0165	0.75	.0295
.009	0.229	.370	9.4	.740	18.8	0.09	.0035	0.43	.0169	0.76	.0299
.010	0.254	.380	9.65	.750	19.05	0.1	.0039	0.44	.0173	0.77	.0303
.020	0.508	.390	9.91	.760	19.3	0.11	.0043	0.45	.0177	0.78	.0307
.030	0.762	.400	10.16	.770	19.56	0.12	.0047	0.46	.0181	0.79	.0311
.040	1.016	.410	10.41	.780	19.81	0.13	.0051	0.47	.0185	0.8	.0315
.050	1.27	.420	10.67	.790	20.07	0.14	.0055	0.48	.0189	0.81	.0319
.060	1.524	.430	10.92	.800	20.32	0.15	.0059	0.49	.0193	0.82	.0323
.070	1.778	.440	11.18	.810	20.57	0.16	.0063	0.5	.0197	0.83	.0327
.080	2.032	.450	11.43	.820	20.83	0.17	.0067	0.51	.0201	0.84	.0331
.090	2.286	.460	11.68	.830	21.08	0.18	.0071	0.52	.0205	0.85	.0335
.100	2.54	.470	11.94	.840	21.34	0.19	.0075	0.53	.0209	0.86	.0339
.110	2.794	.480	12.19	.850	21.59	0.2	.0079	0.54	.0213	0.87	.0343
.120	3.048	.490	12.45	.860	21.84	0.21	.0083	0.55	.0217	0.88	.0346
.130	3.302	.500	12.7	.870	22.1	0.22	.0087	0.56	.0220	0.89	.0350
.140	3.56	.510	12.95	.880	22.35	0.23	.0091	0.57	.0224	0.9	.0354
.150	3.81	.520	13.21	.890	22.61	0.24	.0094	0.58	.0228	0.91	.0358
.160	4.06	.530	13.46	.900	22.86	0.25	.0098	0.59	.0232	0.92	.0362
.170	4.32	.540	13.72	.910	23.11	0.26	.0102	0.6	.0236	0.93	.0366
.180	4.57	.550	13.97	.920	23.37	0.27	.0106	0.61	.0240	0.94	.0370
.190	4.83	.560	14.22	.930	23.62	0.28	.0110	0.62	.0244	0.95	.0374
.200	5.08	.570	14.48	.940	23.88	0.29	.0114	0.63	.0248	0.96	.0378
.210	5.33	.580	14.73	.950	24.13	0.3	.0118	0.64	.0252	0.97	.0382
.220	5.59	.590	14.99	.960	24.38	0.31	.0122	0.65	.0256	0.98	.0386
.230	5.84	.600	15.24	.970	24.64	0.32	.0126	0.66	.0260	0.99	.0390
.240	6.1	.610	15.49	.980	24.89	0.33	.0130	0.67	.0264	1	.0394
.250	6.35	.620	15.75	.990	25.15	0.34	.0134	. . . .	. . . .	. . . .	. . . .
.260	6.66	.630	16	1.000	25.4						
.270	6.86	.640	16.26	. . . .	. . . .						
.280	7.11	.650	16.51	. . . .	. . . .						

Table 3 Fractional sizes expressed in millimeters

25.4 mm equals 1 inch

Fractional sizes		1 In.	2 In.	3 In.	4 In.	5 In.	6 In.	Fractional sizes		1 In.	2 In.	3 In.	4 In.	5 In.	6 In.
		25.4	50.8	76.2	101.6	127.	152.4	1/2	12.7	38.1	63.5	88.9	114.3	139.7	165.1
1/64	0.4	25.80	51.20	76.60	102.	127.39	152.79	33/64	13.1	38.49	63.90	89.3	114.69	140.09	165.49
1/32	0.79	26.19	51.59	76.99	102.39	127.79	153.19	17/32	13.49	38.89	64.29	89.69	115.09	140.49	165.89
3/64	1.19	26.59	51.99	77.39	102.79	128.19	153.59	35/64	13.89	39.29	64.69	90.09	115.49	140.89	166.29
1/16	1.59	26.99	52.39	77.79	103.19	128.59	153.98	9/16	14.29	39.69	65.09	90.49	115.89	141.29	166.68
5/64	1.98	27.38	52.78	78.18	103.58	128.98	154.38	37/64	14.68	40.08	65.48	90.88	116.28	141.68	167.08
3/32	2.38	27.78	53.18	78.58	103.98	129.38	154.78	19/32	15.08	40.48	65.88	91.28	116.68	142.08	167.48
7/64	2.77	28.17	53.58	78.98	104.37	129.78	155.18	39/64	15.48	40.88	66.28	91.68	117.08	142.48	167.88
1/8	3.17	28.57	53.97	79.37	104.77	130.17	155.57	5/8	15.87	41.27	66.67	92.07	117.47	142.87	168.27
9/64	3.57	28.97	54.37	79.77	105.17	130.57	155.97	41/64	16.27	41.67	67.07	92.47	117.87	143.27	168.67
5/32	3.97	29.37	54.77	80.17	105.57	130.97	156.37	21/32	16.67	42.07	67.47	92.87	118.27	143.67	169.07
11/64	4.37	29.76	55.16	80.56	105.96	131.36	156.76	43/64	17.07	42.46	67.86	93.26	118.66	144.06	169.46
3/16	4.76	30.16	55.56	80.96	106.36	131.76	157.16	11/16	17.46	42.86	68.26	93.66	119.06	144.46	169.86
13/64	5.16	30.56	55.96	81.36	106.76	132.16	157.56	45/64	17.86	43.26	68.66	94.06	119.46	144.86	170.26
7/32	5.56	30.96	56.36	81.75	107.16	132.55	157.95	23/32	18.26	43.66	69.05	94.45	119.85	145.25	170.65
15/64	5.95	31.35	56.75	82.15	107.55	132.95	158.35	47/64	18.65	44.05	69.45	94.85	120.25	145.65	171.05
1/4	6.35	31.75	57.15	82.55	107.95	133.35	158.75	3/4	19.05	44.45	69.85	95.25	120.65	146.05	171.45
17/64	6.75	32.15	57.55	82.95	108.34	133.74	159.14	49/64	19.45	44.85	70.25	95.65	121.04	146.44	171.84
9/32	7.14	32.54	57.94	83.34	108.74	134.14	159.54	25/32	19.84	45.24	70.64	96.04	121.44	146.84	172.24
19/64	7.54	32.94	58.34	83.74	109.14	134.54	159.94	51/64	20.24	45.64	71.04	96.44	121.84	147.24	172.64
5/16	7.94	33.34	58.74	84.14	109.54	134.94	160.33	13/16	20.64	46.04	71.44	96.84	122.24	147.63	173.03
21/64	8.33	33.73	59.13	84.53	109.93	135.33	160.73	53/64	21.03	46.43	71.83	97.23	122.63	148.03	173.43
11/32	8.73	34.13	59.53	84.93	110.33	135.73	161.13	27/32	21.43	46.83	72.23	97.63	123.03	148.43	173.83
23/64	9.13	34.53	59.93	85.33	110.73	136.13	161.53	55/64	21.83	47.23	72.63	98.03	123.43	148.83	174.22
3/8	9.52	34.92	60.32	85.72	111.12	136.52	161.92	7/8	22.22	47.62	73.02	98.42	123.82	149.22	174.62
25/64	9.92	35.32	60.72	86.12	111.52	136.92	162.32	57/64	22.62	48.02	73.42	98.82	124.22	149.62	175.02
13/32	10.32	35.72	61.12	86.52	111.92	137.32	162.72	29/32	23.02	48.42	73.82	99.22	124.62	150.02	175.42
27/64	10.72	36.11	61.51	86.91	112.31	137.71	163.11	59/64	23.42	48.81	74.21	99.61	125.01	150.41	175.81
7/16	11.11	36.51	61.91	87.31	112.71	138.11	163.51	15/16	23.81	49.21	74.61	100.01	125.41	150.81	176.21
29/64	11.51	36.91	62.31	87.71	113.11	138.51	163.91	61/64	24.21	49.61	75.01	100.41	125.81	151.21	176.61
15/32	11.91	37.31	62.71	88.1	113.5	138.9	164.3	31/32	24.61	50.01	75.4	100.8	126.2	151.6	177.
31/64	12.3	37.7	63.1	88.5	113.9	139.3	164.7	63/64	25	50.4	75.8	101.2	126.6	152.	177.4

*To use, read down appropriate inch column to the desired fraction line. The number indicated is the size in millimeters.

Table 4   Metric-English conversion table

Multiply	By	To get equivalent number of	Multiply	By	To get equivalent number of
Length			**Acceleration**		
Inch	25.4	Millimeters (mm)	Foot/second^2	0.304 8	Meter per second^2 (m/s^2)
Foot	0.304 8	Meters (m)	Inch/second^2	0.025 4	Meter per second^2
Yard	0.914 4	Meters	**Torque**		
Mile	1.609	Kilometers (km)	Pound-inch	0.112 98	Newton-meters (N-m)
Area			Pound-foot	1.355 8	Newton-meters
Inch^2	645.2	Millimeters^2 (mm^2)	**Power**		
	6.45	Centimeters^2 (cm^2)	Horsepower	0.746	Kilowatts (kW)
Foot^2	0.092 9	Meters^2 (m^2)	**Pressure or stress**		
Yard^2	0.836 1	Meters^2	Inches of water	0.249 1	Kilopascals (kPa)
Volume			Pounds/square inch	6.895	Kilopascals
Inch^3	16 387.	mm^3	**Energy or work**		
	16.387	cm^3	BTU	1 055.	Joules (J)
	0.016 4	Liters (l)	Foot-pound	1.355 8	Joules
Quart (U.S.)	0.946 4	Liters	Kilowatthour	3 600 000.	Joules (J = one W's)
Quart (imperial)	1.136	Liters		or 3.6 × 10^6	
Gallon (U.S.)	3.785 4	Liters	**Light**		
Gallon (imperial)	4.459	Liters	Footcandle	1.076 4	Lumens per meter^2 (lm/m^2)
Yard^3	0.764 6	Meters^3 (m^3)			
Mass			**Fuel performance**		
Pound	0.453 6	Kilograms (kg)	Miles per gallon	0.425 1	Kilometers per liter (km/l)
Ton	907.18	Kilograms (kg)	Gallons per mile	2.352 7	Liters per kilometers (l/km)
Ton	0.907	Tonne (t)	**Velocity**		
Force			Miles per hour	1.609 3	Kilometers per hr (km/h)
Kilogram	9.807	Newtons (N)			
Ounce	0.278 0	Newtons			
Pound	4.448	Newtons			
Temperature					
Degree Fahrenheit	(°F − 32) ÷ 1.8	Degree Celsius (C)			
Degree Celsius	(°C × 1.8) + 32	Degree Fahrenheit (F)			

Table 5   Common Formulas

Code		
c.p.t. = Chip per tooth CS = Cutting speed D = Large diameter d = Small diameter G.L. = Guide bar length	N = Number of threads per inch = Number of strokes per minute = Number of teeth in cutter O.L. = Overall length of work P = Pitch	T.D.S. = Tap drill size T.L. = Taper length tpf = Taper per foot tpmm = Taper per millimeter T.O. = Tallstock offset

Inch	Metric
$\text{T.D.S.} = D - \left(\dfrac{1}{N}\right)$	$\text{T.D.S.} = D - P$
$\text{r/min} = \dfrac{\text{CS (ft)} \times 4}{D \text{ (in.)}}$	$\text{r/min} = \dfrac{\text{CS (m)} \times 320}{D \text{ (mm)}}$
$\text{tpf} = \dfrac{(D - d) \times 12}{\text{T.L.}}$	$\text{tpmm} = \dfrac{(D - d)}{\text{T.L.}}$
$\text{T.O.} = \dfrac{\text{t/ft} \times \text{O.L.}}{24}$	$\text{T.O.} = \dfrac{\text{t/mm} \times \text{O.L.}}{2}$
$\text{Guide bar setover} = \dfrac{(D - d) \times 12}{\text{T.L.}}$	$\text{Guide bar setover} = \dfrac{(D - d)}{2} \times \dfrac{\text{G.L.}}{\text{T.L.}}$
Milling feed (in./min) = $N \times$ c.p.t. $\times$ r/min	Milling feed (mm/min) = $N \times$ c.p.t. $\times$ r/min

Table 6   Formula Shortcuts

For the correct formula, block out (cover) the unknown; the remainder is the formula. In each diagram the horizontal line is the division line; the vertical line(s) is the multiplication line.

Code

A = Area	L = Length	b = Base
C = Circumference	R = Radius	h = Height
CS = Cutting speed	r/min = Revolutions/minute	m = Meters
D = Diameter	S = Strokes/minute	mm = Millimeters

Circle

Division line ⟶ ⟵ Multiplication line

$$C = \pi \times D$$

$$D = \frac{C}{\pi}$$

Four-element formulas

1. Block out unknown.
2. Cross-multiply diagonally opposite elements.
3. Divise by remaining element.

Triangles

$$A = \frac{b \times h}{2}$$

$$b = \frac{A \times 2}{h}$$

$$h = \frac{A \times 2}{b}$$

Area

Squares and rectangles

$$A = L \times W$$

$$L = \frac{A}{W}$$

$$W = \frac{A}{L}$$

Circles

$$A = \pi \times R^2$$

$$R^2 = \frac{A}{\pi}$$

Revolutions per minute (r/min)
(Lathe, drill, mill, grinder)

Inch

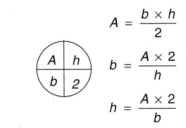

$$r/min = \frac{CS\ (ft) \times 4}{D\ (in.)}$$

$$CS = \frac{r/min \times D}{4}$$

$$D = \frac{CS \times 4}{r/min}$$

Metric

$$r/min = \frac{CS\ (m) \times 320}{D\ (mm)}$$

$$CS = \frac{r/min \times D}{320}$$

$$D = \frac{CS \times 320}{r/min}$$

Table 7 Letter Drill Sizes

Letter	in.	mm	Letter	in.	mm	Letter	in.	mm	Letter	in.	mm
A	.234	5.9	H	.266	6.7	N	.302	7.7	T	.358	9.1
B	.238	6	I	.272	6.9	O	.316	8	U	.368	9.3
C	.242	6.1	J	.277	7	P	.323	8.2	V	.377	9.5
D	.246	6.2	K	.281	7.1	Q	.332	8.4	W	.386	9.8
E	.250	6.4	L	.290	7.4	R	.339	8.6	X	.397	10.1
F	.257	6.5	M	.295	7.5	S	.348	8.8	Y	.404	10.3
G	.261	6.6							Z	.413	10.5

Table 8 Number Drill Sizes

No.	inch	mm	No.	inch	mm	No.	inch	mm
1	.228	5.8	34	.111	2.81	66	.033	0.84
2	.221	5.6	35	.110	2.79	67	.032	0.81
3	.213	5.4	36	.1065	2.7	68	.031	0.79
4	.209	5.3	37	.104	2.65	69	.0292	0.74
5	.2055	5.22	38	.1015	2.6	70	.028	0.71
6	.204	5.18	39	.0995	2.55	71	.026	0.66
7	.201	5.1	40	.098	2.5	72	.025	0.64
8	.199	5.05	41	.096	2.45	73	.024	0.61
9	.196	5	42	.0935	2.4	74	.0225	0.57
10	.1935	4.912	43	.089	2.25	75	.021	0.53
11	.191	4.85	44	.086	2.2	76	.020	0.51
12	.189	4.8	45	.082	2.1	77	.018	0.46
13	.185	4.7	46	.081	2.05	78	.016	0.41
14	.182	4.62	47	.0785	2	79	.0145	0.37
15	.180	4.57	48	.076	1.95	80	.0135	0.34
16	.177	4.5	49	.073	1.85	81	.013	0.33
17	.173	4.4	50	.070	1.8	82	.0125	0.32
18	.1695	4.3	51	.067	1.7	83	.012	0.31
19	.166	4.2	52	.0635	1.6	84	.0115	0.29
20	.161	4.1	53	.0595	1.5	85	.011	0.28
21	.159	4.03	54	.055	1.4	86	.0105	0.27
22	.157	4	55	.052	1.3	87	.010	0.25
23	.154	3.91	56	.0465	1.2	88	.0095	0.24
24	.152	3.86	57	.043	1.1	89	.0091	0.23
25	.1495	3.8	58	.042	1.06	90	.0087	0.22
26	.147	3.73	59	.041	1.04	91	.0083	0.21
27	.144	3.65	60	.040	1	92	.0079	0.2
28	.1405	3.6	61	.039	0.99	93	.0075	0.19
29	.136	3.5	62	.038	0.97	94	.0071	0.18
30	.1285	3.3	63	.037	0.94	95	.0067	0.17
31	.120	3	64	.036	0.92	96	.0063	0.16
32	.116	2.95	65	.035	0.89	97	.0059	0.15
33	.113	2.85						

Table 9   Tap Drill Sizes (75% of thread depth)

American National Unified Form Thread					
NC National Coarse			**NF National Fine**		
Tap Size	Threads per inch	Tap Drill Size	Tap Size	Threads per inch	Tap Drill Size
# 5	40	#38	# 5	44	#37
# 6	32	#36	# 6	40	#33
# 8	32	#29	# 8	36	#29
#10	24	#25	#10	32	#21
#12	24	#16	#12	28	#14
1/4	20	# 7	1/4	28	# 3
5/16	18	F	5/16	24	I
3/8	16	5/16	3/8	24	Q
7/16	14	U	7/16	20	25/64
1/2	13	27/64	1/2	20	29/64
9/16	12	31/64	9/16	18	33/64
5/8	11	17/32	5/8	18	37/64
3/4	10	21/32	3/4	16	11/16
7/8	9	49/64	7/8	14	13/16
1	8	7/8	1	14	15/16
1-1/8	7	63/64	1-1/8	12	1-3/64
1-1/4	7	1-7/64	1-1/4	12	1-11/64
1-3/8	6	1-7/32	1-3/8	12	1-19/64
1-1/2	6	1-11/32	1-1/2	12	1-27/64
1-3/4	5	1-9/16			
2	4-1/2	1-25/32			
NPT National Pipe Thread					
1/8	27	11/32	1	11-1/2	1-5/32
1/4	18	7/16	1-1/4	11-1/2	1-1/2
3/8	18	19/32	1-1/2	11-1/2	1-23/32
1/2	14	23/32	2	11-1/2	2-3/16
3/4	14	15/16	2-1/2	8	2-5/8

The major diameter of an NC or NF number size tap or screw = (N × .013) + .060

Example: The major diameter of a #5 tap equals (5 × .013) + .060 = .125 diameter

Table 10   Metric Tap Drill Sizes

Nominal Diameter mm	Thread Pitch mm	Tap Drill Size mm	Nominal Diameter mm	Thread Pitch mm	Tap Drill Size mm
1.6	0.35	1.2	20	2.5	17.5
2	0.4	1.6	24	3	21
2.5	0.45	2.05	30	3.5	26.5
3	0.5	2.5	36	4	32
3.5	0.6	2.9	42	4.5	37.5
4	0.7	3.3	48	5	43
5	0.8	4.2	56	5.5	50.5
6	1	5.3	64	6	58
8	1.25	6.8	72	6	66
10	1.5	8.5	80	6	74
12	1.75	10.2	90	6	84
14	2	12	100	6	94
16	2	14			

Table 11   ISO Metric Pitch and Diameter Combinations

Nominal Diameter mm	Thread Pitch mm	Nominal Diameter mm	Thread Pitch mm
1.6	0.35	20	2.5
2	0.4	24	3
2.5	0.45	30	3.5
3	0.5	36	4
3.5	0.6	42	4.5
4	0.7	48	5
5	0.8	56	5.5
6	1	64	6
8	1.25	72	6
10	1.5	80	6
12	1.75	90	6
14	2	100	6
16	2		

Table 12 Three Wire Thread Measurement

**Three Wire Thread Measurement
(60° Metric Thread)**

M = PD + C   PD = M − C
M = Measurement over wires
PD = Pitch diameter
C = Constant

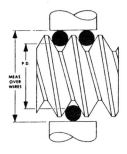

Pitch		Best Wire Size		Constant	
Inches	mm	Inches	mm	Inches	mm
.00787	0.2	.00455	0.1155	.00682	0.1732
.00886	0.225	.00511	0.1299	.00767	0.1949
.00934	0.25	.00568	0.1443	.00852	0.2165
.01181	0.3	.00682	0.1732	.01023	0.2598
.01378	0.35	.00796	0.2021	.01193	0.3031
.01575	0.4	.00909	0.2309	.01364	0.3464
.01772	0.45	.01023	0.2598	.01534	0.3897
.01969	0.5	.01137	0.2887	.01705	0.433
.02362	0.6	.01364	0.3464	.02046	0.5196
.02756	0.7	.01591	0.4041	.02387	0.6062
.02953	0.75	.01705	0.433	.02557	0.6495
.03150	0.8	.01818	0.4619	.02728	0.6928
.03543	0.9	.02046	0.5196	.03069	0.7794
.03937	1	.02273	0.5774	.03410	0.866
.04921	1.25	.02841	0.7217	.04262	1.0825
.05906	1.5	.03410	0.866	.05114	1.299
.06890	1.75	.03978	1.0104	.05967	1.5155
.07874	2	.04546	1.1547	.06819	1.7321
.09843	2.5	.05683	1.4434	.08524	2.1651
.11811	3	.06819	1.7321	.10229	2.5981
.13780	3.5	.07956	2.0207	.11933	3.0311
.15748	4	.09092	2.3094	.13638	3.4641
.17717	4.5	.10229	2.5981	.15343	3.8971
.19685	5	.11365	2.8868	.17048	4.3301
.21654	5.5	.12502	3.1754	.18753	4.7631
.23622	6	.13638	3.4641	.20457	5.1962
.27559	7	.15911	4.0415	.23867	6.0622
.31496	8	.18184	4.6188	.27276	6.9282
.35433	9	.20457	5.1962	.30686	7.7942
.39370	10	.22730	5.7735	.34095	8.6603

Table 13   Morse Tapers

ANGLE OF KEY 8° 19' = TAPER 1-3/4 IN 12

Number of taper	Diameter of plug at small end	Diameter at end of socket	Whole length of shank	Shank depth	Depth of hole	Standard plug depth	Thickness of tongue	Length of tongue	Diameter of tongue	Width of keyway	Length of keyway	End of socket to keyway	Taper per foot
	D	A	B	S	H	P	t	T	d	w	L	K	
0	.252	.356	2-11/32	2-7/32	2-1/32	2	5/32	1/4	.235	.160	9/16	1-15/16	.6246
1	.369	.475	2-9/16	2-7/16	2-3/16	2-1/8	13/64	3/8	.343	.213	3/4	2-1/16	.5986
2	.572	.700	3-1/8	2-15/16	2-5/8	2-9/16	1/4	7/16	17/32	.260	7/8	2-1/2	.5994
3	.778	.938	3-7/8	3-11/16	3-1/4	3-3/16	5/16	9/16	23/32	.322	1-3/16	3-1/16	.6023
4	1.020	1.231	4-7/8	4-5/8	4-1/8	4-1/16	15/32	5/8	31/32	.478	1-1/4	3-7/8	.6232
5	1.475	1.748	6-1/8	5-7/8	5-1/4	5-3/16	5/8	3/4	1-13/32	.635	1-1/2	4-15/16	.6315
6	2.116	2.494	8-9/16	8-1/4	7-3/8	7-1/4	3/4	1-1/8	2	.760	1-3/4	7	.6256
7	2.750	3.270	11-1/4	11-5/8	10-1/8	10	1-1/8	1-3/8	2-5/8	1.135	2-5/8	9-1/2	.6240

Note: All measurements are in inches

Table 14 Standard Milling Machine Taper

Milling Machine Spindles	Milling Machine Arbors

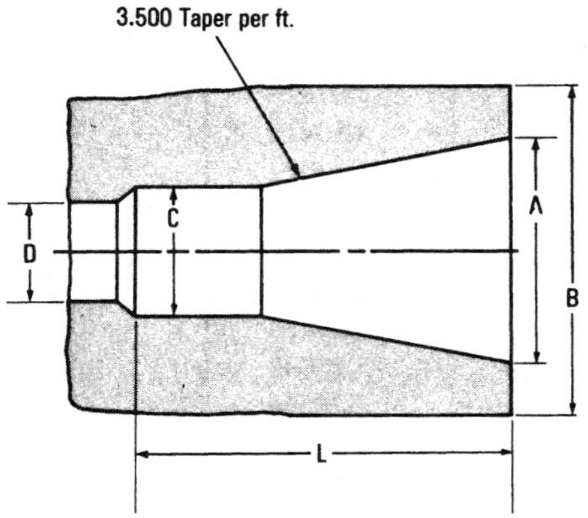
3.500 Taper per ft.
D
C
A
B
L

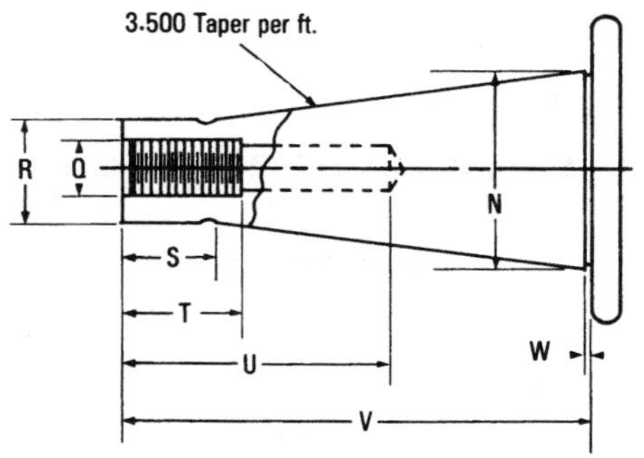
3.500 Taper per ft.
R
Q
S
T
U
V
N
W

Taper No.	A	B	C	D	L	N	Q	R	S	T	U	V	W
30	1.250	2.7493	.685 .692	21/32	2-7/8	1.250	1/2 – 13	.673 .675	13/16	1	2	2-3/4	1/16
40	1.750	3.4993	.997 1.005	21/32	3-7/8	1.750	5/8 – 11	.985 .987	1	1-1/8	2-5/16	3-3/4	1/16
50	2.750	5.0618	1.559 1.568	1-1/16	5-1/2	2.750	1 – 8	1.547 1.549	1	1-3/4	3-1/2	5-1/8	1/8
60	4.250	8.718	2.371 2.381	1-3/8	8-5/8	4.250	1-1/4 – 7	2.359 2.361	1-3/4	2-1/4	4-1/4	8-5/16	1/8

Note: All measurements are in inches

Table 15 Hardness Conversion Chart

10 mm Ball 3000 kg	120° Cone 150 kg	1/16 in. Ball 100 kg	Model C	Mpa	10 mm Ball 3000 kg	120° Cone 150 kg	1/16 in. Ball 100 kg	Model C	Mpa
Brinell	Rockwell C	Rockwell B	Shore Scleroscope	Tensile Strength	Brinell	Rockwell C	Rockwell B	Shore Scleroscope	Tensile Strength
800	72	.....	100	....	276	30	105	42	938
780	71	.....	99	....	269	29	104	41	910
760	70	.....	98	....	261	28	103	40	889
745	68	.....	97	2530	258	27	102	39	876
725	67	.....	96	2460	255	26	102	39	862
712	66	.....	95	2413	249	25	101	38	848
682	65	.....	93	2324	245	24	100	37	820
668	64	.....	91	2248	240	23	99	36	807
652	63	.....	89	2193	237	23	99	35	793
626	62	.....	87	2110	229	22	98	34	779
614	61	.....	85	2062	224	21	97	33	758
601	60	.....	83	2013	217	20	96	33	738
590	59	.....	81	2000	211	19	95	32	717
576	57	.....	79	1937	206	18	94	32	703
552	56	.....	76	1862	203	17	94	31	689
545	55	.....	75	1848	200	16	93	31	676
529	54	.....	74	1786	196	15	92	30	662
514	53	120	72	1751	191	14	92	30	648
502	52	119	70	1703	187	13	91	29	634
495	51	119	69	1682	185	12	91	29	627
477	49	118	67	1606	183	11	90	28	621
461	48	117	66	1565	180	10	89	28	614
451	47	117	65	1538	175	9	88	27	593
444	46	116	64	1510	170	7	87	27	579
427	45	115	62	1441	167	6	87	27	565
415	44	115	60	1407	165	5	86	26	558
401	43	114	58	1351	163	4	85	26	552
388	42	114	57	1317	160	3	84	25	538
375	41	113	55	1269	156	2	83	25	524
370	40	112	54	1255	154	1	82	25	517
362	39	111	53	1234	152	.....	82	24	510
351	38	111	51	1193	150	.....	81	24	510
346	37	110	50	1172	147	.....	80	24	496
341	37	110	49	1158	145	.....	79	23	490
331	36	109	47	1124	143	.....	79	23	483
323	35	109	46	1089	141	.....	78	23	476
311	34	108	46	1055	140	.....	77	22	476
301	33	107	45	1020	135	.....	75	22	462
293	32	106	44	993	130	.....	72	22	448
285	31	105	43	965	.....	.....	.....	....	....

Table 16   Solutions for Right-Angle Triangles

Sine $\angle$ = $\dfrac{\text{Side opposite}}{\text{Hypotenuse}}$	Cosecant $\angle$ = $\dfrac{\text{Hypotenuse}}{\text{Side opposite}}$
Cosine $\angle$ = $\dfrac{\text{Side adjacent}}{\text{Hypotenuse}}$	Secant $\angle$ = $\dfrac{\text{Hypotenuse}}{\text{Side adjacent}}$
Tangent $\angle$ = $\dfrac{\text{Side opposite}}{\text{Side adjacent}}$	Cotangent $\angle$ = $\dfrac{\text{Side adjacent}}{\text{Side opposite}}$

Knowing	Formulas to find	
Sides a & b	$c = \sqrt{a^2 - b^2}$	$\sin B = \dfrac{b}{a}$
Side a & angle B	$b = a \times \sin B$	$c = a \times \cos B$
Sides a & c	$b = \sqrt{a^2 - c^2}$	$\sin C = \dfrac{c}{a}$
Side a & angle C	$b = a \times \cos C$	$c = a \times \sin C$
Sides b & c	$a = \sqrt{b^2 + c^2}$	$\tan B = \dfrac{b}{c}$
Side b & angle B	$a = \dfrac{b}{\sin B}$	$c = b \times \cot B$
Side b & angle C	$a = \dfrac{b}{\cos C}$	$c = b \times \tan C$
Side c & angle B	$a = \dfrac{c}{\cos B}$	$b = c \times \tan B$
Side c & angle C	$a = \dfrac{c}{\sin C}$	$b = c \times \cot C$

Table 17 Sine Bar Constants (5" bar) (Multiply constants by two for a 10" sine bar)

Min.	0°	1°	2°	3°	4°	5°	6°	7°	8°	9°	10°	11°	12°	13°	14°	15°	16°	17°	18°	19°	Min.
0	.00000	.08725	.17450	.26170	.34880	.43580	.52265	.60935	.69585	.78215	.86825	.95405	1.0395	1.1247	1.2096	1.2941	1.3782	1.4618	1.5451	1.6278	0
2	.00290	.09015	.17740	.26460	.35170	.43870	.52555	.61225	.69875	.78505	.87110	.95690	.0424	.1276	.2124	.2969	.3810	.4646	.5478	.6306	2
4	.00580	.09310	.18030	.26750	.35460	.44155	.52845	.61510	.70165	.78790	.87395	.95975	.0452	.1304	.2152	.2997	.3838	.4674	.5506	.6333	4
6	.00875	.09600	.18320	.27040	.35750	.44445	.53130	.61800	.70450	.79080	.87685	.96260	.0481	.1332	.2181	.3025	.3865	.4702	.5534	.6361	6
8	.01165	.09890	.18615	.27330	.36040	.44735	.53420	.62090	.70740	.79365	.87970	.96545	.0509	.1361	.2209	.3053	.3893	.4730	.5561	.6388	8
10	.01455	.10180	.18905	.27620	.36330	.45025	.53710	.62380	.71025	.79655	.88255	.96830	1.0538	1.1389	1.2237	1.3081	1.3921	1.4757	1.5589	1.6416	10
12	.01745	.10470	.19195	.27910	.36620	.45315	.54000	.62665	.71315	.79940	.88540	.97115	.0566	.1417	.2265	.3109	.3949	.4785	.5616	.6443	12
14	.02035	.10760	.19485	.28200	.36910	.45605	.54290	.62955	.71600	.80230	.88830	.97405	.0594	.1446	.2293	.3137	.3977	.4813	.5644	.6471	14
16	.02325	.11055	.19775	.28490	.37200	.45895	.54580	.63245	.71890	.80515	.89115	.97690	.0623	.1474	.2322	.3165	.4005	.4841	.5672	.6498	16
18	.02620	.11345	.20065	.28780	.37490	.46185	.54865	.63530	.72180	.80800	.89400	.97975	.0651	.1502	.2350	.3193	.4033	.4868	.5699	.6525	18
20	.02910	.11635	.20355	.29070	.37780	.46475	.55155	.63820	.72465	.81090	.89685	.98260	1.0680	1.1531	1.2378	1.3221	1.4061	1.4896	1.5727	1.6553	20
22	.03200	.11925	.20645	.29365	.38070	.46765	.55445	.64110	.72755	.81375	.89975	.98545	.0708	.1559	.2406	.3250	.4089	.4924	.5755	.6580	22
24	.03490	.12215	.20940	.29655	.38360	.47055	.55735	.64400	.73040	.81665	.90260	.98830	.0737	.1587	.2434	.3278	.4117	.4952	.5782	.6608	24
26	.03780	.12505	.21230	.29945	.38650	.47345	.56025	.64685	.73330	.81950	.90545	.99115	.0765	.1615	.2462	.3306	.4145	.4980	.5810	.6635	26
28	.04070	.12800	.21520	.30235	.38940	.47635	.56315	.64975	.73615	.82235	.90830	.99400	.0793	.1644	.2491	.3334	.4173	.5007	.5837	.6663	28
30	.04365	.13090	.21810	.30525	.39230	.47925	.56600	.65265	.73905	.82525	.91120	.99685	1.0822	1.1672	1.2519	1.3362	1.4201	1.5035	1.5865	1.6690	30
32	.04655	.13380	.22100	.30815	.39520	.48210	.56890	.65550	.74190	.82810	.91405	.99970	.0850	.1700	.2547	.3390	.4228	.5063	.5893	.6718	32
34	.04945	.13670	.22390	.31105	.39810	.48500	.57180	.65840	.74480	.83100	.91690	1.0016	.0879	.1729	.2575	.3418	.4256	.5091	.5920	.6745	34
36	.05235	.13960	.22680	.31395	.40100	.48790	.57470	.66130	.74770	.83385	.91975	1.0054	.0907	.1757	.2603	.3446	.4284	.5118	.5948	.6772	36
38	.05525	.14250	.22970	.31685	.40390	.49080	.57760	.66415	.75055	.83670	.92260	1.0082	.0935	.1785	.2631	.3474	.4312	.5146	.5975	.6800	38
40	.05820	.14540	.23265	.31975	.40680	.49370	.58045	.66705	.75345	.83960	.92545	1.0110	1.0964	1.1813	1.2660	1.3502	1.4340	1.5174	1.6003	1.6827	40
42	.06110	.14835	.23555	.32265	.40970	.49660	.58335	.66995	.75630	.84245	.92835	.0139	.0992	.1842	.2688	.3530	.4368	.5201	.6030	.6855	42
44	.06400	.15125	.23845	.32555	.41260	.49950	.58625	.67280	.75920	.84530	.93120	.0168	.1020	.1870	.2716	.3558	.4396	.5229	.6058	.6882	44
46	.06690	.15415	.24135	.32845	.41550	.50240	.58915	.67570	.76205	.84820	.93405	.0196	.1049	.1898	.2744	.3586	.4423	.5257	.6085	.6909	46
48	.06980	.15705	.24425	.33135	.41840	.50530	.59200	.67860	.76495	.85105	.93690	.0225	.1077	.1926	.2772	.3614	.4451	.5285	.6113	.6937	48
50	.07270	.15995	.24715	.33425	.42130	.50820	.59490	.68145	.76780	.85390	.93975	1.0253	1.1106	1.1955	1.2800	1.3642	1.4479	1.5312	1.6141	1.6964	50
52	.07565	.16285	.25005	.33715	.42420	.51105	.59780	.68435	.77070	.85680	.94260	.0281	.1134	.1983	.2828	.3670	.4507	.5340	.6168	.6991	52
54	.07855	.16580	.25295	.34010	.42710	.51395	.60070	.68720	.77355	.85965	.94550	.0310	.1162	.2011	.2856	.3698	.4535	.5368	.6196	.7019	54
56	.08145	.16870	.25585	.34300	.43000	.51685	.60355	.69010	.77645	.86250	.94835	.0338	.1191	.2039	.2884	.3726	.4563	.5395	.6223	.7046	56
58	.08435	.17160	.25875	.34590	.43290	.51975	.60645	.69300	.77930	.86540	.95120	.0367	.1219	.2068	.2913	.3754	.4591	.5423	.6251	.7073	58
60	.08725	.17450	.26170	.34880	.43580	.52265	.60935	.69585	.78215	.86825	.95405	1.0395	1.1247	1.2096	1.2941	1.3782	1.4618	1.5451	1.6278	1.7101	60

Table 17 (Continued)

Min.	20°	21°	22°	23°	24°	25°	26°	27°	28°	29°	30°	31°	32°	33°	34°	35°	36°	37°	38°	39°	Min.
0	1.7101	1.7918	1.8730	1.9536	2.0337	2.1131	2.1918	2.2699	2.3473	2.4240	2.5000	2.5752	2.6496	2.7232	2.7959	2.8679	2.9389	3.0091	3.0783	3.1466	0
2	.7128	.7945	.8757	.9563	.0363	.1157	.1944	.2725	.3499	.4266	.5025	.5777	.6520	.7256	.7984	.8702	.9413	.0114	.0806	.1488	2
4	.7155	.7972	.8784	.9590	.0390	.1183	.1971	.2751	.3525	.4291	.5050	.5802	.6545	.7280	.8008	.8726	.9436	.0137	.0829	.1511	4
6	.7183	.8000	.8811	.9617	.0416	.1210	.1997	.2777	.3550	.4317	.5075	.5826	.6570	.7305	.8032	.8750	.9460	.0160	.0852	.1534	6
8	.7210	.8027	.8838	.9643	.0443	.1236	.2023	.2803	.3576	.4342	.5100	.5851	.6594	.7329	.8056	.8774	.9483	.0183	.0874	.1556	8
10	1.7237	1.8054	1.8865	1.9670	2.0469	2.1262	2.2049	2.2829	2.3602	2.4367	2.5126	2.5876	2.6619	2.7354	2.8080	2.8798	2.9507	3.0207	3.0897	3.1579	10
12	.7265	.8081	.8892	.9697	.0496	.1289	.2075	.2855	.3627	.4393	.5151	.5901	.6644	.7378	.8104	.8821	.9530	.0230	.0920	.1601	12
14	.7292	.8108	.8919	.9724	.0522	.1315	.2101	.2881	.3653	.4418	.5176	.5926	.6668	.7402	.8128	.8845	.9554	.0253	.0943	.1624	14
16	.7319	.8135	.8946	.9750	.0549	.1341	.2127	.2906	.3679	.4444	.5201	.5951	.6693	.7427	.8152	.8869	.9577	.0276	.0966	.1646	16
18	.7347	.8162	.8973	.9777	.0575	.1368	.2153	.2932	.3704	.4469	.5226	.5976	.6717	.7451	.8176	.8893	.9600	.0299	.0989	.1669	18
20	1.7374	1.8189	1.8999	1.9804	2.0602	2.1394	2.2179	2.2958	2.3730	2.4494	2.5251	2.6001	2.6742	2.7475	2.8200	2.8916	2.9624	3.0322	3.1012	3.1691	20
22	.7401	.8217	.9026	.9830	.0628	.1420	.2205	.2984	.3755	.4520	.5276	.6025	.6767	.7499	.8224	.8940	.9647	.0345	.1034	.1714	22
24	.7428	.8244	.9053	.9857	.0655	.1447	.2232	.3010	.3781	.4545	.5301	.6050	.6791	.7524	.8248	.8964	.9671	.0369	.1057	.1736	24
26	.7456	.8271	.9080	.9884	.0681	.1473	.2258	.3036	.3807	.4570	.5327	.6075	.6816	.7548	.8272	.8988	.9694	.0392	.1080	.1759	26
28	.7483	.8298	.9107	.9911	.0708	.1499	.2284	.3061	.3832	.4596	.5352	.6100	.6840	.7572	.8296	.9011	.9718	.0415	.1103	.1781	28
30	1.7510	1.8325	1.9134	1.9937	2.0734	2.1525	2.2310	2.3087	2.3858	2.4621	2.5377	2.6125	2.6865	2.7597	2.8320	2.9035	2.9741	3.0438	3.1125	3.1804	30
32	.7537	.8352	.9161	.9964	.0761	.1552	.2336	.3113	.3883	.4646	.5402	.6149	.6889	.7621	.8344	.9059	.9764	.0461	.1148	.1826	32
34	.7565	.8379	.9188	.9991	.0787	.1578	.2362	.3139	.3909	.4672	.5427	.6174	.6914	.7645	.8368	.9082	.9788	.0484	.1171	.1849	34
36	.7592	.8406	.9215	2.0017	.0814	.1604	.2388	.3165	.3934	.4697	.5452	.6199	.6938	.7669	.8392	.9106	.9811	.0507	.1194	.1871	36
38	.7619	.8433	.9241	.0044	.0840	.1630	.2414	.3190	.3960	.4722	.5477	.6224	.6963	.7694	.8416	.9130	.9834	.0530	.1216	.1893	38
40	1.7646	1.8460	1.9268	2.0070	2.0867	2.1656	2.2440	2.3216	2.3985	2.4747	2.5502	2.6249	2.6987	2.7718	2.8440	2.9153	2.9858	3.0553	3.1239	3.1916	40
42	.7673	.8487	.9295	.0097	.0893	.1683	.2466	.3242	.4011	.4773	.5527	.6273	.7012	.7742	.8464	.9177	.9881	.0576	.1262	.1938	42
44	.7701	.8514	.9322	.0124	.0920	.1709	.2492	.3268	.4036	.4798	.5552	.6298	.7036	.7766	.8488	.9200	.9904	.0599	.1285	.1961	44
46	.7728	.8541	.9349	.0150	.0946	.1735	.2518	.3293	.4062	.4823	.5577	.6323	.7061	.7790	.8512	.9224	.9928	.0622	.1307	.1983	46
48	.7755	.8568	.9376	.0177	.0972	.1761	.2544	.3319	.4087	.4848	.5602	.6348	.7085	.7815	.8535	.9248	.9951	.0645	.1330	.2005	48
50	1.7782	1.8595	1.9402	2.0204	2.0999	2.1787	2.2570	2.3345	2.4113	2.4874	2.5627	2.6372	2.7110	2.7839	2.8559	2.9271	2.9974	3.0668	3.1353	3.2028	50
52	.7809	.8622	.9429	.0230	.1025	.1814	.2596	.3371	.4138	.4899	.5652	.6397	.7134	.7863	.8583	.9295	.9997	.0691	.1375	.2050	52
54	.7837	.8649	.9456	.0257	.1052	.1840	.2621	.3396	.4164	.4924	.5677	.6422	.7158	.7887	.8607	.9318	3.0021	.0714	.1398	.2072	54
56	.7864	.8676	.9483	.0283	.1078	.1866	.2647	.3422	.4189	.4949	.5702	.6446	.7183	.7911	.8631	.9342	.0044	.0737	.1421	.2095	56
58	.7891	.8703	.9510	.0310	.1104	.1892	.2673	.3448	.4215	.4975	.5727	.6471	.7207	.7935	.8655	.9365	.0067	.0760	.1443	.2117	58
60	1.7918	1.8730	1.9536	2.0337	2.1131	2.1918	2.2699	2.3473	2.4240	2.5000	2.5752	2.6496	2.7232	2.7959	2.8679	2.9389	3.0091	3.0783	3.1466	3.2139	60

Table 17   Sine Bar Constants (5" bar) (Multiply constants by two for a 10" sine bar) (Continued)

Min.	40°	41°	42°	43°	44°	45°	46°	47°	48°	49°	50°	51°	52°	53°	54°	55°	56°	57°	58°	59°	Min.
0	3.2139	3.2803	3.3456	3.4100	3.4733	3.5355	3.5967	3.6567	3.7157	3.7735	3.8302	3.8857	3.9400	3.9932	4.0451	4.0957	4.1452	4.1933	4.2402	4.2858	0
2	.2161	.2825	.3478	.4121	.4754	.5376	.5987	.6587	.7176	.7754	.8321	.8875	.9418	.9949	.0468	.0974	.1468	.1949	.2418	.2873	2
4	.2184	.2847	.3499	.4142	.4774	.5396	.6007	.6607	.7196	.7773	.8339	.8894	.9436	.9967	.0485	.0991	.1484	.1965	.2433	.2888	4
6	.2206	.2869	.3521	.4163	.4795	.5417	.6027	.6627	.7215	.7792	.8358	.8912	.9454	.9984	.0502	.1007	.1500	.1981	.2448	.2903	6
8	.2228	.2890	.3543	.4185	.4816	.5437	.6047	.6647	.7235	.7811	.8377	.8930	.9472	.0001	.0519	.1024	.1517	.1997	.2464	.2918	8
10	3.2250	3.2912	3.3564	3.4206	3.4837	3.5458	3.6068	3.6666	3.7254	3.7830	3.8395	3.8948	3.9490	4.0019	4.0536	4.1041	4.1533	4.2012	4.2479	4.2933	10
12	.2273	.2934	.3586	.4227	.4858	.5478	.6088	.6686	.7274	.7850	.8414	.8967	.9508	.0036	.0553	.1057	.1549	.2028	.2494	.2948	12
14	.2295	.2956	.3607	.4248	.4879	.5499	.6108	.6706	.7293	.7869	.8433	.8985	.9525	.0054	.0570	.1074	.1565	.2044	.2510	.2963	14
16	.2317	.2978	.3629	.4269	.4900	.5519	.6128	.6726	.7312	.7887	.8451	.9003	.9543	.0071	.0587	.1090	.1581	.2060	.2525	.2978	16
18	.2339	.3000	.3650	.4291	.4921	.5540	.6148	.6745	.7332	.7906	.8470	.9021	.9561	.0089	.0604	.1107	.1597	.2075	.2540	.2992	18
20	3.2361	3.3022	3.3672	3.4312	3.4941	3.5560	3.6168	3.6765	3.7351	3.7925	3.8488	3.9039	3.9579	4.0106	4.0621	4.1124	4.1614	4.2091	4.2556	4.3007	20
22	.2384	.3044	.3693	.4333	.4962	.5581	.6188	.6785	.7370	.7944	.8507	.9058	.9596	.0123	.0638	.1140	.1630	.2107	.2571	.3022	22
24	.2406	.3065	.3715	.4354	.4983	.5601	.6208	.6805	.7390	.7963	.8525	.9076	.9614	.0141	.0655	.1157	.1646	.2122	.2586	.3037	24
26	.2428	.3087	.3736	.4375	.5004	.5621	.6228	.6824	.7409	.7982	.8544	.9094	.9632	.0158	.0672	.1173	.1662	.2138	.2601	.3052	26
28	.2450	.3109	.3758	.4396	.5024	.5642	.6248	.6844	.7428	.8001	.8562	.9112	.9650	.0175	.0689	.1190	.1678	.2154	.2617	.3066	28
30	3.2472	3.3131	3.3779	3.4417	3.5045	3.5662	3.6268	3.6864	3.7448	3.8020	3.8581	3.9130	3.9667	4.0193	4.0706	4.1206	4.1694	4.2169	4.2632	4.3081	30
32	.2494	.3153	.3801	.4439	.5066	.5683	.6288	.6883	.7467	.8039	.8599	.9148	.9685	.0210	.0722	.1223	.1710	.2185	.2647	.3096	32
34	.2516	.3174	.3822	.4460	.5087	.5703	.6308	.6903	.7486	.8058	.8618	.9166	.9703	.0227	.0739	.1239	.1726	.2201	.2662	.3111	34
36	.2538	.3196	.3844	.4481	.5107	.5723	.6328	.6923	.7505	.8077	.8636	.9184	.9720	.0244	.0756	.1255	.1742	.2216	.2677	.3125	36
38	.2561	.3218	.3865	.4502	.5128	.5744	.6348	.6942	.7525	.8096	.8655	.9202	.9738	.0262	.0773	.1272	.1758	.2232	.2692	.3140	38
40	3.2583	3.3240	3.3886	3.4523	3.5149	3.5764	3.6368	3.6962	3.7544	3.8114	3.8673	3.9221	3.9756	4.0279	4.0790	4.1288	4.1774	4.2247	4.2708	4.3155	40
42	.2605	.3261	.3908	.4544	.5169	.5784	.6388	.6981	.7563	.8133	.8692	.9239	.9773	.0296	.0807	.1305	.1790	.2263	.2723	.3170	42
44	.2627	.3283	.3929	.4565	.5190	.5805	.6408	.7001	.7582	.8152	.8710	.9257	.9791	.0313	.0823	.1321	.1806	.2278	.2738	.3184	44
46	.2649	.3305	.3950	.4586	.5211	.5825	.6428	.7020	.7601	.8171	.8729	.9275	.9809	.0331	.0840	.1337	.1822	.2294	.2753	.3199	46
48	.2671	.3326	.3972	.4607	.5231	.5845	.6448	.7040	.7620	.8190	.8747	.9293	.9826	.0348	.0857	.1354	.1838	.2309	.2768	.3213	48
50	3.2693	3.3348	3.3993	3.4628	3.5252	3.5866	3.6468	3.7060	3.7640	3.8208	3.8765	3.9311	3.9844	4.0365	4.0874	4.1370	4.1854	4.2325	4.2783	4.3228	50
52	.2715	.3370	.4014	.4649	.5273	.5886	.6488	.7079	.7659	.8227	.8784	.9329	.9861	.0382	.0891	.1386	.1870	.2340	.2798	.3243	52
54	.2737	.3391	.4036	.4670	.5293	.5906	.6508	.7099	.7678	.8246	.8802	.9347	.9879	.0399	.0907	.1403	.1886	.2356	.2813	.3257	54
56	.2759	.3413	.4057	.4691	.5314	.5926	.6528	.7118	.7697	.8265	.8820	.9364	.9896	.0416	.0924	.1419	.1902	.2371	.2828	.3272	56
58	.2781	.3435	.4078	.4712	.5335	.5947	.6548	.7138	.7716	.8283	.8839	.9382	.9914	.0433	.0941	.1435	.1917	.2387	.2843	.3286	58
60	3.2803	3.3456	3.4100	3.4733	3.5355	3.5967	3.6567	3.7157	3.7735	3.8302	3.8857	3.9400	3.9932	4.0451	4.0957	4.1452	4.1933	4.2402	4.2858	4.3301	60

Table 18A Coordinate Factors and Angles

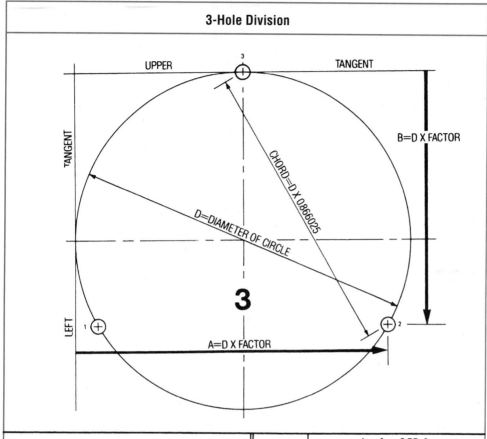

→	Factor For A		Factor For B	↓		Angle of Hole		
						Deg.	Min.	Sec.
1	.066987	1	.750000		1	120	0	0
2	.933013	2	.750000		2	240	0	0
3	.500000	3	.000000		3	360	0	0

Courtesy W. J. Woodworth and J. D. Woodworth

Table 18B Coordinate Factors and Angles

4-Hole Division

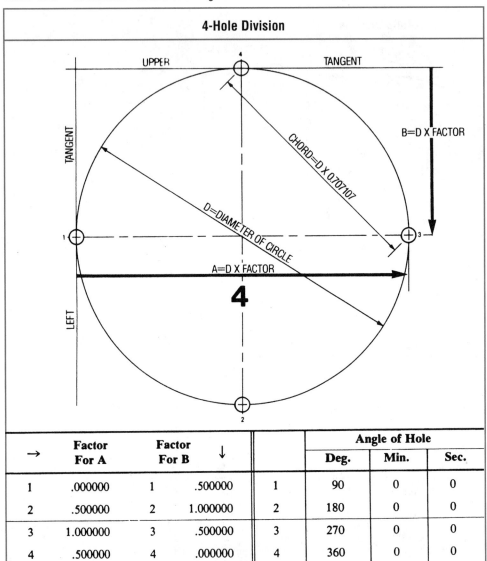

	Factor For A →		Factor For B ↓			Angle of Hole		
						Deg.	Min.	Sec.
1	.000000	1	.500000		1	90	0	0
2	.500000	2	1.000000		2	180	0	0
3	1.000000	3	.500000		3	270	0	0
4	.500000	4	.000000		4	360	0	0

Courtesy W. J. Woodworth and J. D. Woodworth

Table 18C Coordinate Factors and Angles

5-Hole Division

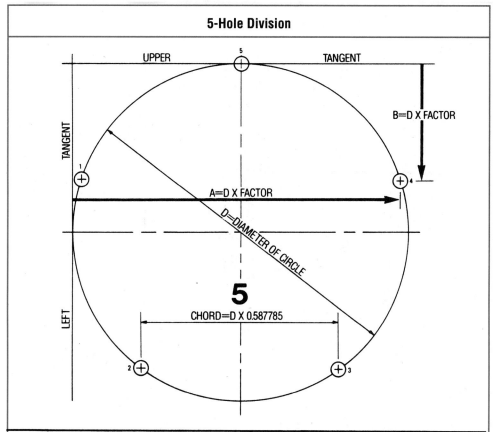

UPPER TANGENT

TANGENT

B=D X FACTOR

A=D X FACTOR

D=DIAMETER OF CIRCLE

LEFT

5

CHORD=D X 0.587785

→	Factor For A		Factor For B ↓			Angle of Hole		
						Deg.	Min.	Sec.
1	.024472	1	.345492		1	72	0	0
2	.206107	2	.904508		2	144	0	0
3	.793893	3	.904508		3	216	0	0
4	.975528	4	.345492		4	288	0	0
5	.500000	5	.000000		5	360	0	0

Courtesy W. J. Woodworth and J. D. Woodworth

Table 18D Coordinate Factors and Angles

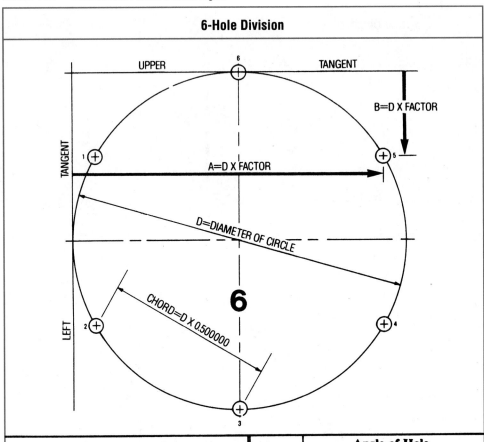

→	Factor For A		Factor For B ↓		Angle of Hole		
					Deg.	Min.	Sec.
1	.066987	1	.250000	1	60	0	0
2	.066987	2	.750000	2	120	0	0
3	.500000	3	1.000000	3	180	0	0
4	.933013	4	.750000	4	240	0	0
5	.933013	5	.250000	5	300	0	0
6	.500000	6	.000000	6	360	0	0

Courtesy W. J. Woodworth and J. D. Woodworth

Table 18E   Coordinate Factors and Angles

7-Hole Division

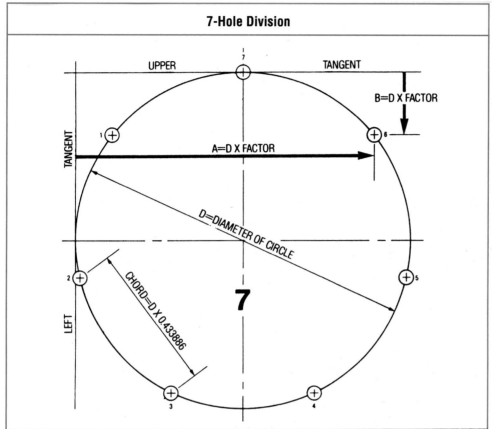

→	Factor For A		Factor For B ↓			Angle of Hole		
						Deg.	Min.	Sec.
1	.109084	1	.188255		1	51	25	42-6/7
2	.012536	2	.611261		2	102	51	23-5/7
3	.283058	3	.950484		3	154	17	8-4/7
4	.716942	4	.950484		4	205	42	51-3/7
5	.987464	5	.611261		5	257	8	34-2/7
6	.890916	6	.188255		6	308	34	17-1/7
7	.500000	7	.000000		7	360	0	0

Courtesy W. J. Woodworth and J. D. Woodworth

Table 18F   Coordinate Factors and Angles

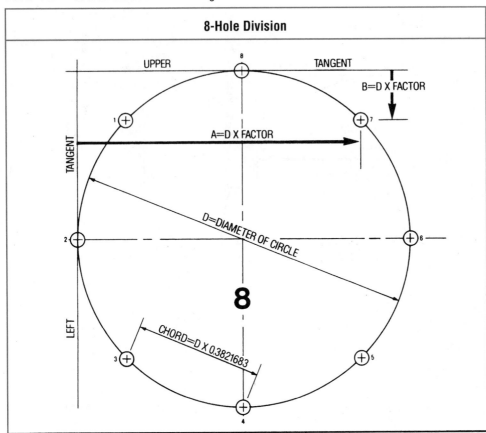

8-Hole Division							

→	Factor For A		Factor For B			Angle of Hole		
				↓		Deg.	Min.	Sec.
1	.146447	1	.146447		1	45	0	0
2	.000000	2	.500000		2	90	0	0
3	.146447	3	.853553		3	135	0	0
4	.500000	4	1.000000		4	180	0	0
5	.853553	5	.853553		5	225	0	0
6	1.000000	6	.500000		6	270	0	0
7	.853553	7	.146447		7	315	0	0
8	.500000	8	.000000		8	360	0	0

Courtesy W. J. Woodworth and J. D. Woodworth

Table 18G Coordinate Factors and Angles

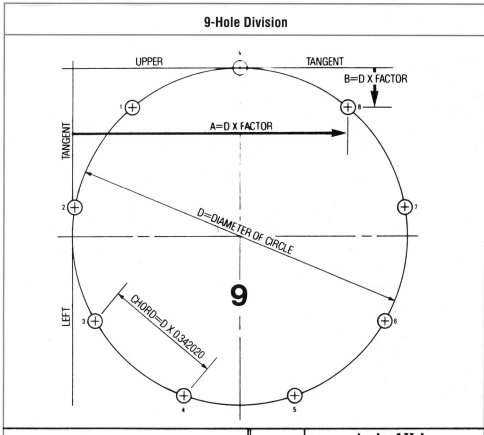

9-Hole Division

→	Factor For A		Factor For B ↓		Angle of Hole		
					Deg.	Min.	Sec.
1	.178606	1	.116978	1	40	0	0
2	.007596	2	.413176	2	80	0	0
3	.066987	3	.750000	3	120	0	0
4	.328990	4	.969846	4	160	0	0
5	.671010	5	.969846	5	200	0	0
6	.933013	6	.750000	6	240	0	0
7	.992404	7	.413176	7	280	0	0
8	.821394	8	.116978	8	320	0	0
9	.500000	9	.000000	9	360	0	0

Courtesy W. J. Woodworth and J. D. Woodworth

Table 18H   Coordinate Factors and Angles

10-Hole Division

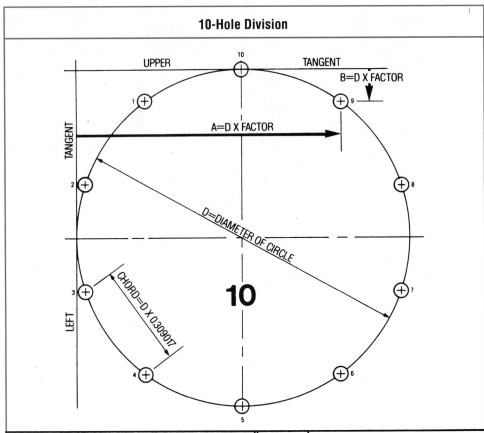

→	Factor For A		Factor For B ↓		Angle of Hole		
					Deg.	**Min.**	**Sec.**
1	.206107	1	.095492	1	36	0	0
2	.024472	2	.345492	2	72	0	0
3	.024472	3	.654508	3	108	0	0
4	.206107	4	.904508	4	144	0	0
5	.500000	5	1.000000	5	180	0	0
6	.793893	6	.904508	6	216	0	0
7	.975528	7	.654508	7	252	0	0
8	.975528	8	.345492	8	288	0	0
9	.793893	9	.095492	9	324	0	0
10	.500000	10	.000000	10	360	0	0

Courtesy W. J. Woodworth and J. D. Woodworth

Table 18I Coordinate Factors and Angles

5-Hole Division

→	Factor For A		Factor For B ↓		Angle of Hole		
					Deg.	Min.	Sec.
1	.229680	1	.079373	1	32	43	38-2/11
2	.045184	2	.292293	2	65	27	16-4/11
3	.005089	3	.571157	3	98	10	54-6/11
4	.122125	4	.827430	4	130	54	32-8/11
5	.359134	5	.979746	5	163	38	10-10/11
6	.640866	6	.979746	6	196	21	49-1/11
7	.877875	7	.827430	7	229	5	27-3/11
8	.994911	8	.571157	8	261	49	5-5/11
9	.954816	9	.292293	9	294	32	43-7/11
10	.770320	10	.079373	10	327	16	21-9/11
11	.500000	11	.000000	11	360	0	0

Courtesy W. J. Woodworth and J. D. Woodworth

Table 19 Natural Trigonometric Functions

	0°						1°					
′	sin	cos	tan	cot	sec	cosec	sin	cos	tan	cot	sec	cosec
0	.00000	1.0000	.00000	Infinite	1.0000	Infinite	.01745	.99985	.01745	57.290	1.0001	57.299
1	.00029	.0000	.00029	3437.7	.0000	3437.7	.01774	.99984	.01775	56.350	.0001	56.359
2	.00058	.0000	.00058	1718.9	.0000	1718.9	.01803	.99984	.01804	55.441	.0001	55.450
3	.00087	.0000	.00087	1145.9	.0000	1145.9	.01832	.99983	.01833	54.561	.0002	54.570
4	.00116	.0000	.00116	859.44	.0000	859.44	.01861	.99983	.01862	53.708	.0002	53.718
5	.00145	1.0000	.00145	687.55	.0000	687.55	.01891	.99982	.01891	52.882	.0002	52.891
6	.00174	.0000	.00174	572.96	.0000	572.96	.01920	.99981	.01920	52.081	.0002	52.090
7	.00204	.0000	.00204	491.11	.0000	491.11	.01949	.99981	.01949	51.303	.0002	51.313
8	.00233	.0000	.00233	429.72	.0000	429.72	.01978	.99980	.01978	50.548	.0002	50.558
9	.00262	.0000	.00262	381.97	.0000	381.97	.02007	.99980	.02007	49.816	.0002	49.826
10	.00291	0.99999	.00291	343.77	.0000	343.77	.02036	.99979	.02036	49.104	.0002	49.114
11	.00320	.99999	.00320	312.52	.0000	312.52	.02065	.99979	.02066	48.412	.0002	43.422
12	.00349	.99999	.00349	286.48	.0000	286.48	.02094	.99978	.02095	47.739	.0002	47.750
13	.00378	.99999	.00378	264.44	.0000	264.44	.02123	.99977	.02124	47.085	.0002	47.096
14	.00407	.99999	.00407	245.55	.0000	245.55	.02152	.99977	.02153	46.449	.0002	46.460
15	.00436	.99999	.00436	229.18	.0000	229.18	.02181	.99976	.02182	45.829	.0002	45.840
16	.00465	.99999	.00465	214.86	.0000	214.86	.02210	.99975	.02211	45.226	.0002	45.237
17	.00494	.99999	.00494	202.22	.0000	202.22	.02240	.99975	.02240	44.638	.0002	44.650
18	.00524	.99999	.00524	190.98	.0000	190.99	.02269	.99974	.02269	44.066	.0002	44.077
19	.00553	.99998	.00553	180.93	.0000	180.93	.02298	.99974	.02298	43.508	.0003	43.520
20	.00582	.99998	.00582	171.88	.0000	171.89	.02326	.99973	.02327	42.964	.0003	42.976
21	.00611	.99998	.00611	163.70	.0000	163.70	.02356	.99972	.02357	42.433	.0003	42.445
22	.00640	.99998	.00640	156.26	.0000	156.26	.02385	.99971	.02386	41.916	.0003	41.928
23	.00669	.99998	.00669	149.46	.0000	149.47	.02414	.99971	.02415	41.410	.0003	41.423
24	.00698	.99997	.00698	143.24	.0000	143.24	.02443	.99970	.02444	40.917	.0003	40.930
25	.00727	.99997	.00727	137.51	.0000	137.51	.02472	.99969	.02473	40.436	.0003	40.448
26	.00756	.99997	.00756	132.22	.0000	132.22	.02501	.99969	.02502	39.965	.0003	39.978
27	.00785	.99997	.00785	127.32	.0000	127.32	.02530	.99968	.02531	39.506	.0003	39.518
28	.00814	.99997	.00814	122.77	.0000	122.78	.02559	.99967	.02560	39.057	.0003	39.069
29	.00843	.99996	.00844	118.54	.0000	118.54	.02589	.99966	.02589	38.618	.0003	38.631
30	.00873	.99996	.00873	114.59	.0000	114.59	.02618	.99966	.02618	38.188	.0003	38.201
31	.00902	.99996	.00902	110.89	.0000	110.90	.02647	.99965	.02648	37.769	.0003	37.782
32	.00931	.99996	.00931	107.43	.0000	107.43	.02676	.99964	.02677	37.358	.0003	37.371
33	.00960	.99995	.00960	104.17	.0000	104.17	.02705	.99963	.02706	36.956	.0004	36.969
34	.00989	.99995	.00989	101.11	.0000	101.11	.02734	.99963	.02735	36.563	.0004	36.576
35	.01018	.99995	.01018	98.218	.0000	98.223	.02763	.99962	.02764	36.177	.0004	36.191
36	.01047	.99994	.01047	95.489	.0000	95.495	.02792	.99961	.02793	35.800	.0004	35.814
37	.01076	.99994	.01076	92.908	.0000	92.914	.02821	.99960	.02822	35.431	.0004	35.445
38	.01105	.99994	.01105	90.463	.0001	90.469	.02850	.99959	.02851	35.069	.0004	35.084
39	.01134	.99993	.01134	88.143	.0001	88.149	.02879	.99958	.02880	34.715	.0004	34.729
40	.01163	.99993	.01164	85.940	.0001	85.946	.02908	.99958	.02910	34.368	.0004	34.382
41	.01193	.99993	.01193	83.843	.0001	83.849	.02937	.99957	.02939	34.027	.0004	34.042
42	.01222	.99992	.01222	81.847	.0001	81.853	.02967	.99956	.02968	33.693	.0004	33.708
43	.01251	.99992	.01251	79.943	.0001	79.950	.02996	.99955	.02997	33.366	.0004	33.381
44	.01280	.99992	.01280	78.126	.0001	78.133	.03025	.99954	.03026	33.045	.0004	33.060
45	.01309	.99991	.01309	76.390	.0001	76.396	.03054	.99953	.03055	32.730	.0005	32.745
46	.01338	.99991	.01338	74.729	.0001	74.736	.03083	.99952	.03084	32.421	.0005	32.437
47	.01367	.99991	.01367	73.139	.0001	73.146	.03112	.99951	.03113	32.118	.0005	32.134
48	.01396	.99990	.01396	71.615	.0001	71.622	.03141	.99951	.03143	31.820	.0005	31.836
49	.01425	.99990	.01425	70.153	.0001	70.160	.03170	.99950	.03172	31.528	.0005	31.544
50	.01454	.99989	.01454	68.750	.0001	68.757	.03199	.99949	.03201	31.241	.0005	31.257
51	.01483	.99989	.01484	67.402	.0001	67.409	.03228	.99948	.03230	30.960	.0005	30.976
52	.01512	.99988	.01513	66.105	.0001	66.113	.03257	.99947	.03259	30.683	.0005	30.699
53	.01542	.99988	.01542	64.858	.0001	64.866	.03286	.99946	.03288	30.411	.0005	30.428
54	.01571	.99988	.01571	63.657	.0001	63.664	.03315	.99945	.03317	30.145	.0005	30.161
55	.01600	.99987	.01600	62.499	.0001	62.507	.03344	.99944	.03346	29.882	.0005	29.899
56	.01629	.99987	.01629	61.383	.0001	61.391	.03374	.99943	.03375	29.624	.0006	29.641
57	.01658	.99987	.01658	60.306	.0001	60.314	.03403	.99942	.03405	29.371	.0006	29.388
58	.01687	.99986	.01687	59.266	.0001	59.274	.03432	.99941	.03434	29.122	.0006	29.139
59	.01716	.99985	.01716	58.261	.0001	58.270	.03461	.99940	.03463	28.877	.0006	28.894
60	.01745	.99985	.01745	57.290	.0001	57.299	.03490	.99939	.03492	28.636	.0006	28.654
′	cos	sin	cot	tan	cosec	sec	cos	sin	cot	tan	cosec	sec
	89°						88°					

	2°						3°						
′	sin	cos	tan	cot	sec	cosec	sin	cos	tan	cot	sec	cosec	′
0	.03490	.99939	.03492	28.636	1.0006	28.654	.05234	.99863	.05241	19.081	1.0014	19.107	60
1	.03519	.99938	.03521	28.399	.0006	28.417	.05263	.99861	.05270	18.975	.0014	19.002	59
2	.03548	.99937	.03550	28.166	.0006	28.184	.05292	.99860	.05299	18.871	.0014	18.897	58
3	.03577	.99936	.03579	27.937	.0006	27.955	.05321	.99858	.05328	18.768	.0014	18.794	57
4	.03606	.99935	.03608	27.712	.0006	27.730	.05350	.99857	.05357	18.665	.0014	18.692	56
5	.03635	.99934	.03638	27.490	.0007	27.508	.05379	.99855	.05387	18.564	.0014	18.591	55
6	.03664	.99933	.03667	27.271	.0007	27.290	.05408	.99854	.05416	18.464	.0015	18.491	54
7	.03693	.99932	.03696	27.056	.0007	27.075	.05437	.99852	.05445	18.365	.0015	18.393	53
8	.03722	.99931	.03725	26.845	.0007	26.864	.05466	.99850	.05474	18.268	.0015	18.295	52
9	.03751	.99930	.03754	26.637	.0007	26.655	.05495	.99849	.05503	18.171	.0015	18.198	51
10	.03781	.99928	.03783	26.432	.0007	26.450	.05524	.99847	.05532	18.075	.0015	18.103	50
11	.03810	.99927	.03812	26.230	.0007	26.249	.05553	.99846	.05562	17.980	.0015	18.008	49
12	.03839	.99926	.03842	26.031	.0007	26.050	.05582	.99844	.05591	17.886	.0016	17.914	48
13	.03868	.99925	.03871	25.835	.0007	25.854	.05611	.99842	.05620	17.793	.0016	17.821	47
14	.03897	.99924	.03900	25.642	.0008	25.661	.05640	.99841	.05649	17.701	.0016	17.730	46
15	.03926	.99923	.03929	25.452	.0008	25.471	.05669	.99839	.05678	17.610	.0016	17.639	45
16	.03955	.99922	.03958	25.264	.0008	25.284	.05698	.99837	.05707	17.520	.0016	17.549	44
17	.03984	.99921	.03987	25.080	.0008	25.100	.05727	.99836	.05737	17.431	.0016	17.460	43
18	.04013	.99919	.04016	24.898	.0008	24.918	.05756	.99834	.05766	17.343	.0017	17.372	42
19	.04042	.99918	.04045	24.718	.0008	24.739	.05785	.99832	.05795	17.256	.0017	17.285	41
20	.04071	.99917	.04075	24.542	.0008	24.562	.05814	.99831	.05824	17.169	.0017	17.198	40
21	.04100	.99916	.04104	24.367	.0008	24.388	.05843	.99829	.05853	17.084	.0017	17.113	39
22	.04129	.99915	.04133	24.196	.0008	24.216	.05872	.99827	.05883	16.999	.0017	17.028	38
23	.04158	.99913	.04162	24.026	.0009	24.047	.05902	.99826	.05912	16.915	.0017	16.944	37
24	.04187	.99912	.04191	23.859	.0009	23.880	.05931	.99824	.05941	16.832	.0018	16.861	36
25	.04217	.99911	.04220	23.694	.0009	23.716	.05960	.99822	.05970	16.750	.0018	16.779	35
26	.04246	.99910	.04249	23.532	.0009	23.553	.05989	.99820	.05999	16.668	.0018	16.698	34
27	.04275	.99908	.04279	23.372	.0009	23.393	.06018	.99819	.06029	16.587	.0018	16.617	33
28	.04304	.99907	.04308	23.214	.0009	23.235	.06047	.99817	.06058	16.507	.0018	16.538	32
29	.04333	.99906	.04337	23.058	.0009	23.079	.06076	.99815	.06087	16.428	.0018	16.459	31
30	.04362	.99905	.04366	22.904	.0009	22.925	.06105	.99813	.06116	16.350	.0019	16.380	30
31	.04391	.99903	.04395	22.752	.0010	22.774	.06134	.99812	.06145	16.272	.0019	16.303	29
32	.04420	.99902	.04424	22.602	.0010	22.624	.06163	.99810	.06175	16.195	.0019	16.226	28
33	.04449	.99901	.04453	22.454	.0010	22.476	.06192	.99808	.06204	16.119	.0019	16.150	27
34	.04478	.99900	.04483	22.308	.0010	22.330	.06221	.99806	.06233	16.043	.0019	16.075	26
35	.04507	.99898	.04512	22.164	.0010	22.186	.06250	.99804	.06262	15.969	.0019	16.000	25
36	.04536	.99897	.04541	22.022	.0010	22.044	.06279	.99803	.06291	15.894	.0020	15.926	24
37	.04565	.99896	.04570	21.881	.0010	21.904	.06308	.99801	.06321	15.821	.0020	15.853	23
38	.04594	.99894	.04599	21.742	.0010	21.765	.06337	.99799	.06350	15.748	.0020	15.780	22
39	.04623	.99893	.04628	21.606	.0011	21.629	.06366	.99797	.06379	15.676	.0020	15.708	21
40	.04652	.99892	.04657	21.470	.0011	21.494	.06395	.99795	.06408	15.605	.0020	15.637	20
41	.04681	.99890	.04687	21.337	.0011	21.360	.06424	.99793	.06437	15.534	.0021	15.566	19
42	.04711	.99889	.04716	21.205	.0011	21.228	.06453	.99791	.06467	15.464	.0021	15.496	18
43	.04740	.99888	.04745	21.075	.0011	21.098	.06482	.99790	.06496	15.394	.0021	15.427	17
44	.04769	.99886	.04774	20.946	.0011	20.970	.06511	.99788	.06525	15.325	.0021	15.358	16
45	.04798	.99885	.04803	20.819	.0011	20.843	.06540	.99786	.06554	15.257	.0021	15.290	15
46	.04827	.99883	.04832	20.693	.0012	20.717	.06569	.99784	.06583	15.189	.0022	15.222	14
47	.04856	.99882	.04862	20.569	.0012	20.593	.06598	.99782	.06613	15.122	.0022	15.155	13
48	.04885	.99881	.04891	20.446	.0012	20.471	.06627	.99780	.06642	15.056	.0022	15.089	12
49	.04914	.99879	.04920	20.325	.0012	20.350	.06656	.99778	.06671	14.990	.0022	15.023	11
50	.04943	.99878	.04949	20.205	.0012	20.230	.06685	.99776	.06700	14.924	.0022	14.958	10
51	.04972	.99876	.04978	20.087	.0012	20.112	.06714	.99774	.06730	14.860	.0023	14.893	9
52	.05001	.99875	.05007	19.970	.0012	19.995	.06743	.99772	.06759	14.795	.0023	14.829	8
53	.05030	.99873	.05037	19.854	.0013	19.880	.06772	.99770	.06788	14.732	.0023	14.765	7
54	.05059	.99872	.05066	19.740	.0013	19.766	.06801	.99768	.06817	14.668	.0023	14.702	6
55	.05088	.99870	.05095	19.627	.0013	19.653	.06830	.99766	.06846	14.606	.0023	14.640	5
56	.05117	.99869	.05124	19.515	.0013	19.541	.06859	.99764	.06876	14.544	.0024	14.578	4
57	.05146	.99867	.05153	19.405	.0013	19.431	.06888	.99762	.06905	14.482	.0024	14.517	3
58	.05175	.99866	.05182	19.296	.0013	19.322	.06918	.99760	.06934	14.421	.0024	14.456	2
59	.05204	.99864	.05212	19.188	.0013	19.214	.06947	.99758	.06963	14.361	.0024	14.395	1
60	.05234	.99863	.05241	19.081	.0014	19.107	.06976	.99756	06993	14.301	.0024	14.335	0
′	cos	sin	cot	tan	cosec	sec	cos	sin	cot	tan	cosec	sec	′
	87°						86°						

Table 19 Natural Trigonometric Functions (Continued)

	4°						5°						6°						7°						
′	sin	cos	tan	cot	sec	cosec	sin	cos	tan	cot	sec	cosec	sin	cos	tan	cot	sec	cosec	sin	cos	tan	cot	sec	cosec	′
0	.06976	.99756	.06993	14.301	1.0024	14.335	.08715	.99619	.08749	11.430	1.0038	11.474	.10453	.99452	.10510	9.5144	1.0055	9.5668	.12187	.99255	.12278	8.1443	1.0075	8.2055	60
1	.07005	.99754	.07022	14.241	.0025	14.276	.08744	.99617	.08778	11.392	.0038	11.436	.10482	.99449	.10540	.4878	.0055	.5404	.12216	.99251	.12308	.1248	.0075	.1861	59
2	.07034	.99752	.07051	14.182	.0025	14.217	.08773	.99614	.08807	11.354	.0039	11.398	.10511	.99446	.10569	.4614	.0056	.5141	.12245	.99247	.12337	.1053	.0076	.1668	58
3	.07063	.99750	.07080	14.123	.0025	14.159	.08802	.99612	.08837	11.316	.0039	11.360	.10540	.99443	.10599	.4351	.0056	.4880	.12273	.99244	.12367	.0860	.0076	.1476	57
4	.07092	.99748	.07110	14.065	.0025	14.101	.08831	.99609	.08866	11.279	.0039	11.323	.10568	.99440	.10628	.4090	.0056	.4620	.12302	.99240	.12396	.0667	.0076	.1285	56
5	.07121	.99746	.07139	14.008	.0025	14.043	.08860	.99607	.08895	11.242	.0039	11.286	.10597	.99437	.10657	9.3831	.0057	9.4362	.12331	.99237	.12426	8.0476	.0077	8.1094	55
6	.07150	.99744	.07168	13.951	.0026	13.986	.08889	.99604	.08925	11.205	.0040	11.249	.10626	.99434	.10687	.3572	.0057	.4105	.12360	.99233	.12456	.0285	.0077	.0905	54
7	.07179	.99742	.07197	13.894	.0026	13.930	.08918	.99601	.08954	11.168	.0040	11.213	.10655	.99431	.10716	.3315	.0057	.3850	.12389	.99229	.12485	.0095	.0078	.0717	53
8	.07208	.99740	.07226	13.838	.0026	13.874	.08947	.99599	.08983	11.132	.0040	11.176	.10684	.99428	.10746	.3060	.0057	.3596	.12418	.99226	.12515	7.9906	.0078	.0529	52
9	.07237	.99738	.07256	13.782	.0026	13.818	.08976	.99596	.09013	11.095	.0040	11.140	.10713	.99424	.10775	.2806	.0058	.3343	.12447	.99222	.12544	.9717	.0078	.0342	51
10	.07266	.99736	.07285	13.727	.0026	13.763	.09005	.99594	.09042	11.059	.0041	11.104	.10742	.99421	.10805	9.2553	.0058	9.3092	.12476	.99219	.12574	7.9530	.0079	8.0156	50
11	.07295	.99733	.07314	13.672	.0027	13.708	.09034	.99591	.09071	11.024	.0041	11.069	.10771	.99418	.10834	.2302	.0058	.2842	.12504	.99215	.12603	.9344	.0079	7.9971	49
12	.07324	.99731	.07343	13.617	.0027	13.654	.09063	.99588	.09101	10.988	.0041	11.033	.10800	.99415	.10863	.2051	.0059	.2593	.12533	.99211	.12633	.9158	.0079	.9787	48
13	.07353	.99729	.07373	13.563	.0027	13.600	.09092	.99586	.09130	10.953	.0041	10.998	.10829	.99412	.10893	.1803	.0059	.2346	.12562	.99208	.12662	.8973	.0080	.9604	47
14	.07382	.99727	.07402	13.510	.0027	13.547	.09121	.99583	.09159	10.918	.0042	10.963	.10858	.99409	.10922	.1555	.0059	.2100	.12591	.99204	.12692	.8789	.0080	.9421	46
15	.07411	.99725	.07431	13.457	.0027	13.494	.09150	.99580	.09189	10.883	.0042	10.929	.10887	.99406	.10952	9.1309	.0060	9.1855	.12620	.99200	.12722	7.8606	.0080	7.9240	45
16	.07440	.99723	.07460	13.404	.0028	13.441	.09179	.99578	.09218	10.848	.0042	10.894	.10916	.99402	.10981	.1064	.0060	.1612	.12649	.99197	.12751	.8424	.0081	.9059	44
17	.07469	.99721	.07490	13.351	.0028	13.389	.09208	.99575	.09247	10.814	.0043	10.860	.10944	.99399	.11011	.0821	.0060	.1370	.12678	.99193	.12781	.8243	.0081	.8879	43
18	.07498	.99718	.07519	13.299	.0028	13.337	.09237	.99572	.09277	10.780	.0043	10.826	.10973	.99396	.11040	.0579	.0061	.1129	.12706	.99189	.12810	.8062	.0082	.8700	42
19	.07527	.99716	.07548	13.248	.0028	13.286	.09266	.99570	.09306	10.746	.0043	10.792	.11002	.99393	.11069	.0338	.0061	.0890	.12735	.99186	.12840	.7882	.0082	.8522	41
20	.07556	.99714	.07577	13.917	.0029	13.235	.09295	.99567	.09335	10.712	.0043	10.758	.11031	.99390	.11099	9.0098	.0061	9.0651	.12764	.99182	.12869	7.7703	.0082	7.8344	40
21	.07585	.99712	.07607	13.146	.0029	13.184	.09324	.99564	.09365	10.678	.0044	10.725	.11060	.99386	.11128	8.9860	.0062	.0414	.12793	.99178	.12899	.7525	.0083	.8168	39
22	.07614	.99710	.07636	13.096	.0029	13.134	.09353	.99562	.09394	10.645	.0044	10.692	.11089	.99383	.11158	.9623	.0062	.0179	.12822	.99174	.12928	.7348	.0083	.7992	38
23	.07643	.99707	.07665	13.046	.0029	13.084	.09382	.99559	.09423	10.612	.0044	10.659	.11118	.99380	.11187	.9387	.0062	8.9944	.12851	.99171	.12958	.7171	.0084	.7817	37
24	.07672	.99705	.07694	12.996	.0029	13.034	.09411	.99556	.09453	10.579	.0044	10.626	.11147	.99377	.11217	.9152	.0063	.9711	.12879	.99167	.12988	.6996	.0084	.7642	36
25	.07701	.99703	.07724	12.947	.0030	12.985	.09440	.99553	.09482	10.546	.0045	10.593	.11176	.99373	.11246	8.8918	.0063	8.9479	.12908	.99163	.13017	7.6821	.0084	7.7469	35
26	.07730	.99701	.07753	12.898	.0030	12.937	.09469	.99551	.09511	10.514	.0045	10.561	.11205	.99370	.11276	.8686	.0063	.9248	.12937	.99160	.13047	.6646	.0085	.7296	34
27	.07759	.99698	.07782	12.849	.0030	12.888	.09498	.99548	.09541	10.481	.0045	10.529	.11234	.99367	.11305	.8455	.0064	.9018	.12966	.99156	.13076	.6473	.0085	.7124	33
28	.07788	.99696	.07812	12.801	.0030	12.840	.09527	.99545	.09570	10.449	.0046	10.497	.11262	.99364	.11335	.8225	.0064	.8790	.12995	.99152	.13106	.6300	.0085	.6953	32
29	.07817	.99694	.07841	12.754	.0031	12.793	.09556	.99542	.09599	10.417	.0046	10.465	.11291	.99360	.11364	.7996	.0064	.8563	.13024	.99148	.13136	.6129	.0086	.6783	31
30	.07846	.99692	.07870	12.706	.0031	12.745	.09584	.99540	.09629	10.385	.0046	10.433	.11320	.99357	.11393	8.7769	.0065	8.8337	.13053	.99144	.13165	7.5957	.0086	7.6613	30
31	.07875	.99689	.07899	12.659	.0031	12.698	.09613	.99537	.09658	10.354	.0046	10.402	.11349	.99354	.11423	.7542	.0065	.8112	.13081	.99141	.13195	.5787	.0087	.6444	29
32	.07904	.99687	.07929	12.612	.0031	12.652	.09642	.99534	.09688	10.322	.0047	10.371	.11378	.99350	.11452	.7317	.0065	.7888	.13110	.99137	.13224	.5617	.0087	.6276	28
33	.07933	.99685	.07958	12.566	.0032	12.606	.09671	.99531	.09717	10.291	.0047	10.340	.11407	.99347	.11482	.7093	.0066	.7665	.13139	.99133	.13254	.5449	.0087	.6108	27
34	.07962	.99682	.07987	12.520	.0032	12.560	.09700	.99528	.09746	10.260	.0047	10.309	.11436	.99344	.11511	.6870	.0066	.7444	.13168	.99129	.13284	.5280	.0088	.5942	26
35	.07991	.99680	.08016	12.474	.0032	12.514	.09729	.99525	.09776	10.229	.0048	10.278	.11465	.99341	.11541	8.6648	.0066	8.7223	.13197	.99125	.13313	7.5113	.0088	7.5776	25
36	.08020	.99678	.08046	12.429	.0032	12.469	.09758	.99523	.09805	10.199	.0048	10.248	.11494	.99337	.11570	.6427	.0067	.7004	.13226	.99121	.13343	.4946	.0089	.5611	24
37	.08049	.99675	.08075	12.384	.0032	12.424	.09787	.99520	.09834	10.168	.0048	10.217	.11523	.99334	.11600	.6208	.0067	.6786	.13254	.99118	.13372	.4780	.0089	.5446	23
38	.08078	.99673	.08104	12.339	.0033	12.379	.09816	.99517	.09864	10.138	.0048	10.187	.11551	.99330	.11629	.5989	.0067	.6569	.13283	.99114	.13402	.4615	.0089	.5282	22
39	.08107	.99671	.08134	12.295	.0033	12.335	.09845	.99514	.09893	10.108	.0049	10.157	.11580	.99327	.11659	.5772	.0068	.6353	.13312	.99110	.13432	.4451	.0090	.5119	21
40	.08136	.99668	.08163	12.250	.0033	12.291	.09874	.99511	.09922	10.078	.0049	10.127	.11609	.99324	.11688	8.5555	.0068	8.6138	.13341	.99106	.13461	7.4287	.0090	7.4957	20
41	.08165	.99666	.08192	12.207	.0033	12.248	.09903	.99508	.09952	10.048	.0049	10.098	.11638	.99320	.11718	.5340	.0068	.5924	.13370	.99102	.13491	.4124	.0090	.4795	19
42	.08194	.99664	.08221	12.163	.0034	12.204	.09932	.99505	.09981	10.019	.0050	10.068	.11667	.99317	.11747	.5126	.0069	.5711	.13399	.99098	.13520	.3961	.0091	.4634	18
43	.08223	.99661	.08251	12.120	.0034	12.161	.09961	.99503	.10011	9.9893	.0050	10.039	.11696	.99314	.11777	.4913	.0069	.5499	.13427	.99094	.13550	.3800	.0091	.4474	17
44	.08252	.99659	.08280	12.077	.0034	12.118	.09990	.99500	.10040	9.9601	.0050	10.010	.11725	.99310	.11806	.4701	.0069	.5289	.13456	.99090	.13580	.3639	.0092	.4315	16
45	.08281	.99656	.08309	12.035	.0034	12.076	.10019	.99497	.10069	9.9310	.0050	9.9812	.11754	.99307	.11836	8.4489	.0070	8.5079	.13485	.99086	.13609	7.3479	.0092	7.4156	15
46	.08310	.99654	.08339	11.992	.0035	12.034	.10048	.99494	.10099	9.9021	.0051	9.9525	.11783	.99303	.11865	.4279	.0070	.4871	.13514	.99083	.13639	.3319	.0092	.3998	14
47	.08339	.99652	.08368	11.950	.0035	11.992	.10077	.99491	.10128	9.8734	.0051	9.9239	.11811	.99300	.11895	.4070	.0070	.4663	.13543	.99079	.13669	.3160	.0093	.3840	13
48	.08368	.99649	.08397	11.909	.0035	11.950	.10106	.99488	.10158	9.8448	.0051	9.8955	.11840	.99296	.11924	.3862	.0071	.4457	.13571	.99075	.13698	.3002	.0093	.3683	12
49	.08397	.99647	.08426	11.867	.0035	11.909	.10134	.99485	.10187	9.8164	.0052	9.8672	.11869	.99293	.11954	.3655	.0071	.4251	.13600	.99071	.13728	.2844	.0094	.3527	11
50	.08426	.99644	.08456	11.826	.0036	11.868	.10163	.99482	.10216	9.7882	.0052	9.8391	.11898	.99290	.11983	8.3449	.0071	8.4046	.13629	.99067	.13757	7.2687	.0094	7.3372	10
51	.08455	.99642	.08485	11.785	.0036	11.828	.10192	.99479	.10246	9.7601	.0052	9.8112	.11927	.99286	.12013	.3244	.0072	.3843	.13658	.99063	.13787	.2531	.0094	.3217	9
52	.08484	.99639	.08514	11.745	.0036	11.787	.10221	.99476	.10275	9.7322	.0053	9.7834	.11956	.99283	.12042	.3040	.0072	.3640	.13687	.99059	.13817	.2375	.0095	.3063	8
53	.08513	.99637	.08544	11.704	.0036	11.747	.10250	.99473	.10305	9.7044	.0053	9.7558	.11985	.99279	.12072	.2837	.0073	.3439	.13716	.99055	.13846	.2220	.0095	.2909	7
54	.08542	.99634	.08573	11.664	.0037	11.707	.10279	.99470	.10334	9.6768	.0053	9.7283	.12014	.99276	.12101	.2635	.0073	.3238	.13744	.99051	.13876	.2066	.0096	.2757	6
55	.08571	.99632	.08602	11.625	.0037	11.668	.10308	.99467	.10363	9.6493	.0053	9.7010	.12042	.99272	.12131	8.2434	.0073	8.3039	.13773	.99047	.13906	7.1912	.0096	7.2604	5
56	.08600	.99629	.08632	11.585	.0037	11.628	.10337	.99464	.10393	9.6220	.0054	9.6739	.12071	.99269	.12160	.2234	.0074	.2840	.13802	.99043	.13935	.1759	.0097	.2453	4
57	.08629	.99627	.08661	11.546	.0037	11.589	.10366	.99461	.10422	9.5949	.0054	9.6469	.12100	.99265	.12190	.2035	.0074	.2642	.13831	.99039	.13965	.1607	.0097	.2302	3
58	.08658	.99624	.08690	11.507	.0038	11.550	.10395	.99458	.10452	9.5679	.0054	9.6200	.12129	.99262	.12219	.1837	.0074	.2446	.13860	.99035	.13995	.1455	.0097	.2152	2
59	.08687	.99622	.08719	11.468	.0038	11.512	.10424	.99455	.10481	9.5411	.0055	9.5933	.12158	.99258	.12249	.1640	.0075	.2250	.13888	.99031	.14024	.1304	.0098	.2002	1
60	.08715	.99619	.08749	11.430	.0038	11.474	.10453	.99452	.10510	9.5144	.0055	9.5668	.12187	.99255	.12278	8.1443	.0075	8.2055	.13917	.99027	.14054	7.1154	.0098	7.1853	0
′	cos	sin	cot	tan	cosec	sec	cos	sin	cot	tan	cosec	sec	cos	sin	cot	tan	cosec	sec	cos	sin	cot	tan	cosec	sec	′
	85°						84°						83°						82°						

Table 19 Natural Trigonometric Functions (Continued)

	8°						9°						10°						11°						
′	sin	cos	tan	cot	sec	cosec	sin	cos	tan	cot	sec	cosec	sin	cos	tan	cot	sec	cosec	sin	cos	tan	cot	sec	cosec	′
	cos	sin	cot	tan	cosec	sec	cos	sin	cot	tan	cosec	sec	cos	sin	cot	tan	cosec	sec	cos	sin	cot	tan	cosec	sec	
		81°						80°						79°						78°					

Table 19 Natural Trigonometric Functions (Continued)

′	12°						13°						14°						15°						
	sin	cos	tan	cot	sec	cosec	sin	cos	tan	cot	sec	cosec	sin	cos	tan	cot	sec	cosec	sin	cos	tan	cot	sec	cosec	′
0	.20791	.97815	.21256	4.7046	1.0223	4.8097	.22495	.97437	.23087	4.3315	1.0263	4.4454	.24192	.97029	.24933	4.0108	1.0306	4.1336	.25882	.96592	.26795	3.7320	1.0353	3.8637	60
1	.20820	.97809	.21286	.6979	.0224	.8032	.22523	.97430	.23117	.3257	.0264	.4398	.24220	.97022	.24964	.0058	.0307	.1287	.25910	.96585	.26826	.7277	.0353	.8595	59
2	.20848	.97803	.21316	.6912	.0225	.7966	.22552	.97424	.23148	.3200	.0264	.4342	.24249	.97015	.24995	.0009	.0308	.1239	.25938	.96577	.26857	.7234	.0354	.8553	58
3	.20876	.97797	.21347	.6845	.0225	.7901	.22580	.97417	.23179	.3143	.0265	.4287	.24277	.97008	.25025	3.9959	.0308	.1191	.25966	.96570	.26888	.7191	.0355	.8512	57
4	.20905	.97790	.21377	.6778	.0226	.7835	.22608	.97411	.23209	.3086	.0266	.4231	.24305	.97001	.25056	.9910	.0309	.1144	.25994	.96562	.26920	.7147	.0356	.8470	56
5	.20933	.97784	.21408	4.6712	.0226	4.7770	.22637	.97404	.23240	4.3029	.0266	4.4176	.24333	.96994	.25087	3.9861	.0310	4.1096	.26022	.96555	.26951	3.7104	.0357	3.8428	55
6	.20962	.97778	.21438	.6646	.0227	.7706	.22665	.97398	.23270	.2972	.0267	.4121	.24361	.96987	.25118	.9812	.0311	.1048	.26050	.96547	.26982	.7062	.0358	.8387	54
7	.20990	.97772	.21468	.6580	.0228	.7641	.22693	.97391	.23301	.2916	.0268	.4065	.24390	.96980	.25149	.9763	.0311	.1001	.26078	.96540	.27013	.7019	.0358	.8346	53
8	.21019	.97766	.21499	.6514	.0228	.7576	.22722	.97384	.23332	.2859	.0268	.4011	.24418	.96973	.25180	.9714	.0312	.0953	.26107	.96532	.27044	.6976	.0359	.8304	52
9	.21047	.97760	.21529	.6448	.0229	.7512	.22750	.97378	.23363	.2803	.0269	.3956	.24446	.96966	.25211	.9665	.0313	.0906	.26135	.96524	.27076	.6933	.0360	.8263	51
10	.21076	.97754	.21560	4.6382	.0230	4.7448	.22778	.97371	.23393	4.2747	.0270	4.3901	.24474	.96959	.25242	3.9616	.0314	4.0859	.26163	.96517	.27107	3.6891	.0361	3.8222	50
11	.21104	.97748	.21590	.6317	.0230	.7384	.22807	.97364	.23424	.2691	.0271	.3847	.24502	.96952	.25273	.9568	.0314	.0812	.26191	.96509	.27138	.6848	.0362	.8181	49
12	.21132	.97741	.21621	.6252	.0231	.7320	.22835	.97358	.23455	.2635	.0271	.3792	.24531	.96944	.25304	.9520	.0315	.0765	.26219	.96502	.27169	.6806	.0362	.8140	48
13	.21161	.97735	.21651	.6187	.0232	.7257	.22863	.97351	.23485	.2579	.0272	.3738	.24559	.96937	.25335	.9471	.0316	.0718	.26247	.96494	.27201	.6764	.0363	.8100	47
14	.21189	.97729	.21682	.6122	.0232	.7193	.22892	.97344	.23516	.2524	.0273	.3684	.24587	.96930	.25366	.9423	.0317	.0672	.26275	.96486	.27232	.6722	.0364	.8059	46
15	.21218	.97723	.21712	4.6057	.0233	4.7130	.22920	.97338	.23547	4.2468	.0273	4.3630	.24615	.96923	.25397	3.9375	.0317	4.0625	.26303	.96479	.27263	3.6679	.0365	3.8018	45
16	.21246	.97717	.21742	.5993	.0234	.7067	.22948	.97331	.23577	.2413	.0274	.3576	.24643	.96916	.25428	.9327	.0318	.0579	.26331	.96471	.27294	.6637	.0366	.7978	44
17	.21275	.97711	.21773	.5928	.0234	.7004	.22977	.97324	.23608	.2358	.0275	.3522	.24672	.96909	.25459	.9279	.0319	.0532	.26359	.96463	.27326	.6596	.0367	.7937	43
18	.21303	.97704	.21803	.5864	.0235	.6942	.23005	.97318	.23639	.2303	.0276	.3469	.24700	.96901	.25490	.9231	.0320	.0486	.26387	.96456	.27357	.6554	.0367	.7897	42
19	.21331	.97698	.21834	.5800	.0235	.6879	.23033	.97311	.23670	.2248	.0276	.3415	.24728	.96894	.25521	.9184	.0320	.0440	.26415	.96448	.27388	.6512	.0368	.7857	41
20	.21360	.97692	.21864	4.5736	.0236	4.6817	.23061	.97304	.23700	4.2193	.0277	4.3362	.24756	.96887	.25552	3.9136	.0321	4.0394	.26443	.96440	.27419	3.6470	.0369	3.7816	40
21	.21388	.97686	.21895	.5673	.0237	.6754	.23090	.97298	.23731	.2139	.0278	.3309	.24784	.96880	.25583	.9089	.0322	.0348	.26471	.96433	.27451	.6429	.0370	.7776	39
22	.21417	.97680	.21925	.5609	.0237	.6692	.23118	.97291	.23762	.2084	.0278	.3256	.24813	.96873	.25614	.9042	.0323	.0302	.26499	.96425	.27482	.6387	.0371	.7736	38
23	.21445	.97673	.21956	.5546	.0238	.6631	.23146	.97284	.23793	.2030	.0279	.3203	.24841	.96865	.25645	.8994	.0323	.0256	.26527	.96417	.27513	.6346	.0371	.7697	37
24	.21473	.97667	.21986	.5483	.0239	.6569	.23175	.97277	.23823	.1976	.0280	.3150	.24869	.96858	.25676	.8947	.0324	.0211	.26556	.96409	.27544	.6305	.0372	.7657	36
25	.21502	.97661	.22017	4.5420	.0239	4.6507	.23202	.97271	.23854	4.1921	.0280	4.3098	.24897	.96851	.25707	3.8900	.0325	4.0165	.26584	.96402	.27576	3.6263	.0373	3.7617	35
26	.21530	.97655	.22047	.5357	.0240	.6446	.23231	.97264	.23885	.1867	.0281	.3045	.24925	.96844	.25738	.8853	.0326	.0120	.26612	.96394	.27607	.6222	.0374	.7577	34
27	.21559	.97648	.22078	.5294	.0241	.6385	.23260	.97257	.23916	.1814	.0282	.2993	.24953	.96836	.25769	.8807	.0327	.0074	.26640	.96386	.27638	.6181	.0375	.7538	33
28	.21587	.97642	.22108	.5232	.0241	.6324	.23288	.97250	.23946	.1760	.0283	.2941	.24982	.96829	.25800	.8760	.0327	.0029	.26668	.96378	.27670	.6140	.0376	.7498	32
29	.21615	.97636	.22139	.5169	.0242	.6263	.23316	.97244	.23977	.1706	.0283	.2838	.25010	.96822	.25831	.8713	.0328	3.9984	.26696	.96371	.27701	.6100	.0376	.7459	31
30	.21644	.97630	.22169	4.5107	.0243	4.6201	.23344	.97237	.24008	4.1653	.0284	4.2836	.25038	.96815	.25862	3.8667	.0329	3.9939	.26724	.96363	.27732	3.6059	.0377	3.7420	30
31	.21672	.97623	.22200	.5045	.0243	.6142	.23373	.97230	.24039	.1600	.0285	.2785	.25066	.96807	.25893	.8621	.0330	.9894	.26752	.96355	.27764	.6018	.0378	.7380	29
32	.21701	.97617	.22230	.4983	.0244	.6081	.23401	.97223	.24069	.1546	.0285	.2733	.25094	.96800	.25924	.8574	.0330	.9850	.26780	.96347	.27795	.5977	.0379	.7341	28
33	.21729	.97611	.22261	.4921	.0245	.6021	.23429	.97216	.24100	.1493	.0286	.2681	.25122	.96793	.25955	.8528	.0331	.9805	.26808	.96340	.27826	.5937	.0380	.7302	27
34	.21757	.97604	.22291	.4860	.0245	.5961	.23458	.97210	.24131	.1440	.0287	.2630	.25151	.96785	.25986	.8482	.0332	.97[illegible]	.26836	.96332	.27858	.5896	.0381	.7263	26
35	.21786	.97598	.22322	4.4799	.0246	4.5901	.23486	.97203	.24162	4.1388	.0288	4.2579	.25179	.96778	.26017	3.8436	.0333	3.9716	.26864	.96324	.27889	3.5856	.0382	3.7224	25
36	.21814	.97592	.22353	.4737	.0247	.5841	.23514	.97196	.24192	.1335	.0288	.2527	.25207	.96771	.26048	.8390	.0334	.9672	.26892	.96316	.27920	.5816	.0382	.7186	24
37	.21843	.97575	.22383	.4676	.0247	.5782	.23542	.97189	.24223	.1282	.0289	.2476	.25235	.96763	.26079	.8345	.0334	.9627	.26920	.96308	.27952	.5776	.0383	.7147	23
38	.21871	.97579	.22414	.4615	.0248	.5722	.23571	.97182	.24254	.1230	.0290	.2425	.25263	.96756	.26110	.8299	.0335	.9583	.26948	.96301	.27983	.5736	.0384	.7108	22
39	.21899	.97573	.22444	.4555	.0249	.5663	.23599	.97175	.24285	.1178	.0291	.2375	.25291	.96749	.26141	.8254	.0336	.9539	.26976	.96293	.28014	.5696	.0385	.7070	21
40	.21928	.97566	.22475	4.4494	.0249	4.5604	.23627	.97169	.24316	4.1126	.0291	4.2324	.25319	.96741	.26172	3.8208	.0337	3.9495	.27004	.96285	.28046	3.5656	.0386	3.7031	20
41	.21956	.97560	.22505	.4434	.0250	.5545	.23655	.97162	.24346	.1073	.0292	.2273	.25348	.96734	.26203	.8163	.0338	.9451	.27032	.96277	.28077	.5616	.0387	.6993	19
42	.21985	.97553	.22536	.4373	.0251	.5486	.23684	.97155	.24377	.1022	.0293	.2223	.25376	.96727	.26234	.8118	.0338	.9408	.27060	.96269	.28109	.5576	.0387	.6955	18
43	.22013	.97547	.22566	.4313	.0251	.5428	.23712	.97148	.24408	.0970	.0293	.2173	.25404	.96719	.26266	.8073	.0339	.9364	.27088	.96261	.28140	.5536	.0388	.6917	17
44	.22041	.97541	.22597	.4253	.0252	.5369	.23740	.97141	.24439	.0918	.0294	.2122	.25432	.96712	.26297	.8027	.0340	.9320	.27116	.96253	.28171	.5497	.0389	.6878	16
45	.22070	.97534	.22628	4.4194	.0253	4.5311	.23768	.97134	.24470	4.0867	.0295	4.2072	.25460	.96704	.26328	3.7983	.0341	3.9277	.27144	.96245	.28203	3.5457	.0390	3.6840	15
46	.22098	.97528	.22658	.4134	.0253	.5253	.23797	.97127	.24501	.0815	.0296	.2022	.25488	.96697	.26359	.7938	.0341	.9234	.27172	.96238	.28234	.5418	.0391	.6802	14
47	.22126	.97521	.22689	.4074	.0254	.5195	.23825	.97120	.24531	.0764	.0296	.1972	.25516	.96690	.26390	.7893	.0342	.9190	.27200	.96230	.28266	.5378	.0392	.6765	13
48	.22155	.97515	.22719	.4015	.0255	.5137	.23853	.97113	.24562	.0713	.0297	.1923	.25544	.96682	.26421	.7848	.0343	.9147	.27228	.96222	.28297	.5339	.0393	.6727	12
49	.22183	.97508	.22750	.3956	.0255	.5079	.23881	.97106	.24593	.0662	.0298	.1873	.25573	.96675	.26452	.7804	.0344	.9104	.27256	.96214	.28328	.5300	.0393	.6689	11
50	.22211	.97502	.22781	4.3897	.0256	4.5021	.23910	.97099	.24624	4.0611	.0299	4.1824	.25601	.96667	.26483	3.7759	.0345	3.9061	.27284	.96206	.28360	3.5261	.0394	3.6651	10
51	.22240	.97495	.22811	.3838	.0257	.4964	.23938	.97092	.24655	.0560	.0299	.1774	.25629	.96660	.26514	.7715	.0345	.9018	.27312	.96198	.28391	.5222	.0395	.6614	9
52	.22268	.97489	.22842	.3779	.0257	.4907	.23966	.97086	.24686	.0509	.0300	.1725	.25657	.96652	.26546	.7671	.0346	.8976	.27340	.96190	.28423	.5183	.0396	.6576	8
53	.22297	.97483	.22872	.3721	.0258	.4850	.23994	.97079	.24717	.0458	.0301	.1676	.25685	.96645	.26577	.7627	.0347	.8933	.27368	.96182	.28454	.5144	.0397	.6539	7
54	.22325	.97476	.22903	.3662	.0259	.4793	.24023	.97072	.24747	.0408	.0302	.1627	.25713	.96638	.26608	.7583	.0348	.8890	.27396	.96174	.28486	.5105	.0398	.6502	6
55	.22353	.97470	.22934	4.3604	.0260	4.4736	.24051	.97065	.24778	4.0358	.0302	4.1578	.25741	.96630	.26639	3.7539	.0349	3.8848	.27424	.96166	.28517	3.5066	.0399	3.6464	5
56	.22382	.97463	.22964	.3546	.0260	.4679	.24079	.97058	.24809	.0307	.0303	.1529	.25769	.96623	.26670	.7495	.0349	.8805	.27452	.96158	.28549	.5028	.0399	.6427	4
57	.22410	.97457	.22995	.3488	.0261	.4623	.24107	.97051	.24840	.0257	.0304	.1481	.25798	.96615	.26701	.7451	.0350	.8763	.27480	.96150	.28580	.4989	.0400	.6390	3
58	.22438	.97450	.23025	.3430	.0262	.4566	.24136	.97044	.24871	.0207	.0305	.1432	.25826	.96608	.26732	.7407	.0351	.8721	.27508	.96142	.28611	.4951	.0401	.6353	2
59	.22467	.97443	.23056	.3372	.0262	.4510	.24164	.97037	.24902	.0157	.0305	.1384	.25854	.96600	.26764	.7364	.0352	.8679	.27536	.96134	.28643	.4912	.0402	.6316	1
60	.22495	.97437	.23087	4.3315	.0263	4.4454	.24192	.97029	.24933	4.0108	.0306	4.1336	.25882	.96592	.26795	3.7320	.0353	3.8637	.27564	.96126	.28674	3.4874	.0403	3.6279	0
′	cos	sin	cot	tan	cosec	sec	cos	sin	cot	tan	cosec	sec	cos	sin	cot	tan	cosec	sec	cos	sin	cot	tan	cosec	sec	′
	77°						76°						75°						74°						

Table 19 Natural Trigonometric Functions (Continued)

	16°						17°						18°						19°						
′	sin	cos	tan	cot	sec	cosec	sin	cos	tan	cot	sec	cosec	sin	cos	tan	cot	sec	cosec	sin	cos	tan	cot	sec	cosec	′
0	.27564	.96126	.28674	3.4874	1.0403	3.6279	.29237	.95630	.30573	3.2708	1.0457	3.4203	.30902	.95106	.32492	3.0777	1.0515	3.2361	.32557	.94552	.34433	2.9042	1.0576	3.0715	60
1	.27592	.96118	.28706	.4836	.0404	.6243	.29265	.95622	.30605	.2674	.0458	.4170	.30929	.95097	.32524	.0746	.0516	.2332	.32584	.94542	.34465	.9015	.0577	.0690	59
2	.27620	.96110	.28737	.4798	.0405	.6206	.29293	.95613	.30637	.2640	.0459	.4138	.30957	.95088	.32556	.0716	.0517	.2303	.32612	.94533	.34498	.8987	.0578	.0664	58
3	.27648	.96102	.28769	.4760	.0406	.6169	.29321	.95605	.30668	.2607	.0460	.4106	.30985	.95079	.32588	.0686	.0518	.2274	.32639	.94523	.34530	.8960	.0579	.0638	57
4	.27675	.96094	.28800	.4722	.0406	.6133	.29348	.95596	.30700	.2573	.0461	.4073	.31012	.95070	.32621	.0655	.0519	.2245	.32667	.94514	.34563	.8933	.0580	.0612	56
5	.27703	.96086	.28832	3.4684	.0407	3.6096	.29376	.95588	.30732	3.2539	.0461	3.4041	.31040	.95061	.32653	3.0625	.0520	3.2216	.32694	.94504	.34595	2.8905	.0581	3.0586	55
6	.27731	.96078	.28863	.4646	.0408	.6060	.29404	.95579	.30764	.2505	.0462	.4009	.31068	.95051	.32685	.0595	.0521	.2188	.32722	.94495	.34628	.8878	.0582	.0561	54
7	.27759	.96070	.28895	.4608	.0409	.6024	.29432	.95571	.30796	.2472	.0463	.3977	.31095	.95042	.32717	.0565	.0522	.2159	.32749	.94485	.34661	.8851	.0584	.0535	53
8	.27787	.96062	.28926	.4570	.0410	.5987	.29460	.95562	.30828	.2438	.0464	.3945	.31123	.95033	.32749	.0535	.0523	.2131	.32777	.94476	.34693	.8824	.0585	.0509	52
9	.27815	.96054	.28958	.4533	.0411	.5951	.29487	.95554	.30859	.2405	.0465	.3913	.31150	.95024	.32782	.0505	.0524	.2102	.32804	.94466	.34726	.8797	.0586	.0484	51
10	.27843	.96045	.28990	3.4495	.0412	3.5915	.29515	.95545	.30891	3.2371	.0466	3.3881	.31178	.95015	.32814	3.0475	.0525	3.2074	.32832	.94457	.34758	2.8770	.0587	3.0458	50
11	.27871	.96037	.29021	.4458	.0413	.5879	.29543	.95536	.30923	.2338	.0467	.3849	.31206	.95006	.32846	.0445	.0526	.2045	.32859	.94447	.34791	.8743	.0588	.0433	49
12	.27899	.96029	.29053	.4420	.0413	.5843	.29571	.95528	.30955	.2305	.0468	.3817	.31233	.94997	.32878	.0415	.0527	.2017	.32887	.94438	.34824	.8716	.0589	.0407	48
13	.27927	.96021	.29084	.4383	.0414	.5807	.29598	.95519	.30987	.2271	.0469	.3785	.31261	.94988	.32910	.0385	.0528	.1989	.32914	.94428	.34856	.8689	.0590	.0382	47
14	.27955	.96013	.29116	.4346	.0415	.5772	.29626	.95511	.31019	.2238	.0470	.3754	.31289	.94979	.32943	.0356	.0529	.1960	.32942	.94418	.34889	.8662	.0591	.0357	46
15	.27983	.96005	.29147	3.4308	.0416	3.5736	.29654	.95502	.31051	3.2205	.0471	3.3722	.31316	.94970	.32975	3.0326	.0530	3.1932	.32969	.94409	.34921	2.8636	.0592	3.0331	45
16	.28011	.95997	.29179	.4271	.0417	.5700	.29682	.95493	.31083	.2172	.0472	.3690	.31344	.94961	.33007	.0296	.0531	.1904	.32996	.94399	.34954	.8609	.0593	.0306	44
17	.28039	.95989	.29210	.4234	.0418	.5665	.29710	.95485	.31115	.2139	.0473	.3659	.31372	.94952	.33039	.0267	.0532	.1876	.33024	.94390	.34987	.8582	.0594	.0281	43
18	.28067	.95980	.29242	.4197	.0419	.5629	.29737	.95476	.31146	.2106	.0474	.3627	.31399	.94942	.33072	.0237	.0533	.1848	.33051	.94380	.35019	.8555	.0595	.0256	42
19	.28094	.95972	.29274	.4160	.0420	.5594	.29765	.95467	.31178	.2073	.0475	.3596	.31427	.94933	.33104	.0208	.0534	.1820	.33079	.94370	.35052	.8529	.0596	.0231	41
20	.28122	.95964	.29305	3.4124	.0420	3.5559	.29793	.95459	.31210	3.2041	.0476	3.3565	.31454	.94924	.33136	3.0178	.0535	3.1792	.33106	.94361	.35085	2.8502	.0598	3.0206	40
21	.28150	.95956	.29337	.4087	.0421	.5523	.29821	.95450	.31242	.2008	.0477	.3534	.31482	.94915	.33169	.0149	.0536	.1764	.33134	.94351	.35117	.8476	.0599	.0181	39
22	.28178	.95948	.29368	.4050	.0422	.5488	.29848	.95441	.31274	.1975	.0478	.3502	.31510	.94906	.33201	.0120	.0537	.1736	.33161	.94341	.35150	.8449	.0600	.0156	38
23	.28206	.95940	.29400	.4014	.0423	.5453	.29876	.95433	.31306	.1942	.0478	.3471	.31537	.94897	.33233	.0090	.0528	.1708	.33189	.94332	.35183	.8423	.0601	.0131	37
24	.28234	.95931	.29432	.3977	.0424	.5418	.29904	.95424	.31338	.1910	.0479	.3440	.31565	.94888	.33265	.0061	.0539	.1681	.33216	.94322	.35215	.8396	.0602	.0106	36
25	.28262	.95923	.29463	3.3941	.0425	3.5383	.29932	.95415	.31370	3.1877	.0480	3.3409	.31592	.94878	.33298	3.0032	.0540	3.1653	.33243	.94313	.35248	2.8370	.0603	3.0081	35
26	.28290	.95915	.29495	.3904	.0426	.5348	.29959	.95407	.31402	.1845	.0481	.3378	.31620	.94869	.33330	.0003	.0541	.1625	.33271	.94303	.35281	.8344	.0604	.0056	34
27	.28318	.95907	.29526	.3868	.0427	.5313	.29987	.95398	.31434	.1813	.0482	.3347	.31648	.94860	.33362	2.9974	.0542	.1598	.33298	.94293	.35314	.8318	.0605	.0031	33
28	.28346	.95898	.29558	.3832	.0428	.5279	.30015	.95389	.31466	.1780	.0483	.3316	.31675	.94851	.33395	.9945	.0543	.1570	.33326	.94283	.35346	.8291	.0606	.0007	32
29	.28374	.95890	.29590	.3795	.0428	.5244	.30043	.95380	.31498	.1748	.0484	.3286	.31703	.94841	.33427	.9916	.0544	.1543	.33353	.94274	.35379	.8265	.0607	2.9982	31
30	.28401	.95882	.29621	3.3759	.0429	3.5209	.30070	.95372	.31530	3.1716	.0485	3.3255	.31730	.94832	.33459	2.9887	.0545	3.1515	.33381	.94264	.35412	2.8239	.0608	2.9957	30
31	.28429	.95874	.29653	.3723	.0430	.5175	.30098	.95363	.31562	.1684	.0486	.3224	.31758	.94823	.33492	.9858	.0546	.1488	.33408	.94254	.35445	.8213	.0609	.9933	29
32	.28457	.95865	.29685	.3687	.0431	.5140	.30126	.95354	.31594	.1652	.0487	.3194	.31786	.94814	.33524	.9829	.0547	.1461	.33435	.94245	.35477	.8187	.0611	.9908	28
33	.28485	.95857	.29716	.3651	.0432	.5106	.30154	.95345	.31626	.1620	.0488	.3163	.31813	.94805	.33557	.9800	.0548	.1433	.33463	.94235	.35510	.8161	.0612	.9884	27
34	.28513	.95849	.29748	.3616	.0433	.5072	.30181	.95337	.31658	.1588	.0489	.3133	.31841	.94795	.33589	.9772	.0549	.1406	.33490	.94225	.35543	.8135	.0613	.9859	26
35	.28541	.95840	.29780	3.3580	.0434	3.5037	.30209	.95328	.31690	3.1556	.0490	3.3102	.31868	.94786	.33621	2.9743	.0550	3.1379	.33518	.94215	.35576	2.8109	.0614	2.9835	25
36	.28569	.95832	.29811	.3544	.0435	.5003	.30237	.95319	.31722	.1524	.0491	.3072	.31896	.94777	.33654	.9714	.0551	.1352	.33545	.94206	.35608	.8083	.0615	.9810	24
37	.28597	.95824	.29843	.3509	.0436	.4969	.30265	.95310	.31754	.1492	.0492	.3042	.31923	.94767	.33686	.9686	.0552	.1325	.33572	.94196	.35641	.8057	.0616	.9786	23
38	.28624	.95816	.29875	.3473	.0437	.4935	.30292	.95301	.31786	.1460	.0493	.3011	.31951	.94758	.33718	.9657	.0553	.1298	.33600	.94186	.35674	.8032	.0617	.9762	22
39	.28652	.95807	.29906	.3438	.0438	.4901	.30320	.95293	.31818	.1429	.0494	.2981	.31978	.94749	.33751	.9629	.0554	.1271	.33627	.94176	.35707	.8006	.0618	.9738	21
40	.28680	.95799	.29938	3.3402	.0438	3.4867	.30348	.95284	.31850	3.1397	.0495	3.2951	.32006	.94740	.33783	2.9600	.0555	3.1244	.33655	.94167	.35739	2.7980	.0619	2.9713	20
41	.28708	.95791	.29970	.3367	.0439	.4833	.30375	.95275	.31882	.1366	.0496	.2921	.32034	.94730	.33816	.9572	.0556	.1217	.33682	.94157	.35772	.7954	.0620	.9689	19
42	.28736	.95782	.30001	.3332	.0440	.4799	.30403	.95266	.31914	.1334	.0497	.2891	.32061	.94721	.33848	.9544	.0557	.1190	.33709	.94147	.35805	.7929	.0622	.9665	18
43	.28764	.95774	.30033	.3296	.0441	.4766	.30431	.95257	.31946	.1303	.0498	.2861	.32089	.94712	.33880	.9515	.0558	.1163	.33737	.94137	.35838	.7903	.0623	.9641	17
44	.28792	.95765	.30065	.3261	.0442	.4732	.30459	.95248	.31978	.1271	.0499	.2831	.32116	.94702	.33913	.9487	.0559	.1137	.33764	.94127	.35871	.7878	.0624	.9617	16
45	.28820	.95757	.30096	3.3226	.0443	3.4698	.30486	.95239	.32010	3.1240	.0500	3.2801	.32144	.94693	.33945	2.9459	.0560	3.1110	.33792	.94118	.35904	2.7852	.0625	2.9593	15
46	.28847	.95749	.30128	.3191	.0444	.4665	.30514	.95231	.32042	.1209	.0501	.2772	.32171	.94684	.33978	.9431	.0561	.1083	.33819	.94108	.35936	.7827	.0626	.9569	14
47	.28875	.95740	.30160	.3156	.0445	.4632	.30542	.95222	.32074	.1177	.0502	.2742	.32199	.94674	.34010	.9403	.0562	.1057	.33846	.94098	.35969	.7801	.0627	.9545	13
48	.28903	.95732	.30192	.3121	.0446	.4598	.30569	.95213	.32106	.1146	.0503	.2712	.32226	.94665	.34043	.9375	.0563	.1030	.33874	.94088	.36002	.7776	.0628	.9521	12
49	.28931	.95723	.30223	.3087	.0447	.4565	.30597	.95204	.32138	.1115	.0504	.2683	.32254	.94655	.34075	.9347	.0565	.1004	.33901	.94078	.36035	.7751	.0629	.9497	11
50	.28959	.95715	.30255	3.3052	.0448	3.4532	.30625	.95195	.32171	3.1084	.0505	3.2653	.32282	.94646	.34108	2.9319	.0566	3.0977	.33928	.94068	.36068	2.7725	.0630	2.9474	10
51	.28987	.95707	.30287	.3017	.0448	.4498	.30653	.95186	.32203	.1053	.0506	.2624	.32309	.94637	.34140	.9291	.0567	.0951	.33956	.94058	.36101	.7700	.0632	.9450	9
52	.29014	.95698	.30319	.2983	.0449	.4465	.30680	.95177	.32235	.1022	.0507	.2594	.32337	.94627	.34173	.9263	.0568	.0925	.33983	.94049	.36134	.7675	.0633	.9426	8
53	.29042	.95690	.30350	.2948	.0450	.4432	.30708	.95168	.32267	.0991	.0508	.2565	.32364	.94618	.34205	.9235	.0569	.0898	.34011	.94039	.36167	.7650	.0634	.9402	7
54	.29070	.95681	.30382	.2914	.0451	.4399	.30736	.95159	.32299	.0960	.0509	.2535	.32392	.94608	.34238	.9208	.0570	.0872	.34038	.94029	.36199	.7625	.0635	.9379	6
55	.29098	.95673	.30414	3.2879	.0452	3.4366	.30763	.95150	.32331	3.0930	.0510	3.2506	.32419	.94599	.34270	2.9180	.0571	3.0846	.34065	.94019	.36232	2.7600	.0636	2.9355	5
56	.29126	.95664	.30446	.2845	.0453	.4334	.30791	.95141	.32363	.0899	.0511	.2477	.32447	.94590	.34303	.9152	.0572	.0820	.34093	.94009	.36265	.7575	.0637	.9332	4
57	.29154	.95656	.30478	.2811	.0454	.4301	.30819	.95132	.32395	.0868	.0512	.2448	.32474	.94580	.34335	.9125	.0573	.0793	.34120	.93999	.36298	.7550	.0638	.9308	3
58	.29181	.95647	.30509	.2777	.0455	.4268	.30846	.95124	.32428	.0838	.0513	.2419	.32502	.94571	.34368	.9097	.0574	.0767	.34147	.93989	.36331	.7525	.0639	.9285	2
59	.29209	.95639	.30541	.2742	.0456	.4236	.30874	.95115	.32460	.0807	.0514	.2390	.32529	.94561	.34400	.9069	.0575	.0741	.34175	.93979	.36364	.7500	.0641	.9261	1
60	.29237	.95630	.30573	3.2708	.0457	3.4203	.30902	.95106	.32492	3.0777	.0515	3.2361	.32557	.94552	.34433	2.9042	.0576	3.0715	.34202	.93969	.36397	2.7475	.0642	2.9238	0
′	cos	sin	cot	tan	cosec	sec	cos	sin	cot	tan	cosec	sec	cos	sin	cot	tan	cosec	sec	cos	sin	cot	tan	cosec	sec	′
	73°						72°						71°						70°						

Table 19 Natural Trigonometric Functions (Continued)

′	20°						21°						22°						23°						′
′	sin	cos	tan	cot	sec	cosec	sin	cos	tan	cot	sec	cosec	sin	cos	tan	cot	sec	cosec	sin	cos	tan	cot	sec	cosec	′
0	.34202	.93969	.36397	2.7475	1.0642	2.9238	.35837	.93358	.38386	2.6051	1.0711	2.7904	.37461	.92718	.40403	2.4751	1.0785	2.6695	.39073	.92050	.42447	2.3558	1.0864	2.5593	60
1	.34229	.93959	.36430	.7450	.0643	.9215	.35864	.93348	.38420	.6028	.0713	.7883	.37488	.92707	.40436	.4730	.0787	.6675	.39100	.92039	.42482	.3539	.0865	.5575	59
2	.34257	.93949	.36463	.7425	.0644	.9191	.35891	.93337	.38453	.6006	.0714	.7862	.37514	.92696	.40470	.4709	.0788	.6656	.39126	.92028	.42516	.3520	.0866	.5558	58
3	.34284	.93939	.36496	.7400	.0645	.9168	.35918	.93327	.38486	.5983	.0715	.7841	.37541	.92686	.40504	.4689	.0789	.6637	.39153	.92016	.42550	.3501	.0868	.5540	57
4	.34311	.93929	.36529	.7376	.0646	.9145	.35945	.93316	.38520	.5960	.0716	.7820	.37568	.92675	.40538	.4668	.0790	.6618	.39180	.92005	.42585	.3482	.0869	.5523	56
5	.34339	.93919	.36562	2.7351	.0647	2.9122	.35972	.93306	.38553	2.5938	.0717	2.7799	.37595	.92664	.40572	2.4647	.0792	2.6599	.39207	.91993	.42619	2.3463	.0870	2.5506	55
6	.34366	.93909	.36595	.7326	.0648	.9098	.36000	.93295	.38587	.5916	.0719	.7778	.37622	.92653	.40606	.4627	.0793	.6580	.39234	.91982	.42654	.3445	.0872	.5488	54
7	.34393	.93899	.36628	.7302	.0650	.9075	.36027	.93285	.38620	.5893	.0720	.7757	.37649	.92642	.40640	.4606	.0794	.6561	.39260	.91971	.42688	.3426	.0873	.5471	53
8	.34421	.93889	.36661	.7277	.0651	.9052	.36054	.93274	.38654	.5871	.0721	.7736	.37676	.92631	.40673	.4586	.0795	.6542	.39287	.91959	.42722	.3407	.0874	.5453	52
9	.34448	.93879	.36694	.7252	.0652	.9029	.36081	.93264	.38687	.5848	.0722	.7715	.37703	.92620	.40707	.4565	.0797	.6523	.39314	.91948	.42757	.3388	.0876	.5436	51
10	.34475	.93869	.36727	2.7228	.0653	2.9006	.36108	.93253	.38720	2.5826	.0723	2.7694	.37730	.92609	.40741	2.4545	.0798	2.6504	.39341	.91936	.42791	2.3369	.0877	2.5419	50
11	.34502	.93859	36760	.7204	.0654	.8983	.36135	.93243	.38754	.5804	.0725	.7674	.37757	.92598	.40775	.4525	.0799	.6485	.39367	.91925	.42826	.3350	.0878	.5402	49
12	.34530	.93849	.36793	.7179	.0655	.8960	.36162	.93232	.38787	.5781	.0726	.7653	.37784	.92587	.40809	.4504	.0801	.6466	.39394	.91913	.42860	.3332	.0880	.5384	48
13	.34557	.93839	.36826	.7155	.0656	.8937	.36189	.93222	.38821	.5759	.0727	.7632	.37811	.92576	.40843	.4484	.0802	.6447	.39421	.91902	.42894	.3313	.0881	.5367	47
14	.34584	.93829	.36859	.7130	.0658	.8915	.36217	.93211	.38854	.5737	.0728	.7611	.37838	.92565	.40877	.4463	.0803	.6428	.39448	.91891	.42929	.3294	.0882	.5350	46
15	.34612	.93819	.36892	2.7106	.0659	2.8892	.36244	.93201	.38888	2.5715	.0729	2.7591	.37865	.92554	.40911	2.4443	.0804	2.6410	.39474	.91879	.42963	2.3276	.0884	2.5333	45
16	.34639	.93809	.36925	.7082	.0660	.8869	.36271	.93190	.38921	.5693	.0731	.7570	.37892	.92543	.40945	.4423	.0806	.6391	.39501	.91868	.42998	.3257	.0885	.5316	44
17	.34666	.93799	.36958	.7058	.0661	.8846	.36298	.93180	.38955	.5671	.0732	.7550	.37919	.92532	.40979	.4403	.0807	.6372	.39528	.91856	.43032	.3238	.0886	.5299	43
18	.34693	.93789	.36991	.7033	.0662	.8824	.36325	.93169	.38988	.5649	.0733	.7529	.37946	.92521	.41013	.4382	.0808	.6353	.39554	.91845	.43067	.3220	.0888	.5281	42
19	.34721	.93779	.37024	.7009	.0663	.8801	.36352	.93158	.39022	.5627	.0734	.7509	.37972	.92510	.41047	.4362	.0810	.6335	.39581	.91833	.43101	.3201	.0889	.5264	41
20	.34748	.93769	.37057	2.6985	.0664	2.8778	.36379	.93148	.39055	2.5605	.0736	2.7488	.37999	.92499	.41081	2.4342	.0811	2.6316	.39608	.91822	.43136	2.3183	.0891	2.5247	40
21	.34775	.93758	.37090	.6961	.0666	.8756	.36406	.93137	.39089	.5583	.0737	.7468	.38026	.92488	.41115	.4322	.0812	.6297	.39635	.91810	.43170	.3164	.0892	.5230	39
22	.34803	.93748	.37123	.6937	.0667	.8733	.36433	.93127	.39122	.5561	.0738	.7447	.38053	.92477	.41149	.4302	.0813	.6279	.39661	.91798	.43205	.3145	.0893	.5213	38
23	.34830	.93738	.37156	.6913	.0668	.8711	.36460	.93116	.39156	.5539	.0739	.7427	.38080	.92466	.41183	.4282	.0815	.6260	.39688	.91787	.43239	.3127	.0895	.5196	37
24	.34857	.93728	.37190	.6889	.0669	.8688	.36488	.93105	.39189	.5517	.0740	.7406	.38107	.92455	.41217	.4262	.0816	.6242	.39715	.91775	.43274	.3109	.0896	.5179	36
25	.34884	.93718	.37223	2.6865	.0670	2.8666	.36515	.93095	.39223	2.5495	.0742	2.7386	.38134	.92443	.41251	2.4242	.0817	2.6223	.39741	.91764	.43308	2.3090	.0897	2.5163	35
26	.34912	.93708	.37256	.6841	.0671	.8644	.36542	.93084	.39257	.5473	.0743	.7366	.38161	.92432	.41285	.4222	.0819	.6205	.39768	.91752	.43343	.3072	.0899	.5146	34
27	.34939	.93698	.37289	.6817	.0673	.8621	.36569	.93074	.39290	.5451	.0744	.7346	.38188	.92421	.41319	.4202	.0820	.6186	.39795	.91741	.43377	.3053	.0900	.5129	33
28	.34966	.93687	.37322	.6794	.0674	.8599	.36596	.93063	.39324	.5430	.0745	.7325	.38214	.92410	.41353	.4182	.0821	.6168	.39821	.91729	.43412	.3035	.0902	.5112	32
29	.34993	.93677	.37355	.6770	.0675	.8577	.36623	.93052	.39357	.5408	.0747	.7305	.38241	.92399	.41387	.4162	.0823	.6150	.39848	.91718	.43447	.3017	.0903	.5095	31
30	.35021	.93667	.37388	2.6746	.0676	2.8554	.36650	.93042	.39391	2.5386	.0748	2.7285	.38268	.92388	.41421	2.4142	.0824	2.6131	.39875	.91706	.43481	2.2998	.0904	2.5078	30
31	.35048	.93657	.37422	.6722	.0677	.8532	.36677	.93031	.39425	.5365	.0749	.7265	.38295	.92377	.41455	.4122	.0825	.6113	.39901	.91694	.43516	.2980	.0906	.5062	29
32	.35075	.93647	.37455	.6699	.0678	.8510	.36704	.93020	39458	.5343	.0750	.7245	.38322	.92366	.41489	.4102	.0826	.6095	.39928	.91683	.43550	.2962	.0907	.5045	28
33	.35102	.93637	.37488	.6675	.0679	.8488	.36731	.93010	.39492	.5322	.0751	.7225	.38349	.92354	.41524	.4083	.0828	.6076	.39955	.91671	.43585	.2944	.0908	.5028	27
34	.35130	.93626	.37521	.6652	.0681	.8466	.36758	.92999	.39525	.5300	.0753	.7205	.38376	.92343	.41558	.4063	.0829	.6058	.39981	.91659	.43620	.2925	.0910	.5011	26
35	.35157	.93616	.37554	2.6628	.0682	2.8444	.36785	.92988	.39559	2.5278	.0754	2.7185	.38403	.92332	.41592	2.4043	.0830	2.6040	.40008	.91648	.43654	2.2907	.0911	2.4995	25
36	.35184	.93606	.37587	.6604	.0683	.8422	.36812	.92978	.39593	.5257	.0755	.7165	.38429	.92321	.41626	.4023	.0832	.6022	.40035	.91636	.43689	.2889	.0913	.4978	24
37	.35211	.93596	.37621	.6581	.0684	.8400	.36839	.92967	.39626	.5236	.0756	.7145	.38456	.92310	.41660	.4004	.0833	.6003	.40061	.91625	.43723	.2871	.0914	.4961	23
38	.35239	.93585	.37654	.6558	.0685	.8378	.36866	.92956	.39660	.5214	.0758	.7125	.38483	.92299	.41694	.3984	.0834	.5985	.40088	.91613	.43758	.2853	.0915	.4945	22
39	.35266	.93575	.37687	.6534	.0686	.8356	.36893	.92945	.39694	.5193	.0759	.7105	.38510	.92287	.41728	.3964	.0836	.5967	.40115	.91601	.43793	.2835	.0917	.4928	21
40	.35293	.93565	.37720	2.6511	.0688	2.8334	.36921	.92935	.39727	2.5171	.0760	2.7085	.38537	.92276	.41762	2.3945	.0837	2.5949	.40141	.91590	.43827	2.2817	.0918	2.4912	20
41	.35320	.93555	.37754	.6487	.0689	.8312	.36948	.92924	.39761	.5150	.0761	.7065	.38564	.92265	.41797	.3925	.0838	.5931	.40168	.91578	.43862	.2799	.0920	.4895	19
42	.35347	.93544	.37787	.6464	.0690	.8290	.36975	.92913	.39795	.5129	.0763	.7045	.38591	.92254	.41831	.3906	.0840	.5913	.40195	.91566	.43897	.2781	.0921	.4879	18
43	.35375	.93534	.37820	.6441	.0691	.8269	.37002	.92902	.39828	.5108	.0764	.7026	.38617	.92242	.41865	.3886	.0841	.5895	.40221	.91554	.43932	.2763	.0922	.4862	17
44	.35402	.93524	.37853	.6418	.0692	.8247	.37029	.92892	.39862	.5086	.0765	.7006	.38644	.92231	.41899	.3867	.0842	.5877	.40248	.91543	.43966	.2745	.0924	.4846	16
45	.35429	.93513	.37887	2.6394	.0694	2.8225	.37056	.92881	.39896	2.5065	.0766	2.6986	.38671	.92220	.41933	2.3847	.0844	2.5958	.40205	.91531	.44001	2.2727	.0925	2.4829	15
46	.35456	.93503	.37920	.6371	.0695	.8204	.37083	.92870	.39930	.5044	.0768	.6967	.38698	.92209	.41968	.3828	.0845	.5841	.40301	.91519	.44036	.2709	.0927	.4813	14
47	.35483	.93493	.37953	.6348	.0696	.8182	.37110	.92859	.39963	.5023	.0769	.6947	.38725	.92197	.42002	.3808	.0846	.5823	.40328	.91508	.44070	.2691	.0928	.4797	13
48	.35511	.93482	.37986	.6325	.0697	.8160	.37137	.92848	.39997	.5002	.0770	.6927	.38751	.92186	.42036	.3789	.0847	.5805	.40354	.91496	.44105	.2673	.0929	.4780	12
49	.35538	.93472	.38020	.6302	.0698	.8139	.37164	.92838	.40031	.4981	.0771	.6908	.38778	.92105	.42070	.3770	.0849	.5787	.40381	.91484	.44140	.2655	.0931	.4764	11
50	.35565	.93462	.38053	2.6279	.0699	2.8117	.37191	.92827	.40065	2.4960	.0773	2.6888	.38805	.92164	.42105	2.3750	.0850	2.5770	.40408	.91472	.44175	2.2637	.0932	2.4748	10
51	.35592	.93451	.38086	.6256	.0701	.8096	.37218	.92816	.40098	.4939	.0774	.6869	.38832	.92152	.42139	.3731	.0851	.5752	.40434	.91461	.44209	.2619	.0934	.4731	9
52	.35619	.93441	.38120	.6233	.0702	.8074	.37245	.92805	.40132	.4918	.0775	.6849	.38859	.92141	.42173	.3712	.0853	.5734	.40461	.91449	.44244	.2602	.0935	.4715	8
53	.35647	.93431	.38153	.6210	.0703	.8053	.37272	.92794	.40166	.4897	.0776	.6830	.38886	.92130	.42207	.3692	.0854	.5716	.40487	.91437	.44279	.2584	.0936	.4699	7
54	.35674	.93420	.38186	.6187	.0704	.8032	.37299	.92784	.40200	.4876	.0778	.6810	.38912	.92118	.42242	.3673	.0855	.5699	.40514	.91425	.44314	.2566	.0938	.4683	6
55	.35701	.93410	.38220	2.6164	.0705	2.8010	.37326	.92773	.40233	2.4855	.0779	2.6791	.38939	.92107	.42276	2.3654	.0857	2.5681	.40541	.91414	.44349	2.2548	.0939	2.4666	5
56	.35728	.93400	.38253	.6142	.0707	.7989	.37353	.92762	.40267	.4834	.0780	.6772	.38966	.92096	.42310	.3635	.0858	.5663	.40567	.91402	.44383	.2531	.0941	.4650	4
57	.35755	.93389	.38286	.6119	.0708	.7968	.37380	.92751	.40301	.4813	.0781	.6752	.38993	.92084	.42344	.3616	.0859	.5646	.40594	.91390	.44418	.2513	.0942	.4634	3
58	.35782	.93379	.38320	.6096	.0709	.7947	.37407	.92740	.40335	.4792	.0783	.6733	.39019	.92073	.42379	.3597	.0861	.5628	.40620	.91378	.44453	.2495	.0943	.4618	2
59	.35810	.93368	.38353	.6073	.0710	.7925	.37434	.92729	.40369	.4772	.0784	.6714	.39046	.92062	.42413	.3577	.0862	.5610	.40647	.91366	.44488	.2478	.0945	.4602	1
60	.35837	.93358	.38386	2.6051	.0711	2.7904	.37461	.92718	.40403	2.4751	.0785	2.6695	.39073	.92050	.42447	2.3558	.0864	2.5593	.40674	.91354	.44523	2.2460	.0946	2.4586	0
′	cos	sin	cot	tan	cosec	sec	cos	sin	cot	tan	cosec	sec	cos	sin	cot	tan	cosec	sec	cos	sin	cot	tan	cosec	sec	′
	69°						68°						67°						66°						

Table 19 Natural Trigonometric Functions (Continued)

′	24°						25°					
′	sin	cos	tan	cot	sec	cosec	sin	cos	tan	cot	sec	cosec
0	.40674	.91354	.44523	2.2460	1.0946	2.4586	.42262	.90631	.46631	2.1445	1.1034	2.3662
1	.40700	.91343	.44558	.2443	.0948	.4570	.42288	.90618	.46666	.1429	.1035	.3647
2	.40727	.91331	.44593	.2425	.0949	.4554	.42314	.90606	.46702	.1412	.1037	.3632
3	.40753	.91319	.44627	.2408	.0951	.4538	.42341	.90594	.46737	.1396	.1038	.3618
4	.40780	.91307	.44662	.2390	.0952	.4522	.42367	.90581	.46772	.1380	.1040	.3603
5	.40806	.91295	.44697	2.2373	.0953	2.4506	.42394	.90569	.46808	2.1364	.1041	2.3588
6	.40833	.91283	.44732	.2355	.0955	.4490	.42420	.90557	.46843	.1348	.1043	.3574
7	.40860	.91271	.44767	.2338	.0956	.4474	.42446	.90544	.46879	.1331	.1044	.3559
8	.40886	.91260	.44802	.2320	.0958	.4458	.42473	.90532	.46914	.1315	.1046	.3544
9	.40913	.91248	.44837	.2303	.0959	.4442	.42499	.90520	.46950	.1299	.1047	.3530
10	.40939	.91236	.44872	2.2286	.0961	2.4426	.42525	.90507	.46985	2.1283	.1049	2.3515
11	.40966	.91224	.44907	.2268	.0962	.4418	.42552	.90495	.47021	.1267	.1050	.3501
12	.40992	.91212	.44942	.2251	.0963	.4395	.42578	.90483	.47056	.1251	.1052	.3486
13	.41019	.91200	.44977	.2234	.0965	.4379	.42604	.90470	.47092	.1235	.1053	.3472
14	.41045	.91188	.45012	.2216	.0966	.4363	.42630	.90458	.47127	.1219	.1055	.3457
15	.41072	.91176	.45047	2.2199	.0968	2.4347	.42657	.90445	.47163	2.1203	.1056	2.3443
16	.41098	.91164	.45082	.2182	.0969	.4332	.42683	.90433	.47199	.1187	.1058	.3428
17	.41125	.91152	.45117	.2165	.0971	.4316	.42709	.90421	.47234	.1171	.1059	.3414
18	.41151	.91140	.45152	.2147	.0972	.4300	.42736	.90408	.47270	.1155	.1061	.3399
19	.41178	.91128	.45187	.2130	.0973	.4285	.42762	.90396	.47305	.1139	.1062	.3385
20	.41204	.91116	.45222	2.2113	.0975	2.4269	.42788	.90383	.47341	2.1123	.1064	2.3371
21	.41231	.91104	.45257	.2096	.0976	.4254	.42815	.90371	.47376	.1107	.1065	.3356
22	.41257	.91092	.45292	.2079	.0978	.4238	.42841	.90358	.47412	.1092	.1067	.3342
23	.41284	.91080	.45327	.2062	.0979	.4222	.42867	.90346	.47448	.1076	.1068	.3328
24	.41310	.91068	.45362	.2045	.0981	.4207	.42893	.90333	.47483	.1060	.1070	.3313
25	.41337	.91056	.45397	2.2028	.0982	2.4191	42920	.90321	.47519	2.1044	.1072	2.3299
26	.41363	.91044	.45432	.2011	.0984	.4176	.42946	.90308	.47555	.1028	.1073	.3285
27	.41390	.91032	.45467	.1994	.0985	.4160	.42972	.90296	.47590	.1013	.1075	.3271
28	.41416	.91020	.45502	.1977	.0986	.4145	.42998	.90283	.47626	.0997	.1076	.3256
29	.41443	.91008	.45537	.1960	.0988	.4130	.43025	.90271	.47662	.0981	.1078	.3242
30	.41469	.90996	.45573	2.1943	.0989	2.4114	.43051	.90258	.47697	2.0965	.1079	2.3228
31	.41496	.90984	.45608	.1926	.0991	.4099	.43077	.90246	.47733	.0950	.1081	.3214
32	.41522	.90972	.45643	.1909	.0992	.4083	.43104	.90233	.47769	.0934	.1082	.3200
33	.41549	.90960	.45678	.1892	.0994	.4068	.43130	.90221	.47805	.0918	.1084	.3186
34	.41575	.90948	.45713	.1875	.0995	.4053	.43156	.90208	.47840	.0903	.1085	.3172
35	.41602	.90936	.45748	2.1859	.0997	2.4037	.43182	.90196	.47876	2.0887	.1087	2.3158
36	.41628	.90924	.45783	.1842	.0998	.4022	.43208	.90183	.47912	.0872	.1088	.3143
37	.41654	.90911	.45819	.1825	.1000	.4007	.43235	.90171	.47948	.0856	.1090	.3129
38	.41681	.90899	.45854	.1808	.1001	.3992	.43261	.90158	.47983	.0840	.1092	.3115
39	.41707	.90887	.45889	.1792	.1003	.3976	.43287	.90145	.48019	.0825	.1093	.3101
40	.41734	.90875	.45924	2.1775	.1004	2.3961	.43313	.90133	.48055	2.0809	.1095	2.3087
41	.41760	.90863	.45960	.1758	.1005	.3946	.43340	.90120	.48091	.0794	.1096	.3073
42	.41787	.90851	.45995	.1741	.1007	.3931	.43366	.90108	.48127	.0778	.1098	.3059
43	.41813	.90839	.46030	.1725	.1008	.3916	.43392	.90095	.48162	.0763	.1099	.3046
44	.41839	.90826	.46065	.1708	.1010	.3901	.43418	.90082	.48198	.0747	.1101	.3032
45	.41866	.90814	.46101	2.1692	.1011	2.3886	.43444	.90070	.48234	2.0732	.1102	2.3018
46	.41892	.90812	.46136	.1675	.1013	.3871	.43471	.90057	.48270	.0717	.1104	.3004
47	.41919	.90790	.46171	.1658	.1014	.3856	.43497	.90044	.48306	.0701	.1106	.2990
48	.41945	.90778	.46206	.1642	.1016	.3841	.43523	.90032	.48342	.0686	.1107	.2976
49	.41972	.90765	.46242	.1625	.1017	.3826	.43549	.90019	.48378	.0671	.1109	.2962
50	.41998	.90753	.46277	2.1609	.1019	2.3811	.43575	.90006	.48414	2.0655	.1110	2.2949
51	.42024	.90741	.46312	.1592	.1020	.3796	.43602	.89994	.48449	.0640	.1112	.2935
52	.42051	.90729	.46348	.1576	.1022	.3781	.43628	.89981	.48485	.0625	.1113	.2921
53	.42077	.90717	.46383	.1559	.1023	.3766	.43654	.89968	.48521	.0609	.1115	.2907
54	.42103	.90704	.46418	.1543	.1025	.3751	.43680	.89956	.48557	.0594	.1116	.2894
55	.42130	.90692	.46454	2.1527	.1026	2.3736	.43706	.89943	.48593	2.0579	.1118	2.2880
56	.42156	.90680	.46489	.1510	.1028	.3721	.43732	.89930	.48629	.0564	.1120	.2866
57	.42183	.90668	.46524	.1494	.1029	.3706	.43759	.89918	.48665	.0548	.1121	.2853
58	.42209	.90655	.46560	.1478	.1031	.3691	.43785	.89905	.48701	.0533	.1123	.2839
59	.42235	.90643	.46595	.1461	.1032	.3677	.43811	.89892	.48737	.0518	.1124	.2825
60	.42262	.90631	.46631	2.1445	.1034	2.3662	.43837	.89879	.48773	2.0503	.1126	2.2812
′	cos	sin	cot	tan	cosec	sec	cos	sin	cot	tan	cosec	sec
	65°						64°					

′	26°						27°						′
′	sin	cos	tan	cot	sec	cosec	sin	cos	tan	cot	sec	cosec	′
0	.43837	.89879	.48773	2.0503	1.1126	2.2812	.45399	.89101	.50952	1.9626	1.1223	2.2027	60
1	.43863	.89867	.48809	.0488	.1127	.2798	.45425	.89087	.50989	.9612	.1225	.2014	59
2	.43889	.89854	.48845	.0473	.1129	.2784	.45451	.89074	.51026	.9598	.1226	.2002	58
3	.43915	.89841	.48881	.0458	.1131	.2771	.45477	.89061	.51062	.9584	.1228	.1989	57
4	.43942	.89828	.48917	.0443	.1132	.2757	.45503	.89048	.51099	.9570	.1230	.1977	56
5	.43968	.89815	.48953	2.0427	.1134	2.2744	.45528	.89034	.51136	1.9556	.1231	2.1964	55
6	.43994	.89803	.48989	.0412	.1135	.2730	.45554	.89021	.51172	.9542	.1233	.1952	54
7	.44020	.89790	.49025	.0397	.1137	.2717	.45580	.89008	.51209	.9528	.1235	.1939	53
8	.44046	.89777	.49062	.0382	.1139	.2703	.45606	.88995	.51246	.9514	.1237	.1927	52
9	.44072	.89764	.49098	.0367	.1140	.2690	.45632	.88981	.51283	.9500	.1238	.1914	51
10	.44098	.89751	.49134	2.0352	.1142	2.2676	.45658	.88968	.51319	1.9486	.1240	2.1902	50
11	.44124	.89739	.49170	.0338	.1143	.2663	.45684	.88955	.51356	.9472	.1242	.1889	49
12	.44150	.89726	.49206	.0323	.1145	.2650	.45710	.88942	.51393	.9458	.1243	.1877	48
13	.44177	.89713	.49242	.0308	.1147	.2636	.45736	.88928	.51430	.9444	.1245	.1865	47
14	.44203	.89700	.49278	.0293	.1148	.2623	.45761	.88915	.51466	.9430	.1247	.1852	46
15	.44229	.89687	.49314	2.0278	.1150	2.2610	.45787	.88902	.51503	1.9416	.1248	2.1840	45
16	.44255	.89674	.49351	.0263	.1151	.2596	.45813	.88888	.51540	.9402	.1250	.1828	44
17	.44281	.89661	.49387	.0248	.1153	.2583	.45839	.88875	.51577	.9388	.1252	.1815	43
18	.44307	.89649	.49423	.0233	.1155	.2570	.45865	.88862	.51614	.9375	.1253	.1803	42
19	.44333	.89636	.49459	.0219	.1156	.2556	.45891	.88848	.51651	.9361	.1255	.1791	41
20	.44359	.89623	.49495	2.0204	.1158	2.2543	.45917	.88835	.51687	1.9347	.1257	2.1778	40
21	.44385	.89610	.49532	.0189	.1159	.2530	.45942	.88822	.51724	.9333	.1258	.1766	39
22	.44411	.89597	.49568	.0174	.1161	.2517	.45968	.88808	.51761	.9319	.1260	.1754	38
23	.44437	.89584	.49604	.0159	.1163	.2503	.45994	.88795	.51798	.9306	.1262	.1742	37
24	.44463	.89571	.49640	.0145	.1164	.2490	.46020	.88781	.51835	.9292	.1264	.1730	36
25	.44489	.89558	.49677	2.0130	.1166	2.2477	.46046	.88768	.51872	1.9278	.1265	2.1717	35
26	.44516	.89545	.49713	.0115	.1167	.2464	.46072	.88755	.51909	.9264	.1267	.1705	34
27	.44542	.89532	.49749	.0101	.1169	.2451	.46097	.88741	.51946	.9251	.1269	.1693	33
28	.44568	.89519	.49785	.0086	.1171	.2438	.46123	.88728	.51983	.9237	.1270	.1681	32
29	.44594	.89506	.49822	.0071	.1172	.2425	.46149	.88714	.52020	.9223	.1272	.1669	31
30	.44620	.89493	.49858	2.0057	.1174	2.2411	.46175	.88701	.52057	1.9210	.1274	2.1657	30
31	.44646	.89480	.49894	.0042	.1176	.2398	.46201	.88688	.52094	.9196	.1275	.1645	29
32	.44672	.89467	.49931	.0028	.1177	.2385	.46226	.88674	.52131	.9182	.1277	.1633	28
33	.44698	.89454	.49967	.0013	.1179	.2372	.46252	.88661	.52168	.9169	.1279	.1620	27
34	.44724	.89441	.50003	1.9998	.1180	.2359	.46278	.88647	.52205	.9155	.1281	.1608	26
35	.44750	.89428	.50040	1.9984	.1182	2.2346	.46304	.88634	.52242	1.9142	.1282	2.1596	25
36	.44776	.89415	.50076	.9969	.1184	.2333	.46330	.88620	.52279	.9128	.1284	.1584	24
37	.44802	.89402	.50113	.9955	.1185	.2320	.46355	.88607	.52316	.9115	.1286	.1572	23
38	.44828	.89389	.50149	.9940	.1187	.2307	.46381	.88593	.52353	.9101	.1287	.1560	22
39	.44854	.89376	.50185	.9926	.1189	.2294	.46407	.88580	.52390	.9088	.1289	.1548	21
40	.44880	.89363	.50222	1.9912	.1190	2.2282	.46433	.88566	.52427	1.9074	.1291	2.1536	20
41	.44906	.89350	.50258	.9897	.1192	.2269	.46458	.88553	.52464	.9061	.1293	.1525	19
42	.44932	.89337	.50295	.9883	.1193	.2256	.46484	.88539	.52501	.9047	.1294	.1513	18
43	.44958	.89324	.50331	.9868	.1195	.2243	.46510	.88526	.52538	.9034	.1296	.1501	17
44	.44984	.89311	.50368	.9854	.1197	.2230	.46536	.88512	.52575	.9020	.1298	.1489	16
45	.45010	.89298	.50404	1.9840	.1198	2.2217	.46561	.88499	.52612	1.9007	.1299	2.1477	15
46	.45036	.89285	.50441	.9825	.1200	.2204	.46587	.88485	.52650	.8993	.1301	.1465	14
47	.45062	.89272	.50477	.9811	.1202	.2192	.46613	.88472	.52687	.8980	.1303	.1453	13
48	.45088	.89258	.50514	.9797	.1203	.2179	.46639	.88458	.52724	.8967	.1305	.1441	12
49	.45114	.89245	.50550	.9782	.1205	.2166	.46664	.88444	.52761	.8953	.1306	.1430	11
50	.45140	.89232	.50587	1.9768	.1207	2.2153	.46690	.88431	.52798	1.8940	.1308	2.1418	10
51	.45166	.89219	.50623	.9754	.1208	.2141	.46716	.88417	.52836	.8927	.1310	.1406	9
52	.45191	.89206	.50660	.9739	.1210	.2128	.46741	.88404	.52873	.8913	.1312	.1394	8
53	.45217	.89193	.50696	.9725	.1212	.2115	.46767	.88390	.52910	.8900	.1313	.1382	7
54	.45243	.89180	.50733	.9711	.1213	.2103	.46793	.88376	.52947	.8887	.1315	.1371	6
55	.45269	.89166	.50769	1.9697	.1215	2.2090	.46819	.88363	.52984	1.8873	.1317	2.1359	5
56	.45295	.89153	.50806	.9683	.1217	.2077	.46844	.88349	.53022	.8860	.1319	.1347	4
57	.45321	.89140	.50843	.9668	.1218	.2065	.46870	.88336	.53059	.8847	.1320	.1335	3
58	.45347	.89127	.50879	.9654	.1220	.2052	.46896	.88322	.53096	.8834	.1322	.1324	2
59	.45373	.89114	.50916	.9640	.1222	.2039	.46921	.88308	.53134	.8820	.1324	.1312	1
60	.45399	.89101	.50952	1.9626	.1223	2.2027	.46947	.88295	.53171	1.8807	.1326	2.1300	0
′	cos	sin	cot	tan	cosec	sec	cos	sin	cot	tan	cosec	sec	′
	63°						62°						

Table 19 Natural Trigonometric Functions (Continued)

	28°						29°						30°						31°						
′	sin	cos	tan	cot	sec	cosec	sin	cos	tan	cot	sec	cosec	sin	cos	tan	cot	sec	cosec	sin	cos	tan	cot	sec	cosec	′
0	.46947	.88295	.53171	1.8807	1.1326	2.1300	.48481	.87462	.55431	1.8040	1.1433	2.0627	.50000	.86603	.57735	1.7320	1.1547	2.0000	.51504	.85717	.60086	1.6643	1.1666	1.9416	60
1	.46973	.88281	.53208	.8794	.1327	.1289	.48506	.87448	.55469	.8028	.1435	.0616	.50025	.86588	.57774	.7309	.1549	1.9990	.51529	.85702	.60126	.6632	.1668	.9407	59
2	.46998	.88267	.53245	.8781	.1329	.1277	.48532	.87434	.55507	.8016	.1437	.0605	.50050	.86573	.57813	.7297	.1551	.9980	.51554	.85687	.60165	.6621	.1670	.9397	58
3	.47024	.88254	.53283	.8768	.1331	.1266	.48557	.87420	.55545	.8003	.1439	.0594	.50075	.86559	.57851	.7286	.1553	.9970	.51578	.85672	.60205	.6610	.1672	.9388	57
4	.47050	.88240	.53320	.8754	.1333	.1254	.48583	.87405	.55583	.7991	.1441	.0583	.50101	.86544	.57890	.7274	.1555	.9960	.51603	.85657	.60244	.6599	.1674	.9378	56
5	.47075	.88226	.53358	1.8741	.1334	2.1242	.48608	.87391	.55621	1.7979	.1443	2.0573	.50126	.86530	.57929	1.7262	.1557	1.9950	.51628	.85642	.60284	1.6588	.1676	1.9369	55
6	.47101	.88213	.53395	.8728	.1336	.1231	.48633	.87377	.55659	.7966	.1445	.0562	.50151	.86515	.57968	.7251	.1559	.9940	.51653	.85627	.60324	.6577	.1678	.9360	54
7	.47127	.88199	.53432	.8715	.1338	.1219	.48659	.87363	.55697	.7954	.1446	.0551	.50176	.86500	.58007	.7239	.1561	.9930	.51678	.85612	.60363	.6566	.1681	.9350	53
8	.47152	.88185	.53470	.8702	.1340	.1208	.48684	.87349	.55735	.7942	.1448	.0540	.50201	.86486	.58046	.7228	.1562	.9920	.51703	.85597	.60403	.6555	.1683	.9341	52
9	.47178	.88171	.53507	.8689	.1341	.1196	.48710	.87335	.55774	.7930	.1450	.0530	.50226	.86471	.58085	.7216	.1564	.9910	.51728	.85582	.60443	.6544	.1685	.9332	51
10	.47204	.88158	.53545	1.8676	.1343	2.1185	.48735	.87320	.55812	1.7917	.1452	2.0519	.50252	.86457	.58123	1.7205	.1566	1.9900	.51753	.85566	.60483	1.6534	.1687	1.9322	50
11	.47229	.88144	.53582	.8663	.1345	.1173	.48760	.87306	.55850	.7905	.1454	.0508	.50277	.86442	.58162	.7193	.1568	.9890	.51778	.85551	.60522	.6523	.1689	.9313	49
12	.47255	.88130	.53619	.8650	.1347	.1162	.48786	.87292	.55888	.7893	.1456	.0498	.50302	.86427	.58201	.7182	.1570	.9880	.51803	.85536	.60562	.6512	.1691	.9304	48
13	.47281	.88117	.53657	.8637	.1349	.1150	.48811	.87278	.55926	.7881	.1458	.0487	.50327	.86413	.58240	.7170	.1572	.9870	.51827	.85521	.60602	.6501	.1693	.9295	47
14	.47306	.88103	.53694	.8624	.1350	.1139	.48837	.87264	.55964	.7868	.1459	.0476	.50352	.86398	.58279	.7159	.1574	.9860	.51852	.85506	.60642	.6490	.1695	.9285	46
15	.47332	.88089	.53732	1.8611	.1352	2.1127	.48862	.87250	.56003	1.7856	.1461	2.0466	.50377	.86383	.58318	1.7147	.1576	1.9850	.51877	.85491	.60681	1.6479	.1697	1.9276	45
16	.47357	.88075	.53769	.8598	.1354	.1116	.48887	.87235	.56041	.7844	.1463	.0455	.50402	.86369	.58357	.7136	.1578	.9840	.51902	.85476	.60721	.6469	.1699	.9267	44
17	.47383	.88061	.53807	.8585	.1356	.1104	.48913	.87221	.56079	.7832	.1465	.0444	.50428	.86354	.58396	.7124	.1580	.9830	.51927	.85461	.60761	.6458	.1701	.9258	43
18	.47409	.88048	.53844	.8572	.1357	.1093	.48938	.87207	.56117	.7820	.1467	.0434	.50453	.86339	.58435	.7113	.1582	.9820	.51952	.85446	.60801	.6447	.1703	.9248	42
19	.47434	.88034	.53882	.8559	.1359	.1082	.48964	.87193	.56156	.7808	.1469	.0423	.50478	.86325	.58474	.7101	.1584	.9811	.51977	.85431	.60841	.6436	.1705	.9239	41
20	.47460	.88020	.53919	1.8546	.1361	2.1070	.48989	.87178	.56194	1.7795	.1471	2.0413	.50503	.86310	.58513	1.7090	.1586	1.9801	.52002	.85416	.60881	1.6425	.1707	1.9230	40
21	.47486	.88006	.53957	.8533	.1363	.1059	.49014	.87164	.56232	.7783	.1473	.0402	.50528	.86295	.58552	.7079	.1588	.9791	.52026	.85400	.60920	.6415	.1709	.9221	39
22	.47511	.87992	.53995	.8520	.1365	.1048	.49040	.87150	.56270	.7771	.1474	.0392	.50553	.86281	.58591	.7067	.1590	.9781	.52051	.85385	.60960	.6404	.1712	.9212	38
23	.47537	.87979	.54032	.8507	.1366	.1036	.49065	.87136	.56309	.7759	.1476	.0381	.50578	.86266	.58630	.7056	.1592	.9771	.52076	.85370	.61000	.6293	.1714	.9203	37
24	.47562	.87965	.54070	.8495	.1368	.1025	.49090	.87121	.56347	.7747	.1478	.0370	.50603	.86251	.58670	.7044	.1594	.9761	.52101	.85355	.61040	.6383	.1716	.9193	36
25	.47588	.87951	.54107	1.8482	.1370	2.1014	.49116	.87107	.56385	1.7735	.1480	2.0360	.50628	.86237	.58709	1.7033	.1596	1.9752	.52126	.85340	.61080	1.6372	.1718	1.9184	35
26	.47613	.87937	.54145	.8469	.1372	.1002	.49141	.87093	.56424	.7723	.1482	.0349	.50653	.86222	.58748	.7022	.1598	.9742	.52151	.85325	.61120	.6361	.1720	.9175	34
27	.47639	.87923	.54183	.8456	.1373	.0991	.49166	.87078	.56462	.7711	.1484	.0339	.50679	.86207	.58787	.7010	.1600	.9732	.52175	.85309	.61160	.6350	.1722	.9166	33
28	.47665	.87909	.54220	.8443	.1375	.0980	.49192	.87064	.56500	.7699	.1486	.0329	.50704	.86192	.58826	.6999	.1602	.9722	.52200	.85294	.61200	.6340	.1724	.9157	32
29	.47690	.87895	.54258	.8430	.1377	.0969	.49217	.87050	.56539	.7687	.1488	.0318	.50729	.86178	.58865	.6988	.1604	.9713	.52225	.85279	.61240	.6329	.1726	.9148	31
30	.47716	.87882	.54295	1.8418	.1379	2.0957	.49242	.87035	.56577	1.7675	.1489	2.0308	.50754	.86163	.58904	1.6977	.1606	1.9703	.52250	.85264	.61280	1.6318	.1728	1.9139	30
31	.47741	.87868	.54333	.8405	.1381	.0946	.49268	.87021	.56616	.7663	.1491	.0297	.50779	.86148	.58944	.6965	.1608	.9693	.52275	.85249	.61320	.6308	.1730	.9130	29
32	.47767	.87854	.54371	.8392	.1382	.0935	.49293	.87007	.56654	.7651	.1493	.0287	.50804	.86133	.58983	.6954	.1610	.9683	.52299	.85234	.61360	.6297	.1732	.9121	28
33	.47792	.87840	.54409	.8379	.1384	.0924	.49318	.86992	.56692	.7639	.1495	.0276	.50829	.86118	.59022	.6943	.1612	.9674	.52324	.85218	.61400	.6286	.1734	.9112	27
34	.47818	.87826	.54446	.8367	.1386	.0912	.49343	.86978	.56731	.7627	.1497	.0266	.50854	.86104	.59061	.6931	.1614	.9664	.52349	.85203	.61440	.6276	.1737	.9102	26
35	.47844	.87812	.54484	1.8354	.1388	2.0901	.49369	.86964	.56769	1.7615	.1499	2.0256	.50879	.86089	.59100	1.6920	.1616	1.9654	.52374	.85188	.61480	1.6265	.1739	1.9093	25
36	.47869	.87798	.54522	.8341	.1390	.0890	.49394	.86949	.56808	.7603	.1501	.0245	.50904	.86074	.59140	.6909	.1618	.9645	.52398	.85173	.61520	.6255	.1741	.9084	24
37	.47895	.87784	.54559	.8329	.1391	.0879	.49419	.86935	.56846	.7591	.1503	.0235	.50929	.86059	.59179	.6898	.1620	.9635	.52423	.85157	.61560	.6244	.1743	.9075	23
38	.47920	.87770	.54597	.8316	.1393	.0868	.49445	.86921	.56885	.7579	.1505	.0224	.50954	.86044	.59218	.6887	.1622	.9625	.52448	.85142	.61601	.6233	.1745	.9066	22
39	.47946	.87756	.54635	.8303	.1395	.0857	.49470	.86906	.56923	.7567	.1507	.0214	.50979	.86030	.59258	.6875	.1624	.9616	.52473	.85127	.61641	.6223	.1747	.9057	21
40	.47971	.87742	.54673	1.8291	.1397	2.0846	.49495	.86892	.56962	1.7555	.1508	2.0204	.51004	.86015	.59297	1.6864	.1626	1.9606	.52498	.85112	.61681	1.6212	.1749	1.9048	20
41	.47997	.87728	.54711	.8278	.1399	.0835	.49521	.86877	.57000	.7544	.1510	.0194	.51029	.86000	.59336	.6853	.1628	.9596	.52522	.85096	.61721	.6202	.1751	.9039	19
42	.48022	.87715	.54748	.8265	.1401	.0824	.49546	.86863	.57039	.7532	.1512	.0183	.51054	.85985	.59376	.6842	.1630	.9587	.52547	.85081	.61761	.6191	.1753	.9030	18
43	.48048	.87701	.54786	.8253	.1402	.0812	.49571	.86849	.57077	.7520	.1514	.0173	.51079	.85970	.59415	.6831	.1632	.9577	.52572	.85066	.61801	.6181	.1756	.9021	17
44	.48073	.87687	.54824	.8240	.1404	.0801	.49596	.86834	.57116	.7508	.1516	.0163	.51104	.85955	.59454	.6820	.1634	.9568	.52597	.85050	.61842	.6170	.1758	.9013	16
45	.48099	.87673	.54862	1.8227	.1406	2.0790	.49622	.86820	.57155	1.7496	.1518	2.0152	.51129	.85941	.59494	1.6808	.1636	1.9558	.52621	.85735	.61882	1.6160	.1760	1.9004	15
46	.48124	.87659	.54900	.8215	.1408	.0779	.49647	.86805	.57193	.7484	.1520	.0142	.51154	.85926	.59533	.6797	.1638	.9549	.52646	.85020	.61922	.6149	.1762	.8995	14
47	.48150	.87645	.54937	.8202	.1410	.0768	.49672	.86791	.57232	.7473	.1522	.0132	.51179	.85911	.59572	.6786	.1640	.9539	.52671	.85004	.61962	.6139	.1764	.8986	13
48	.48175	.87631	.54975	.8190	.1411	.0757	.49697	.86776	.57270	.7461	.1524	.0122	.51204	.85896	.59612	.6775	.1642	.9530	.52695	.84989	.62003	.6128	.1766	.8977	12
49	.48201	.87617	.55013	.8177	.1413	.0746	.49723	.86762	.57309	.7449	.1526	.0111	.51229	.85881	.59651	.6764	.1644	.9520	.52720	.84974	.62043	.6118	.1768	.8968	11
50	.48226	.87603	.55051	1.8165	.1415	2.0735	.49748	.86748	.57348	1.7437	.1528	2.0101	.51254	.85866	.59691	1.6753	.1646	1.9510	.52745	.84959	.62083	1.6107	.1770	1.8959	10
51	.48252	.87588	.55089	.8152	.1417	.0725	.49773	.86733	.57386	.7426	.1530	.0091	.51279	.85851	.59730	.6742	.1648	.9501	.52770	.84943	.62123	.6097	.1772	.8950	9
52	.48277	.87574	.55127	.8140	.1419	.0714	.49798	.86719	.57425	.7414	.1531	.0081	.51304	.85836	.59770	.6731	.1650	.9491	.52794	.84928	.62164	.6086	.1775	.8941	8
53	.48303	.87560	.55165	.8127	.1421	.0703	.49823	.86704	.57464	.7402	.1533	.0071	.51329	.85821	.59809	.6720	.1652	.9482	.52819	.84912	.62204	.6076	.1777	.8932	7
54	.48328	.87546	.55203	.8115	.1422	.0692	.49849	.86690	.57502	.7390	.1535	.0061	.51354	.85806	.59849	.6709	.1654	.9473	.52844	.84897	.62244	.6066	.1779	.8924	6
55	.48354	.87532	.55241	1.8102	.1424	2.0681	.49874	.86675	.57541	1.7379	.1537	2.0050	.51379	.85791	.59888	1.6698	.1656	1.9463	.52868	.84882	.62285	1.6055	.1781	1.8915	5
56	.48379	.87518	.55279	.8090	.1426	.0670	.49899	.86661	.57580	.7367	.1539	.0040	.51404	.85777	.59928	.6687	.1658	.9454	.52893	.84866	.62325	.6045	.1783	.8906	4
57	.48405	.87504	.55317	.8078	.1428	.0659	.49924	.86646	.57619	.7355	.1541	.0030	.51429	.85762	.59967	.6676	.1660	.9444	.52918	.84851	.62366	.6034	.1785	.8897	3
58	.48430	.87490	.55355	.8065	.1430	.0648	.49950	.86632	.57657	.7344	.1543	.0020	.51454	.85747	.60007	.6665	.1662	.9435	.52942	.84836	.62406	.6024	.1787	.8888	2
59	.48455	.87476	.55393	.8053	.1432	.0637	.49975	.86617	.57696	.7332	.1545	.0010	.51479	.85732	.60046	.6654	.1664	.9425	.52967	.84820	.62446	.6014	.1790	.8879	1
60	.48481	.87462	.55431	1.8040	.1433	2.0627	.50000	.86603	.57735	1.7320	.1547	2.0000	.51504	.85717	.60086	1.6643	.1666	1.9416	.52992	.84805	.62487	1.6003	.1792	1.8871	0
′	cos	sin	cot	tan	cosec	sec	cos	sin	cot	tan	cosec	sec	cos	sin	cot	tan	cosec	sec	cos	sin	cot	tan	cosec	sec	′
	61°						60°						59°						58°						

Table 19 Natural Trigonometric Functions (Continued)

′	32°						33°					
′	sin	cos	tan	cot	sec	cosec	sin	cos	tan	cot	sec	cosec
0	.52992	.84805	.62487	1.6003	1.1792	1.8871	.54464	.83867	.64941	1.5399	1.1924	1.8361
1	.53016	.84789	.62527	.5993	.1794	.8862	.54488	.83851	.64982	.5389	.1926	.8352
2	.53041	.84774	.62568	.5983	.1796	.8853	.54513	.83835	.65023	.5379	.1928	.8344
3	.53066	.84758	.62608	.5972	.1798	.8844	.54537	.83819	.65065	.5369	.1930	.8336
4	.53090	.84743	.62649	.5962	.1800	.8836	.54561	.83804	.65106	.5359	.1933	.8328
5	.53115	.84728	.62689	1.5952	.1802	1.8827	.54586	.83788	.65148	1.5350	.1935	1.8320
6	.53140	.84712	.62730	.5941	.1805	.8818	.54610	.83772	.65189	.5340	.1937	.8311
7	.53164	.84697	.62770	.5931	.1807	.8809	.54634	.83756	.65231	.5330	.1939	.8303
8	.53189	.84681	.62811	.5921	.1809	.8801	.54659	.83740	.65272	.5320	.1942	.8295
9	.53214	.84666	.62851	.5910	.1811	.8792	.54683	.83724	.65314	.5311	.1944	.8287
10	.53238	.84650	.62892	1.5900	.1813	1.8783	.54708	.83708	.65355	1.5301	.1946	1.8279
11	.53263	.84635	.62933	.5890	.1815	.8775	.54732	.83692	.65397	.5291	.1948	.8271
12	.53288	.84619	.62973	.5880	.1818	.8766	.54756	.83676	.65438	.5282	.1951	.8263
13	.53312	.84604	.63014	.5869	.1820	.8757	.54781	.83660	.65480	.5272	.1953	.8255
14	.53337	.84588	.63055	.5859	.1822	.8749	.54805	.83644	.65521	.5262	.1955	.8246
15	.53361	.84573	.63095	1.5849	.1824	1.8740	.54829	.83629	.65563	1.5252	.1958	1.8238
16	.53386	.84557	.63136	.5839	.1826	.8731	.54854	.83613	.65604	.5234	.1960	.8230
17	.53411	.84542	.63177	.5829	.1828	.8723	.54878	.83597	.65646	.5233	.1962	.8222
18	.53435	.84526	.63217	.5818	.1831	.8714	.54902	.83581	.65688	.5223	.1964	.8214
19	.53460	.84511	.63258	.5808	.1833	.8706	.54926	.83565	.65729	.5214	.1967	.8206
20	.53484	.84495	.63299	1.5798	.1835	1.8697	.54951	.83549	.65771	1.5204	.1969	1.8198
21	.53509	.84479	.63339	.5788	.1837	.8688	.54975	.83533	.65813	.5195	.1971	.8190
22	.53533	.84464	.63380	.5778	.1839	.8680	.54999	.83517	.65854	.5185	.1974	.8182
23	.53558	.84448	.63421	.5768	.1841	.8671	.55024	.83501	.65896	.5175	.1976	.8174
24	.53583	.84433	.63462	.5757	.1844	.8663	.55048	.83485	.65938	.5166	.1978	.8166
25	.53607	.84417	.63503	1.5747	.1846	1.8654	.55002	.83469	.65980	1.5156	.1980	1.8158
26	.53632	.84402	.63543	.5737	.1848	.8646	.55090	.83453	.66021	.5147	.1983	.8150
27	.53656	.84386	.63584	.5727	.1850	.8637	.55121	.83437	.66063	.5137	.1985	.8142
28	.53681	.84370	.63625	.5717	.1852	.8629	.55145	.83421	.66105	.5127	.1987	.8134
29	.53705	.84355	.63666	.5707	.1855	.8620	.55169	.83405	.66147	.5118	.1990	.8126
30	.53730	.84339	.63707	1.5697	.1857	1.8611	.55194	.83388	.56188	1.5108	.1992	1.8118
31	.53754	.84323	.63748	.5687	.1859	.8603	.55218	.83372	.66230	.5099	.1994	.8110
32	.53779	.84308	.63789	.5677	.1861	.8595	.55242	.83356	.66272	.5089	.1997	.8102
33	.53803	.84292	.63830	.5667	.1863	.8586	.55266	.83340	.66314	.5080	.1999	.8094
34	.53828	.84276	.63871	.5657	.1866	.8578	.55291	.83324	.66356	.5070	.2001	.8086
35	.53852	.84261	.63912	1.5646	.1868	1.8569	.55315	.83308	.66398	1.5061	.2004	1.8078
36	.53877	.84245	.63953	.5636	.1870	.8561	.55339	.83292	.66440	.5051	.2006	.8070
37	.53901	.84229	.63994	.5626	.1872	.8552	.55363	.83276	.66482	.5042	.2008	.8062
38	.53926	.84214	.64035	.5616	.1874	.8544	.55388	.83260	.66524	.5032	.2010	.8054
39	.53950	.84198	.64076	.5606	.1877	.8535	.55412	.83244	.66566	.5023	.2013	.8047
40	.53975	.84182	.64117	1.5596	.1879	1.8527	.55436	.83228	.66608	1.5013	.2015	1.8039
41	.53999	.84167	.64158	.5586	.1881	.8519	.55460	.83211	.66650	.5004	.2017	.8031
42	.54024	.84151	.64199	.5577	.1883	.8510	.55484	.83195	.66692	.4994	.2020	.8023
43	.54048	.84135	.64240	.5567	.1886	.8502	.55509	.83179	.66734	.4985	.2022	.8015
44	.54073	.84120	.64281	.5557	.1888	.8493	.55533	.83163	.66776	.4975	.2024	.8007
45	.54097	.84104	.64322	1.5547	.1890	1.8485	.55557	.83147	.66818	1.4966	.2027	1.7999
46	.54122	.84088	.64363	.5537	.1892	.8477	.55581	.83131	.66860	.4957	.2029	.7992
47	.54146	.84072	.64404	.5527	.1894	.8468	.55605	.83115	.66902	.4947	.2031	.7984
48	.54171	.84057	.64446	.5517	.1897	.8460	.55629	.83098	.66944	.4938	.2034	.7976
49	.54195	.84041	.64487	.5507	.1899	.8452	.55654	.83082	.66986	.4928	.2036	.7968
50	.54220	.84025	.64528	1.5497	.1901	1.8443	.55678	.83066	.67028	1.4919	.2039	1.7960
51	.54244	.84009	.64569	.5487	.1903	.8435	.55702	.83050	.67071	.4910	.2041	.7953
52	.54268	.83993	.64610	.5477	.1906	.8427	.55726	.83034	.67113	.4900	.2043	.7945
53	.54293	.83978	.64652	.5467	.1908	.8418	.55750	.83017	.67155	.4891	.2046	.7937
54	.54317	.83962	.64693	.5458	.1910	.8410	.55774	.83001	.67197	.4881	.2048	.7929
55	.54342	.83946	.64734	1.5448	.1912	1.8402	.55799	.82985	.67239	1.4872	.2050	1.7921
56	.54366	.83930	.64775	.5438	.1915	.8394	.55823	.82969	.67282	.4863	.2053	.7914
57	.54391	.83914	.64817	.5428	.1917	.8385	.55847	.82952	.67324	.4853	.2055	.7906
58	.54415	.83899	.64858	.5418	.1919	.8377	.55871	.82936	.67366	.4844	.2057	.7898
59	.54439	.83883	.64899	.5408	.1921	.8369	.55895	.82920	.67408	.4835	.2060	.7891
60	.54464	.83867	.64941	1.5399	.1922	1.8361	.55919	.82904	.67451	1.4826	.2062	1.7883
′	cos	sin	cot	tan	cosec	sec	cos	sin	cot	tan	cosec	sec
	57°						56°					

′	34°						35°						′
′	sin	cos	tan	cot	sec	cosec	sin	cos	tan	cot	sec	cosec	′
0	.55919	.82904	.67451	1.4826	1.2062	1.7883	.57358	.81915	.70021	1.4281	1.2208	1.7434	60
1	.55943	.82887	.67493	.4816	.2064	.7875	.57381	.81898	.70064	.4273	.2210	.7427	59
2	.55967	.82871	.67535	.4807	.2067	.7867	.57405	.81882	.70107	.4264	.2213	.7420	58
3	.55992	.82855	.67578	.4798	.2069	.7860	.57429	.81865	.70151	.4255	.2215	.7413	57
4	.56016	.82839	.67620	.4788	.2072	.7852	.57453	.81848	.70194	.4246	.2218	.7405	56
5	.56040	.82822	.67663	1.4779	.2074	1.7844	.57477	.81832	.70238	1.4237	.2220	1.7398	55
6	.56064	.82806	.67705	.4770	.2076	.7837	.57500	.81815	.70281	.4228	.2223	.7391	54
7	.56088	.82790	.67747	.4761	.2079	.7829	.57524	.81798	.70325	.4220	.2225	.7384	53
8	.56112	.82773	.67790	.4751	.2081	.7821	.57548	.81781	.70368	.4211	.2228	.7377	52
9	.56136	.82757	.67832	.4742	.2083	.7814	.57572	.81765	.70412	.4202	.2230	.7369	51
10	.56160	.82741	.67875	1.4733	.2086	1.7806	.57596	.81748	.70455	1.4193	.2233	1.7362	50
11	.56184	.82724	.67917	.4724	.2088	.7798	.57619	.81731	.70499	.4185	.2235	.7355	49
12	.56208	.82708	.67960	.4714	.2091	.7791	.57643	.81714	.70542	.4176	.2238	.7348	48
13	.56232	.82692	.68002	.4705	.2093	.7783	.57667	.81698	.70586	.4167	.2240	.7341	47
14	.56256	.82675	.68045	.4696	.2095	.7776	.57691	.81681	.70629	.4158	.2243	.7334	46
15	.56280	.82659	.68087	1.4687	.2098	1.7768	.57714	.81664	.70673	1.4150	.2245	1.7327	45
16	.56304	.82643	.68130	.4678	.2100	.7760	.57738	.81647	.70717	.4141	.2248	.7319	44
17	.56328	.82626	.68173	.4669	.2103	.7753	.57762	.81630	.70760	.4132	.2250	.7312	43
18	.56353	.82610	.68215	.4659	.2105	.7745	.57786	.81614	.70804	.4123	.2253	.7305	42
19	.56377	.82593	.68258	.4650	.2107	.7738	.57809	.81597	.70848	.4115	.2255	.7298	41
20	.56401	.82577	.68301	1.4641	.2110	1.7730	.57833	.81580	.70891	1.4106	.2258	1.7291	40
21	.56425	.82561	.68343	.4632	.2112	.7723	.57857	.81563	.70935	.4097	.2260	.7284	39
22	.56449	.82544	.68386	.4623	.2115	.7715	.57881	.81546	.70979	.4089	.2263	.7277	38
23	.56473	.82528	.68429	.4614	.2117	.7708	.57904	.81530	.71022	.4080	.2265	.7270	37
24	.56497	.82511	.68471	.4605	.2119	.7700	.57928	.81513	.71066	.4071	.2268	.7263	36
25	.56521	.82495	.68514	1.4595	.2122	1.7693	.57952	.81496	.71110	1.4063	.2270	1.7256	35
26	.56545	.82478	.68557	.4586	.2124	.7685	.57975	.81479	.71154	.4054	.2273	.7249	34
27	.56569	.82462	.68600	.4577	.2127	.7678	.57999	.81462	.71198	.4045	.2276	.7242	33
28	.56593	.82445	.68642	.4568	.2129	.7670	.58023	.81445	.71241	.4037	.2278	.7234	32
29	.56617	.82429	.68685	.4559	.2132	.7663	.58047	.81428	.71285	.4028	.2281	.7227	31
30	.56641	.82413	.68728	1.4550	.2134	1.7655	.58070	.81411	.71329	1.4019	.2283	1.7220	30
31	.56664	.82396	.68771	.4541	.2136	.7648	.58094	.81395	.71373	.4011	.2286	.7213	29
32	.56688	.82380	.68814	.4532	.2139	.7640	.58118	.81378	.71417	.4002	.2288	.7206	28
33	.56712	.82363	.68857	.4523	.2141	.7633	.58141	.81361	.71461	.3994	.2291	.7199	27
34	.56736	.82347	.68899	.4514	.2144	.7625	.58165	.81344	.71505	.3985	.2293	.7192	26
35	.56760	.82330	.68942	1.4505	.2146	1.7618	.58189	.81327	.71549	1.3976	.2296	1.7185	25
36	.56784	.82314	.68985	.4496	.2149	.7610	.58212	.81310	.71593	.3968	.2298	.7178	24
37	.56808	.82297	.69028	.4487	.2151	.7603	.58236	.81293	.71637	.3959	.2301	.7171	23
38	.56832	.82280	.69071	.4478	.2153	.7596	.58259	.81276	.71681	.3951	.2304	.7164	22
39	.56856	.82264	.69114	.4469	.2156	.7588	.58283	.81259	.71725	.3942	.2306	.7157	21
40	.56880	.82247	.69157	1.4460	.2158	1.7581	.58307	.81242	.71769	1.3933	.2309	1.7151	20
41	.56904	.82231	.69200	.4451	.2161	.7573	.58330	.81225	.71813	.3925	.2311	.7144	19
42	.56928	.82214	.69243	.4442	.2163	.7566	.58354	.81208	.71857	.3916	.2314	.7137	18
43	.56952	.82198	.69286	.4433	.2166	.7559	.58378	.81191	.71901	.3908	.2316	.7130	17
44	.56976	.82181	.69329	.4424	.2168	.7551	.58401	.81174	.71945	.3899	.2319	.7123	16
45	.57000	.82165	.69372	1.4415	.2171	1.7544	.58425	.81157	.71990	1.3891	.2322	1.7116	15
46	.57023	.82148	.69415	.4406	.2173	.7537	.58448	.81140	.72034	.3882	.2324	.7109	14
47	.57047	.82131	.69459	.4397	.2175	.7529	.58472	.81123	.72078	.3874	.2327	.7102	13
48	.57071	.82115	.69502	.4388	.2178	.7522	.58496	.81106	.72122	.3865	.2329	.7095	12
49	.57095	.82098	.69545	.4379	.2180	.7514	.58519	.81089	.72166	.3857	.2332	.7088	11
50	.57119	.82082	.69588	1.4370	.2183	1.7507	.58543	.81072	.72211	1.3848	.2335	1.7081	10
51	.57143	.82065	.69631	.4361	.2185	.7500	.58566	.81055	.72255	.3840	.2337	.7075	9
52	.57167	.82048	.69674	.4352	.2188	.7493	.58590	.81038	.72299	.3831	.2340	.7068	8
53	.57191	.82032	.69718	.4343	.2190	.7485	.58614	.81021	.72344	.3823	.2342	.7061	7
54	.57214	.82015	.69761	.4335	.2193	.7478	.58637	.81004	.72388	.3814	.2345	.7054	6
55	.57238	.81998	.69804	1.4326	.2195	1.7471	.58661	.80987	.72432	1.3806	.2348	1.7047	5
56	.57262	.81982	.69847	.4317	.2198	.7463	.58684	.80970	.72477	.3797	.2350	.7040	4
57	.57286	.81965	.69891	.4308	.2200	.7456	.58708	.80953	.72521	.3789	.2353	.7033	3
58	.57310	.81948	.69934	.4299	.2203	.7449	.58731	.80936	.72565	.3781	.2355	.7027	2
59	.57334	.81932	.69977	.4290	.2205	.7442	.58755	.80919	.72610	.3772	.2358	.7020	1
60	.57358	.81915	.70021	1.4281	.2208	1.7434	.58778	.80902	.72654	1.3764	.2361	1.7013	0
′	cos	sin	cot	tan	cosec	sec	cos	sin	cot	tan	cosec	sec	′
	55°						54°						

Table 19 Natural Trigonometric Functions (Continued)

36°

′	sin	cos	tan	cot	sec	cosec
0	.58778	.80902	.72654	1.3764	1.2361	1.7013
1	.58802	.80885	.72699	.3755	.2363	.7006
2	.58825	.80867	.72743	.3747	.2366	.6999
3	.58849	.80850	.72788	.3738	.2368	.6993
4	.58873	.80833	.72832	.3730	.2371	.6986
5	.58896	.80816	.72877	1.3722	.2374	1.6979
6	.58920	.80799	.72921	.3713	.2376	.6972
7	.58943	.80782	.72966	.3705	.2379	.6965
8	.58967	.80765	.73010	.3697	.2382	.6959
9	.58990	.80747	.73055	.3688	.2384	.6952
10	.59014	.80730	.73100	1.3680	.2387	1.6945
11	.59037	.80713	.73144	.3672	.2389	.6938
12	.59060	.80696	.73189	.3663	.2392	.6932
13	.59084	.80679	.73234	.3655	.2395	.6925
14	.59107	.80662	.73278	.3647	.2397	.6918
15	.59131	.80644	.73323	1.3638	.2400	1.6912
16	.59154	.80627	.73368	.3630	.2403	.6905
17	.59178	.80610	.73412	.3622	.2405	.6898
18	.59201	.80593	.73457	.3613	.2408	.6891
19	.59225	.80576	.73502	.3605	.2411	.6885
20	.59248	.80558	.73547	1.3597	.2413	1.6878
21	.59272	.80541	.73592	.3588	.2416	.6871
22	.59295	.80524	.73637	.3580	.2419	.6865
23	.59318	.80507	.73681	.3572	.2421	.6858
24	.59342	.80489	.73726	.3564	.2424	.6851
25	.59365	.80472	.73771	1.3555	.2427	1.6845
26	.59389	.80455	.73816	.3547	.2429	.6838
27	.59412	.80437	.73861	.3539	.2432	.6831
28	.59435	.80420	.73906	.3531	.2435	.6825
29	.59459	.80403	.73951	.3522	.2437	.6818
30	.59482	.80386	.73996	1.3514	.2440	1.6812
31	.59506	.80368	.74041	.3506	.2443	.6805
32	.59529	.80351	.74086	.3498	.2445	.6798
33	.59552	.80334	.74131	.3489	.2448	.6792
34	.59576	.80316	.74176	.3481	.2451	.6785
35	.59599	.80299	.74221	1.3473	.2453	1.6779
36	.59622	.80282	.74266	.3465	.2456	.6772
37	.59646	.80264	.74312	.3457	.2459	.6766
38	.59669	.80247	.74357	.3449	.2461	.6759
39	.59692	.80230	.74402	.3440	.2464	.6752
40	.59716	.80212	.74447	1.3432	.2467	1.6746
41	.59739	.80195	.74492	.3424	.2470	.6739
42	.59762	.80177	.74538	.3416	.2472	.6733
43	.59786	.80160	.74583	.3408	.2475	.6726
44	.59809	.80143	.74628	.3400	.2478	.6720
45	.59832	.80125	.74673	1.3392	.2480	1.6713
46	.59856	.80108	.74019	.3383	.2483	.6707
47	.59879	.80090	.74064	.3375	.2486	.6700
48	.59902	.80073	.74809	.3367	.2488	.6694
49	.59926	.80056	.74855	.3359	.2491	.6687
50	.59949	.80038	.74900	1.3351	.2494	1.6681
51	.59972	.80021	.74946	.3343	.2497	.6674
52	.59995	.80003	.74991	.3335	.2499	.6668
53	.60019	.79986	.75037	.3327	.2502	.6661
54	.60042	.79968	.75082	.3319	.2505	.6655
55	.60065	.79951	.75128	1.3311	.2508	1.6648
56	.60088	.79933	.75173	.3303	.2510	.6642
57	.60112	.79916	.75219	.3294	.2513	.6636
58	.60135	.79898	.75264	.3286	.2516	.6629
59	.60158	.79881	.75310	.3278	.2519	.6623
60	.60181	.79863	.75355	1.3270	.2521	1.6616
′	cos	sin	cot	tan	cosec	sec

53°

37°

′	sin	cos	tan	cot	sec	cosec
0	.60181	.79863	.75355	1.3270	1.2521	1.6616
1	.60205	.79846	.75401	.3262	.2524	.6610
2	.60228	.79828	.75447	.3254	.2527	.6603
3	.60251	.79811	.75492	.3246	.2530	.6597
4	.60274	.79793	.75538	.3238	.2532	.6591
5	.60298	.79776	.75584	1.3230	.2535	1.6584
6	.60320	.79758	.75629	.3222	.2538	.6578
7	.60344	.79741	.75675	.3214	.2541	.6572
8	.60367	.79723	.75721	.3206	.2543	.6565
9	.60390	.79706	.75767	.3198	.2546	.6559
10	.60413	.79688	.75812	1.3190	.2549	1.6552
11	.60437	.79670	.75858	.3182	.2552	.6546
12	.60460	.79653	.75904	.3174	.2554	.6540
13	.60483	.79635	.75950	.3166	.2557	.6533
14	.60506	.79618	.75996	.3159	.2560	.6527
15	.60529	.79600	.76042	1.3151	.2563	1.6521
16	.60552	.79582	.76088	.3143	.2565	.6514
17	.60576	.79565	.76134	.3135	.2568	.6508
18	.60599	.79547	.76179	.3127	.2571	.6502
19	.60622	.79530	.76225	.3119	.2574	.6496
20	.60645	.79512	.76271	1.3111	.2577	1.6489
21	.60668	.79494	.76317	.3103	.2579	.6483
22	.60691	.79477	.76364	.3095	.2582	.6477
23	.60714	.79459	.76410	.3087	.2585	.6470
24	.60737	.79441	.76456	.3079	.2588	.6464
25	.60761	.79424	.76502	1.3071	.2591	1.6458
26	.60784	.79406	.76548	.3064	.2593	.6452
27	.60807	.79388	.76594	.3056	.2596	.6445
28	.60830	.79371	.76640	.3048	.2599	.6439
29	.60853	.79353	.76686	.3040	.2602	.6433
30	.60876	.79335	.76733	1.3032	.2605	1.6427
31	.60899	.79318	.76779	.3024	.2607	.6420
32	.60922	.79300	.76825	.3016	.2610	.6414
33	.60945	.79282	.76871	.3009	.2613	.6408
34	.60968	.79264	.76918	.3001	.2616	.6402
35	.60991	.79247	.76964	1.2993	.2619	1.6396
36	.61014	.79229	.77010	.2985	.2622	.6389
37	.61037	.79211	.77057	.2977	.2624	.6383
38	.61061	.79193	.77103	.2970	.2627	.6377
39	.61084	.79176	.77149	.2962	.2630	.6371
40	.61107	.79158	.77196	1.2954	.2633	1.6365
41	.61130	.79140	.77242	.2946	.2636	.6359
42	.61153	.79122	.77289	.2938	.2639	.6352
43	.61176	.79104	.77335	.2931	.2641	.6346
44	.61199	.79087	.77382	.2923	.2644	.6340
45	.61222	.79069	.77428	1.2915	.2647	1.6334
46	.61245	.79051	.77475	.2907	.2650	.6328
47	.61268	.79033	.77521	.2900	.2653	.6322
48	.61290	.79015	.77568	.2892	.2656	.6316
49	.61314	.78998	.77614	.2884	.2659	.6309
50	.61337	.78980	.77661	1.2876	.2661	1.6303
51	.61360	.78962	.77708	.2869	.2664	.6297
52	.61383	.78944	.77754	.2861	.2667	.6291
53	.61405	.78926	.77801	.2853	.2670	.6285
54	.61428	.78908	.77848	.2845	.2673	.6279
55	.61451	.78890	.77895	1.2838	.2676	1.6273
56	.61474	.78873	.77941	.2830	.2679	.6267
57	.61497	.78855	.77988	.2822	.2681	.6261
58	.61520	.78837	.78035	.2815	.2684	.6255
59	.61543	.78819	.78082	.2807	.2687	.6249
60	.61566	.78801	.78128	1.2799	.2690	1.6243
′	cos	sin	cot	tan	cosec	sec

52°

38°

′	sin	cos	tan	cot	sec	cosec
0	.61566	.78801	.78128	1.2799	1.2690	1.6243
1	.61589	.78783	.78175	.2792	.2693	.6237
2	.61612	.78765	.78222	.2784	.2696	.6231
3	.61635	.78747	.78269	.2776	.2699	.6224
4	.61658	.78729	.78316	.2769	.2702	.6218
5	.61681	.78711	.78363	1.2761	.2705	1.6212
6	.61703	.78693	.78410	.2753	.2707	.6206
7	.61726	.78675	.78457	.2746	.2710	.6200
8	.61749	.78657	.78504	.2738	.2713	.6194
9	.61772	.78640	.78551	.2730	.2716	.6188
10	.61795	.78622	.78598	1.2723	.2719	1.6182
11	.61818	.78604	.78645	.2715	.2722	.6176
12	.61841	.78586	.78692	.2708	.2725	.6170
13	.61864	.78568	.78739	.2700	.2728	.6164
14	.61886	.78550	.78786	.2692	.2731	.6159
15	.61909	.78532	.78834	1.2685	.2734	1.6153
16	.61932	.78514	.78881	.2677	.2737	.6147
17	.61955	.78496	.78928	.2670	.2739	.6141
18	.61978	.78478	.78975	.2662	.2742	.6135
19	.62001	.78460	.79022	.2655	.2745	.6129
20	.62023	.78441	.79070	1.2647	.2748	1.6123
21	.62046	.78423	.79117	.2639	.2751	.6117
22	.62069	.78405	.79164	.2632	.2754	.6111
23	.62092	.78387	.79212	.2624	.2757	.6105
24	.62115	.78369	.79259	.2617	.2760	.6099
25	.62137	.78351	.79306	1.2609	.2763	1.6093
26	.62160	.78333	.79354	.2602	.2766	.6087
27	.62183	.78315	.79401	.2594	.2769	.6081
28	.62206	.78297	.79449	.2587	.2772	.6077
29	.62229	.78279	.79496	.2579	.2775	.6070
30	.62251	.78261	.79543	1.2572	.2778	1.6064
31	.62274	.78243	.79591	.2564	.2781	.6058
32	.62297	.78224	.79639	.2557	.2784	.6052
33	.62320	.78206	.79686	.2549	.2787	.6046
34	.62342	.78188	.79734	.2542	.2790	.6040
35	.62365	.78170	.79781	1.2534	.2793	1.6034
36	.62388	.78152	.79829	.2527	.2795	.6029
37	.62411	.78134	.79876	.2519	.2798	.6023
38	.62433	.78116	.79924	.2512	.2801	.6017
39	.62456	.78097	.79972	.2504	.2804	.6011
40	.62479	.78079	.80020	1.2497	.2807	1.6005
41	.62501	.78061	.80067	.2489	.2810	.6000
42	.62524	.78043	.80115	.2482	.2813	.5994
43	.62547	.78025	.80163	.2475	.2816	.5988
44	.62570	.78007	.80211	.2467	.2819	.5982
45	.62592	.77988	.80258	1.2460	.2822	1.5976
46	.62615	.77970	.80306	.2452	.2825	.5971
47	.62638	.77952	.80354	.2445	.2828	.5965
48	.62660	.77934	.80402	.2437	.2831	.5959
49	.62683	.77915	.80450	.2430	.2834	.5957
50	.62706	.77897	.80498	1.2423	.2837	1.594
51	.62728	.77879	.80546	.2415	.2840	.5942
52	.62751	.77861	.80594	.2408	.2843	.5936
53	.62774	.77842	.80642	.2400	.2846	.5930
54	.62796	.77824	.80690	.2393	.2849	.5924
55	.62819	.77806	.80738	1.2386	.2852	1.5919
56	.62841	.77788	.80786	.2378	.2855	.5913
57	.62864	.77769	.80834	.2371	.2858	.5907
58	.62887	.77751	.80882	.2364	.2861	.5901
59	.62909	.77733	.80930	.2356	.2864	.5896
60	62932	.77715	.80978	1.2349	.2867	1.5890
′	cos	sin	cot	tan	cosec	sec

51°

39°

′	sin	cos	tan	cot	sec	cosec	′
0	.62932	.77715	.80978	1.2349	1.2867	1.5890	60
1	.62955	.77696	.81026	.2342	.2871	.5884	59
2	.62977	.77678	.81075	.2334	.2874	.5879	58
3	.63000	.77660	.81123	.2327	.2877	.5873	57
4	.63022	.77641	.81171	.2320	.2880	.5867	56
5	.63045	.77623	.81219	1.2312	.2883	1.5862	55
6	.63067	.77605	.81268	.2305	.2886	.5856	54
7	.63090	.77586	.81316	.2297	.2889	.5850	53
8	.63113	.77568	.81364	.2290	.2892	.5854	52
9	.63135	.77549	.81413	.2283	.2895	.5839	51
10	.63158	.77531	.81461	1.2276	.2898	1.5833	50
11	.63180	.77513	.81509	.2268	.2901	.5828	49
12	.63203	.77494	.81558	.2261	.2904	.5822	48
13	.63225	.77476	.81606	.2254	.2907	.5816	47
14	.63248	.77458	.81655	.2247	.2910	.5811	46
15	.63270	.77439	.81703	1.2239	.2913	1.5805	45
16	.63293	.77421	.81752	.2232	.2916	.5799	44
17	.63315	.77402	.81800	.2225	.2919	.5794	43
18	.63338	.77384	.81849	.2218	.2922	.5788	42
19	.63360	.77365	.81898	.2210	.2926	.5783	41
20	.63383	.77347	.81946	1.2203	.2929	1.5777	40
21	.63405	.77329	.81995	.2196	.2932	.5771	39
22	.63428	.77310	.82043	.2189	.2935	.5766	38
23	.63450	.77292	.82092	.2181	.2938	.5760	37
24	.63473	.77273	.82141	.2174	.2941	.5755	36
25	.63495	.77255	.82190	1.2167	.2944	1.5749	35
26	.63518	.77236	.82238	.2160	.2947	.5743	34
27	.63540	.77218	.82287	.2152	.2950	.5738	33
28	.63563	.77199	.82336	.2145	.2953	.5732	32
29	.63585	.77181	.82385	.2138	.2956	.5727	31
30	.63608	.77162	.82434	1.2131	.2960	1.5721	30
31	.63630	.77144	.82482	.2124	.2963	.5716	29
32	.63653	.77125	.82531	.2117	.2966	.5710	28
33	.63675	.77107	.82580	.2109	.2969	.5705	27
34	.63697	.77088	.82629	.2102	.2972	.5699	26
35	.63720	.77070	.82678	1.2095	.2975	1.5694	25
36	.63742	.77051	.82727	.2088	.2978	.5688	24
37	.63765	.77033	.82776	.2081	.2981	.5683	23
38	.63787	.77014	.82825	.2074	.2985	.5677	22
39	.63810	.76996	.82874	.2066	.2988	.5672	21
40	.63832	.76977	.82923	1.2059	.2991	1.5666	20
41	.63854	.76958	.82972	.2052	.2994	.5661	19
42	.63877	.76940	.83022	.2045	.2997	.5655	18
43	.63899	.76921	.83071	.2038	.3000	.5650	17
44	.63921	.76903	.83120	.2031	.3003	.5644	16
45	.63944	.76884	.83169	1.2024	.3006	1.5639	15
46	.63966	.76865	.83218	.2016	.3010	.5633	14
47	.63989	.76847	.83267	.2009	.3013	.5628	13
48	.64011	.76828	.83317	.2002	.3016	.5622	12
49	.64033	.76810	.83366	.1995	.3019	.5617	11
50	.64056	.76791	.83415	1.1988	.3022	1.5611	10
51	.64078	.76772	.83465	.1981	.3025	.5606	9
52	.64100	.76754	.83514	.1974	.3029	.5600	8
53	.64123	.76735	.83563	.1967	.3032	.5595	7
54	.64145	.76716	.83613	.1960	.3035	.5590	6
55	.64160	.76698	.83662	1.1953	.3038	1.5584	5
56	.64189	.76679	.83712	.1946	.3041	.5579	4
57	.64212	.76660	.83761	.1939	.3044	.5573	3
58	.64234	.76642	.83811	.1932	.3048	.5568	2
59	.64256	.76623	.83860	.1924	.3051	.5563	1
60	.64279	.76604	.83910	1.1917	.3054	1.5557	0
′	cos	sin	cot	tan	cosec	sec	′

50°

Table 19 Natural Trigonometric Functions (Continued)

	40°						41°						42°						43°						
′	sin	cos	tan	cot	sec	cosec	sin	cos	tan	cot	sec	cosec	sin	cos	tan	cot	sec	cosec	sin	cos	tan	cot	sec	cosec	′
0	.64279	.76604	.83910	1.1917	1.3054	1.5557	.65606	.75471	.86929	1.1504	1.3250	1.5242	.66913	.74314	.90040	1.1106	1.3456	1.4945	.68200	.73135	.93251	1.0724	1.3673	1.4663	60
1	.64301	.76586	.83959	.1910	.3057	.5552	.65628	.75452	.86980	.1497	.3253	.5237	.66935	.74295	.90093	.1100	.3460	.4940	.68221	.73115	.93306	.0717	.3677	.4658	59
2	.64323	.76567	.84009	.1903	.3060	.5546	.65650	.75433	.87031	.1490	.3257	.5232	.66956	.74276	.90146	.1093	.3463	.4935	.68242	.73096	.93360	.0711	.3681	.4654	58
3	.64345	.76548	.84059	.1896	.3064	.5541	.65672	.75414	.87082	.1483	.3260	.5227	.66978	.74256	.90198	.1086	.3467	.4930	.68264	.73076	.93415	.0705	.3684	.4649	57
4	.64368	.76530	.84108	.1889	.3067	.5536	.65694	.75394	.87133	.1477	.3263	.5222	.66999	.74236	.90251	.1080	.3470	.4925	.68285	.73056	.93469	.0699	.3688	.4644	56
5	.64390	.76511	.84158	1.1882	.3070	1.5530	.65716	.75375	.87184	1.1470	.3267	1.5217	.67021	.74217	.90304	1.1074	.3474	1.4921	.68306	.73036	.93524	1.0692	.3692	1.4640	55
6	.64412	.76492	.84208	.1875	.3073	.5525	.65037	.75356	.87235	.1463	.3270	.5212	.67043	.74197	.90357	.1067	.3477	.4916	.68327	.73016	.93578	.0686	.3695	.4635	54
7	.64435	.76473	.84257	.1868	.3076	.5520	.65759	.75337	.87287	.1456	.3274	.5207	.67064	.74178	.90410	.1061	.3481	.4911	.68349	.72996	.93633	.0680	.3699	.4631	53
8	.64457	.76455	.84307	.1861	.3080	.5514	.65781	.75318	.87338	.1450	.3277	.5202	.67086	.74158	.90463	.1054	.3485	.4906	.68370	.72976	.93687	.0674	.3703	.4626	52
9	.64479	.76436	.84357	.1854	.3083	.5509	.65803	.75299	.87389	.1443	.3280	.5197	.67107	.74139	.90515	.1048	.3488	.4901	.68391	.72956	.93742	.0667	.3707	.4622	51
10	.64501	.76417	.84407	1.1847	.3086	1.5503	.65285	.75280	.87441	1.1436	.3284	1.5192	.67129	.74119	.90568	1.1041	.3492	1.4897	.68412	.72937	.93797	1.0661	.3710	1.4617	50
11	.64523	.76398	.84457	.1840	.3089	.5498	.65847	.75261	.87492	.1430	.3287	.5187	.67150	.74100	.90621	.1035	.3495	.4892	.68433	.72917	.93851	.0655	.3714	.4613	49
12	.64546	.76380	.84506	.1833	.3092	.5493	.65869	.75241	.87543	.1423	.3290	.5182	.67172	.74080	.90674	.1028	.3499	.4887	.68455	.72897	.93906	.0649	.3718	.4608	48
13	.64568	.76361	.84556	.1826	.3096	.5487	.65891	.75222	.87595	.1416	.3294	.5177	.67194	.74061	.90727	.1022	.3502	.4882	.68476	.72877	.93961	.0643	.3722	.4604	47
14	.64590	.76342	.84606	.1819	.3099	.5482	.65913	.75203	.87646	.1409	.3297	.5171	.67215	.74041	.90780	.1015	.3506	.4877	.68497	.72857	.94016	.0636	.3725	.4599	46
15	.64612	.76323	.84656	1.1812	.3102	1.5477	.65934	.75184	.87698	1.1403	.3301	1.5166	.67237	.74022	.90834	1.1009	.3509	1.4873	.68518	.72837	.94071	1.0630	.3729	1.4595	45
16	.64635	.76304	.84706	.1805	.3105	.5471	.65956	.75165	.87749	.1396	.3304	.5161	.67258	.74002	.90887	.1003	.3513	.4868	.68539	.72817	.94125	.0624	.3733	.4590	44
17	.64657	.76286	.84056	.1798	.3109	.5466	.65978	.75146	.87801	.1389	.3307	.5156	.67280	.73983	.90940	.0996	.3517	.4863	.68561	.72797	.94180	.0618	.3737	.4586	43
18	.64679	.76267	.84806	.1791	.3112	.5461	.66000	.75126	.87852	.1383	.3311	.5151	.67301	.73963	.90993	.0990	.3520	.4858	.68582	.72777	.94235	.0612	.3740	.4581	42
19	.64701	.76248	.84856	.1785	.3115	.5456	.66022	.75107	.87904	.1376	.3314	.5146	.67323	.73943	.91046	.0983	.3524	.4854	.68603	.72757	.94290	.0605	.3744	.4577	41
20	.64723	.76229	.84906	1.1778	.3118	1.5450	.66044	.75088	.87955	1.1369	.3318	1.5141	.67344	.73924	.91099	1.0977	.3527	1.4849	.68624	.72737	.94345	1.0599	.3748	1.4572	40
21	.64745	.76210	.84956	.1771	.3121	.5445	.66066	.75069	.88007	.1363	.3321	.5136	.67366	.73904	.91153	.0971	.3531	.4844	.68645	.72717	.94400	.0593	.3752	.4568	39
22	.64768	.76191	.85006	.1764	.3125	.5440	.66087	.75049	.88058	.1356	.3324	.5131	.67387	.73885	.91206	.0964	.3534	.4839	.68666	.72697	.94455	.0587	.3756	.4563	38
23	.64790	.76173	.85056	.1757	.3128	.5434	.66109	.75030	.88110	.1349	.3328	.5126	.67409	.73865	.91259	.0958	.3538	.4835	.68688	.72677	.94510	.0581	.3759	.4559	37
24	.64812	.76154	.85107	.1750	.3131	.5429	.66131	.75011	.88162	.1343	.3331	.5121	.67430	.73845	.91312	.0951	.3542	.4830	.68709	.72657	.94565	.0575	.3763	.4554	36
25	.64834	.76135	.85157	1.1743	.3134	1.5424	.66153	.74992	.88213	1.1336	.3335	1.5116	.67452	.73826	.91366	1.0945	.3545	1.4825	.68730	.72637	.94620	1.0568	.3767	1.4550	35
26	.64856	.76116	.85207	.1736	.3138	.5419	.66175	.74973	.88265	.1329	.3338	.5111	.67473	.73806	.91419	.0939	.3549	.4821	.68751	.72617	.94675	.0562	.3771	.4545	34
27	.64878	.76097	.85257	.1729	.3141	.5413	.66197	.74953	.88317	.1323	.3342	.5106	.67495	.73787	.91473	.0932	.3552	.4816	.68772	.72597	.94731	.0556	.3774	.4541	33
28	.64900	.76078	.85307	.1722	.3144	.5408	.66218	.74934	.88369	.1316	.3345	.5101	.67516	.73767	.91526	.0926	.3556	.4811	.68793	.72577	.94786	.0550	.3778	.4536	32
29	.64923	.76059	.85358	.1715	.3148	.5403	.66240	.74915	.88421	.1309	.3348	.5096	.67537	.73747	.91580	.0919	.3560	.4806	.68814	.72557	.94841	.0544	.3782	.4532	31
30	.64945	.76041	.85408	1.1708	.3151	1.5398	.66262	.74895	.88472	1.1303	.3352	1.5092	.67559	.73728	.91633	1.0913	.3563	1.4802	.68835	.72537	.94896	1.0538	.3786	1.4527	30
31	.64967	.76022	.85458	.1702	.3154	.5392	.66284	.74876	.88524	.1296	.3355	.5087	.67580	.73708	.91687	.0907	.3567	.4797	.68856	.72517	.94952	.0532	.3790	.4523	29
32	.64989	.76003	.85509	.1695	.3157	.5387	.66305	.74857	.88576	.1290	.3359	.5082	.67602	.73688	.91740	.0900	.3571	.4792	.68878	.72497	.95007	.0525	.3794	.4518	28
33	.65011	.75984	.85559	.1688	.3161	.5382	.66327	.74838	.88628	.1283	.3362	.5077	.67623	.73669	.91794	.0894	.3574	.4788	.68899	.72477	.95062	.0519	.3797	.4514	27
34	.65033	.75965	.85609	.1681	.3164	.5377	.66349	.74818	.88680	.1276	.3366	.5072	.67645	.73649	.91847	.0888	.3578	.4783	.68920	.72457	.95118	.0513	.3801	.4510	26
35	.65055	.75946	.85660	1.1674	.3167	1.5371	.66371	.74799	.88732	1.1270	.3369	1.5067	.67666	.73629	.91901	1.0881	.3581	1.4778	.68941	.72437	.95173	1.0507	.3805	1.4505	25
36	.65077	.75927	.85710	.1667	.3170	.5366	.66393	.74780	.88784	.1263	.3372	.5062	.67688	.73610	.91955	.0875	.3585	.4774	.68962	.72417	.95229	.0501	.3809	.4501	24
37	.65100	.75908	.85761	.1660	.3174	.5361	.66414	.74760	.88836	.1257	.3376	.5057	.67709	.73590	.92008	.0868	.3589	.4769	.68983	.72397	.95284	.0495	.3813	.4496	23
38	.65121	.75889	.85811	.1653	.3177	.5356	.66436	.74741	.88888	.1250	.3379	.5052	.67730	.73570	.92062	.0862	.3592	.4764	.69004	.72377	.95340	.0489	.3816	.4492	22
39	.65144	.75870	.85862	.1647	.3180	.5351	.66458	.74722	.88940	.1243	.3383	.5047	.67752	.73551	.92116	.0856	.3596	.4760	.69025	.72357	.95395	.0483	.3820	.4487	21
40	.65166	.75851	.85912	1.1640	.3184	1.5345	.66479	.74702	.88992	1.1237	.3386	1.5042	.67773	.73531	.92170	1.0849	.3600	1.4755	.69046	.72337	.95451	1.0476	.3824	1.4483	20
41	.65188	.75832	.85963	.1633	.3187	.5340	.66501	.74683	.89044	.1230	.3390	.5037	.67794	.73511	.92223	.0843	.3603	.4750	.69067	.72317	.95506	.0470	.3828	.4479	19
42	.65210	.75813	.86013	.1626	.3190	.5335	.66523	.74664	.89097	.1224	.3393	.5032	.67816	.73491	.92277	.0837	.3607	.4746	.69088	.72297	.95562	.0464	.3832	.4474	18
43	.65232	.75794	.86064	.1619	.3193	.5330	.66545	.74644	.89149	.1217	.3397	.5027	.67837	.73472	.92331	.0830	.3611	.4741	.69109	.72277	.95618	.0458	.3836	.4470	17
44	.65254	.75775	.86115	.1612	.3197	.5325	.66566	.74625	.89201	.1211	.3400	.5022	.67859	.73452	.92385	.0824	.3614	.4736	.69130	.72256	.95673	.0452	.3839	.4465	16
45	.65276	.75756	.86165	1.1605	.3200	1.5319	.66588	.74606	.89253	1.1204	1.3404	1.5018	.67880	.73432	.92439	1.0818	.3618	1.4732	.69151	.72236	.95729	1.0446	.3843	1.4461	15
46	.65298	.75737	.86216	.1599	.3203	.5314	.66610	.74586	.89306	.1197	.3407	.5013	.67901	.73412	.92493	.0812	.3622	.4727	.69172	.72216	.95785	.0440	.3847	.4457	14
47	.65320	.75718	.86267	.1592	.3207	.5309	.66631	.74567	.89358	.1191	.3411	.5008	.67923	.73393	.92547	.0805	.3625	.4723	.69193	.72196	.95841	.0434	.3851	.4452	13
48	.65342	.75700	.86318	.1585	.3210	.5304	.66653	.74548	.89410	.1184	.3414	.5003	.67944	.73373	.92601	.0799	.3629	.4718	.69214	.72176	.95896	.0428	.3855	.4448	12
49	.65364	.75680	.86368	.1578	.3213	.5299	.66675	.74528	.89463	.1178	.3418	.4998	.67965	.73353	.92655	.0793	.3633	.4713	.69235	.72156	.95952	.0422	.3859	.4443	11
50	.65386	.75661	.86419	1.1571	.3217	1.5294	.66697	.74509	.89515	1.1171	.3421	1.4993	.67987	.73333	.92709	1.0786	.3636	1.4709	.69256	.72136	.96008	1.0416	.3863	1.4439	10
51	.65408	.75642	.86470	.1565	.3220	.5289	.66718	.74489	.89567	.1165	.3425	.4988	.68008	.73314	.92763	.0780	.3640	.4704	.69277	.72115	.96064	.0410	.3867	.4435	9
52	.65430	.75623	.86521	.1558	.3223	.5283	.66740	.74470	.89620	.1158	.3428	.4983	.68029	.73294	.92817	.0774	.3644	.4699	.69298	.72095	.96120	.0404	.3870	.4430	8
53	.65452	.75604	.86572	.1551	.3227	.5278	.66762	.74450	.89672	.1152	.3432	.4979	.68051	.73274	.92871	.0767	.3647	.4695	.69319	.72075	.96176	.0397	.3874	.4426	7
54	.65474	.75585	.86623	.1544	.3230	.5273	.66783	.74431	.89725	.1145	.3435	.4974	.68072	.73254	.92926	.0761	.3651	.4690	.69340	.72055	.96232	.0391	.3878	.4422	6
55	.65496	.75566	.86674	1.1537	.3233	1.5268	.66805	.74412	.89777	1.1139	.3439	1.4969	.68093	.73234	.92980	1.0755	.3655	1.4686	.69361	.72035	.96288	1.0385	.3882	1.4417	5
56	.65518	.75547	.86725	.1531	.3237	.5263	.66826	.74392	.89830	.1132	.3442	.4964	.68115	.73215	.93034	.0749	.3658	.4681	.69382	.72015	.96344	.0379	.3886	.4413	4
57	.65540	.75528	.86775	.1524	.3240	.5258	.66848	.74373	.89882	.1126	.3446	.4959	.68136	.73195	.93088	.0742	.3662	.4676	.69403	.71994	.96400	.0373	.3890	.4408	3
58	.65562	.75509	.86826	.1517	.3243	.5253	.66870	.74353	.89935	.1119	.3449	.4954	.68157	.73175	.93143	.0736	.3666	.4672	.69424	.71974	.96456	.0367	.3894	.4404	2
59	.65584	.75490	.86878	.1510	.3247	.5248	.66891	.74334	.89988	.1113	.3453	.4949	.68178	.73155	.93197	.0730	.3669	.4667	.69445	.71954	.96513	.0361	.3898	.4400	1
60	.65606	.75401	.86929	1.1504	.3250	1.5242	.66913	.74314	.90040	1.1106	.3456	1.4945	.68200	.73135	.93251	1.0724	.3673	1.4663	.69466	.71934	.96569	1.0355	.3902	1.4395	0
′	cos	sin	cot	tan	cosec	sec	cos	sin	cot	tan	cosec	sec	cos	sin	cot	tan	cosec	sec	cos	sin	cot	tan	cosec	sec	′
	49°						48°						47°						46°						

Table 19 Natural Trigonometric Functions (Continued)

′	sin	cos	tan	cot	sec	cosec	′
0	.69466	.71934	.96569	1.0355	1.3902	1.4395	60
1	.69487	.71914	.96625	.0349	.3905	.4391	59
2	.69508	.71893	.96681	.0343	.3909	.4387	58
3	.69528	.71873	.96738	.0337	.3913	.4382	57
4	.69549	.71853	.96794	.0331	.3917	.4378	56
5	.69570	.71833	.96850	1.0325	.3921	1.4374	55
6	.69591	.71813	.96907	.0319	.3925	.4370	54
7	.69612	.71792	.96963	.0313	.3929	.4365	53
8	.69633	.71772	.97020	.0307	.3933	.4361	52
9	.69654	.71752	.97076	.0301	.3937	.4357	51
10	.69675	.71732	.97133	1.0295	.3941	1.4352	50
11	.69696	.71711	.97189	.0289	.3945	.4348	49
12	.69716	.71691	.97246	.0283	.3949	.4344	48
13	.69737	.71671	.97302	.0277	.3953	.4339	47
14	.69758	.71650	.97359	.0271	.3957	.4335	46
15	.69779	.71630	.97416	1.0265	.3960	1.4331	45
16	.69800	.71610	.97472	.0259	.3964	.4327	44
17	.69821	.71589	.97529	.0253	.3968	.4322	43
18	.69841	.71569	.97586	.0247	.3972	.4318	42
19	.69862	.71549	.97643	.0241	.3976	.4314	41
20	.69883	.71529	.97700	1.0235	.3980	1.4310	40
21	.69904	.71508	.97756	.0229	.3984	.4305	39
22	.69925	.71488	.97813	.0223	.3988	.4301	38
23	.69945	.71468	.97870	.0218	.3992	.4297	37
24	.69966	.71447	.97927	.0212	.3996	.4292	36
25	.69987	.71427	.97984	1.0206	.4000	1.4288	35
26	.70008	.71406	.98041	.0200	.4004	.4284	34
27	.70029	.71386	.98098	.0194	.4008	.4280	33
28	.70049	.71366	.98155	.0188	.4012	.4276	32
29	.70070	.71345	.98212	.0182	.4016	.4271	31
30	.70091	.71325	.98270	1.0176	.4020	1.4267	30
31	.70112	.71305	.98327	.0170	.4024	.4263	29
32	.70132	.71284	.98384	.0164	.4028	.4259	28
33	.70153	.71264	.98441	.0158	.4032	.4254	27
34	.70174	.71243	.98499	.0152	.4036	.4250	26
35	.70194	.71223	.98556	1.0146	.4040	1.4246	25
36	.70215	.71203	.98613	.0141	.4044	.4242	24
37	.70236	.71182	.98671	.0135	.4048	.4238	23
38	.70257	.71162	.98728	.0129	.4052	.4233	22
39	.70277	.71141	.98786	.0123	.4056	.4229	21
40	.70298	.71121	.98843	1.0117	.4060	1.4225	20
41	.70319	.71100	.98901	.0111	.4065	.4221	19
42	.70339	.71080	.98958	.0105	.4069	.4217	18
43	.70360	.71059	.99016	.0099	.4073	.4212	17
44	.70381	.71039	.99073	.0093	.4077	.4208	16
45	.70401	.71018	.99131	1.0088	.4081	1.4204	15
46	.70422	.70998	.99189	.0082	.4085	.4200	14
47	.70443	.70977	.99246	.0076	.4089	.4196	13
48	.70463	.70957	.99304	.0070	.4093	.4192	12
49	.70484	.70936	.99362	.0064	.4097	.4188	11
50	.70505	.70916	.99420	1.0058	.4101	1.4183	10
51	.70525	.70895	.99478	.0052	.4105	.4179	9
52	.70546	.70875	.99536	.0017	.4109	.4175	8
53	.70566	.70854	.99593	.0011	.4113	.4171	7
54	.70587	.70834	.99651	.0035	.4117	.4167	6
55	.70608	.70813	.99709	1.0029	.4122	1.4163	5
56	.70628	.70793	.99767	.0023	.4126	.4159	4
57	.70649	.70772	.99826	.0017	.4130	.4154	3
58	.70669	.70752	.99884	.0012	.4134	.4150	2
59	.70690	.70731	.99942	.0006	.4138	.4146	1
60	.70711	.70711	1.00000	1.0000	.4142	1.4142	0
′	cos	sin	cot	tan	cosec	sec	′

$45°$

Glossary

Abrasion Resistance The property of a cutting tool material to resist the wear and abrasion that occurs during a metal-removal process.

Abrasive The material used in making grinding wheels or abrasive cloth; it may be either natural or artificial. The natural abrasives are emery and corundum. The artificial abrasives are silicon carbide and aluminum oxide.

Abrasive Cloth, Coated Abrasives A strong flexible backing to which aluminum oxide or silicon carbide grains are cemented. Used to improve the surface finish of a workpiece.

Abrasive Grain A part of a grinding wheel that can vary from coarse to fine, depending on the job at hand.

Absolute Positioning Positioning where all dimensions are taken from one fixed point or location.

Absolute System A CNC system where all positional dimensions are taken from the same common datum point.

Active Cutting Oils These oils contain sulfur that is not firmly attached to the oil, therefore it is released during the machining operation and reacts with the work surface.

Agile Manufacturing Combines state-of-the-art fabrication and product delivery techniques to custom-made products suiting each customer's taste, specifications, and budget.

Alignment Linear accuracy, uniformity, or coincidence of the centers of a lathe; a straight line of adjustment through two or more points. Setting the lathe in alignment means adjusting the tailstock in line with the headstock spindle.

Allowance Allowance refers to a difference in dimensions prescribed in order to secure classes of fits. It is the minimum or maximum interference intentionally permitted between mating parts.

Alloy A mixture of two or more metals melted together. As a rule, when two or more metals are melted together to form an alloy, the substance formed is a new metal.

Aluminum A very light silvery-white metal used independently or in alloys with copper and other metals.

Annealing The process of relieving internal stresses and softening steel by heating it above its critical temperature and allowing it to cool slowly in a closed furnace.

Apprentice A person who is bound by an agreement to learn a trade or business under the guidance of a skilled craftsperson.

Arbor A short shaft or spindle on which an object may be mounted. Spindles or supports for milling machine cutters and saws are called arbors.

Artificial Intelligence (AI) A new manufacturing tool that combines the use of artificial vision, expert systems, robotics, natural language understanding, and voice recognition.

Automated-Guided Vehicle (AGV) A computer controlled material-handling device used in Flexible Manufacturing Systems to supply materials and tools to machines, and to remove the finished product.

Automatic Identification System A system in which data is collected automatically and produced by advanced technology, such as laser-read bar coding, machine vision, and radio frequency identification.

Automatic Tool Changing A system in which preset tools necessary to machine a part are changed automatically as required by a command in the computer program.

Automation A manufacturing approach in which all or part of a machining or manufacturing process is accomplished by setting in motion a sequence of events that completes the process without further human intervention.

Bar Coding A data-collection system that can be used to monitor quality control on assembly lines and collect information through multiple work stations.

Basic-Oxygen Furnace A cylindrical brick-lined furnace that uses a water-cooled oxygen lance to create a turbulent, churning action to burn out undesirable elements in the steel.

Basic Size The size from which limits of size are based when applying allowances and tolerances.

Bastard A coarse-cut file, not as rough as a first-cut, used to remove material quickly.

Bauxite A white to red earthy aluminum hydroxide largely used in the preparation of aluminum and alumina and for the lining of furnaces that are exposed to intense heat.

Belt Grinder See Grinder (Belt).

Bench Grinder See Grinder (Bench).

Blast The volume of air forced into furnaces where combustion is hastened artificially.

Blind Hole A hole that does not go through a workpiece.

Bore The internal diameter of a pipe, cylinder, or hole.

Boring The operation of making or finishing circular holes in metal, usually done with a boring tool.

Buffing The operation of finishing a surface using a soft cotton wheel or a belt with some very fine abrasive.

Burr A thin edge on a machined or ground surface left by the cutting tool.

Bushing A sleeve or liner for a bearing. Some bushings can be adjusted to compensate for wear.

CAD (Computer Aided Design) The use of computers in various stages in the design of a product or component.

Caliper A tool for measuring the inside and outside diameter of cylindrical work or the thickness of flat work.

CAM (Computer Aided Manufacturing) The use of computers to control machining and manufacturing.

Cam-Lock Spindle A type of spindle nose that has holes to receive the notched holding pins of a lathe spindle accessory. A chuck wrench is used to tighten the cam-lock screws on the spindle taper.

Canned Cycle A preset sequence of machine movements that are initiated by a single command in the CNC program. For example, G81 will perform a drilling cycle.

Carburizing The process of increasing the carbon content of low-carbon steel by heating the metal below the melting point while it is in contact with carbonaceous material.

Cartesian Coordinate System A dimensional-locating system in which the position of any point can be defined with reference to a set of axes (X and Y) at right angles to each other.

Casehardening A process of forming a thin, hard film on the surface of low-carbon steel by heating it while it is in contact with some carburizing material.

Cast Iron A term covering a large family of cast ferrous alloys containing at least 2% carbon, which exceeds the solubility of carbon in austenite at the eutectic temperature.

CBN (Cubic Boron Nitride) A super-hard, super wear-resistant abrasive used to machine or grind hard, abrasive, difficult-to-cut ferrous metals.

Cellular Manufacturing A process where machinery, machining processes, and people are grouped together to manufacture a specific family of parts.

Cemented Carbide A very hard metal carbide cemented together with a little cobalt as a binder, to form a cutting edge nearly as hard as a diamond.

Center Drill A short drill, used for centering work so it may be supported by lathe centers. Center drills are usually made in combination with a countersink, that permits a double operation with one tool.

Center Gage A gage used to align a threading toolbit for thread cutting in a lathe.

Center Head A tool used for finding the center of a circle or an arc of a circle. It is most frequently used to find the center of a cylindrical piece of metal.

Center-Point Programming A method of programming arcs and circles in which the arc is defined by the end point of the arc and the arc center point coordinates (I and J).

Centrifugal Force The force used to prevent a body from trying to move away from the center of its path. Centrifugal force tries to move chuck jaws outward as the speed is increased.

Cermet A cutting tool material consisting of ceramic particles bonded with a metal. Cermets, more shock-resistant than ceramics, are used for high-speed machining.

Chamfer A bevelled edge or a cut-off corner.

Chatter Caused by lack of rigidity in the cutting tools or in machine bearings or parts while machining work.

Chip Pressure The force exerted on a cutting tool when removing material during machining.

Chip-Producing Machine Tools Tools that produce parts to size and shape by cutting away metal in the form of chips. These types of machines, usually called standard or conventional machine tools, can include some types of Computer Numerical Control (CNC) machines.

Chip-Tool Interface The area where the cutting tool contacts the workpiece and a chip is produced.

Chisel Edge The end of a drill, formed at the web where the two cutting edges meet, that does not have a very efficient cutting action.

Chuck A holding device, used on many machine tools to hold workpieces on a lathe spindle, or hold cutting tools in a tailstock or on a drill press spindle.

Chucking Center A chucking center is similar to a turning center, but workpieces are held in some form of chuck or holding fixture.

Circular Interpolation A mode of contouring control that allows arcs or circles up to 360° to be produced using only one block of programmed information.

Climb Milling A milling operation in which the cutter rotation and the machine table (work) feed are in the same direction.

CNC (Computer Numerical Control) A system consisting of letters, numbers, and symbols used to program the basic motions of CNC machines.

Coarseness A term referring to the pitch or spacing of the saw teeth. A coarse pitch blade will cut faster than a fine pitch blade.

Cold Chisel An all-steel chisel without a handle, used for chipping metals.

Collets Tools used to hold and drive cutting tools such as end mills. The two most common types are the spring collet and the solid collet.

Compensation The control function that adjusts the tool's programmed path to make allowance for differences in cutter diameter and tool-nose radii for cuts using linear or circular interpolation on the X and Y axes.

Computer-Aided Design See CAD.

Computer-Aided Manufacturing See CAM.

Computer-Aided Process Planning (CAPP) An approach to using a computer to perform the computational work involved in programming tools to perform manufacturing tasks.

Computer-Integrated Manufacturing (CIM) A computer system controlling all phases of manufacturing from product design to product shipment.

Computer Numerical Control See CNC.

Continuous Casting A casting technique where liquid metal is continuously fed through a mold where it is formed into shape and solidifies.

Continuous Corporate Renewal A management of change and vision, that includes continuous change, rapid response, and evolving quality standards.

Continuous-Path Control A system on almost all modern CNC controls in which the drive motors of the X and Y axes can operate at different rates of speed.

Continuous-Path Machining An operation in which there is no pause for data reading, and the rate and direction of movement of machine slides is under continuous CNC control.

Continuous-Path Positioning The ability to control motions on two or more machine axes simultaneously to keep a constant cutter-workpiece relationship.

Contour Bandsaw A vertical sawing machine, using an endless saw band, that can cut intricate shapes and contours on a workpiece.

Contouring A CNC system for controlling the path of a machine tool that results from the coordinated movement of more than one axis of motion.

Conventional Milling A milling operation where the cutter rotation and the machine table (work) feed are going in opposite directions.

Coordinate Measuring Machine (CMM) An inspection machine capable of making many very accurate measurements of three-dimensional objects in a short period of time.

Copper A soft, ductile, malleable metal second only to silver for electrical conductivity. It is also the basis of all the alloys known as brass and bronze.

Corundum A natural abrasive material used in place of emery.

Counterbore A tool used to enlarge a hole through part of its length.

Countersink A tool used to recess a hole conically for the head of a screw or rivet.

Critical Temperature The temperature at which certain changes take place in the chemical composition of steel during heating and cooling.

Crossfeed A transverse feed. In a lathe, that which usually operates at right angles to the axis of the work.

Cubic Boron Nitride (CBN) A material harder than any other material except diamond, manufactured from hexagonal boron nitride under high-pressure and high-temperature conditions.

Customized Manufacturing Same as Agile Manufacturing.

Cutting Fluids Various fluids used in machining and grinding operations to assist the cutting action by cooling and lubricating the cutting tool and workpiece.

Cutting Off The operation of using a narrow blade parting tool in a lathe to cut off work held in a chuck or to produce narrow grooves.

Cutting Speed The speed in feet or meters per minute at which the cutting tool passes the work, or vice versa.

Dial Caliper A caliper with a dial indicator that provides a direct reading in inches or millimeters.

Dial Indicator A comparison measuring instrument where any movement of the spindle or plunger is magnified on a dial usually graduated in .001 in. (0.02 mm) or less.

Diamond (manufactured) Developed by the General Electric Co. in 1954, by putting a form of carbon and a catalyst under high pressure and high temperature to form diamond.

Die An internal screw used for cutting outside thread on cylindrical work.

Dielectric Fluid The fluid used in EDM to control the arc discharge in the erosion gap.

Digital-Readout System Calibrated machined spars attached to the main machine tool slides and a reading head that transfers any slide movement to the digital readout display.

Digitizing The operation of tracing a drawing with a cursor or light pen that is then automatically entered into the computer memory.

Direct Computer Control An automated inspection process where the computer makes all the necessary calculations and inspects the part as per preprogrammed instructions.

Direct Indexing See Indexing (Direct).

Direct Steelmaking A steelmaking process designed to bypass the blast furnace and cooking ovens to produce steel directly from iron ore.

Dog A clamp-type device that is fastened to work held between centers to connect the work to a positive drive in a milling machine or an engine lathe.

Draw Chisel A pointed cold chisel, usually diamond shaped, used for shifting the center of a hole being drilled to the correct location.

Dressing The operation of removing dull abrasive grains and metal particles from a grinding wheel to make it cut better.

Drift A strip of steel, rectangular in section, wedge-shaped in its length, used for driving drill sockets from their spindles or sleeves.

Drill Chuck A holding device, generally with three jaws, used to hold and drive straight-shank cutting tools.

Drill Press A machine tool used for drilling operations available in a wide variety of types and sizes, in different types and sizes of workpieces.

Drilling The operation of using a twist drill to produce a hole where none existed before.

Drill Sleeve A tool having an internal and external taper that can adapt a cutting tool shank to fit into a machine spindle.

Drill Socket The socket that receives the tapered shank of a drill or reamer.

Dynamic Gripping Force The chuck jaw force exerted on the workpiece when the machine spindle is running.

Electrical Discharge Machining (EDM) A process that removes metal by controlled electric spark erosion.

Electric-Arc Furnace A controlled-atmospheric furnace that uses carbon electrodes to generate the heat required to produce fine alloy and tool steels.

Electro-Optical Measuring Tools Tools used for a variety of workpiece measurements, such as dimensions, angles, contours, and surface conditions. They can be one of two types: contact measuring tools or non-contact measuring tools.

Emery Cloth Powdered emery glued onto cloth, used for removing file marks and for polishing metallic surfaces.

Emulsifier Any liquid that mixes with water to form a cutting fluid used to improve machinability, prolong tool life, and cool the cutting tool and workpiece.

End Mill A milling cutter, usually smaller than 1 in.

(25.4 mm) in diameter, with straight or tapered shanks. The cutting portion is cylindrical in shape, made so that it can cut both on the sides and the end.

Engine Lathe A basic machine tool that uses a stationary cutting tool to machine round forms on a revolving workpiece.

Expert Systems A form of AI that uses a knowledge base to solve problems formerly requiring some form of human expertise.

Extreme Pressure Oils Cutting fluid additives (chlorine, sulfur, or phosphorous compounds) that react with the work material to reduce built-up edges and are good for high-speed machining.

Faceplate A circular plate for attachment to the spindle in the headstock of a lathe. Work can be clamped or bolted to its surface for machining.

Facing The operation of machining surfaces at 90° to the centerline of the lathe spindle axis.

Feed The longitudinal movement of a tool in inches per minute or thousandths of an inch per revolution. Metric feeds are stated in millimeters per minute or hundredths of a millimeter per revolution.

Feeler A gage, consisting of a number of leaves in varying thicknesses of thousandths of an inch or hundredths of a millimeter, used for setting tools and measuring narrow openings.

Ferrous Metals Metals containing ferrite or iron.

File A hand-cutting tool with many teeth used to remove burrs, sharp edges, and surplus metal, and produce finished surfaces.

Fillet A concave or radius surface joining two adjacent faces of an article to strengthen the joint, as between two diameters on a shaft.

Finishing Machining a surface to size with a fine feed produced in a lathe, milling machine, or grinder.

Fit The range of tightness between two mating parts, generally classified as clearance fit and interference fit.

Fixture A special device designed and built for holding a particular piece of work for machining operations.

Flame Hardening A process in which an intense flame is used to bring the surface of hardenable ferrous alloys above their upper critical temperature, followed by rapid cooling.

Flexible Manufacturing System (FMS) A manufacturing system consisting of a number of CNC machine tools, serviced by a material-handling system, under the control of one or more dedicated computers.

Flycutter A tool consisting of a shank that fits into a milling machine spindle and a body into which are fastened a number of single point or indexable tools used for machining flat surfaces.

Gage A tool used for checking dimensions of a job. A surface gage can also be used to lay out for machining.

Gage Block The accepted world physical standard of measurement, gage blocks give the inch or millimeter physical form and allow them to be used as a means of calibration or measurement.

Gage Height A predetermined reference or partial retraction point along the Z axis that allows the cutting tool to retract above the work surface to allow clearance for travel in the X Y axes.

General Shop A shop that is usually associated with a manufacturing plant, a school lab, or a foundry.

Goto Command A programming technique that causes the program to return to a certain line to repeat the same commands.

Grade Refers to the degree of strength with which a bond holds the abrasive particles in a grinding wheel. The grade symbols range from A (softest) to Z (hardest).

Graduated Collars Devices used on machine tools to accurately locate the workpiece or cutting tool in a horizontal or longitudinal position and to set vertically for a depth of cut.

Grinder (Belt) A finishing machine, equipped with an endless aluminum oxide or silicon carbide abrasive belt, used primarily for finishing metal surfaces.

Grinder (Bench) A bench or pedestal machine, usually equipped with a coarse and fine wheel, used for offhand grinding of cutting tools and the rough grinding of metal.

Grinder (Surface) A precision machine used to finish the surface of hardened metal pieces to close tolerances and a high surface finish.

Gripper (End Effector) The device on the end of a robot arm, which may be some form of gripping finger or vacuum cup, that is controlled by the computer program to pick up and move material.

Grooving The operation of machining a square, round, or V-shaped groove at the end of a shoulder or thread.

Group Technology A system for classifying parts into families on the basis of their similarities in shape, holes, threads, and machining operations to improve manufacturing productivity.

Hacksaw A hand tool used to cut metal.

Hammer A hand tool used to strike metal. Examples of hammers used in the shop are the ball-peen hammer, the layout hammer, and the soft-faced hammer.

Hardening The process of heating steel above its lower critical temperature and quickly quenching it in water, oil, or air to produce a martensitic grain structure.

Headstock The part of a lathe containing the driving mechanism for the main spindle.

Heat Treating A process using controlled heating and cooling of metals and their alloys to produce certain properties in the metal.

Height Gage A precision instrument used in toolrooms and inspection rooms to lay out and measure vertical distances to .001 in. (0.025 mm) accuracy.

Helical Interpolation A control function that combines the use of two-axes circular interpolation with a linear movement in the third axis to form a helical path.

Histogram A quality-control tool used as a final inspection for lot-size parts to see if a process is holding tolerance and where it lies in relation to the minimum and maximum limits.

Horizontal Bandsaw A horizontal sawing machine that uses an endless saw band to cut workpieces to length.

Horizontal Milling Machine See Milling Machine (Horizontal).

Inactive Cutting Oils These oils contain sulfur that is firmly attached to the oil so very little is released during machining to react with the work surface.

Inch Micrometer A precision instrument used to measure accurately to within .001 in.

Inch System The standard of measurement for North American industry.

Inch Taper A uniform increase or decrease in the diameter of a piece of work measured along its length, expressed in taper per foot or taper per inch.

Incremental Positioning Postioning where the dimension for the next location is given from the previous point.

Incremental System A CNC dimensional-control system where any coordinate or dimension is given from the previous point.

Index Grating A unit with an index grating having the same line structure as the reference scale, which is set on a slight angle on top of the reference scale. As the unit moves along the scale it produces an integrated interference or Moire fringe pattern.

Index Head A device, also known as a dividing head, used on milling machines to divide the circumference of a workpiece into a number of equal divisions for cutting gear teeth, sprockets, etc.

Indexing (Direct) A form of indexing in which a plunger pin fits into numbered slots of an index plate for dividing the circumference of a part into 2 to 36 equal divisions.

Indexing (Simple) A form of indexing that engages a 40 tooth spindle gear with a worm wheel to accurately divide the circumference of a part into a wide range of divisions.

Induction Hardening The rapid heating of a part surrounded by a high-frequency electric coil to above its critical temperature for a few seconds, followed by rapid cooling.

In-Process Gaging A system using probes, lasers, optical devices, etc., to measure or inspect work while it is being machined.

Jig A device that holds and locates a piece of work and guides the tools that operate upon it.

Job Shop A shop generally equipped with a variety of standard and CNC machine tools. It generally does not have a product of its own to manufacture; it does a variety of work for other companies.

Just-In-Time A system where materials are available at the time they are needed for production. Its purpose is to improve productivity, reduce costs, reduce scrap and rework, overcome the shortage of machines, reduce inventory and work-in-process, and use manufacturing space efficiently.

Keyway A groove, usually rectangularly cut in a shaft or hub, keyed and fitted for a driving purpose.

Knowledge-Base Building Blocks The base of an Expert System consisting of a large body of information that has been assembled, widely shared, accepted by experts, and a diagnostic computer software package.

Knurling The operation of impressing diamond or straight-line patterns on a diameter to improve the appearance and to provide a better grip.

Laser The acronym of Light Amplification by Stimulated Emission of Radiation used in applications of material processing, identification and tracking, communications, and precision measurement.

Laser (Excimer) The fastest-growing class of lasers using a mixture of noble gas and halogen gas to remove material quickly from the machining area.

Lasercaving A process similar to milling using lasers to produce cavities in material too hard to mill.

Lathe Centers 60° high speed steel or carbide insert centers used in the headstock and tailstock to support work held between lathe centers.

Laying Out An important first step before any machining operations take place. The two main types are the semi-precision layout and the precision layout.

Leadscrew A threaded shaft running longitudinally in front of the lathe bed.

Limits of Tolerance The limits of accuracy, over or under size, in which a part must be kept to be acceptable.

Linear Interpolation A control function where data points are generated between coordinate positions to allow two or more axes of motion in a straight (linear) path.

Linear Logic This is when a program goes from the beginning command and follows each program step in sequence to the end.

Load Balancing A means of distributing the work load so that manufacturing takes place efficiently on the available machine tools.

Loop A program designed to repeat a group of commands as often as they are required in the program. Dumb loops repeat the same motions until stopped manually; smart loops repeat themselves only a specific number of times.

Machinability The ease at which a metal is machined or ground.

Machine Control Unit A system of hardware and software that controls the operation of a CNC machine.

Machining Center A machine tool, similar to a milling machine, that is CNC controlled to perform a wide variety of machining operations automatically under the control of the part program.

Macro Same as subroutine.

Mainframe Computer A large-scale data processing system that can store billions of bytes of information

in the main memory and can process millions of instructions per second. It is a company's main computer and may have many individual keyboards using its information.

Mandrel A shaft or spindle on which an object may be fixed for rotation, such as that used when a piece is to be machined in a lathe between centers.

Manufacturing Cell A group of machines combined to perform all operations on a part before it leaves the cell.

Mass Production A manufacturing process that can include specialized and single-purpose machines to produce a great many identical parts.

Material-Handling System Various accessories such as a bar feeder, loader/unloader, robot, etc., that are used to handle material to increase the machine's productivity.

Mesh The engagement of teeth or gears of a sprocket and chain.

Metallurgy The art or science of separating metals from ores by smelting or alloying. The study of metals.

Metal Stamp A stencil used to mark or identify workpieces.

Metric Micrometer A precision instrument used to measure accurately to within 0.002 mm.

Metric System The standard of length for most countries.

Metric Taper A uniform increase or decrease in the diameter of a piece of work measured along its length, expressed as a ratio of 1 mm per unit of length.

Microcrystalline The structure of an abrasive grain where each crystal is composed of many micron or submicrometer crystallites.

Micrometer Collar The graduated collars provided on machine slide screws to make accurate movements in inches or millimeters.

Microstructure A term referring to the structural characteristics such as size, shape, and arrangement of the crystals present in a metal or alloy.

Milling Cutter A rotary cutting tool, with a number of equally-spaced teeth around its periphery, generally used on a horizontal milling machine for a variety of milling operations.

Milling Machine (Horizontal) A machine tool used to produce flats, contours, gears, cams, sprocket teeth, etc., with a rotary cutter mounted on a horizontal arbor.

Milling Machine (Vertical) A machine tool, with a cutting tool mounted in a vertical spindle, used for machining flat surfaces, contours, and face and end milling operations.

Minicomputer A small machine that has all the basic processing and storage functions of a computer with limited capabilities.

Minimills These are smaller, faster, and more efficient steel mills that bypass the coke and iron-making steps of large steel mills.

Modal Code A code that stays in effect in the program

until it is replaced or changed by another code of the same group number.

Modular Tooling A complete tooling system combining the flexibility and versatility to build a series of tools necessary to produce a part. Modular systems combine accuracy and quick-change capabilities to increase productivity.

Monocrystalline A single abrasive crystal with a regular arrangement of atoms in all three directions.

Natural Language Generally refers to whether a machine can understand and interpret a natural language.

New-Generation Machine Tools Machine tools, such as electro-discharge and laser machines, that form or cut parts to size and shape using either chemical or electrical energy.

Non-Chip Producing Machine Tools Machines that form metal to size and shape by a pressing, drawing, bending, extruding, or shearing action. Examples of these machines are the punch press, forming press, hobbing press, etc.

Nonferrous Alloys A combination of two or more nonferrous metals completely dissolved in each other, having the properties of the original metals.

Nonferrous Metals Metals containing little or no ferrite or iron, resistant to corrosion, and nonmagnetic.

Nonmodal Code A code that stays in effect for only one operation and must be replaced in the program whenever required.

Normalizing Heating steel to just above its critical temperature and cooling it in still air to improve the grain structure and remove stresses and strains.

Numerical Control A system composed of a control program, a control unit, and a machine tool that can control the actions of a machine through coded-command instructions.

Offhand Grinding The use of bench or pedestal grinders for the sharpening of cutting tools and the rough grinding of metal. The work is usually held in the hand.

Orthographic View A view used by a draftsperson to show an object as seen from three different views: the top, front, and right-side.

Parabolic Interpolation A method used to program complex contour forms especially free-form designs used in mold work and automotive die sculpturing.

Parallel A straight, rectangular bar of uniform thickness or width, used for setting up work in the same plane as a fixed surface.

Part Families An approach to classifying standard parts based on their physical characteristics so they can be grouped together with similar parts that could be produced using the same manufacturing process.

Pawl A pin having a pointed edge or hook, made to engage with ratchet teeth.

Pedestal Grinder A grinder used for the sharpening of cutting tools and the rough grinding of metals.

Periphery The line bounding a round surface, such as the circumference of a wheel.

Pilot Hole A small hole drilled to guide and allow free passage for the thickness of the web of a twist drill.

Pinion The smaller of two gears in mesh, generally the driving gear.

Pitch The distance from the center of one thread or gear tooth to the corresponding point on the next thread or tooth. For threads, it is measured parallel to the axis. For gear teeth, it is measured on the pitch circle.

Point-to-Point Positioning A system that moves the machine spindle (tool) rapidly to a location while the tool is not in contact with the workpiece.

Polar Coordinates A control function used to calculate and define rotary and angular movements when the arc radius and the angle of movement are provided.

Polar Reference Line The base line, usually the X+ axis, that divides the top and bottom as well as the left and right portions of the cartesian coordinate system.

Polishing The operation of improving the surface finish on a workpiece by using abrasive cloth or some other media.

Production Shop A shop, usually associated with a large plant or factory, that makes many types of machined parts.

Productivity The output realized when human and material resources are used most efficiently.

Quadrant Any one of the four sections divided by the rectangular coordinate axes (X and Y) in that plane.

Quality Control A list of procedures, programs, and activities designed to gather and analyze data so that the manufacturing process does not produce defective products.

Quick-Change Gearbox A lathe part containing a number of different-sized gears to provide the feed rod and leadscrew with various feed rates for turning and thread-cutting operations.

Rack A straight strip of metal having teeth to engage with those of a gear wheel, as in a rack and pinion.

Radius Programming A method of programming arcs and circles that requires only the coordinates for the arc/circle center and the required radius.

Rake Angle The angle that provides keenness to the tool's cutting edge.

Rapid Protoyping and Manufacturing (RP&M) Same as stereolithography.

Rapid Response The ability of a company to respond quickly to any manufacturing challenge to suit customer needs or changes in the market conditions.

Reamer A tool used to enlarge, smooth, and size a hole that has been drilled or bored.

Reference Scale A part of a coordinate measuring machine containing a graduated spar; movements of which are measured to a high degree of accuracy by a fibre-optic laser system.

Rivet A soft metal pin with a head at one end used to permanently fasten metal parts together.

Robot Generally a single-arm device that can be programmed to automatically move tools, parts, materials, or to perform a variety of tasks.

Robotics The study of robots.

Roughing An operation done before finishing to remove surplus stock rapidly where fine surface finish is not important.

Run-Out The amount a saw blade will run out of square as it cuts through a material. Dull blades will have more run-out than sharp blades.

S.A.E. (Society of Automotive Engineers) These letters are used to indicate that the article or measurement is approved by the Society of Automotive Engineers.

Safety Glasses Protective eye devices that must be worn in the shop at all times.

Safety Shoes Shoes with steel-reinforced toes. Many companies require employees to wear safety shoes.

Scale A thin surface on castings or rolled metal caused by burning, oxidizing, or cooling.

Scale Size Used on most shop or engineering drawings to indicate the ratio of the drawing size to the actual size of the part.

Section View A view that involves an imaginary cut through an object to produce a view of the cross section.

Sensors Devices used in a robot gripper to obtain information about the part it is handling or the work environment. With this information it can modify its movements as necessary, as in the alignment or placement of parts.

Sequence of Operations A carefully planned set of procedures used to produce a part quickly and accurately.

Serration A series of grooves produced in metal to provide a grip or locking action. Vise jaws are often serrated.

Servomechanism A control mechanism used in EDM machines to feed the electrode into the part, maintain the work/electrode gap, and slow or speed up the drive motors as required.

Shoulder Turning The operation of machining a square, round, or angular form on a workpiece where one diameter meets another.

Side Milling A method often used to machine a vertical surface on the sides or the ends of a workpiece.

Simple Indexing See Indexing (Simple).

Spark Erosion A material-removal process that uses an electrical discharge of short duration and high-current density between the electrode and the material to be cut.

Soaking Heating metal at a uniform heat for a period of time for complete penetration.

Spark Testing A process of identifying the composition of a metal by the characteristics of the sparks produced during a grinding process.

Spindle Speed The number of r/min. that is made by the spindle (cutting tool) of a machine.

Spotting An operation done on a lathe with a toolbit or a center drill.

Static Gripping Force The chuck-jaw force exerted on a workpiece when the machine spindle is stopped.

Statistical Control Chart A chart containing points plotted in a time sequence of individual measurements or summary values obtained from process data which define the state of statistical control of a process.

Statistical Process Control (SPC) A Quality Assurance method of using performance data to identify product and process errors that lead to the production of faulty goods. Correct analysis of this data should lead to correcting the errors and produce only acceptable goods.

Stereolithography (SL) The process that takes a CAD design and makes a solid three-dimensional prototype (model) using a combination of laser, photochemistry, optical scanning, and computer software technology.

Strategic Planning A listing of existing skills, resources, and capabilities that can be compared to the manufacturing needs of the future.

Subroutine A series of computer programming statements or commands that perform frequently used operations. (Same as macro).

Superabrasives A term referring to manufactured diamond and cubic boron nitride abrasives used to make super-hard, super wear-resistant cutting tools and grinding wheels.

Surface Finish The surface characteristics of a piece, including waviness, roughness, lay, and flaw.

Surface Gage A machinist's gage consisting of a heavy base and a scriber for marking in layout for machining.

Surface Grinder See Grinder (Surface).

Surface Plate A cast iron scraped plate used for layout work. Granite plates are also used.

Tailstock The lathe part that fits on the bed (ways) and is used for supporting long workpieces.

Tap An accurate, hardened cutting tool used to produce internal threads.

Taper A uniform change in diameter of a workpiece measured along its length.

Taper Spindle Nose This generally refers to the tapered section on the end of the headstock spindle on which accessories such as chucks, faceplates, drive plates, etc., are mounted.

Tensile Strength The resistance of steel or iron to a lengthwise pull.

Tempering The process of heating hardened steel to below its critical temperature, followed by a rate of slow cooling to reduce the brittleness and toughness.

Thermal Conductivity The characteristics of a material defining its capabilities to transfer heat.

Thread A helical form of uniform section formed on the inside or outside of a cylinder or cone.

Thread-Chasing Dial A revolving dial, fastened to the

lathe carriage, indicating when to engage the split-nut lever to take successive cuts for thread cutting.

Threading The act of producing internal or external threads on a workpiece.

Through Hole A hole that goes through a workpiece.

Tolerance The amount of interference required for two or more parts that are in contact. The amount of variation, over or under the required size, permitted on a part.

Tombstone Fixture Vertical or horizontal workholding fixtures, that may be square, hexagonal or vertical, used on machining centers to hold parts for machining.

Toolbit A single-point cutting tool, made of high speed steel, cemented carbide, ceramic, or superabrasive material, used to machine metal.

Tool-Monitoring System The most common tool-monitoring system measures the force or load on the machine spindle required for a machining operation. Once the load is exceeded, which indicates a dull tool, the feed and speed are generally reduced automatically.

Toolpost The accessory that fits on a lathe compound rest to hold toolholders, toolbits, or other types of cutting tools.

Truing The operation of making the periphery of a job or tool concentric with the axis of rotation. When applied to a grinding wheel, it makes the wheel round, true to its axis, and forms it to a shape.

Turning Center A machine tool, similar to a lathe, that is CNC controlled to perform a wide variety of machining operations automatically under the control of the part program.

Two-Axes Programming Positioning programming locations only along the X Y axes.

VARIAX A type of machining center that has a lower platform to hold the workpiece, and an upper platform containing a free-floating spindle to drive the cutting tool. The two platforms are connected by six legs that form triangles.

Vector Angle The angle formed by the vector and the polar reference line.

Vector Radius The length of the line from the center of the circle to any point on the circle.

Vernier Caliper Precision measuring instruments used to make internal and external measurements that would be difficult to make with a micrometer or other measuring instrument.

Vernier Micrometer Basically the same as the standard inch micrometer with the addition of a vernier scale on the sleeve.

Vertical Head The vertical head on a vertical milling machine that must be set at 90° to the table, or else any surfaces machined or holes drilled will not be square with the work surface.

Vertical Milling Machine One of the most versatile and useful machines in a school or manufacturing shop. The many types of cutters and attachments that are available for this machine allow machining operations such as end and surface milling, radius and cam milling, drilling, reaming, boring, cutting slots and keyways, etc. to be performed.

Vise A work-holding device used to hold work for such operations as sawing, filing, chipping, tapping, threading, etc.

Vision Systems Artificial Intelligence (AI) technology combined with computers, software, television cameras, and optical sensors, that allows machines to perform jobs normally done by humans.

Work-in-Process The components that are in the process of being manufactured before the final product is assembled.

Zero-Defect Manufacturing A quality-improvement program whose goal is to continually improve manufacturing to where no defects are produced during manufacturing operations.

Zero-Reference Point A function of the CNC that allows the zero point on an axis to be quickly moved to another location within a specific range; also called zero shift.

Index